CHEM, Second Edition

Product Director: Mary Finch

Sr. Product Manager: Maureen Rosener

Content Developer: Liz Woods

Product Assistant: Jessica Wang

Executive Brand Manager: Nicole Hamm

Market Development Manager: Janet del Mundo

Associate Product Manager, 4LTR Press: Pierce Denny

Sr. Content Project Manager: Tanya Nigh

Sr. Art Director: Maria Epes

Manufacturing Planner: Judy Inouye

Rights Acquisitions Specialist: Don Schlotman

Production Service and Compositor: Lachina Publishing Services

Photo and Text Researcher: Q2A/Bill Smith

Copy Editor: Lachina Publishing Services

Text and Cover Designer: Riezebos Holzbaur

Cover Image: Shutterstock

Inside Front Cover: © iStockphoto.com/sdominick; © iStockphoto.com/alexsl; © iStockphoto.com/A-Digit

Page I: © iStockphoto.com/CostinT; © iStockphoto.com/photovideostock; © iStockphoto.com/Leontura

Back Cover: © iStockphoto.com/Leontura

For product information and technology assistance, contact us at **Cengage Learning Customer & Sales Support, 1-800-354-9706**.

For permission to use material from this text or product, submit all requests online at **www.cengage.com/permissions**. Further permissions questions can be e-mailed to **permissionrequest@cengage.com**.

Library of Congress Control Number: 2013936571

Student Edition:

ISBN-13: 978-1-133-96298-4

ISBN-10: 1-133-96298-X

Cengage Learning
200 First Stamford Place, 4th Floor
Stamford, CT 06902
USA

Cengage Learning is a leading provider of customized learning solutions with office locations around the globe, including Singapore, the United Kingdom, Australia, Mexico, Brazil, and Japan. Locate your local office at **www.cengage.com/global**.

Cengage Learning products are represented in Canada by Nelson Education, Ltd.

To learn more about Cengage Learning Solutions, visit **www.cengage.com**.

Purchase any of our products at your local college store or at our preferred online store **www.cengagebrain.com**.

Printed in the United States of America
3 4 5 6 7   17 16 15

CHEM Brief Contents

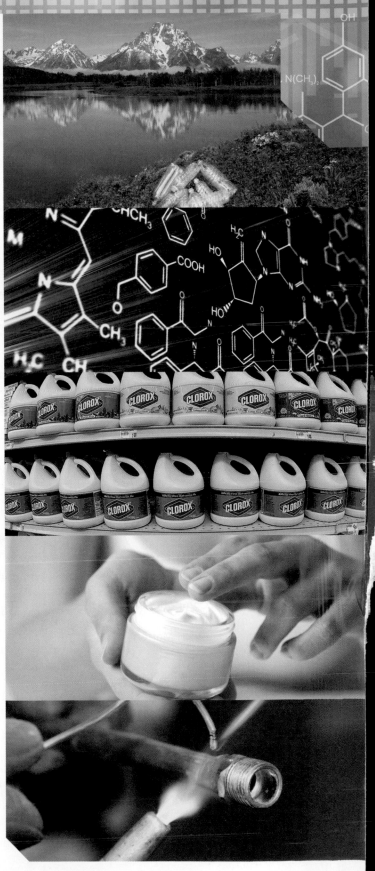

Contents

CHEM Contents

Contents

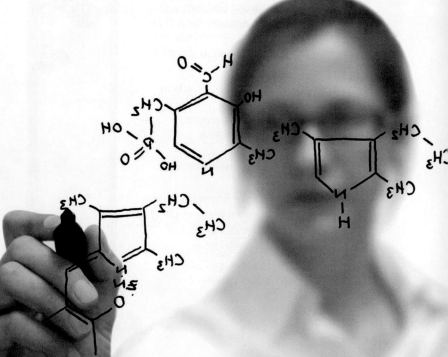

CHEM Contents

Contents

4LTR Press solutions are designed for today's learners through the continuous feedback of students like you. Tell us what you think about CHEM2 and help us improve the learning experience for future students.

YOUR FEEDBACK MATTERS.

Living in a World of Chemistry

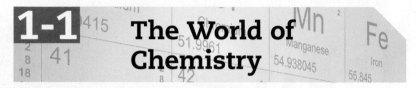

1-1 The World of Chemistry

On the first day of class each semester, professors often try to gather some general impressions about student backgrounds, aspirations, and career goals. Typical reasons for enrolling in a chemistry course designed for nonmajors range from "I'm taking this course to fulfill a graduation requirement" to "I really liked chemistry in high school." Most students do not plan to pursue careers that they perceive to be heavily dependent on a knowledge of chemistry. The professor then sets out on a semester-long journey to increase students' awareness of the impact of chemistry on their daily lives. His or her goal is to help students to develop an appreciation of the fact that some level of understanding of chemistry is critical to an understanding of life and its associated activities.

Chemistry can be defined as the study of matter and the changes it undergoes. We all make daily use, consciously or not, of a knowledge of chemical phenomena. Try to imagine how your favorite activity might directly or indirectly depend on chemistry. Cooking? Playing golf? Word processing on a computer? Art? If you can readily see how chemistry, or some product or device dependent on chemistry, influences these and other activities, then you may already have an appreciation of the importance of chemistry. Even if you don't see the connection, you should know that every activity in which we are involved on a daily basis depends on chemicals in some way. It is the purpose of this book to help you to see the connection. However, you may still question how much fundamental chemistry you actually need to know.

Many argue that chemistry is uniquely positioned at the crossroad between the biological and physical sciences. By exploiting their understanding of both realms, chemists and other professionals with strong backgrounds in chemistry have made, and continue to make, valuable contributions to our global society. Major technological and biological discoveries almost always depend on a fundamental understanding of chemistry. The pursuit of these discoveries, as a way to improve the world in which we live, drives those who seek to be a part of the process. While you probably aspire to improve the world, at least to some extent, you probably don't imagine your effort to revolve around chemistry. A little analysis

NASA Goddard Space Flight Center Image by Reto Stöckli (land surface, shallow water, clouds). Enhancements by Robert Simmon (ocean color, compositing, 3D globes, animation). Data and technical support: MODIS Land Group; MODIS Science Data Support Team; MODIS Atmosphere Group; MODIS Ocean Group Additional data: USGS EROS Data Center (topography); USGS Terrestrial Remote Sensing Flagstaff Field Center (Antarctica); Defense Meteorological Satellite Program (city lights).

of even the most mundane of your daily activities, however, reveals that chemistry's influence is pervasive in modern society.

Consider the following. Why are some consumer products touted as being "chemical-free," and why are so many people drinking bottled water these days? Does that mean that chemicals are bad? Should you be concerned about this as you shop for consumer goods and food products? We'll answer those questions many times as we proceed through this course.

Many of the issues on which a conscientious citizen votes depend heavily on the ability to evaluate scientific information. For instance, should you vote to elect a representative who is convinced that global warming is a myth or one who is convinced that it is a serious issue about which something needs to be done?

The well-being of our planet and the health of its inhabitants are dependent on chemistry. Chemistry is all around us.

Should you object to legislation to allow irradiation of foods because you think this will render the foods radioactive? Should you really understand something about carbon dioxide and water, the combustion products formed by burning gasoline, when you decide to purchase a car? These are just a few of the social issues facing us today that are heavily dependent on some knowledge of chemistry.

Such issues often make it seem as if chemistry only causes problems. We also want to focus on the contributions that chemistry has made to the development of the modern conveniences, wonderful medicines, and consumer goods we enjoy. These contributions have certainly led to an increased life span and a higher quality of life for many of the people of the world. However, not everyone yet enjoys these benefits.

There is an old saying that "every cloud has a silver lining." Could it possibly be that there are problems that may arise, or have already arisen, as a result of our quest for this wonderful lifestyle? Could it be that almost every silver lining has a little tarnish on it? As wonderful as some of our chemical and technological advances have been, they all carry with them some not-so-desirable aspects. A study of chemistry will leave you better prepared to understand that benefit must be balanced against risk many times daily. Often you do this analysis without even thinking, as you've already established that the benefits of a certain activity or use of a certain product outweigh any risks. New information often calls these decisions into question, however, so you must be capable of making informed decisions as you encounter these dilemmas.

Let's briefly consider how you might encounter "chemistry" in the future. You may be surprised to know that third graders are required to know the difference between chemical and physical changes and between atoms and molecules. Would you, as a future third-grade teacher, feel comfortable teaching these concepts to your students? As a future farmer, would you understand enough chemistry to utilize the chemicals necessary for modern farming, or would you feel that "organic farming without chemicals"

Are these foods "chemical-free"?

is the way to go? (Is such farming even possible?) Would you, as a stockbroker, feel comfortable recommending that your clients invest in a company seeking to develop automobile engines that run on water instead of gasoline? Is it worth the extra expense to purchase an audio cable that has gold-plated contacts? Would you, as a parent, feel comfortable when considering unconventional medicines or medical treatments recommended by a friend, or would you feel more comfortable with advice from your physician? We could discuss hypothetical situations like these forever, but hopefully it is apparent by now that maybe, just maybe, you do need to know a little more chemistry than you do now.

Making you a little more comfortable with and knowledgeable of chemistry is a major goal of this course. Let's begin.

Maybe *chemistry* means *chemicals* to you. And perhaps you think this word should be used only with the adjective *toxic*. That belief wouldn't be surprising, because you have probably heard of "toxic chemical spills" or warnings about "toxins" in the environment. Indeed, some chemicals are toxic—very toxic: the arsenic of mystery stories, the poisonous gases of World War I, and the botulinum toxin released by the microorganism that grows in badly canned food and causes severe food poisoning are a few examples.

Learning about chemistry doesn't, however, mean learning only about harmful substances. Nor does it mean learning only about the wonders of our modern world provided by chemistry, such as the medications that cure once incurable diseases; the synthetic fabrics that are beautiful, durable, and inexpensive; or the colors on the television screen.

We suggest you come to the subject with a "What's in it for me?" attitude. The first photographs of planet Earth taken from the moon provided a forceful reminder to many of us that this planet and the materials on it are finite. The world of chemistry really has but one concern: the materials provided by our planet and what we do with them.

In a global view, some understanding of chemistry will be helpful in dealing with major social issues that lie ahead in the 21st century. The present world population of nearly 7 billion is expanding at an ever-increasing rate and is expected to double in the next 50 years. How can everyone be fed, clothed, and housed? How can we prevent spoiling what planet Earth has provided us? How can we reverse some of the damage done when we knew less than we do now about the consequences of our activities? How can we prevent doing new damage to our environment? None of these questions can be fully addressed without some serious applied chemistry.

Knowing something about chemistry adds a new dimension to everyday life, too. If an advertisement proclaims that a product "contains no chemicals," what does that tell you about the producers of that product? Is it worth paying twice the price for something that proclaims itself to be "all natural"? (Sometimes it is and sometimes it isn't.) Do you know that you shouldn't mix household ammonia and chlorine bleach because they react to form a very toxic gas?

Lying ahead in the chapters of this book is a look into the chemical view of the world. It is based on observations and facts. A *scientific fact* results from repeated observations that produce the same result every time. (Water boils at 100°C at sea level—that's a scientific fact.) The chemical world is also based on models, theories, and experiments. Scientists use models and theories to organize knowledge and make predictions. The predictions must then be tested by experiments. If experimental results disagree with the predictions, the reason must be explored. Possibly the theory is wrong.

You will also see that the chemical view of the world requires learning to look at materials in two ways. The first is what direct observation provides; the second is the mental image of what scientists have learned about the world that cannot be observed directly.

To continue this introduction to the world of chemistry, we offer a glimpse of three chemistry-related topics that are sufficiently newsworthy to be mentioned often in the news—the DNA that governs heredity for all plants and animals, the disappearance of ozone in our atmosphere, and the connection between fossil fuel use and global warming. As you progress through this and subsequent chapters of this book, you will learn more about the connection between chemistry-related topics that make the news and the chemists' view of matter.

1-2 DNA, Biochemistry, and Science

DNA (*deoxyribonucleic acid*) is present in many cells throughout our bodies. About 55 years ago, DNA was identified as the carrier of genetic information from one generation to the next, determining how we will be the same or different from our ancestors. In the intervening years, the chemical composition of DNA and the shape of the molecule came to be understood.

In 2000, the door opened on a revolution in our knowledge of DNA. Two research endeavors, one in a for-profit company (Celera Genomics) and one in a not-for-profit consortium (the Human Genome Project) jointly announced their first results in mapping human DNA. The map of human DNA was completed in April 2003. The location of every segment of human DNA had been identified.

In their original publication, the leaders of the Human Genome Project made the following prophetic statement: "It has not escaped our notice that the more we learn about the human genome, the more there is to explore." Among these ongoing explorations are the following:

- Development of products that improve the health of humans, other animals, and plants

- Understanding of hereditary diseases

- Development of drugs to cure hereditary diseases

- Alteration of an individual's genetic makeup to prevent or cure a disease

- The study of why a drug can be effective in some individuals but not in others

- Exact matching of drugs to an individual's genetic makeup

- Development of improved agricultural crops and animals

- Creation of genetically modified bacteria that will mass-produce desirable chemical products

- Use of genetic information for a better understanding of evolution

The composition of an individual's DNA provides the equivalent of a fingerprint in identifying that individual. Because methods have now become available for analyzing DNA, such analysis has played an important role in forensic science. In response to arguments about the validity of courtroom use of DNA identification, the National Academy of Sciences conducted a review. They concluded that there is no reason to question the reliability of properly analyzed DNA evidence.

Comparison of the DNA analysis of a sample of blood or other biological material found at the scene of a crime with that of a sample from a suspect can weigh heavily as evidence for guilt or innocence. In the early days of the courtroom presentation of DNA evidence, however, there was concern over the use of competing scientific "experts" by opposing sides. Judges were often put in the position of ruling on scientific matters about which they knew very little. In 1995, the U.S. Supreme Court decided that federal judges, and presumably also state judges, have the power to hold pretrial hearings to validate the competence of experts before allowing them to testify. One goal is to prevent "junk science"—science not widely accepted and

Chemistry: The study of matter and the changes it undergoes

not developed with the necessary care—from reaching the courtroom. As a result, an effort was begun to educate judges about DNA. In 1995, the first group of 35 judges spent six days studying the subject; they even sequenced their own DNA in the laboratory.

The results of establishing the validity of DNA testing have been widespread. By 2011, more than 280 individuals in the United States, including 17 people awaiting execution, had been proven innocent by DNA tests. Proof of innocence or guilt by DNA testing is expected to expand rapidly, thanks to the DNA Index System established by the FBI. This system links the DNA evidence collected in almost every state and makes it available for searching by law enforcement officials in all states.

The study of DNA is one aspect of *biochemistry*, the natural science that unites chemistry, a physical science, with the biological sciences. Chemistry is defined as the study of matter and the changes it undergoes. A moment's thought should convince you that this makes an understanding of chemistry fundamental to an understanding of everything in nature. And, of course, with an understanding of natural matter, chemists are able to modify matter and synthesize new kinds of matter.

Historically, the natural sciences have been associated with observations of nature—our physical and biological environment. A traditional classification of the natural sciences is represented in Figure 1.1, with an emphasis on the relationship of chemistry to other sciences. As each science grows more sophisticated, the boundaries become more blurred, however. The dynamic character of science is illustrated by the emergence of new disciplines. There are now

FIGURE 1.1 The natural sciences, with some of their subdivisions.
The number of subdivisions continues to increase as the boundaries between the sciences disappear.

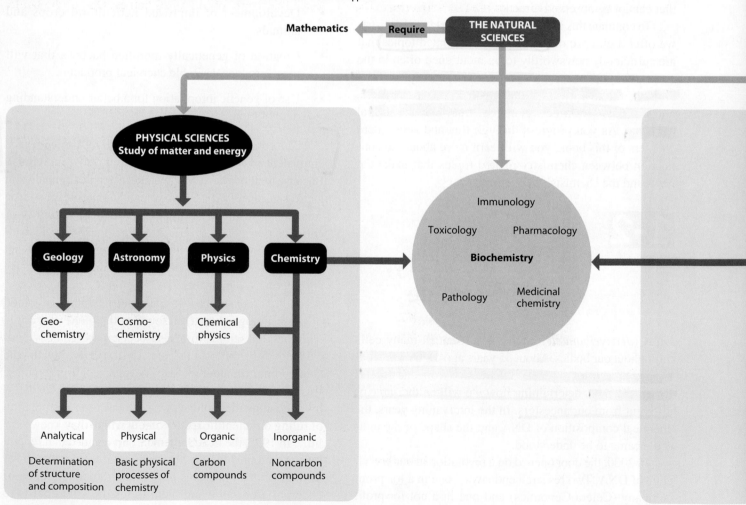

individuals who refer to themselves as biophysicists, bioinorganic chemists, geochemists, chemical physicists, and even molecular paleontologists (they use modern chemical methods to analyze ancient artifacts).

The study of DNA is one of the frontier areas of science. Hardly a week passes without the report of some new development, which might be the unraveling of the cause of an inherited disease or the introduction of a new product of the technology known as *genetic engineering*. Because these developments often relate to human health or behavior, they make the newspapers. DNA is one of the topics found in Chapter 15, "The Chemistry of Life."

1-3 Air-Conditioning, the Ozone Hole, and Technology

Next, we consider something that directly affects more of us than DNA testing: the chemicals used as refrigerants. The development of refrigerants is rooted in some interesting history of chemistry. Refrigeration as we know it has been in use since the late 1800s. The process requires a fluid that absorbs heat as it evaporates, releases heat when it condenses, and can be continuously cycled through evaporation and condensation without breaking down.

In the early 1900s, the fluids used as refrigerants were mostly flammable or toxic. One day, Charles Kettering, then director of research at General Motors, passed along a challenge to Thomas Midgley, a young engineer: "The refrigeration industry needs a new refrigerant if they ever expect to get anywhere." Midgley and a colleague vigorously dug into the problem. They hunted for candidates in the extensive tables of data kept at hand in every chemistry laboratory and surveyed the systematic variations in the properties of the chemical elements. Although fluorine had a reputation for being toxic, they decided that in some cases it might not be. To begin with, they decided to make a new fluorine-containing substance in the laboratory. Because one of the necessary raw materials was a fluorine chemical that was scarce, they quickly purchased the only available supply of five small bottles. One of the bottles

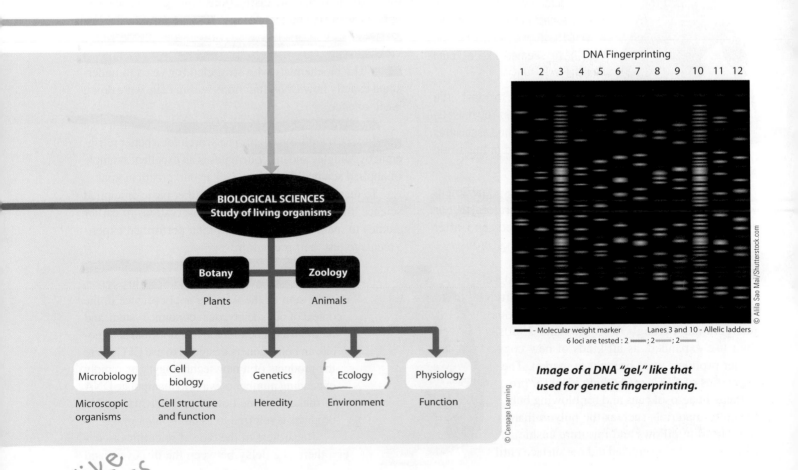

Image of a DNA "gel," like that used for genetic fingerprinting.

was used to make the first sample of the new chemical. In Midgley's own words:

> A guinea pig was placed under a bell jar with it and, much to the surprise of the physician in charge, didn't suddenly gasp and die. In fact, it wasn't even irritated. Our predictions were fulfilled.

A second sample made from the starting material in a different bottle did, however, kill the guinea pig. Here was a problem to be solved. Some investigation showed that only the first bottle contained pure material. It was a contaminant that had killed the guinea pig, not Midgley's newly prepared chemical.

> Of five bottles . . . one had really contained good material. We had chosen that one by accident for our first trial. Had we chosen any one of the other four, the animal would have died as expected by everyone else in the world except ourselves. I believe we would have given up what would then have seemed like a "bum hunch."
>
> And the moral of this last little story is simply this: You must be lucky as well as have good associates and assistants to succeed in this world of applied chemistry.

The outcome of this experiment was the proliferation of refrigerators in our homes. Food spoilage diminished, and after World War II the food processing industry expanded to produce frozen foods.

So far, this is a success story—the development of chlorofluorocarbons, often referred to as CFCs, as much less hazardous refrigerants. Through the 1950s and 1960s, their use expanded, as all kinds of new consumer products were brought to market. The properties of CFCs made them ideal for propellants in aerosol cans and for blowing bubbles into materials such as the polyurethane foam used in pillows and furniture cushions. The problem being created did not surface until

the 1970s, because no one was aware of the fate of these very stable compounds as they were released into the environment.

By the 1980s, it had become clear that CFCs drift unchanged into the stratosphere, where they interact with ozone and destroy it. The result is the "ozone hole" and the potential for damage from the resulting increase in solar radiation that reaches Earth's surface.

Ozone in the stratosphere acts as a filter that absorbs ultraviolet radiation. With ozone depletion comes a corresponding increase in the radiation that reaches us and an increase in its harmful effects, which include human skin cancer and cataracts that can lead to blindness. Increased ultraviolet exposure is also harmful throughout the environment. In the oceans, the phytoplankton population is decreased, resulting in a loss of the fish that feed on phytoplankton.

In 1987, nations that produce CFCs drafted a plan of action designed to reduce CFC use. Later amendments called for complete phaseout of all chemicals that harm the ozone layer and accelerated the schedule for that phaseout. (These chemicals and the ozone layer are further discussed in Chapter 7.)

It is helpful in reading about refrigerants and similar topics to recognize the distinctions among basic science, applied science, and technology. Basic science, or basic research, is the pursuit of knowledge about the universe with no short term practical objectives for application in mind. The biochemists who struggled for years to understand exactly how DNA functions within cells were doing basic science.

Applied science has the well-defined, short-term goal of solving a specific problem. The search for a better refrigerant by Midgley and his colleagues is an excellent example of applied science: They had a clear-cut, practical goal.

To reach the goal, as is done in either basic or applied science, they utilized the recorded observations of earlier studies to make a prediction and then performed experiments to test their prediction.

Technology, also an application of scientific knowledge, is a bit more difficult to define. In essence, it is the sum of the way we apply science in the context of our society, our economic system, and our industry. The first refrigerators and automobile air conditioners designed to use CFCs were the products of a new technology. The rapidly expanding number of ways to manipulate DNA to make new medicines or other marketable products is referred to as *biotechnology*.

Regardless of the type of scientific discovery, there is a delay between the discovery and

© Tetra Images/Jupiterimages

AP Images/Wilfredo Lee

it all gets kinda hazy when your heart's fucked up

TABLE 1.1 Time Needed to Develop Technology

Innovation	Conception	Realization	Incubation Interval (Years)
Photography	1782	1838	56
Antibiotics	1910	1940	30
Zipper	1883	1913	30
Nylon	1927	1939	12
Roomba vacuuming robot	1990	2002	12
Roll-on deodorant	1948	1955	7
Blu-ray disc	2000	2006	6
Internet	1964	1969	5
Post-it note	1974	1977	3
Medical X-ray	12/1895	01/1896	0.08

its technological application. The incubation intervals for a number of practical applications of various types are given in Table 1.1.

The important point is that technology, like science, is a human activity. Decisions about the uses of technology and priorities for technological developments are made by men and women. How scientific knowledge is used to promote technology depends on the people who have the authority to make the decisions. Sometimes those people are all of us—when we go to the polls in a democratic society, we can influence decisions about technology. When we have this chance, it is important to be well enough informed to critically evaluate the societal issues related to the technology.

As the history of CFCs illustrates, science and technology, like social conditions, are constantly evolving. When CFCs were introduced as refrigerants in 1930, they were a great advance for the economy and replaced hazardous materials. At that time, the number of technologically advanced nations and the world population were smaller. People were in the habit of assuming that natural processes would keep the environment healthy, and to a greater extent than today that was true. Moreover, some of the sophisticated instruments that have since revealed ozone depletion were not available then. Only by staying informed can we be ready to adjust to changing times.

1-4 Fossil Fuel Use and Global Warming

The world has become increasingly dependent on the use of *fossil fuels* (coal, oil, and natural gas) to support consumer-driven societies and their associated lifestyles. Enormous quantities of these fuels are used daily. The fuels may be burned in a variety of vehicles to propel them down the road, through the air, or across the oceans. The fuels may be used to drive industrial manufacturing processes directly, such as by powering an engine, or they may be used indirectly to generate electricity to power the economy by producing the consumer products upon which we've become so dependent. Our ability to heat and cool our homes; grow, harvest, transport, and cook our food; provide health care and medicines; and entertain ourselves during our free time all depend on fossil fuels in one way or another. Some activities depend on burning the fuels, whereas others depend on utilizing the fossil fuels as raw materials in the chemical industry. Fossil fuels are nonrenewable resources, however, and the supply will eventually run out.

It is becoming increasingly apparent that the products of fossil fuel combustion may be causing Earth to warm

there's got to be more than going back & forth

beyond what is natural and desirable. It is a known fact that both carbon dioxide and water vapor absorb and trap heat radiation. It is also known that the levels of carbon dioxide and some other heat-trapping substances have increased significantly in the atmosphere over the past hundred or so years. The debate among scientists, politicians, and concerned citizens centers on whether the warming of Earth is due to the increase in concentration of these substances. If so, what will be the consequences of this warming? And is there anything we can or should do about it?

Some people believe this *global warming* will be good for Earth, some think it will be inconsequential, some think we will adjust to it, and some think it threatens life as we know it. Some people think there is nothing that can or should be done about it, whereas others feel it is imperative that we do everything within our power to halt this phenomenon. The average citizen must become knowledgeable enough about this issue to make informed decisions concerning factors affecting global warming. Such knowledge could influence the type of automobile you purchase, the size of the house you live in, and your lifestyle in general. It could eventually influence where and how you live or, possibly, even if you live. Knowledge is power, and only by obtaining, properly assessing, and using this knowledge can we adjust to changing times.

1-5 Benefits/Risks and the Law

A glance through the stories in any daily newspaper or weekly news magazine or a half hour spent watching almost any television newscast is bound to demonstrate that the potential for risk to public health and safety is newsworthy.

The risk associated with an activity is determined by two main factors, in combination: hazard and exposure. An activity can be low-risk if it is associated with low levels of hazard and there is little opportunity for exposure. An example of such an activity would be drinking a glass of purified water. There is little hazard associated with this activity, and exposing yourself to this activity involves little risk. However, if you were to increase your exposure to purified water, say by swimming in it, then the risk of overexposure to purified water can increase. You may be interested to know that even an activity

that involves a high level of hazard can be considered safe, as long as the possibility of exposure is small. Here, the use of electricity comes to mind. As long as there is little possibility of coming into contact with live wires, there is little risk associated with plugging in a lamp.

The risk that accompanies an activity is often weighed against the benefit derived from the activity, and it is important to consider how society weighs the benefit of some activity, the use of some chemical, or the use of some new technology against the potential risks. For example, automobiles offer many transportation advantages over horses, yet automobiles are responsible for thousands of deaths and considerable pollution each year. Clearly, we've accepted these risks while, at the same time, making efforts to minimize or eliminate them.

The Boston-based Harvard Center for Risk Analysis (HCRA) points out that in the days when measles and polio were prevalent, most people agreed that the benefits of vaccination against these conditions far outweighed the risk associated with the vaccines and opted for immunization. However, now these diseases are quite rare in most parts of the world. Many parents of small children have decided that the benefit to be gained from vaccinations does not outweigh the risks from side effects and are opting out of voluntary vaccination programs. As concern has grown that terrorists might release the smallpox virus into a world in which smallpox has essentially been eliminated, some have called for renewed vaccination against this deadly disease. However, many physicians and health care providers who would be first responders in such a terrorist attack have been reluctant to voluntarily participate in renewed vaccination programs because of the small risk of side effects. The current perception is frequently that

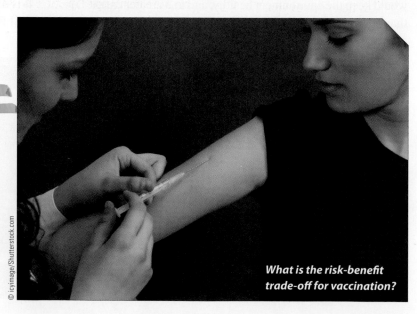

What is the risk-benefit trade-off for vaccination?

the benefit (protection against an unlikely smallpox attack) does not outweigh the risks. On the other hand, when one of the authors of this book was a young boy, people were routinely vaccinated against smallpox because the risk of side effects was small compared to the benefit: protection against a deadly killer. Less clear are situations in which there are small benefits and small risks or large benefits with large risks.

In any benefit versus risk analysis, several factors come into play. An excellent discussion of these factors is presented in the June 2003 HCRA publication *Risk in Perspective* (http://www.hcra.harvard.edu). The authors identify the following major risk perception factors to consider:

- Level of dread: Which do you dread more, for example, heart attack or cancer?

- Control: We feel safer when we are in control.

- Natural or human-made risk: Natural risks seem less worrisome.

- Choice: Do you have a choice in the risk activity?

- Involvement of children: Risks that are acceptable for adults may not be seen in the same light when children are involved.

- Newness of risk: Newly perceived risks seem worse than "old" risks.

- Awareness: Simply becoming more aware of a risk makes it seem more serious than it may really be. The popular press may contribute to this problem in its effort to keep us well informed.

- Can it happen to you: Not long ago, American citizens didn't worry much about terrorism because "it doesn't happen here." September 11, 2001, certainly changed that.

- Trust: We tend to be more willing to accept risks imposed on us by those we trust than by those we do not.

- Risk-benefit trade-off: Benefits tend to make risks more acceptable.

Consider the following risk comparisons of "natural" versus "human-made" risks. Many people are opposed to the use of "toxic chemicals" around food products, yet toxicologists estimate that the typical American diet contains ten thousand times more naturally occurring cancer-causing chemicals than those of the synthetic variety. The Centers for Disease Control and Prevention estimates that about nine thousand Americans die each year from natural food poisoning, whereas worst-case estimates of deaths due to pesticide residues in foods are generally less than a few thousand. Suppose legislation were proposed that banned the sale of any foodstuff containing a human-made chemical for which a reasonable rate of consumption would lead to one additional cancer death in ten thousand people. Most people probably would be in favor of banning the use of the foodstuff containing this chemical. However, the following activities are estimated by experts to carry this same risk of cancer due to naturally occurring carcinogens, shown in parentheses, in the foodstuff: drinking a glass of orange juice (*d*-limonene) every other week; consuming a head of lettuce (caffeic acid) every other year or one carrot (caffeic acid) per week; or drinking one glass of wine (ethyl alcohol) every three years. Does this place a different perspective on such legislation? See Table 1.2 for some activities estimated to increase the probability of one additional death per one million people involved in the activity.

It is rather ironic that as life expectancies have increased dramatically because of chemical and technological progress, our level of acceptance of risk is declining in areas associated with chemicals. Can there be a rational and logical explanation for this? Almost every chemical, natural or human-made, has some risk associated with it at some level

TABLE 1.2 Activities with a Probability of One Additional Death per One Million People Exposed to the Risk

Activity	Cause of Death
Smoking 1.4 cigarettes	Cancer, lung disease
Living 2 months with a cigarette smoker	Cancer, lung disease
Eating 100 charcoal-broiled steaks	Cancer
Traveling 6 minutes by canoe	Accident
Traveling 10 minutes by bicycle	Accident
Traveling 300 miles by car	Accident
Traveling 1000 miles by jet aircraft	Accident
Drinking Miami tap water for 1 year	Cancer from chloroform
Living 2 months in Denver	Cancer caused by cosmic radiation
Living 5 years at the boundary of a typical nuclear power plant	Cancer

© Brian Hagiwara/FoodPix/Jupiterimages/© Cengage Learning

of exposure. The risk associated with a chemical is a function of its toxicity and a person's level of exposure to it. The toxicity of a given chemical is an inherent property of the particular chemical, but exposure to that toxicity is something we may have some control over. For instance, water is inherently nontoxic, but overexposure to it under certain circumstances may lead to death by drowning. Peanut butter contains a potent, naturally occurring toxic compound known as aflatoxin, yet most of us are not concerned about its presence because its concentration, and hence our exposure to it, is very low. As our ability grows to detect increasingly lower concentrations of such chemicals in our foods and environments, the level of anxiety over their presence appears to have risen.

Over time, our ability to detect the presence of toxic substances has changed from levels in the parts per hundred (%) to levels below parts per million (ppm), parts per billion (ppb), and now parts per trillion (ppt). It is not necessarily true that the concentrations of these toxic substances have increased over the years. Surely there must be some balance between a certain level of informed, serious concern and chemical or technical paranoia. We must assess the risk from the use of a certain chemical or the pursuit of a certain technology or activity and then balance that against the benefit to be obtained. Some individuals seem to believe that a certain amount of public activism and government regulation can guarantee risk-free living. Is that possible?

$1\% = 1/100$
$1\ ppm = 1/1{,}000{,}000$
$1\ ppb = 1/1{,}000{,}000{,}000$
$1\ ppt = 1/1{,}000{,}000{,}000{,}000$

No absolute answer can be provided to the question, "How safe is safe enough?" The determination of acceptable levels of risk requires value judgments that are difficult and complex, involving the consideration of scientific, social, economic, and political factors. Over the years, a number of laws designed to protect human health and the environment have been enacted.

The Safe Drinking Water Act, the Toxic Substances Control Act, and the Clean Air Act are laws of this type. The federal Environmental Protection Agency (EPA) is required to balance regulatory costs and benefits in its decision-making activities. Risk assessments are used here. Chemicals are regulated or banned when they pose "unreasonable risks" to or have "adverse effects" on human health or the environment. Parts of the Clean Air Act and the Clean Water Act impose pollution controls based on the best economically available technology or the best practical technology. Such laws assume that complete elimination of the discharge of human and industrial wastes into water or air is not feasible. Controls are imposed to reduce exposure, but true balancing is not attempted. The goal is to provide an "ample margin of safety" to protect the public. As technology advances and the cost of the technology decreases, the margin of safety can be adjusted.

Risk management requires value judgments that integrate social, economic, and political issues with risk assessment. Risk assessment is the province of scientists, but determination of the acceptability of the risk is a societal issue. It is up to all of us to weigh the benefits against the risks in an intelligent and competent manner.

1-6 What Is Your Attitude Toward Chemistry?

Before proceeding with this study of chemistry and its relationship to our society, you might want to examine your prejudices (if any) and attitudes about chemistry, science, and technology. Many nonscientists regard science and its various branches as a mystery and think that they cannot possibly comprehend the basic concepts and their relation to societal issues. Many also have what has been christened *chemophobia* (an unreasonable fear of chemicals) and a feeling of hopelessness about the environment. These attitudes are most likely the result of news stories about the harmful effects of technology. Some of these harmful effects are indeed tragic.

What is needed, however, is a full realization of both the benefits and the harmful effects of science and technology. In the analysis of these pluses and minuses, we need to determine why the harmful effects occurred and

Cottonwood Lake near Buena Vista, Colorado. It's up to us to keep such places beautiful.

© altrendo nature/Altrendo/Getty Images; © Jamie Grill/Photographer's Choice/Getty Images

if the risk can be reduced for future generations. This book will give you the basics in chemistry along with some insight into the role of chemistry in both positive and negative aspects of the world today. We hope this knowledge will afford you a healthier and more satisfying life by allowing you to make wise decisions about personal problems and problems of concern to society as a whole.

Your Resources

In the back of the textbook:

→ *Review Card on Living in a World of Chemistry*

 In OWL for CHEM2 at www.cengagebrain.com

→ *Review Key Terms with Flash Cards (printable or digital)*

→ *Complete Interactive Practice Quizzes to prepare for tests*

→ *Submit Assigned Homework and Exercises*

Applying Your Knowledge

1. Which of the following is not one of the physical sciences?
   (a) Physics   (b) Chemistry
   (c) Zoology   (d) Geology

2. A scientist who is trying to make an insecticide that is more toxic to mosquitoes than to humans would be involved in

   _____.

   (a) basic research   (b) applied research
   (c) technology   (d) serendipity

3. An appropriate risk assessment regarding the use of a toxic chemical for control of household pests would include a consideration of the inherent toxicity of the chemical and what other factor?

4. Define chemistry.

5. In which branch of the biological sciences would the study of DNA fall?

6. Is it possible that scientists in other branches of biological and physical sciences might also study DNA? If so, explain what they might be looking for.

7. Dallas, Houston, and San Antonio are three of the ten largest cities in the United States. They have certainly benefited from the technology associated with air-conditioning and the internal combustion engine. Have there been risks or downsides associated with the application of this technology? Do you think the benefits have outweighed the risks?

8. List as many risk-perception factors as you can that you believe most people consider, consciously or subconsciously, when evaluating a new activity, product, medicine, or technology. Refer to Section 1-5.

9. How would you assess the risks and benefits of eating only organically grown fruits and vegetables (that is, those grown without the use of human-made chemicals) compared to the risks and benefits of eating fruits and vegetables grown using human-made chemicals allowed by federal and state laws? Does your level of *chemophobia* (fear of chemicals) enter into this?

10. For each of the following activities, list the hazards involved and the ways in which exposure could occur. Then, decide if the benefit of the activity outweighs the risk. Propose ways in which the overall risk might be lessened.
    (a) installing a smoke detector that uses a radioactive capsule to detect particles in the air
    (b) eating food that has been treated with radiation to kill harmful bacteria
    (c) eating food derived from genetically modified organisms or plants
    (d) drinking raw (unpasteurized) milk
    (e) smoking cigarettes
    (f) playing rugby

The Chemical View
of Matter

Practically everything we use has been changed from its natural state of little or no utility to one of very different appearance and much greater utility. Some of these changes are mechanical, such as turning trees into the lumber used to build houses. Many others are chemical, such as heating sand to make glass. Exploring, understanding, and managing the processes by which natural materials can be changed are basic to the science of chemistry. These activities require close examination of the composition and structure of matter, the stuff of which everything is made, which we begin to do in this chapter.

Many times a day, you expect things to behave in certain ways, and you act on these expectations. If you are cooking spaghetti, you know that you must first boil some water and that after the spaghetti has been in the boiling water for a while, it will soften. If you are pouring gasoline into the fuel tank of your lawn mower, you know that you must protect the gasoline from sparks or it will explode into fire. To cook the spaghetti or safely fill the gas tank, you don't need to think about why these expectations are correct.

Chemistry, however, has taken on the job of understanding why matter behaves as it does. Answering questions such as, "Why does water boil at 100°C?" or "Why does gasoline burn, but water doesn't?" or "Why does spaghetti get soft in boiling water?" requires a different approach than direct observation. We have to look more deeply into the nature of matter than we can actually see, at least under everyday conditions.

Samples of matter large enough to be seen, felt, and handled—and thus large enough for ordinary laboratory experiments—are called **macroscopic** samples. In contrast, **microscopic** samples are so small that they have to be viewed with the aid of a microscope. The structure of matter that really interests chemists, however, is at the **nanoscopic** level. Our senses have limited access into this small world of structure, although new kinds of instruments are beginning to change this condition.

In the chemical model of matter, everything around us, matter of every possible kind, is pictured as collections of very small particles. We'll begin our introduction to this way of thinking with a look at the ways in which matter is described, then turn our attention to the nanoscopic scale.

Macroscopic: Large enough to be seen, felt, and handled

Microscopic: Visible only with the aid of a microscope

Nanoscopic: In the range of the nanometer (0.000000001 m)

2-1 States of Matter, Mixtures, and Pure Substances

Matter is anything that has mass and occupies volume. Solids, liquids, and gases are the three common states of matter. The fourth state, plasma, occurs in flames, stars, and the outer atmosphere of Earth. Even though you can easily tell the difference between the three states, it is useful to describe their differences accurately. For example, the chemical substance we all call water is commonly found as a solid (ice),

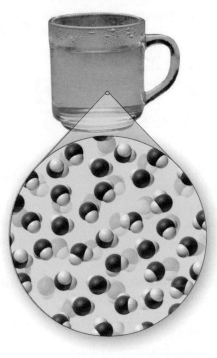

FIGURE 2.1 A macroscopic and nanoscopic view of the three states of matter in which water is commonly found.

Heterogeneous mixture: Matter that is not uniform in composition

Homogeneous mixture: Matter that is uniform in composition

Solution: A homogeneous mixture that may be in the solid, liquid, or gaseous state

Pure substance: Matter with a uniform and fixed composition at the nanoscopic level

Element: A pure substance composed of only one kind of atom

Atom: The smallest particle of an element

a liquid (water), or a gas (steam, water vapor). Unlike water, however, most other substances do not have unique names for each of their three states. What are the differences between the phases of matter for these other substances that can allow us to avoid confusing them?

From a scientific point of view, a solid is matter that retains its shape and volume, regardless of the container in which it is placed. While an ice cube might be the same shape as the tray in which it was formed (Figure 2.1), its shape and volume do not change when you place the cube into a drinking glass. A liquid is matter that retains its volume, but not its shape. When you pour water from a small glass into a larger glass, the liquid does not expand, but it does have a different shape in the larger glass. A gas is matter that does not retain its shape or volume. When a boiling tea kettle releases steam, the gas is not shaped like the kettle, and it quickly expands into the room. Figure 2.2 illustrates the behaviors of the states of matter for another chemical substance, bromine.

Most samples of matter as they occur in nature are not pure substances, but mixtures. Some are **heterogeneous mixtures**—their nonuniform composition is clearly visible and uneven, as in chocolate chip cookies or the different kinds of crystals in many rocks. **Homogeneous mixtures,** which are uniform in composition, are referred to as **solutions**. No amount of optical magnification will reveal a solution to be heterogeneous, because it is a uniform mixture of particles too small to be seen using ordinary methods. In addition, different portions of the same homogeneous mixture have the same composition. Examples of solutions are *clean* air, which consists mostly of nitrogen and oxygen;

freshly brewed tea; and some brass alloys, which are homogeneous mixtures of the elements copper and zinc. As these examples show, *solutions can be found in the gaseous, liquid, and solid states.*

Some mixtures may appear to be homogeneous even if they are not. For example, the air in your classroom appears homogeneous until a beam of light enters the room and reveals floating dust particles. Blood appears homogeneous as it flows from a cut in your finger, but a microscope shows that it is quite a heterogeneous mixture (Figure 2.3).

When any mixture is separated into its components, the components are said to be *purified*. Efforts at separation are usually incomplete in a single step, and repetition of the process is necessary to produce a purer substance. Ultimately, the goal is to arrive at substances that cannot be purified further.

Separation of mixtures is done by taking advantage of the different properties of substances in the mixture. For example, many people want to separate certain substances from their tap water before they drink it. One way to do this is based on the forces of attraction between particles at the nanoscopic level. The water is passed through a material that attracts the undesirable substances and holds them back.

2-2 Elements: The Simplest Kind of Matter

The chemical model of matter starts with the following three concepts. A **pure substance** is something with a uniform and fixed composition at the nanoscopic level.

FIGURE 2.2 The three states of matter for bromine: (a) bromine as a solid, (b) bromine as a liquid, and (c) bromine as a gas.

(a)   (b)   (c)

Pure substances can be recognized by the unchanging nature of their properties. An **element** is a pure substance composed of only one kind of atom. An **atom** is the smallest particle of an element, and different elements have different atoms.

Most of the materials you handle every day are pretty complex in their structure, and many are mixtures. Only occasionally do you encounter elements not combined with something else. One example is the helium used to fill a birthday balloon. The gas is helium, an element. From the chemical point of view, what does this mean? First, it means that helium cannot be separated or broken down into any other kind of matter that can still be recognized as helium. Second, it means that the balloon contains the very, very small particles known as helium atoms. Third, it means that helium has a unique and consistent set of properties by which it can be identified.

Another element you may have seen around your house is mercury. It is used to conduct electricity in switches like those in some thermostats and as the liquid in some thermometers. Mercury is certainly very different from helium in its properties. It is a liquid, not a gas, and it conducts electricity, which helium cannot do. What mercury and helium have in common is that they are both elements and each is composed of a single kind of atom. The liquid in the mercury thermometer is made of mercury atoms. Because the properties of helium and mercury are so different, we can safely conclude that helium and mercury atoms are not the same (Figure 2.4).

2-3 Chemical Compounds: Atoms in Combination

If a pure substance is not an element, it is a chemical compound. To take our most common example, what does it mean that water is a chemical compound? You probably know that

FIGURE 2.3 Blood, another heterogeneous mixture.
Color has been added to show the different kinds of blood cells. The doughnut-shaped ones are red blood cells, the yellow ones are immune system cells, and the brown cells are platelets.

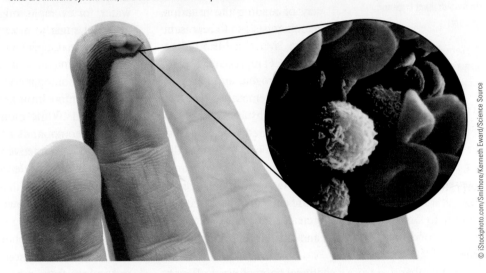

© iStockphoto.com/Smithore/Kenneth Eward/Science Source

FIGURE 2.4 Mercury atoms are larger and much heavier than helium atoms.
With 3 times the diameter and 50 times the weight, the difference in size between the two atoms is about the same as the difference between a 16-pound bowling ball and a baseball.

Mercury atom
3 times the diameter
50 times the weight

Helium atom

© iStockphoto.com/billnoll/© iStockphoto.com/NickyBlade

water is referred to as "h-2-oh" even if you have never studied chemistry. The "h-2-oh" is a way of reading the notation that chemists use to represent water: H_2O. In this kind of notation, H represents the element hydrogen, and O represents the element oxygen.

A **chemical compound** is a pure substance composed of atoms of one or more elements combined in fixed ratios. We know that water is a pure substance because it has the same properties no matter where it comes from. We know that water is a chemical compound because passing energy through it in the form of an electric current causes decomposition to the elements hydrogen and oxygen. Also, under the right conditions, hydrogen and oxygen combine to form water.

Remember that from a chemical point of view, all matter consists of tiny particles. In water, the smallest particle that can still be recognized as water is a water **molecule**. Each water molecule is composed of two hydrogen atoms (shown by the subscript "2" in H_2O) and only one oxygen atom (shown by the absence of a subscript after the O in H_2O). Experimentally then, pure substances are classified into two categories: (1) *chemical compounds*, which can be broken down into simpler substances called elements, and (2) *elements*, which cannot be broken down into any smaller particle still recognizable as that element.

Once elements are combined to form compounds, the original, characteristic properties of the elements are replaced by the characteristic properties of the compounds. Consider the difference between table sugar, a white crystalline substance that is soluble in water, and its constituent elements: carbon, which is usually a black powder and is not water soluble; hydrogen, the lightest gas known; and oxygen, the atmospheric gas needed for respiration.

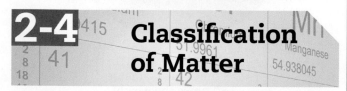

2-4 Classification of Matter

The kinds of matter we have described—elements, compounds, and mixtures—can be classified according to their composition and how they can be separated into other substances, as shown in Figure 2.5.

Heterogeneous samples of matter are all mixtures and can be physically separated into various kinds of homogeneous matter. Homogeneous matter can be a pure substance or a mixture. If it is a mixture, it is described as a solution and has the same composition throughout. Different solutions of the same substances, however, can vary in composition, sometimes over a very wide range. A teacup full of water, for example, might dissolve anywhere from a few grains of sugar to more than one measuring cup of sugar. The water and sugar can be separated by the physical process of evaporating the water.

If the homogeneous matter is a pure substance, it has a fixed composition and must be either an element or a compound. While elements contain only atoms of that element, compounds contain atoms of different elements combined in distinctive ways. At the nanoscopic level, each table sugar molecule contains 12 carbon (C) atoms, 22 hydrogen (H) atoms, and 11 oxygen (O) atoms, represented by $C_{12}H_{22}O_{11}$. At the macroscopic level, every 100 g of table sugar contains 42.1 g of carbon, 6.5 g of hydrogen, and 51.4 g of oxygen in chemical combination. Only chemical reactions could be used to produce pure carbon, hydrogen, and oxygen from sugar.

2-4a Why Study Pure Substances?

Perhaps by now you are wondering why we should be interested in elements and compounds and their properties. There are three basic reasons.

1. Only by studying pure matter can we understand how to utilize its properties, design new kinds of matter, and make desirable changes in the nature of everyday life. A long time ago, for example, it was known that swallowing extracts from the willow tree could relieve pain, although there were some unpleasant side effects. Once chemists identified the pain-relieving substance in the mixture of extracts, they were able to create a molecule that was similar enough to relieve pain but different enough to have fewer side effects. We know this pure substance as aspirin, which has the chemical formula $C_9H_8O_4$.

   At one time, most natural materials could be changed only by physical means. Only a few useful materials, such as pottery and iron, were the products of chemical change. Otherwise, the design of everyday objects was limited to using the properties of natural substances. Today, we have evidence all around us that the chemical modification of materials has indeed changed the quality of life. Synthetic fibers, plastics, lifesaving drugs, latex paints, new and better fuels, photographic films, compact discs, and DVDs are but a few of the materials produced by controlled chemical change. It is equally important that understanding the properties

FIGURE 2.5 A classification of matter.

Once elements have combined into compounds, only chemical reactions can separate them.

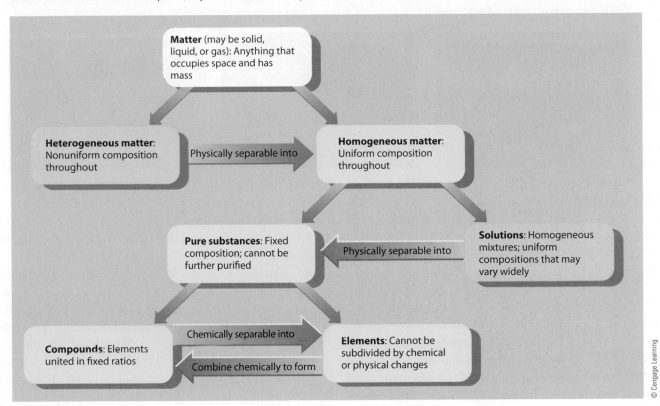

Matter (may be solid, liquid, or gas): Anything that occupies space and has mass

Heterogeneous matter: Nonuniform composition throughout

Physically separable into

Homogeneous matter: Uniform composition throughout

Pure substances: Fixed composition; cannot be further purified

Physically separable into

Solutions: Homogeneous mixtures; uniform compositions that may vary widely

Compounds: Elements united in fixed ratios

Chemically separable into

Combine chemically to form

Elements: Cannot be subdivided by chemical or physical changes

© Cengage Learning

of substances allows us to prevent *undesirable* chemical changes, such as the corrosion of metal objects, the spoiling of food, and the effects of hereditary diseases.

2. Studying the properties of matter helps us deal intelligently with our environment, both in everyday life and in the social and political arena. Knowing something about the properties of fertilizers or pesticides helps us decide which ones to use, which ones to avoid, and when their use is or is not necessary. In the context of community life, many decisions are enlightened by some knowledge of the properties of matter. When you pour waste down a sewer drain,

what happens to it? Does it matter whether the waste is motor oil or the water from washing your car? Does your town or city really need to spend millions of dollars to upgrade its sewage treatment plant to keep nitrogen-containing compounds out of natural waters or to remove asbestos from the high school auditorium ceiling?

3. The third reason for studying the properties of elements and compounds is simple curiosity. Chemicals and chemical change are a part of nature that is open to investigation. For some people, this is as big a challenge as a high mountain is to a climber. If we hope to understand matter, the first steps are to discover the simplest forms of matter and study their interactions. Curiosity draws many chemists to basic research.

2-4b The Structure of Matter Explains Chemical and Physical Properties

An essential part of basic research is investigation into the *structure of matter*—how atoms of the elements are connected in larger units of matter and how these units are arranged in

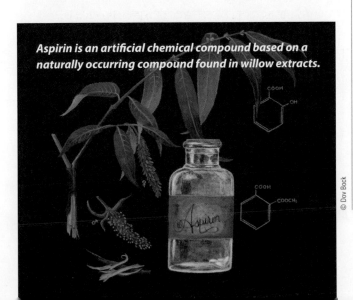

Aspirin is an artificial chemical compound based on a naturally occurring compound found in willow extracts.

© Dow Bock

© iStockphoto.com/AVTG/© iStockphoto.com/ milehightraveler/Clive Freeman/Biosym Technologies/Science Source

FIGURE 2.6 Three ways to look at snow.

(a) A snowy landscape. (b) The beautiful patterns of snowflakes as they can be seen under a microscope. (c) The orderly arrangement, in ice and snow, of water molecules represented by molecular models.

(a)

(b)

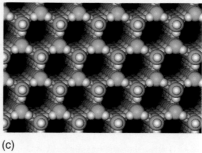

(c)

Diatomic molecules:
Molecules composed of two atoms

samples of matter large enough to be seen (Figure 2.6).

The properties of a sample of matter are determined by the nature of its parts, just as the capabilities of a computer are determined by the parts that have been assembled. If we hope to understand the nature of matter, it is absolutely necessary that we understand the minute parts and how they are related to each other. Indeed, *the basic theme of this text and of chemistry itself is the relationship between the structure of matter and its properties.* Armed with an understanding of this relationship, chemists are developing an ever-increasing ability to create new substances and predict the properties of these substances.

2-5 The Chemical Elements

Iron, gold, sulfur, tin, and a few other substances that occur in Earth's crust were recognized as elements long before the modern meaning of the term developed. Other naturally occurring elements were identified only after long and laborious efforts to separate them from the compounds in which they are found. Although it isn't always easy, chemists can now separate and identify the elements present in the most complex mixtures and compounds.

The names of the elements are listed on the periodic table review card. About 18 of these elements are not found anywhere in nature but have only been produced artificially in laboratories, usually in extremely small quantities.

The elements range widely in their properties. As is discussed in Chapter 3 and represented by the arrangement

of the elements in the periodic table (on the periodic table review card), the properties of the elements are largely determined by the composition of their atoms.

Under everyday conditions, some elements, including helium, hydrogen, nitrogen, oxygen, and chlorine, are gases. Only two elements—mercury and bromine—are ordinarily liquids. Most of the elements are solids, and of these most are silvery metals. Two distinguishing properties of metals are that they conduct electric current and have a shiny appearance. A second major class of elements, the nonmetals, do not conduct electricity and are not lustrous. The familiar nonmetals include all of the elemental gases, and carbon, sulfur, phosphorus, and iodine, which are solids.

2-5a Symbols for the Elements

Sometimes people are discouraged about understanding chemistry when they encounter chemical formulas and names such as CH_3CH_2OH, H_2SO_4, acetylsalicylic acid, nitric oxide. . . . There is nothing unique about chemistry, however, in needing its own vocabulary. Surely the symbols

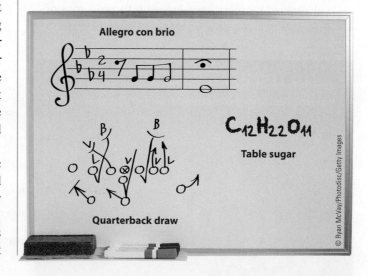

and language on the whiteboard below are equally mysterious to anyone who has never studied chemistry, been a musician, or played football. Chemistry, just like music or football, needs special symbols and language. Without them, communication would be impossible.

Since elements are the building blocks of all matter, symbols for the elements are fundamental to communicating about chemistry. The symbols for the elements are listed next to their names on the periodic table review card. Some symbols are single letters that are the first letter of the name. Others are two letters from the name, always with the first letter capitalized and the second lowercase. For example, *i*odine is represented by I and *in*dium is represented by In; *n*itrogen is represented by N and *ni*ckel by Ni; *m*agnesium is represented by Mg and *ma*nganese by Mn. You can see that it is important to recognize the symbols and know how they differ for different elements.

Eleven metals are represented by symbols not derived from their modern names but instead from older names in other languages, often Latin or Greek. Most of these metals, listed in Table 2.1, were known in ancient times because they are found free in nature or are easily obtained from their naturally occurring compounds.

Of the known elements, only 18 are nonmetals. Most of the common nonmetals have single-letter symbols:

H	C	N	O
Hydrogen	Carbon	Nitrogen	Oxygen
P	S	F	I
Phosphorus	Sulfur	Fluorine	Iodine

Earlier, we noted that a sample of gaseous helium (also a nonmetal) contains helium atoms. The symbol He is therefore used to represent a single He atom or a large collection of helium atoms. One of the first milestones in understanding the structure of matter was the discovery that a number of nonmetallic elements are composed not of atoms but of molecules. A sample of pure hydrogen gas does contain only hydrogen atoms as expected for an element, but the atoms are joined together in two-atom hydrogen molecules, represented by H_2. Seven nonmetals exist under everyday conditions as two-atom molecules, known as **diatomic molecules**:

Element	Symbol	Properties
Hydrogen	H_2	Colorless, odorless, occurs as a very light gas, burns in air
Oxygen	O_2	Colorless, odorless gas, reactive, constituent of air
Nitrogen	N_2	Colorless, odorless gas, rather unreactive
Chlorine	Cl_2	Greenish-yellow gas, very sharp choking odor, poisonous
Fluorine	F_2	Pale yellow, highly reactive gas
Bromine	Br_2	Dark red liquid, vaporizes readily, very corrosive
Iodine	I_2	Dark purple solid that changes directly to gas when heated gently

© Cengage Learning

TABLE 2.1 Elements with Symbols Not Based on Their Modern Names

Modern Name	Symbol	Origin of Symbol
Antimony	Sb	*stibium:* Latin name for *antimony*, which comes from the Greek *anti + monos*, or element not found alone
Copper	Cu	*cuprum:* Latin name, meaning from the island of Cyprus
Gold	Au	*aurum:* Latin name, meaning shining dawn
Iron	Fe	*ferrum:* Latin name for the element
Lead	Pb	*plumbum:* Latin name for the element
Mercury	Hg	*hydrargyrum:* Greek name, meaning liquid silver or quick silver
Potassium	K	*kalium:* Latin name, meaning alkali
Silver	Ag	*argentum:* Latin name for the element
Sodium	Na	*natrium:* Latin name for the element
Tin	Sn	*stannum:* Latin name for the element
Tungsten	W	*wolfram:* Swedish name, meaning devourer of tin (because it interferes with purifying tin)

© Cengage Learning

2-6 Using Chemical Symbols

Chemical formula:
Written combination of element symbols that represents the atoms combined in a chemical compound

Subscripts: In chemical formulas, numbers written below the line (for example, the "2" in H_2O) to show numbers or ratios of atoms in a compound

Structural formula:
A chemical formula that illustrates the connections between atoms in molecules as straight lines

Molecular formula:
A chemical formula that illustrates molecules with atomic symbols and subscripts

Organic compound:
A derivative of the carbon and hydrogen compound

Inorganic compound:
Any compound other than an organic compound

2-6a Formulas for Chemical Compounds

We have already used a few **chemical formulas**, combinations of the symbols for the elements that represent the stable combinations of atoms in molecules: H_2 for elemental hydrogen, H_2O for water, and $C_{12}H_{22}O_{11}$ for table sugar. The principle is the same no matter how many atoms are combined. The symbols represent the elements, and the **subscripts** (for example, the "2" in H_2O) indicate the relative numbers of atoms of each kind. The formulas and properties of a few simple molecular compounds are given in Table 2.2. Such formulas can represent one molecule or a large sample of the compound.

Sometimes a line or lines are drawn between symbols to indicate which atoms are connected in molecules:

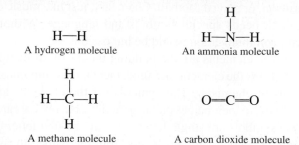

A hydrogen molecule   An ammonia molecule

A methane molecule   A carbon dioxide molecule

Formulas that show the connections in this way are known as **structural formulas**, whereas formulas that give just one symbol for each element present are called **molecular formulas**. The molecular formula for methane is CH_4, for example. Methane and the many millions of compounds that contain carbon combined with hydrogen—and often also with nitrogen, oxygen, phosphorus, or sulfur—are known as **organic compounds**.

Compounds not based on carbon are referred to as **inorganic compounds**. A few in everyday use include the following:

	NaCl	NH_4NO_3
Chemical name:	Sodium chloride	Ammonium nitrate
Use:	Table salt	Fertilizer
	H_2SO_4	$Mg(OH)_2$
Chemical name:	Sulfuric acid	Magnesium hydroxide
Use:	Battery acid	A laxative

When it is necessary to explore the shapes of molecules beyond what can be shown on a flat piece of paper, chemists resort to physical models or computer-drawn pictures, such as those shown in Figure 2.7.

TABLE 2.2 Some Simple Molecular Compounds

Compound	Formula	Properties
Water	H_2O	Odorless liquid
Carbon monoxide	CO	Odorless, flammable, toxic gas
Carbon dioxide	CO_2	Odorless, nonflammable, suffocating gas
Sulfur dioxide	SO_2	Nonflammable gas, suffocating odor
Ammonia	NH_3	Colorless, nonflammable gas, pungent odor
Methane	CH_4	Odorless, flammable gas
Carbon tetrachloride	CCl_4	Nonflammable, dense liquid
Nitrogen dioxide	NO_2	Reddish-brown gas, irritant

© Cengage Learning

FIGURE 2.7 The water molecule, as represented in print (a and b), in physical models that can be handled (c and d), and on a computer screen (e).

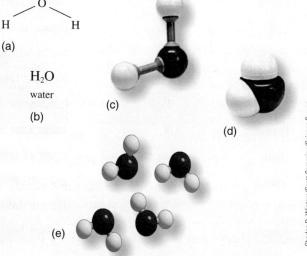

(a)

H_2O
water
(b)

(c)

(d)

(e)

Charles D. Winters/Scott Camazine/Science Source

EXAMPLE 2.1 Chemical Symbols

 Which of the following represent elements and which represent compounds?

(a) KI (b) Co (c) Ag (d) NO (e) Cl₂

SOLUTION

The chemical symbols in (a) and (d) represent compounds—the second letter in an element symbol is never a capital letter; (b) and (c) represent the elements cobalt and silver; and (e) represents the element chlorine, which exists as diatomic molecules (page 21).

TRY IT 2.1

Write the symbols or formulas for (a) lead, (b) phosphorus, (c) a molecule containing one hydrogen atom and one chlorine atom, (d) a molecule containing one aluminum atom and three bromine atoms, and (e) elemental fluorine.

2-6b Chemical Equations

A chemical reaction is a process in which one or more pure substances are transformed into one or more different pure substances. Here again, chemists use symbols to communicate. To concisely represent chemical reactions, symbols and formulas are arranged in **chemical equations**. For example, in words, a chemical reaction could be expressed as "carbon reacts with oxygen to form carbon monoxide," a reaction that happens whenever something containing carbon burns incompletely. Like most solid elements, carbon is represented just by its symbol, C; oxygen must be represented by the molecular formula for its diatomic molecules, O_2; and carbon monoxide molecules, which contain two atoms, one each of carbon and oxygen, are represented by their molecular formula, CO.

Although atoms are rearranged to form new substances in a chemical reaction, the number of atoms stays the same. To represent this correctly, we must have **balanced chemical equations**—the number of atoms of each kind in the reactants and products must be the same. In this case, one oxygen molecule will combine with two carbon atoms to form two carbon monoxide molecules. All of this information is contained in the following equation:

$$2\,C(s) + O_2(g) \longrightarrow 2\,CO(g)$$

We have also included here the information that carbon is a solid (s) and that oxygen and carbon monoxide are gases (g). The arrow ($\longrightarrow$) is often read as "yields," and represents the change that is taking place in going from the left of the equation (the reactant side, which can include elements or compounds, or both) to the right (the product side, which can also contain elements or compounds, or both). The equation then states the following:

At the macroscopic level: Carbon, a solid, plus oxygen gas yields carbon monoxide gas.

At the nanoscopic level: Two atoms of carbon plus one diatomic molecule of oxygen yield two molecules of carbon monoxide.

The number written before a formula in an equation, the **coefficient**, gives the relative amount of the substance involved. The subscripts give the composition of the pure substances. Changing the coefficient changes only the amount of the element or compound involved, whereas changing a subscript would change the identity of the reactant or product. For example, 2 CO represents two molecules of carbon monoxide, whereas CO_2 represents a molecule of carbon dioxide, a very different substance formed in the complete combustion of carbon-containing materials:

$$C(s) + O_2(g) \longrightarrow CO_2(g)$$

We'll talk more about the process of balancing a chemical equation in Chapter 8.

> **Chemical equation:** A representation of a chemical reaction by the formulas of reactants and products
>
> **Balanced chemical equation:** A chemical equation in which the total number of atoms of each kind is the same in reactants and products
>
> **Reactant:** A substance that undergoes chemical change
>
> **Product:** A substance produced as a result of chemical change
>
> **Coefficient:** In a chemical equation, a number written before a formula to balance the equation

EXAMPLE 2.2 Chemical Equations

 Nitrogen dioxide (NO_2) is a red-brown gas often visible in the haze during a period of air pollution over a city.

(a) Interpret in words the information given in the equation for formation of nitrogen dioxide from nitrogen monoxide (NO):

$$2\,NO(g) + O_2(g) \longrightarrow 2\,NO_2(g)$$

(b) Explain how the equation is balanced.

SOLUTION

(a) Nitrogen monoxide, a gas, reacts with oxygen gas to yield nitrogen dioxide, also a gas. The coefficients show that two molecules of NO react with one molecule of O_2 to give two molecules of NO_2.

(b) *Reactants:* The coefficient 2 in 2 NO shows the presence of two molecules of NO, which means two N atoms and two O atoms. The O_2 has no coefficient; thus, there is one O_2 molecule, but the subscript shows that the single molecule has two O atoms. Therefore, there are two N atoms and four O atoms on the reactants side. *Products:* The coefficient in $2\ NO_2$ shows two molecules of NO_2, which gives a total of two N atoms (N has no subscript) and four O atoms (two in each of the two molecules, as shown by the subscript on O). The equation is balanced.

TRY IT 2.2A

Write a balanced equation, including the state designations, for the reaction of hydrogen gas with oxygen gas to yield water.

TRY IT 2.2B

Repeat the question of Example 2.2 for the equation for the formation of hydrogen chloride (HCl):

$$H_2(g) + Cl_2(g) \longrightarrow 2\ HCl(g)$$

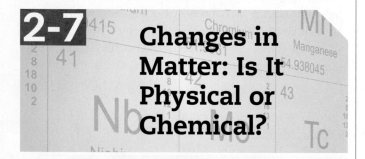

2-7 Changes in Matter: Is It Physical or Chemical?

Mass: A measure of the quantity of matter in an object

Physical property: A property that can be observed without changing the identity of a substance

Density: Mass per unit volume

Chemical property: A property that is observable in chemical reactions and changes in the identity of one or more reactants

Sulfur is yellow; iron is magnetic; water boils at 100°C. These are different properties of matter, but with something in common. We can observe the color of a substance, pick it up with a magnet, measure its boiling point, or determine its **mass**, its quantity of matter, without changing its chemical identity. Such properties are classified as **physical properties**. In chemistry, the word *physical* is used to refer to processes that do not change chemical identities.

Separating sulfur and iron with a magnet is a *physical* separation. Boiling water is a *physical* change—the steam that forms is still water (H_2O).

Whenever possible, the physical properties of substances are pinpointed by making numerical measurements. You might describe lead as heavier than aluminum—if you pick up same-sized pieces of these metals, lead will definitely feel heavier. However, the difference is better represented by giving the numerical value of the **density** of each metal, which takes into account the need to compare pieces of equal size—that is, of equal volume. One cubic centimeter (cm^3) of lead weighs 11.3 grams (g), and 1 cubic centimeter of aluminum weighs 2.7 g. Lead is *denser* than aluminum. Stated in the usual way:

	Densities (at 20°C)
Lead	11.3 g/cm³
Aluminum	2.7 g/cm³

By contrast to physical changes such as boiling, there are processes that do result in changes of identity, and these are chemical changes. When gasoline burns, it is converted to a mixture of carbon dioxide, carbon monoxide, and water. Burning in air, a property that gasoline, kerosene, and similar substances have in common, is classified as a **chemical property**.

As an adjective, the word *chemical* is used to describe processes that result in a change in identity. A chemical reaction produces a new arrangement of atoms. The number and kinds of atoms in the reactants and products remain the same, but the reactants and products are different substances that can be recognized by their different properties. To distinguish between a hydrogen–oxygen mixture and a compound composed of hydrogen and oxygen, we might say that in water, hydrogen and oxygen are *chemically* combined. The meaning is that the hydrogen and oxygen atoms are held together strongly enough to form the individual units we call water molecules.

Some easily observed results of chemical reactions are the rusting of iron, the change of leaf color in the fall, and the formation of carbon dioxide bubbles by an antacid tablet. Often, though not always, the occurrence of a chemical reaction can be detected because there is some observable change (Figure 2.8).

The word *chemical* is also commonly used as a noun. In this sense, every substance in the universe is a chemical. You might read of polluted waters that contain chemicals. Used this way, the word really stands for *chemical compounds*. The chemical compounds in polluted water may be unhealthful or harmful. Anyone who has studied even

FIGURE 2.8 Observable indications of chemical reactions.
Chances are that a chemical reaction has occurred when (a) a flame is visible, (b) mixing substances produces a glow, (c) a solid appears when two solutions are mixed, or (d) there is a color change.

(a)   (b)   (c)   (d)

© Jeffrey L. Rotman/CORBIS/© Julian Smith/ Corbis/Charles D. Winters/© iStockphoto.com/ alexkotlov/© iStockphoto.com/belterz

a little chemistry understands that useful substances—penicillin, table salt, nylon fabrics, laundry detergents, and all natural herbs and spices—are also made of chemical compounds. So are all plants and animals. Pure water is also a chemical compound.

Some physical changes and almost all chemical reactions are accompanied by changes in energy. Frequently, the energy is taken up or released in the form of heat. Heat must be added for the physical changes of melting or boiling to occur. Heat is released when the chemical reaction of combustion, or burning, takes place. But many chemical reactions, such as those of photosynthesis, require energy from their surroundings in order to take place.

Energy is the ability to cause change or, in the formal terms of physics, to do work. Energy in storage, waiting to be used, is **potential energy**. There is potential energy in gasoline, known as *chemical energy*—it is released as heat and light when gasoline burns. Chemical energy can also be released in the form of electrical energy or mechanical energy (Table 2.3). Energy in use, rather than in storage, is **kinetic energy**, the energy associated with motion.

The diver has potential energy because of his position above the surface of the water.

As the diver falls, potential energy is converted to kinetic energy.

© Cengage Learning

> **Energy:** The capacity for doing work or causing change
>
> **Potential energy:** Energy in storage by virtue of position or arrangement
>
> **Kinetic energy:** The energy of objects in motion

2-8 The Quantitative Side of Science

Several times in the preceding sections, we used the numerical results of measurements of the boiling points, masses, or densities of pure substances. These and hundreds of other kinds of measurements are fundamental to chemistry and every other science. The result of a measurement is what

TABLE 2.3 Some Examples of Conversion of Chemical Potential Energy to Other Forms of Energy

Conversion to—	Electrical Energy	Heat	Light	Mechanical Energy
Done by—	Batteries	Combustion	Burning candle, logs in fireplace	Rocket
	Fuel cells	Digestion of food	Luminescence in firefly	Animal muscles
		Many kinds of chemical reactions	Luminescence from chemical reaction	Dynamite explosion

© Cengage Learning

we refer to as **quantitative** information—it uses numbers. Weighing yourself is a quantitative experiment. By contrast, there is **qualitative** information that does not deal with numbers. For example, sticking your hand into a tub of water and determining it to be "hot" is a qualitative observation.

The result of a measurement is recorded as a number plus a unit. "A boiling point of 52" doesn't mean anything without the unit. Was the temperature measured in Fahrenheit degrees, as is done by the usual household thermometer in the United States, or in Celsius degrees, as would be done in the rest of the world?

2-8a Unit Conversions

Being able to express a measured quantity using units different from those obtained in the initial measurement can be very useful. The process by which this can take place is often called **unit conversion**. To convert between units of measurement, you need to know some equalities beforehand. An **equality** is a way of expressing one quantity using two different units.

For example, consider what would happen if you ran a bakery and needed 48 eggs to complete the day's business: you would go to the store and purchase four dozen eggs. Perhaps you would do this almost without thinking . . . but how is this the same as unit conversion? The answer to this question is that you were aware of an important equality that relates the number of individual eggs to a dozen eggs:

$$12 \text{ eggs} = 1 \text{ dozen eggs}$$

Mathematically, this can also be written as a **conversion factor**, which is a ratio in which the numerator refers to the same quantity as the denominator. Another way of saying this is that a conversion factor is a ratio in which the numerator and denominator are the two measurements that refer to the same amounts. A conversion factor always holds true, even if it is inverted, since the amounts are the same. One could say "there are 1 dozen eggs for every 12 eggs" or "there are 12 eggs for every dozen eggs":

$$\frac{1 \text{ dozen eggs}}{12 \text{ eggs}} \quad \text{OR} \quad \frac{12 \text{ eggs}}{1 \text{ dozen eggs}}$$

To use a conversion factor, begin with the initial quantity, expressed in the original units. Then, multiply by the conversion factor that has the original unit in the denominator. When you carry out the calculation, you'll notice that one of the units is canceled, leaving the new unit of measurement:

$$48 \cancel{\text{ eggs}} \times \frac{1 \text{ dozen eggs}}{12 \cancel{\text{ eggs}}} = 4 \text{ dozen eggs}$$

The same equality can be used to convert dozens of eggs into the number of individual eggs. If you knew instead that you required four dozen eggs:

$$4 \cancel{\text{ dozen eggs}} \times \frac{12 \text{ eggs}}{1 \cancel{\text{ dozen eggs}}} = 48 \text{ eggs}$$

If an equality needed for unit conversion is not known, it is possible that more than one equality can be used. Consider, for example, the following equalities, which can be used to convert a measurement of 5 inches to one in yards:

$$12 \text{ inches} = 1 \text{ foot}$$
$$3 \text{ feet} = 1 \text{ yard}$$

We can calculate

$$5 \cancel{\text{ inches}} \times \frac{1 \cancel{\text{ foot}}}{12 \cancel{\text{ inches}}} \times \frac{1 \text{ yard}}{3 \cancel{\text{ feet}}} = 0.14 \text{ yard}$$

Take a moment to be sure you understand how units are included and canceled in this unit conversion. An excellent way to keep track of what you are doing is to include units in setting up the problem and then to cancel them like numbers. You will find that this method applies to far more than unit conversion or the mathematics of chemistry. By including the units with numbers, you can often figure out what is needed to find a mathematical answer; when the units cancel to give the desired unit with the answer, chances are that the answer will be the right one (if, of course, your arithmetic is correct).

The establishment of scientific facts and laws is dependent on accurate observations and measurements. Although measurements can be reported as precisely in one system of measurement as another, there has been an effort since the time of the French Revolution in the late 1700s to have all scientists embrace the same simple system. The hope was and is to facilitate communication. The metric system, which was born of this effort, has two advantages. First, it is easy to convert from one unit to another, since smaller and larger units for the same physical quantity differ only by multiples of 10. Consequently, to change millimeters to meters, the decimal point need only be moved three places to the left:

TABLE 2.4 Common Prefixes for Multiples and Fractions of SI Units

Prefix	Abbreviation	Meaning	Example
mega-	M	10^6 (1 million)	1 megaton $= 1 \times 10^6$ tons
kilo-	k	10^3 (1 thousand)	1 kilogram (kg) $= 1 \times 10^3$ g
deci-	d	10^{-1} (1 tenth)	1 decimeter (dm) $= 0.1$ m
centi-	c	10^{-2} (1 one-hundredth)	1 centimeter (cm) $= 0.01$ m
milli-	m	10^{-3} (1 one-thousandth)	1 millimeter (mm) $= 0.001$ m
micro-	μ*	10^{-6} (1 one-millionth)	1 micrometer (μm) $= 1 \times 10^{-6}$ m
nano-	n	10^{-9} (1 one-billionth)	1 nanometer (nm) $= 1 \times 10^{-9}$ m
pico-†	p	10^{-12} (1 one-trillionth)	1 picometer (pm) $= 1 \times 10^{-12}$ m

© Cengage Learning

* This is the Greek letter mu (pronounced "mew").

† This prefix is pronounced "peako."

$$1 \text{ millimeter} = 0.001 \text{ meter} = 1 \times 10^{-3} \text{ meter}$$

$$\text{therefore, } 5.0 \text{ millimeters} = 0.0050 \text{ meter}$$

$$= 5.0 \times 10^{-3} \text{ meter}$$

Compare this to the problem of changing inches to yards. Notice, too, that we've also included the use of scientific notation. For a review of this method of representing very small or very large numbers, see the significant figures and scientific notation review card at the end of this text.

The second advantage of the modern metric system is that standards for most fundamental units are defined by reproducible phenomena of nature. For example, the metric unit for time—the second—is now defined in terms of a specific number of cycles of radiation from a radioactive cesium atom, a time period believed never to vary.

The units still in everyday use in the United States have evolved from the English system of measurement, which began with the decrees of various English monarchs. The *yard* started out as the length of the waist sash worn by Saxon kings and obviously varied with the girth of the king. A step toward standardization came when King Henry I decreed that the yard should be the distance from the tip of his nose to the end of his thumb. Standards for the English units in use in the United States today are designated by the National Institute of Standards and Technology (NIST).

Since its introduction in 1790, the metric system has been continually modified and improved. The current version—the International System of Units, abbreviated SI (from the French, Système International d'Unités)—was adopted by the International Bureau of Weights and Measures in 1960. The SI system defines seven fundamental units from which other units are derived. For example, the fundamental SI unit for length is the meter (abbreviated m), which then gives the SI unit for area as the square meter

(m²) and the SI unit for volume as the cubic meter (m³). Multiples and fractions of SI units are designated by adding prefixes (Table 2.4). Therefore, when a length unit smaller than the meter is needed, we can choose the *centi*meter (cm), the *milli*meter (mm), or the *nano*meter (nm); when a longer length unit is needed, we can use the *kilo*meter (km). For mass units, prefixes are added to gram, such as the *kilo*gram (kg), the *milli*gram (mg), the *micro*gram (μg, where μ is the Greek letter mu), or the *nano*gram (ng).

Because of their convenient sizes, the five units listed in Table 2.5 are commonly used in chemistry and indeed are in everyday use in the rest of the world. We will use the unit symbols listed in Table 2.5 in this text. The volume unit of liters is preferable to the SI-derived unit of cubic meters, which is much too large (1 cubic meter = 1000 liters).

TABLE 2.5 Some Common Units in Chemistry

Name of Unit	Symbol	Common Equivalent
meter	m	39.4 inches
liter	L	1.06 quarts
gram	g	0.0352 ounce
degree Celsius	°C	Pure water boils at 100°C and freezes at 0°C
calorie	cal	Energy required to raise the temperature of 1 g of water by 1°C

© Cengage Learning

EXAMPLE 2.3 Metric Unit Prefixes

 How many meters are in 2 km?

SOLUTION

The prefix *kilo-* (k) means 1000 times, meaning that 1000 m = 1 km. It is hardly worth the trouble to write anything down in this solution. One just thinks 1000 m for every kilometer as one thinks 10 dimes for every one dollar bill. The best way to write this problem, or any more complicated problem, however, is to include the units and cancel them out (like numbers in algebra). This way, if you are left with the correct unit, you have probably done the problem correctly.

$$2 \text{ km} \times \frac{1000 \text{ m}}{1 \text{ km}} = 2000 \text{ m}$$

TRY IT 2.3

How much in dollars is 20 *mega*bucks?

Your Resources

Located at back of the textbook

→ *Review Card on The Chemical View of Matter*

In OWL for CHEM2 at www.cengagebrain.com

→ *Review Key Terms with Flash Cards (printable or digital)*

→ *Complete Interactive Practice Quizzes to prepare for tests*

→ *Submit Assigned Homework and Exercises*

Applying Your Knowledge

1. Name as many materials as you can that you have used during the past day that were not chemically changed from their natural states.

2. Identify the following as physical or chemical changes. Explain your choices.
   (a) Formation of snowflakes
   (b) Rusting of a piece of iron
   (c) Ripening of fruit
   (d) Fashioning a table leg from a piece of wood
   (e) Fermenting grapes
   (f) Boiling a potato

3. Would it be possible for two pure substances to have exactly the same set of properties? Give reasons for your answer.

4. List three physical properties that can be used to identify pure substances. Give a specific example of each property.

5. Nitroglycerine, the chemical compound that led to Alfred Nobel's fortune, is also used for the treatment of heart conditions. What does this illustrate about the versatility

of many chemical compounds? Are the risks and benefits associated with the use of nitroglycerin as an explosive the same as for using it to treat angina, a heart condition?

6. Classify each of the following as a physical property or a chemical property. Explain your answers.
   (a) Density
   (b) Melting temperature
   (c) Decomposition of a substance into two elements on heating
   (d) Electrical conductivity
   (e) The failure of a substance to react with sulfur
   (f) The ignition temperature of a piece of paper

7. Chemical changes can be both useful and destructive to humanity's purposes. Cite a few examples of each kind of change from your own experience. Also give evidence from observation that each is indeed a chemical change and not a physical change.

8. Classify each of the following as an element, a compound, or a mixture. Justify each answer.
   (a) Mercury
   (b) Milk
   (c) Pure water
   (d) A piece of lumber
   (e) Ink
   (f) Iced tea
   (g) Pure ice
   (h) Carbon
   (i) Antimony

9. Use the periodic table review card for help in answering this question. What is the state (solid, liquid, gas) in which each of the following elements is commonly encountered?
   (a) tin
   (b) bromine
   (c) dysprosium
   (d) xenon
   (e) samarium
   (f) lithium
   (g) mercury
   (h) iodine

10. Are the following observations examples of chemical or physical changes?
    (a) The antifreeze in a car does not function below −40°F.
    (b) Paper bursts into flame at high temperature.
    (c) Bubbles of gas form when vinegar and baking soda are combined.
    (d) A layer of ice forms on the surface of a pond during the winter.
    (e) Your body digests food.

11. This question may seem a bit odd, but it will help you learn the symbols of the elements and their names. Take each of the following common words, identify the elemental symbols that can be combined to spell the words, and then write out the full names of the elements. Example: The word *chosen* can be spelled either C-H-O-Se-N or C-Ho-Se-N: carbon-hydrogen-oxygen-selenium-nitrogen *or* carbon-holmium-selenium-nitrogen.
    (a) sine
    (b) cry
    (c) virus (two possibilities)
    (d) resistance (two possibilities)
    (e) crossbow (four possibilities)
    (f) fender
    (g) accuse (two possibilities)

12. Write your last name. How many element symbols can you produce from the letters in your name?

13. Choose a common word and try to spell it using elemental symbols. This may not be possible in all cases (only 27,000 words can be spelled using elemental symbols). Which symbols or letters of the alphabet would you need to complete your word?

14. As was pointed out in Section 2-5, many of the elements have symbols based on older names in languages other than English. Many other elements are named for people, qualities, origins, or geographic locations. For each of the following elemental symbols, provide the name of the element and use this text or another resource to learn about the origin of the name.
    (a) Md
    (b) K
    (c) Cf
    (d) Bh
    (e) Ir
    (f) Y
    (g) Cm

15. The elements in the following sets of elemental names or symbols are often confused. Provide the names of the elements, or the symbols for the elements.
    (a) copper, copernicium
    (b) copper, chromium, cerium
    (c) tungsten, titanium, tin
    (d) thallium, thorium
    (e) nitrogen, nickel
    (f) C, Ca
    (g) iron, fluorine
    (h) nitrogen, nickel, neon

16. Is it possible for the properties of iron to change? What about the properties of steel, which is an alloy and a homogeneous mixture? Explain your answer.

17. An element is the smallest part of a compound that may still be identified as the compound.
    (a) True
    (b) False

18. An atom is the smallest part of an element that may still be identified as the element.
    (a) True
    (b) False

19. An atom is the smallest part of a compound that may still be identified as the compound.
    (a) True
    (b) False

20. A pure compound may be broken down into two or more different kinds of pure elements.
    (a) True
    (b) False

21. You have a mixture of sand (SiO_2) and salt (NaCl). How would you separate these two substances? When you have them as separate substances, how can you prove which is which?

22. Consider the following five elements: nitrogen, sulfur, chlorine, magnesium, and cobalt. By using this text or any other source available at the library, find the major source for these elements and at least one compound that uses the element in combined form.

23. Atrazine is a selective herbicide that has the molecular formula $C_8H_{14}N_5Cl$. This compound is used for season-long weed control in corn, sorghum, and certain other crops. What elements are present in atrazine?

24. Cytoxan, also known as cyclophosphamide, is widely used alone or in combination in the treatment of certain kinds of cancer. It interferes with protein synthesis and in the process kills rapidly replicating cells, particularly malignant ones. Cytoxan has the molecular formula $C_7H_{15}O_2N_2PCl_2$.
    (a) How many atoms are in one molecule of Cytoxan?
    (b) What elements are present in Cytoxan?
    (c) What is the ratio of hydrogen atoms to nitrogen atoms in Cytoxan?
    (d) Would Cytoxan be classified as an organic compound?

25. By reading labels, identify a commercial product that contains each of the following compounds:
    (a) Calcium carbonate
    (b) Phosphoric acid
    (c) Water
    (d) Fructose
    (e) Sodium chloride
    (f) Potassium sorbate
    (g) Potassium iodide
    (h) Glycerol
    (i) Aluminum
    (j) Butylated hydroxytoluene (BHT)

26. There are three common states of matter: gas, liquid, and solid. Name a material that is a pure substance for each state of matter. Do not use water, oxygen, or salt. Name a material that is a mixture for each state of matter. Do not use air, gasoline, or brass.

27. Which of the following statements is true of a balanced chemical equation?
    (a) The total number of molecules of reactants has to be the same as the total number of product molecules.
    (b) The total number of atoms in the reactant molecules doesn't have to equal the total number of atoms in the product molecules.
    (c) The identity of the atoms in the reactants has to be the same as the identity of the atoms in the products.
    (d) The physical state of the reactants has to be the same as the physical state of the products.

28. Is it possible to have a mixture of two elements and also to have a compound of the same two elements? Explain. Can you think of an example?

29. Given the following sentence, write a chemical equation using chemical symbols that convey the same information: "One nitrogen molecule reacts with three hydrogen molecules to produce two ammonia molecules, each containing one nitrogen and three hydrogen atoms."

30. Chemical equations are a shorthand way of describing how species known as _____ are converted into _____.
    (a) reactants; solvents
    (b) products; solutes
    (c) solutes; solutions
    (d) reactants; products

31. Name four kinds of energy.

32. Describe in words the chemical processes represented by the following equations:
    (a) $2\,Na(s) + Cl_2(g) \longrightarrow 2\,NaCl(s)$
    (b) $N_2(g) + 3\,Cl_2(g) \longrightarrow 2\,NCl_3(g)$ [named as nitrogen trichloride]
    (c) $CO_2(g) + H_2O(\ell) \longrightarrow H_2CO_3(aq)$ [carbonic acid]
    (d) $2\,H_2O_2(aq)$ [hydrogen peroxide] $\longrightarrow O_2(g) + 2\,H_2O(\ell)$

33. Prove that each of the equations in question 32 is balanced.

34. For equations **b** and **d** in question 32, identify the reactant(s) and the product(s).

35. Are the following equations balanced?
    (a) $AgNO_3(aq) + Na_2SO_4(aq) \longrightarrow Ag_2SO_4(s) + NaNO_3(aq)$
    (b) $AgNO_3(aq) + HCl(aq) \longrightarrow AgCl(s) + HNO_3(aq)$

36. Write a chemical equation for the following descriptions:
    (a) Two atoms of potassium metal react with two molecules of liquid water to produce two molecules of solid potassium hydroxide (KOH) and one molecule of hydrogen gas.
    (b) One molecule of carbon dioxide gas reacts with one molecule of liquid water to produce one molecule of liquid carbonic acid (H_2CO_3).

37. Is the tea in tea bags a pure substance? Use the process of making tea to make an argument for your answer. How would your argument apply to instant tea?

38. Find and list as many pure substances as you can in a kitchen.

39. (a) How many milligrams are there in 1 g?
    (b) How many meters are there in 1 km?
    (c) How many centigrams are there in 1 g?

40. What are the most common units in chemistry for mass, length, and volume?

41. Which of the following quantities is a density?
    (a) 9 cal/g   (b) 100 cm/m
    (c) 1.5 g/mL   (d) 454 g/lb

42. A cook wants to pour 1.5 L of batter into a 2-quart bowl. Will it fit?

43. A rancher needs one acre of grazing land for 10 cows. How many acres are needed for 55 cows? Solve this problem (and others where appropriate) by including units. In this case, the answer will have the "units" of acres per cow.

44. There are 200 mg of ibuprofen in an Advil tablet. How many grams is this? How many micrograms?

45. How many meters does a runner cover in a 10-km race?

46. If a 1-ounce portion of cereal contains 3.0 g of protein, how many milligrams of protein does it contain?

47. Complete the following:
    (a) 4 cm = _____ m
    (b) 0.043 g = _____ mg
    (c) 15.5 m = _____ mm
    (d) 328 mL = _____ L
    (e) 0.98 kg = _____ g

48. An average African gorilla has a mass of 163 kg. What is this mass in grams?

49. An average adult man has a mass of 70 kg. Convert this mass to milligrams.

50. An aspirin tablet usually contains 325 mg of aspirin. What is this in grams?

51. What are the results for the following operations in exponential notation? You may need to refer to the periodic table review card for questions 51 and 52.
    (a) $(4.0 \times 10^5) (2.0 \times 10^2)$
    (b) $(6.0 \times 10^7)/(2.0 \times 10^2)$
    (c) $(7.4 \times 10^4)/(4.6 \times 10^9)$
    (d) $(2.3 \times 10^{-3}) (4.2 \times 10^4)$

52. Express the following in exponential (scientific) notation:
    (a) 8,000,000   (b) 0.000075
    (c) 23,600,000,000   (d) 37,000
    (e) 6492   (f) 0.000000028

53. An electric power plant produces 450 megawatts.
    (a) What is this expressed in watts?
    (b) How many 100-watt lightbulbs will this light?

54. A computer has a 750-gigabyte hard drive. How many bytes is this?

55. What is the mass in grams for a 1000-mL volume (1 liter) of iridium if it has a density of 22.42 g/mL?

56. Lead has a density of 11.4 g/cm³. What is the mass in grams of a cube of lead that has the dimensions 10 cm × 10 cm × 10 cm?

57. The densities for aluminum and iron are 2.7 g/cm³ and 7.86 g/cm³, respectively. What is the ratio of the mass of an object made of aluminum to that of the same object made of iron?

58. There are 8 fluid ounces in 1 cup, and there are 2 cups in 1 pint. Write the equalities described, then use unit conversion to calculate the number of fluid ounces in 1 pint.

59. There are 8 fluid ounces in one cup, 2 cups in 1 pint, 2 pints in 1 quart, and 4 quarts in 1 gallon. Write the equalities described, then use unit conversion to calculate the number of fluid ounces in 1 gallon.

60. One kilometer is equal to 0.621 miles. There are 5280 feet in 1 mile. How many feet are there in 0.5 kilometer?

61. One system of measurement used to describe amounts of pharmaceutical compounds uses units of grains, drams, and scruples. Some equivalencies from this system are shown here:

    20 grains = 1 scruple

    3 scruples = 1 dram

    8 dram = 1 ounce

    1 dram = 3.888 g

    If an aspirin tablet contains 5.01×10^2 mg of aspirin, how many grains of aspirin does the tablet contain?

3

Atoms and
the Periodic Table

W hy does an element or compound have the properties it has? Why does one element or compound undergo a change that another element or compound will not undergo? Inanimate matter behaves the way it does because of the nature of its parts. The use of atoms to represent these parts dates back to about 400 BCE, when the Greek philosopher Leucippus and his student Democritus (460–370 BCE) argued for a limit to the divisibility of matter, which was counter to the prevailing view of Greek philosophers that matter is endlessly divisible. Democritus used the Greek word *atomos*, which literally means "uncuttable," to describe the ultimate particles of matter, particles that could not be divided further. However, it wasn't until John Dalton (1766–1844) introduced his atomic theory in 1803 that the importance of using atoms to explain properties of matter was recognized.

Dalton's atomic theory and the development of the periodic table by Mendeleev in 1869 led to the rapid growth of chemistry as a science. In particular, the influence of the location and number of electrons in atoms on the properties of elements has become one of the essential ideas of chemistry. In this chapter, we will use current knowledge about the atom together with the periodic table as the basis for our understanding of the chemical view of matter.

3-1 John Dalton's Atomic Theory

John Dalton, drawing from his own quantitative experiments and those of earlier scientists, proposed the following in 1803:

1. All matter is made up of indivisible and indestructible particles called atoms.

2. All atoms of a given element are identical, both in mass and in properties. By contrast, atoms of different elements have different masses and different properties.

3. Compounds form when atoms of different elements combine in ratios of small whole numbers.

4. Elements and compounds are composed of definite arrangements of atoms. Chemical change occurs when the atomic arrays are rearranged.

5. In chemical reactions, atoms are combined, separated, or rearranged.

Although it has been modified over time, the main principles of this theory hold true today. John Dalton's atomic theory was accepted because it could be used to explain several scientific laws that Dalton and other scientists of the time had established. Two of these laws are (1) the law of conservation of matter and (2) the law of definite proportions.

Some years before Dalton proposed his atomic theory, Antoine Lavoisier (1743–1794) had carried out a series of experiments in which the reactants were carefully weighed before a chemical reaction and the products were carefully weighed afterward. He found no change in mass when a reaction occurred, proposed that this was true for every reaction, and called his proposal the **law of conservation of matter**: *Matter is neither lost nor gained during a chemical reaction.* Others verified

Law of conservation of matter: Matter is neither lost nor gained during a chemical reaction

The "Cloud Gate" sculpture in Chicago was inspired by the structure of a droplet of mercury.

his results, and the law became accepted. The second and fourth points in Dalton's theory imply the same thing: If each kind of atom has a particular characteristic mass, and if there are exactly the same numbers of each kind of atom before and after a reaction, the masses before and after must also be the same.

Another chemical law known in Dalton's time had been proposed by Joseph Louis Proust (1754–1826) as a result of his analyses of minerals. Proust found that a particular compound, once purified, always contained the same elements in the same ratio by mass. One such study, which Proust made in 1799, involved copper carbonate. Proust discovered that regardless of how copper carbonate was prepared in the laboratory or how it was isolated from nature, it always contained the same proportions by mass— five parts copper, four parts oxygen, and one part carbon. Careful analyses of this and other compounds led Proust to propose the **law of definite proportions**: *In a compound, the constituent elements are always present in a definite proportion by mass.* For example, pure water, a compound, is always made up of 11.2% hydrogen and 88.8% oxygen by weight. Pure table sugar, another compound, always contains 42.1% carbon, 6.5% hydrogen, and 51.4% oxygen by weight. The source of the pure substance is irrelevant.

dropped object falls downward, not up. A **scientific model** is something that represents a part of the world around us. Examples of scientific models can include a small-scale rocket that mimics a larger design, or the mathematical equation that describes a circle.

Many important scientific theories and models have been developed by reducing an object of study into its constituent pieces, then identifying the pieces and learning how they fit together. This is an approach that is as useful in chemistry as it is in biology, physics, accounting, engineering, and psychology. For example, an engineer seeking to understand ways in which to make cars safer during collisions may first begin by crashing the car into a wall (in a laboratory!), then looking at the effect of the crash on the individual pieces that make up the vehicle.

In this section, we'll see just how it is that our model of atomic structure came about from taking a close look at the pieces formed when an atom is broken into its subatomic parts.

Henri Becquerel (1852–1908) discovered natural radioactivity in uranium and radium ores in 1896 (Section 13-1). In 1898, Marie Sklodowska Curie (1867–1934)—a student of Becquerel and the first person honored with two Nobel Prizes—and her husband, Pierre, discovered two radioactive elements, *radium* and *polonium*. In 1899, Marie Curie suggested that atoms of radioactive substances disintegrate when they emit these unusual rays. She named this phenomenon **radioactivity**. A given radioactive element gives off exactly the same type of radioactive particles or rays regardless of

3-2 Structure of the Atom

Almost 95 years after Dalton's atomic theory was published, it became apparent that there are particles of matter smaller than an atom. The discovery of these subatomic particles was a result of research into a phenomenon that was unknown to Dalton: radioactivity.

3-2a Natural Radioactivity

A **scientific theory** is a set of ideas that, together, aid in understanding the behavior of the world around us. One example of a scientific theory with which you are familiar is gravity; because of this theory, we understand why a

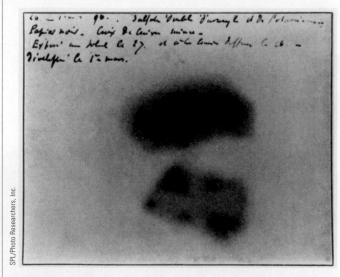

Becquerel discovered radioactivity by noticing that uranium ore could be used to expose a photographic plate in the absence of sunlight. The top exposure is from a sample of ore; the lower exposure is from a sample of ore placed on top of a Maltese cross.

whether it is found in its pure state or combined with other elements. About 25 elements exist only in radioactive forms. Marie Curie's suggestion that atoms disintegrate contradicted Dalton's idea that atoms are indivisible.

Ernest Rutherford (1871–1937) began studying the radiation emitted from radioactive elements soon after experiments by the Curies and others had shown that three types of radiation may be spontaneously emitted by radioactive elements (Section 13-1). These are referred to as **alpha** (α) **particles**, **beta** (β) **particles**, and **gamma** (γ) **rays**. They behave differently when they pass between electrically charged plates, as shown in Figure 3.1. Alpha and beta particles are deflected, while gamma rays pass straight through undeflected. This implies that alpha and beta particles are electrically charged particles, because particles with a charge would be attracted or repelled by the charged plates. Even though an alpha particle has an electrical charge (+2) twice as large as that of a beta particle (−1), alpha particles are deflected less; hence, alpha particles must be heavier than beta particles. Careful studies by Rutherford showed that alpha particles are equivalent to helium atoms that have lost two electrons and thus have a +2 charge (He^{2+}). Beta particles were found to be negatively charged particles identical to electrons. Gamma rays have no detectable charge or mass—they behave like rays of light.

3-2b The Nucleus of the Atom

Rutherford's experiments with alpha particles led him to consider using them in experiments on the structure of the atom. In 1909, he suggested to two of his coworkers that they bombard a piece of gold foil with alpha particles. Hans Geiger (1882–1945), a German physicist, and Ernest Marsden (1889–1970), an undergraduate student, set up an apparatus like that diagrammed in Figure 3.2 and observed what happened when alpha particles hit the thin gold foil. Most passed straight through, but Geiger and Marsden were amazed to find that a few alpha particles were deflected through large angles, and some came almost straight back. Rutherford later described this unexpected result by saying, "It was about as incredible as if you had fired a 15-inch [artillery] shell at a piece of paper and it came back and hit you."

What allowed most of the alpha particles to pass through the gold foil in a rather straight path? According to Rutherford's interpretation, the atom is mostly empty space and therefore offers little resistance to the alpha particles (Figure 3.2).

What caused a few alpha particles to be deflected? According to Rutherford's model of the atom, all of the positive charge and most of the mass of the atom must be concentrated in a very small volume at the center of the atom. He named this part of the atom, which contained most of the

> **Alpha (α) particle:** A positively charged particle emitted by certain radioactive isotopes
>
> **Beta (β) particle:** An electron ejected at high speeds from the nuclei of certain radioactive isotopes
>
> **Gamma (γ) ray:** A high-energy electromagnetic radiation emitted from radioactive isotopes

Marie Sklodowska Curie

© Photo Researchers/Alamy

FIGURE 3.1 Separation of alpha and beta particles and gamma rays by an electric field.

Alpha particles are deflected toward the negative plate, beta particles are attracted toward the positive plate, and gamma rays are not deflected. Additional studies showed that alpha particles are high-energy helium nuclei, beta particles are high-energy electrons, and gamma rays are high-energy electromagnetic radiation.

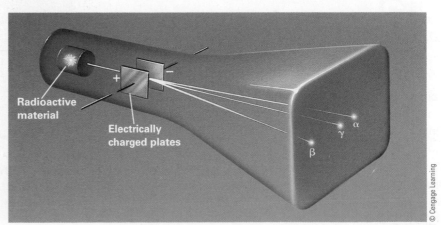

Radioactive material

Electrically charged plates

© Cengage Learning

FIGURE 3.2 Rutherford's scattering experiment (left) and the model of atomic structure he proposed (right).

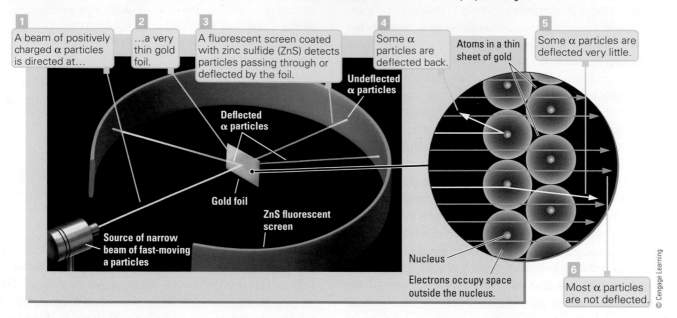

① A beam of positively charged α particles is directed at...

② ...a very thin gold foil.

③ A fluorescent screen coated with zinc sulfide (ZnS) detects particles passing through or deflected by the foil.

④ Some α particles are deflected back.

Atoms in a thin sheet of gold

⑤ Some α particles are deflected very little.

Undeflected α particles

Deflected α particles

Gold foil

ZnS fluorescent screen

Source of narrow beam of fast-moving α particles

Nucleus

Electrons occupy space outside the nucleus.

⑥ Most α particles are not deflected.

© Cengage Learning

mass of the atom and all of the positive charge, the nucleus. When an alpha particle passes near the nucleus, the positive charge of the nucleus repels the positive charge of the alpha particle; the path of the smaller alpha particle is consequently deflected. The closer an alpha particle comes to a target nucleus, the more it is deflected. Those alpha particles that meet a nucleus head-on bounce back toward the source as a result of the strong positive–positive repulsion, since the alpha particles do not have enough energy to penetrate the nucleus.

Rutherford's calculations, based on the observed deflections, indicated that the nucleus is a very small part of an atom, and an important scientific model of atomic structure could be proposed: *the diameter of an atom is about 100,000 times the diameter of its nucleus.*

Truly, Rutherford's model of the atom was one of the most dramatic interpretations of experimental evidence to come out of this period of significant discoveries.

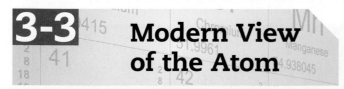

3-3 Modern View of the Atom

Scientists now have experimental evidence for the existence of more than 60 subatomic particles, which are small portions of matter that combine to form atoms. However, only three of these are important to our understanding of the chemical view of matter: protons, neutrons, and electrons.

Early experiments on the structure of the atom clearly showed that the three primary constituents of atoms are electrons, protons, and neutrons. The nucleus, or core, of the atom is made up of protons with a positive electrical charge and neutrons with no charge. The electrons, with a negative electrical charge, are found in the space surrounding the nucleus (Figure 3.3). The properties of these small particles, of which atoms are made, are shown in Table 3.1. For an atom, which always has no net electrical charge, *the number*

FIGURE 3.3 Model of an atom.

All atoms consist of one or more protons (positively charged) and usually at least as many neutrons (no charge) packed into an extremely small nucleus. Electrons (negatively charged) are arranged in a cloud around the nucleus.

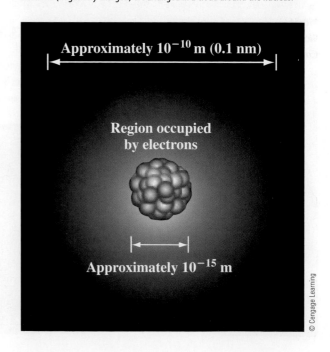

Approximately 10^{-10} m (0.1 nm)

Region occupied by electrons

Approximately 10^{-15} m

© Cengage Learning

TABLE 3.1 Summary of Properties of Electrons, Protons, and Neutrons

	Relative Charge	Mass (g)	Mass (amu)*	Approximate Relative Mass (amu)*	Location
Electron	−1	0.000911×10^{-24}	0.00055	~0	Outside the nucleus
Proton	+1	1.672623×10^{-24}	1.00727	~1	Nucleus
Neutron	0	1.674929×10^{-24}	1.00867	~1	Nucleus

*amu is the abbreviation for *atomic mass unit* (page 37); 1 amu = 1.6605×10^{-24} g

of negatively charged electrons around the nucleus equals the number of positively charged protons in the nucleus.

Atoms are extremely small—far too small to be seen with even the most powerful optical microscopes. The diameters of most atoms range from 1×10^{-8} cm to 5×10^{-8} cm. For example, the diameter of a gold atom is 3×10^{-8} cm. To visualize how small this is, consider that it would take approximately 517 million gold atoms to run the length (15.5 cm) of a dollar bill. If this weren't hard enough to imagine, remember that Rutherford's experiments provided evidence that the diameter of the nucleus is 1/100,000 the diameter of the atom. For example, if a gold atom were scaled upward in size so that the nucleus was the size of a small marble, the diameter of the scaled-up atom would be more than 1200 meters (3/4 mile) wide, and most of the space between the outer edge and the nucleus would be empty. Because the nucleus carries most of the mass of the atom (the neutrons and protons) in such a small volume, a small matchbox full of atomic nuclei would weigh more than 2.5 billion tons. The interior of a collapsed star is made up of nuclear material estimated to be nearly this dense.

So, if atoms are so small that they cannot be seen by ordinary methods, how do we know that they exist? This important question was finally answered by the invention of two important imaging techniques. Scientists have been able to obtain computer-enhanced images of the outer surface of atoms (Figure 3.4) using the scanning tunneling microscope (STM) and the atomic force microscope (AFM).

> **Atomic number:**
> The number of protons in the nucleus of an element

3-3a Atomic Number

Knowing the makeup of an atomic nucleus allows us to identify the element to which it belongs. The **atomic number** of an element indicates the number of protons in the nucleus of the atom. *All atoms of the same element have*

FIGURE 3.4 Scanning tunneling microscope (STM).

(a) When an electric current passes through a tungsten needle with a narrow tip (atom's width) into the atoms on the surface of the sample being examined, the electron flow between the tip and the surface changes in relation to the electron clouds around the atoms. By adjusting the position of the needle to maintain a constant current, the positions of the atoms are measured. (b) An STM image depicting a field of xenon (Xe) atoms. The sample of matter being imaged is 600 billionths of an inch (16.8 nm) wide.

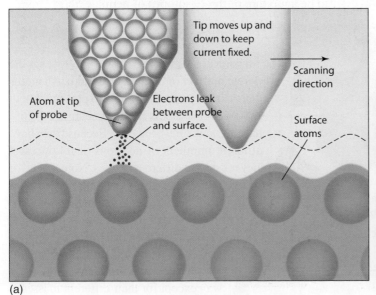

(a)

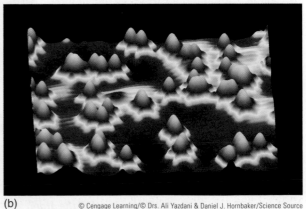

(b)

Mass number: The number of neutrons and protons in the nucleus of an atom

Isotopes: Atoms of an element with different mass numbers due to different numbers of neutrons

the same number of protons in the nucleus. In the periodic table (Figure 3.9), the atomic number for each element is given above the element's symbol. There is a different atomic number for each element, beginning with the atomic number 1 for hydrogen. Phosphorus, for example, has an atomic number of 15; thus, the nucleus of a phosphorus atom contains 15 protons.

It is the number of protons, and nothing more, that determines the identity of an atom. In an atom, the number of protons is equal to the number of electrons; therefore, the atomic number also gives the number of electrons, all of which exist outside the nucleus in an atom of an element.

3-3b Mass Number

The **mass number** of a particular atom is the total number of neutrons and protons present in the nucleus of an atom. Because the atomic number gives the number of protons in the nucleus, the difference between the mass number and the atomic number equals the number of neutrons in the nucleus.

Mass number = # of protons + # of neutrons

A notation frequently used for showing the mass number and atomic number of an atom places subscripts and superscripts to the left of the symbol.

Mass number $\longrightarrow$ $^{19}_{9}\text{F}$ $\longleftarrow$ Symbol of element
Atomic number $\longrightarrow$

The subscript giving the atomic number is optional because the element symbol tells you what the atomic number must be. For example, the fluorine atom can have the notation $^{19}_{9}\text{F}$ or simply ^{19}F. For an atom of fluorine, $^{19}_{9}\text{F}$, the number of protons is 9, the number of electrons is also 9, and the number of neutrons is $19 - 9 = 10$. To represent isotopes with words instead of symbols, the mass number is added to the name—for example, neon-20, neon-21, and neon-22.

Outfielder Manny Ramirez, while suspended from playing in Major League Baseball games, played with a Triple-A team—the Albuquerque Isotopes. The Isotopes, whose mascot Orbit is a large fuzzy electron, sport a logo based on Rutherford's model of the atom.

AP Photo/Craig Fritz

EXAMPLE 3.1 Atomic Composition

How many protons, neutrons, and electrons are in an atom of gold (Au) with a mass number of 197?

SOLUTION

The atomic number of an element gives the number of protons and electrons. If the element is known, its atomic number can be obtained from the periodic table (on the periodic table review card). The atomic number of gold is 79. Gold has 79 protons and 79 electrons. The number of neutrons is obtained by subtracting the atomic number from the mass number:

$$197 - 79 = 118 \text{ neutrons}$$

TRY IT 3.1A

How many protons, neutrons, and electrons are there in a $^{59}_{28}\text{Ni}$ atom?

TRY IT 3.1B

An atom has a mass number of 165 and contains 88 neutrons in its nucleus. What is the identity of the element?

3-3c Isotopes

When a natural sample of almost any element is analyzed, the element is found to be composed of atoms with different mass numbers. Atoms of the same element having different mass numbers are called **isotopes** of that element.

How can this be true? The element neon is a good example to consider. A natural sample of neon gas is found to be a mixture of three isotopes of neon:

$$^{20}_{10}\text{Ne} \qquad ^{21}_{10}\text{Ne} \qquad ^{22}_{10}\text{Ne}$$

The fundamental difference between isotopes is the different number of neutrons per atom. All atoms of neon have 10 electrons and 10 protons. About 90.92% of the atoms have 10 neutrons, 0.26% have 11, and 8.82% have 12. Because they have different numbers of neutrons, it follows that they must have different masses. Note that all of the isotopes have the same atomic number because they are all neon. An analogy may be helpful here. Consider three brand-new, identical automobiles that have had different numbers of 50-lb. weights hidden in their trunks. The automobiles are identical in every way except for their different masses.

More than 100 elements are known, and more than 1000 isotopes have been identified, many of them produced artificially (see Section 13-5). Some elements have many isotopes; tin, for example, has 10 natural isotopes. Keep in mind that not all isotopes are radioactive. Hydrogen has three isotopes, which are the only known isotopes generally referred to by different names: ^1_1H is commonly called hydrogen or protium, ^2_1H is called deuterium, and ^3_1H is called tritium. Only tritium is radioactive.

EXAMPLE 3.2 Isotopes

→ *Carbon has seven known isotopes. Three of these have six, seven, and eight neutrons, respectively. Write the complete chemical notation for these three isotopes, giving mass number, atomic number, and symbol.*

SOLUTION

The atomic number of carbon is 6. The mass number for the three isotopes is 6 plus the number of neutrons, which equal 12, 13, and 14, respectively.

$$^{12}_6C \quad ^{13}_6C \quad ^{14}_6C$$

TRY IT 3.2

Silver has two natural isotopes, one with 60 neutrons and the other with 62 neutrons. Give the complete chemical notation for these isotopes.

3-3d Atomic Mass and Atomic Weight

Although Dalton knew nothing about subatomic particles, he proposed that atoms of different elements have different masses. Eventually, it was found that an oxygen atom is about 16 times as heavy as a hydrogen atom. This fact, however, does not tell us the mass of either atom. These are relative masses in the same way that a grapefruit may weigh twice as much as an orange. This information gives neither the mass of the grapefruit nor that of the orange. Nevertheless, if a specific number is assigned as the mass of any particular atom, this fixes the numbers assigned to the masses of all other atoms. The present atomic mass scale, adopted by scientists worldwide in 1961, is based on assigning the mass of a particular isotope of the carbon atom, the carbon-12 isotope, as exactly 12 **atomic mass units (amu)**. Therefore, one atomic mass unit is a very small number and is *exactly* 1/12 the mass of a carbon-12 atom.

Information related to the mass of each element is listed below the symbol for each element in the periodic table. So

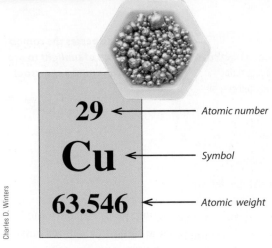

Atomic number

Symbol

Atomic weight

The periodic table entry for copper. (The complete periodic table and its significance are discussed in Section 3-6.)

why, then, is the atomic mass of copper (above) not an integer number? The answer is that the value shown represents an average mass, which takes into account the relative abundances of the different isotopes as found in nature. This average is often referred to as the **atomic weight** of the element. For example, boron has two naturally occurring isotopes, $^{10}_5B$ and $^{11}_5B$, with natural percent abundances of 19.91% and 80.09%, respectively. The masses in atomic mass units are 10.0129 and 11.0093, respectively. The atomic weight listed in the periodic table is the weighted average mass of a natural sample of atoms, expressed in atomic mass units.

$$(10.0129 \times 0.1991) + (11.0093 \times 0.8009) = 10.81$$

The average mass of 10.81 for B is closer to the mass of $^{11}_5B$ because of its higher abundance.

Atomic mass unit (amu): The unit for elements' relative atomic masses

Atomic weight: The average atomic mass of an element's isotopes weighted by percentage abundance

3-4 Building the Atomic Model: Where Are the Electrons in Atoms?

Rutherford's experiment helped identify the composition of the atomic nucleus but did not indicate precisely *where* the electrons are located, other than being outside the nucleus. It is important to know this with a bit more detail, because the way in which atoms combine to form chemical

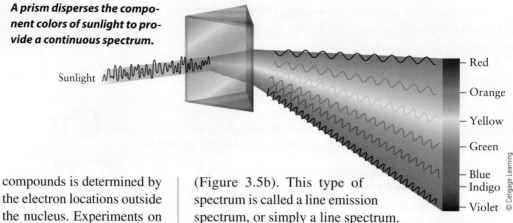

A prism disperses the component colors of sunlight to provide a continuous spectrum.

Sunlight

Red
Orange
Yellow
Green
Blue
Indigo
Violet

© Cengage Learning

compounds is determined by the electron locations outside the nucleus. Experiments on the interactions of light with matter provide important information about the energy and location of electrons in atoms.

3-4a Continuous and Line Spectra

We are familiar with the spectrum of colors that makes up visible light. The spectrum of white light is a display of separated colors. This type of spectrum is referred to as a **continuous spectrum** and is obtained by passing sunlight or light from an incandescent lightbulb through a glass prism. When we see a rainbow, we are looking at the continuous visible spectrum that forms when raindrops act as prisms and disperse the sunlight.

If a high voltage is applied to an element in the gas phase in a partially evacuated tube, the atoms absorb energy and are said to be "excited." When the excited atoms "relax," they emit light. An example of this is a neon advertising sign in which excited neon atoms emit orange-red light (Figure 3.5a). When the light emitted by the atoms passes through a prism, a different type of spectrum is obtained, one that is not continuous but has characteristic lines that correspond to light of specific wavelengths

(Figure 3.5b). This type of spectrum is called a line emission spectrum, or simply a line spectrum.

3-4b The Theory of Light

To understand how light is related to the locations of electrons outside the atomic nucleus, you will need to know a bit more about how light is described.

Microwaves, gamma rays, and X-rays are light, even though they cannot be seen with the naked eye. The light that you can see is called **visible radiation** (not radioactive!) and represents only a small fraction of the light that surrounds us. Visible radiation can easily be described by its color, and many of us are familiar with the mnemonic ROY G BIV to describe the colors of light present in a rainbow. But how can we begin to describe the other light that surrounds us, which we cannot see?

The English physicist Thomas Young discovered in 1801 that light behaved as though it were a wave and could be described by a wavelength. A **wavelength** is the distance between similar points in a wave and is represented by the Greek letter lambda (λ). For light, the units used to describe wavelength are nanometers (nm).

Blue light has a wavelength of about 400 nm, shorter than that of red light (700 nm). Other forms of light that are not visible, such as microwaves, have very long wave-

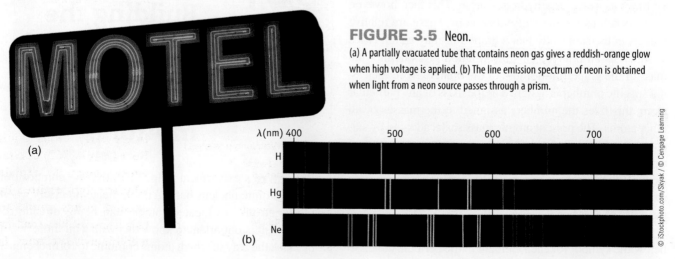

(a)

(b)

λ(nm) 400   500   600   700

H

Hg

Ne

© iStockphoto.com/Skvak / © Cengage Learning

FIGURE 3.5 Neon.

(a) A partially evacuated tube that contains neon gas gives a reddish-orange glow when high voltage is applied. (b) The line emission spectrum of neon is obtained when light from a neon source passes through a prism.

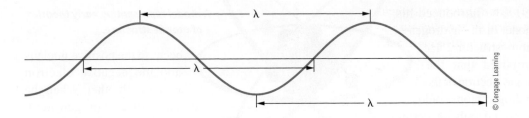

Water wave (ripple): The wavelength (λ) is the distance between any two adjacent identical points of a wave, such as two adjacent peaks or two adjacent troughs.

Frequency: The number of waves that pass a fixed point in one unit of time

Energy: For light, energy is determined by the frequency and wavelength

lengths (measured in centimeters), and X-rays have very short wavelengths.

Another way in which light is described is by its frequency. The **frequency**, represented by the Greek letter nu (ν), is the number of wavelengths that pass a point in a specific amount of time. To understand this, it may be useful to think of a more macroscopic example. If you were to drop a small stone into the center of a pond, you would notice that a series of small waves would begin to travel toward the shore. If you focused on one point on the surface of the pond and counted the number of times a wave passed that point in one second, you would be able to determine the frequency of waves traveling across the pond surface. The unit used to describe frequency of all waves, including light, is the hertz (Hz, also represented as s^{-1}, "per second of time").

Last, light can be described by its **energy** (E). What is interesting is that the energy of light is related to both its frequency and its wavelength. These relationships were discovered by Max Planck near the beginning of the 20th century and include a constant now known as Planck's constant (h):

$$E = h\nu$$
$$E = hc/\lambda$$

The first of these equations implies that light with a high energy also has a high frequency. This must be true, since if we increase the energy of light by a small increment, the frequency must also increase in order for the equality to remain true. But the second equation indicates that light with a high energy must have a shorter wavelength if the equality is to remain true. Perhaps this seems wrong, but it has been observed to be true over and over. The relationships between wavelength, frequency, and energy are shown in Figure 3.6, along with a continuous spectrum of all forms of light and the terms used to describe them. You'll make use of some of this knowledge in Chapter 7, where the importance of the ozone layer is presented. For now, we are ready to understand how light can be used to determine where the electrons are located outside of the atomic nucleus. Keep the following in mind as you continue: *The energy of light is directly related to its wavelength. If the light is visible, then the energy of the light is directly related to its wavelength and therefore its color.*

FIGURE 3.6 The electromagnetic spectrum.

Visible light (enlarged section) is but a small part of the entire spectrum. The energy of electromagnetic radiation increases from the radio wave end to the gamma ray end. The frequency of electromagnetic radiation is related to the wavelength by $\nu\lambda = c$, where ν = frequency; λ = wavelength; and c = speed of light, 3.00×10^8 meters (m)/second (s). The higher the frequency, the lower the wavelength and the larger the energy. The energy (E) of a *photon* (or *quantum*) of light is given by the expression $E_{photon} = h\nu$, where h is Planck's constant (6.6262×10^{-34} J · s).

It is important to understand the relationships among λ, ν, and E

As $\downarrow$, $\nu\uparrow E\uparrow$

As $\lambda\uparrow$, $\nu\downarrow E\downarrow$

Wavelength and frequency (or energy) are inversely related.

3-4c Bohr Model of the Atom

In the early 1900s, Niels Bohr (1885–1962) as elsewhere sought to explain why the light emitted by excited atoms produced only a line spectrum and not a continuous spectrum. In

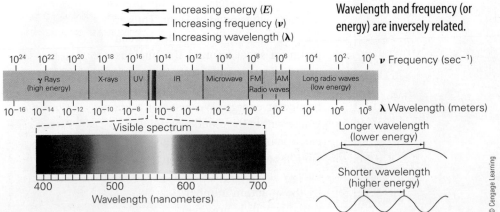

1913, he introduced his model of the hydrogen atom structure. He proposed that the single electron of the hydrogen atom could be found only at certain locations outside the atomic nucleus. He referred to these locations as *orbits* and called the energy difference of an electron located in one orbit or its closest neighboring orbit a **quantum** of energy. When the hydrogen electron in an orbit close to the nucleus absorbs a quantum of energy, the energy of the electron increases and it moves to a higher energy level in an orbit farther from the nucleus. When this electron returns to the lower, more stable energy level, the quantum of energy is emitted as light. The energy of the light is the same as the energy difference between the two orbits. Because the energy and wavelength of light are related, it is possible to measure the wavelength of light released as a way to measure the difference in energy levels, or orbits.

Because there are many possible orbits, many possible wavelengths of light can be released. This is the origin of the line spectra shown in Section 3-4.

In Bohr's model, each allowed orbit is assigned an integer, n, known as the principal quantum number. The values of n for the orbits range from 1 to infinity. Any atom with its electrons in their normal, lowest energy level is said to be in the **ground state**. Energy must be supplied to move the electron farther away from the nucleus

The colors of fireworks are a result of atoms in an excited state releasing their energy as light. The color can be selected by using difference elements: sodium for yellow, magnesium for white, strontium for red, and barium for green.

A model representing early theories of atomic structure.

because the positive nucleus and the negative electron attract each other. When the electron of a hydrogen atom occupies a higher energy level with n greater than 1, the atom has more energy than in its ground state and is said to be in an **excited state**. The excited state is an unstable state, and the extra energy is emitted when the electron returns to the ground state. According to Bohr, the light forming the lines in the spectrum of hydrogen comes from electrons moving toward the nucleus, having first been excited to energy levels farther from the nucleus. Because the energy levels have only certain energies, the emitted light has only certain wavelengths.

With brilliant imagination, Bohr applied a little algebra and some classic mathematical equations of physics to his tiny solar system model of the hydrogen atom and was able to calculate the wavelengths of the lines in the hydrogen spectrum. By 1900, scientists had measured the wavelengths of lines for hydrogen in the ultraviolet, visible, and infrared regions, and Bohr's calculated values agreed with the measured values. Niels Bohr had tied the unseen (the interior of the atom) with the seen (the observable lines in the hydrogen spectrum)—a fantastic achievement. The concepts of quantum number and energy level are valid for all atoms and molecules.

The Bohr model was accepted almost immediately after its presentation, and Bohr was awarded the Nobel Prize in Physics in 1922 for his contribution to the understanding of the hydrogen atom. However, his model gave only approximate agreement with line spectra of atoms having more than one electron. Later models of the atom have been more successful by considering electrons as having both particle and wave characteristics. This led to mathematical treatment of the locations of electrons as probabilities instead of as the precise locations envisioned by Bohr. Thus, in the modern view of the atom, we can picture a space around the nucleus occupied by each electron, somewhat like a cloud of electrical charge of a particular energy. We just don't know where within the cloud a particular electron is at any given instant. Bohr's concept of the main energy levels represented by the quantum number n remains valid, however;

Niels Bohr

TABLE 3.2 Electron Arrangements of the First 20 Elements*

Element	Atomic Number	Number of Electrons in Each Energy Level			
		1st	2nd	3rd	4th
Hydrogen (H)	1	1 e			
Helium (He)	2	2 e			
Lithium (Li)	3	2 e	1 e		
Beryllium (Be)	4	2 e	2 e		
Boron (B)	5	2 e	3 e		
Carbon (C)	6	2 e	4 e		
Nitrogen (N)	7	2 e	5 e		
Oxygen (O)	8	2 e	6 e		
Fluorine (F)	9	2 e	7 e		
Neon (Ne)	10	2 e	8 e		
Sodium (Na)	11	2 e	8 e	1 e	
Magnesium (Mg)	12	2 e	8 e	2 e	
Aluminum (Al)	13	2 e	8 e	3 e	
Silicon (Si)	14	2 e	8 e	4 e	
Phosphorus (P)	15	2 e	8 e	5 e	
Sulfur (S)	16	2 e	8 e	6 e	
Chlorine (Cl)	17	2 e	8 e	7 e	
Argon (Ar)	18	2 e	8 e	8 e	
Potassium (K)	19	2 e	8 e	8 e	1 e[†]
Calcium (Ca)	20	2 e	8 e	8 e	2 e

© Cengage Learning

* Valence electrons are shown in blue.

[†] The discussion on page 42, Beyond the Bohr Atom, more clearly explains why the 19th and 20th electrons do not add to the third energy level as might have been predicted using the $2n^2$ guide.

for our purposes, this concept is all that we need to discuss the location of electrons in atoms.

3-4d Atom Building Using the Bohr Model

Recall that the atomic number is the number of electrons (or protons) per atom of an element. Imagine building a model of an atom by adding one electron to an orbit as another proton is added to the nucleus. As part of his theory, Bohr proposed that only a fixed number of electrons could be accommodated in any one orbit, and he calculated that this number was given by the formula $2n^2$, where n equals the number of the energy

© Bettmann/CORBIS

Erwin Schrödinger

level. For the lowest energy level (first level, closest to the atomic nucleus), n equals 1, and the maximum number of electrons allowed is $2(1)^2$, or 2. For the second energy level, the maximum number of electrons is $2(2)^2$, or 8. By using $2n^2$, the maximum number of electrons allowed for levels 3, 4, and 5 are 18, 32, and 50, respectively. A general overriding rule to the preceding numbers is that the highest energy level can have no more than eight electrons, except for transition elements, for a stable atom. Table 3.2 lists the Bohr electron arrangements for the first 20 elements.

Electrons in the highest occupied energy level listed for the elements in Table 3.2 are at the greatest stable distance from the nucleus. These are the most important electrons in the study of chemistry because they are the ones that interact when atoms react with each other. G. N. Lewis first proposed that each energy level could hold only a characteristic number of electrons and that only those electrons in the highest occupied level were involved when one atom combined with another to form a chemical compound. These outermost electrons came to be known as **valence electrons**. The number of valence electrons an atom has is important in determining how the atom combines with other atoms.

> **Valence electrons:** The outermost electrons in an atom

Energy Level (n)	$2n^2$	Maximum Number of Electrons
1	$2(1)^2$	2
2	$2(2)^2$	8
3	$2(3)^2$	18
4	$2(4)^2$	32
5	$2(5)^2$	50
6	$2(6)^2$	72

© Cengage Learning

For example, look at phosphorus (P). The Bohr arrangement of electrons is 2–8–5. This means that the stable state of the phosphorus atom has electrons in three energy levels. The one closest to the nucleus has two electrons, the second energy level has eight electrons, and the highest energy level has five electrons. The energy level with five electrons is farthest from the nucleus; thus, P has *five valence electrons*, and these are the electrons that are most available for interactions with valence electrons of other atoms in chemical reactions. This interaction results

Shell: A principal energy level defined by a given value of *n*

Orbital: A region of three-dimensional space around an atom within which there is a significant probability that a given electron will be found

Subshell: A more specific energy level (orbital) within a given shell

in the formation of bonds, which will be discussed in Chapter 5.

3-4e Beyond the Bohr Atom

The simple Bohr model of the atom, in which each **shell** is capable of holding $2n^2$ electrons, can be used to explain many properties of atoms and their associated electrons. However, more sophisticated treatments of atoms depend on the fact that electrons moving at very high speeds, up to 2 million meters per second (or roughly 1 percent the speed of light), exhibit properties of waves. Although it is appropriate at some level of understanding to envision electrons as particles moving around the nucleus like planets moving around the sun, electron locations around the nucleus can be more accurately described by using wave theory. This theory defines electron locations in terms of regions of probability known as **orbitals**. A full treatment of highly mathematical wave theory, developed by an Austrian scientist named Erwin Schrödinger, is well beyond the scope of this text, but his results are nonetheless useful.

In essence, Schrödinger concluded that the volume of space in which each pair of electrons could most likely be found defines an orbital and that each orbital can accommodate a maximum of two electrons. Furthermore, he

concluded that Bohr's planetary shells can contain more than one such orbital and the orbitals within a given shell may be of different types. Quantum theory shows that the two electrons residing in the first Bohr shell occupy a single, spherical orbital called a 1*s* orbital (Figure 3.7).

The second Bohr shell can accommodate a total of eight electrons (2×2^2), but quantum theory shows that these eight electrons occupy two different types of orbitals. This second energy shell thus contains two **subshells** comprising a single spherical 2*s* orbital and three separate, dumbbell-shaped orbitals oriented at right angles with respect to each other and known as 2*p* orbitals. All of these orbitals, like all atomic orbitals regardless of type, are centered about the nucleus of the atom. The 2*s* orbital is higher in energy than the 1*s* orbital, but it is slightly lower in energy than the three 2*p* orbitals. The 2*s* orbital holds two electrons, and the three 2*p* orbitals each hold two electrons, for a total of eight electrons in the second energy shell.

A continuation of this treatment showed that the third energy level, which can accommodate a maximum of 2×3^2 or 18 electrons, is actually composed of three subshells. The lowest energy orbital in this shell is the spherical 3*s* orbital, capable of holding two electrons. The next-lowest energy orbitals in this shell are three dumbbell-shaped 3*p* orbitals (Figure 3.7) capable of holding a total of six electrons. Lastly, the third energy shell contains five orbitals known as 3*d* orbitals. These five 3*d* orbitals can accommodate a total of 10 electrons. Summarizing, we see that the third shell contains a single 3*s* orbital (2-electron capacity), three 3*p* orbitals (6-electron capacity), and five 3*d* orbitals (10-electron capacity) for a maximum of 18 electrons in the third shell.

FIGURE 3.7 Atomic orbitals.

Boundary surface diagrams for electron densities of 1*s*, 2*s*, 2*p*, 3*s*, 3*p*, and 3*d* orbitals. For the *p* orbitals, the subscript letter on the orbital notation (*x, y, z*) indicates the cartesian axis along which the orbital lies.

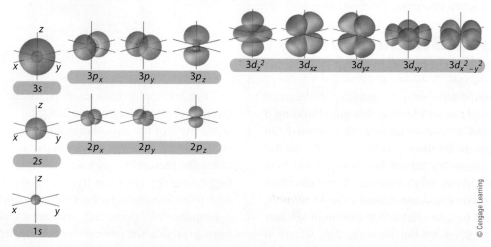

© Cengage Learning

TABLE 3.3 Electron Configurations of the First 20 Elements*

Element	Atomic Number		Element	Atomic Number	
Hydrogen (H)	1	$1s^1$	Sodium (Na)	11	$1s^22s^22p^6\ 3s^1$
Helium (He)	2	$1s^2$	Magnesium (Mg)	12	$1s^22s^22p^6\ 3s^2$
Lithium (Li)	3	$1s^2\ 2s^1$	Aluminum (Al)	13	$1s^22s^22p^6\ 3s^23p^1$
Beryllium (Be)	4	$1s^2\ 2s^2$	Silicon (Si)	14	$1s^22s^22p^6\ 3s^23p^2$
Boron (B)	5	$1s^2\ 2s^22p^1$	Phosphorus (P)	15	$1s^22s^22p^6\ 3s^23p^3$
Carbon (C)	6	$1s^2\ 2s^22p^2$	Sulfur (S)	16	$1s^22s^22p^6\ 3s^23p^4$
Nitrogen (N)	7	$1s^2\ 2s^22p^3$	Chlorine (Cl)	17	$1s^22s^22p^6\ 3s^23p^5$
Oxygen (O)	8	$1s^2\ 2s^22p^4$	Argon (Ar)	18	$1s^22s^22p^6\ 3s^23p^6$
Fluorine (F)	9	$1s^2\ 2s^22p^5$	Potassium (K)	19	$1s^22s^22p^63s^23p^6\ 4s^1$
Neon (Ne)	10	$1s^2\ 2s^22p^6$	Calcium (Ca)	20	$1s^22s^22p^63s^23p^6\ 4s^2$

* Valence electrons are shown in blue.

The fourth energy shell contains, after the $4s$ orbital, three $4p$ orbitals, five $4d$ orbitals, and seven $4f$ orbitals (not pictured) that can accommodate a maximum of 14 electrons. Thus, the fourth shell or energy level can accommodate a total of 2×4^2 (or 32) electrons. The trend continues in this fashion until no additional orbitals are required to accommodate the electrons.

Although the Bohr model correctly predicts the number of electrons in each energy shell according to the $2n^2$ calculation with n equaling the number of the energy shell, the quantum wave model more accurately describes the location probability of the electrons in three-dimensional space.

As one progresses from one atom to the atom of next higher atomic number in the periodic table, electrons are added singly to the lowest-energy orbital in a given subshell that does not already have an electron. Once each orbital in a subshell has a single electron, additional electrons are then added to the singly occupied orbitals in that subshell until they are each doubly occupied. At that point, additional electrons are then added to the orbitals with the next highest energy until one exhausts the number of electrons associated with the atom.

Figure 3.8 provides an order-of-filling chart that allows the exact specification of orbitals occupied by electrons in a given atom. Simply follow the arrowed pathway shown in Figure 3.8 and place the appropriate number of electrons in each orbital or set of orbitals in the order specified. Remember that there are 2 electrons in the $2s$ orbitals, 6 electrons in the $2p$ orbitals, 10 electrons in the $3d$ orbitals, and 14 electrons in the $4f$ orbitals.

Using this method, we can write the electron configuration of each atom as shown in Table 3.3, where superscripts are used to indicate the number of electrons in the orbitals. All electrons that occupy a subshell with the same main shell number (for example, where the second shell is signified by using a 2 in front of s and p orbitals) are valence electrons.

FIGURE 3.8 Subshell filling order.

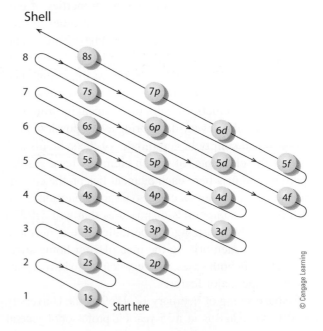

This discussion helps explain why the last electron added when going from $_{18}$Ar to $_{19}$K goes in the fourth energy level (as shown in Table 3.2) and not in the third energy level, as one might have predicted using the simple Bohr model. This happens because according to the filling model, electrons are added to the 4s orbital before being added to the 3d orbital. Following this model, the next electron added to arrive at $_{20}$Ca goes into the 4s orbital; the 3d orbitals are filled next, before the 4p orbitals. If one were to follow the filling method across the entire periodic table, the transition metals (Section 3-6) would occupy the region of the periodic table where the d and f orbitals come into play.

In summary, quantum theory's more sophisticated treatment of electrons allows us to more accurately describe the positions of electrons around a given nucleus. This can help us understand chemical bonding in a more detailed way.

3-5 Development of the Periodic Table

So far, in exploring the chemical view of matter, you have seen that everything is made of atoms. Atoms of elements combined in specific ratios are present in chemical compounds. All matter consists of elements, compounds, or mixtures of elements or compounds. You have also been introduced to the subatomic composition of atoms—the electrons, protons, and neutrons. The next part of the story is to see how atomic structure and the properties of elements and compounds are related to each other. Fortunately, the chemistry of the elements can be organized and classified in a way that helps both chemists and nonchemists. The periodic table is the single most important classification system in chemistry because it summarizes, correlates, and predicts a wealth of chemical information. Chemists consult it every day during every possible kind of work. It can simply be a reminder of the symbols and names of the elements, of which elements have similar properties, and of where each element lies on the continuum of atomic numbers. It can also be an inspiration in the search for new compounds or mixtures that will fulfill a specific need. Memorizing the periodic table is no more necessary than memorizing the map of your home state. Nevertheless, in both cases, it's very helpful to have a general idea of the major features.

On the evening of February 17, 1869, at the University of St. Petersburg in Russia, a 35-year-old professor of general chemistry—Dmitri Ivanovich Mendeleev (1834–1907)—was writing a chapter of his soon-to-be-famous textbook on chemistry. He had the properties of each element written on cards, with a separate card for each element. While he was shuffling the cards trying to gather his thoughts before writing his manuscript, Mendeleev realized that if the elements were arranged in the order of their atomic weights, there was a trend in properties that repeated itself several times. He arranged the elements into groups that had similar properties and used the resulting periodic chart to predict the properties and places in the chart of as yet undiscovered elements.

Thus, the periodic law and the periodic table were born, although only 63 elements had been discovered by 1869 (for example, the noble gases were not discovered until after 1893), and the clarifying concept of the atomic number was not known until 1913. Mendeleev's idea and textbook achieved great success, and he rose to a position of prestige and fame while he continued to teach at the University of St. Petersburg.

Mendeleev aided the discovery of new elements by predicting their properties with remarkable accuracy, and he even suggested the geographical regions in which minerals containing the elements could be found. The properties of a missing element were predicted by consideration of the properties of its neighboring elements in the table. For example, for the element we now know as germanium, which falls below silicon in the modern periodic table (Figure 3.9), Mendeleev predicted a gray element of atomic weight 72 with a density of 5.5 g/cm^3. Germanium, once discovered, proved to be a gray element of atomic weight 72.59 with a density of 5.36 g/cm^3.

The empty spaces in the table and Mendeleev's predictions of the properties of missing elements stimulated a flurry of prospecting for elements in the 1870s and 1880s, and eight more were discovered by 1886.

Mendeleev found that a few elements did not fit under other elements with similar chemical properties when arranged according to increasing atomic weight. Eventually, it was found that the atomic weight is *not* the property that governs the similarities and differences among the elements. This was discovered in 1913 by H. G. J. Moseley (1888–1915), a young scientist working with Ernest Rutherford. Moseley found that the wavelengths of X-rays emitted by a particular element are related in a precise way to the atomic number of that element. He quickly realized that other atomic properties may be similarly related to atomic number and not, as Mendeleev had believed, to atomic weight.

By building on the work of Mendeleev and others, and by using the concept of the atomic number, we are now

FIGURE 3.9 Modern periodic table of elements.

Legend (example):
Uranium — Name
92 — Atomic number
U — Symbol
238.0289 — Atomic weight

Legend color key:
- MAIN GROUP METALS
- TRANSITION METALS
- METALLOIDS
- NONMETALS

Period	1A (1)	2A (2)	3B (3)	4B (4)	5B (5)	6B (6)	7B (7)	8B (8)	8B (9)	8B (10)	1B (11)	2B (12)	3A (13)	4A (14)	5A (15)	6A (16)	7A (17)	8A (18)
1	Hydrogen 1 H 1.0079																	Helium 2 He 4.0026
2	Lithium 3 Li 6.941	Beryllium 4 Be 9.0122											Boron 5 B 10.811	Carbon 6 C 12.011	Nitrogen 7 N 14.0067	Oxygen 8 O 15.9994	Fluorine 9 F 18.9984	Neon 10 Ne 20.1797
3	Sodium 11 Na 22.9898	Magnesium 12 Mg 24.3050											Aluminum 13 Al 26.9815	Silicon 14 Si 28.0855	Phosphorus 15 P 30.9738	Sulfur 16 S 32.066	Chlorine 17 Cl 35.4527	Argon 18 Ar 39.948
4	Potassium 19 K 39.0983	Calcium 20 Ca 40.078	Scandium 21 Sc 44.9559	Titanium 22 Ti 47.867	Vanadium 23 V 50.9415	Chromium 24 Cr 51.9961	Manganese 25 Mn 54.9380	Iron 26 Fe 55.845	Cobalt 27 Co 58.9332	Nickel 28 Ni 58.6934	Copper 29 Cu 63.546	Zinc 30 Zn 65.38	Gallium 31 Ga 69.723	Germanium 32 Ge 72.61	Arsenic 33 As 74.9216	Selenium 34 Se 78.96	Bromine 35 Br 79.904	Krypton 36 Kr 83.80
5	Rubidium 37 Rb 85.4678	Strontium 38 Sr 87.62	Yttrium 39 Y 88.9059	Zirconium 40 Zr 91.224	Niobium 41 Nb 92.9064	Molybdenum 42 Mo 95.96	Technetium 43 Tc (97.907)	Ruthenium 44 Ru 101.07	Rhodium 45 Rh 102.9055	Palladium 46 Pd 106.42	Silver 47 Ag 107.8682	Cadmium 48 Cd 112.411	Indium 49 In 114.818	Tin 50 Sn 118.710	Antimony 51 Sb 121.760	Tellurium 52 Te 127.60	Iodine 53 I 126.9045	Xenon 54 Xe 131.29
6	Cesium 55 Cs 132.9055	Barium 56 Ba 137.327	Lanthanum 57 La 138.9055	Hafnium 72 Hf 178.49	Tantalum 73 Ta 180.9479	Tungsten 74 W 183.84	Rhenium 75 Re 186.207	Osmium 76 Os 190.23	Iridium 77 Ir 192.22	Platinum 78 Pt 195.084	Gold 79 Au 196.9666	Mercury 80 Hg 200.59	Thallium 81 Tl 204.3833	Lead 82 Pb 207.2	Bismuth 83 Bi 208.9804	Polonium 84 Po (208.98) Discovered 1999	Astatine 85 At (209.99)	Radon 86 Rn (222.02)
7	Francium 87 Fr (223.02)	Radium 88 Ra (226.0254)	Actinium 89 Ac (227.0278)	Rutherfordium 104 Rf (267)	Dubnium 105 Db (268)	Seaborgium 106 Sg (271)	Bohrium 107 Bh (272)	Hassium 108 Hs (270)	Meitnerium 109 Mt (276)	Darmstadtium 110 Ds (281)	Roentgenium 111 Rg (280)	Copernicium 112 Cn (285)	Ununtrium 113 Uut Discovered 2004	Ununquadium 114 Uuq Discovered 1999	Ununpentium 115 Uup Discovered 2004	Ununhexium 116 Uuh Discovered 1999	Ununseptium 117 Uus Discovered 2010	Ununoctium 118 Uuo Discovered 2002

Lanthanides

Cerium 58 Ce 140.116	Praseodymium 59 Pr 140.9076	Neodymium 60 Nd 144.242	Promethium 61 Pm (144.91)	Samarium 62 Sm 150.36	Europium 63 Eu 151.964	Gadolinium 64 Gd 157.25	Terbium 65 Tb 158.9254	Dysprosium 66 Dy 162.50	Holmium 67 Ho 164.9303	Erbium 68 Er 167.26	Thulium 69 Tm 168.9342	Ytterbium 70 Yb 173.054	Lutetium 71 Lu 174.9668

Actinides

Thorium 90 Th 232.0381	Protactinium 91 Pa 231.0359	Uranium 92 U 238.0289	Neptunium 93 Np (237.0482)	Plutonium 94 Pu (244.664)	Americium 95 Am (243.061)	Curium 96 Cm (247.07)	Berkelium 97 Bk (247.07)	Californium 98 Cf (251.08)	Einsteinium 99 Es (252.08)	Fermium 100 Fm (257.10)	Mendelevium 101 Md (258.10)	Nobelium 102 No (259.10)	Lawrencium 103 Lr (262.11)

Note: Atomic masses are 2007 IUPAC values (up to four decimal places). Numbers in parentheses are atomic masses or mass numbers of the most stable isotope of an element.

Standard Colors for Atoms in Molecular Models
- carbon atoms
- hydrogen atoms
- oxygen atoms
- nitrogen atoms
- chlorine atoms

able to state the modern **periodic law:** *When elements are arranged in the order of their atomic numbers, their chemical and physical properties show repeatable, or periodic, trends.* Other familiar periodic phenomena include the changing of the seasons and the orbits of the planets, which are periodic with time. To illustrate this idea on a simple level, a shingle roof, which has the same pattern over and over, is periodic.

Thus, to build up a periodic table according to the periodic law, the elements are lined up in a horizontal row in the order of their atomic numbers, as in Figure 3.9. At an element with similar properties to one already previously placed in the row, a new row is started. *Each column then contains elements with similar properties.* Some chemical and physical properties of the first 20 elements are summarized in Table 3.4. Do you see any trends and similarities among the elements in Table 3.4? For example, lithium (Li) is a soft metal with low density that is very reactive. It combines with chlorine gas to form lithium chloride with the formula LiCl. The other elements in Table 3.4 that have properties similar to those of lithium are sodium (Na) and potassium (K). According to the periodic law, lithium, sodium, and potassium should be in the same group—and they are. Look for similarities among other elements listed in Table 3.4 and check your grouping with that shown in the periodic table in Figure 3.9.

Dmitri Mendeleev

© SSPL/The Image Works

In the modern **periodic table** (Figure 3.9), the elements are arranged in order of their atomic numbers so that elements with similar chemical and physical properties fall together in vertical columns. These vertical columns are called **groups**. The first two and last six groups (Groups 1, 2, and 13 to 18) are the representative or main-group elements. The middle 10 groups (Groups 3 to 12) are the transition elements that link the two areas of representative elements. The inner transition elements are the *lanthanide series* and the *actinide series*. They are placed at the bottom of the periodic table because the similarity of properties within the two series would require their placement between lanthanum and hafnium, and between actinium and rutherfordium, respectively, which would make the table inconveniently wide.

The horizontal rows are called **periods**. These periods or rows are related to energy levels for electrons in atoms (Figure 3.9). The length of a row is linked to the maximum number of electrons, $2n^2$, that can fit into an energy level. The periods are not equal in size because the maximum number of electrons per energy level increases as the distance of the energy level from the nucleus increases. Periods one through seven have 2, 8, 8, 18, 18, 32, and 23 (incomplete) elements, respectively. Larger periods, seen as the atoms of elements get larger, are similar to the longer rows and more seats per row in a stadium as you proceed up from field level in the stands.

Most of the elements are **metals**. These are found in Groups 1 and 2 and parts of Groups 13 to 16 (light blue in Figure 3.9). All of the transition elements are metals (dark blue in Figure 3.9). Characteristic physical properties of metals include malleability (ability to be beaten into thin sheets such as aluminum foil), ductility (ability to be stretched or drawn into wire such as copper), and good conduction of heat and electricity.

Eighteen elements are **nonmetals** (in yellow in Figure 3.9), and except for hydrogen they are found on the right side of the periodic table. Hydrogen is shown above Group 1 because its atoms have one electron. However, hydrogen is a nonmetal and probably should be in a group by itself, although you may see H in both Group 1 and Group 17 in some periodic tables. Hydrogen forms compounds with formulas similar to those of the Group 1 elements but with vastly different properties. For example, compare NaCl (table salt) with HCl (a strong acid), or compare Na_2O (an active metal

TABLE 3.4 Some Properties of the First 20 Elements

Element	Atomic Number	Description	Compound Formation* With Cl (or Na)	Compound Formation* With O (or Mg)
Hydrogen (H)	1	Colorless gas, reactive	HCl	H_2O
Helium (He)	2	Colorless gas, unreactive	None	None
Lithium (Li)	3	Soft metal, low density, very reactive	LiCl	Li_2O
Beryllium (Be)	4	Harder metal than Li, low density, less reactive than Li	$BeCl_2$	BeO
Boron (B)	5	Both metallic and nonmetallic, very hard, not very reactive	BCl_3	B_2O_3
Carbon (C)	6	Brittle nonmetal, unreactive at room temperature	CCl_4	CO_2
Nitrogen (N)	7	Colorless gas, nonmetallic, not very reactive	NCl_3	N_2O_5
Oxygen (O)	8	Colorless gas, nonmetallic, reactive	Na_2O, Cl_2O	MgO
Fluorine (F)	9	Greenish-yellow gas, nonmetallic, extremely reactive	NaF, ClF	MgF_2, OF_2
Neon (Ne)	10	Colorless gas, unreactive	None	None
Sodium (Na)	11	Soft metal, low density, very reactive	NaCl	Na_2O
Magnesium (Mg)	12	Harder metal than Na, low density, less reactive than Na	$MgCl_2$	MgO
Aluminum (Al)	13	Metal as hard as Mg, less reactive than Mg	$AlCl_3$	Al_2O_3
Silicon (Si)	14	Brittle nonmetal, not very reactive	$SiCl_4$	SiO_2
Phosphorus (P)	15	Nonmetal, low melting point, white solid, reactive	PCl_3	P_2O_5
Sulfur (S)	16	Yellow solid, nonmetallic, low melting point, moderately reactive	Na_2S, SCl_2	MgS
Chlorine (Cl)	17	Yellow-green gas, nonmetallic, extremely reactive	NaCl	$MgCl_2$, Cl_2O
Argon (Ar)	18	Colorless gas, unreactive	None	None
Potassium (K)	19	Soft metal, low density, very reactive	KCl	K_2O
Calcium (Ca)	20	Harder metal than K, low density, less reactive than K	$CaCl_2$	CaO

* The chemical formulas shown are lowest ratios. The molecular formula for $AlCl_3$ is Al_2Cl_6, and that for P_2O_5 is P_4O_{10}.

oxide) with H_2O (water). Hydrogen also forms compounds similar to those of the Group 17 elements: NaCl and NaH (sodium hydride), and $CaBr_2$ and CaH_2 (calcium hydride).

The physical and chemical properties of nonmetals are opposite those of metals. For example, nonmetals are **insulators**; that is, they are extremely poor conductors of heat and electricity.

Elements that border the staircase in Figure 3.9—those between metals and nonmetals—are six **metalloids** (in green). Their properties are intermediate between those of metals and nonmetals. For example, silicon (Si), germanium (Ge), and arsenic (As) are **semiconductors** and are the elements that form the basic components of computer chips. Semiconductors conduct electricity less well than metals, such as silver and copper, but better than insulators, such as sulfur. The six **noble gases** in Group 18 have little tendency to undergo chemical reactions and at one time were considered inert (totally unreactive). The classifications of metals, nonmetals, and metalloids will enable you to predict the kinds of compounds formed between elements.

Insulator: A poor conductor of heat and electricity

Metalloid: An element with properties intermediate between those of metals and nonmetals

Semiconductor: A material with electrical conductivity intermediate between those of metals and insulators

Noble gas: An element in Group 18 of the periodic table

EXAMPLE 3.3 Periodic Table

For the elements with atomic numbers 17, 33, and 82, give the names and symbols and identify the elements as metals, metalloids, or nonmetals.

SOLUTION

Chlorine (Cl) is the element with atomic number 17. It is in Group 17. Chlorine and all the other elements in Group 17 are nonmetals.

Arsenic (As) is the element with atomic number 33. It is in Group 15. Because it lies along the line between metals and nonmetals, arsenic is a metalloid.

Lead (Pb) is the element with atomic number 82. It is in Group 14. Like other elements at the bottom of Groups 13 to 16, lead is a metal.

TRY IT 3.3

List the main groups in the periodic table that **(a)** consist entirely of metals, **(b)** consist entirely of nonmetals, and **(c)** include metalloids. Identify the numbers of valence electrons in atoms from groups listed under **(a)**, **(b)**, and **(c)**.

3-7 Periodic Trends

Why do elements in the same group in the periodic table have similar chemical behavior? Why do metals and nonmetals have different properties? G. N. Lewis was seeking answers to these questions during his development of the concept of valence electrons. He assumed that each noble gas atom had a completely filled outermost shell, which he regarded as a stable configuration because of the lack of reactivity of noble gases. He also assumed that the reactivity of other elements was influenced by their numbers of valence electrons.

3-7a Lewis Dot Symbols

Lewis used the element's symbol to represent the atomic nucleus together with all but the outermost electrons. The valence electrons, which are the outermost electrons of an atom, are represented by dots. The dots are placed around the symbol one at a time until they are used up or until all four sides are occupied; any remaining electron dots are paired with the ones already there. Lewis dot symbols for atoms of the first 20 elements are shown

TABLE 3.5 Lewis Dot Symbols for Atoms

1	2	13	14	15	16	17	18
H·							He:
Li·	·Be·	·Ḃ·	·Ċ·	·Ṅ·	:Ö·	:Ḟ·	:Ṅe:
Na·	·Mg·	·Äl·	·Ṡi·	·Ṗ·	:Ṡ·	:Ċl·	:Är:
K·	·Ca·						

© Cengage Learning

in Table 3.5. Notice that all atoms of elements in a given main group have the same number of valence electrons. *The importance of valence electrons in the study of chemistry cannot be overemphasized.* The identical number of valence electrons primarily accounts for the similar properties of elements in the same group. *The chemical view of matter is primarily concerned with what valence electrons are doing in the course of chemical reactions.*

Lewis dot symbols will be used extensively in Chapter 5 in the discussions of bonding.

3-7b Describing Atoms: Trends in Atomic Properties

From left to right across each period, metallic character gives way to nonmetallic character (Figure 3.9). The elements with the most metallic character are at the lower-left part of the periodic table near cesium (Cs). The elements with the most nonmetallic character are at the upper-right portion of the periodic table near fluorine. The six metalloid elements (in green) that begin with boron and move down like a staircase to astatine (At) roughly separate the metals and the nonmetals.

Atomic radii show periodicity by increasing with atomic number down the main groups of the periodic table (Figure 3.10). Why do atoms get larger from the top to the bottom of a group? The larger atoms simply have more energy levels occupied by electrons than do the smaller atoms; the more electrons there are surrounding the atomic nucleus, the more space they will occupy.

Atomic radii decrease across a period from left to right (Figure 3.10). You may see a paradox in adding electrons and getting smaller atoms, but protons also are being added. The increasing positive charge of the nucleus pulls electrons closer to the nucleus and causes contraction of the atom. Overall, atomic radii increase as one proceeds downward and to the left in the periodic table.

We can also use the trends in size of atomic radii to predict trends in reactivity. The valence electrons in larger atoms are farther from the nucleus. The larger the atom, the easier it is to remove the valence electrons because the attractive forces between protons in the nucleus and valence electrons decrease with increasing size of the atom.

In ionic compounds, atoms have gained and lost electrons to form **ions**, which have positive or negative charges. Metal atoms lose valence electrons to form positive ions. The larger the metal atom, the greater the tendency to lose valence electrons and the more reactive the metal. Therefore, we would predict that the most reactive metal in Figure 3.10 is cesium (Cs), the metal with the largest radius, and this is correct. The resulting *increase* in reactivity of metals down a given group in the periodic table is dramatically illustrated by lithium, sodium, and potassium—the first three metals in Group 1. Their atoms increase in size in this order down the group. Each element reacts with water—lithium quietly and smoothly, sodium more vigorously, and potassium much more quickly. The reactions of both lithium and potassium give off enough heat to ignite the hydrogen gas produced by the reaction, but as shown in Figure 3.11, potassium reacts with explosive violence. For elements at the bottom of Group 1, exposure to moist air produces a vigorous explosion.

Nonmetal atoms gain electrons from metals to form negative ions. The smaller the nonmetal atom, the higher the reactivity of the nonmetal. For example, fluorine atoms are the smallest of Group 17 elements, and fluorine is the most reactive nonmetal. It reacts with all other elements except three noble gases—helium, neon, and argon. The reaction of Group 17 elements with hydrogen illustrates how the reactivity of nonmetals decreases down the group. Fluorine reacts explosively with hydrogen, but the reaction

FIGURE 3.10 Atomic radii of the main-group elements (in picometers).
A picometer (pm) is 1×10^{-12} m (100 nm). Remember that the radius is one-half of the diameter.

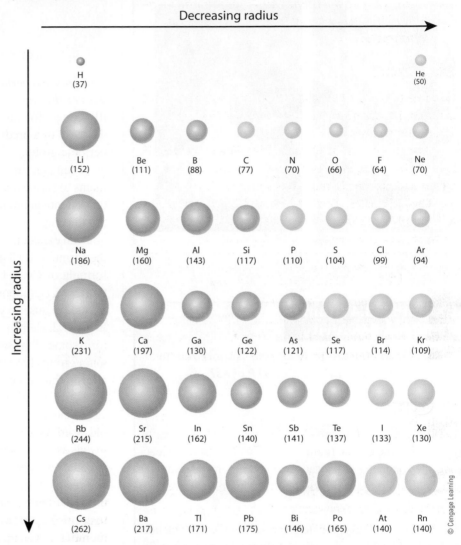

© Cengage Learning

with hydrogen is less violent for chlorine and is very slow for iodine.

Why are there repeatable patterns of properties across the periods in the periodic table? Again, it is because there is a repeatable pattern in atomic structure, and properties depend on atomic structure. *Each period begins with one valence electron for atoms of the elements in Group 1. Each period builds up to eight valence electrons, and the period ends. This pattern repeats across periods two through six.* As more elements are synthesized by chemists using nuclear accelerators (Chapter 13), it is possible that period seven will be completed someday.

Ion: An atom or group of atoms with a positive or negative charge

FIGURE 3.11 Reaction of alkali metals with water: (a) lithium, and (b) potassium.

© Cengage Learning/Charles D. Winters/© Cengage Learning/Charles D. Winters

EXAMPLE 3.4 Atomic Radii and Reactivity

→ **Which element in each pair is more reactive: (a) O or S, (b) Be or Ca, or (c) P or As?**

SOLUTION

(**a**) Oxygen and sulfur are Group 16 nonmetals. Generally, the smaller the atomic radius of a nonmetal, the more reactive the nonmetal is. Because oxygen atoms are smaller than sulfur atoms, oxygen should be more reactive than sulfur.

(**b**) Beryllium (Be) and calcium (Ca) are Group 2 metals. Reactivity of metals in a given group increases down the group as the atomic radius increases. Therefore, calcium should be more reactive than beryllium.

(**c**) Phosphorus and arsenic are Group 15 nonmetals; thus, phosphorus, which has a smaller atomic radius, is predicted to be more reactive than arsenic.

TRY IT 3.4A

Which element in each pair has the larger radius—that is, which is the larger atom in each pair: (**a**) Ca or Ba, (**b**) S or Se, (**c**) Si or S, or (**d**) Ga or Br?

TRY IT 3.4B

Which element in each pair is more reactive: (**a**) Mg or Sr, (**b**) Cl or Br, or (**c**) Rb or Cs?

3-8 Properties of Main-Group Elements

Elements in a periodic table group have similar properties but not the same properties. Some properties, as already illustrated for atomic radii and reactivity, increase or decrease in a predictable fashion from top to bottom of a periodic group.

Elements in a group generally react with other elements to form similar compounds, a fact accounted for by their identical numbers of valence electrons. For example, the formula for the compound of Li and Cl is LiCl; thus, you can expect there to be a compound of Rb and Cl with the formula RbCl and a compound of Cs and Cl with the formula of CsCl. Likewise, the formula Na_2O is known; therefore, a compound with the formula Na_2S predictably exists, because oxygen and sulfur are in the same group. In general, elements in the same group of the periodic table form some of the same types of compounds. In fact, several main groups have group names because of the distinctive and similar properties of the group elements.

The Group 1 elements (Li, Na, K, Rb, Cs, and Fr) are called the **alkali metals**. The name *alkali* derives from an old word meaning "ashes of burned plants." All the alkali metals are soft enough to be cut with a knife. None is found in nature as a free element, because all combine rapidly and completely with virtually all of the nonmetals and, as illustrated in Figure 3.11, with water. These elements form positive ions (**cations**) with a +1 charge. Francium, the last member of Group 1, is found only in trace amounts in nature, and all of its 21 isotopes are naturally radioactive.

The Group 2 elements (Be, Mg, Ca, Sr, Ba, and Ra) are called the **alkaline earth metals**. Compared with the alkali metals, the alkaline earth metals are harder, are more dense, and melt at higher temperatures. These elements form cations with a +2 charge. Because the valence electrons of alkaline earths are held more tightly, they are less reactive than their alkali metal neighbors. All of the alkaline earth metals react with oxygen to form an oxide MO, where M represents any of the alkaline earth metals.

$$2\,M(s) + O_2(g) \longrightarrow 2\,MO(s)$$

The **halogens** (F, Cl, Br, I, and At) are Group 17 elements. In the elemental state, each of these elements exists as diatomic molecules (X_2). Fluorine (F_2) and chlorine (Cl_2) are gases at room temperature, whereas bromine (Br_2) is a liquid and iodine (I_2) is a solid. (Recall that the only other diatomic elements are H_2, N_2, and O_2.) This illustrates a typical trend within a group—an increase in melting point and boiling point in going down a group. All isotopes of astatine (At) are naturally radioactive and disintegrate quickly. The name *halogen* comes from a Greek word and means "salt producing." The best-known salt containing a halogen is sodium chloride (NaCl), table salt. However, there are many other halogen salts, such as calcium fluoride (CaF_2), a natural source of fluorine; potassium iodide (KI), an additive to table salt that prevents goiter; and silver bromide (AgBr), the active photosensitive component of photographic film. The halogen salts contain negatively charged ions known as **anions**.

The *noble gases* (He, Ne, Ar, Kr, Xe, and Rn) are Group 18 elements. They are all colorless gases composed of single atoms at room temperature. They are referred to as *noble* because they lack chemical reactivity and generally do not react with "common" elements. Neon is the gas that glows orange-red in tubes of neon lights. Other gases and color-tinted tubes are used to give different colors. Radon (Rn) is naturally radioactive. Problems associated with indoor radon pollution are discussed in Chapter 13.

Alkali metal: An element in Group 1 of the periodic table

Cation: An atom or group of bonded atoms that has a positive charge

Alkaline earth metal: An element in Group 2 of the periodic table

Halogen: An element in Group 17 of the periodic table

Anion: An atom or group of bonded atoms that has a negative charge

Your Resources

In the back of the textbook:

→ *Review Card on Atoms and the Periodic Table*

In OWL for CHEM2 at www.cengagebrain.com

→ *Review Key Terms with Flash Cards (printable or digital)*

→ *Complete Interactive Practice Quizzes to prepare for tests*

→ *Submit Assigned Homework and Exercises*

Applying Your Knowledge

1. What is the law of conservation of matter? Give an example of the law in action.

2. State the law of definite proportions. Give an example to illustrate what it means.

3. Which early scientist is given credit for much of the development of the atomic theory of atoms, molecules, elements, and compounds?

4. What kinds of evidence did Dalton have for atoms that the early Greeks (Democritus, Leucippus) did not have?

5. How does Dalton's atomic theory explain these?
   (a) The law of conservation of matter
   (b) The law of definite proportions

6. Describe the most significant observation to be derived from Ernest Rutherford's experiments in which he bombarded gold foil with alpha particles.

7. One experiment we did not write about was one conducted by James Chadwick, who was the first to identify the neutron. This took place several years after the discovery of protons and electrons. Based on the experiments described in this chapter, why do you think that it took longer to identify the neutron?

8. The fact that carbon dioxide always contains 27.27% carbon and 72.73% oxygen by mass is a manifestation of the:
   (a) Law of mass action.
   (b) Law of definite proportions.

(c) Law of conservation of matter.

(d) Periodic law.

9. What is meant by the term *subatomic particles*? Give two examples.

10. Give short definitions for the following terms:
    (a) Atomic number   **(b)** Mass number
    (c) Atomic weight   **(d)** Isotope
    (e) Natural abundance   **(f)** Atomic mass unit

11. There are more than 1000 kinds of atoms, each with a different weight, yet only 118 elements were known as this book was being written. How does one explain this in terms of subatomic particles?

12. Some of the elements that appear in the periodic table are not found in nature but are instead prepared by fusing the nuclei of existing atoms with each other through collision at high speed and energy. Based on what you know about atomic structure, why are these high energies required?

13. Dmitri Mendeleev was the last born of 17 children in his family. Can you write the atomic symbol and give the total number of electrons and the number of valence electrons for each of the first 17 elements in his periodic table?

14. Which of the following pairs are isotopes? Explain your answers.
    (a) ^{50}Ti and ^{50}V   **(b)** ^{12}C and ^{14}C
    (c) ^{40}Ar and ^{40}K

15. What do all the atoms of an element have in common?

16. Correct the following statements related to the atomic structure of the elements, and explain.
    (a) An atom with 78 protons in its nucleus is a palladium atom.
    (b) All atoms of a given element contain the same number of neutrons in their nuclei.
    (c) All isotopes are radioactive.

17. If $^{37}_{17}$Cl is used to designate a certain isotope of chlorine, what would $^{35}_{17}$Cl represent?
    (a) An isotope with two fewer neutrons
    (b) An isotope with two fewer protons
    (c) A different element
    (d) An isotope with two more electrons
    (e) An isotope with one less proton and one less neutron

18. A common isotope of Li has a mass of 7. The atomic number of Li is 3. How can this information be used to determine the number of protons and neutrons in the nucleus?

19. The element iodine (I) occurs naturally as a single isotope of atomic mass 127; its atomic number is 53. How many protons and how many neutrons does it have in its nucleus?

20. You'll see in Chapter 13 that each radioactive isotope decays over time. The amount of time required for one-half of a sample to decay is called the half-life. For example, the half-life of carbon-14 (^{14}C) is 5740 years. If you had a 1-g sample of carbon-14, how many grams of carbon-14 atoms would remain after 5740 years? What about 11,480 years?

21. Which of the following elements would need only to gain two electrons to have a full outer shell (energy level) of electrons equivalent to the noble gas of the next highest atomic number?
    (a) $_7$N   **(b)** $_{20}$Ca   **(c)** $_{50}$Sn   **(d)** $_{35}$Br   **(e)** $_{34}$Se

22. Complete the following table so that it contains information about five different atoms.

	Number of Protons	Number of Neutrons	Number of Electrons	Atomic Number	Mass Number
(a)	32				73
(b)		14	14		
(c)				28	59
(d)	48	64			
(e)		115		77	

23. Identify the elements in question 22.

24. What number is most important in identifying an atom?

25. The atomic weight listed in the periodic table for magnesium is 24.305 amu. Someone said that there wasn't a single magnesium atom on the entire Earth with a weight of 24.305 amu. Is this statement correct? Why or why not?

26. The extent to which an isotope is present in a pure sample of an element is called the *abundance* of that isotope. The abundance and atomic mass of the three isotopes of silicon are shown in the following table. Use this information to calculate the atomic weight of silicon.

Isotope	Abundance	Atomic Mass
^{28}Si	92.23%	27.97693
^{29}Si	4.683%	28.97649
^{30}Si	3.087%	29.97377

27. Complete the following table for bromine, which has an atomic weight of 79.904.

Isotope	Abundance	Atomic Mass
^{79}Br	50.69%	78.91833
^{81}Br	?	80.91629

28. Complete the following table:

Isotope	Atomic No.	Mass No.	No. of Protons	No. of Neutrons	No. of Electrons
Bromine-81		81			
Boron-11	5				
^{35}Cl	17				
^{52}Cr		52			
Ni-60					
Sr-90					
Lead-206					

29. The longest wavelength for visible light is approximately 700 nm. What is the frequency for this type of light? Recall: $c = \lambda\nu$; $c = 3.00 \times 10^8$ m/s.

30. Which has a higher frequency, light having a wavelength of 700 nm or light having a wavelength of 600 nm?

31. Which of the wavelengths of light in question 30 has the higher energy?

32. Krypton is the name of Superman's home planet and also that of an element. Look up the element krypton and list its symbol, atomic number, atomic weight, and electron arrangement.

33. If the frequency of a given type of electromagnetic radiation increases, the:
    (a) Speed of light increases.
    (b) Energy decreases.
    (c) Wavelength increases.
    (d) Energy increases.

34. Some ultraviolet (UV) radiation has a wavelength of 350 nm. What is this equivalent to according to the following analysis?

$$350 \text{ nm} \times \frac{10^{-9} \text{ m}}{1 \text{ nm}} \times \frac{10^2 \text{ cm}}{1 \text{ m}} = \underline{\hspace{1cm}}$$

    (a) 350×10^{-7} cm   (b) 350×10^9 m
    (c) 35 cm   (d) 350×10^{-11} mm

35. Infrared (IR) radiation (wavelengths in the range of 10^{-4} cm) is significantly less energetic than UV radiation. This means that:
    (a) Infrared radiation has a higher frequency than UV radiation.
    (b) Ultraviolet radiation has a longer wavelength than IR radiation.
    (c) Infrared radiation has a lower frequency than UV radiation.
    (d) Infrared radiation has a shorter wavelength than UV radiation.

36. Which equation represents the maximum number of electrons that can occupy a given energy level (shell) in an element if n is an integer representing the energy level?
    (a) n^2 (b) $2n^2$ (c) $2n$ (d) n (e) $n/2$

37. Write the placement of electrons in their ground-state energy levels according to the Bohr theory for atoms having 6, 10, 13, and 20 electrons.

38. Write the Bohr electron notation for atoms of the elements sodium through argon.

39. How many valence electrons do atoms of each of the elements in question 38 contain?

40. Ultraviolet light has a wavelength of approximately 200 nm. What is the frequency for this form of UV light? Recall: $c = \lambda\nu$; $c = 3.00 \times 10^8$ m/s.

41. State the periodic law.

42. Give definitions for the following terms:
    (a) Group   (b) Period
    (c) Chemical properties   (d) Transition element
    (e) Inner transition element
    (f) Representative element

43. How did the discovery of the periodic law lead to the discovery of elements?

44. Describe the relative locations of metals, nonmetals, and metalloids in the periodic table.

45. How do metals differ from nonmetals?

46. Identify each of the following elements as either a metal, nonmetal, or metalloid.
    (a) Nitrogen   (b) Arsenic
    (c) Argon   (d) Calcium
    (e) Uranium

47. Answer the following questions about the periodic table.
    (a) How many periods are there?
    (b) How many representative groups or families are there?
    (c) How many groups consist of all metals?
    (d) How many groups consist of all nonmetals?
    (e) Is there a period that consists of all metals?

48. What do the electron structures of alkali metals have in common?

49. The diameter of an atom is approximately 10^{-10} m. The diameter of an atomic nucleus is about 10^{-15} m. What is the ratio of the atomic diameter to the nuclear diameter?

50. Give the number of valence electrons for each of the following:
    (a) Ba   (b) Al
    (c) P   (d) Se
    (e) Br   (f) K

51. Give the symbol for an element that has
    (a) 3 valence electrons   (b) 4 valence electrons
    (c) 7 valence electrons   (d) 1 valence electron

52. Draw the Lewis dot symbols for Be, Cl, K, As, and Kr.

53. From their position in the periodic table, predict which will be more metallic.
    (a) Be or B   (b) Be or Ca
    (c) As or Ge   (d) As or Bi

54. Which atom in the following pairs is more metallic?
    (a) Li or F   (b) Li or Cs
    (c) Be or Ba   (d) C or Pb
    (e) B or Al   (f) Na or Ar

55. What general electron arrangement is conducive to chemical inactivity?

56. Use the information in the periodic table to locate the following details.
    (a) The charge of a cadmium (Cd) nucleus
    (b) The atomic number of As
    (c) The atomic mass (or mass number) of an isotope of Br having 46 neutrons

(d) The number of electrons in an atom of Ba

(e) The number of protons in an isotope of Zn

(f) The number of protons and neutrons in an isotope of Sr, atomic mass (or mass number) of 88

(g) An element forming compounds similar to those of Ga

57. Complete the following table.

Atomic Number	Name of Element	Number of Valence Electrons	Period	Metal or Nonmetal
6				
12				
17				
37				
42				
54				

58. Write the symbols of the halogen family in the order of increasing size of their atoms.

59. Why does Cs have larger atoms than Li?

60. How does the atomic radius for a metal atom relate to the reactivity of the metal?

61. How does the atomic radius for a nonmetal atom relate to reactivity?

USE THE TOOLS.

- Rip out the Review Cards in the back of your book to study.

Or Visit OWL to:

- Read, search, highlight, and take notes in the Interactive eBook
- Review Flashcards to master key terms
- Test yourself with Mastery Questions and End-of-Chapter Questions
- Bring concepts to life with Tutorials and Animations!

Go to OWL for CHEM2 to begin using these tools.
Access at **www.cengagebrain.com**

Complete the Speak Up
survey in CourseMate at
www.cengagebrain.com

f **Follow us at**
www.facebook.com/4ltrpress

The Air
We Breathe

From its formation roughly 4.5 billion years ago to about 1.5 billion years ago, Earth's atmosphere was quite different than it is today. There was gaseous nitrogen (N_2), water vapor (H_2O), carbon dioxide (CO_2), methane (CH_4), and ammonia (NH_3), but little or no oxygen (O_2). Not all life required oxygen, and life was limited to deep in the oceans.

Scientists believe that by about 2.5 billion years ago, cyanobacteria capable of photosynthesis had evolved and that these bacteria began a process of changing our atmosphere. They produced oxygen as a waste product of their metabolism! The oxygen built up very slowly in our atmosphere because Earth contained abundant amounts of materials that are very reactive with oxygen. Iron and other materials found both in the oceans and on land, and reactive toward oxygen, consumed most of the waste oxygen at first. Slowly, however, these materials exhausted their capacity to consume more oxygen, and the concentration of oxygen in our atmosphere began to increase. By about 1.75 billion years ago, the oxygen content of our atmosphere had increased to near its present value of about 21%. The increase in oxygen concentration played two key roles in the development of oxygen-dependent life. First, oxygen was needed for respiration. Second, the increased oxygen concentration allowed the ozone layer to form in our stratosphere. Without the protection from high-energy ultraviolet radiation afforded by this ozone layer, oxygen-dependent organisms, of which humans are but one example, would not have survived or, if they had, would have been dramatically different than they are today.

Although we've just begun our exploration of chemistry, our goal here is to illustrate that one can use a relatively limited knowledge of elements, atoms, molecules, and chemical reactions to understand some pretty important phenomena. In this chapter, we will focus on air, a precious commodity that, as has become increasingly apparent, is at considerable risk. Have we begun to pollute our atmosphere to such an extent that our health and way of life are endangered? We will look at our air, that life-sustaining chemical canopy that surrounds Earth, and address some of the issues related to its composition and our influence on it.

Charles D. Winters

4-1 The Lower Atmospheric Regions and Their Composition

Earth is enveloped by a few vertical miles of chemicals that compose the gaseous medium in which we exist. We refer to that region of space as our *atmosphere*. The gases in Earth's atmosphere are held in place by gravitational force. The thermal energy that keeps the molecules in motion ensures that these gases do not simply sink to Earth's surface, while gravity ensures that they do not simply escape into outer space. The percentage composition of our atmosphere remains essentially constant with increasing altitude, but the density of the atmosphere drops significantly as the altitude increases and the temperature decreases. The

The atmosphere varies from ground level to the stratosphere.

Atmospheric pressure: The pressure exerted by the weight of the atmosphere at a given altitude

Troposphere: The region of Earth's atmosphere that extends from sea level up to about 10 km

Stratosphere: The region of Earth's atmosphere that extends from the troposphere to an altitude of about 50 km

Ozone layer: The stratospheric layer of gaseous ozone that protects life on Earth by filtering out most of the harmful ultraviolet radiation emitted by the Sun

Parts per million (ppm): 1 part out of every million parts

Parts per billion (ppb): 1 part out of every billion parts

atmospheric pressure decreases from 14.7 lbs/in² = 1 kg/cm² (1 atm) at sea level, to about 12 lbs/in² (~0.8 atm) in Denver, the "mile-high city," and finally to about 4.4 lbs/in² (~0.3 atm) at the top of Mt. Everest—8850 meters above sea level.

Even though the air at the top of Mt. Everest is still about 21% oxygen, the fact that the air there is only about one-third as dense as that at sea level means that a mountain climber would have to take in three times the normal volume of air to take in the same amount of oxygen. It is little wonder that most climbers use supplemental oxygen at high altitudes, where even slight exertion can leave one gasping for breath.

The region of Earth's atmosphere that is closest to Earth is known as the **troposphere** (Figure 4.1). The troposphere extends to an altitude of roughly 10 km (~6 mi). This is the region in which our weather occurs. Propeller and jet aircraft and balloons routinely travel in the troposphere. Denver sits at about 1.6 km above sea level, while Mt. Everest reaches almost 9 km into the troposphere. At the upper level of the troposphere, the temperature of the air is about –60°C (–76°F), much colder than Earth's average surface temperature of 15°C (59°F). Most of the pollution discussed in this chapter occurs near the surface of Earth. We will refer to this as *ground-level air pollution*, even though it may extend for several kilometers into our atmosphere.

Above the troposphere is the **stratosphere**, which extends to about 50 km (~30 mi). The temperature of the stratosphere rises gradually to about 0°C (32°F) at 50 km. This temperature increase is primarily due to the absorption of ultraviolet light (UV light) by the protective **ozone layer**, which is a band of higher-level ozone concentrations at about 15 to 30 km. Ozone, O_3, although a pollutant in the troposphere, plays a protective role in the stratosphere. Only some supersonic planes and some high-flying weather balloons venture into this region, although spacecraft travel through this region on their voyages beyond Earth's atmosphere.

FIGURE 4.1 Lower layers of Earth's atmosphere.

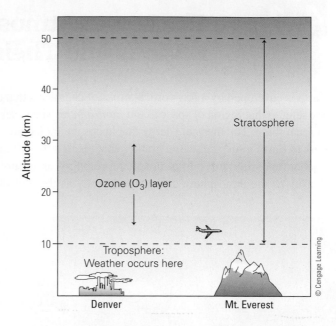

Before we consider atmospheric pollutants, let's examine the normal composition of dry air in more detail. As shown in Table 4.1, nitrogen and oxygen make up about 99% of Earth's atmosphere. There are lesser amounts of many other gaseous substances.

The concentration values in Table 4.1 are listed as percentages and as parts per million. All of us are familiar with percentages. For instance, in a class composed of 136 females and 64 males, we can easily determine the percentage of females by dividing the number of females by the total class membership (200) and multiplying by 100 to get 68%. This means that out of every 100 students, there are 68 females. Thus, 68% is 68 parts per hundred. Many of the constituents of air are present at concentrations far less than 1%. It gets quite awkward to deal with such small numbers; for example, neon is present at a level of 0.00182%. It is easier to express such low-level concentrations in **parts per million (ppm)** or **parts per billion (ppb)**. You will need to become comfortable with these units in your role as a chemically informed citizen.

To convert 0.00182% to ppm, we are really converting parts per hundred to ppm. In other words, we are asking the following question: If 0.00182 out of every 100 particles in the air are Ne atoms, what number would that be out of 1,000,000 particles? A simple proportional equation can be set up as shown here to solve for this value (represented by x):

$$\frac{0.00182}{100} = \frac{x}{1,000,000}$$

TABLE 4.1 Composition of Dry Air at Sea Level

Composition (by volume)		
Gas	Percentage	ppm
Nitrogen (N_2)	78.084	780,840
Oxygen (O_2)	20.948	209,480
Argon (Ar)	0.934	9340
Carbon dioxide (CO_2)	0.037	370
Neon (Ne)	0.00182	18.2
Helium (He)	0.00052	5.2
Methane (CH_4)	0.0002	2
Hydrogen (H_2)	0.0001	1
Krypton (Kr)	0.0001	1
Carbon monoxide (CO)	0.00001	0.1
Xenon (Xe)	0.000008	0.08
Ozone (O_3)	0.000002	0.02
Ammonia (NH_3)	0.000001	0.01
Nitrogen dioxide (NO_2)	0.0000001	0.001
Sulfur dioxide (SO_2)	0.00000002	0.0002

© Cengage Learning

By cross multiplying, we get $100x = 1820$ or $x = 18.2$ ppm. This is the same as saying that if we could reach into the air and grab 1,000,000 particles, 18.2 would, on average, be Ne atoms. Of course, we can't actually have two-tenths of a neon atom, but this is a statistical average. With a little practice, you should be easily able to convert among percent, ppm, and ppb in either direction.

EXAMPLE 4.1 Conversions Between Percent, Parts per Million, and Parts per Billion

(a) On a particularly humid day, the water vapor concentration in the air might be 0.85%. How would this concentration of water vapor be expressed in ppm and ppb?

SOLUTION

$$\frac{0.85}{100} = \frac{x}{1,000,000}$$

so $100x = 850,000$ and $x = 8500$ or 8500 ppm

and

$$\frac{0.85}{100} = \frac{x}{1,000,000,000}$$

so $100x = 850,000,000$ and $x = 8,500,000$ ppb

By solving the preceding proportional equations, you can see that 0.85% is the same as 8500 ppm or 8,500,000 ppb.

(b) Breathing air that contains 1000 ppm of carbon monoxide for about four hours will probably cause death in most adults. To what percentage of carbon monoxide does 1000 ppm correspond?

SOLUTION

The following equation is another way of stating that, if 1000 out of every 1,000,000 particles in the air are CO, (x) of every 100 particles are CO.

$$\frac{1000}{1,000,000} = \frac{x}{100}$$

so $1,000,000x = 100,000$ and $x = 0.1$ or 0.1%

In this example, in every 100 particles in the air, an average of 0.1 is CO. Of course, we can't have 0.1 molecules. To make this statistical average more realistic, we might choose to convert 0.1% to parts per thousand (ppt) to see that this corresponds to 1 molecule per thousand.

4-2 Air: A Source of Pure Gases

Air is a homogenous mixture of gases, and the gases can be separated from each other to provide pure chemical compounds. Before air can be separated into pure nitrogen, oxygen, and other gases by the process known as *fractionation of air*, water vapor and carbon dioxide must be removed. This is usually done by precooling the air so that frozen water vapor and carbon dioxide can be removed. Afterward, the air is compressed to more than 100 times normal atmospheric pressure, which causes the air to heat up. It is then cooled to room temperature and allowed to expand into a chamber. Overcoming the forces that attract molecules to each other requires energy (Section 5-7), so as the gas expands, the chamber in which they are held cools, and the gas gets cooler. If the compression and expansion are repeated many times and controlled properly, the expanding air cools to the point

FIGURE 4.2 (a) A diagram of the fractionation process. (b) Liquid nitrogen, which boils at −196°C.

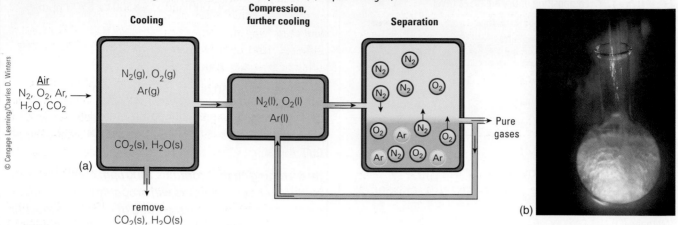

© Cengage Learning/Charles D. Winters

Cooling

Compression, further cooling

Separation

Air
N₂, O₂, Ar, H₂O, CO₂

$N_2(g)$, $O_2(g)$
$Ar(g)$

$CO_2(s)$, $H_2O(s)$

$N_2(l)$, $O_2(l)$
$Ar(l)$

(a)

remove
$CO_2(s)$, $H_2O(s)$

Pure gases

(b)

---

of liquefaction (also called condensation). Once air has been liquefied, its pure gases can be separated by taking advantage of their different boiling points.

When liquid air is allowed to warm up, nitrogen (boiling point −196°C) vaporizes first, and the remaining liquid becomes more concentrated in oxygen and argon. Further processing allows separation of high-purity oxygen (bp −183°C) and argon (bp −189°C). The less-abundant gases (neon, bp −246°C; xenon, bp −108°C; and krypton, bp −153°C) are also separated from liquid air.

Helium (bp −271°C) is not commercially recovered from air because it is cheaper to isolate it from natural gas, where it is sometimes present in as much as 7% by volume in the natural gas fields of West Texas.

4-2a Oxygen

Most oxygen produced by the fractionation of liquid air is used in steelmaking. Some is also used in rocket propulsion and in other controlled reactions with oxygen.

Liquid oxygen (LOX) can be stored and shipped at its boiling point of −183°C under atmospheric pressure. Substances this cold are called *cryogens* (from the Greek *kryos*, meaning "icy cold"). Cryogens present special hazards, since contact produces instantaneous frostbite, and structural materials such as plastics, rubber gaskets, and some metals become brittle and fracture easily at these low temperatures. Because liquid oxygen can react explosively with some substances, contact must be prevented between it and substances that can ignite and burn in air.

4-2b Nitrogen

Because nitrogen gas is so chemically unreactive, it is used as an inert atmosphere for applications such as welding

and other high-temperature metallurgical processes. If air is not excluded from these processes, unwanted reactions of hot metals with oxygen would occur.

Liquid nitrogen is used in medicine (*cryosurgery*)—for example, in cooling a localized area of skin prior to removal of a wart or other unsightly or pathogenic tissue. Because of its low temperature and inertness, liquid nitrogen is also widely used in frozen food preparation and preservation during transit. Trucks or railroad boxcars with nitrogen atmospheres present health hazards, since they contain little (if any) oxygen to support life. Workers must either enter such areas with breathing apparatus or first allow fresh air to enter.

Nitrogen is an essential element for plants, but they cannot derive it from gaseous nitrogen. It must first be "fixed." **Nitrogen fixation** is the process of changing atmospheric nitrogen into compounds that dissolve in water and can be absorbed through plant roots (Section 19-3).

4-2c Noble Gases

Approximately 250,000 tons of argon, the most abundant of the noble gases in the air, are recovered each year in the United States. Most of the argon is used to provide inert atmospheres for metallurgical processes. It is also used as a filler gas in incandescent lightbulbs to prolong the life of the hot filament. Neon is used in many "neon" signs, but argon, krypton, and xenon are also used for this purpose.

A "neon" sign using several noble gases.

© iStockphoto.com/Graffizone

4-3 Natural Versus Anthropogenic Air Pollution

The gases we have mentioned thus far are not the only substances found in the air we breathe. Some of these additional substances are present as the result of natural processes, while others are the result of human activity. When these substances are deemed undesirable or harmful, they are called pollutants.

Nature pollutes the air on a massive scale with ash, mercury vapor, hydrogen chloride, and hydrogen sulfide released during volcanic eruptions. Carbon dioxide and chlorinated organic compounds are released during forest and grassland fires. Many reactive organic compounds are naturally released into the atmosphere from coniferous and deciduous plants. We can do little to control these natural events, so our main concern in this chapter is pollution that results from human activity, known as **anthropogenic pollution**.

> **Anthropogenic pollution:** Pollution that is attributable to human activity

Human activity, especially in heavily populated (urban) areas, seems to have the most noticeable effects on the quality of the air we breathe. Automobiles, fossil fuel–burning power plants, smelting plants, other metallurgical plants, and petroleum refineries add significant quantities of polluting chemicals to the atmosphere. These atmospheric pollutants, especially in the concentrations found in urban areas, and increasingly in rural areas, can cause people to experience such symptoms as burning eyes, coughing, breathing difficulties, and even death. Air pollution is nothing new; Shakespeare wrote about it in the seventeenth century.

Prior to 1960, there was little concern about air pollution and little effort toward its control in the United States, in spite of some dramatic episodes in which many people suffered as a direct result of polluted air. For example, for five days in October 1948, the town of Donora, Pennsylvania, was overcome by air pollution that caused almost 6000 residents to become ill and 18 to die. In the past, smoke, carbon monoxide, sulfur dioxide, nitrogen oxides, and organic vapors were emitted into the air from industrial facilities with little apparent thought about their harmful nature as long as they were scattered into the atmosphere and away from human smell and sight.

In the 1960s, people in the United States began to realize that air pollution was a growing problem. The result was passage of the first Clean Air Act (CAA) in 1970. This law helped control air pollution from sources such as industry and automobiles, but it was not very comprehensive. The Clean Air Act was amended in 1977 to add stricter requirements.

(a)

(b)

(a) The eruption of Mt. St. Helens in 1980 ejected 2/3 cubic miles of material into the atmosphere. (b) The "Donora Smog" darkened the sky at noon. The main source of the pollution was zinc-smelting factories.

Courtesy USGS/Cascades Volcano Observatory/© Everett Collection Inc/Alamy

Hydrocarbon: A compound that contains only carbon and hydrogen atoms

Particulate: Either a solid particle or liquid droplet suspended in the atmosphere

Aerosol: A pollutant particle so small that it remains suspended in the atmosphere for a long period

Thermal inversion: An atmospheric condition in which warmer air is on top of cooler air

In June 1989, President George H. W. Bush signed into law the 1990 Clean Air Act amendments, a major overhaul of the earlier Clean Air Act. These amendments affect almost everything that is manufactured in this country. The substances regulated by the 1990 amendments include particulates, ozone, carbon monoxide, oxides of nitrogen and sulfur, **hydrocarbons** (compounds that contain only carbon and hydrogen atoms), volatile toxic substances, carbon dioxide, and stratospheric ozone-depleting chemicals. In the next section, we look first at the particles that obscure our vision, aggravate respiratory illnesses, and cause regional and global cooling by scattering sunlight. Subsequent sections will cover the other major pollutants.

4-4 Air Pollutants: Particle Size Makes a Difference

The effect that air pollutants can have is partly determined by their size. Air pollutant particles range in size from fly ash particles, which are big enough to see, down to individual molecules, ions, or atoms. Many pollutants are attracted into the water droplets of fog. Solids and liquid droplets suspended in the atmosphere are collectively known as **particulates**. The solids may be metal oxides, soil particles, sea salt, fly ash from electric generating plants and incinerators, elemental carbon, or even small metal particles. **Aerosol** particles range upward from a diameter of 1 nanometer (nm) to about 10,000 nm and may contain as many as a trillion atoms, ions, or small molecules. Particles in the 2000-nm range are largely responsible for the deterioration of visibility.

Aerosol particles are small enough to remain suspended in the atmosphere for long periods. Such particles are easily inhaled and can cause lung diseases. They may also contain mutagenic or carcinogenic compounds. Because of their relatively large surface area, aerosol particles have great capacities to *adsorb*, or concentrate chemicals on their surfaces. Liquid aerosols or particles covered with a thin coating of water may *absorb* air pollutants into the water, thereby concentrating them and providing a medium in which reactions may occur.

Millions of tons of soot, dust, and smoke particles are emitted into the atmosphere of the United States each year. The average suspended particulate concentrations in the United States vary from about 0.00001 g/m^3 of air in rural areas to about six times as much in urban locations. In heavily polluted areas, concentrations of particulates may increase to 0.002 g/m^3.

Particulates in the atmosphere can cool Earth by scattering and partially reflecting light from the Sun. Large volcanic eruptions such as those from Mt. St. Helens in 1980 and Mt. Pinatubo in 1991 had measurable cooling effects on Earth.

Particulates and aerosols are removed naturally from the atmosphere by gravitational settling and by rain and snow. Industrial emissions of particulates can be prevented by treating the emissions with one or more of a variety of physical methods such as filtration, centrifugal separation, and scrubbing. Another method often used is electrostatic precipitation, which is more than 98% effective in removing aerosols and dust particulates even smaller than 1 micrometer (μm) from exhaust gases. The effects of an efficient electrostatic precipitator can be quite dramatic, as Figure 4.3 shows.

4-5 Smog

The poisonous mixture of smoke, fog, air, and other chemicals was first called *smog* in 1911 by Dr. Harold de Voeux in his report on a London air pollution disaster that caused the deaths of 1150 people. Through the years, smog has been a technological plague in many communities and industrial regions.

What general conditions are necessary to produce smog? Although the chemical ingredients of smogs vary depending on the sources of the pollutants, certain geographical and meteorological conditions exist in nearly every instance of smog.

First, there must be a period of windlessness so that pollutants can collect without being dispersed vertically or horizontally. This sets the conditions for a **thermal inversion**, which is an abnormal temperature arrangement for air masses (Figure 4.4). Normally, warmer air is on the bottom, nearer the warm Earth, and this warmer, less dense air rises and transports most of the pollutants to the upper troposphere, where they are dispersed. In a thermal inversion, the warmer air overlays cooler, denser air near ground

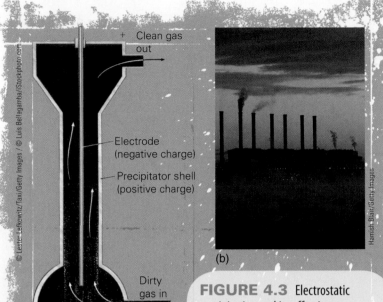

+ Clean gas out

Electrode (negative charge)

Precipitator shell (positive charge)

Dirty gas in

Dust falls off wall into collector

(a)

(b)

FIGURE 4.3 Electrostatic precipitation and its effectiveness.

(a) An electrostatic precipitator. The central electrode is negatively charged and imparts a negative charge to particles in smoke that passes over it. The particles are attracted to the positively charged walls and fall into the collector. (b) Of the stacks shown, one does not have a functioning precipitator, one requires cleaning, and the others have functioning precipitators.

Industrial smog: Smog generally caused by industrial activity such as coal burning

and partially oxidized organic compounds. Because it was first characterized in and around the city of London, it is sometimes called "London" smog. Smog of this kind, also called **industrial smog** because of its association with industrial activity, generally diminishes in intensity and frequency as less coal is burned and more controls are installed on industrial emissions.

The main ingredient in industrial smog is sulfur dioxide. Laboratory experiments have shown that sulfur dioxide increases aerosol formation, particularly in the presence of mixtures of hydrocarbons and nitrogen oxides. For example, mixtures of 3 ppm hydrocarbons, 1 ppm NO_2, and 0.5 ppm SO_2 at 50% relative humidity (H_2O) form aerosols inside which a chemical reaction occurs, producing sulfuric acid (H_2SO_4) as a major product. Breathing a sulfuric acid acrosol is very harmful, especially to people with respiratory diseases such as asthma or emphysema. At a concentration of 5 ppm for 1 hour, this kind of aerosol can cause constriction of bronchial tubes. A level of 10 ppm for 1 hour can cause severe breathing distress.

level. The air becomes stagnated. If the land is bowl shaped (surrounded by mountains, cliffs, or the like), a stagnant air mass can remain in place for quite some time. When these atmospheric conditions exist, the pollutants supplied by combustion in automobiles, electric power plants, and industrial plants accumulate to form smog.

Two general kinds of smog have been identified. One is derived largely from the combustion of coal and oil and contains sulfur dioxide mixed with soot, fly ash, smoke,

A notorious case of London smog occurred in 1952. On Thursday, December 4, of that year, a thermal inversion, a particularly foggy night, and freezing temperatures along the Thames River valley prompted Londoners to stoke their house fires with large amounts of coal. As was later revealed, low-quality coal had been supplied to the citizenry so that the higher-quality coal could be exported. This low-quality coal had a high sulfur content, which contributed to the resultant problems.

FIGURE 4.4 A thermal inversion.

Normally, air that is warmed near the surface rises, carrying pollutants with it. During a thermal inversion, a blanket of warm air becomes stationary over a layer of cooler, denser air. The result is that pollutants are trapped near the surface.

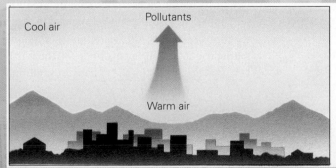

Cool air

Pollutants

Warm air

Cool air

Warm air layer (inversion layer)

Polluted (cooler) air

Photochemical smog:
Smog formed by the action of sunlight on photoreactive pollutants in the air

Primary pollutant:
A pollutant emitted directly into the atmosphere

Secondary pollutant:
A pollutant formed by chemical reactions in the atmosphere

Photodissociation:
Decomposition of a reactant, caused by the energy of light

In response to the continuing cold, fires continued to burn throughout the weekend, adding their smoke and chemical fumes to an increasingly polluted atmosphere. Accounts of the incident indicated that no sunlight shone through the smog for miles around London as early as Saturday. As the weekend wore on and weather conditions persisted, things only got worse. Respiratory problems mounted, and people resorted to sleeping in upright positions to ease their breathing. Hospitals were flooded with people gasping for breath, and more than 100 people died from respiration-related heart conditions on Monday.

Weather conditions began to improve on Tuesday, December 9, but much damage had been done by that time. An estimated 4000 deaths were initially attributed to the smog, but some recent (2001) estimates, based on a new assessment of the data, put the total death count, including belated deaths directly related to the smog, at 11,000 to 12,000 people. There have been other deadly cases of London smog since then, but none has ever matched the magnitude of the 1952 disaster. For comparison, consider that the London smog death total from this single incident almost doubled the combined tolls of the Japanese attack on Pearl Harbor in 1941 and the collapse of New York City's World Trade Center twin towers following the terrorist attacks in 2001. And yet, this is an event that most people have never even heard of. Pollution may not be as acutely newsworthy, but it can be more deadly.

The second type of smog is typical of Los Angeles and other urban centers where exhaust fumes from internal combustion engines are highly concentrated in the atmosphere. This predominantly urban smog is called **photochemical smog** because light—sunlight in this instance—is important in initiating several chemical reactions that together make the smog harmful. Photochemical smog is practically free of sulfur dioxide but contains substantial amounts of nitrogen oxides, ozone, oxygenated and ozonated hydrocarbons, and organic peroxide compounds, together with unreacted hydrocarbons of varying complexity. The automobile is a direct or indirect source of many of the components of photochemical smog. Consider Figure 4.5. It shows how several components of photochemical smog increase during rush hour in an urban area.

Many of the chemical reactions that create photochemical smog take place in aerosol particles. In this way, pollutants emitted directly into the atmosphere (**primary pollutants**) are converted into **secondary pollutants**—pollutants that are not directly released from some source but are formed by reactions with other components in the air. A few of these pollutants are summarized in Table 4.2.

One process by which primary pollutants form the secondary pollutants of photochemical smog begins with the absorption of light energy by a molecule of nitrogen dioxide. Nitrogen dioxide reacts with light with a wavelength between 280 nm and 430 nm. This **photodissociation** reaction (*photo*, light; *dissociation*, breaking apart) produces nitric oxide and free oxygen atoms that can react with a molecule of oxygen to produce a molecule of ozone (O_3), which is an important secondary pollutant.

$$NO_2(g) \xrightarrow{\text{light}} NO(g) + O(g)$$
$$O_2(g) + O(g) \longrightarrow O_3(g)$$

Atomic oxygen can also react with hydrocarbons to form other chemicals that are toxic and also impart an odor to the air. On a sunny day, only about 0.2 ppm of nitrogen oxides and 1 ppm of reactive hydrocarbons are sufficient to initiate these photochemical smog reactions. The hydrocarbons involved in these reactions come mostly from unburned petroleum products such as gasoline, and the nitrogen oxides come from the exhausts of internal combustion engines.

In the following sections, we shall look at the major ingredients of photochemical smog—the nitrogen oxides, ozone, sulfur dioxide, and hydrocarbons.

FIGURE 4.5 The concentrations of pollutants during rush hour.
High levels of primary pollutants are produced early in the day, followed by the formation of secondary pollutants.

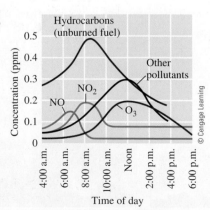

TABLE 4.2 Primary and Secondary Pollutants

Primary Pollutants	Secondary Pollutants
Nitric oxide (NO)	Nitrate (NO_3^-)
Nitrogen dioxide (NO_2)	Nitric acid (HNO_3)
Sulfur dioxide (SO_2)	Sulfur trioxide (SO_3)
Carbon monoxide (CO)	Sulfuric acid (H_2SO_4)
Carbon dioxide (CO_2)	Sulfate (SO_4^{2-})
Hydrocarbons	Organic peroxides
Ash	Ozone (O_3)

© Cengage Learning

NO$_x$ gases = ↓

4-6 Nitrogen Oxides

There are eight known oxides of nitrogen, three of which are recognized as important components of the atmosphere: dinitrogen oxide (N_2O), nitric oxide (NO), and nitrogen dioxide (NO_2). These oxides of nitrogen are collectively known as "NO_x." About 97% of the nitrogen oxides in the atmosphere are naturally produced; only 3% result from human activity. Certain bacteria can produce N_2O, so this oxide of nitrogen is also commonly found in the atmosphere in trace amounts.

Normally, the atmospheric concentration of NO_2 is a few parts per billion or less; most of the nitrogen oxides formed during lightning storms are washed out by rain. This is one of the ways nitrogen is made available to plants. Looking at all the sources of oxides of nitrogen (Table 4.3), it is apparent that combustion processes are their primary sources. In the United States, most oxides of nitrogen from sources other than nature are produced from fossil fuel combustion in vehicles, industry, and power plants.

TABLE 4.3 Emissions of NO_x

Source	Emissions (millions of tons) United States
Fossil fuel combustion	66
Biomass burning	1.1
Lightning	3.3
Microbial activity in soil	3.3
Input from the stratosphere	0.3
Total (uncertainty in estimates)	74 ($\pm$1)

Source: Stanford Research Institute.

© Cengage Learning

One of the key culprits in the formation of photochemical smog is NO (nitric oxide), a colorless but reactive gas. Its nominal concentration in our atmosphere is quite low, and there are natural sources of nitric oxide about which we can do little. It is the anthropogenic nitric oxide about which we are most concerned, as it is a pollutant over which we may have some control. Where does this nitric oxide come from? You are probably aware that an internal combustion engine takes in huge amounts of air as it runs. This air comprises about 78% nitrogen gas, which is unreactive under normal circumstances. About 21% of the air sucked into the engine is oxygen gas. Nitrogen and oxygen do not react with each other at normal air temperatures, but the combustion chamber of an engine may reach temperatures as high as 2500°C (the average is probably around 400–600°C). Under these conditions, the reaction that forms nitric oxide, which requires heat, takes place quite readily. The same reaction is triggered in our atmosphere by lightning strikes.

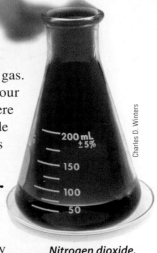

Nitrogen dioxide.

Charles D. Winters

$$N_2(g) + O_2(g) \xrightarrow{\text{energy}} 2\,NO(g)$$

Nitric oxide itself is a toxic pollutant, and the U.S. Environmental Protection Agency (EPA) includes it among the primary toxic pollutants it attempts to regulate. If this nitric oxide is able to escape the exhaust system of the engine, it reacts quite rapidly with atmospheric oxygen to produce nitrogen dioxide, a reddish-brown, highly toxic gas.

$$2\,NO(g) + O_2(g) \xrightarrow{\text{energy}} 2\,NO_2(g)$$

nitrogen dioxide

Almost anyone who has ever observed the skyline of a major city such as Houston, Los Angeles, or Mexico City has seen a reddish-brown haze hanging over the city at times. This haze is composed largely of nitrogen dioxide, an ocular and respiratory irritant. The EPA has set the allowable yearly average exposure to all nitrogen oxides at 53 ppb. A city or state whose level of nitrogen oxides exceeds this number is subject to fine and government intervention.

A primary role of nitrogen dioxide as a pollutant is in the formation of the secondary pollutant ozone (Section 4-7) and photochemical smog. Nitrogen dioxide can also react with water to form nitric acid and nitrous acid. This

The buildup of nitrogen dioxide in an urban area.

Ed Pritchard/Stone/Getty Images

reaction takes place readily in aqueous aerosols, producing acids that help stabilize the droplet.

$$2\ NO_2(g) + H_2O(\ell) \longrightarrow HNO_3(aq) + HNO_2(aq)$$

Nitric acid   Nitrous acid

Of course, breathing air containing these aerosol droplets is harmful because of the corrosive nature of the acids. The acids in turn can react with ammonia or metallic particles in the atmosphere to produce nitrate or nitrite salts. For example,

$$NH_3(g) + HNO_3\ (aq) \longrightarrow NH_4NO_3(aq)$$

Ammonia   Ammonium nitrate
(a salt)

Both the acids and the salts stabilize the aerosol particles, which eventually settle from the air or dissolve in larger raindrops. Nitrogen dioxide, besides causing the formation of ozone, is a primary cause of haze in urban or industrial atmospheres because of its participation in the process of aerosol formation.

4-7 Ozone as a Pollutant

Ozone consists of three oxygen atoms bound together in a molecule with the formula O_3. It is an **allotrope** of molecular oxygen. It has a pungent odor that we often smell near sparking electrical appliances or after a thunderstorm when rainfall washes lightning-produced ozone out of the troposphere.

As you've seen, any nitric oxide that escapes into our atmosphere can react with oxygen to form nitrogen dioxide. Nitrogen dioxide can, under the intensity of the bright summer sun, be photochemically degraded into nitric oxide and very reactive oxygen atoms (O). As stated earlier, the reaction between these oxygen atoms and oxygen molecules leads to the formation of ozone.

Because it is generated by chemical reactions in the atmosphere, ozone is one of the most difficult pollutants to control. One can observe the daily course of ozone formation in cities throughout the United States by accessing the EPA's ozone monitoring web site, http://www.airnow.gov. You may even find that your local newspaper includes color-coded ozone alerts on its weather page.

Table 4.4 shows the relationship between the local ozone concentration in the atmosphere, air quality indices, and air quality descriptors in common use. A period of unhealthy air may be quite short or quite lengthy depending on weather conditions, intensity of sunlight, prevailing winds, topography, and the level of nitrogen dioxides emitted into the air in a given area by automobiles, buses, trucks, trains, ships, industry, and even lawn mowers.

Although there has been an overall improvement in air quality in the United States in recent years, there are still problems with ozone in many urban areas and even in some remote areas such as Big Bend National Park deep in southwest Texas. Federal, state, and local efforts to control emissions of nitrogen oxides are constantly discussed and debated in the local newspapers and by political action groups in many areas of the United States. The popularity of the automobile in American culture has had a dramatic impact on air quality. In an approach not yet tried in the United States, there have been efforts in some major cities in other parts of the world to limit the number of automobiles allowed into those cities on a given day or during a given time period through the use of permits or taxes. There have been noticeable improvements in the air quality of these cities during the times these programs have been in effect, but as you can imagine, the programs have met some opposition.

In the United States, the EPA's efforts to control ozone concentrations in our ground-level air have had some success. In 2004, thanks to their emission control programs, the EPA reported a continuous decrease in ozone levels. The result was the lowest national ozone level since 1980. However, concerns that the downward trend in ozone levels may be slowing have led regulators to propose additional national rules designed to decrease vehicle exhaust and industrial emissions.

TABLE 4.4 Air Quality Index and Air Quality Descriptors for Ozone

Ozone Concentration, ppm (8-hr average unless noted)	Air Quality Index Values	Air Quality Descriptor
0.00 to 0.064	0 to 50	Good
0.065 to 0.084	51 to 100	Moderate
0.085 to 0.104	101 to 150	Unhealthy for sensitive groups
0.105 to 0.124	151 to 200	Unhealthy
0.125 (8 hr) to 0.404 (1 hr)	201 to 300	Very unhealthy

© Cengage Learning

© Miriam Maslo/Science Photo Library/Corbis

4-8 Sulfur Dioxide: A Major Primary Pollutant

Sulfur dioxide is produced when sulfur or compounds containing sulfur are burned in air.

$$S(s) + O_2(g) \longrightarrow SO_2(g)$$

Although volcanoes put large amounts of SO_2 into the atmosphere annually, human activities probably account for up to 70% of all emissions on a global basis. Once formed, SO_2 generally becomes distributed in aerosol droplets, which are numerous enough to contribute to significantly reduced visibility, and can affect both global and regional climate by causing the scattering of sunlight that would otherwise warm Earth. Emissions of SO_2 cause the mean temperature in the United States to be about 1°C cooler than it would be otherwise.

Coal, a heterogenous solid mixture, contains the mineral pyrite.

© schankz/Shutterstock.com

Most of the coal burned in the United States contains sulfur in the form of the mineral pyrite (FeS_2). The percentage (by weight) of sulfur in this coal ranges from 1% to 4%. The pyrite reacts with oxygen as the coal burns, forming SO_2.

$$4\ FeS_2(s) + 11\ O_2(g) \longrightarrow 2\ Fe_2O_3(s) + 8\ SO_2(g)$$

Large amounts of coal are burned in this country to generate electricity. A 1000-megawatt (MW) coal-fired generating plant can burn about 700 tons of coal an hour. If the coal contains 4% sulfur, that equals 56 tons of SO_2 an hour, or 490,560 tons of SO_2 every year. About 800 million tons of coal are burned each year to produce electricity.

Oil-burning electric generating plants can also produce comparable amounts of SO_2 because some fuel oils can contain up to 4% sulfur. The sulfur in the oil is in the form of compounds in which sulfur atoms are bound to carbon and hydrogen atoms. Gasoline contains relatively low concentrations of sulfur-containing compounds. Even so, in 1999 the EPA mandated that the maximum concentration of sulfur-containing compounds be reduced from 120 ppm to 90 ppm by 2005 and to 15 ppm by 2010, leading to a decrease in emissions of more than 90%.

States that rely mainly on coal for their electricity production and industrial furnaces have the highest SO_2 emissions in the United States. Operators of all coal-fired burners are under EPA orders to eliminate most of the SO_2 before it reaches the exhaust stack.

The removal of sulfur from high-sulfur coal is costly and incomplete. One method is to pulverize the coal to the consistency of talcum powder and remove the pyrite (FeS_2) by magnetic separation. Reducing the sulfur content of fuel oil is also costly. It involves the formation of hydrogen sulfide (H_2S) by bubbling hydrogen through the oil in the presence of metal catalysts.

At present, most sulfur-containing coal is burned without prior treatment, and SO_2 is removed from the exhaust gases after it is formed. In one method, lime (calcium oxide) reacts with SO_2 to form calcium sulfite, a solid particulate, which can be removed from an exhaust stack by an electrostatic precipitator. Lime is prepared by heating limestone, which is mainly calcium carbonate.

$$CaCO_3(s) \xrightarrow{\text{Heat}} CaO(s) + CO_2(g)$$
$$\underset{\text{Limestone}}{} \qquad \underset{\text{Lime}}{}$$

$$CaO(s) + SO_2(g) \longrightarrow CaSO_3(s)$$
$$\underset{\text{Calcium sulfite}}{}$$

In another method, exhaust gases containing SO_2 are passed through molten sodium carbonate and solid sodium sulfite is formed (Figure 4.6).

Notice how both of these methods of SO_2 removal emit additional CO_2, a major contributor to global warming (Section 6-3), into the atmosphere. Newer technology to address this problem continues to be explored.

A less desirable method of lowering the effects of SO_2 emissions, but one still being used, is sending the smoke up very tall exhaust stacks. Tall stacks emit SO_2 at a high elevation and away from the immediate vicinity, which allows the SO_2 to be diluted before it forms aerosol particles. The fact remains, however, that the longer it stays in the air, the greater chance SO_2 has to become sulfuric acid. A 10-year study in Great Britain showed that although SO_2 emissions from power plants increased by 35%, the construction of tall stacks decreased the ground-level concentrations of SO_2 by as much as 30%. The question is, who got the SO_2? In this case, Britain's solution was others' pollution. In the United States, the EPA may have added to a pollution problem unwittingly by implementing rules in 1970 that caused plants to increase the height of smokestacks and caused pollutants to be carried longer distances by winds.

Most of the SO_2 that does get into the atmosphere reacts with oxygen to form sulfur trioxide (SO_3). This compound has a strong affinity for water and dissolves in aqueous aero- sol particles, forming sulfuric acid, which in turn contributes to acid rain (Section 11-2).

$$SO_3(g) + H_2O(\ell) \longrightarrow H_2SO_4(aq)$$

4-9 Hydrocarbons and Air Pollution

Hydrocarbons enter the atmosphere from both natural sources and human activities. Certain natural hydrocarbons are produced in large quantities by both coniferous and deciduous trees. In 1988, William Chameides of Georgia Tech in Atlanta published a report in *Science* magazine in which he stated that in some cities, trees may account for more hydrocarbons in the atmosphere than those produced from human activities. The EPA has since found this to be true. Methane gas (CH_4) is produced by such diverse sources as termites, ants, rice growing, ruminant animals such as cows, and decay-causing bacteria acting on dead plants and animals. Human activities such as the use of industrial solvents, petroleum refining and distribution, and the release of unburned gasoline and diesel fuel components account for a large amount of hydrocarbons in the atmosphere.

In addition to simpler hydrocarbons, some larger hydrocarbon molecules (more carbon atoms than in CH_4) are released into the atmosphere, primarily from motor vehicle exhaust. The greatest danger of some of these pollutants and the organic derivatives formed from them is their toxicity. Hydrocarbons are known to react with

FIGURE 4.6 Removal of SO_2 from flue gas by reaction with molten sodium carbonate.

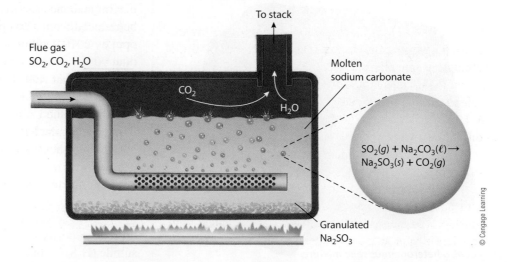

To stack

Flue gas
SO_2, CO_2, H_2O

CO_2

H_2O

Molten
sodium carbonate

$$SO_2(g) + Na_2CO_3(\ell) \rightarrow$$
$$Na_2SO_3(s) + CO_2(g)$$

Granulated
Na_2SO_3

© Cengage Learning

TABLE 4.5 Emission Rates for Hydrocarbons (HC), Carbon Monoxide (CO), and Nitrogen Oxides (NO_x)

1960 (Precontrol—no catalytic mufflers installed)		1993 (Catalytic mufflers required on all U.S. automobiles)		2012 Standards*,†	
HC	10.6 g/mi	HC	0.41 g/mi	NMHC‡	0.075 g/mi
CO	84.0 g/mi	CO	3.4 g/mi	CO	3.4 g/mi
NO_x	4.1 g/mi	NO_x	1.0 g/mi	NO_x	0.2 g/mi

© Cengage Learning

* Source: EPA Office of Transportation and Air Quality.

† These standards were enacted in 2004.

‡ NMHC stands for nonmethane hydrocarbons.

ground-level ozone to form a class of compounds called organic peroxides, which are severe irritants.

Although it is practically impossible to control hydrocarbon emissions from living plants and other natural sources, hydrocarbon emissions from automobiles can be controlled. Two means of control are in use at present. First, the spouts and hoses on gasoline pumps have been redesigned to prevent gasoline from entering the air. Second, catalytic converters that reduce emissions of hydrocarbons, CO, and NO_x are now part of every U.S. automobile's exhaust system.

The effectiveness of these catalytic converters, which have been on automobiles sold in the United States since the mid-1970s, can be seen by comparing the emissions of hydrocarbons, CO, and NO_x in grams per vehicle-mile in 1960, before there were controls, to the same values now (Table 4.5).

4-10 Carbon Monoxide

At least 10 times as much carbon monoxide enters the atmosphere from natural sources as from all industrial and automotive sources combined. Of the 3.8 billion tons of carbon monoxide emitted every year, about 3 billion tons are emitted by the oxidation of decaying organic matter in the topsoil. In spite of this fact, carbon monoxide is considered an air pollutant, primarily because so much of it is produced by human activities in the urban environment.

Like ozone, carbon monoxide is one of the most difficult pollutants to control. Cities such as Los Angeles, Houston, and other highly populated urban centers with high densities of automobiles tend to be repeatedly cited

by the EPA for not attaining the required ambient air quality for carbon monoxide.

Some carbon monoxide is always produced when carbon from coal or carbon-containing compounds like those found in petroleum are burned using an insufficient quantity of oxygen.

$$2\ C(s) + O_2(g) \longrightarrow 2\ CO(g)$$

Gasoline engines are notorious sources of CO. This is because the rapid combustion inside the combustion chamber does not burn all the fuel to CO_2 before the exhaust gases are swept out. Modern catalytic converters convert much of this carbon monoxide to carbon dioxide, but the amounts that are not converted make being near a heavily traveled street dangerous because of the carbon monoxide concentrations. At peak traffic times, concentrations as high as 50 ppm can occur. In the countryside, carbon monoxide levels are closer to the global average of 0.1 ppm. The only effective means of controlling carbon monoxide concentrations in urban air is to control the major emitters—automobiles. Of course, transportation that does not depend on burning hydrocarbon fuels does not emit *any* CO.

A pine tree, which can release hydrocarbons into the air.

© iStockphoto.com/ironika

You likely know that carbon monoxide is dangerous, but why and how? Recall from Chapter 1 that the risk associated with any activity depends on the likelihood of exposure and the hazard associated with that exposure. Carbon monoxide is a colorless, odorless gas at room temperature. There is no common observation you can make to detect its presence, and exposure can occur without warning. It is also hazardous because of the effect it can have on the ability of your body to use the oxygen you take in with each breath. Overall, the risks associated with carbon monoxide make it a quite dangerous pollutant.

The interference of carbon monoxide with oxygen transport in the blood is one of the best understood kinds of metabolic poisoning. Because it is a gas, carbon monoxide mixes with air, is inhaled, and comes into contact with the blood while in the lungs. There, it is picked up by the hemoglobin in red blood cells and transported throughout the body, just as oxygen would be. This transport can take place because hemoglobin contains iron atoms that bind (form a bond to) either of these two molecules. The chemical reactions for the binding of O_2 or CO by hemoglobin are shown here. (We've drawn the hemoglobin molecule in a simplified form in order to focus on the topic at hand. See Chapter 15 for more detail.)

$$O_2 + Fe_{hemoglobin} \rightleftharpoons O_2-Fe_{hemoglobin}$$

$$CO + Fe_{hemoglobin} \rightleftharpoons CO-Fe_{hemoglobin}$$

Both of these reactions are reversible, as indicated by the double arrow ($\rightleftharpoons$). That is, once hemoglobin picks up a molecule, it can also release it in the reverse process (that's what it does when it transports oxygen to a cell in the body). Here, there is an important difference between CO and O_2. Laboratory tests show that carbon monoxide is held 140 times as strongly by hemoglobin as oxygen is held. Since hemoglobin is so effectively tied up by carbon monoxide, hemoglobin molecules that contain a CO molecule cannot perform their vital function of transporting oxygen, as they normally would. So, even though there is plenty of oxygen in the air we breathe, carbon monoxide impairs our bodies' ability to use that oxygen.

You do not have to breathe pure carbon monoxide in order to be affected. Because the reactions of both carbon monoxide and oxygen with hemoglobin are reversible, the concentrations of the two gases also determine how much hemoglobin will be combined with either molecule. In air that contains 0.1% CO or less, oxygen molecules outnumber the CO molecules by at least 200 to 1. Under these conditions, there is little danger from carbon monoxide poisoning. However, breathing air that contains higher concentrations of CO allows it to effectively compete with the oxygen in binding to hemoglobin, and this causes problems. Breathing air with a concentration of 30 ppm of CO for 8 hours is sufficient to cause headache and nausea for most people. Breathing air that is 0.1%

In some occupations, it is difficult to avoid long-term exposure to carbon monoxide.

(1000 ppm) CO for 4 hours ties up approximately 60% of the hemoglobin of an average adult, and death is likely to result unless the attached CO molecules can be freed. Because the amount of hemoglobin bound to CO is related to the amount of CO in the air, increasing the level of O_2 an affected person breathes can be an effective treatment. A victim of carbon monoxide poisoning can be saved if he or she is quickly exposed to fresh air or, still better, pure oxygen.

An analogy is perhaps appropriate. Imagine that instead of air, hemoglobin, O_2, and CO, we are observing a cafeteria full of college students where only two meal options are available: a big, healthy salad or greasy (but very tasty) pizza. As students pass through the line it is likely that more will choose the pizza even if few pieces are displayed, although some will still choose the salad. Because the students have a higher affinity for pizza than salad, a greater number will return to their tables with an unhealthy meal. If even more pizza is displayed, then the percentage of students receiving an unhealthy meal will increase. One way to undo this unhealthy balance would be to set two salads in front of each student. Then, after eating the pizza they would choose to eat a salad. Hemoglobin responds in much the same way as these students; it has a strong affinity for CO (unhealthy) and "prefers" it over oxygen (healthy). Overcoming this preference would require limiting the choice between O_2 and CO (less carbon monoxide), or by offering more O_2 once the initial choice was made.

© iStockphoto.com/RelaxFoto.de

4-11   A Look Ahead

The quality of the air we breathe can be dramatically affected by the presence of the air pollutants we've discussed in this chapter. Both legislative and technological efforts to reduce the concentrations of these pollutants have shown positive results, but there is clearly still much to be done. As we progress through the text, we will frequently revisit some of the chemistry involved in these efforts. For instance, in Chapter 6 we will focus on the combustion of fossil fuels and the production of, among other things, carbon dioxide. We don't normally think of carbon dioxide as a pollutant, but it is having a major impact on our environment in the form of **global warming**. Although the concentration of CO_2 in our atmosphere is low, as is its toxicity, its ability to trap heat near the surface of Earth may be of even more concern than the pollutants discussed in this chapter.

In Chapter 12, we will discuss, in more detail, some of the technology associated with reducing air pollution from internal combustion engines.

You will also see that some pollutants (sulfur oxides and nitrogen oxides) have a major impact on our environment as primary contributors to acid rainfall. You will need some knowledge of water chemistry and of the pH scale to fully appreciate this problem, so we will defer that discussion until Chapter 11.

Lastly, some chemicals known as chlorofluorocarbons (CFCs) are pollutants of major concern in an entirely different region of our atmosphere. These chemicals, whose effect at tropospheric levels is minimal, play havoc with the protective ozone layer of our stratosphere.

Air pollution can take many forms, come from many sources, have its impact in different regions of our atmosphere, and require different forms of personal, legislative, and technological action to control. You will decide what contribution, if any, you as an individual can make to these efforts. Alternatively, you may decide that no action is required. Only by being informed will you be able to make these decisions.

> **Global warming:**
> Warming of Earth above a level that is desirable from a human viewpoint as a result of increased concentrations of one or more greenhouse gases

In the back of the textbook:

→ Review Card on *The Air We Breathe*.

   In OWL for CHEM2 at www.cengagebrain.com

→ Review Key Terms with Flash Cards (printable or digital)

→ Complete Interactive Practice Quizzes to prepare for tests

→ Submit Assigned Homework and Exercises

Applying Your Knowledge

1. Define the following terms:
   (**a**) Polluted air   (**b**) Aerosol
   (**c**) Photochemical   (**d**) Secondary
        smog   pollutant

2. Define the following terms:
   (**a**) Photodissociation   (**b**) Particulate
   (**c**) Thermal inversion   (**d**) Free radical

3. Name three major air pollutants and give a source for each.

4. Which of these would not be considered a primary pollutant in our atmosphere—nitric oxide (NO), carbon monoxide (CO), ozone (O_3), hydrocarbons, or sulfur dioxide (SO_2)?

5. What federal legislation has the abbreviation CAA? Describe how this legislation has changed over the years.

6. Describe what a thermal inversion is and explain how it can aggravate the problems caused by smog.

7. How is an industrial-type smog different from a photochemical smog?

8. Explain how ozone is formed in the troposphere.

9. Place the following four reactions in a sequence that explains the production of ozone in a typical urban environment.
   (**a**) $NO_2 + sunlight \longrightarrow NO + O$
   (**b**) $N_2 + O_2 \longrightarrow 2\ NO$
   (**c**) $2\ NO + O_2 \longrightarrow 2\ NO_2$
   (**d**) $O + O_2 \longrightarrow O_3$

10. Which reaction in question 9 involves an atom combining with a diatomic molecule to produce a triatomic molecule?

11. What are the main sources of nitrogen oxides in the atmosphere?

12. What is an aerosol and how does it play a role in air pollution?

13. Name two sources of hydrocarbons in the atmosphere. Which one is more readily controlled?

14. Write the equation for the reaction that occurs when lightning causes nitrogen to react with oxygen.

15. Nitrogen dioxide plays a role in the formation of what secondary pollutant?

16. What are the names of the two regions of the atmosphere discussed in this chapter? Which one is closer to the surface of Earth?

17. What happens when a photon of light strikes a molecule of NO_2? Write the reaction.

18. Describe how ozone can be harmful when it is present in the air we breathe.

19. How are the harmful effects of SO_2-containing aerosols, NO_2, and ozone similar?

20. Describe two sources of carbon monoxide.

21. Describe how pure oxygen can be obtained from air.

22. How is the control of ozone in the lower atmosphere connected to the control of NO_x emissions?

23. What health risks are posed by breathing air contaminated with particulates?

24. What is the product of the combustion of sulfur or sulfur-containing compounds in air? Write the reaction.

25. Pick a source of SO_2 emissions and describe two ways SO_2 emissions from that source can be controlled.

26. Write the chemical equation for the reaction between carbon and oxygen when an insufficient amount of oxygen is present. What is the name of the product of this reaction?

27. A human forced to breathe air that is 7% carbon dioxide would probably become unconscious, suffer brain damage, and die within a few minutes. However, we are worried that Earth's atmosphere may approach 400 ppm of carbon dioxide within the next few years. Which of the following statements is correct?
    (a) 400 ppm of carbon dioxide is greater than 7% carbon dioxide.
    (b) 7% is equal to 7000 ppm.
    (c) 400 ppm is equal to 0.04%.
    (d) 400 ppm of carbon dioxide in our air means the percentage of oxygen will drop by 4%, making it harder to breathe.
    (e) 400 ppm is equal to 40,000 ppb.

28. A December 2002 ice storm in the southeastern United States knocked out electricity for millions of people. About 150 people were hospitalized and treated in hyperbaric chambers where they could breathe much higher-than-normal concentrations of oxygen; several people died. Use your chemical literacy to decide which of the following scenarios probably led to this.
    (a) People purchased large blocks of dry ice (solid carbon dioxide) to keep the items in their refrigerators from spoiling, and the passage of the dry ice from the solid state to the gaseous state caused carbon dioxide poisoning. The increased amounts of carbon dioxide also made it more difficult to keep fires burning, so many people experienced hypothermia.
    (b) Massive quantities of broken tree limbs released huge amounts of carbon dioxide into the atmosphere, leading to severe breathing difficulties for those with respiratory problems.
    (c) Downed electrical wires sparked, and this led to the production of significant amounts of ozone from atmospheric oxygen in nearby areas.
    (d) Trying to keep warm without electricity, people resorted to poorly ventilated indoor fires that produced significant amounts of carbon monoxide due to incomplete combustion. This led to many cases of CO poisoning.

5

Chemical Bonding
and States of Matter

A toms are the fundamental building blocks of all matter. Everything that exists is made up of atoms held together in innumerable combinations that range from the simplest molecules,
composed of only two atoms, to the most complex molecules, composed
of thousands of atoms. Atoms of the 90 naturally occurring elements,
and occasionally the 25 or so nonnatural elements, combine to make
everything that we can see, touch, and smell, and lots of things that we
can't so easily observe.

What makes atoms stick together? Valence electrons form the
glue, but how? Atoms form a few different types of bonds when they
combine with other atoms. In this chapter, we describe two of them—
ionic bonds and *covalent bonds*. Transfer of valence electrons from
an atom of a metal to an atom of a nonmetal produces an ionic bond.
Sharing of electrons between atoms of nonmetals produces covalent
bonds. These simple ideas about valence electrons are the basis for
understanding the bonding in two major classes of chemical compounds—ionic compounds and molecular compounds. In addition,
once you understand how valence electrons function, you can better
understand how the states of matter—gas, liquid, and solid—are
formed.

The concept of valence electrons developed by G. N. Lewis is
useful for understanding how atoms of different elements interact
and why elements in the same group have similar properties (Section 3-7). Lewis assumed that each noble gas atom had a completely
filled outermost shell, which he regarded as a stable configuration
because of the lack of reactivity of noble gases. Since all noble
gases (except He) have eight valence electrons, the observation
came to be known as the **octet rule**: *When atoms of elements react,
they tend to lose, gain, or share electrons to achieve the same
electron arrangement as the noble gas nearest them in the periodic
table.* Metals can achieve a noble gas
electron arrangement by giving up electrons, and nonmetals can achieve a
noble gas electron arrangement by adding or sharing electrons. This is the
basis for our discussion of the two
major classes of bonding—ionic
bonding and covalent bonding.

Octet rule: In forming
bonds, main-group elements
gain, lose, or share electrons
to achieve a stable electron
configuration with eight
valence electrons

5-1 Electronegativity and Bonding

The tendency of an atom to participate in bonding with another atom is related directly to a property of the atom known as its **electronegativity**. The electronegativity of the elements is shown in the periodic table in Figure 5.1. The electronegativity values range from a low of 0.8 for Cs to a high of 4.0 for F. Atoms with high values are said to be electronegative and attract the valence electrons of other atoms quite strongly. Atoms with lower electronegativity values are said to be electropositive, and they donate their valence electrons to a significantly more electronegative atom.

> **Electronegativity:**
> A measure of the relative tendency of an atom to attract electrons to itself

FIGURE 5.1 Electronegativity values.

Nonmetals have high electronegativity values (e.g., O, 3.5), metalloids have intermediate values (e.g., Ge, 2.0), and metals have low values (e.g., Mg,1.3).

1	2	3	4	5	6	7	8	9	10	11	12	13	14	15	16	17
																H 2.2
Li 1.0	Be 1.6											B 2.0	C 2.5	N 3.0	O 3.5	F 4.0
Na 0.9	Mg 1.3											Al 1.6	Si 1.9	P 2.2	S 2.6	Cl 3.2
K 0.8	Ca 1.0	Sc 1.4	Ti 1.5	V 1.6	Cr 1.7	Mn 1.5	Fe 1.8	Co 1.9	Ni 1.9	Cu 1.9	Zn 1.6	Ga 1.8	Ge 2.0	As 2.2	Se 2.6	Br 3.0
Rb 0.8	Sr 1.0	Y 1.2	Zr 1.3	Nb 1.6	Mo 2.2	Tc 1.9	Ru 2.2	Rh 2.3	Pd 2.2	Ag 1.9	Cd 1.7	In 1.8	Sn 2.0	Sb 1.9	Te 2.1	I 2.7
Cs 0.8	Ba 0.9	La 1.1	Hf 1.3	Ta 1.5	W 2.4	Re 1.9	Os 2.2	Ir 2.2	Pt 2.3	Au 2.5	Hg 2.0	Tl 1.6	Pb 2.3	Bi 2.0	Po 2.0	At 2.2

© Cengage Learning

■ <1.0   □ 1.5–1.9   ■ 2.5–2.9
□ 1.0–1.4   ■ 2.0–2.4   ■ 3.0–4.0

Anion: An atom (or molecule) that has a negative charge

Cation: An atom (or molecule) that has a positive charge

Lewis dot symbol: A representation of a molecule, ion, or formula unit by atomic symbols with the valence electrons shown as dots

Inspection of the electronegativity values in Figure 5.1 leads to the observation that nonmetals, the elements in the upper right corner of the periodic table, are the most electronegative, while the metals in the lower left corner are the least electronegative. The general trend is that electronegativity increases across a period in the periodic table and decreases down a group of the periodic table. A key to predicting the type of bonding interactions between atoms is the difference in the electronegativity of the bonding atoms. As you will see, bonding interactions between atoms on opposite sides of the periodic table, where the difference in electronegativity values is greatest, result from the tendency of the less electronegative metal atom to transfer its valence electrons to the more electronegative nonmetal atom. This type of bonding interaction is referred to as an *ionic bond* (Section 5-2). In contrast, interactions between atoms of similar electronegativities result in the formation of a bond referred to as a *covalent bond* (Section 5-3). The following analysis of both types of bonding interactions illustrates the magnitude of electronegativity differences that leads to ionic or covalent bonding and helps us predict bonding interactions in other compounds.

5-2 Ionic Bonds and Ionic Compounds

The type of bonding between sodium (Group 1) and chlorine (Group 17) atoms can be predicted using the octet rule and the difference in electronegativities of the two atoms. Sodium, with a single valence electron, has an electronegativity of 0.9, and chlorine, with seven valence electrons, has an electronegativity of 3.2. The large difference in these values (ΔEN), 2.3, indicates a situation in which one atom (chlorine, in this case) is capable of taking a valence electron away from the other atom (sodium, in this case) to complete its octet of outer shell electrons. As a general rule of thumb, ΔEN values of 2.0 or greater indicate the likelihood of transfer of electrons from the less electronegative atom to the more electronegative atom. In this way, both atoms attain full outer shells of valence electrons, one by gaining electrons and the

other by losing electrons. Atoms are called **anions** after they gain electron(s) and become negatively charged; atoms are called **cations** after they lose electron(s) and become positively charged. The chlorine atom becomes a chlor*ide* anion (Cl^-), and the sodium atom becomes a sodium cation (Na^+).

Let's look at this situation more closely. A sodium atom (atomic number 11) has a total of 11 electrons surrounding the 11 protons in its nucleus; the numbers of positive and negative charges cancel and the atom is uncharged, or electrically neutral. Only one of these electrons is a valence electron (Table 3.3). By giving its single valence electron away, the sodium atom becomes a sodium cation with a net +1 charge because it now has 11 protons but only 10 electrons. Furthermore, the sodium cation now has exactly the same number of outer shell electrons (valence electrons) as the nearest noble gas, Ne.

Sodium atom: $1s^2\,2s^2\,2p^6\,\boxed{3s^1} \longrightarrow$

(11 protons)   (11 electrons)

Sodium cation: $1s^2\,\boxed{2s^2\,2p^6}\, + 1$ electron

(11 protons)   (10 electrons)

Neon atom: $1s^2\ \ \boxed{2s^2\,2p^6}$

(11 protons)   (11 electrons)

The chlorine atom (atomic number 17) has a total of 17 electrons and 17 protons. Seven of these electrons are outer shell or valence electrons. By gaining the electron from a sodium atom, the chlorine atom becomes a chloride anion with a net −1 charge since it now has 17 protons and 18 electrons. Furthermore, the chloride anion now has exactly the same number of outer shell (valence electrons) as the nearest noble gas, Ar.

Chlorine atom: $1s^2\,2s^2\,2p^6\,\boxed{3s^2\,3p^5}\, + 1$ electron $\longrightarrow$

(17 protons)   (17 electrons)

Chloride anion: $1s^2\,2s^2\,2p^6\,\boxed{3s^2\,3p^6}$

(17 protons)   (18 electrons)

Argon atom: $1s^2\,2s^2\,2p^6\,\boxed{3s^2\,3p^6}$

(18 protons)   (18 electrons)

The reaction between sodium and chlorine can be illustrated using the **Lewis dot symbols**, in which valence electrons in lone pairs are represented as dots:

$$Na\cdot + \cdot\ddot{\underset{..}{Cl}}: \longrightarrow Na^+ + :\ddot{\underset{..}{Cl}}:^-$$

The less electronegative atom transfers its valence electron(s) to the more electronegative atom when the difference in electronegativity is large. By this process, both atoms obtain the same number of outer shell electrons as the nearest noble gas and satisfy the octet rule. This is a generalization that can be successfully used to predict the nature of bonding between atoms of many different elements.

The strong electrostatic attraction between the positive and negative ions is known as an **ionic bond**. Compounds that are held together by ionic bonds are known as **ionic compounds**. Because ionic compounds are overall electrically neutral (the sum of positive and negative charges cancels), sodium chloride must be composed of one sodium ion for every chloride ion. To show this composition, the formula of the compound is written as NaCl. (Note that the ionic charges are not indicated in the formula.) In ionic compounds, the *simplest* ratio of oppositely charged ions that gives an electrically neutral unit is represented in the formula and is called a **formula unit**. The formula unit for sodium chloride is NaCl, or one sodium ion and one chloride ion, but not Na_2Cl_2.

Ionic crystals are made up of large numbers of formula units that form a regular three-dimensional crystalline **lattice**. A model of the sodium chloride crystalline lattice is shown in Figure 5.2a. Note that each Na^+ ion has six Cl^- ions around it. Similarly, each Cl^- ion has six Na^+ ions around it (Figures 5.2b and c). In this way, the one-to-one ratio of the singly charged ions is preserved. There is no *unique molecule* in ionic structures; no particular ion is attached exclusively to another ion, but each ion is attracted to all the oppositely charged ions surrounding it.

When atoms become ions, properties are drastically altered. For example, a collection of Br_2 molecules is red, but bromide ions (Br^-) contribute no color to a crystal of a compound such as NaBr. A chunk of sodium atoms (Figure 5.3, *left*, on page 80) is soft, metallic, and violently reactive with water, but Na^+ ions are stable in water. A large collection of Cl_2 molecules (Figure 5.3, *center*) constitutes a greenish-yellow poisonous gas, but chloride ions (Cl^-) produce no color in compounds and are not poisonous. We have everyday evidence of this because we use NaCl (Figure 5.3, *right*) to season food. When atoms become ions, atoms change their nature.

Metals in Groups 1, 2, and 13 are capable of donating one, two, or three valence electrons, respectively, to the significantly more electronegative nonmetals in Groups 15, 16, and 17. These nonmetal atoms can accept three electrons, two electrons, or one electron, respectively. Thus, cations of +1, +2, or +3 charge are formed from Group 1, 2, and 13 metals, respectively, and anions of −3, −2, or −1 charge are formed from Group 15, 16, and 17 nonmetals, respectively. Carbon, in Group 14, represents a sort of crossover point from elements that tend to form cations to ones more likely to form anions. Carbon itself rarely forms the predicted +4 or −4 ions, since losing four electrons to another atom is simply too energetically unfavorable, as is gaining four electrons from another atom. Thus, carbon tends to be found mostly in molecules and not in ionic compounds.

Main-group elements form ions of a single type as shown in Figure 5.4. Transition metal elements form multiple ions in many cases. For instance, iron (atomic number 26) readily forms both +2 and +3 ions, whereas silver (atomic number 47) forms only the +1 ion. Therefore, it is necessary to learn some of the most important transition metal ions shown in Figure 5.4.

> **Ionic bond:** The attraction between positive and negative ions
>
> **Ionic compound:** A compound composed of positive and negative ions
>
> **Formula unit:** In ionic compounds, the simplest ratio of oppositely charged ions that gives an electrically neutral unit
>
> **Lattice:** A regular arrangement

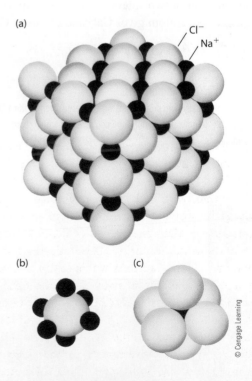

(a)

Cl⁻

Na⁺

(b)   (c)

© Cengage Learning

FIGURE 5.2 Structure of sodium chloride.

FIGURE 5.3 Sodium (stored under mineral oil to protect it from oxygen in the air), chlorine, and sodium chloride (which we know commonly as "salt").

Larry Cameron

5-2a Predicting Formulas of Ionic Compounds

Because opposite charges are attracted to each other, it stands to reason that cations of the metals of Groups 1, 2, and 13 tend to form ionic compounds with the anions of the nonmetals of Groups 15, 16, and 17. There are, of course, ionic compounds formed by combination of the cations of the transition metal elements with various non-metal anions. The formula units for ionic compounds are always represented by using *the simplest ratio of cation and anion necessary to give an electrically neutral formula unit*. The transfer of two valence electrons from a calcium atom to an oxygen atom gives Ca^{+2} and O^{-2} ions. Calcium

oxide, formed by the combination of these two ions, is represented as CaO and not Ca_2O_2. Similarly, magnesium and bromine atoms would combine to give the formula unit $MgBr_2$, since magnesium forms a $+2$ ion and bromine forms a -1 ion.

EXAMPLE 5.1 Predicting Formulas of Ionic Compounds

→ *What is the formula of aluminum oxide, the ionic compound formed by the combination of aluminum and oxygen?*

SOLUTION

1. Aluminum (Al) is in Group 13, so remove three electrons from an Al atom to give the Al^{3+} ion.

2. Oxygen (O) is in Group 16, so add two electrons to an O atom to give eight. The resulting ion is O^{2-}.

3. To make this formula electrically neutral: Al^{3+} ion is $+3$, and two ions would give a total charge of $+6$; O^{2-} ion is -2, and three ions would give a total charge of -6; $+6$ and $-6 = 0$; the formula is Al_2O_3.

TRY IT 5.1*a*

What is the formula of calcium fluoride, the ionic compound formed by the combination of calcium (Ca) and fluorine (F)?

FIGURE 5.4 Common ions.

Main-group metals usually form positive ions with a charge given by the group number (blue). For transition metals, the positive charge is variable (red), and ions other than those shown are possible. Nonmetals generally form negative ions with a charge equal to 18 minus the group number (yellow). The Group 18 elements, known as the noble gases, form very few compounds.

1	2	3	4	5	6	7	8	9	10	11	12	13	14	15	16	17	18
H^+																H^-	N
Li^+														N^{3-}	O^{2-}	F^-	O B L
Na^+	Mg^{2+}											Al^{3+}		P^{3-}	S^{2-}	Cl^-	E
K^+	Ca^{2+}		Ti^{2+}		Cr^{2+} Cr^{3+}	Mn^{2+}	Fe^{2+} Fe^{3+}	Co^{2+} Co^{3+}	Ni^{2+}	Cu^+ Cu^{2+}	Zn^{2+}				Se^{2-}	Br^-	G
Rb^+	Sr^{2+}									Ag^+	Cd^{2+}		Sn^{2+}		Te^{2-}	I^-	A S
Cs^+	Ba^{2+}										Hg_2^{2+} Hg^{2+}		Pb^{2+}	Bi^{3+}			E S

© Cengage Learning

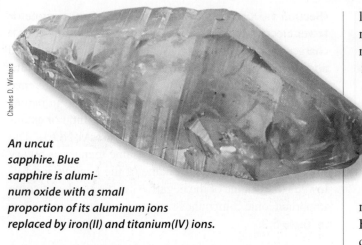

An uncut
sapphire. Blue
sapphire is alumi-
num oxide with a small
proportion of its aluminum ions
replaced by iron(II) and titanium(IV) ions.

lowed by the name of the negative ion. The element name is used for the positive ion. If the compound is made up of only one type

> **Binary compound:**
> A chemical compound composed of two elements

of metal and one type of nonmetal (a **binary compound**), the name of the negative ion ends in *-ide*. Examples we have already used besides sodium chloride include calcium oxide (CaO), aluminum oxide (Al_2O_3), and lithium nitride (Li_3N). (Note that *binary* means only two elements are present, but the number of atoms in the formula can be more than two.) For transition metals that form ions with different charges, Roman numerals are used with the names to indicate the charge. For example, iron(II) ion refers to Fe^{2+}, and iron(III) ion means Fe^{3+}. The name of $FeCl_2$ is iron(II) chloride, and the name of $FeCl_3$ is iron(III) chloride.

TRY IT 5.1*b*

How many ionic compounds can be formed between cobalt and chlorine? What are their formulas?

EXAMPLE 5.2 Predicting More Ionic Compound Formulas

 What is the formula of lithium nitride, the ionic compound formed by the combination of lithium and nitrogen?

SOLUTION

1. Lithium (Li) is in Group 1, so remove one electron from the Li atom to give the Li^+ ion.

2. Nitrogen (N) is in Group 15, so add three electrons to the N atom to give eight outer shell electrons and form an N^{3-} ion.

3. The electrically neutral compound would be formed from the combination of three Li^+ ions with one N^{3-} ion to give Li_3N.

TRY IT 5.2

Write the formula units for the ionic compounds formed by the combination of (**a**) cesium (Cs) and iodine (I), (**b**) strontium (Sr) and chlorine (Cl), and (**c**) barium (Ba) and sulfur (S).

5-2b Naming Binary Ionic Compounds

By starting once again with table salt, we can use the formula and chemical name of this compound, NaCl (sodium chloride), to help remember the rules for naming ionic compounds. The positive ion is named first, fol-

EXAMPLE 5.3 Naming Binary Ionic Compounds

 Name the following ionic compounds: (a) K_2S, (b) $BaBr_2$, (c) Li_2O, and (d) Fe_2O_3.

SOLUTION

All of these compounds are binary compounds—made up of ions of two elements. The positive ion is named first, followed by the negative ion. The positive ion is named as the element. The negative ion is named by adding *-ide* to the stem of the name of the element. The correct names of the first three are (**a**) potassium sulfide, (**b**) barium bromide, and (**c**) lithium oxide. Because iron can be a +2 or +3 cation, compounds of iron require the use of a Roman numeral after *iron* to identify its charge. The charge on the iron ion in Fe_2O_3 is determined by calculating the total negative charge (–6 from 3 O^{2-}) and dividing by 2 (because there are two iron ions). This gives a charge of +3; thus, the correct name of Fe_2O_3 is iron(III) oxide.

TRY IT 5.3

Name the following compounds: (**a**) RbCl, (**b**) Ga_2O_3, (**c**) $CaBr_2$, and (**d**) Fe_3N_2.

EXAMPLE 5.4 Writing Formulas for Binary Ionic Compounds

 Write formulas for the following compounds: (a) cesium bromide, (b) cobalt(III) chloride, and (c) barium oxide.

SOLUTION

(a) The correct formula is CsBr since Cs (Group 1) forms Cs^+ and Br (Group 17) forms Br^-. (b) The Roman numeral III indicates the Co^{3+} ion, and Cl (Group 17) forms Cl^-; thus, the correct formula is $CoCl_3$. (c) The correct formula is BaO since Ba (Group 2) forms Ba^{2+} and O (Group 16) forms O^{2-}.

TRY IT 5.4

What are the formulas of (a) cobalt(II) sulfide, (b) magnesium fluoride, and (c) potassium iodide?

5-2c Ionic Compounds with Polyatomic Ions

Polyatomic ion:
A positive or negative ion composed of two or more atoms

Atoms of two or more elements can also combine to form a **polyatomic ion**, a chemically distinct species with an electric charge. Communication in the world of chemistry and understanding many applications require one to know the names, formulas, and charges of the common polyatomic ions listed in Table 5.1. Most common polyatomic ions are negatively charged; the ammonium ion (NH_4^+) is the major exception.

The bonding between the atoms within polyatomic ions is just like the bonding within molecular compounds (Section 5-3), but the group of atoms has either more or fewer electrons than protons and therefore has an overall charge. Compounds that contain polyatomic ions are ionic, and their formulas are written by the same procedure described for binary ionic compounds. The only difference is that the polyatomic ion formula is enclosed in parentheses when more than one such ion is present. For example, the formula of aluminum nitrate is $Al(NO_3)_3$. The compounds are also named in the same manner as binary ionic compounds, with the name of the positive ion followed by the name of the negative ion. Examples of some important ionic compounds with polyatomic ions are given in Table 5.2.

EXAMPLE 5.5 Writing Formulas for Compounds of Polyatomic Ions

 Write the formulas of (a) magnesium sulfate and (b) calcium hydrogen sulfite.

SOLUTION

(a) The sulfate ion is SO_4^{2-} and the charge on the magnesium ion is $+2$, so the formula is $MgSO_4$. No parentheses are needed around the sulfate ion because only one SO_4^{2-} ion is present. (b) The hydrogen sulfite ion is HSO_3^- and the charge on the calcium ion is $+2$; thus, the formula is $Ca(HSO_3)_2$.

TABLE 5.1 Names and Composition of Some Common Polyatomic Ions

Cation (Positive Ion)			
NH_4^+	Ammonium ion		
Anions (Negative Ions)			
OH^-	Hydroxide ion	CO_3^{2-}	Carbonate ion
$CH_3CO_2^-$	Acetate ion	HCO_3^-	Hydrogen carbonate ion (or bicarbonate ion)
NO_2^-	Nitrite ion	PO_4^{3-}	Phosphate ion
NO_3^-	Nitrate ion	HPO_4^{2-}	Hydrogen phosphate ion
SO_3^{2-}	Sulfite ion	$H_2PO_4^-$	Dihydrogen phosphate ion
HSO_3^-	Hydrogen sulfite ion	ClO^-	Hypochlorite ion
SO_4^{2-}	Sulfate ion	ClO_3^-	Chlorate ion
HSO_4^-	Hydrogen sulfate ion (or bisulfate ion)	ClO_4^-	Perchlorate ion
CN^-	Cyanide ion		

TABLE 5.2 Some Commercially Important Ionic Compounds with Polyatomic Ions

Formula	Name (Common Name)	Uses
NH_4NO_3	Ammonium nitrate	Fertilizers and explosives
KNO_3	Potassium nitrate	Gunpowder and matches
$NaOH$	Sodium hydroxide (lye)	Extracting aluminum from ore; preparing soaps, detergents, and rayon; pulp and paper industry
$Mg(OH)_2$	Magnesium hydroxide	Milk of magnesia
Na_2CO_3	Sodium carbonate (washing soda, soda ash)	Water softening; detergents and cleansers; pulp and paper industry; glass and ceramics
$NaHCO_3$	Sodium bicarbonate (baking soda)	Household use; food industry; fire extinguishers
Na_3PO_4	Sodium phosphate	Food additive
$Ca(H_2PO_4)_2$	Calcium dihydrogen phosphate	Fertilizer
$CaSO_4$	Calcium sulfate	Gypsum, drywall (wallboard)
$Al_2(SO_4)_3$	Aluminum sulfate	Water purification

© Cengage Learning

A few minerals formed from ionic compounds

Limestone
$CaCO_3$

Gypsum
$CaSO_4 \cdot 2 H_2O$

Fluorite
CaF_2

Common minerals of Group 2A elements.

© Cengage Learning/Charles D. Winters

5-3 Covalent Bonds

Most chemical compounds are not ionic. The atoms in most chemical compounds are held together in units

> **Covalent bond:** A bond in which two atoms share electrons

known as *molecules* (Section 2-3). In the chemical bonds in molecules, one or more pairs of electrons are shared between two atoms in what are known as **covalent bonds**. The shared electrons feel the positive attraction of the protons in the nucleus of each of the two atoms connected by the bond, but it is possible that the sharing is not equal. Most molecules are composed of atoms of nonmetals, as in the elements that exist naturally as diatomic molecules (i.e., H_2, F_2, C_{12}, Br_2, I_2, O_2, and N_2). Here, the electrons are shared equally between the two atoms. In other molecules, like HCl, the electrons are shared, but unequally because of the difference in electronegativity between H and Cl atoms. In ionic compounds the large difference in electronegativity between a metal and nonmetal results in little or no sharing. Ionic compounds are composed of metal cations and nonmetal anions.

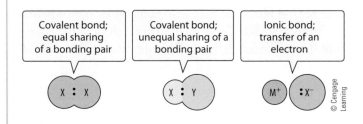

Covalent bond; equal sharing of a bonding pair

Covalent bond; unequal sharing of a bonding pair

Ionic bond; transfer of an electron

© Cengage Learning

The Lewis dot structures discussed for atoms in Chapter 3 can also be used to represent the covalent bonding in molecules. This is a simple process, but does require a little practice. A few steps are involved:

1. Place the symbols of each atom near each other. If there are more than two atoms in the molecule, place the one with the lowest group number (farthest to the left in the periodic table) in the middle of the other atoms. The only exception to this rule is hydrogen.

2. Use the periodic table to determine the number of valence electrons for each atom. Draw these around the symbol for each element.

Bonding pair: A pair of electrons shared between two atoms

Lone pair: A pair of valence electrons that is not shared between two atoms but instead is associated with a single atom in the structure

3. Connect the atoms using bonds (shared pairs of electrons). To do this, use one electron from each adjacent atom. You can use two dots (:) to represent the shared pair, or a line (—).

4. Check to make sure that the octet rule is satisfied for all atoms. (H cannot obey the octet rule.)

5. If the octet rule is not obeyed for every atom, you must share a second (or third) pair of electrons between the atom that does not have an octet and one to which it is attached. This will give rise to multiple bonds between two atoms.

6. Check the structure you've drawn to make sure that each atom (except for H) has an octet of electrons. Also, add up the number of electrons in the structure you've drawn to make sure that the total is the same as the sum of valence electrons for all of the atoms that make up the molecule.

You will see that shared electrons are involved in covalent bonds, but there may also sometimes be unshared electrons associated with an individual atom in the structure. The electrons in these Lewis structures are classified as **bonding pairs** or **lone pairs**.

bonding pair,
2 electrons
$$H\overset{..}{O}—H$$
lone pair

For many molecules or ions, the actual connectivity of the atoms may not be obvious, and more information may be needed to choose the best among several possible Lewis structures. The remainder of this section is devoted to examples that will help you learn how to draw Lewis structures. Compounds containing single, double, and triple bonds are described with consideration given to choosing among several possible Lewis structures.

5-3a Single Covalent Bonds

A single covalent bond is formed when two atoms share a single pair of electrons. The simplest case in which this can occur is the hydrogen molecule (H_2). Since each hydrogen atom (Group 1) has a single valence electron, the two hydrogen atoms are connected by a single pair of electrons that you can represent in shorthand fashion with a line, as shown here:

$$H\cdot + H\cdot \longrightarrow H:H \quad \text{or} \quad H—H$$
Lewis structure

In a molecule such as F_2, each fluorine atom has seven valence electrons because fluorine is in Group 17. Thus, the Lewis structure will contain 14 electrons. All of the "other" electrons are still present, but since they are not involved in bonding they are not shown. Only bonding pairs and lone pairs of electrons are shown in Lewis structures. Connect the two F atoms with a pair of electrons, and you arrive at the structure that follows.

$$\cdot\overset{..}{\underset{..}{F}}:+\cdot\overset{..}{\underset{..}{F}}:\longrightarrow :\overset{..}{\underset{..}{F}}:\overset{..}{\underset{..}{F}}: \quad \text{or} \quad :\overset{..}{\underset{..}{F}}—\overset{..}{\underset{..}{F}}:$$

You can use this process to arrive at Lewis structures for many molecules. For instance, a Lewis structure for water (H_2O) should contain a total of eight valence electrons (one from each of the two hydrogen atoms and six from the oxygen atom).

$$H\cdot \ \ \cdot\overset{..}{O}\cdot \ \ \cdot H \longrightarrow H:\overset{..}{O}:H \quad \text{or} \quad H—\overset{..}{O}—H$$

EXAMPLE 5.6 Writing Lewis Dot Structures

➡ *Use your knowledge of valence electrons to draw a Lewis dot structure for ammonia (NH_3).*

SOLUTION

1. Place the Lewis dot structures of each atom near each other, with nitrogen in the middle.

$$H\cdot \ \ \cdot\overset{..}{N}\cdot \ \ \cdot H$$
$$\overset{\cdot}{H}$$

2. Connect the atoms together using pairs of electrons to arrive at the following structure.

$$H:\overset{..}{N}:H$$
$$\overset{\cdot}{H}$$

3. For ease of representation, you may replace the bonding pairs of electrons with lines to arrive at the following structure:

$$H-\overset{\displaystyle \cdot\cdot}{N}-H$$
$$|$$
$$H$$

TRY IT 5.6

Use the process just discussed to draw Lewis dot structures for the following molecules: HI, Br_2, CH_4, and $HOCl$.

5-3b Single Covalent Bonds in Hydrocarbons

Hydrocarbons, compounds that contain only carbon and hydrogen atoms, represent a large class of compounds. They are extremely important molecules in the fuel industry, as will be seen later. Hydrocarbons that contain only $C-H$ and $C-C$ single bonds are known as **alkanes**. In Example 5.6, you drew a Lewis dot structure for the simplest alkane, known as methane (CH_4). Methane is the major component of natural gas. One could imagine bringing four hydrogen atoms, each with a single valence electron, together with a carbon atom with four unpaired valence electrons to form the methane molecule.

$$H:\overset{\displaystyle H}{\underset{\displaystyle H}{\overset{\cdot\cdot}{C}}}:H \quad \text{or} \quad H-\overset{\displaystyle H}{\underset{\displaystyle H}{\overset{|}{C}}}-H$$

Two ways to represent methane.

As will be seen over and over again, carbon has a tendency to form four bonds. In alkanes, these four bonds will be either single bonds to H or single bonds to C. Another simple alkane is propane, C_3H_8, which is represented by the following Lewis structure:

$$H-\overset{\displaystyle H}{\underset{\displaystyle H}{\overset{|}{C}}}-\overset{\displaystyle H}{\underset{\displaystyle H}{\overset{|}{C}}}-\overset{\displaystyle H}{\underset{\displaystyle H}{\overset{|}{C}}}-H$$

© iStockphoto.com/lucato

TRY IT 5.7

(a) Draw a Lewis dot structure for the alkane intermediate between methane and propane. It is known as ethane and has the formula C_2H_6.

(b) There are two different ways to connect the fourteen atoms of butane (C_4H_{10}). See if you can draw Lewis dot structures for both of these. This should lead you to see that as the number of carbon atoms increases, the number of ways to connect the atoms together increases as well.

Alkanes are often referred to as **saturated hydrocarbons** because they have the highest possible ratio of hydrogen to carbon atoms that can be bonded in a molecule. The general formula for saturated hydrocarbons is C_nH_{2n+2} where n is the number of carbon atoms.

Alkane: A hydrocarbon with carbon–carbon single bonds

Saturated hydrocarbon: A hydrocarbon that is an alkane

Double bond: A bond in which two pairs of electrons are shared between two atoms

5-3c Multiple Covalent Bonds

Frequently, one encounters molecules in which more than one pair of electrons is shared between two atoms. An atom with fewer than seven valence electrons can form covalent bonds in two ways. The atom can contribute one electron to each of several single covalent bonds, or it can share two (or three) pairs of electrons with a single other atom in forming multiple bonds to that atom.

For instance, we can arrive at the Lewis dot structure for carbon dioxide (CO_2) in the following manner. First, write the dot structures for each atom with C, which has the lowest group number, in the middle. Join the atoms to each other with a single bond (structure a). Next, notice that neither carbon (6 electrons surrounding) nor oxygen (7 electrons surrounding) has an octet of electrons. To remedy this, pair an electron from each oxygen atom with an electron from the carbon atom to arrive at structure c, in which the octet rule is satisfied for all atoms. The carbon dioxide molecule has two **double bonds**.

$$:\overset{\cdot\cdot}{O}\cdot\cdot\overset{\cdot\cdot}{C}\cdot\cdot\overset{\cdot\cdot}{O}: \qquad :\overset{\cdot\cdot}{O}::C::\overset{\cdot\cdot}{O}: \quad \text{or} \quad :\overset{\cdot\cdot}{O}=C=\overset{\cdot\cdot}{O}:$$

$$\quad a \qquad\qquad\qquad\qquad b$$

In the N_2 molecule, the initial joining of the two atoms gives a, in which neither of the atoms has an octet of electrons. Sharing of an additional electron from each atom as in the previous example gives b, but again there is no octet for either atom. Sharing of yet another electron from each atom gives *c*, a molecule in which each atom now satisfies the octet rule. These atoms are connected by a **triple bond**.

:N··N: :N::N: :N:::N: or   :N≡N:

a   b   c

EXAMPLE 5.7 Molecules with Triple Bonds

 Show that the best Lewis dot structure for acety-lene (C_2H_2) contains a triple bond between the carbon atoms and a single bond between the carbon and hydrogen atoms.

SOLUTION

Connect the basic skeleton of the molecule using pairs of electrons to arrive at structure *a*. Sharing of an additional two electrons between the C atoms gives structure *b*. A second sharing gives the final structure *c*.

H··C··C··H   H··C::C··H

a   b

H:C:::C:H   or   H—C≡C—H

c

Triple bond: A bond in which three pairs of electrons are shared between two atoms

Resonance structures: Two or more possible Lewis dot representations of a molecule or ion in which the only difference is the position of the valence electrons

Bond energy: The amount of energy required to break one mole of bonds between a specific pair of atoms

Alkene: A hydrocarbon with one or more carbon–carbon double bonds

Unsaturated hydrocarbon: A hydrocarbon that contains fewer than the maximum number of hydrogen atoms

Ozone (O_3) is an interesting molecule for which to consider Lewis dot structures, as it illustrates another important feature of such structures. After following the rules for drawing molecular structure, it is possible to arrive at two different results, structures *a* and *b*. The two structures differ only in the position of electrons and not in the position of atoms and are known as **resonance structures**. So, which one is correct? In some cases, like that of ozone, we run into one limitation of using Lewis structures to represent molecular structures. Experiments show that the actual structure of ozone is neither *a* nor *b*. It is a hybrid structure of the two, in which each oxygen–oxygen bond is about halfway between a single bond and a double bond. The two resonance structures are the limiting Lewis dot structures that can be drawn to represent the actual structure of ozone, a compound for which the bonding cannot accurately be represented by a single Lewis dot structure.

(a) :O::O:O: or   :O=O—O:

(b) :O:O::O: or   :O—O=O:

Single, double, and triple bonds differ in length and strength. Triple bonds are shorter than double bonds, which in turn are shorter than single bonds. Bond energies normally increase with decreasing bond length. **Bond energy** is defined as the amount of energy required to break a defined number of the bonds (this number, a *mole*, is discussed in Section 8-2). Some typical bond lengths and energies are listed in Table 5.3.

5-3d Multiple Covalent Bonds in Hydrocarbons

Ethylene (Figure 5.6) contains a double bond between the carbon atoms and single bonds between the hydrogen atoms and the carbon atoms. Ethylene is the first member of the **alkene** series of hydrocarbons, compounds that have one or more C=C bonds—that is, carbon–carbon double bonds and the general formula C_nH_{2n}.

H   H
 \   /
  C = C
 /   \
H   H

Ethylene (C_2H_4)

The structural formula of ethylene illustrates why alkenes are said to be **unsaturated hydrocarbons**: they contain fewer hydrogen atoms than the corresponding alkanes and react with hydrogen to form alkanes.

TABLE 5.3 Some Bond Lengths and Bond Energies

Bond type	C—C	C=C	C≡C	N—N	N=N	N≡N
Bond length (nm)	0.154	0.134	0.120	0.140	0.124	0.109
Bond energy (kcal/mol)*	83	147	200	40	100	225

* kcal/mol (kilocalories per mole) = thousands of calories necessary to break 6.02×10^{23} bonds (Section 8-2).

Acetylene, the gas that produces a flame hot enough to cut steel when it is mixed with oxygen and burned, has a carbon–carbon triple bond, C≡C.

$$H—C≡C—H$$

5-4 Shapes of Molecules

Lewis dot structures can be drawn only in two dimensions. Since the shapes of very few molecules can be adequately represented using two dimensions, it is necessary to go beyond the Lewis structure to infer the shapes of molecules.

The *valence shell electron pair repulsion (VSEPR) model* is quite useful in this regard. According to this model, the valence electrons represented in a Lewis dot structure are key to predicting the geometry of the molecule. Valence electrons that are not shared in the same bond repel each other. This repulsion leads to arrangements in which this repulsion is minimized. The result is a geometry, or shape, in which the electrons around a central atom are as far away from each other as possible while maintaining their association with the central atom. In molecules with many atoms, more than one atom may be considered as a central atom as one focuses in turn on the geometry of different parts of a molecule.

Imagine differing numbers of identical balloons attached to a central point. If the balloons each have identical electrostatic charges, they repel each other. The final geometry adopted by the balloons will be one of the following: The balloons will be either 180° apart (two balloons) to define a *linear geometry*, 120° apart (three balloons) to define a *trigonal planar geometry*, or 109.5° apart (four balloons) to define a *tetrahedral geometry* (Figure 5.5). These are some of the more common situations found in chemical bonding. Five- and six-electron pairs assume *trigonal bipyramidal* and *octahedral geometries*.

If we now extend this idea to the regions of high electron density around a central atom (bonds and lone pairs),

we can predict both the **molecular geometry** and the **electron pair geometry** of a molecule.

If all of the regions of high electron density are bonds, then the electron pair geometry is the same as the molecular geometry. This is what is seen for the structures of methane, carbon dioxide, and acetylene. Methane (CH_4), the simplest hydrocarbon molecule, has a tetrahedral shape, with the carbon atom at the center of the tetrahedron and the four hydrogen atoms at the corners. This arrangement

> **Molecular geometry:** The arrangement of atomic nuclei in a molecule
>
> **Electron pair geometry:** The arrangement of bonds and lone pairs in a molecule

FIGURE 5.5 Balloon models of electron-pair geometries for two- to six-electron pairs.
Balloons of similar size and shape, when tied together, naturally assume the arrangements shown.

Linear

Trigonal planar

Tetrahedral

Trigonal bipyramidal

Octahedral

Charles D. Winters /© Thomson Learning/Charles D. Winters

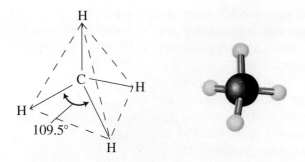

allows the hydrogen atoms bonded to carbon atoms to be located as far as possible from one another. The angle between any two bonds, and therefore the hydrogen atoms, is 109.5°.

Carbon dioxide and acetylene (pages 85 and 86) each have two regions of electron density about the central atom, and their electron pair and molecular geometries are linear with a 180° bond angle. These two examples illustrate the four-balloon and two-balloon analogies, respectively.

Let's look at another simple molecule, formaldehyde, to illustrate the three-balloon analogy. Formaldehyde has the formula H_2CO. The central carbon atom has three regions of electron density (composed of two $C-H$ bonds and one $C=O$ bond). The VSEPR model suggests that these three groups assume a trigonal planar geometry with approximately 120° bond angles. That is what is found. The four atoms lie in a plane, and the bond angles are very close to the predicted angle. This is another case in which the electron pair geometry and the molecular geometry are the same. Note that the geometry at the carbon atoms of ethylene ($H_2C=CH_2$), shown in Figure 5.6, is the same as that about the central carbon atom in formaldehyde.

FIGURE 5.6 Structure of an ethylene molecule.
All of the atoms in this molecule lie in the same plane.

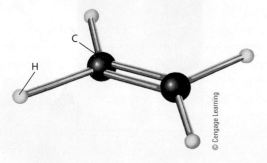

© Cengage Learning

If a central atom has unshared pairs, the electron pair geometry will not be the same as the molecular geometry. This situation can be illustrated using water and ammonia. A Lewis dot structure for water reveals that there are two bonds to hydrogen atoms and two unshared pairs on the central oxygen atom. Thus, if we take into account the actual geometric relationship between the atoms and unshared pairs, we can predict that water has a tetrahedral electron pair geometry with an $H-O-H$ bond angle of about 109.5°. (The actual bond angle is slightly compressed to 104.5° due to excess repulsion by the two unshared pairs; the lone pairs exert a larger influence than do bonded pairs in determining repulsive interactions.) However, *the molecular geometry of water is said to be "bent"* because the three *atoms* must all lie in a plane. Thus, the electron pair geometry of water is tetrahedral, but the molecular geometry is bent.

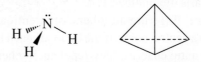

A similar analysis of ammonia (NH_3) reveals that there are four regions of electron density around the central nitrogen atom. Thus, we predict the *electron pair geometry of ammonia to be tetrahedral* with an $H-N-H$ bond angle of about 109.5°. The actual bond angle is again slightly compressed to 107° due to excess repulsion by the single unshared pair. (Compare the amount of repulsion and angle compression from one unshared pair in ammonia to two unshared pairs in water.) The *molecular geometry of ammonia is said to be trigonal pyramidal* since the nitrogen atom is at the apex of a shallow trigonal pyramid with the three hydrogen atoms forming the base of the pyramid.

Let's compare methane, ammonia, and water.

Molecule	Bonds	Lone Pairs	Bond Angle (degrees)	Molecular Geometry	Electron Pair Geometry
CH_4	4	0	109.5	tetrahedral	tetrahedral
NH_3	3	1	107	trigonal pyramidal	tetrahedral
H_2O	2	2	104.5	bent	tetrahedral

© Cengage Learning

5-4a Names for Binary Compounds

In Section 5-2, we showed you how to name binary ionic compounds. The naming of binary molecular compounds is a little different, but just as descriptive. Hydrogen forms binary compounds with all of the nonmetals except the noble gases. For compounds of oxygen, sulfur, and the halogens, the H atom is generally written first in the formula and is named first using the element name. The other nonmetal is named as if it were a negative ion. For example, HF is hydrogen fluoride, and H_2S is hydrogen sulfide.

When there is more than one possible combination of two elements, the number of atoms of a given type in the compound is designated with a prefix such as *mono-*, *di-*, *tri-*, *tetra-*, and so on. Table 5.4 lists the prefixes for up to 10 atoms, and some common molecular compounds and their names are given in Table 5.5.

Many molecular compounds were discovered years ago and have names so common that they continue to be used. Examples include water (H_2O), ammonia (NH_3), nitric oxide (NO), and nitrous oxide (N_2O).

EXAMPLE 5.8 Naming Molecular Compounds

 What is the name of N_2O_4?

SOLUTION

The prefix for two N atoms is *di-* and the prefix for four O atoms is *tetra-*; thus, the name of N_2O_4 is dinitrogen tetraoxide.

TRY IT 5.8

What is the name of P_4S_3?

TABLE 5.4 Prefixes for Number of Atoms in a Compound

Number of Atoms	Prefix	Number of Atoms	Prefix
1	mono-	6	hexa-
2	di-	7	hepta-
3	tri-	8	octa-
4	tetra-	9	nona-
5	penta-	10	deca-

© Cengage Learning

TABLE 5.5 Common Molecular Compounds

Compound	Name	Use
CO	Carbon monoxide	Preparation of methanol and other organic chemicals
CO_2	Carbon dioxide	Carbonated beverages; fire extinguisher; inert atmosphere; dry ice
NO	Nitrogen monoxide (nitric oxide)	Preparation of nitric acid
NO_2	Nitrogen dioxide	Preparation of nitric acid
N_2O	Dinitrogen oxide (nitrous oxide)	Spray-can propellant; anesthetic
SO_2	Sulfur dioxide	Preparation of sulfuric acid; food preservative; metal refining
SO_3	Sulfur trioxide	Preparation of sulfuric acid
CCl_4	Carbon tetrachloride	Solvent
SF_6	Sulfur hexafluoride	Insulator in electric transformers
P_4O_{10}	Tetraphosphorus decaoxide	Preparation of phosphoric acid

© Cengage Learning

5-5 Polar and Non-polar Bonding

The way in which atoms combine to form chemical compounds and the geometries that result have an important effect on the behavior of chemical compounds. In this and the following two sections we'll take a close look at what this means.

In a molecule like H_2 or F_2, where both atoms are alike, there is equal sharing of the electron pair. Where two unlike atoms are bonded, however, the sharing of the electron pair is unequal and results in a shift of electric charge toward one partner. The more electronegative an element is, the more that element attracts electrons.

When two atoms are bonded covalently and their abilities to attract electrons are the same, there is an equal sharing

Nonpolar: Describes a bond or molecule with no or symmetrically oriented (and thus canceled) polar bonds

Polar: Describes a bond or molecule with positive and negative regions

of the bonding electrons, and the bond is a **nonpolar** covalent bond. The bonds in H_2, F_2, and NCl_3 (N and Cl have equal abilities to attract electrons) are nonpolar.

Two atoms with different abilities to attract electrons bonded covalently form a **polar** covalent bond. The bonds in HF, NO, SO_2, H_2O, CCl_4, and BeF_2 are polar. In a molecule of HF, for example, the bonding pair of electrons is drawn more toward the fluorine atom and away from the hydrogen atom (Figure 5.7). The unequal sharing of electrons makes the fluorine end of the molecule more negative than the hydrogen end.

Polar bonds fall between the extremes of nonpolar covalent bonds and ionic bonds. In a covalent bond between two atoms of the same element, there is no charge separation—that is, the negative charge of the electrons is evenly distributed over the bond. In ionic bonds there is complete separation of the charges, and in polar bonds the separation falls somewhere in between.

Polar bonds in molecules can result in the molecule itself being polar when the shape of the molecule allows a permanent separation of charge. A linear diatomic molecule containing a polar bond, such as HF, is polar because there is a separation between the positive end of the molecule and the negative end (Figure 5.7b). Some linear molecules can be nonpolar because the separated charges cancel each other. The carbon dioxide (CO_2) molecule is an example. The oxygen atoms attract electrons to themselves, leaving the carbon atom with a partial positive charge, but because the oxygen atoms are opposite one another, the molecule itself is not polar.

$$\overset{\leftrightarrow}{O}=C=\overset{\leftrightarrow}{O}$$

The water molecule, on the other hand, is an example of a polar molecule. The shape of the water molecule is bent; thus, the partial charges of the polar bonds between the hydrogen atoms and the oxygen atom do not cancel one another.

5-6 Properties of Molecular and Ionic Compounds Compared

The general properties of ionic and molecular compounds are summarized in Table 5.6. All of these properties can be interpreted using the chemical view of matter. Ionic compounds form hard, brittle, crystalline solids with high melting points. Their crystalline nature is explained by the regular arrangement of positive and negative ions needed to place each ion in contact with those of the opposite charge. Their hardness is accounted for by the strong electrostatic forces of attraction that hold the ions in place. The differences in melting and boiling points (high for ionic compounds and low for molecular compounds) are largely accounted for by the strength of the electrostatic attraction between ions, which is greater than that between molecules. Melting requires adding enough energy to overcome these forces.

Whether a molecular compound is a gas, liquid, or solid is a function of the strength of the attractions between molecules. Only when there are huge molecules, with attractions between many polar regions in the molecule, as in polymers (Figure 14.9), do we find very hard and strong molecular substances.

Another major difference between ionic and molecular compounds is in their ability to conduct electricity, an ability that depends on the presence of mobile carriers of positive or negative charge. A solid ionic compound cannot conduct electricity, because the ions are held in fixed positions.

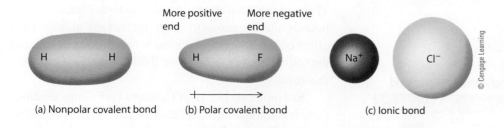

© Cengage Learning

FIGURE 5.7 Nonpolar, polar, and ionic bonds.

(a) H_2, a nonpolar molecule. (b) In a polar molecule like HF, the valence electron density is shifted toward the fluorine atom. An arrow is used to show the direction of molecular polarity, with the arrowhead pointing toward the negative end of the molecule and the plus sign at the positive end of the molecule. (c) In ionic compounds such as NaCl, the valence electron or electrons of the metal are transferred completely to the nonmetal to give ions.

TABLE 5.6 Properties of Ionic and Molecular Compounds

Ionic Compounds	Molecular Compounds
Examples: NaCl, CaF$_2$	Examples: CH$_4$, CO$_2$, NH$_3$, CH$_3$CH$_2$CH$_3$
Many are formed by combination of reactive metals with reactive nonmetals	Many are formed by combination of nonmetals with other nonmetals or with less reactive metals
Crystalline solids	Gases, liquids, and solids
Hard and brittle	Solids are brittle and weak, or soft and waxy
High melting points	Low melting points
High boiling points (700°C to 3500°C)	Low boiling points (−250°C to 600°C)
Good conductors of electricity when molten; poor conductors of heat and electricity when solid	Poor conductors of heat and electricity
Many are soluble in water	Many are insoluble in water but soluble in organic solvents

© Cengage Learning

When melted or dissolved in water, the situation is different—the ions are free to move and current can flow. A compound that conducts electricity under these conditions is referred to as an **electrolyte**. Ionic compounds, to whatever extent they dissolve, are electrolytes in water solution because the ions separate from the crystal and can move about in the solution. Molecules, whether in pure compounds or dissolved in water or any other liquid, have no overall charge and therefore do not carry current; they are referred to as **nonelectrolytes**.

5-7 Intermolecular Forces

Although atoms within molecules are held together by strong chemical bonds, the attractive forces between two separate molecules are much weaker. For example, only about 1% of the energy required to break one of the C—H bonds in a methane (CH$_4$, the major component of natural gas) molecule is required to pull two methane molecules away from one another. These small forces between molecules are called **intermolecular forces**. Even though these attractions are small, they give samples of matter exceedingly important properties. For example, intermolecular attractions are responsible for the fact that all gases condense to form liquids and solids. Both pressure and temperature play a role in the liquefaction of a gas. When the pressure is high, the volume decreases and the molecules of a gas are close together. This condition allows the effect of attractive forces to be appreciable (Section 5-8).

For polar molecules (Section 5-5), intermolecular forces act between the positive end of one polar molecule and the negative end of an adjacent polar molecule. Molecules of SO$_2$ (Figure 5.8) are polar, with the partially negative region of one molecule (represented by δ$^-$) being attracted to the partially positive region (represented by δ$^+$) of an adjacent molecule. This allows the SO$_2$ molecule to interact strongly with similar molecules, such as H$_2$O, and leads to the formation of aerosols that contain sulfuric acid (Chapter 4).

Even nonpolar molecules can have momentary unequal distribution of their electrons, which results in weak intermolecular forces. For example, why would nitrogen—composed of nonpolar molecules—liquefy? Why would carbon dioxide—also composed of nonpolar molecules—form a solid ("dry ice")? When the molecules get close enough, one molecule will cause an uneven distribution of charge in its neighbor. The two molecules become momentarily polar and are attracted to each other. These induced attractive forces are weak but become more pronounced when the molecules are larger and contain more electrons.

Electrolyte: A compound that conducts electricity when melted or dissolved in water

Nonelectrolyte: A compound that does not conduct electricity when melted or dissolved in water

Intermolecular force: An attractive force that acts between molecules

FIGURE 5.8 Polar sulfur dioxide molecules attracted to one another (oxygen atoms attract electrons more than do sulfur atoms).

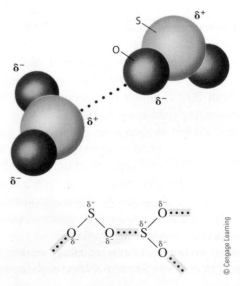

© Cengage Learning

Hydrogen bonding:
Attraction between a hydrogen atom bonded to a highly electronegative atom and the lone pair on an electronegative atom in another or the same molecule

5-7a Hydrogen Bonding

An especially strong intermolecular force called **hydrogen bonding** acts between molecules in which a hydrogen atom is covalently bonded to a strongly electronegative atom with nonbonding electron pairs. The electronegative atom may be fluorine, oxygen, or nitrogen. The hydrogen bond is the attraction between such a hydrogen atom with a partial positive charge on one molecule and an electronegative atom (F, O, or N) of another molecule that has a partial negative charge and one or more unshared electron pairs. The greater the electron attraction of the atom connected to H, the greater the partial positive charge on the H, hence, the stronger the hydrogen bond is between it and a partially negative atom on another molecule. Hydrogen bonds are typically shown as dotted lines between atoms.

$$\overset{\delta^+}{H}-\overset{\delta^-}{F}\cdots\overset{\delta^+}{H}-\overset{\delta^-}{F}\cdots\overset{\delta^+}{H}-\overset{\delta^-}{F}$$

Hydrogen bond ↑   Covalent bond ↑

Water provides the most common example of hydrogen bonding. Hydrogen compounds of oxygen's neighbors and family members in the periodic table are all gases at room temperature. However, water is a liquid at room temperature, and this indicates a strong degree of intermolecular attraction. Figure 5.9 illustrates that the boiling point of H_2O is about 200°C higher than would be predicted if hydrogen bonding were not present.

Because each hydrogen atom in one water molecule can form a hydrogen bond to an oxygen atom in another water molecule, and each oxygen atom has two nonbonding electron pairs, each water molecule can form a maximum of four hydrogen bonds to four other water molecules (Figure 5.10). The result is a tetrahedral cluster of water molecules around the central water molecule.

Although hydrogen bonds are much weaker (with a bond energy of only ~3–5 kcal/mol) than ordinary covalent

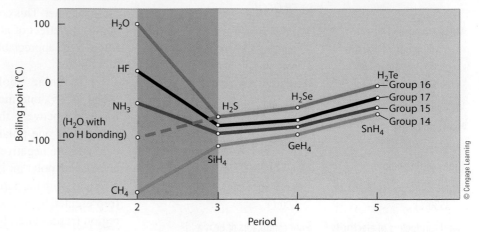

FIGURE 5.9 Boiling points of simple hydrogen-containing compounds.

Lines connect molecules in which hydrogen combines with atoms from the same periodic table group. As shown for water, the temperature at which the compounds would boil if there were no hydrogen bonding is found by following the lines on the right down to the left.

© Cengage Learning

bonds (O — H bond energies are ~111 kcal/mol), they play a key role in the chemistry of life (Chapter 15). For instance, all of the water in your body would simply boil away if the hydrogen bonds were all suddenly broken.

5-8 The States of Matter

By now you know enough about bonding to understand the states of matter: gas, liquid, and solid. Matter in any state is composed of atoms, molecules, or ions in constant motion. In gases (Figure 5.11a), the particles are in rapid, random motion relative to one another. In addition, they are, on average, far apart from one another. In liquids (Fig-

FIGURE 5.10 The four hydrogen bonds between one water molecule and its neighbors.

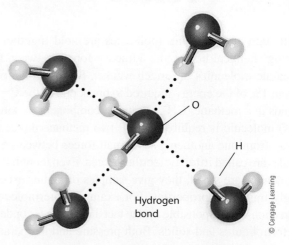

© Cengage Learning

FIGURE 5.11 The three states of matter.

The particles represented by the circles can be atoms, molecules, or ions (in liquids and solids). (a) In a gas, particles are very far apart and move rapidly in straight lines. (b) In a liquid, the particles move about at random, alone or in clusters. (c) In a solid, the particles or ions are in fixed positions and can only vibrate in place.

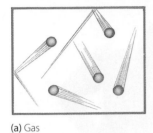
(a) Gas

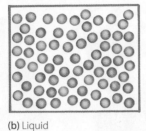
(b) Liquid

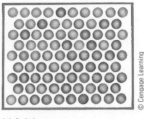
(c) Solid

© Cengage Learning

Condensation: The change of molecules from the gaseous state to the liquid state

Fluid: A state of matter that is capable of flowing; a gas or a liquid

ure 5.11b), the particles are also in motion, but they come close enough to touch one another. In solids (Figure 5.11c), the particles are also close to one another. They do not, however, move appreciable distances, but instead vibrate in fixed positions.

The particles in a sample of gas move at high speeds when they are at or near room temperature. For example, the molecules found in the air in your room are moving at over 1000 miles per hour. They travel only about 400 times their own diameter before colliding with another gas molecule, a short distance. If you cool a gas to a very low temperature, the molecules move much less rapidly. On further cooling, the molecules are moving so slowly that they attract one another and begin to form tiny droplets of liquid, a phenomenon called **condensation**. If that liquefied sample of gas is cooled even further, it will become a solid. All gases behave this way. For gases consisting of nonpolar molecules, the condensation and solidification temperatures are quite low. However, for gases whose molecules are polar, the temperatures are much higher. In this table, the condensation and solidification temperatures for SO_2, which is a polar molecule, and H_2O, which can engage in hydrogen bonding, are given for comparison:

In contrast to solids, whose particles are in rigid arrangements, liquids and gases have the property of being **fluid**—that is, they flow—because their atoms, ions, or molecules are not as strongly attracted to each other as they are in solids. Not being confined to specific locations, the particles in a liquid can move past one another. For most substances, the particles are a little farther apart in a liquid than in the corresponding solid; thus, the volume occupied by a given mass of the liquid is a little larger than the volume occupied by the same mass of the solid. This means that the liquid is less dense than the solid, and the solid form of a sample of matter sinks in its liquid (Figure 5.12). *There is a rather important exception to this rule: Solid water floats on liquid water.* The importance of this property of water is discussed in Section 11-1.

For solids, liquids, or gases, the higher the temperature, the faster the particles move. A solid melts when its temperature is raised to the point at which the particles vibrate fast enough and far enough to get away from the attraction of their neighbors and move out of their regularly spaced positions. As the temperature rises, the particles move even faster, until finally they escape their neighbors and become independent; the substance becomes a gas.

Condensation and solidification temperatures for some gases at 1 atm pressure:		
	Condenses (°C)	Solidifies (°C)
He	−269.0	—
Ar	−185.9	−189
O_2	−183.0	−218.4
N_2	−195.8	−209.8
Cl_2	−34.6	−101.0
SO_2	−10.1	−72.7
H_2O	100.0	0

© Cengage Learning

Charles D. Winters

FIGURE 5.12 Water (H_2O) and benzene (C_6H_6).
(a) Ice floats on water. (b) Solid benzene (melting point 5.5°C) sinks in liquid benzene.

5-9 Gases

Pressure: Force per unit area

Miscibility: Ability to mix in all proportions

Volatile: Easily vaporized

Gases surround us in our atmosphere. We breathe a mixture of nitrogen, oxygen, and other gases. Every breath (~0.5 liter in volume) we take carries with it the oxygen gas we need to burn the foods we eat. When we exhale, the carbon dioxide gas that is produced by the food-burning processes in our cells leaves our bodies, as does all of the inhaled nitrogen. This is one of the ways we rid ourselves of waste materials.

Interestingly, all gases possess a set of common physical properties. At constant temperature, all gases expand when the surrounding pressure decreases and contract when the pressure increases. At constant pressure, all gases expand with increasing temperature and contract with decreasing temperature. All gases have the ability to mix in any proportion with other gases. These properties are explained by the fact that the gas particles are far apart, move fast, and have very little chance to interact. Their molecular properties, such as size, number of electrons, and shape, have virtually no effect on the properties of the gas as a whole.

A sample of gas confined inside a container exerts a **pressure**, which is caused by the individual particles of the gas sample striking the walls of the container. Earth's atmosphere, a mixture of gases, exhibits a pressure that is dependent on the altitude relative to sea level and temperature.

Another general property of gases is *compressibility*. All gases can be compressed by applying a pressure on a confined sample (Figure 5.13). This property of all gases, known as *Boyle's law* after Robert Boyle who discovered it in 1661, is explained by the great distances between gas particles—applying pressure only confines the particles in a smaller space. Of course, a gas will always expand if the pressure is reduced. If a sample of gas is released into deep space, where the pressure is effectively zero, the molecules would begin randomly moving away from one another and eventually would become attracted to some nearby star system.

Perhaps one of the more interesting properties of all gases is that of **miscibility**—the ability to mix in all proportions with other gases. The miscibility of gases is explained by the great distances between gas molecules. In effect, there is always room for more molecules.

To illustrate how gases mix with one another, consider what happens when a person wearing a strong perfume

FIGURE 5.13 A bicycle pump.
The effects described by Boyle's law in action. The pump compresses air into a smaller volume. You experience Boyle's law because you can feel the increasing pressure of the gas as you press down on the plunger.

© Photos.com/Jupiterimages

enters a room. The smell of the perfume is noticed immediately by those close by and eventually by everyone in the room. The perfume contains **volatile**, meaning easily vaporized, compounds that gradually mix with the other gases in the room's atmosphere. Even if there were no apparent movement of the air in the room, the aroma of perfume would eventually reach everywhere in the room. This mixing of two or more gases due to random molecular motion is called *gaseous diffusion*. Given time, the molecules of one component in a gas mixture will thoroughly and completely mix with all of the other components to form a homogeneous mixture.

5-10 Liquids

In a liquid, the molecules are close enough together that, unlike a gas, a liquid is only slightly compressible. However, the molecules remain mobile enough that the liquid flows. Because they are difficult to compress and their molecules are moving in all directions, confined liquids can transmit applied pressure equally in all directions. This property is used in the *hydraulic fluids* that operate automotive brakes and airplane wing surfaces, tail flaps, and rudders.

Every liquid has a **vapor pressure**, which is the pressure in the gaseous state of molecules that have escaped from the liquid at a given temperature. As you would expect, the vapor pressure of a liquid increases with increasing temperature because more molecules escape the liquid (vaporize) and enter the gaseous state. The higher the temperature, the greater the volatility, because a larger fraction of the molecules have sufficient energy to overcome the attractive forces at the liquid's surface. Everyday experiences, such as heating water on a stove or spilling a liquid on a hot pavement in the summer, tell us that raising the temperature of the liquid makes vaporization take place more readily. Conversely, at lower temperatures, the volatility of a liquid will be lower. The same amount of water spilled on the pavement on a winter day will remain there much longer.

The temperature at which the vapor pressure equals the atmospheric pressure is the **boiling point** of the liquid. If the atmospheric pressure is 1 atm, the boiling temperature is called the **normal boiling point**. The normal boiling point of a compound is a reflection of the extent of intermolecular interactions (e.g., hydrogen bonding) and the molecular weight of the compound. In general, compounds with the ability to form hydrogen bonds have higher boiling points than those that do not. Higher molecular weight compounds also have higher boiling points. Water and ethanol, both capable of hydrogen bonding, boil at 100°C and 78.5°C, respectively. Ethane, which is incapable of hydrogen bonding, boils at 288°C, and octane, a higher molecular weight hydrocarbon also incapable of hydrogen bonding, boils at 125.6°C.

5-11 Solutions

Solutions, as explained in Section 2-1, are homogeneous mixtures that can be in the gaseous, liquid, or solid state. Most commonly, however, we encounter liquid solutions. How many liquid solutions are familiar to you? How about sugar or salt dissolved in water, oil paints dissolved in turpentine, or grease dissolved in gasoline? In each of these solutions, the substance present in the greater amount, the liquid, is the **solvent**; and the substance dissolved in the liquid, the one present in a smaller amount, is the **solute**. For example, in a glass of iced tea, water is the solvent, while sugar, lemon juice, and the components extracted from the tea leaves that impart taste and color are solutes.

A substance's **solubility** is defined as the quantity of solute that will dissolve in a given amount of solvent at a given temperature. Solubility is determined by the strengths of the forces of attraction between solvent molecules; between solute atoms, molecules, or ions; and between solute and solvent. The forces acting between the solvent and solute particles must be greater than those within the solute for the solute to dissolve. In other words, the solute and solvent must like each other more than they like themselves. If the forces between a liquid solvent and a solute are strong enough, then the solute will dissolve (Figure 5.14).

Although some solutes are so much like the solvent they are dissolving in that they dissolve in all proportions (a property called miscibility, like that of gases), there are limits to the solubility of most solutes in a given solvent. When a quantity of solvent has dissolved all of the solute it can, the solution is said to be a **saturated solution**. If more solute can be dissolved in the solution,

---

Vapor pressure: The pressure above a liquid caused by molecules that have escaped from the liquid surface

Boiling point: The temperature at which the vapor pressure of a liquid equals the atmospheric pressure and boiling occurs

Normal boiling point: The boiling temperature if the atmospheric pressure is 1 atm

Solvent: In a solution, the substance present in the greater amount

Solute: In a solution, the substance dissolved in the solvent

Solubility: The maximum quantity of a solute that will dissolve in a given amount of a solvent at a specified temperature

Saturated solution: Solution that has dissolved all the solute that it can dissolve at a given temperature

Solubility of Oxygen in Water at Various Temperatures*	
Temperature (°C)	Solubility of O_2 (g O_2/L H_2O)
0	0.0141
10	0.0109
20	0.0092
25	0.0083
30	0.0077
35	0.0070
40	0.0065

* These data are for water in contact with air at 1 atm pressure.

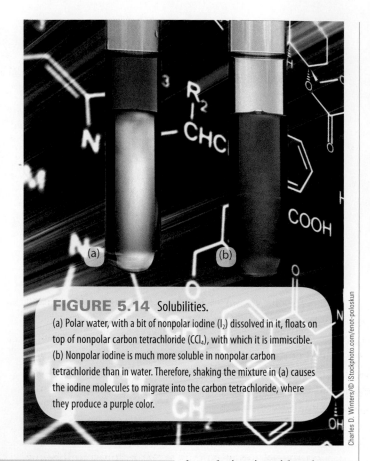

FIGURE 5.14 Solubilities.

(a) Polar water, with a bit of nonpolar iodine (I_2) dissolved in it, floats on top of nonpolar carbon tetrachloride (CCl_4), with which it is immiscible. (b) Nonpolar iodine is much more soluble in nonpolar carbon tetrachloride than in water. Therefore, shaking the mixture in (a) causes the iodine molecules to migrate into the carbon tetrachloride, where they produce a purple color.

Charles D. Winters/© iStockphoto.com/enot-poloskun

Unsaturated solution: Solution in which less than the maximum amount of solute is dissolved at a given temperature

Insoluble: Describes a substance that will not dissolve in a given solvent

Solubilities of gases over a range of temperatures.

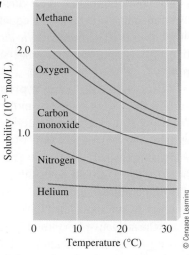

© Cengage Learning

increases. The simpler alcohols are very water soluble because the —OH group is polar and forms strong hydrogen bonds with water. As the hydrocarbon chain lengthens, the influence of the —OH group decreases, and with it, the water solubility of the molecule. A general rule of solubility is that "like dissolves like," which means that structurally similar compounds are more likely to be soluble in each other.

The effect of temperature on the solubility of solutes in various solvents is difficult to predict. Experience with everyday solutions such as sugar in water would lead us to predict that increasing the temperature of the solvent will cause more solute to dissolve. Although this is true for sugar and most other nonionic solutes, it is not true for all ionic compounds. In fact, the solubility of table salt in water is about the same at all temperatures.

Gases dissolve in liquids to an extent dependent on the similarity of the gas molecules and the solvent molecules. Polar gas molecules dissolve to a greater extent in polar solvents than nonpolar gas molecules do. Pressure also affects gas solubility. At higher pressure, more gas dissolves in a given volume of liquid than at lower pressure. When the pressure is lowered, gas will be evolved from a gas-in-liquid solution. The behavior of a carbonated beverage when the cap is removed is a common illustration of this principle.

Temperature has a greater effect on gas solubility in liquids than it has on solids or liquids dissolving in liquids. Without exception, lower temperatures cause more gas to dissolve in a given volume of liquid. Higher temperatures cause less gas to dissolve. As every fisherman knows, fish prefer deeper water in the summer months. This is because more oxygen dissolves in colder water, and the water temperature is generally cooler at greater depths.

the solution is said to be an **unsaturated solution**. Our everyday experiences illustrate this kind of behavior in solutions. For example, if a spoonful of sugar is added to a glass of iced tea, it quickly dissolves with a little stirring, which hastens the mixing process. The resulting solution is unsaturated (it can dissolve more sugar); if you like your iced tea a little sweeter, you just add another spoonful of sugar and stir. However, if sugar is continually added to the solution, a point is reached when no more appears to dissolve and the added sugar simply sinks to the bottom of the glass. At this point, the solution has become saturated in the dissolved sugar.

If almost none of the solute dissolves in a solvent, then it is said to be **insoluble** in that solvent. Metals, except those that react with water, are insoluble in water. Oily substances are also insoluble in water, as is water in oily substances. The low solubility of oily substances in water is caused by a lack of attraction between the highly polar water molecules and the nonpolar molecules of an oily substance. The water solubility of a nonpolar molecule can be increased if a polar part can be added. Conversely, the water solubility of a molecule decreases as the nonpolar portion of the molecule

5-12 Solids

In contrast to gases and liquids, in which molecules are in continual random motion, the movement of atoms, molecules, or ions in solids is restricted to vibration and sometimes rotation around an average position. This leads to an orderly array of particles and to properties different from those of liquids or gases.

Because the particles that make up a solid are very close together, solids are difficult to compress. In this respect, solids are like liquids, the particles of which are also very close together. Unlike liquids, however, solids are rigid, so they cannot transmit pressure in all directions. Solids have definite shapes, occupy fixed volumes, and have varying degrees of hardness. Hardness depends on the kinds of bonds that hold the particles of the solid together. Graphite, one form of elemental carbon, is one of the softest solids known; diamond (also carbon) is one of the hardest. Graphite is used as a lubricant. At the atomic level, graphite consists of layered sheets that contain carbon atoms. The attractive forces between the sheets are very weak; as a result, one sheet can slide along another and be removed easily from the rest. In diamond, on the other hand, each carbon atom is strongly bonded to four neighbors, and each of those neighbors is strongly bonded to three more carbon atoms, and so on throughout the solid. Because of the number and strength of the bonds holding each carbon atom to its neighbors, diamond is so hard that it can scratch or cut almost any other solid. The cutting and abrasive uses of diamonds are far more important commercially than their gemstone uses.

5-12a Properties of Solids

When a solid is heated to a temperature at which molecular motions are violent enough to partially overcome the interparticle forces, the orderliness of the solid's structure collapses and the solid melts. The temperature at which melting occurs is the **melting point** of the solid. Melting requires energy to overcome the attractions between the particles in a solid lattice. The energy required to melt a compound is called the *heat of fusion*. The heat of fusion of ice (solid water) is 80 cal/g (334.7 J/g). In the reverse of melting, **crystallization** (or **solidification**), energy is evolved. Melting points of solids depend on the kinds of forces holding the particles together in the solid, because these forces must be overcome for a solid to become a liquid. Table 5.7 gives some melting points for several types of solids.

Molecules can also escape directly from the solid to enter the gaseous state by a process known as **sublimation**. A common substance that sublimes at normal atmospheric pressure is solid carbon dioxide, which has a vapor pressure of 1 atm at $-78°C$. Solid CO_2 is known as dry ice, because it is cold like ice (although much colder at $-78°C$) and produces no liquid residue because it does not melt. Ice can sublime or melt. Have you ever noticed that ice and snow slowly disappear even if the temperature never gets above freezing? The reason is that ice sublimes readily in dry air (Figure 5.15). Given enough air passing over it, a sample of ice will sublime away, even at temperatures well below its melting point, leaving no trace behind. This is what happens in a frost-free refrigerator. A current of dry air periodically blows across any ice formed in the freezer compartment, taking away water vapor (and hence the ice) without having to warm the freezer compartment to melt the ice.

Melting point: The temperature at which a solid substance turns into a liquid

Crystallization (solidification): Formation of a crystalline solid from a melt or a solution

Sublimation: The process whereby molecules escape from the surface of a solid into the gas phase

5-13 Reversible State Changes

Each of the states of matter we've discussed have been associated with the changes they undergo. Solids melt to give liquids, and liquids vaporize to produce gases. Each of these changes in state requires an energy input. But the reverse processes *release* the same amount of energy. Changes in state like these are said to be reversible; the amount of heat energy needed to cause a change is the same as the heat energy released for the reverse process. Figure 5.16 summarizes the reversible changes of state that matter can undergo.

You are certainly aware that ice turns into liquid water upon melting

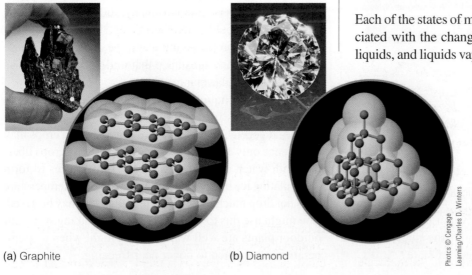

(a) Graphite (b) Diamond

Photos © Cengage Learning/Charles D. Winters

FIGURE 5.15 Ice cubes shrink in the freezer because of sublimation.
Even a solid like ice has a vapor pressure caused by its molecules in the vapor state. Dry air will sweep these molecules away.

Dry air

Air containing water molecules from ice

© Cengage Learning

TABLE 5.7 Melting Points of Some Solids

Solid	Melting Point (°C)
Molecular Solids: Nonpolar Molecules	
H_2	−259
F_2	−219.6
O_2	−218.4
N_2	−209.8
Cl_2	−101.0
Molecular Solids: Polar Molecules	
HCl	−115
HBr	−88
H_2S	−86
HF*	−83.1
NH_3	−77.7
SO_2	−72.7
HI	−51
H_2O	0
Ionic Solids	
NaI	662
NaBr	747
$CaCl_2$	782
NaCl	800
Al_2O_3	>2072
MgO	2852

© Cengage Learning

* Strongly hydrogen-bonded molecules highlighted in blue

and that liquid water turns into water vapor upon evaporation, but did you ever stop to consider the microscopic changes associated with these changes of state? At 0°C,

the water molecules in ice are tightly held in a crystalline lattice in which there is little molecular motion. As the ice absorbs heat and starts to melt, the lattice structure is disrupted and the molecular motion increases. However, the temperature of the liquid water immediately upon melting is still 0°C. If the water is heated even more, the liquid-state molecular motion becomes greater and greater while the water temperature increases from 0°C to 100°C. At 100°C, there is enough thermal energy to overcome all intermolecular interactions between water molecules. The water begins to boil and turn into the vapor state. The temperature of the water molecules in the vapor state is exactly 100°C. It is possible, however, to superheat steam above this point, which is why burns from steam sometimes are worse than those from boiling water.

This process of changing a solid to a liquid and then changing the liquid to a gas involves melting to the liquid state, heating of the liquid state to the boiling point, and vaporization of the liquid state to the gaseous state. Note that in this direction the processes are all *endothermic* (heat input is required). The reverse process (condensing water vapor to liquid water, cooling the water to its freezing point, and allowing the water to freeze) is *exothermic*, meaning that heat is released. By using the heat of fusion, the specific heat, and the heat of vaporization for water, one can calculate the amount of heat required to, for instance, convert 5 grams of ice at 0°C into 5 grams of steam at 100°C. Melting requires 400 cal (5 g × 80 cal/g), increasing the temperature of the water from 0°C to 100°C requires 500 cal (5 g × 1 cal/g-°C × 100°C), and converting the water to steam requires 2700 cal (5 g × 540 cal/g). Thus, the entire process is endothermic by 3600 cal. The reverse of this process would be exothermic by 3600 cal. Assuming we know the appropriate numbers, we can perform these calculations for any substance that undergoes these phase changes in a regular manner.

The fact that liquid water releases heat as it cools down and then freezes is sometimes used advantageously by citrus and vegetable farmers who anticipate that crops may experience only a very light freeze. They spray crops liberally with water, which releases heat as it freezes to form an insulating ice blanket around the crop. If the temperature does not drop much below freezing, the crop may be saved. You might use this to your advantage by making sure your outdoor plants are well watered on a night when the temperature is expected to hover near freezing.

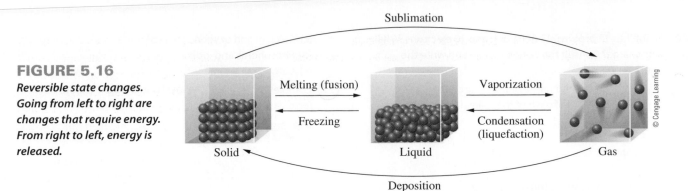

FIGURE 5.16
Reversible state changes.
Going from left to right are
changes that require energy.
From right to left, energy is
released.

Sublimation

Melting (fusion) — Vaporization

Freezing — Condensation (liquefaction)

Solid — Liquid — Gas

Deposition

© Cengage Learning

Your Resources

In the back of the textbook:

→ *Review Card on Chemical Bonding and States of Matter*

   In OWL for CHEM2 at www.cengagebrain.com

→ *Review Key Terms with Flash Cards (printable or digital)*

→ *Complete Interactive Practice Quizzes to prepare for tests*

→ *Submit Assigned Homework and Exercises*

Applying Your Knowledge

1. Give definitions for the following terms:
   (**a**) Cation   (**b**) Anion
   (**c**) Octet rule   (**d**) Formula unit

2. Give definitions for the following terms:
   (**a**) Nonmetal   (**b**) Binary compound

3. Give definitions for the following terms:
   (**a**) Shared pair   (**b**) Double bond
   (**c**) Triple bond   (**d**) Unshared pair
   (**e**) Single bond   (**f**) Multiple bond

4. Give definitions for the following terms:
   (**a**) Covalent bond   (**b**) Polyatomic ion
   (**c**) Ionic bond   (**d**) Binary compound

5. Give definitions for the following terms:
   (**a**) Nonpolar bond   (**b**) Polar bond

6. What is the octet rule?

7. Describe what each of the following terms means:
   (**a**) Hydrocarbon
   (**b**) Saturated hydrocarbon
   (**c**) Unsaturated hydrocarbon
   (**d**) Alkene

8. Is Ca^{3+} a possible ion under normal chemical conditions? Why or why not?

9. Predict the ions that would be formed by the following:
   (**a**) Br   (**b**) Al
   (**c**) Na   (**d**) Ba
   (**e**) Ca   (**f**) Ga
   (**g**) I   (**h**) S
   (**i**) All Group 1 metals
   (**j**) All Group 17 nonmetals

10. An ion has 12 protons, 13 neutrons, and 10 electrons. What is its charge? Consult the periodic table and write the symbol of the ion.

11. Write the formula and name of the ionic compounds formed from atoms of each of the following pairs of elements:
    (a) Al and I
    (b) Sr and Cl
    (c) Ca and N
    (d) K and S
    (e) Al and S
    (f) Li and N

12. What holds ionic solids together?

13. Name the following compounds:
    (a) $CaSO_4$
    (b) Na_3PO_4
    (c) $NaHCO_3$
    (d) K_2HPO_4
    (e) $NaNO_2$
    (f) $Cu(NO_3)_2$

14. The following are known mostly by the names that follow the formula for naming such compounds. Give another acceptable name for these molecules.
    (a) H_2O water or _____
    (b) O_2 oxygen or _____
    (c) H_2O_2 hydrogen peroxide or _____

15. Write correct formulas for the ionic compounds you expect to be formed when the following pairs of elements react:
    (a) Li and Te
    (b) Mg and Br
    (c) Ga and S

16. Provide molecular formulas for the following ionic compounds.
    (a) Calcium chloride, used to melt ice on sidewalks and driveways
    (b) Strontium carbonate, used to produce red fireworks
    (c) Magnesium hydroxide, the active ingredient in many liquid antacids
    (d) Iron(III) oxide, commonly known as rust
    (e) Potassium phosphate, a food additive used to prepare emulsions

17. Bismuth telluride is an ionic compound. When this compound is subjected to chemical analysis, it is found that there are three telluride anions for every two bismuth cations.
    (a) What is the likely formula for bismuth telluride?
    (b) What is the charge of the telluride anion?

18. Ammonium perchlorate has the formula NH_4ClO_4. What is the charge of the perchlorate anion?

19. Provide molecular formulas for the following compounds.
    (a) Ammonium phosphate
    (b) Sodium sulfate
    (c) Copper(II) chloride
    (d) Chromium(III) nitrate
    (e) Potassium bromide
    (f) Calcium carbonate
    (g) Sodium hypochlorite

20. Describe the difference between an ionic bond and a covalent bond.

21. Predict the type of bond formed between each of the following pairs of elements:
    (a) Sodium and sulfur
    (b) Nitrogen and bromine
    (c) Calcium and oxygen
    (d) Phosphorus and iodine
    (e) Carbon and oxygen

22. Complete the following table by writing the predicted formulas for each pair of elements:

	F	O	Cl	S	Br	Se
Na	NaF					
K				K_2S		
B		B_2O_3				
Al						
Ga			$GaCl_3$			
C					CBr_4	
Si						$SiSe_2$

23. Name the following compounds:
    (a) NO
    (b) SO_3
    (c) N_2O
    (d) NO_2

24. For each of the following compounds, first identify the compound as ionic or covalent. Then provide a name for each compound.
    (a) NBr_3
    (b) ZnI_2
    (c) CCl_4
    (d) HBr
    (e) Cr_2O_3

25. Draw Lewis structures for the following:
    (a) CO
    (b) SiF_4
    (c) C_2H_4
    (d) H_2S
    (e) C_2H_2
    (f) C_2H_6
    (g) OH^-
    (h) NF_3

26. Summarize the differences between ionic, polar covalent, and nonpolar covalent bonding.

27. Draw Lewis dot structures for the following molecules and then use your knowledge of the VSEPR model to predict the shapes of the molecules.
    (a) Dichlorodifluoromethane
    (b) Nitrogen trichloride
    (c) Oxygen dichloride

28. Which are the more polar bonds in the following molecules?
    (a) Chloroethane (CH_3CH_2Cl)
    (b) Freon 12 (CCl_2F_2)

29. A compound that has an odd number of total valence electrons can never satisfy the octet rule. Why?

30. Nitric oxide (NO) is a molecule that does not satisfy the octet rule. How many total valence electrons are present in the molecule? Draw the best Lewis structure you can for this molecule.

31. Which of the following compounds has the most polar bonds?
   (a) H — F
   (b) H — Cl
   (c) H — Br

32. Which of the following molecules is (are) not polar? For each polar molecule, which is the negative and which is the positive end of the molecule?
   (a) CO
   (b) GeH$_4$
   (c) BCl$_3$
   (d) HF

33. Which of the following molecules is polar and which is nonpolar? Explain.
   (a) Acetone (CH$_3$COCH$_3$), a common solvent

   (b) Butane (CH$_3$CH$_2$CH$_2$CH$_3$), a common fuel

   (c) Ammonia (NH$_3$)

34. Describe the following states of matter:
   (a) Gas
   (b) Liquid
   (c) Solid

35. The structural formula for ethanol, the alcohol in alcoholic beverages, is

   Give the total number of:
   (a) Valence electrons
   (b) Single bonds
   (c) Bonding pairs of electrons
   How many extra pairs of electrons are left? What are these called, and where should they be placed in the structural formula?

36. Lewis dot structures of two compounds are shown. For each, indicate the expected electron pair geometry and the molecular geometry.
   (a)

   :Cl:
   |
   :Cl — C — Cl:
   |
   :Cl:

   (b)

   H — P̈ — H
   |
   H

37. In which state of matter are the particles in fixed positions?
   (a) Liquid
   (b) Solid
   (c) Gas

38. In which state of matter are the particles the greatest average distance apart?
   (a) Liquid
   (b) Solid
   (c) Gas

39. Explain why the pressure of the atmosphere decreases with increasing altitude.

40. Which state of matter can be described as "particles close together and in constant, random motion"?
   (a) Liquid
   (b) Solid
   (c) Gas

41. Explain why molecules of a perfume can be detected a few feet away from the person wearing the perfume.

42. What happens to the pressure in an automobile tire in cold weather? Explain.

43. Name two properties of water that are unusual because of the presence of hydrogen bonding between adjacent water molecules.

44. Draw a structure showing four water molecules bonded to a central water molecule by means of hydrogen bonding. Indicate all the hydrogen bonds by drawing arrows to them.

45. Whenever a liquid evaporates, heat is required. Use this statement to explain why you get chilled when you emerge from a swimming pool on a windy day.

46. Which of the following compounds would you expect to exhibit hydrogen bonding? Explain your answer.
   (a) CH$_3$OH
   (b) NH$_3$
   (c) SO$_2$
   (d) CO$_2$
   (e) CH$_4$
   (f) HF
   (g) CH$_3$OCH$_3$

47. Unlike sodium chloride, barium sulfate is insoluble in water. Why do you think this might be?

48. Based on Figure 5.9, approximately what would be the boiling point of ammonia if there were no hydrogen bonding between the molecules?

49. Why are gases compressible, whereas liquids and solids are not?

50. In Chapter 4, you learned that as part of the fractionation of air, compression of a gas releases heat. Is the compression of a gas an exothermic or endothermic process?

51. Why is a gas more soluble in a solvent when a higher pressure is applied? Give an example of where you see this behavior of gases dissolving in liquids.

52. What causes surface tension in liquids? Name a compound that has a high surface tension.

53. The solubility of gases in liquids depends on temperature and gas pressure. Henry's law says the concentration of dissolved gas, C, is equal to the pressure of the gas multiplied by a constant, $C = kP$. The solubility of O$_2$ in water at 10°C is 0.0109 g O$_2$/L H$_2$O when the pressure

is 1 atm. What would be the solubility if the pressure is decreased to 0.5 atm?

54. Victims of carbon monoxide poisoning are placed in a hyperbaric chamber where the pressure is raised above 1 atm and a richer oxygen environment exists. If the pressure is raised from 1 atm to 3 atm, what will be the proportional change in oxygen solubility in blood? Use Henry's law and assume blood behaves like water.

55. Gold has a normal melting point of 1063°C and a normal boiling point of 2600°C when the pressure is 1 atm. What state (solid, liquid, or gas) will gold be in at 800°C and 1 atm?

56. A common solvent, acetone, $(CH_3)_2CO$, has a normal melting point of −95°C and a normal boiling point of 56.5°C when the pressure is 1 atm. What state (solid, liquid, or gas) will acetone be in at −50°C and 1 atm?

57. Francium is radioactive; therefore, physical data for francium compounds are not readily available. What melting point do you expect for francium chloride, FrCl, based on the normal melting points for other Group 1 ionic chlorides? Justify your estimate. NaCl, 801°C; KCl, 776°C; CsCl, 646°C; FrCl, _____.

58. Predict the boiling point for octane, C_8H_{18}, using the observed boiling points for the similar hydrocarbons hexane, C_6H_{14}, 68.7°C; heptane, C_7H_{16}, 98.4°C; and nonane, C_9H_{20}, 150.8°C.

WHY CHOOSE?

Every 4LTR Press solution comes complete with a visually engaging textbook in addition to an interactive eBook. Go to OWL for CHEM2 to begin using the eBook. Access at **www.cengagebrain.com**

www.cengagebrain.com

f Follow us at
www.facebook.com/4ltrpress

©iStockphoto.com/A-Digit | © Cengage Learning 2011

Carbon Dioxide
and the Greenhouse Effect

As pollutants go, carbon dioxide might be considered a relatively innocuous component of our atmosphere. After all, every breath we exhale contains about 4% carbon dioxide even though the inhaled air is only about 0.04% carbon dioxide. Fossil fuel combustion and natural decay processes combined with forest and grassland fires release billions of metric tons of carbon dioxide into the atmosphere annually. At the same time, trees, grasses, and other plants remove equivalent quantities of carbon dioxide from our atmosphere each year. Carbon dioxide is constantly being dissolved in and released from the ocean and other bodies of water as their temperatures fluctuate. In other words, carbon dioxide is constantly being added to and removed from our atmosphere by a variety of processes, some natural and some of human origin.

Carbon dioxide is not considered a pollutant in the same sense as are nitrogen oxides, sulfur oxides, ozone, carbon monoxide, and other gaseous pollutants. However, there is much concern that the increasing concentration of carbon dioxide in our lower atmosphere is having a major impact on our climate.

6-1 Atmospheric Carbon Dioxide Concentration over Time

Although the concentration of carbon dioxide in our atmosphere is relatively low, evidence indicates that this concentration has fluctuated quite significantly over the past 200,000 years. This is not a long time, considering Earth has existed for about 4.5 billion years, but this is the time frame for which the concentration can be reliably inferred using available data. Some have suggested that significantly higher concentrations of carbon dioxide may have been, at least in part, responsible for the 10°C to 20°C higher terrestrial temperatures about 100 million years ago when dinosaurs roamed Earth.

Data from ice cores taken from deep below the surface of Antarctica have been used to estimate Earth's temperature in the past. The exact manner in which these determinations have been made depends on the fact that water molecules containing different isotopes of hydrogen (protium, the most common isotope, or deuterium) have slightly different boiling points. By looking at the relative amounts of these isotopes in the ice, one can determine the atmospheric temperature at the time the ice was laid down. There are also bubbles of air trapped within the ice at the time of its formation. Thus, one can compare the temperature-related data with the carbon dioxide concentration to determine if there is a correlation. This correlation is shown in Figure 6.1 on page 106. It reveals that as the amount of carbon dioxide trapped in these ice cores fluctuated, the temperature fluctuated in the same direction; higher carbon dioxide concentrations correlate with higher than average temperatures in a strikingly similar pattern. This has led scientists to believe that the amount of carbon dioxide in our atmosphere is related to Earth's temperature.

© iStockphoto.com/Gannet77

The greenhouse effect results in an increase in temperature. These strawberries can be grown in a cold climate during the winter, thanks to a covering that prevents accumulated heat from escaping.

Greenhouse effect:
A phenomenon of Earth's atmosphere by which solar radiation, trapped by Earth and re-emitted from the surface as infrared radiation, is prevented from escaping by various gases in the atmosphere

Infrared (IR) radiation:
Electromagnetic radiation whose wavelength is in the range between 2.5×10^{-5} m and 2.5×10^{-6} m

This correlation suggests, but does not prove, that the high concentration of carbon dioxide is the cause of the temperature increase. It is theoretically possible, but thought to be less likely, that other phenomena led to higher temperatures which, in turn, caused the carbon dioxide concentration to rise. However, carbon dioxide is known to trap heat, so it is incumbent on us to see if the increasing concentration of carbon dioxide in our atmosphere and the resultant temperature increase in the past two hundred years, a mere blip in time on Figure 6.1, are related to human activity.

Ice core samples from the Greenland Ice Sheet. The samples, stored at −33°F, are examined for evidence of global warming.

© Roger Ressmeyer/CORBIS

FIGURE 6.1 Estimated long-term variations in mean global surface temperature and average tropospheric carbon dioxide levels of the past 160,000 years.

These carbon dioxide levels were obtained by inserting metal tubes deep into Antarctic glaciers, removing the ice, and analyzing bubbles of ancient air trapped in ice at various depths throughout the past. Such analyses reveal that since the last great ice age ended about 10,000 years ago, we have enjoyed a warm interglacial period. The rough correlation between tropospheric carbon dioxide level and temperatures suggests, but does not prove, a connection between the two.

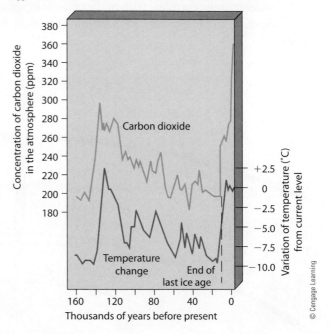

© Cengage Learning

6-2 What Is the Greenhouse Effect?

The **greenhouse effect** is a well-known phenomenon that results in warming of Earth's atmosphere through the absorption of **infrared (IR) radiation** by molecules in the atmosphere such as carbon dioxide, water vapor, and methane. Figure 6.2 shows that infrared radiation accounts for about 53% of the radiation coming from our Sun. About 8% is higher-energy ultraviolet (UV) radiation and 39% is visible (Vis) radiation. What is the impact of this radiation as it travels through our atmosphere and then strikes Earth?

Radiation making its way from the Sun to Earth encounters many types of molecules once it reaches Earth's stratosphere. Certain molecules are more efficient than others at absorbing some types of radiation, and different types of radiation bring about different responses in the molecules they encounter. Both features are critical to understanding the greenhouse effect.

Figure 6.3 illustrates the fate of solar radiation. Much of the radiation coming from the Sun is simply reflected back into space once it encounters our atmosphere. This fact is of little practical consequence to us except as a reminder that Earth would be too hot to support life if this did not happen. Fortunately for us, much of the higher-energy ultraviolet radiation is absorbed by molecules in our atmosphere (primarily oxygen and ozone); thus, only a small portion of this potentially damaging radiation reaches the surface of Earth. In Chapter 7, we will look at the importance of the oxygen–ozone screen as a mechanism for protecting us from dangerous UV radiation.

FIGURE 6.2 A comparison of the intensity of solar radiation and the wavelength of the radiation.

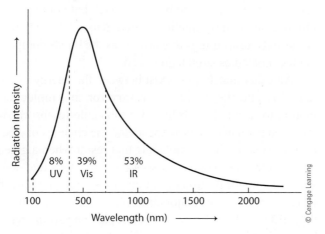

Most of the solar radiation that reaches Earth's surface is in the visible and infrared range. At Earth's surface, this radiation is absorbed by all manner of objects including water, dirt, buildings, sidewalks, ice, snow, grass, and, of course, humans. By and large, the absorbed radiation increases molecular vibrations in whatever object it strikes. Sunlight striking your skin causes the molecules that compose the skin to vibrate, making you feel warm. The same is true of sunlight striking Earth, a building, or a body of water. The energy of this absorbed radiation increases the temperature of the body that has absorbed the radiation. An object, once warmed, will begin to reradiate this energy away to cooler objects with which it is in contact, including the atmosphere. This energy is lost from the warmer body in the form of infrared radiation.

As infrared radiation heads away from Earth back toward space, about 84% of it is reabsorbed by molecules (water vapor, carbon dioxide, and methane) in our atmosphere and then reradiated back to Earth. This "greenhouse blanket" is critical to the balance of incoming and outgoing radiation that keeps Earth's temperature tolerable for life. Thankfully, this phenomenon warms the surface of Earth to a comfortable average temperature of 15°C or about 59°F (compared to about −270°C in outer space) and allows us to flourish here. Without this warming, our oceans would be frozen solid!

The greenhouse effect is desirable only to a certain point. Imagine what would happen if Earth were encased in a glass sphere capable of letting in UV and visible

FIGURE 6.3 The natural greenhouse effect.

Without the atmospheric warming provided by this natural effect, Earth would be a cold and mostly lifeless planet. According to the widely accepted greenhouse theory, when concentrations of greenhouse gases in the atmosphere rise, the average temperature of the troposphere rises. (Source: Modified by permission from Cecie Starr, Biology Concepts and Applications, 4th ed, Brooks/Cole (Wadsworth) 2000.)

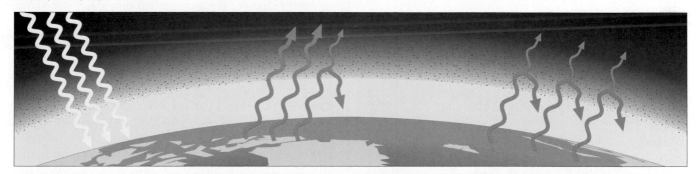

(a) Rays of sunlight penetrate the lower atmosphere and warm Earth's surface.

(b) Earth's surface absorbs much of the incoming solar radiation and degrades it to longer-wavelength infrared (IR) radiation, which rises into the lower atmosphere. Some of this IR radiation escapes into space as heat and some is absorbed by molecules of greenhouse gases and emitted as even longer wavelength IR radiation, which warms the lower atmosphere.

(c) As concentrations of greenhouse gases rise, their molecules absorb and emit more infrared radiation, which adds more heat to the lower atmosphere.

A row of greenhouses.

© iStockphoto.com/thehague

additional concepts have been introduced, we will be able to compute the exact energies in more meaningful terms (Section 8-2).

Energy Type	Wavelength Range	Relative Energy
Infrared	2.5×10^{-5} to 2.5×10^{-6} m	Low
Visible	700×10^{-9} to 400×10^{-9} m	Medium
Ultraviolet	400×10^{-9} to 200×10^{-9} m	High

light but incapable of letting any of the reradiated IR radiation pass back out to space. The result would be catastrophic. (Think of what happens inside a car on a hot, sunny day.) Fortunately for us, Earth's control mechanisms have regulated the natural greenhouse effect for thousands of years. However, increasing concern exists that humans have begun to upset this fine balance, and that is why we now speak of global warming (Section 4-11) as a potential environmental disaster whose time may have come.

TRY IT 6.1

Convert the 0.03% to 0.04% average concentration of carbon dioxide in our atmosphere to ppm and ppb.

TRY IT 6.2

Convert the range of wavelengths for infrared radiation (2.5×10^{-5} m to 2.5×10^{-6} m) to nm, and compare these wavelengths to those for ultraviolet radiation (200–400 nm). Which is longer? Which is more energetic?

6-3 Why Worry About Carbon Dioxide?

By using the relationships among energy, wavelength, and frequency (Figure 3.7), it is possible to calculate the energies associated with the types of electromagnetic radiation discussed earlier. For now, we will just list the relative energies of infrared (IR), visible (Vis), and ultraviolet (UV) radiation and note, once again, that longer-wavelength radiation corresponds to lower-energy radiation. Later, after

Each type of radiation elicits a certain response when it impinges on a molecule or an object. An inspection of the relative energies associated with IR, Vis, and UV radiation reveals that IR radiation has the lowest energy of the three. Infrared radiation is sufficiently energetic to cause certain chemical bonds to bend or stretch but not to break. Infrared radiation is sometimes referred to as heat radiation because it causes things to warm up as the bonds composing the molecules stretch and bend.

An exact match must exist between the energy of the radiation and the energy necessary for this molecular motion to occur. The ability of a molecule to absorb IR radiation is related to its shape and to the changes in electron distribution in the molecule that result when a given bond stretches or bends. Most of the molecules that constitute the major portion of our atmosphere do not absorb IR radiation. The two major atmospheric gases (N_2 and O_2) do not absorb IR radiation and, thus, play no part in greenhouse warming.

That brings us to carbon dioxide, methane, and water. These gases do absorb infrared radiation because the energy associated with certain stretches and bends of the bonds in these molecules are the same as the energy of infrared radiation. Because carbon dioxide, water vapor, and methane are three molecules that are capable of absorbing IR radiation, these molecules play a major role in keeping Earth warm.

6-3a Carbon Dioxide in the Air

Carbon dioxide's effect on temperature can be illustrated by comparing the temperature on Venus (the second planet from the Sun) to that on Earth. Both planets are warmer than can be explained simply by looking at their distance from the Sun. Venus, whose average temperature is 450°C, is estimated to be 300°C warmer than might be predicted

Carbon dioxide is used to make sparkling water fizzy.

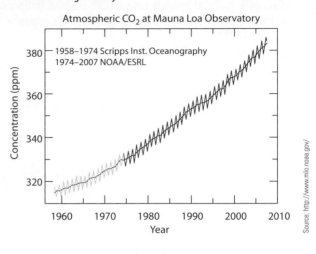

FIGURE 6.4 Atmospheric carbon dioxide concentrations have been rising steadily.

Atmospheric CO$_2$ at Mauna Loa Observatory

1958–1974 Scripps Inst. Oceanography
1974–2007 NOAA/ESRL

Source: http://www.mlo.noaa.gov/

based only on its distance from the Sun. Earth, which is almost 26 million miles farther from the Sun, is still about 33°C higher in average temperature (15°C) than the −18°C predicted. Why is that?

These elevated temperatures are largely attributable to the composition of the atmosphere of the two planets. Venus's atmosphere is 96% CO$_2$ with H$_2$SO$_4$ clouds and an atmospheric pressure of 90 atm, whereas Earth's atmosphere is 78% N$_2$, 21% O$_2$, 0.03% to 0.04% CO$_2$, and 0.1% to 1.0% H$_2$O at 1 atm of pressure at sea level. This difference in composition, as well as the increased pressure (higher pressure means more gas molecules in a given volume), can account for the beneficial greenhouse effect of Earth's atmosphere despite its distance from the Sun.

The percentage of water in Earth's atmosphere and, thus, the other percentages fluctuate over relatively small ranges depending on humidity. The average CO$_2$ concentration in Earth's atmosphere has increased from about 280 ppm in 1860 to 380 ppm in 2004, and it continues to rise (Figure 6.4). It took 150 years for the concentration to increase by 50 ppm from 280 ppm to 330 ppm, but it has only taken 30 years for the most recent 50 ppm increase from 330 to 380 ppm. Earth has experienced an increase in CO$_2$ concentration of over 25% in the last century or so. During this time, Earth's average temperature has increased by about 0.5°C to 0.7°C, as shown in Figure 6.5. The big question is, has the increase in carbon dioxide concentration in our atmosphere been responsible for the increase in temperature? Furthermore, if there is a cause-and-effect relationship here, what will be the consequences of further increases in atmospheric carbon dioxide concentrations?

TRY IT 6.3

Use relationships between the Fahrenheit and Celsius temperature scales to convince yourself that 59°F, Earth's average temperature, corresponds to 15°C. Do the same to show that Venus's average temperature of 450°C is 840°F. [Relationships: °F = (9/5)°C + 32 and °C = (5/9)(°F − 32)]

TRY IT 6.4

The Kelvin scale of absolute temperature is often used by scientists. The relationship between the Celsius and Kelvin scales (abbreviated as K only) is: K = °C + 273.15. What is Earth's average temperature in kelvins?

Although unanimous agreement does not exist regarding the answer to the first question, more and more scientists are becoming convinced that the recent increases in Earth's temperature are largely, or entirely, due to increasing atmospheric concentrations of carbon dioxide. Although difficult to predict exactly, the general estimate is that if the concentration of carbon dioxide reaches around 600 ppm, Earth's temperature will increase by between 1.5°C and 6°C (2.4–10.5°F). This could come as early as 2050. While the magnitude of this predicted temperature change may not seem significant, temperature decreases within this range resulted in the last ice age about 20,000 years ago. (We will return to the possible consequences of an increase in Earth's temperature in Section 6-6.)

FIGURE 6.5 a) Trends in temperature, sea level, and snow cover over the past 150 years. b) Actual and predicted surface warming from 1900–2100, according to a variety of models. c) Geographic distribution of predicted warming.

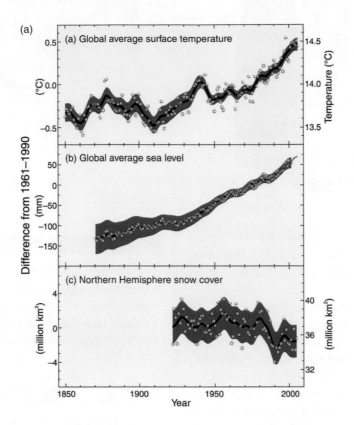

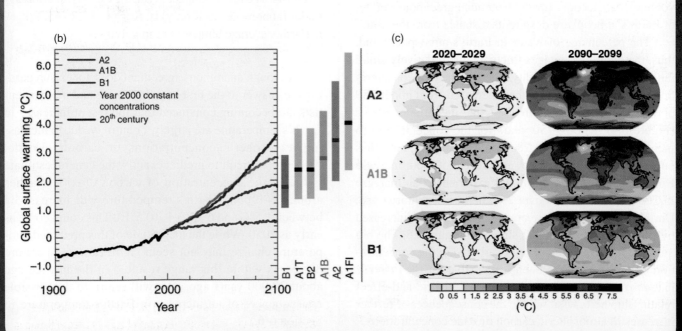

TABLE 6.1 Atmospheric Concentration and Lifetimes of Some Key Greenhouse Gases

Time/Gas	CO_2	CH_4	N_2O	CFC-11 (CCl_3F)	CFC-12 (CCl_2F_2)	HCFC-22 ($CHClF_2$)
Pre-1750 (preindustrial) concentration*	~280 ppm	~700 ppb	270 ppb	0[†]	0[†]	0[†]
Concentration in 2001/2002	372 ppm	~1800 ppb	318 ppb	258 ppt[‡]	546 ppt	146 ppt
Atmospheric lifetime (years)	Variable[§] (50–200)	12	114	45	100	11.9

* Pre-1750 concentrations are assumed to be practically uninfluenced by human activities associated with significant fossil fuel combustion.

[†] Not yet discovered at this time.

[‡] This stands for parts per trillion (not parts per thousand).

[§] The mechanisms by which carbon dioxide is removed have quite different lifetimes for removal, so this is just an estimate.

Source: All data taken from a tabulation compiled by T. J. Blasing and Sonja Jones at http://cdiac.esd.ornl.gov/pns/current_ghg.html, which contains extensive documentation of the original sources.

6-4 Other Greenhouse Gases

Some atmospheric gases other than carbon dioxide also absorb IR radiation. Some, such as **chlorofluorocarbons** (also known as **CFCs**), owe their atmospheric presence to human activities while others, such as water vapor, methane, and nitrous oxides, do not. Most, with the exception of water vapor, are present in relatively low concentrations, as is shown in Table 6.1. Although water vapor is a major greenhouse gas, we clearly can do little about the amount of water vapor in the air. However, one can observe the effect of water vapor on **radiational cooling** on a clear winter night versus a cloudy winter night. The temperature drops more quickly after sunset on clear nights due to the absence of heat-trapping clouds in the lower atmosphere. For this same reason, desert nights are often very cool due to low humidity. Is there anything we can do about potential greenhouse gases other than water? If so, on which gases should we focus?

The efficiency with which these gases absorb IR radiation is not equal, as shown by the greenhouse factors in Table 6.2. Some gases are orders of magnitude more efficient at absorbing IR radiation than others. We are fortunate that the concentration of these efficient absorbers in our atmosphere (the *tropospheric abundance*) is quite low. Since the seriousness of the problem is related both to the efficiency with which the gas absorbs IR and its concentration, we will focus on CO_2 and methane, both present in relatively high and increasing concentrations.

TRY IT 6.5

Convert the tropospheric abundance values listed in Table 6.2 into ppm and ppb.

Chlorofluorocarbons (CFCs): A term used to refer collectively to compounds containing only carbon, chlorine, fluorine, and sometimes hydrogen

Radiational cooling: The phenomenon of enhanced loss of heat from Earth on a clear, cloudless night due to the absence of significant water vapor in the atmosphere

TABLE 6.2 Relative Efficiencies of IR Absorption by Some Key Greenhouse Gases Compared with Their Tropospheric Abundance

Substance	Greenhouse Factor*	Tropospheric Abundance (%)
CO_2	1 (assigned value for comparison)	3.7×10^{-2}
CH_4	30	1.8×10^{-4}
N_2O	160	3.1×10^{-5}
O_3	2000	4.0×10^{-6}
CCl_3F	21,000	2.6×10^{-8}
CCl_2F_2	25,000	5.2×10^{-8}

* The greenhouse factor is a measure of the effectiveness of infrared absorption by various atmospheric gases when compared with carbon dioxide, which is arbitrarily assigned a greenhouse factor of 1.

Source: Intergovernmental Panel on Climate Change, 1996.

6-5 Sources of Greenhouse Gases

Prior to the Industrial Revolution, which began in the late 18th century in Europe, the average citizen of the world was responsible for the release of relatively small amounts of carbon dioxide gas into the atmosphere. This release occurred through the burning of coal and wood for cooking and personal heating. However, the internal combustion engine brought with it an enormous appetite for fossil fuel. This fuel was coal in the early days of the revolution. Oil, gasoline, diesel fuel, and natural gas became major fuels as time progressed, and our dependence on factories, electricity, and motorized transportation increased to its present level.

No way exists to avoid the fact that burning fossil fuels produces carbon dioxide and water vapor, both greenhouse gases. A gallon of gasoline produces about 18 pounds of carbon dioxide and about 1 gallon of water upon combustion (Section 12-1). The increased release of these gases into our atmosphere brings up the question of the rate of change of the concentration of these gases in our atmosphere. In other words, are we putting more carbon dioxide into our atmosphere than is or can be taken out of the atmosphere by natural mechanisms? The answer to this question appears to be "yes," as evidenced by the increasing carbon dioxide concentration.

Spewing volcanoes, natural vents at the bottom of some lakes and the oceans, naturally decaying plant and vegetable materials, forest fires (natural and those of human origin), and a few other mechanisms contribute significantly to the estimated 25 billion tons of carbon dioxide that are added to the atmosphere each year. Of this amount,

22 billion tons is generated by burning fossil fuels! It is this anthropogenic contribution of carbon dioxide to our atmosphere about which we are most concerned.

The two main mechanisms for removing carbon dioxide from the air are photosynthesis and dissolution into water (primarily the oceans).

Plants, trees, and grasses take enormous amounts of carbon dioxide from the air as part of their natural growth cycle. In photosynthesis, carbon dioxide and water are combined in the presence of sunlight with the chlorophyll in plants to produce carbohydrates and oxygen gas. This is one of the main reasons there is so much concern about deforestation and loss of grasslands as the world's population increases.

$$6\ CO_2\ +\ 6\ H_2O\ \xrightarrow[\text{Chlorophyll}]{\text{Sunlight}}\ C_6H_{12}O_6\ +\ 6\ O_2$$

carbohydrates
(sucrose, fructose, cellulose)

Carbon dioxide is absorbed by oceans because it readily dissolves in water and reacts with it to produce carbonic acid (H_2CO_3). The acid can stay dissolved or form bicarbonates and carbonate minerals that precipitate out as solids in the form of limestone and other materials. The temperature of the water greatly affects its ability to dissolve carbon dioxide and other gases. At higher temperatures, less CO_2 will dissolve. (Think about what happens when a carbonated beverage is allowed to warm up—it goes flat.) Warming of the world's oceans could lead to release of more carbon dioxide into the atmosphere, which, in turn, could lead to even more warming and greater release of carbon dioxide. This is not a problem to be taken lightly.

It is estimated that natural mechanisms remove 10 billion tons of carbon dioxide per year from our atmosphere. At present levels of fossil fuel combustion, we put 22 billion tons of carbon dioxide into the atmosphere each year. Thus,

Burning fossil fuels produces carbon dioxide and water vapor, both greenhouse gases.

we are producing a net increase of about 12 billion tons of carbon dioxide in our atmosphere each year! That explains the trend in carbon dioxide concentrations. The carbon dioxide concentration is increasing, and we are almost certainly responsible for that increase.

Compare the lifestyle of a typical U.S. citizen to that of a peasant farmer in India, and imagine the difference in the amounts of carbon dioxide released into the atmosphere as a function of their differing lifestyles. Consider, for example, transportation, housing, food production, electricity use, consumer products, and the energy required for their production. Would it surprise you to learn that the average U.S. citizen is responsible for releasing about 6 metric tons of carbon in the form of carbon dioxide into the atmosphere each year, compared to less than 0.5 metric ton for the Indian farmer? Since carbon dioxide is only 27.3% carbon, that equates to almost 22 metric tons of CO_2 per person in the United States versus 1.8 metric tons per person for the people of India.

Pam and Jeff Riesgraf own and operate a small organic dairy farm in Jordan, Minnesota. Though they may produce less carbon dioxide and methane than a commercial farm, they are still producing more greenhouse gases than Indian peasants.

TRY IT 6.6

Show that 6 metric tons of carbon in the form of carbon dioxide equates to 22 metric tons of carbon dioxide.

Methane, another problematic greenhouse gas, is primarily released into the atmosphere from natural sources such as decaying vegetation in swamps (marsh gas), decaying organic materials in landfills and rice paddies, cows, and leaking natural gas lines and wells.

The digestive systems of cows and other ruminant animals contain bacteria that break down cellulose and produce methane gas. Some large ruminants produce hundreds of liters of methane per day. The large increase in the world population of cattle and sheep has given rise to a large increase in methane emissions from this source.

Scientists agree that energy production, energy distribution, and livestock contribute most to methane emissions, followed by landfill emissions and natural sources. Thus, even though the concentration of methane in our atmosphere is far less than that of carbon dioxide, its role in global warming cannot be totally discounted, because methane is an efficient absorber of IR radiation. However, the major focus is on carbon dioxide due to the enormous quantities of this gas released into the atmosphere by our carbon-based economy.

TRY IT 6.7

A gallon is 3.785 liters. How many gallons of methane would an average cow produce in one day if the cow produces 500 liters of methane per day?

6-6 What Do We Know for Sure About Global Warming?

There are some things about global warming that can be stated with increasing confidence.

• Earth is definitely warming, and most scientists agree that humans are responsible for the warming.

• The warming is likely due, at least in part, to increased anthropogenic emissions of greenhouse gases (primarily carbon dioxide). The Intergovernmental Panel on Climate Change (IPCC) sponsored by the United Nations recently predicted a potentially catastrophic warming of 1.4°C to 5.8°C (2.5°F to 10.4°F) by 2100 if nothing is done to curb these emissions.

- A temperature increase of only 2°C would mean Earth would be warmer than it has been for the past 2 million years.

- As the world's population increases, the rate and amount of greenhouse gases released will increase unless some significant changes are made. The population as of early 2012 is 6.99 billion.

6-7 Consequences of Global Warming— Good or Bad?

If global warming is a reality, as most now agree it is, then its consequences must be considered. The possible consequences of global warming are controversial and difficult to predict. However, most of the scientists who study the atmosphere agree that the following are likely results of increased global warming, although the extent to which each of these phenomena may manifest itself is debatable.

- Oceans may warm and release more carbon dioxide due to the reduced solubility of gases in warm water. This would increase warming even more and exacerbate the problem. However, warmer oceans might allow more phytoplankton to grow, which could consume more carbon dioxide.

- Solid, frozen methane deposits at the bottom of the oceans may become gaseous as the ocean warms, and the gaseous methane would be released into air. This would increase warming even more.

- Ice and snow cover may decrease significantly in parts of the world. This would cut down on reflected light, so Earth would absorb even more radiation and warm up at an accelerated rate.

- Ocean levels may rise as ice and snow at the polar caps melt. The consequences for islands and low-lying areas along coasts could be disastrous. There is already evidence of this occurring. Ice shelves that have been frozen for thousands of years are breaking away. There is evidence that this rise is already taking place (described shortly). The Arctic Climate Impact Assessment, a study by more than 300 scientists released in late 2004, suggested that among other consequences of such warming, polar bears could become extinct by 2099.

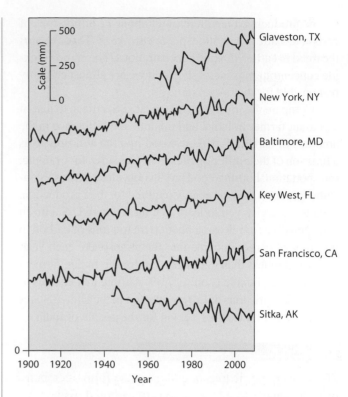

- More water might evaporate from the oceans as they warm. This could increase global warming because water is a greenhouse gas. However, this could be offset by the reflective effect that additional cloud cover might have on incoming solar radiation.

- Warming of the oceans could cause major shifts in ocean currents, leading to unpredictable influences on weather, fisheries, and so on.

- Rainfalls would become much heavier in some areas as a result of increased evaporation. This would lead to increased flooding.

- The tree line would move farther north, causing trees to grow over a broader region. This could consume more carbon dioxide.

- Deserts and arable regions of the world could shift location and change size as a function of shifts in areas of rainfall. The consequences are difficult to predict.

- Tropical diseases might become more prevalent and widespread as Earth warms.

- The intensity of storms and weather systems would probably increase.

- Increased heat may mean worsened air pollution, damaged crops, and depleted resources in some areas.

- Some scientists have speculated that when a certain atmospheric concentration of greenhouse gases is reached, it may trigger a relatively rapid and cataclysmic climate system reorganization.

6-8 The Kyoto Conference Addresses Global Warming

More and more people from less-developed countries are adopting the lifestyle of the United States. Because of this trend, the problems associated with global warming are expected to be dramatically exacerbated. Right now, among industrialized nations, the United States is first worldwide in per capita carbon dioxide emissions, and China is eleventh. However, China is first in total emissions (7.7 billion tons in 2009), and the United States is second (5.4 billion tons in 2009). What will happen as additional countries begin to adopt lifestyles that are characterized by prodigious production of carbon dioxide?

Cognizant of the potential for a global warming disaster and unwilling to wait for utter certainty that global warming is a real problem, many of the world's nations sent representatives to Kyoto, Japan, in 1997 to begin to address the issue. At this conference, 161 countries set greenhouse gas emissions goals for "developed" countries but chose not to set them for "developing" countries. This was a contentious issue and was partly responsible for the fact that, at least as of this writing, the United States—alone among major world powers—has chosen not to support the goals established at Kyoto. Of course, economic issues were addressed, and the United States took the position that it would suffer more economic hardships than the other countries if these goals were enforced. There was also concern that the United States was being asked to bear an unfair share of the burden relative to less-developed countries. Neither Presidents George W. Bush and Barack Obama nor the leading Democrats or Republicans have generally supported these goals, although some individual American politicians have come out in favor of the Kyoto Protocol. Russia ratified the Kyoto Protocol in October 2004. This ratification made carbon dioxide reductions mandatory among all 124 countries that had accepted the accord as of the date when it went into effect as law, February 16, 2005. On that date, carbon emissions became a commodity that can be traded by the signatory countries. The 30 industrialized countries that have signed the protocol will have to meet their protocol emissions targets by 2012. Some countries have vowed to cut emissions even more dramatically than is called for by the protocol. Great Britain plans to cut emissions by 60% by 2050.

Even though the U.S. government has not signed the Kyoto Protocol, many state governments, organizations, and industries have taken action to address the issue. The attorneys general of eight states sued the nation's largest utility companies in July 2004, demanding that they reduce emissions of greenhouse gases. California has proposed a 30% reduction in car emissions by 2015, and other states are considering similar actions. Many companies and other industries responsible for the production of large amounts of greenhouse gases have implemented voluntary reduction plans even in the absence of government dictates. However, even the most ardent supporters of the Kyoto Protocol recognize that it alone will not come close to solving the problem of global climate change. It is, according to UN Environment Program director Klaus Toepfner, "only the first step in a long journey."

Mark Wilson/Getty Images

An atlas of pollution: the world in carbon dioxide emissions

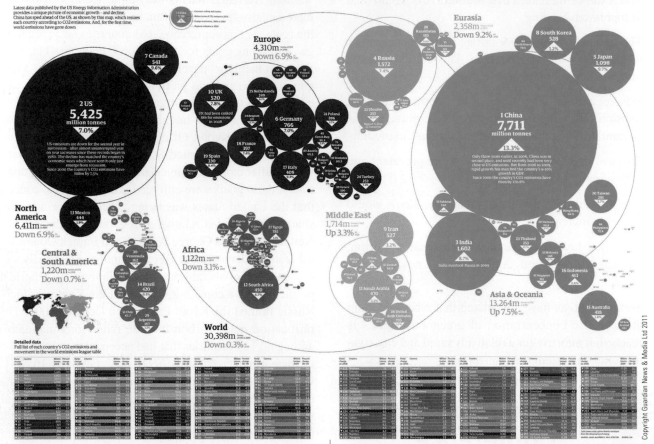

TRY IT 6.8

Comment on the following quotation as it might apply to the United States versus some other countries. "One part of the world cannot live in an orgy of unrestrained consumption while the rest destroys its environment just to survive." *New Scientist*, 17 August 2002, 31.

Gasoline-electric hybrid vehicles, such as Toyota's 2010 Prius hatchback, are becoming increasingly popular among American drivers.

6-9 Possible Responses to Global Warming

Assuming that the problem of global warming is real, scientists have begun to make suggestions about what can be done to either slow its progress or allow us to live with its consequences. Some of the actions suggested are obvious and probably practical. Others are obvious but probably impractical. Still other suggested actions are quite unusual, and their degree of practicality is debatable. However, with a problem as serious as global warming, maybe even the most outrageous of responses deserves some consideration. Engineering feats that at one time would have been considered folly are now routine. Who is to say how the future will judge our response to global warming? For instance, one or more of the following actions might prove effective:

• Investing in low-emission consumer vehicles such as electric, hydrogen-powered, and hybrid cars.

- Decreasing the consumption of fossil fuels and developing and using alternative fuel sources that are not carbon based.

- Developing more efficient internal combustion engines or cars that do not depend so heavily on fossil fuels.

- Developing and requiring the use of more mass transit systems.

- Imposing taxes on goods and services that take into account the amount of greenhouse gases released in their production (a "carbon tax").

- Developing a worldwide emissions trading program wherein countries or other entities could work together to collectively exploit effective reduction programs.

- Developing and using new technologies that do not generate greenhouse gases. For instance, nuclear power plants and wind power generate no greenhouse gases. It is interesting to note that the Freedom Tower in New York City was originally designed to obtain 20% of its power from wind turbines at the top of the tower.

- Using the ocean as a waste dump for carbon dioxide by pumping carbon dioxide from factory stacks to the bottom of the ocean, where it will become liquid at the high pressure there. This might change the chemistry of the ocean, however. Also, problems might occur if the ocean suddenly belches to release large amounts of carbon dioxide.

- Fertilizing the oceans to increase phytoplankton growth. The idea is that this would increase carbon dioxide consumption. Even if this works, what happens when the phytoplankton die and decompose?

Clean coal-burning plant in Florida.

© Larry Lee Photography/Corbis

- Putting a giant parasol between the Sun and Earth to block or reflect part of the Sun's light.

- Transporting large amounts of dust into the stratosphere to reflect sunlight and counteract warming. What happens if we get too much sunlight? Some have suggested an alternative that would involve releasing large numbers of small, reflective spheres into the stratosphere.

- Changing our lifestyles to produce less carbon dioxide per capita.

- Reducing the world's population or at least slowing its growth.

- Banning the use of internal combustion engines and the burning of fossil fuels.

Although some of the actions described here seem sort of like science fiction, all of them have been seriously considered or have been tested on a small scale. Some combination of several of these actions will probably be necessary, although some of the suggestions will never get off the ground.

Finally, even though the evidence for a major anthropogenic contribution to global warming appears to be stronger and stronger with each passing day, we must recognize that there are still different opinions about global warming. Most people will associate themselves with one of the following or a minor variation thereof. Which is most consistent with your opinion about this issue?

- *Wait*—Wait until we're 100% certain that this problem is for real, that it is serious, that we are causing it, and only then take action.

- *Act now*—Make our best scientific analysis of the problem based on the current evidence, and take appropriate action immediately to try to buy time before it's too late to do anything. Continue to collect evidence and work on solutions to the problem, which may require new technology.

- *Ignore it*—If it's natural, just ignore it, because humans cannot control nature.

Only time will tell if we respond appropriately.

In the back of the textbook:

→ *Review Card on Carbon Dioxide and the Greenhouse Effect*

In OWL for CHEM2 at www.cengagebrain.com

→ *Review Key Terms with Flash Cards (printable or digital)*

→ *Complete Interactive Practice Quizzes to prepare for tests*

→ *Submit Assigned Homework and Exercises*

Applying Your Knowledge

1. Define the following terms:
   (a) Greenhouse effect
   (b) Global warming
   (c) CFC
   (d) Kyoto Protocol
   (e) Infrared radiation

2. Most of the radiation coming from the Sun is ultraviolet (UV) radiation.
   (a) True
   (b) False

3. Carbon dioxide is less efficient at absorbing infrared radiation than methane, CFCs, and some other molecules, but its higher atmospheric concentration makes it the global warming gas with which we are most concerned.
   (a) True
   (b) False

4. Photosynthesis removes huge amounts of carbon dioxide from our atmosphere.
   (a) True
   (b) False

5. Which of the following actions has not been proposed as a possible solution or partial solution to global warming?
   (a) Fertilizing the oceans with iron to increase phytoplankton growth
   (b) Burning only gasoline that contains no carbon compounds
   (c) Sequestering CO_2 by pumping it into the ground or the bottom of the oceans
   (d) Putting a giant parasol (a screen) between Earth and the Sun to cut down on the amount of radiation hitting Earth
   (e) Decreasing the rate of deforestation and increasing the planting of trees

6. Visible light coming from the Sun can be absorbed and re-emitted as infrared radiation since visible light is of a higher energy than infrared radiation.
   (a) True
   (b) False

7. If it were not for the greenhouse effect, the temperature of Earth would be about −18°C (0°F) instead of +15°C (+59°F) and the oceans would be frozen.
   (a) True
   (b) False

8. The concentration of CO_2 in our atmosphere has increased from around 315 ppm to more than 370 ppm in the last 40 or so years, and this appears to be related to global warming.
   (a) True
   (b) False

9. IR radiation is sometimes referred to as heat radiation, and all molecules absorb IR radiation.
   (a) True
   (b) False

10. A reasonable political agenda for an environmental group might be to propose legislation aimed at reducing the concentration of all greenhouse gases in our atmosphere to zero.
    (a) True
    (b) False

11. Some of the UV and visible radiation striking Earth's surface is re-emitted from Earth as IR radiation.
    (a) True
    (b) False

12. Which of the following is not a greenhouse gas?
    (a) Methane
    (b) Water
    (c) Carbon dioxide
    (d) Nitrogen
    (e) CFC-12

13. The oceans will be able to dissolve more carbon dioxide when they become warmer.
    (a) True
    (b) False

14. The United States is number 1 in per capita production of carbon dioxide emissions.
    (a) True
    (b) False

15. The U.S. government has not been very supportive of the terms of the Kyoto Protocol.
    (a) True
    (b) False

16. Most greenhouse gases are produced by photochemical reactions in much the same way as photochemical smog is produced.
    (a) True
    (b) False

17. Global warming is mainly associated with which of the following?
    (a) Molecules that absorb UV radiation in the stratosphere
    (b) Molecules that decompose the ozone layer
    (c) Molecules that readily react with ultraviolet light
    (d) The type of electromagnetic radiation that is responsible for stretching and bending chemical bonds
    (e) The huge concentrations of CFCs that have built up in our atmosphere

18. Which of the following is the single major reason that Earth does not continuously lose energy to the surrounding space?
    (a) Earth is close enough to the Sun in the summer to absorb enough extra energy to compensate for that lost during the winter.
    (b) Oxygen and nitrogen in our atmosphere absorb enough reflected IR energy to maintain the present energy balance.
    (c) The "heavy" CFCs added to our atmosphere have trapped a carbon dioxide layer below them in the atmosphere.
    (d) Greenhouse gases like carbon dioxide and water vapor in the atmosphere absorb reflected energy and radiate it back to Earth's surface.
    (e) The ozone layer absorbs the UV emitted from the sun, heating the atmosphere.

19. Every major industrialized country has ratified the Kyoto Protocol.
    (a) True
    (b) False

20. All greenhouse gas molecules are equally efficient at absorbing infrared radiation.
    (a) True
    (b) False

21. The average concentration of carbon dioxide in our atmosphere has increased to around 380 ppm from a preindustrial concentration of around 280 ppm. This is most likely due to which of the following?
    (a) An increase in the animal population of the world
    (b) Increased volcanic activity during the past 150 years
    (c) Increased gasoline efficiency of the internal combustion engine
    (d) A thinning of the ozone layer that allowed more CO to be converted to CO_2 by UV radiation
    (e) Increased use of fossil fuels by an increasingly populous world

22. Some scientists have proposed pumping carbon dioxide to the bottom of the oceans as a partial solution to global warming.
    (a) True
    (b) False

23. Discuss the pros and cons of the various methods that have been proposed to regulate or control global warming.

24. Should industrial processes that raise our standard of living but produce greenhouse gases be exempt from regulation of greenhouse gas emissions?

25. Should developing countries be held to the same standards of greenhouse gas emissions as developed countries if doing so can be shown to slow their rate of development?

26. Would you support a carbon tax added to the cost of consumer goods based on a calculation of how much their production contributed to the greenhouse gas problem?

27. Should people who insist on driving vehicles that get below a certain gas mileage be assessed an additional tax since they contribute more carbon dioxide to the atmosphere?

28. Why do you think it has been so hard to convince the average person that global warming is a real problem?

29. Calculate the energy associated with a photon of infrared radiation that has a wavelength of 2.5×10^{-3} cm and that with a wavelength of 2.5×10^{-4} cm. Use the expressions $\lambda \nu = c$ and $E_{photon} = h\nu$ where $h = 1.58 \times 10^{-34}$ cal.

30. Repeat the calculation of problem 29 using appropriate wavelengths (Section 6-3) for ultraviolet and visible radiation and check to see if the order of these energies when compared with that for infrared radiation agrees with the order listed in Section 6-3.

Chlorofluorocarbons
and the Ozone Layer

The major effects of many air pollutants such as nitrogen oxides, sulfur oxides, carbon monoxide, and ozone are felt at altitudes low enough that they are considered "ground-level" pollutants. These pollutants are so chemically reactive and/or water soluble that they never make it out of our troposphere, the atmospheric region extending from sea level to about 10 km. Many of them react there with other chemicals, including water, to form compounds that are soluble in water and are removed by natural precipitation in a process called **rainout**. When sulfur oxides, nitrogen oxides, and carbon dioxide react with water, they form sulfuric acid, nitric acid, and carbonic acid, respectively.

Humans had done little, if anything, to affect the chemistry of the stratospheric ozone layer until we began to make significant use of a class of chemicals known as *chlorofluorocarbons* (CFCs). CFCs were first synthesized by chemists at DuPont in the 1930s and came to be marketed under the brand name *Freon*. Prior to that time, the little refrigeration available depended on the use of ammonia and sulfur dioxide as heat exchangers. Because of the potential danger associated with these corrosive, noxious, and toxic gases, scientists had been searching for replacement chemicals that would mitigate these problems. As was pointed out in Section 3-1, the development of CFC-12 (dichlorodifluoromethane) presented scientists with a chemical that was nontoxic, extremely stable, and nonflammable and had the right physical properties for use as a refrigerant. Little did anyone realize at the time that some of these properties would one day lead to their widespread use in products including aerosol spray cans and trigger an environmental problem of global proportions.

7-1 The Oxygen–Ozone Screen

rainout: the removal of pollutants from the atmosphere by natural precipitation

As you saw in Chapter 6, our Sun bombards Earth with electromagnetic radiation, of which about 53% is infrared (IR) radiation, about 39% is visible radiation, and about 8% is ultraviolet (UV) radiation. IR radiation is the lowest in energy of these three, and UV radiation is the highest in energy. We have

© Eyebyte/Alamy / © Natphotos/Digital Vision/ Getty Images

focused on the effect that these forms of radiation can have upon reaching Earth and have briefly mentioned that some of this electromagnetic radiation never reaches Earth because it is either absorbed or reflected.

Ultraviolet (UV) radiation has wavelengths in the 200 to 400 nm range. This energy is sufficient to break some chemical bonds and to create havoc with many biological systems. Fortunately for us, much of the UV radiation that streams from the Sun toward Earth never reaches the surface. It is absorbed primarily by two important screening molecules: oxygen and ozone. Let's examine how this happens.

First, we must look at the bonding in oxygen (O_2) and ozone (O_3) molecules. While a classical Lewis dot structure of molecular oxygen indicates that it contains an oxygen–oxygen double bond, it does not. Experimental observations indicate that there is only a single bond between the oxygen atoms, and each atom has an

The use of CFCs as propellants in aerosol cans has contributed to the formation of the ozone hole.

unpaired electron. Oxygen's actual structure is represented below. Ozone can be represented by two different yet equal structures in which there is one oxygen–oxygen double bond and one oxygen–oxygen single bond. The actual structure is one in which both oxygen–oxygen bonds are essentially intermediate between single bonds and double bonds, but that is difficult to represent pictorially. We therefore resort to writing two separate but equal resonance structures (page 86), which differ only in the location of lone pair electrons and multiple bonds. The real structure is a hybrid of the two resonance structures.

$$:\ddot{O}-\ddot{O}: \qquad \ddot{O}=\ddot{O}-\ddot{O}: \longleftrightarrow :\ddot{O}-\ddot{O}=\ddot{O}$$

Oxygen Structure   Ozone Resonance Structures

Although both molecules consist only of bonds between oxygen atoms, the bond between atoms in O_2 is stronger than the oxygen–oxygen bond in ozone (O_3) and requires more energy to break.

If either of these molecules is exposed to energy that matches or exceeds the strength of the bond connecting oxygen atoms within their structures, then the oxygen–oxygen bond will be broken. It turns out that the energy required to break these oxygen–oxygen bonds can be supplied by UV radiation of slightly different wavelengths. UV radiation in the 200 to 400 nm range is generally divided into three separate ranges referred to as *UV-A* (320–400 nm), *UV-B* (280–320 nm), and *UV-C* (200–280 nm), as shown in Table 7.1.

Because the bond in O_2 is stronger, this molecule can absorb higher energy UV-C radiation (shorter wavelength); ozone is capable of absorbing only lower-energy UV-B radiation. Together these two compounds protect us from the harmful effects of these two types of UV radiation. But ozone is not as robust as O_2 and is more prone to photodissociation by lower-energy UV-B light than is oxygen.

7-2 Where Is the Ozone?

The ozone layer is a band of higher-than-average ozone concentration within the stratosphere (Section 4-1). The maximum concentration of ozone in the densest region of the stratospheric ozone layer is slightly less than 10^{19} molecules of ozone per cubic meter. This is a big number. However, it has been estimated that an average human breath (about 0.5 liter) contains 2×10^{22} particles. That equates to 4×10^{22} particles per liter or 4×10^{25} particles per cubic meter (a cubic meter is 1000 liters). In other words, the density of ozone molecules at the densest part of the ozone layer is only 0.000001 times that of the air we breathe. If all of the ozone in a column of air in the atmosphere were compressed to 1 atmosphere pressure at 0°C, the layer would be only 3 mm thick (Figure 7.1). This is about the thickness of two nickels stacked together. However, we must remember that this amount of ozone is spread out over many kilometers of space because the region of maximum ozone concentration is at least 15 km thick.

In the absence of anthropogenic influences, the concentration of ozone in the stratospheric ozone

TABLE 7.1 Comparison of Types of Ultraviolet Radiation

Radiation Type	Wavelength Range	Relative Energy	Comment
UV-A	320–400 nm	Least energetic	Reaches Earth in great amounts but has less potential than UV-B or UV-C to cause biological damage due to its lower energy. Even so, UV-A is suspected of being responsible for premature aging and wrinkling of the skin. It also causes some types of skin cancers that can be fatal.
UV-B	280–320 nm	Intermediate	Most of the UV-B is absorbed by ozone in the atmosphere; some reaches Earth, however. UV-B radiation is very effective at damaging DNA, weakening the human immune system, causing some skin cancers and eye damage, and causing significant crop and marine organism damage.
UV-C	200–280 nm	Most energetic	Completely absorbed by oxygen and/or ozone in the atmosphere; may cause significant biological damage.

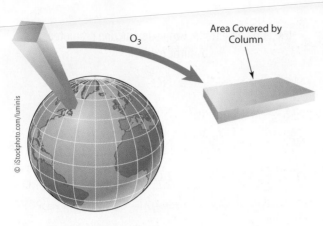

FIGURE 7.1 If all the ozone over a certain area were compressed down to 0°C and 1 atm pressure, it would form a slab 3 mm thick, corresponding to 300 Dobson units (Section 7-3).

(Source: Figure located at http://www.atm.ch.cam.ac.uk/tour/dobson.html. Used by permission of the Centre for Atmospheric Science, Cambridge University.)

layer had maintained a **steady state** (a state in which the net rate of destruction equals the net rate of formation) for millions of years. On average, ozone molecules were destroyed by a variety of chemical reactions and UV-B absorption at the same rate that they were re-formed by other processes. As a result, the net concentration of ozone in the stratosphere remained constant. In other words, a finely balanced system was in place, capable of protecting any life on Earth from an overabundance of high-energy ultraviolet radiation.

We can illustrate a steady state condition by using a bathtub analogy. A bathtub is filled to a certain level with water and then the drain is opened. One can maintain a constant level of water in the tub, even with an open drain, by carefully adding water to the tub with the faucet at exactly the rate at which it is draining away.

A slightly simplified version of the steady state process for ozone formation (reactions 1 and 2) and destruction (reactions 3 and 4) is known as the **Chapman cycle** after the scientist who proposed it.

An inspection of the Chapman cycle reveals that the following chemical reactions are occurring:

- Reaction (1) indicates that molecular oxygen can absorb high-energy UV radiation with wavelengths less than 240 nm to dissociate into two reactive oxygen atoms.

- Reaction (2) shows that an oxygen atom can combine with an oxygen molecule to form an ozone molecule. This is the process by which ozone is generated in the stratosphere.

- Reaction (3) shows that ozone molecules will react with ultraviolet radiation streaming in from the Sun

if the radiation has wavelengths less than about 320 nm. In doing so, the ozone is reconverted to an oxygen atom and an oxygen molecule. Oxygen and ozone protect us from nearly all ultraviolet radiation with wavelengths less than 320 nm. The ozone concentration is maintained by this steady state process.

- Reaction (4) illustrates that it is possible for an oxygen atom to combine with an ozone molecule to generate two oxygen molecules. This reaction leads to a loss of ozone. This is a relatively rare process under natural conditions, due to the exceedingly low concentrations of the species involved. Any chemical process that leads to the promotion of reaction (4) is deleterious to the steady state concentration of the ozone layer.

> **Steady state:** A state in which the concentration of a chemical remains essentially constant even though the chemical may be undergoing numerous chemical reactions
>
> **Chapman cycle:** The set of four reactions that represents the steady state formation and destruction of ozone in the stratosphere

The Chapman cycle for the formation and destruction of stratospheric ozone.

$$< 240 \text{ nm photon} \quad O_2 \xrightarrow{\;1\;} 2 \cdot \ddot{O} \cdot \quad <320 \text{ nm photon}$$

$$\cdot \ddot{O} \cdot + O_2 \underset{3}{\overset{2}{\rightleftharpoons}} O_3$$

$$\cdot \ddot{O} \cdot + O_3 \xrightarrow{\;4\;} 2\, O_2$$

Because the ozone layer developed naturally in the presence of many other perturbing influences such as destruction of ozone by water vapor, these reactions have no real effect on the steady state concentration of ozone. However, as you will soon see, anthropogenic influences have begun to perturb the steady state concentration of ozone.

TRY IT 7.1

For the average person, it is probably more important to be able to use words to describe the process by which ozone is formed in the stratosphere than it is to write out chemical equations describing the process. Can you do this?

Following the logic of Try It 7.1, use words to describe the reactions by which ozone and oxygen protect us from different wavelengths of ultraviolet radiation. Why do oxygen and ozone protect us from different wavelengths of ultraviolet radiation?

Use the bathtub analogy to describe how the steady state concentration of stratospheric ozone could be maintained. Also use the bathtub analogy to describe how anthropogenic influences might perturb the steady state concentration of ozone in the stratosphere.

Dobson unit (DU): A measure of stratospheric ozone concentration

It is known that measured ozone concentrations are lower than can be accounted for by the simple Chapman cycle. This has led scientists to look for other influences on the concentration of ozone. First, let's briefly consider one of the natural reactions that destroys ozone. UV radiation can break the oxygen–hydrogen bond of a water molecule in the stratosphere to generate hydrogen atoms and hydroxyl radicals ($\cdot$OH). These two species are involved in many reactions, some of which actually convert O_3 to O_2. However, this process, which scientists now believe is an efficient process above 50 km, has been occurring since the ozone layer developed, and there is little, if anything, that humans can do about it. The system has obviously attained a steady state that includes this perturbation.

Another natural mechanism for ozone destruction, shown in the next column, involves nitric oxide (NO). The NO can be generated when nitrous oxide (N_2O), produced naturally by microorganisms in soil and water, is released into the lower atmosphere and slowly makes its way into the stratosphere. Once in the stratosphere, one molecule of nitrous oxide can react with oxygen atoms to form two molecules of NO (reaction 1). The NO can then react with ozone to produce NO_2 and O_2 (reaction 2). The NO_2 can then react further with an oxygen atom to regenerate NO and O_2 (reaction 3). The net result of these last two reactions is exactly the same as the last reaction of the Chapman cycle and leads to destruction of ozone. The most significant sources of N_2O are natural biological and oceanic processes, so, once again, the system attains a steady state including this perturbation.

Reaction sequence for ozone destruction promoted by nitric oxide.

$$N_2O + \cdot\ddot{O}\cdot \xrightarrow{\ 1\ } 2\cdot NO$$

$$\cdot NO + O_3 \xrightarrow{\ 2\ } \cdot NO_2 + O_2$$

$$\cdot NO_2 + \cdot\ddot{O}\cdot \xrightarrow{\ 3\ } \cdot NO + O_2$$

$$\overline{\cdot\ddot{O}\cdot + O_3 \xrightarrow{\ 4\ } 2\,O_2 \quad \text{sum of } 2 + 3}$$

7-3 The Ozone Layer Is Disappearing

Scientists, who had no inkling of what they were about to discover, began measuring stratospheric ozone concentrations over 80 years ago using relatively unsophisticated instrumentation. Eventually, as a credit to G. M. B. Dobson, one of the scientists who invented an early instrument used for these measurements, the unit of ozone concentration in the atmosphere became known as a **Dobson unit (DU)**. This unit equates to about 1 ozone molecule for every billion molecules of air, so 1 DU is 1 ppb of ozone. Satellite measurements began to be used as time progressed. Measurements showed that an average value of about 250 DU of ozone was found near the equator, while the value was around 320 in the Northern Hemisphere. The concentration of ozone is actually higher at the North and South Poles.

Beginning in 1979, measurements taken at the South Pole revealed a significant decline in stratospheric ozone, one that was especially prevalent during the months of September to November of each year (Figure 7.2). This was so unexpected that scientists first thought there was something wrong with the data being collected. In fact, computers had been programmed to discard some of the data collected because it was so far outside the expected range. By 1985, the results were unmistakable. Although the expected predictable seasonal fluctuations in ozone concentration over Antarctica were present due to weather conditions and available sunlight, overall decline was obvious (Figure 7.3). This observed decline, which manifests itself as the annual ozone hole (Figure 7.4), could not simply be explained by the previously observed annual drop in late

FIGURE 7.2 Seasonal ozone deviations from pre-ozone-hole (1957–1978) averages over Antarctica show a rapid decline during the southern spring and a lesser decline during the summer.

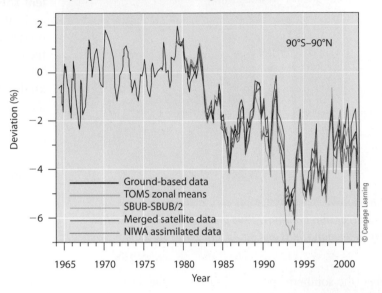

FIGURE 7.3 Antarctic ozone minima, 1979–2003.

The ozone minima are given above the data points, and the dates of recorded minima are given below the data points. The data, in Dobson units (DU), were measured by the Total Ozone Mapping Spectrometer (TOMS) spacecraft. (Source: http:// jwocky.gsfc.nasa.gov/multi/ min_ozone.gif)

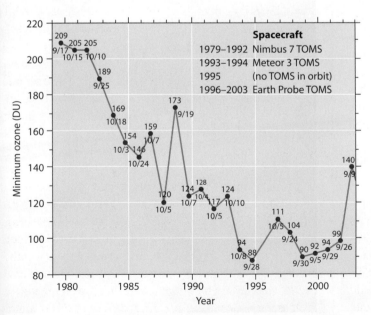

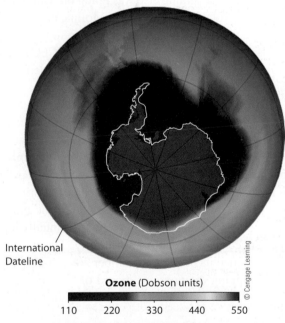

FIGURE 7.4 Ozone hole over Antarctica on October 4, 2008.

The hole in darker blue can be seen to cover the entire South Pole and extend over the southern tip of South America in the upper left quadrant of the image. Higher concentrations of ozone are represented by green and yellow. (Source: http://earthobservatory.nasa.gov/Features/WorldOfChange /ozone.php)

Professor Mario J. Molina (center) gestures during a 1995 press conference with fellow Nobel Prize winners F. Sherwood Rowland (left) and Paul J. Crutzen in Stockholm, Sweden.

ness, low boiling points, and insolubility in water of the CFCs made them popular as propellants in aerosol cans and as agents used to expand foam plastics. Because of their excellent solvent properties for greases and oils, many of the CFCs were used as degreasers during the manufacture of printed computer circuit boards, television sets, and other kinds of appliances. In effect, the increased use of CFCs paralleled the growth of modern, urbanized industrial society.

During the time CFCs were in common use, little effort was made to prevent them from escaping into the atmosphere. For example, if your automobile air conditioner needed repair, the common practice was to vent all of the CFC refrigerant into the atmosphere before any work was done. In fact, most of the CFCs ever manufactured have probably been released into the atmosphere.

September or early October (the southern spring) of each year. Clearly, something else was going on. But what?

F. Sherwood Rowland, Mario Molina, and Paul Crutzen made the most dramatic contribution to the puzzle of the missing ozone when they proposed that CFCs were the key ingredient. They subsequently were awarded the 1995 Nobel Prize in chemistry for their work.

Scientific observations over the past 45 years or so reveal that the ozone concentration over Antarctica shows some seasonal variation (Figure 7.2), with the greatest and most rapid decline occurring from September to November. The interval corresponds to spring in the Southern Hemisphere. Furthermore, detailed observations from 1980 to the present show that the value of the minimum ozone concentration observed fluctuates annually and has experienced a great deal of variability. The hole in the ozone reached its peak in September 2006 at 29.7 million sq km, topping the previous record of 29 million sq km observed in 2003. It has been declining in size slightly since then.

It is obvious that the CFCs used as refrigerants, propellants, and solvents for the past half century were not released into the atmosphere at the South Pole, so why did the ozone hole first show up there? CFCs were, for years, released into our lower atmosphere by a variety of mecha-

7-4 Why CFCs and Why at the Poles?

After the discovery of the refrigerant gas properties of CFCs, their use became widespread in applications such as automobile and home air-conditioning. Soon the inert-

*The campy 2007 film **Hairspray** parodies the culture and popular styles of the early 1960s, which glorified the use of CFC-propelled aerosols.*

nisms, including venting of air-conditioning systems when recharging, spraying of aerosol sprays, evaporation of solvents, and decomposition of blown foams in which CFCs were trapped. At lower altitudes, these chemicals were essentially unreactive and hung around for days, months, or years until the natural air currents and weather patterns transported them into the stratosphere. Although many have argued that these heavier-than-air molecules could never make their way into the stratosphere, that is simply not true. Vertical mixing of such chemicals within the troposphere has been shown to occur within weeks, and such chemicals released into one hemisphere can mix between hemispheres within about a year. Although it may take years for vertical mixing of such chemicals between the troposphere and the stratosphere, it does occur. Most chemicals are removed from the atmosphere by rainout or chemical reactions before they have a chance for this to occur, but CFCs are not water soluble and they are very unreactive until they reach the stratosphere.

Once in the stratosphere, these CFCs are subject to bombardment by intense UV radiation because they no longer are protected by the oxygen–ozone screen. High-energy (<220 nm) radiation is sufficient to break the relatively weak carbon–chlorine bond of CFC molecules, as shown in Figure 7.5, for the common CFC known as trichlorofluoromethane or CFC-11. This generates a very reactive chlorine atom with an odd number of electrons that can then collide with ozone molecules to generate chlorine monoxide (ClO) and an oxygen molecule (O_2) (reaction 2). Reaction 3 shows how the chlorine monoxide can react with an oxygen atom to regenerate the destructive chlorine atom. Reaction 4, which is the sum of reactions 2 and 3, is exactly the same net reaction as the last reaction of the Chapman cycle (p. 123).

An additional destruction pathway in which two chlorine monoxide species combine (reaction 1) can occur is shown below. The resultant ClOOCl undergoes two subsequent reactions (2 and 3), triggered by the UV light from

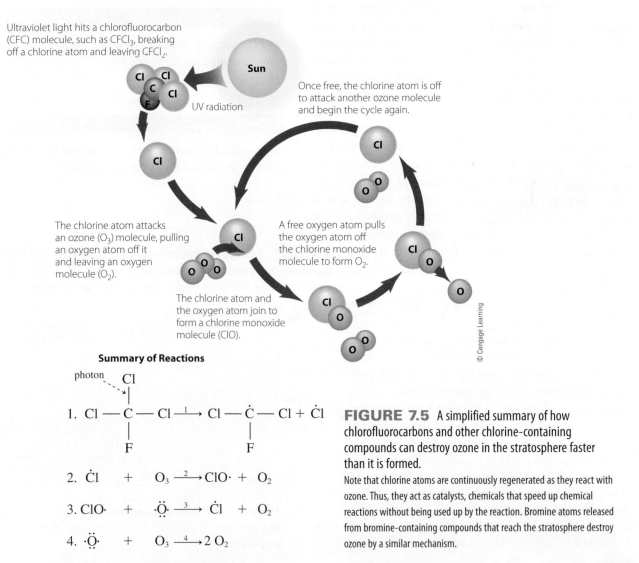

Ultraviolet light hits a chlorofluorocarbon (CFC) molecule, such as $CFCl_3$, breaking off a chlorine atom and leaving $CFCl_2$.

Sun

UV radiation

Once free, the chlorine atom is off to attack another ozone molecule and begin the cycle again.

The chlorine atom attacks an ozone (O_3) molecule, pulling an oxygen atom off it and leaving an oxygen molecule (O_2).

A free oxygen atom pulls the oxygen atom off the chlorine monoxide molecule to form O_2.

The chlorine atom and the oxygen atom join to form a chlorine monoxide molecule (ClO).

© Cengage Learning

Summary of Reactions

1. $Cl \overset{photon}{-} C - Cl \xrightarrow{1} Cl - \dot{C} - Cl + \dot{Cl}$
   (with F below each carbon)

2. $\dot{Cl} + O_3 \xrightarrow{2} ClO\cdot + O_2$

3. $ClO\cdot + \cdot\ddot{O}\cdot \xrightarrow{3} \dot{Cl} + O_2$

4. $\cdot\ddot{O}\cdot + O_3 \xrightarrow{4} 2\,O_2$

FIGURE 7.5 A simplified summary of how chlorofluorocarbons and other chlorine-containing compounds can destroy ozone in the stratosphere faster than it is formed.

Note that chlorine atoms are continuously regenerated as they react with ozone. Thus, they act as catalysts, chemicals that speed up chemical reactions without being used up by the reaction. Bromine atoms released from bromine-containing compounds that reach the stratosphere destroy ozone by a similar mechanism.

the Sun, to generate two chlorine atoms that may then combine with ozone (reaction 4) in exactly the same process as shown in reaction 2 in Figure 7.5. This entire sequence converts two ozone molecules to three oxygen molecules, a reaction that consumes ozone.

Ozone destruction triggered by ClOOCl.

$$ClO\cdot + ClO\cdot \xrightarrow{\ 1\ } ClOOCl$$

$$ClOOCl \xrightarrow[\text{sunlight}]{2} ClOO\cdot + \dot{Cl}$$

$$ClOO\cdot \xrightarrow{\ 3\ } \dot{Cl} + O_2$$

$$\dot{Cl} + O_3 \xrightarrow{\ 4\ } ClO\cdot + O_2$$

Although the overall process is certainly more complicated, these simplified schemes provide an understanding of the role of Cl atoms in triggering the ozone depletion that leads to the seasonal ozone hole over Antarctica. A single Cl atom is thought to be capable of destroying 100,000 ozone molecules before it is rendered unreactive by some other chemical process. Collected data (Figure 7.6) also show that there is an anticorrelation between the concentration of ClO and the concentration of ozone in the stratosphere.

It is also likely that the chlorine atoms undergo other reactions to generate somewhat stable compounds that render them temporarily unreactive in the stratosphere, where they are not going to be removed by rainout. In effect, they are "stored" in less reactive forms such as chlorine nitrate ($ClONO_2$), which can be produced when ClO reacts with NO_2. Another chlorine reservoir or storage molecule is HCl, which can form from the reaction between a chlorine atom and naturally occurring methane (CH_4) in the stratosphere.

So, why does this all happen at the South Pole? The lower stratosphere above Antarctica is the coldest spot in Earth's atmosphere, and that, coupled with the relative lack of dispersive wind currents above the pole during the long dark periods of Antarctic winter, means that the conditions in the stratosphere above the South Pole are ripe for ice crystal formation. During the dark, cold ($-80°C$) polar winter, polar stratospheric ice clouds composed of nitric acid trihydrate ($HNO_3\cdot3$ H_2O) form. These ice clouds trigger the breakdown of the normally stable molecules ($ClONO_2$ and HCl) to generate HOCl and Cl_2, which pho-

todissociate when sunlight returns to the South Pole in the spring (September–November). This photodissociation produces Cl atoms, and the destruction of ozone thus triggered is highly effective during this time of the year.

Scientists have also implicated some bromine-containing chemicals in the destruction of the ozone layer. The chief culprit has been CH_3Br, methyl bromide, a commonly used agricultural fumigant. Here again, it is the relatively weak C—Br bond that is broken by UV radiation to produce Br atoms, which act in the same role as Cl atoms to destroy ozone.

TRY IT 7.4

Explain why the ozone hole has occurred over the South Pole even though the molecules that triggered the hole formation were not released at the pole.

TRY IT 7.5

What chemical bond in CFCs is most critical to their ability to destroy the ozone layer?

TRY IT 7.6

Why are CFCs not removed by rainout like many other chemicals released into the atmosphere?

FIGURE 7.6 The anticorrelation between stratospheric ozone and chlorine monoxide concentration versus latitude near the South Pole in September 1987. (Source: http://www.elmhurst.edu/~chm/onlcourse/chm212/images/ozoant.GIF)

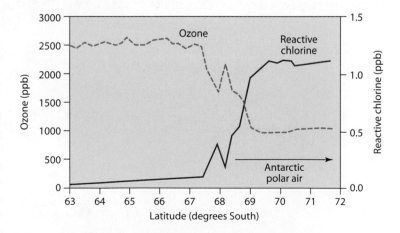

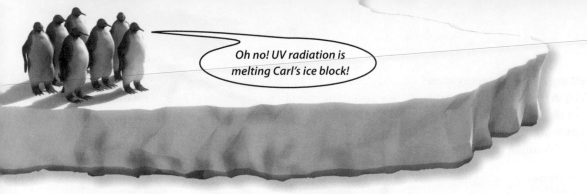

Oh no! UV radiation is melting Carl's ice block!

7-5 What Are the Implications of Increases in the Size of the Ozone Hole?

Although the decrease in ozone concentrations manifests itself most readily over Antarctica, this is simply a reflection of worldwide decreasing ozone concentrations throughout the ozone layer. There is evidence that the lowered ozone concentrations are already starting to have a negative effect on crop yields in the Southern Hemisphere. There is much concern about increases in skin cancers as the lowered ozone concentrations and ozone hole start to spread over populated areas. Australian officials have started to warn citizens about the enhanced danger associated with these lowered levels. Australian schoolchildren are not allowed out at recess without long sleeves and hats with visors and neck flaps to protect them from the Sun. In the fall of 2000, the ozone hole extended over a population center for the first time and exposed thousands of Chileans in the city of Punta Arenas to extremely high levels of UV radiation. Citizens in Punta Arenas are now warned not to go out into the sunlight between the hours of 11 A.M. and 3 P.M. during the height of the seasonal reduction. If the size of the hole expands, the number of people living under the hole and subjected to such restrictions will increase.

Even people living in regions of the world not affected by the ozone hole are increasingly warned about the dangers of UV radiation exposure. Most daily newspapers and television weather reports routinely report a UV index indicating the expected danger associated with outdoor activities. Skin cancer (melanoma) rates are on the rise. A 1% decrease in ozone concentration has been predicted to cause a 2% increase in skin cancers. As early as 1999, the amount of UV radiation reaching New Zealand had increased by over 12% compared to the previous decade.

Not all increases in skin cancer can be attributed directly to the thinning ozone layer, however. Our obsession with bronzed bodies and the increased amount of tanning, whether by natural sunlight or in tanning salons, has also contributed to the dramatic increase in skin cancer. A 2002 report in the *Journal of the National Cancer Institute* reported that people who used tanning devices were anywhere from 1.5 to 2.5 times as likely to develop skin cancer! An earlier study indicated that people under the age of 30 who had used tanning lamps more than 10 times a year faced an eightfold likelihood of developing melanoma (skin cancer). Although many tanning salons argue that they have switched from UV-B lamps to lower-energy UV-A lamps, studies have indicated that this has not

sigh

reduced the problem because longer tanning periods are required to get the same effect. A golden tan may look great, but is the increased risk of skin cancers really worth it? An increasing number of states and municipalities has restricted the use of tanning beds by minors.

TRY IT 7.7

List three possible consequences of a significant decrease in stratospheric ozone concentration.

TRY IT 7.8

Is tanning associated with UV radiation in tanning salons any less dangerous than tanning associated with sunlight? Why or why not?

7-6 Can We Do Anything About the Ozone Hole?

Since most (>80%) of the chlorine and bromine in the stratosphere comes from man-made chemicals, it stands to reason that this would be the place to start addressing the ozone hole problem. As early as 1978, the use of CFCs as propellants in aerosol spray cans was legislatively banned in the United States. By 1990, their use as foaming agents in blown foams was also discontinued in the United States when scientists began to realize the severity of the problems posed by the release of these chemicals. The subsequent release of CFCs from discarded foam products so prevalently used in our society was also recognized as a major problem. The CFC gases trapped in manufactured goods such as couches, car seats, carpet backing, and foam products of every sort begin to decompose with age.

The first major international response to this problem came with the Montreal Protocol on Substances that Deplete the Ozone Layer (1987). At that time, a number of countries agreed to guidelines that called for the reduction of CFC production to half of 1986 levels by 1998. In the intervening years, increased awareness of the seriousness of the problem led nations around the world to agree to accelerated schedules for the elimination of these chemicals. In 1990, one hundred countries agreed to ban the production of CFCs by 2000. Subsequent meetings led to the addition of methyl bromide and some other related compounds to the list, although some developing countries have been given permission to use methyl bromide until 2015. The United States

and 140 other countries agreed to completely halt production of CFCs after December 31, 1995. However, many of these compounds have lifetimes of 100 or more years in the atmosphere, so they'll be around for a while even if no more are being produced. Those already produced will almost certainly find their way into the atmosphere.

These and continuing actions on a worldwide scale have had a significant impact on the amount of chlorine in the atmosphere. Its concentration is known to have peaked in the mid-1990s and is now declining. There is hope that the ozone layer may recover within the next 50 years or so due to this worldwide effort to address a very serious environmental problem. It is anticipated that during this time the Cl concentration will drop to the pre-ozone-hole level of around 2 ppb.

The demand for CFCs did not suddenly disappear once the problem was recognized. International agreements allowed the sale and use of existing stockpiles of these CFCs. Some countries have imposed taxes on CFCs to encourage the switch to alternatives. This has led to price increases and black market smuggling of CFCs into the United States. Some reports have suggested it is more profitable to

smuggle CFCs into the United States from Russia, or China than it is to smuggle cocaine. There is also the concern that some countries may allow or even encourage continued production of banned CFCs to take advantage of the demand for them. One thing is certain: In most places, as a result of increased concern about the consequences, CFCs are no longer being vented into the atmosphere.

7-7 What Will Replace CFCs?

The demand for air-conditioning, aerosol sprays, refrigerants, blown foams, industrial solvents, and agricultural fumigants has not disappeared, so scientists have been working hard to find replacements for CFCs. Since it is the relatively weak C—Cl bonds and C—Br bonds in these molecules that are problematic, a major effort has focused on replacing as many as possible of the Cl or Br atoms with H atoms. Compounds in which some, but not all, of the Cl atoms have been replaced by H atoms are referred to as *hydrochlorofluorocarbons* or HCFCs. The HCFCs have a tendency to break down before they reach the stratosphere and would appear to be acceptable substitutes for CFCs at first glance. Compounds in which all Cl atoms have been replaced by H are *hydrofluorocarbons* or HFCs. HFCs are very unreactive and are not susceptible to UV-induced breakdown in the stratosphere. These compounds would also seem to be acceptable substitutes for the CFCs. However, one must remember that the replacement compounds must still have the necessary physical properties (correct boiling point, low toxicity, high stability, low flammability) that would make them appropriate as refrigerants, if that is the intended use of the compounds. Thus, this has not been a simple task. One cannot simply replace most or all of the Cl atoms with H or F atoms and be guaranteed of having an acceptable replacement compound. Even if one finds such a candidate among the HCFCs and HFCs, one must remember that these compounds are still potential greenhouse gases with varying abilities to absorb infrared radiation. For every solution, we appear to encounter a new problem.

Table 7.2 shows some of the replacement compounds and their properties. Many of these compounds are in current use and have dramatically reduced impacts on the ozone layer. Even though these HCFCs and HFCs also absorb infrared radiation, a report from the United Nations Intergovernmental Panel on Climate Change in the spring of 2005 reported that these replacements generally, with some significant exceptions, have much lower global warming potential than CFCs. The global warming potential of emissions of CFCs, HCFCs, and HFCs was equivalent to 7.5 billion metric tons of carbon dioxide in 1990, compared with the 22 billion metric tons of carbon dioxide released due to fossil fuel combustion. This equivalence had dropped to 2.5 billion metric tons of carbon dioxide by 2000, and the expectation is that the equivalence will drop to 2.3 billion metric tons of carbon dioxide by 2015.

CFCs in aerosol cans and blown foams have been replaced by a variety of compounds such as propane, butane, HCFCs, and HFCs. Most commercial products now routinely advertise that they contain no CFCs, but most fail to acknowledge that the replacement compounds still may cause ozone depletion and global warming. The search for safe and effective alternatives to CFCs continues. Maybe new technology will come to the rescue soon.

© iStockphoto.com/JimLarkin

TABLE 7.2 Alternatives to CFCs and Halons for Common Applications

Application	CFC or Halon Used Previously	Alternatives
Aerosols	CFC-11, CFC-12	hydrocarbons, HFC-152a, dimethyl ether and water, pumps, roll-ons
Refrigerators		
home	CFC-12	HFC-134a
commercial	CFC-11	HCFC-123*
Air conditioners		
home	HCFC-22*	HFC blends
automobiles	CFC-12	HFC-134a
commercial (large buildings)	CFC-11, CFC-12	HCFC-123*, HFC-134a
Foam blowing		
polystyrene (food containers)	CFC-12	hydrocarbons, HCFC-22*, HFC-152a
polyurethane (refrigerator insulation)	CFC-11	HCFC-141b*, HFC-134a, vacuum panels
rigid foams (construction materials)	CFC-11, CFC-12	HCFC-142b*
Cleaning solvents (electronics industry)	CFC-113, CH_3CCl_3	water-based processes, other solvents, supercritical CO_2[†]
Fire extinguishers (aircraft, military, office)	Halon-1301, Halon-1211	CF_3I and other possibilities are being tested

* Alternatives denoted by an asterisk (*) are interim solutions only, as they are gradually being phased out.

[†] Supercritical CO_2 is above the critical temperature (Tc), where gas and liquid coexist with no visible interface.

Source: Institute for Chemical Education.

Your Resources

In the back of the textbook:

→ *Review Card on Chlorofluorocarbons and the Ozone Layer*

    In OWL for CHEM2 at www.cengagebrain.com

→ *Review Key Terms with Flash Cards (printable or digital)*

→ *Complete Interactive Practice Quizzes to prepare for tests*

→ *Submit Assigned Homework and Exercises*

Applying Your Knowledge

1. Define the following terms:
   - (a) Ozone hole
   - (b) CFCs
   - (c) HCFCs
   - (d) Montreal Protocol
   - (e) Resonance structures
   - (f) UV-A
   - (g) UV-B
   - (h) UV-C
   - (i) Chapman cycle
   - (j) HFCs
   - (k) Stratosphere
   - (l) Steady state
   - (m) Dobson unit

2. Explain the different mechanisms by which ozone is formed in the troposphere as a pollutant and in the stratosphere as a protective shield.

3. Classify the following statements as true or false.
   - (a) Increased destruction of the ozone layer would probably lead to an increase in the incidence of skin cancers.
   - (b) Volcanic eruptions produce CFCs, which have been a major factor in the destruction of the ozone layer.
   - (c) The ozone hole that appears in the Southern Hemisphere each fall has now started to spread over some populated areas.
   - (d) Ozone in our stratosphere absorbs primarily UV-B radiation in its protective role, while oxygen absorbs UV-C radiation.
   - (e) UV-A radiation is not absorbed by stratospheric ozone.
   - (f) The ozone molecules in the protective ozone layer of the stratosphere are quite different from the harmful ozone molecules found in ground-level smog and air pollution.
   - (g) The replacement of CFCs by HCFCs has been totally effective at reversing the damage done to the ozone layer since HCFCs do not destroy the ozone layer.
   - (h) The concentration of CFCs in the stratosphere is relatively low. However, the fact that a single chlorine atom generated by the action of UV radiation on the CFC may cause the destruction of as many as 100,000 ozone molecules is cause for much concern.
   - (i) Less than 50% of the chlorine and bromine in the stratosphere comes from man-made chemicals.
   - (j) Oxygen molecules absorb lower-energy ultraviolet radiation than ozone molecules absorb because the oxygen–oxygen double bond is shorter and stronger than the 1.5 bonds of ozone.

4. What is the difference, if any, between "good ozone" and "bad ozone"?

5. Discuss the mechanism by which CFCs and related compounds destroy the ozone layer. What chemical bond in the CFC is critical in this mechanism?

6. What were CFCs primarily used for before their potential as ozone-destroying chemicals was recognized? What compounds have been used as replacements?

7. How thick would the stratospheric ozone layer be if it were compressed to atmospheric pressure at Earth's average temperature?

8. List some of the characteristics that were sought in the chemicals used as replacements for early refrigerants.

9. What type of radiation does ozone shield us from? What wavelengths of this radiation are absorbed by ozone?

10. Oxygen and ozone both protect us from ultraviolet radiation, but there are differences in the wavelength of absorption. What are they?

11. Why do oxygen and ozone absorb UV radiation of different wavelengths? Which absorbs UV radiation of higher energy?

12. Explain why the "ozone hole" has occurred primarily over Antarctica.

13. Give an example of a bromine-containing compound that has been implicated in destroying ozone.

14. Write out the sequence of reactions of the Chapman cycle and explain what is happening in each reaction.

15. Give examples of two common CFCs and draw Lewis structures for them.

16. Discuss the bathtub analogy as it relates to the steady state concentration of ozone in the stratosphere.

17. Why are there seasonal fluctuations in ozone concentration above the South Pole?

18. CFC vapors are fairly heavy, and many people have suggested that this makes it impossible for them to make their way into the stratosphere. Explain how it is possible for that to happen and what factors allow it.

19. Many CFC replacements are greenhouse gases that are worse than CFC-11. How do CFCs and CFC replacements act as greenhouse gases?

20. What might be some of the effects of a significant reduction in stratospheric ozone?

21. Which molecule is primarily responsible for absorbing the UV-C (high-energy UV) emitted by the Sun?
    (a) Nitrogen   (b) Oxygen
    (c) Carbon dioxide   (d) Carbon monoxide
    (e) Ozone

22. The ozone hole has developed primarily over the South Pole because:
    (a) The heavy CFCs sink to the bottom of the world.
    (b) Volcanoes in South America release significant amounts of CFCs into the atmosphere.
    (c) The weather conditions at the South Pole are critical for the chemical reactions that destroy the ozone layer.
    (d) The South Pole is closer to the Sun so there is more UV radiation there.
    (e) Excess CFCs, now banned in the industrialized world, were shipped there for storage and leaked into the atmosphere.

23. Which of the following is a likely explanation for the fact that, even though volcanoes produce HCl gas, the HCl thus produced does not seem to be able to generate chlorine atoms to destroy the ozone layer?
    (a) The chlorine atoms that would be produced by dissociation of HCl are different than the chlorine atoms produced by interaction of CFCs with ultraviolet radiation.
    (b) HCl is a liquid that runs down the sides of the volcano during an eruption.

(c) HCl is a solid and simply falls out of the sky.
(d) HCl is probably removed from the troposphere by being rained out because it reacts readily with water.

24. The Chapman cycle is a series of chemical reactions related to the oxygen–ozone screening process. These reactions indicate that, in the absence of human interference, oxygen and ozone concentrations in our stratosphere would:
    (a) continually increase.
    (b) maintain a steady state.
    (c) continually decrease.
    (d) change with the seasons.
    (e) drop dramatically as global warming increases.

25. As we have progressed from using CFCs to HCFCs to HFCs we have been:
    (a) gradually eliminating the chlorine atoms from these refrigerant molecules.
    (b) replacing many of the carbon atoms with less-harmful chlorine atoms.
    (c) trying to prepare refrigerants that will destroy tropospheric ozone before it contributes too much to global warming.
    (d) trying to produce a refrigerant that is so heavy that it cannot possibly ever make its way into the stratosphere.
    (e) come up with molecules that produce ozone by reaction with oxygen and UV light when they arrive in the stratosphere.

26. The _____ deals with global warming and the _____ deals with the hole in the ozone layer and the phase-out of _____.
    (a) Montreal Protocol; Kyoto Conference; greenhouse gases
    (b) Kyoto Conference; Montreal Protocol; substances that deplete the ozone layer
    (c) Chapman cycle; law of conservation of mass; greenhouse gases
    (d) Kyoto Conference; Montreal Protocol; CFCs
    (e) Kyoto Conference; Montreal Protocol; photons

27. UV-C radiation has wavelengths less than 280 nm while UV-B radiation has wavelengths between 280 nm and 320 nm. Which is higher in energy?

28. If the ozone layer in the stratosphere were to be brought to atmospheric pressure and the average surface temperature of Earth (15°C), the ozone would:
    (a) be about an inch thick.
    (b) be about the thickness of two nickels stacked on top of each other.
    (c) totally block all UV radiation from reaching Earth.
    (d) totally block all UV and IR radiation from reaching Earth.
    (e) would be impervious to damage by Cl and Br atoms.

29. Why do you think that ozone produced at ground level in large urban environments doesn't simply make its way to the stratosphere to replace the ozone that has been destroyed by CFCs?

30. In a previous chapter, we looked at some of the unusual ideas that had been proposed to address global warming. The ozone hole problem does not seem to have generated the same variety of unusual ideas. For instance, no one appears to have suggested putting "sunglasses" between Earth and the Sun. Why do you think that is? Can you suggest differences between global warming and the ozone hole that suggest we may have turned the corner in addressing the ozone hole but have not done so with global warming?

31. Compare and contrast global warming and the ozone hole problem in terms of the following:
    (a) Atmospheric region involved
    (b) Type of radiation involved
    (c) Molecules involved
    (d) Public recognition of the problem
    (e) Legislative action and other action taken
    (f) Public recognition of the problem
    (g) Results of actions taken

8

Chemical Reactivity:
Chemicals in Action

M any disciplines rely on experiments to test the value and accuracy of theories. Psychologists might study rats to determine the effect of diet on memory; an economist might survey households to identify a relation between spending and income; or a sociologist might seek to identify a link between school breakfast programs and children's educational success. However, those who study animals, people, or social interactions are often faced with the difficult task of extending their theories to other groups. Are two plants, two laboratory rats, or two social groups ever identical? Chemists, however, have an advantage: Under *identical conditions*, pure chemicals *always* react with each other in the same way. Sometimes it is difficult, but with effort identical conditions can be achieved.

In this chapter, the conditions that influence the outcome of chemical reactions are explored. To fully investigate a chemical reaction requires answers to the following questions:

What is happening? **How much** matter will react? **How fast** is the reaction? **To what extent** will reactants react or products form? And, **why** do chemical reactions occur?

Some chemists spend a lifetime seeking answers to these questions about a single complex reaction. Others devote themselves to answering one of these questions about many chemical reactions.

In Chapter 2, we introduced chemical reactions and the information needed to answer the "What?" question. What are the reactants and products? In the following reaction, hydrogen and oxygen, the reactants, combine to form water, the product:

$$2 \text{ H}_2(g) + \text{O}_2(g) \longrightarrow 2 \text{ H}_2\text{O}(\ell)$$

$$\text{Hydrogen} \quad \text{Oxygen} \quad \quad \text{Water}$$

Now we're going to focus on the story of chemical reactions and pursue the meaning of the other questions listed here.

SECTIONS

8-1 Balanced Chemical Equations and the "What?" Question

Chemical equations are the best way we have to represent what happens in chemical reactions at the nanoscopic level. Although there are many ways to describe a chemical reaction, only a *balanced* chemical equation can accurately describe the changes that occur. In Chapter 2 you saw that a balanced chemical equation is a chemical equation in which the total number of atoms of each kind is the same in the reactants and products. The importance of having a balanced chemical equation is that it allows chemists

Fireworks provide a dramatic example of chemistry in action.

to understand and predict the amounts of product formed or reactant consumed in a chemical reaction. To determine whether an equation is balanced requires counting up the atoms of each kind in the reactants and products (Section 2-6). Doing this, of course, requires knowing the identity and correct formulas of the reactants and products. Without prior knowledge of the reaction or knowledge of similar reactions, this can be difficult to do. With experience you will be able to use previous examples and chemical intuition to predict the products of many, but not all, chemical reactions. To see how an equation can be balanced, consider the following example.

The products of the complete burning, or complete combustion, of any hydrocarbon are always carbon dioxide and water. Thus, for the burning of propane (C_3H_8), the *unbalanced* equation is

$$C_3H_8(g) + O_2(g) \longrightarrow CO_2(g) + H_2O(g)$$

Notice that the number of C atoms on the reactant side of the equation is three, but that there is only one C atom on the product side. Because matter is not created or destroyed in a chemical reaction (Chapter 3), this equation is not an accurate description of the combustion of propane. It is not balanced.

To balance this and all other chemical equations, we start by identifying one element that is part of only one reactant compound and one product compound. Then, a *coefficient* is added in front of the reactant or product compound that has the lesser number of atoms of the element. This should result in a balance for the element in the equation. A coefficient changes the number of formula units without changing the compound's identity. An equation can *never* be balanced by changing the subscript in a chemical formula; this changes the identity of the chemical compound. Only coefficients can be changed to achieve balance.

In this example, either C or H appears in a single compound on either side of the equation, so it is possible to begin with either of these elements. If we focus on carbon, adding a coefficient of 3 in front of CO_2 balances the number of C atoms:

$$C_3H_8(g) + O_2(g) \longrightarrow 3\ CO_2(g) + H_2O(g)$$

However, the equation is not yet fully balanced. Notice, now, that the number of H atoms is not equal, with eight on the reactant side and only two on the product side. This can be remedied by placing a coefficient of 4 in front of $H_2O(g)$:

$$C_3H_8(g) + O_2(g) \longrightarrow 3\ CO_2(g) + 4\ H_2O(g)$$

There are now equal numbers of C and H atoms on each side of the equation. Last, you can see that the oxygen atoms are not balanced. On the reactant side there is a total of 10 oxygen atoms (3 CO_2 molecules with 2 each plus 4 H_2O molecules with 1 each: $[3 \times 2] + [4 \times 1] = 10$). Placing a coefficient of 5 in front of O_2 on the reactant side of the equation now brings oxygen into balance:

$$C_3H_8(g) + 5\ O_2(g) \longrightarrow 3\ CO_2(g) + 4\ H_2O(g)$$

The chemical equation is now fully balanced. There are 3 carbon atoms, 8 hydrogen atoms, and 10 oxygen atoms on each side of the equation.

To summarize, the steps needed to balance any chemical equation are as follows:

1. Identify and count the types of each atom on each side of the equation. If the numbers are the same for each element, the equation is already balanced.

2. Identify one element that is present in only one reactant and one product compound. Add a coefficient so that the number of atoms of this element on both sides is equal. Never change subscripts!

3. Repeat with another type of atom.

4. Continue until all atoms are balanced.

EXAMPLE 8.1 Equation Balancing

 Balance the following equation for the reaction of hydrofluoric acid [HF(aq)] with glass, which can be represented as calcium silicate ($CaSiO_3$). Decorative glass is etched using this reaction:

$$CaSiO_3(s) + HF(aq) \longrightarrow CaF_2(s) + SiF_4(g) + H_2O(\ell)$$

SOLUTION

The Ca and Si atoms are balanced. To balance the three O atoms on the left requires three H_2O molecules on the right. There must then be six H atoms on the left. Putting in both of these coefficients gives

$$CaSiO_3(s) + 6\ HF(aq) \longrightarrow$$
$$CaF_2(s) + SiF_4(g) + 3\ H_2O(\ell)$$

There are now six F atoms on each side of the equation, and it is fully balanced.

On the left:	1 Ca	1 Si	6 H	6 F	3 O
On the right:	1 Ca	1 Si	6 H	6 F	3 O

TRY IT 8.1A

Balance the following equation for the preparation of aluminum trichloride, which is an ingredient in some antiperspirants.

$$Al(s) + Cl_2(g) \longrightarrow AlCl_3(s)$$

TRY IT 8.1B

Are the following equations balanced? If not, balance them.

(a) $CaO(s) + H_2O(\ell) \longrightarrow Ca(OH)_2$

(b) $SiO_2(s) + C(s) \longrightarrow Si(s) + CO(g)$

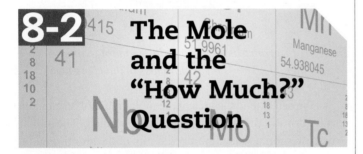

8-2 The Mole and the "How Much?" Question

8-2a Moles and Molar Masses

Eventually, anyone who wants to carry out a chemical reaction must figure the amount of the reactants needed to make a desired amount of product. Somehow a connection must be made between atoms, molecules, and ions at the nanoscopic level and amounts of chemicals that can be easily measured. Here is where balanced equations are essential. The relative atomic weight scale and balanced chemical equations together make it possible to answer "How much?" questions.

To understand the situation, consider the balanced equation for hydrogen burning in chlorine to form hydrogen chloride:

Don't get stuck in a hole—measure your atoms in moles!

$$H_2(g) + Cl_2(g) \longrightarrow 2\ HCl(g)$$

The equation shows that one molecule of hydrogen and one molecule of chlorine combine to form two molecules of hydrogen chloride. Is this information of any help in figuring out, for instance, how many hydrogen and chlorine molecules would be needed to make 100 g of hydrogen chloride? The essential problem is that molecules are very small—so small that it is impossible to count them one by one.

The solution to this problem is to use a counting unit called the **mole**.

A counting unit is a single unit of measurement used to represent a larger number of objects. For example, the counting unit "pair" refers to two objects, and "dozen" refers to twelve objects. The counting unit of "mole" refers to a somewhat larger number of objects, 6.022×10^{23}, but is still a counting unit nonetheless, and is also known as **Avogadro's number**. So, one mole of eggs is the same as 6.022×10^{23} individual eggs, and one mole of neon atoms is the same as 6.022×10^{23} neon atoms.

The use of the mole as a counting unit in chemistry allows chemists to measure out large numbers of atoms or molecules by simply measuring their mass. Recall from Chapter 3 that atoms of different elements have different mass, mostly because the number of protons and neutrons in the nucleus differs from one element to another. Therefore, the mass of one carbon atom is different from the mass of one neon atom, and the mass of one mole of carbon atoms is different from the mass of one mole of neon atoms. By knowing the mass of one atom of each of these elements, you should be able to use a scale to determine how many atoms are contained in a sample of each pure element.

But the exact mass of a single atom of any element is not particularly useful. For instance, who wants to remember that the mass of a single carbon atom is 1.993×10^{-23} grams? Such numbers would be incredibly difficult to remember and even more difficult to measure.

The mole is a useful counting unit in chemistry because one mole of, say, carbon atoms has a mass of 12.01 grams. *The atomic weights listed below each element symbol in the periodic table are both a weighted average of the mass of a single atom in amu and the mass in grams of one mole of atoms of the element.* The mass of one mole of any type of object, especially atoms, molecules, and ions, is known as the **molar mass**. This means

Mole: The counting unit used for atoms

Avogadro's number: 6.022×10^{23}; the number of objects in one mole

Molar mass: Mass in grams of one mole of any substance

that while one mole of C atoms has a mass of about 12.01 grams, one mole of calcium atoms has a mass of 40.08 grams, one mole of iodine atoms has a mass of 126.9 grams, and so on.

Because of this relationship between mass, chemical identity, and the mole, *it is possible to relate the number of atoms in a sample to the mass of the sample* (Figure 8.1). The following example is meant to help you understand this.

FIGURE 8.1 One-mole quantities of some elements.
The cylinders (*left to right*) hold mercury (201 g), lead (207 g), and copper (64 g). The two Erlenmeyer flasks hold sulfur (*left*, 32 g) and magnesium (*right*, 24 g). All rest on 1 mole of aluminum in the form of foil (27 g) and also on the watch glass. Each sample contains 6.022×10^{23} atoms.

EXAMPLE 8.2 Molar Mass

 Show that the mass of Avogadro's number of carbon-12 atoms is 12.0 g. A single carbon atom weighs 12 amu, and one amu is equal to 1.661 $\times$ 10^{-24} g.

SOLUTION

The mass of a single carbon-12 atom in grams can be calculated as 12 amu $\times$ 1.661 $\times$ 10^{-24} g/amu = 1.993 $\times$ 10^{-23} g. Therefore, Avogadro's number of carbon atoms has a mass of (1.993 $\times$ 10^{-23} g/C atom)(6.022 $\times$ 10^{23} C atom/mol) = 12.0 g/mol.

TRY IT 8.2A

What is the mass of one mole of selenium atoms?

SOLUTION

Because the mass of one mole of any element is equal to the atomic weighted average of the element, in grams, the mass of one mole of selenium atoms is 78.96 g.

TRY IT 8.2B

How many magnesium atoms are contained within a 24.3-g sample of magnesium? How about in a 12.15-g sample?

SOLUTION

Consulting the periodic table, we can see that the mass of 1 mole of Mg atoms is 24.3 g. Since this is the same as the mass of the first magnesium sample, that sample must contain 6.022 $\times$ 10^{23} Mg atoms. For the second sample, which weighs half as much as the first, there must be half as many atoms. So, there are 6.022 $\times$ 10^{23}/2 = 3.011 $\times$ 10^{23} Mg atoms.

This discussion and these exercises have shown that *one mole of atoms of any element is simply the atomic*

weight in grams listed below the element symbol in the periodic table. This is the molar mass of the element (Figure 8.1).

We can extend the idea of the molar mass of an element to determine the molar mass of compounds. The molar mass of a compound is simply the sum of the atomic weight of each element in grams multiplied by the subscript on the element in the formula of the compound. For instance, one mole of methane molecules (CH_4) is made of one mole of carbon atoms (molar mass = 12.01 g) and four moles of hydrogen atoms (4 $\times$ 1.0 g/mol = 4.0 g). Thus, the molar mass of CH_4 is 12.01 + 4.0 = 16.01 g. To summarize, *to find the molar mass for a compound, multiply the number of atoms of each element in the formula by the atomic weight of the element and add them together.*

EXAMPLE 8.3 Determining Molar Mass

 Determine the molar mass of rubbing alcohol (C_3H_8O).

SOLUTION

$$3 \text{ C } (3 \text{ mol})(12.01 \text{ g/mol}) = 36.03 \text{ g}$$
$$8 \text{ H } (8 \text{ mol})(1.0 \text{ g/mol}) = 8.0 \text{ g}$$
$$1 \text{ O } (1 \text{ mol}) (16.0 \text{ g/mol}) = \underline{16.0 \text{ g}}$$
$$60.03 \text{ g}$$

TRY IT 8.3A

Determine the molar mass of common table salt (NaCl).

Determine the molar mass of the simple sugar known as glucose ($C_6H_{12}O_6$).

Because the mole is a counting unit, every balanced chemical equation can be interpreted in terms of moles—moles of atoms, moles of molecules, moles of ions, or anything else. For example,

$$H_2(g) \quad + \quad Cl_2(g) \quad \longrightarrow \quad 2\,HCl(g)$$
1 H_2 molecule   1 Cl_2 molecule   2 HCl molecules

also means

$$H_2(g) \quad + \quad Cl_2(g) \quad \longrightarrow \quad 2\,HCl(g)$$
1 mol of H_2 1 mol of Cl_2 2 mol of HCl

The molar mass of a substance provides a conversion factor that can be used to find the number of moles equal to a known mass of a substance. For example, 10.0 g of carbon is equal to 0.830 mol of carbon.

$$(10.0 \text{ g C})(1 \text{ mol}/12.01 \text{ g C}) = 0.830 \text{ mol C}$$

Also, the molar mass allows conversion of a known number of moles to the equivalent mass of a substance. For example,

$$(2 \text{ mol C})(12.01 \text{ g}/1 \text{ mol C}) = 24.02 \text{ g C}$$

EXAMPLE 8.4 Moles and Masses

 What is the equivalent in moles of 3.7 g of water, roughly the amount in a teaspoonful?

SOLUTION

The molar mass of water, H_2O, is 18 g. Molar mass provides the conversion factors that connect mass and numbers of moles. Therefore,

$$3.7 \text{ g } H_2O \times \frac{1 \text{ mol } H_2O}{18 \text{ g } H_2O} = 0.21 \text{ mol}$$

What is the mass equivalent to 50 mol of barium nitrate, $Ba(NO_3)_2$, which gives a green color to fireworks?

8-2b Masses of Reactants and Products

The mole provides us with the means to answer the "How much?" question about any chemical reaction. If all of the reactants are converted to product, 2.0 g of H_2 (1 mol) should react with 71 g of Cl_2 (1 mol) to produce 73 g of HCl (2 mol).

$$H_2(g) \quad + \quad Cl_2(g) \quad \longrightarrow \quad 2\,HCl(g)$$
1 H_2 molecule   1 Cl_2 molecule   2 HCl molecules
1 mol of H_2 1 mol of Cl_2 2 mol of HCl
2.0 g of H_2 71.0 g of Cl_2 2 mol × 36.5 g/mol = 73.0 g of HCl

Have you noticed that these masses adhere to the law of conservation of matter? The 2.0 g + 71.0 g of reactants produce 73.0 g of product.

EXAMPLE 8.5 Information About Masses from a Chemical Equation

 Interpret the equation for a reaction that produces copper from copper ore in terms of moles, molar masses, and masses of reactants and products.

$$Cu_2S(s) + 2\,Cu_2O(s) \xrightarrow{\text{Heat}} 6\,Cu(s) + SO_2(g)$$
Copper(I)   Copper(I)   Copper   Sulfur dioxide
sulfide   oxide

SOLUTION

The equation in terms of moles shows that 1 mol of copper(I) sulfide reacts with 2 mol of copper(I) oxide to produce 6 mol of metallic copper and 1 mol of sulfur dioxide. The molar masses of the reactants and products are found from the molar masses of the elements combined in each reactant and product. Using the molar masses for copper (Cu, 63.55 g/mol), sulfur (S, 32.07 g/mol), and oxygen (O, 16 g/mol) gives for Cu_2S, 159.17 g/mol [(2 × 63.55 g) + 32.07 g]; for SO_2, 64.07 g/mol [32.07 g + (2 × 16 g)]; and for Cu_2O, 143.1 g/mol [(2 × 63.55 g) + 16 g]. Thus, the equation gives the following information about reactants and products:

$$Cu_2S(g) \quad + \quad 2\,Cu_2O(g) \xrightarrow{\text{Heat}} 6\,Cu(s) \quad + \quad SO_2(g)$$
1 mol Cu_2S 2 mol Cu_2O 6 mol Cu   1 mol SO_2
159.17 g Cu_2S 2 × 143.1 g Cu_2O/mol   6 × 63.55 g Cu/mol   64.07 g SO_2
      = 286.2 g Cu_2O = 381.3 g Cu

The number of moles (or the number of molecules) is not necessarily conserved in a chemical reaction. The number of atoms and their mass is conserved, however, according to the law of conservation of matter.

Interpret the equation for making methanol in terms of moles, molar masses, and masses of reactants and products.

$$CO(g) + 2\,H_2(g) \longrightarrow CH_3OH(\ell)$$

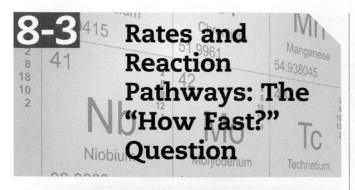

8-3 Rates and Reaction Pathways: The "How Fast?" Question

8-3a Reaction Pathways

The chemical reactions we notice in the world around us are usually fast. The color change, bubbles of gas, or explosion happens right away as visible proof that a reaction has occurred. Many reactions are naturally slow, however. At everyday conditions of temperature and pressure, the conversion of carbon monoxide to carbon dioxide is slow.

$$2\,CO(g) + O_2(g) \longrightarrow 2\,CO_2(g)$$

If this reaction were faster, perhaps cases of carbon monoxide poisoning would not be as prevalent as they are; the gas would not accumulate but would instead be converted into the less hazardous carbon dioxide prior to exposure.

Out of curiosity and also for practical reasons, it is interesting to discover what makes reactions fast or slow. Ideally, to do this a chemist would like to watch the pathway of each atom from its position in the reactants to its position in the products.

What is the connection between the rate of a process and its pathway? Suppose that there are several hundred books in a storeroom on the first floor and that you have been hired to move them to the new third-floor library. Depending on the conditions, there are a variety of possible pathways. One pathway might require the following steps: (1) put ten books (the maximum number you can lift) into a carton, (2) carry the carton up one flight of stairs to the second floor, (3) rest a bit, (4) carry the carton up another flight of stairs to the third floor, (5) empty the carton, and (6) return for another load. Another pathway might involve a different series of steps: (1) fill four cartons with books, (2) pile the cartons onto a dolly, (3) push the dolly onto the elevator, (4) ride to the third floor, (5) push the dolly off the elevator, (6) empty the four cartons, and (7) return for another load. The second pathway would probably be faster than the first. To compare them quantitatively,

the rate for each pathway could be measured in books moved per hour.

The rate at which a chemical reaction takes place is determined by the way in which chemical compounds react. One way to think of what happens during a chemical reaction, at the atomic level, is to imagine what would be required to produce two broken eggs by throwing them at each other. First, and most obvious, it would be necessary for them to collide with each other. However, if thrown too gently, they might simply touch and fall to the floor; no broken egg would be produced as a result of the collision. This is because the speed (and energy) with which they collide is important to producing broken eggs by this method. But what if they barely touch as they go past each other? Even though they might have collided and been thrown at a great speed, this also would result in no broken eggs being produced as a result of the collision. The eggs must therefore also collide with the proper orientation in order for a change to occur.

Now, extend this to the analogous process for two molecules. Molecules fly about at random. Occasionally, two molecules speed directly toward each other. Unlike eggs, as they get closer together, they begin to repel each other through their negatively charged electrons. If the energy of their motion is not great enough to overcome this repulsion, the molecules just veer away from each other and no chemical reaction occurs (Figure 8.2a). But, if the energy (speed) of the approaching molecules is great enough to overcome the repulsion, then they will collide. At this instant, a chemical reaction *may* occur if the collision takes place with the proper orientation. If some of the bonds in the colliding molecules break so that new bonds

Two chemical reactions. (a) Dissolving Alka-Seltzer in water leads to a fast reaction. (b) The combination of iron with oxygen is a slow reaction.

Charles D. Winters

FIGURE 8.2 Collisions and chemical reactions.

Some collisions are successful and some are not. The difference is determined to a great extent by the kinetic energy of the particles.

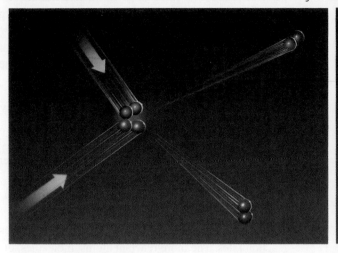

(a) Unsuccessful collision

(b) Successful collision

© Cengage Learning

can form, the collision causes a chemical change and yields the product (Figure 8.2b).

In summary, the following factors are involved in any chemical reaction:

- Atoms, molecules, or ions must collide for a reaction to occur. If no collision is possible, then no reaction can occur.

- The collisions between the atoms, molecules, or ions must be sufficiently energetic for reaction to occur.

- In some reactions, collisions are productive only if the reactants are properly oriented relative to each other upon collision.

- Bonds are made or broken in most chemical reactions. Some reactions may involve only electron transfer, however.

- Energy is absorbed during bond breakage and released during bond formation.

8-3b An Energy Hill to Climb

Why are chemists interested in how fast a reaction takes place? One practical answer is that knowledge of the rate of a reaction can be useful in predicting the time required for a reaction to occur. This is true of all chemical reactions, whether they take place in a laboratory, in a living organism, or in the environment. **Reaction rates** are usually expressed as the amount of reactant converted to product in a specific unit of time. For fast reactions, the time unit might be seconds; for slow reactions, it might be days or

even years. The number of successful collisions per second, minute, hour, or day determines the reaction rate.

For each reaction, there is a threshold, or minimum, quantity of energy needed for successful collisions, known as the **activation energy** (E_{act}). Activation energy is like a steep mountain that must be climbed to reach the valley on the other side. If the activation energy of a reaction is high, only a small number of reactant particles can collide with the energy needed to form product molecules, and the reaction will take place slowly. In the opposite condition, low activation energies result in fast reactions (Figure 8.3). Some reactions do not take place at all unless enough energy is supplied.

Consider the following sets of reactions in which one reaction does not occur (or occurs very slowly) at room temperature while the other occurs much more rapidly at higher temperatures.

> **Reaction rate:** Amount of reactant converted to product in a specific amount of time
>
> **Activation energy (E_{act}):** Quantity of energy needed for successful collision of reactants

a.
$$N_2(g) + O_2(g) \xrightarrow{\text{room temp.}} \text{no reaction}$$
$$N_2(g) + O_2(g) \xrightarrow{\text{very high temp.}} 2\,NO$$

b.
$$2\,H_2(g) + O_2(g) \xrightarrow{\text{room temp.}} \text{no reaction}$$
$$2\,H_2(g) + O_2(g) \xrightarrow{\text{spark}} 2\,H_2O(g)$$

c.
$$4\,Fe(s) + 3\,O_2(g) \xrightarrow{\text{room temp.}} 2\,Fe_2O_3(s) \quad \text{(slow)}$$
$$4\,Fe(s) + 3\,O_2(g) \xrightarrow{\text{heat}} 2\,Fe_2O_3(s) \quad \text{(faster)}$$

Nitrogen and oxygen, the major components of our atmosphere, do not react with each other at room temperature, yet these same two gases react quite quickly to produce nitric oxide at the high temperatures found within a gasoline engine. Similarly, hydrogen and oxygen gas are unreactive toward each other until a spark is added, in which case there is an explosion. Iron will react with oxygen at room temperature to slowly produce rust (Fe_2O_3), but this process is much more rapid at elevated temperatures. The rate at which iron rusts can also be increased by increasing the surface area contact between the iron and oxygen. This fact makes it possible to manufacture small packets containing iron dust that, when opened, generate a small amount of heat as a result of the reaction between iron and oxygen. These packets are used as hand warmers by hikers and backpackers during the winter.

8-3c Controlling Reaction Rates

Understanding what is necessary for a chemical reaction to occur makes it possible for chemists to increase or decrease the rate of a reaction. Four strategies are available: (1) adjust the temperature, (2) adjust the concentration of the reactants, (3) increase contact between the reactants, or (4) add a catalyst.

1. *Effect of temperature on reaction rate.* At higher temperatures, molecules (on average) move faster, so more of them have enough energy to get over the activation energy hill and react. We make use of this principle in cooking by raising the temperature to speed up roasting a piece of meat and in preserving foods by lowering the temperature to slow down the reactions that spoil the food.

2. *Effect of concentration on reaction rate.* The quantity of a substance in a given quantity of a mixture is its *concentration*. For example, a solution might have a concentration of 5 g of sodium chloride in 1 L of water (written as 5 g/L, or 5 grams per liter). A solution of 10 g of sodium chloride in 1 L of water (10 g/L) has a higher concentration.

   Increasing the concentration of reactants increases reaction rates. The more reactant atoms, molecules, or ions, the more frequent the collisions between reactants. (What if, in the earlier example, many eggs, not just two, were thrown toward each other?) There is, for example, a dramatic increase in the rate of combustion of propane in pure oxygen compared with that in air, which is about 20% oxygen.

3. *Effect of contact on reaction rate.* An effect similar to a concentration increase occurs if a solid reactant is very finely divided, which essentially increases concentration by increasing the surface area at which reactions can occur. We don't think of flour as an explosive

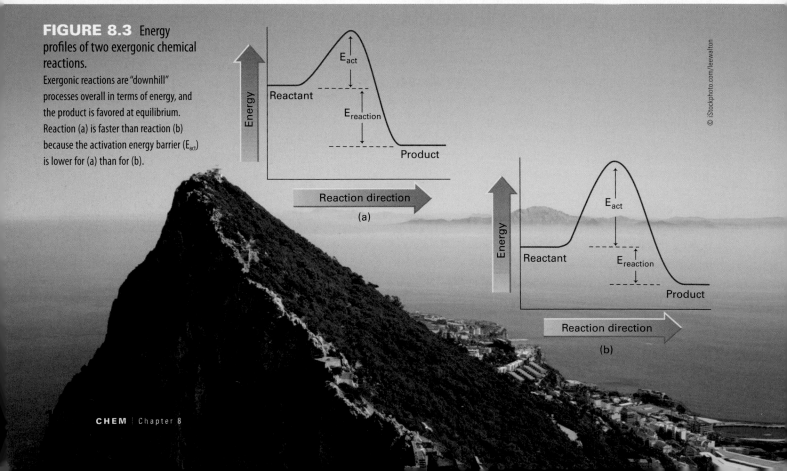

FIGURE 8.3 Energy profiles of two exergonic chemical reactions.

Exergonic reactions are "downhill" processes overall in terms of energy, and the product is favored at equilibrium. Reaction (a) is faster than reaction (b) because the activation energy barrier (E_{act}) is lower for (a) than for (b).

© iStockphoto.com/leewalton

FIGURE 8.4 Effect of a catalyst.

All a catalyst does is lower the activation energy. Because more reactants have enough energy to overcome the reaction energy barrier, the reaction rate increases.

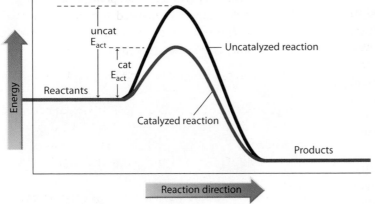

© Cengage Learning

substance, but a spark can set off the explosion of flour suspended in the air. Dust explosions are a hazard in grain storage, clothing factories, and coal mines.

4. **Effect of a catalyst on reaction rate.** Sometimes a reaction speeds up dramatically when a substance other than the reactants is added to the mixture. Substances that increase the rate of a chemical reaction without being changed themselves are called **catalysts**.

Catalysts can be used in a laboratory as well as in other settings. One example is the kitchen trick used to ripen tomatoes quickly. An unripe tomato is placed in a paper bag with a ripe banana. The banana will release a very small amount of ethylene, which is a catalyst for the ripening reaction in many fruits. The result is a "force-ripened" tomato. This has been used to great effect in agriculture and is partly responsible for the availability of ripened fruits during the winter months.

A catalyst functions by allowing a chemical reaction to take place through a pathway that requires less energy and lowers the activation (minimum) energy needed for the reaction to occur. The earlier example in which books were moved included an example of a catalyst, the elevator, which sped up the process but did not have an effect on the overall change. When a catalyst is used, more collisions are successful because less energy is required for success (Figure 8.4). What makes catalysts so practical is that many times they can be recovered after the reaction is over and used again and again. Many industrial processes rely on rare and expensive metals as catalysts, making recovery of the catalyst an economic necessity.

Living systems are even more dependent on catalysts than the chemical industry. Although engineers can manipulate temperature and concentrations in an industrial process to control reaction rates, our bodies cannot. If body temperature varies far from 37°C or if the concentrations of chemicals in body fluids vary much from the normal values, we are in

serious trouble. Catalysis is the major strategy available for controlling biochemical reactions. The amazing constancy of our internal chemistry is maintained by biological catalysts known as *enzymes* (Section 15-8). There is a good reason why blood and the liver are well supplied with a fast-acting catalyst for the decomposition of hydrogen peroxide (Figure 8.5). Because it is highly reactive, hydrogen peroxide must be destroyed before it has the chance to damage essential substances in its surroundings.

> **Catalyst:** A substance that increases the rate of a chemical reaction without being changed in identity

Enzymes are also responsible for converting our food into nutrients, removing waste such as carbon dioxide from our bodies, and building important molecules needed to maintain health and wellness. A number of human metabolic diseases are due to a lack of or defect in important enzymes. In the absence of these important catalysts, necessary metabolic chemical reactions take place slowly or not at all.

8-4 Chemical Equilibrium and the "To What Extent?" Question

Although it is tempting to think that combining reactants together will always result in their complete conversion into

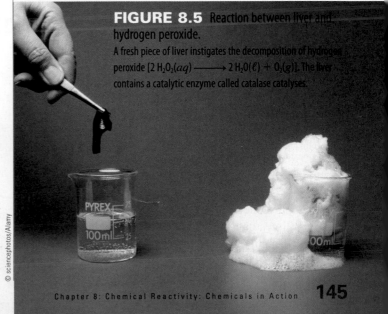

FIGURE 8.5 Reaction between liver and hydrogen peroxide.

A fresh piece of liver instigates the decomposition of hydrogen peroxide [2 H$_2$O$_2$(aq) $\longrightarrow$ 2 H$_2$O(ℓ) + O$_2$(g)]. The liver contains a catalytic enzyme called catalase catalyses.

© sciencephotos/Alamy

products, this is not true for all chemical reactions. Some chemical reactions exist in **chemical equilibrium**, a state in which the amount of reactant molecules and product molecules are in balance.

For a chemical reaction, equilibrium is a balance between the reaction and its reverse, both occurring at equal rates. Studies of many, many chemical reactions have shown that reactions reach equilibrium at a characteristic and predictable point. Understanding why this is provides an answer to the "To what extent?" question for a reaction.

The concept of equilibrium is not unique to chemistry. It is simply a balance between opposite changes that take place at the same rate. In a closed container partly filled with water, the space above the liquid begins to fill with water vapor. As the air starts to become saturated with water vapor, some of the water vapor starts to condense. Eventually, the evaporation and condensation of the water establish an equilibrium—a state of balance between exactly opposite changes occurring at the same rate (Figure 8.6). To indicate this equilibrium, a double arrow is placed between the symbols for water in the liquid and vapor states

$$H_2O(l) \rightleftharpoons H_2O(g)$$

Many chemical reactions can take place in a forward direction (left to right in a chemical equation) and a reverse direction (right to left). These reactions are said to be *reversible*. Whenever the point is reached at which the forward process is proceeding at the same rate as the reverse reaction, equilibrium is established and the amounts of reactants and products remain unchanged. But because equilibrium is a *dynamic* condition, the forward and reverse reactions are still happening so that each reactant or product is replaced as soon as it is consumed.

For equilibrium to be reached, it is important that none of the reactants or products can escape. An example is provided by the conversion of limestone (calcium carbonate) to lime (calcium oxide). By heating limestone in open pits, early U.S. settlers produced lime for mortar:

$$CaCO_3(s) \xrightarrow{\text{Heat}} CaO(s) + CO_2(g)$$

Calcium carbonate (limestone) Calcium oxide (lime) Carbon dioxide

Because the CO_2 gas escapes from the open pit, the reaction keeps going until all of the calcium carbonate in the limestone is converted to lime. If instead some dry limestone is sealed in a closed container and heated, the result is different. As soon as some CO_2 accumulates in the container, the

FIGURE 8.6 A system at equilibrium.
Water has reached equilibrium with its vapor in this ecosphere. Equilibrium also has been reached by the food and waste products of the inhabitants—a carefully balanced community of plants, shrimp, and a hundred or so kinds of microorganisms.

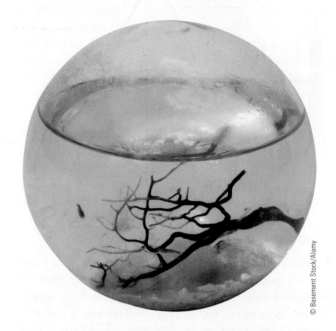

© Basement Stock/Alamy

reverse reaction starts to occur. Once the concentrations of CO_2 and CaO reach the appropriate values, equilibrium is established.

$$CaCO_3(s) \rightleftharpoons CaO(s) + CO_2(g)$$

If there aren't any changes in the pressure or the temperature, the forward and reverse reactions continue to take

Because it stands up well to exposure, the sedimentary rock limestone is used commonly in masonry and architecture. The Great Pyramid of Giza, the only one of the Seven Wonders of the Ancient World still standing, is made entirely from limestone.

© Michael Kelley/Stone+/Getty Images

place at the same rates, and the concentration of CO_2 and the amounts of the solid $CaCO_3$ and CaO in the container remain unchanged.

Many reactions in which all reactants and products are dissolved in water occur in nature or are carried out by chemists. How far such a reaction goes before equilibrium is reached is always a matter of interest. Acids are a very important class of water-soluble chemical compounds, about which we'll have more to say in the next chapter. Acetic acid provides a good example of a reaction that reaches equilibrium with just a small amount of reactant converted to product.

$$CH_3COOH(aq) + H_2O(\ell) \rightleftharpoons CH_3COO^-(aq) + H_3O^+(aq)$$

Acetic acid   Water   Acetate ion   Hydronium ion

Since the hydronium ion (H_3O^+), which is present in all acid solutions (Section 9-1) is what makes vinegar sour, it's a good thing that this reaction doesn't go to completion. If it did, vinegar would be of little use on a salad unless it was diluted with more water to make it less acidic.

For an illustration of the importance of constant conditions to maintaining equilibrium, we can return to the decomposition of limestone that has reached equilibrium in a closed container. If the container is opened briefly to let out some of the CO_2 and then sealed again, the forward reaction will outpace the reverse reaction, and the CO_2 concentration will increase until equilibrium is again established. This type of change illustrates a very important principle that applies to all systems at equilibrium: *If a stress is applied to a system at equilibrium, the system will adjust to relieve the stress.* Known as **Le Chatelier's principle** for the French scientist Henri Le Chatelier, who

first stated it in 1884, this principle means that whatever the disruption, the reaction will shift in the direction that re-establishes equilibrium. For a chemical reaction, the stresses might be adding or taking away a reactant or product, changing the temperature, or, in some cases, changing the pressure.

Le Chatelier's principle: When a stress is applied to a system at equilibrium, the equilibrium shifts to relieve the stress

As another example, consider a reaction that takes place entirely in the gaseous state—the synthesis of ammonia. Because of the need for ammonia in the production of fertilizers, this reaction is of great commercial significance (Section 19-4).

$$N_2(g) + 3 H_2(g) \rightleftharpoons 2 NH_3(g)$$

There are more total reactant gas molecules (4) than product gas molecules (2), which means that changing the pressure will stress the reaction (Figure 8.7). If the pressure is increased, the equilibrium will shift in the direction that decreases the pressure. To do this, the number of gas molecules must be decreased, which for ammonia synthesis means the forward reaction will be favored, and more ammonia will form.

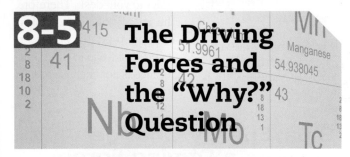

8-5 The Driving Forces and the "Why?" Question

Why does one reaction produce more product at equilibrium than another? Why do some chemicals react when they are mixed, while others have absolutely no tendency to react unless the conditions are changed? In general, *why* is one chemical reaction favorable and another not favorable?

These differences are due to driving forces. A driving force is a force that causes a change to take place. In chemistry, many driving forces are the result of changes in energy. Examining the driving forces that account for such differences between chemical reactions, including equilibria, requires looking at two kinds of change: changes in energy as heat and changes in the amount of order that accompany chemical reactions. We experience such energy changes every day. For example, when heat is removed from liquid water, ice is formed. Even processes that do not involve adding or removing heat involve changes in energy. Creating disorder on your desk while rushing to finish an assignment

FIGURE 8.7 Ammonia synthesis.

In a demonstration of Le Chatelier's principle, pressure shifts the equilibrium of reaction with different amounts of gaseous reactants and products. Higher pressure shifts the reaction toward smaller amounts, and therefore smaller volumes, of gas. To increase the amounts of ammonia produced, the synthesis is done under pressure.

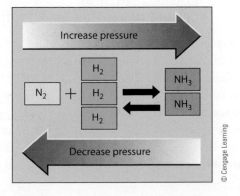

© Cengage Learning

ENERGY

© FoodPix/Jupiterimages/Getty Images

will later require energy (in the form of work) to restore order to your desk. Therefore, the energy of a tidy desk is different from that of a disordered desk. Clearly, this difference is not because of heat loss!

8-5a Energy Change as a Driving Force

First, it is important to note that the rate of a chemical reaction and the energy taken up or released by the reaction are in no way connected. The difference between the energy stored in the reactants and products is the same, no matter how quickly the reaction takes place. Put another way, the energy change of a chemical reaction is not related to the activation energy of the reaction.

Also, how fast or how slowly a reaction occurs has absolutely nothing to do with the position of the equilibrium for the reaction. The position of equilibrium is determined only by the energy difference between the reactants and products.

The natural direction of chemical reactions is much like the natural direction of more familiar changes in everyday life. We all know that water going over a waterfall is favorable, and that water going back up the waterfall is not

favorable. To move water from the bottom of the waterfall back to the top would require energy and work—electricity could drive a mechanical pump that could move the water.

Water held back behind a dam has stored energy—potential energy that can be converted to kinetic energy if the gates in the dam are opened. Like the water behind a dam, chemical compounds have potential energy; it is stored in chemical bonds. In many reactions, this energy is released as heat. The reaction of sodium with water is a good example. When the reactants are mixed, the mixture bursts into flame. Any reaction that releases energy is described as **exergonic**, and most exergonic reactions, once they get going, are favorable—no input of energy is required to keep them going.

The opposite condition is found in an **endergonic** reaction, one that requires energy in order to take place. Most endergonic reactions, such as the decomposition of limestone in an open pit, don't happen at all without a continuous supply of energy.

So far, you've seen evidence for one answer to the "Why?" question. Like water going over a waterfall, chemical reactions are favorable when they release energy, and the products thus have less potential energy than the reactants. Usually the energy is released as heat, although there are chemical reactions that generate light and, under the right conditions, electric current. A few reactions, however, are favorable but endothermic—they absorb heat from their surroundings and keep going without any outside influence. Here is evidence that there must be another driving force for change that does not involve heat energy.

8-5b Entropy Change as a Driving Force

Perhaps you have seen a demonstration of the endothermic but favorable reaction of barium hydroxide and ammonium thiocyanate. This reaction absorbs so much heat from its surroundings that it can freeze water in contact with the reaction flask (Figure 8.8):

$$Ba(OH)_2 \cdot 8\,H_2O(s) + 2\,NH_4SCN(s) \longrightarrow$$
$$Ba(SCN)_2(aq) + 2\,NH_3(aq) + 10\,H_2O(\ell)$$

The reactants are both crystalline solids, and, as such, their components are held together in a repetitive, orderly arrangement. Look at the products—there is liquid water, gaseous ammonia that mostly dissolves in the water, and an ionic compound $Ba(SCN)_2$ that dissolves in the water to give separate ions (Ba^{2+} and 2 SCN^-). What a mixture! And this is the clue to the driving force. The system has moved from an ordered state to a disordered state.

FIGURE 8.8 A favorable but endothermic reaction.

The reaction of barium hydroxide and ammonium thiocyanate is one of the uncommon examples of a favorable reaction that absorbs heat from its surroundings. As a result of this reaction, the wooden block freezes to the bottom of the flask.

(a)

(b)

Charles D. Winters

The melting of ice at room temperature is another change that naturally proceeds from order to disorder. You should have no trouble identifying this tendency on a personal scale. It seems to take no effort to create a random mixture of possessions in your room, but it certainly takes energy to put them back in order.

Physical scientists have given the name **entropy** to the disorder of matter, and it can be measured. Entropy is a form of energy. Gases have a higher entropy than liquids. Liquids have a higher entropy than crystalline solids. Large molecules often have higher entropy than small ones because their atoms can rotate around the bonds in many different ways. The more molecules or the more different kinds of molecules there are in a mixture, the higher the entropy of the mixture.

Entropy is the second factor that determines the answer to the "Why?" question. Reactions are favorable when they result in a *decrease* in energy *and* an *increase* in disorder. When one of these changes is favorable but the other is not, the greater effect controls the favorability of the reaction.

What happens when a process is unfavorable because it requires energy and creates order? Such a process is not forbidden by nature—it can be made to happen when energy is supplied from some other process. Ordering processes must be driven by favorable, disordering processes. The result is always that the net disorder of the universe is increased when an unfavorable process is driven by a favorable one.

Figure 8.9 on the next page illustrates the relationships between the energy of activation (E_{act}) and the overall energy change of a reaction ($E_{reaction}$) for an endergonic reaction in the same way that Figure 8.3 did for an exergonic reaction. In an exergonic reaction, the product is favored at equilibrium since the product has a lower energy content (that is, is more stable) than the reactant. The energy change of a reaction ($E_{reaction}$) is a function of both **enthalpy changes** and **entropy changes**. Enthalpy changes are primarily reflected in the energy required to break bonds and the energy released upon bond formation. Entropy changes primarily reflect the amount of order or disorder created in the chemical change. In an endergonic reaction (Figure 8.9), the reactant is favored at equilibrium since it has a lower energy content (i.e., is more stable) than the product. Such reactions do occur, however, provided there is a source available to provide the additional energy required for product formation. As you will see later (Section 15-10), photosynthesis is such an example, with the Sun serving as the source of the required energy.

Entropy: The disorder of matter

Enthalpy change: A measure of the quantity of heat transferred into or out of a system as it undergoes a chemical or physical change

Entropy change: A measure of the change in the disorder of a system as it undergoes chemical reaction or physical change

FIGURE 8.9 Energy profile of an endergonic reaction. Endergonic reactions are "uphill" processes in terms of energy, and the reactant is favored at equilibrium.

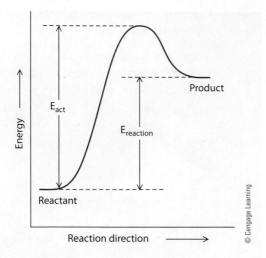

TABLE 8.1 Some Statements of the First and Second Laws of Thermodynamics

The First Law
The energy of the universe is constant.
Energy can be converted from one form to another but cannot be destroyed or created.
You can't get something for nothing.
There's no such thing as a free lunch.

The Second Law
The total entropy of the universe is constantly increasing.
The state of maximum entropy is the most stable state for an isolated system.
Energy is conserved in quantity but not in quality.
Every system that is left to itself will, on the average, change toward a condition of maximum probability.
You can't break even.

© Cengage Learning

8-5c The First Law of Thermodynamics

Thermodynamics: The science of energy and its transformations

First law of thermodynamics: Energy can be converted from one form to another but cannot be destroyed

Second law of thermodynamics: The total entropy of the universe is constantly increasing

The first and second laws of **thermodynamics** summarize the universal conditions for changes in energy and entropy (Table 8.1). Because they have such broad application and meaning beyond science, they are often referred to familiarly as "the first law" and "the second law." Here is a formal statement of the **first law of thermodynamics**, sometimes known as the *law of conservation of energy*:

Energy can be converted from one form to another but cannot be destroyed or created.

When gasoline burns in an automobile engine, *all* of the energy released could be accounted for if one could measure the resulting mechanical energy, the friction of moving parts, the energy that leaves the car in the exhaust, the energy converted to electrical and then chemical potential energy in the battery, and the increase in temperature of the engine and everything surrounding it.

Looked at another way, the first law means that *the total quantity of energy in the universe is constant*. Creation of new energy is not possible. All we can do is change it from one form to another. The first law shows up in conversation whenever someone says, "Oh well, you can't get something for nothing."

In 1900, Albert Einstein created an extension of the first law when he recognized that matter and energy are interconvertible:

The total amount of matter and energy in the universe is constant.

8-5d The Second Law of Thermodynamics

Even more ways have been found to express the **second law of thermodynamics** than the first (Table 8.1). Each statement reflects the observation that although the energy of the universe is constant, once energy is converted to entropy it is never again available for useful purposes.

In chemistry textbooks, the usual statement of the second law is

The total entropy of the universe is constantly increasing.

The universal truth of this law may be hard to accept at first. The formation of the stars

© iStockphoto.com/timsa

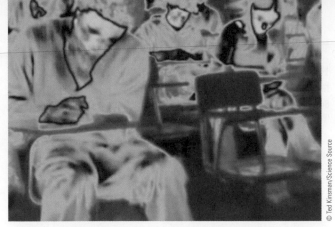

Visualizing body heat. Thermogram photos show heat in reds and yellow, illustrating that the body heats up when energy is expelled. This released energy leads to greater molecular motion in the air and an increase in entropy.

and planets, the formation of continents and oceans, the formation of crystalline mineral deposits, and the growth of plants and animals are all ordering processes. Life itself is a constant struggle against entropy. The essential connection is that every ordering process in one small corner of the universe creates disorder somewhere else in the universe. As we are eating, breathing, and producing new biomolecules, we are emitting disordered waste products and contributing disorder to the atmosphere by giving off heat.

In fact, *every* time energy is generated and used to do work, some of the energy is converted to heat. Consider the release of energy by burning coal, petroleum, or wood. The principal products of combustion, carbon dioxide and water, will not burn and release more energy. The energy from the reactants is dispersed as heat into the random molecular motion of the surroundings, where it is not available to do more work. In the burning process, matter and energy are conserved. However, the products and the energy converted to heat are much less useful than the

reactants and their stored energy. Observations of this type are the basis of one of the alternative statements of the second law:

**Energy is conserved in quantity
but not quality.**

The implications of the second law are wide ranging. Our economy is based on extracting raw materials from our surroundings, using energy to process them, and, in marketplace terms, "adding value" to the raw materials. The second law reminds us that no matter how carefully a process is designed, energy is lost at each manufacturing step.

8-6 Recycling: New Metal for Old

Many communities in the United States now recycle metals, paper, and plastic. There is a long way to go, however, before we can congratulate ourselves too heartily on reducing the quantity of material that is simply dumped somewhere. Between 1960 and 2007 the generation of municipal solid waste almost doubled, whereas the U.S. population increased by only one-third. While recycling rates have soared over the past 25 years, each of us is still producing more garbage than ever before (Figure 8.10). The best remedy remains to generate less waste rather than to expend time, energy, and money in collecting, processing, and recycling it.

Also, recycling alone is not the best answer. Consider what the laws of thermodynamics mean for recycled materials. Recycling counteracts the natural direction of increasing entropy and can be done only with the expenditure of energy in collecting, transporting, and remanufacturing the waste to produce newly useful materials. Remember the second law—some of the energy used at each step is lost forever to entropy.

A different and more direct approach to conserving resources is to increase the useful lifetime of our household materials and objects. Another is to diminish excess use of materials, even those that

are recyclable. The three Rs of waste prevention, in order of their importance, are shown in the following diagram:

1. Reduce *the amount of waste as much as possible.*

3. Recycle *materials as much as possible.*

2. Reuse *products as much as possible.*

To get the whole picture of the amount of waste we generate and its impact on the environment, it's important to keep in mind that municipal waste is only a small fraction of the total. Waste is generated at each stage of a product's life cycle, meaning, of course, that there are opportunities for waste reduction at each stage (Figure 8.11).

Metals are not incinerated but are transported either to waste dumps or to recycling plants. Many factors determine the extent to which a metal is recycled.

Lead is the most recycled metal for several reasons: It is too toxic to go to landfills; the major use is in automobile batteries, which have a predictable life span; and used batteries are collected at legally designated locations in most states. Iron and steel are second in percentage recycled, most of which is recycled within the industry rather than from consumer products. Iron and steel are used in vastly greater quantities than all other metals combined, resulting

FIGURE 8.10 Municipal solid waste generation and recycling rates from 1960 to 2007.

Both waste generation and recycling are trending upwards. (Source: http://www.epa.gov/waste/nonhaz/municipal/pubs/msw07-fs.pdf)

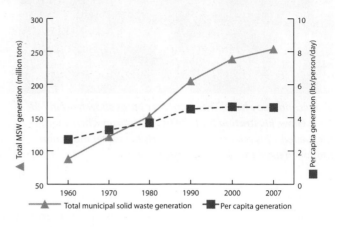

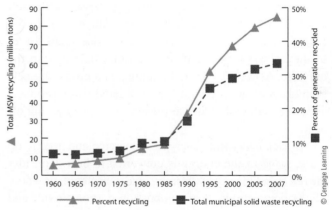

in a huge pool of scrap to be dealt with. Not recovering it would be a financial as well as an environmental burden. Also, all steel is potentially recyclable without separation of pure metals from the mixture.

FIGURE 8.11 Stages in the production, use, and disposal of manufactured products.

At each stage energy is used, there is a possibility of air or water pollution, and there is generation of waste. Therefore, there is also the possibility for conservation and recycling at each stage.

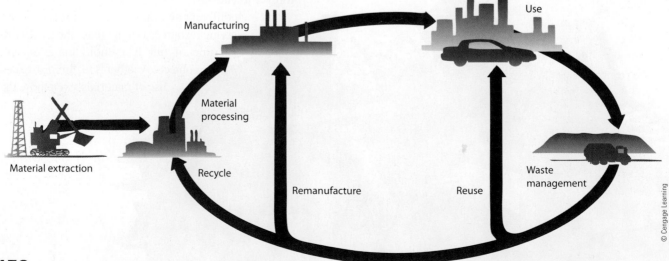

Your Resources

In the back of the textbook:

→ Review Card on Chemical Reactivity: Chemicals in Action

In OWL for CHEM2 at www.cengagebrain.com

→ Review Key Terms with Flash Cards (printable or digital)

→ Complete Interactive Practice Quizzes to prepare for tests

→ Submit Assigned Homework and Exercises

Applying Your Knowledge

1. Look at the following balanced chemical equation for burning ethanol:

$$CH_3CH_2OH(g) + 3\,O_2(g) \longrightarrow 2\,CO_2(g) + 3\,H_2O(g)$$

   (a) How many hydrogen atoms are on the product side of the equation? How many are on the reactant side of the equation?

   (b) How many oxygen atoms are on the product side of the equation? How many are on the reactant side of the equation?

   (c) What are the balancing *coefficients* in the reaction?

   (d) In your own words explain how this balanced equation obeys the law of conservation of matter.

2. Equations can be balanced only by adjusting the coefficients of the reactants and products, while the subscripts within the formulas of the reactants and products cannot be changed and still keep the original sense of the equation. Explain why this is so.

3. In the reaction $2\,H_2O_2 \longrightarrow 2\,H_2O + O_2$, two moles of a reactant produce three moles of products. How is this possible?

4. Balance the following chemical equations:
   (a) $Al + Cl_2 \longrightarrow AlCl_3$
   (b) $Mg + N_2 \longrightarrow Mg_3N_2$
   (c) $NO + O_2 \longrightarrow NO_2$
   (d) $SO_2 + O_2 \longrightarrow SO_3$
   (e) $H_2 + N_2 \longrightarrow NH_3$

5. Balance the following chemical equations:
   (a) $CH_3OH(\ell) + O_2(g) \longrightarrow CO_2(g) + H_2O(g)$
   (b) $C_3H_8(\ell) + O_2(g) \longrightarrow CO_2(g) + H_2O(g)$
   (c) $C_6H_6(\ell) + O_2(g) \longrightarrow CO_2(g) + H_2O(g)$
   (d) $C_3H_8O(\ell) + O_2(g) \longrightarrow CO_2(g) + H_2O(g)$

6. Balance the following equations:
   (a) $Ba(s) + H_2O(\ell) \longrightarrow Ba(OH)_2(aq) + H_2(g)$
   (b) $Fe(s) + H_2O(\ell) \longrightarrow Fe_3O_4(s) + H_2(g)$
   (c) $Na(s) + H_2O(\ell) \longrightarrow NaOH(aq) + H_2(g)$
   (d) $Li(s) + H_2O(\ell) \longrightarrow LiOH(aq) + H_2(g)$

7. Balance the following equations:
   (a) $HBr(aq) + KOH(aq) \longrightarrow KBr(aq) + H_2O(\ell)$
   (b) $H_2S(g) + NaOH(aq) \longrightarrow Na_2S(aq) + H_2O(\ell)$
   (c) $HNO_3(aq) + Ca(OH)_2(aq) \longrightarrow Ca(NO_3)_2(aq) + H_2O(\ell)$
   (d) $HCl(aq) + Al(OH)_3(s) \longrightarrow AlCl_3(aq) + H_2O(\ell)$

8. Balance the equations shown.
   (a) $BaO_2 + H_2SO_4 \longrightarrow BaSO_4 + H_2O_2$
   (b) $BaO_2 + H_3PO_4 \longrightarrow Ba_3(PO_4)_2 + H_2O_2$
   (c) $Na_2S + AlCl_3 + H_2O \longrightarrow Al(OH)_3 + NaCl + H_2S$
   (d) $CaCl_2 + Na_2CO_3 \longrightarrow CaCO_3 + NaCl$
   (e) $AlN + H_2O \longrightarrow Al_2O_3 + NH_3$

9. Ammonia for fertilizers is made from the elements according to this equation:

$$N_2(g) + 3\,H_2(g) \longrightarrow 2\,NH_3(g)$$

   (a) How many molecules of N_2 are needed to react with 30 molecules of H_2?

   (b) How many molecules of N_2 are needed to react with 15 H_2 molecules?

   (c) If 100 N_2 molecules and 300 H_2 molecules are allowed to react, how many NH_3 molecules will be formed?

10. Hydrogen peroxide is formed from the elements according to this equation:

$$H_2(g) + O_2(g) \longrightarrow H_2O_2(\ell)$$

   (a) How many molecules of oxygen are needed to react with 1 molecule of $H_2(g)$?

   (b) How many hydrogen peroxide molecules are expected if 10 oxygen molecules react with 10 hydrogen molecules?

   (c) How many hydrogen peroxide molecules are expected if 100 oxygen molecules are mixed and reacted with 10 hydrogen molecules?

11. For the balanced equation
$$C_6H_{12}O_6(s) + 6\ O_2(g) \longrightarrow 6\ CO_2(g) + 6\ H_2O(g),$$

   (a) how many moles of water will be produced if 0.5 mol of $C_6H_{12}O_6$ reacts?

   (b) how many moles of water will be produced if 180 g of $C_6H_{12}O_6$ react?

   (c) if 90 g of water were produced, how many moles of $C_6H_{12}O_6$ reacted?

   (d) if 72 g of water were produced, how many g of $C_6H_{12}O$ reacted?

12. For the balanced equation $4\ Fe(s) + 3\ O_2(g) \longrightarrow 2\ Fe_2O_3(s),$

   (a) how many moles of Fe_2O_3 will be produced if 4 mol of Fe react?

   (b) how many moles of Fe_2O_3 will be produced if 12.7 mol of Fe react?

   (c) how many moles of Fe reacted if 4 mol of Fe_2O_3 were produced?

   (d) how many moles of Fe reacted if 17 g of Fe_2O_3 were produced?

   (e) How many moles of O_2 are required to react with 13 g of Fe?

13. For the balanced equation $SiO_2(s) + 2\ C(s) \longrightarrow Si(s) + 2\ CO(g),$

   (a) how many moles of Si will be produced when 4 mol of SiO_2 react?

   (b) how many grams of CO will be produced when 4 mol of SiO_2 react?

   (c) how many moles of C reacted if 1 mol of Si was formed?

   (d) how many grams of Si can be produced from the reaction of 150 g of SiO_2?

14. Identify the atom used as the basis of the atomic weight scale.

15. What do 35.5 g of chlorine (Cl) and 12.011 g of carbon (C) have in common?

16. Calculate the number of moles for the given amount of each substance.

   (a) 14.3 g of ethane, C_2H_6

   (b) 300 mg of sodium cyanide, NaCN

   (c) 0.2 g of ibuprofen, $C_{13}H_{18}O_2$

   (d) 74 g of gold

   (e) 1 L (1.00 kg) of water

17. What do 103.5 g of lead (Pb) and 6.006 g of carbon (C) have in common?

18. Look up the unit called the *gross*. How many pairs of jeans are in 4 gross of jeans? In what way is the gross unit like Avogadro's number?

19. What is the numerical value of Avogadro's number?

20. What is meant by the term *molar mass*?

21. Calculate the molar mass for each of the following substances.

   (a) Aspirin, $C_9H_8O_4$

   (b) Water, H_2O

   (c) Octane, C_8H_{18}

   (d) Ammonium phosphate, $(NH_4)_3PO_4$

   (e) Lovastatin (Lipitor), $C_{24}H_{36}O_5$

   (f) Escitalopram (Lexapro), $C_{20}H_{21}FN_2O$

22. What is the definition for the mole?

23. Interpret the following equation for the complete combustion of octane in terms of moles of reactants and products, and their molar masses:

$$2\ C_8H_{18}(\ell) + 25\ O_2(g) \longrightarrow 16\ CO_2(g) + 18\ H_2O(\ell)$$

24. Which is faster, a chemical reaction with a high activation energy or a chemical reaction with a low activation energy?

25. Give at least one reason why some chemical reactions are fast while others are slow. Use your own analogy to explain the reason.

26. How is activation energy related to the change in energy that accompanies some chemical reactions?

27. What influence does temperature usually have on the rate of a chemical reaction? Explain why this is so.

28. What effect does freezing food have on reaction rates? Why is freezing used for preservation of food, tissue samples, and biological samples?

29. In recent years it has become possible to test the DNA of several prehistoric specimens, because it has been preserved. One notable example is the "Ice Man," called Ötzi. Many of these specimens have been discovered in cold climates. Why do you think this is the case?

30. Explain the term *activation energy* as it applies to chemical reactions.

31. How does the magnitude of the activation energy influence the rate of a chemical reaction?

32. What is the effect of a catalyst on the activation energy of a chemical reaction?

33. Explain why hydrogen and oxygen can remain mixed at room temperature without any noticeable reaction, yet they will combine explosively to form water if the mixture is ignited with a tiny spark.

34. Describe how each of the following changes will affect the rate of a reaction:

   (a) Increase in temperature

   (b) Increase in concentration

   (c) Introduction of a catalyst

35. Paper burns fairly rapidly in air. How would you expect paper to burn in pure oxygen? Explain.

36. Use your knowledge of the control of reaction rates to explain why a carbon dioxide fire extinguisher can be used to put out some fires.

37. One trick used by hikers to preserve raw eggs during a long trip is to coat the eggs with vegetable oil. Using your knowledge of the control of reaction rates, how might this method of preservation work?

38. What is the general name for biological catalysts?

39. What is meant by the term *reversible* when describing chemical reactions?

40. What is meant by the term *dynamic equilibrium*?

41. If you heat limestone in a closed vessel, explain which way you would expect the equilibrium to shift if
    (a) you added more CO_2?
    (b) you allowed some of the CO_2 to escape from the vessel
    $$CaCO_3(s) \rightleftharpoons CaO(s) + CO_2(g)?$$

42. In the equilibrium $CO_2 + H_2O \rightleftharpoons HCO_3^- + H^+$, what will be the effect of:
    (a) Increasing the amount of CO_2?
    (b) Adding HCO_3^-?

43. Give a statement of Le Chatelier's principle.

44. What is meant when it is said that the reaction shifts in favor of the products of the reaction?

45. What is meant when it is said that the reaction shifts in favor of the reactants?

46. The reaction between HCl and NaOH is described as going to completion. Explain what this means in terms of how much of the reactants remain unreacted.

47. The reaction between N_2 and H_2 to produce ammonia (NH_3) is described as an equilibrium reaction
    $$N_2(g) + 3 H_2(g) \rightleftharpoons 2 NH_3(g)$$
    After the reaction reaches equilibrium, what is present in the reaction mixture?

48. Define the following terms:
    (a) Potential energy
    (b) Exothermic
    (c) Endothermic
    (d) Entropy
    (e) Favorable reaction

49. State the first law of thermodynamics. What does it mean?

50. State the second law of thermodynamics. What does it mean?

51. Does entropy increase or decrease in the following reactions? Explain your answer.
    (a) Nitrogen reacting with oxygen to form nitrogen dioxide
    $$N_2(g) + 2 O_2(g) \longrightarrow 2 NO_2(g)$$
    (b) Acetylene reacting with oxygen to form carbon dioxide and water
    $$2 HC \equiv CH(g) + 5 O_2(g) \longrightarrow$$
    $$4 CO_2(g) + 2 H_2O(g)$$

52. Calculate the molar mass for each of the following:
    (a) H_2O
    (b) I_2
    (c) KOH
    (d) NH_3

53. Calculate the molar mass for each of the following:
    (a) $C_6H_{12}O_6$
    (b) H_2SO_4
    (c) Na_2HPO_4
    (d) $Ca(NO_3)_2$

54. Which of the following is the largest mass?
    (a) 0.5 mol CO_2
    (b) 2 mol Li
    (c) 12 mol H_2

55. Ethanol in oxygenated gasoline burns as a fuel according to this equation:
    $$CH_3CH_2OH(\ell) + 3 O_2(g) \longrightarrow 2 CO_2(g) + 3 H_2O(g)$$
    (a) What is the mole ratio of ethanol to oxygen?
    (b) How many grams of oxygen are needed to burn 1 mol of ethanol?
    (c) How many grams of oxygen are needed to completely burn 500 g of ethanol?

56. Propane in gas-fired barbecues burns ideally according to this equation:
    $$CH_3CH_2CH_3(g) + 5 O_2(g) \longrightarrow 3 CO_2(g) + 4 H_2O(g)$$
    (a) How many moles of oxygen are needed to burn 100 mol of propane?
    (b) How many grams of oxygen are needed to burn 4400 g of propane?
    (c) How many pounds of oxygen are needed to burn 88 lb of propane?

57. This equation shows how water is formed from the elements:
    $$2 H_2(g) + O_2(g) \longrightarrow 2 H_2O(\ell)$$
    (a) How many grams of H_2 are needed to react with every 32 g of O_2?
    (b) How many pounds of H_2 are needed to react with 32 lb of O_2?
    (c) How many tons of H_2 are needed to react with 32 tons of O_2?

58. Based on the balanced equation
    $$Cu(s) + Cl_2(g) \longrightarrow CuCl_2(s)$$
    (a) How many moles of copper will be required to react with 0.5 mol of chlorine?
    (b) How many moles of $CuCl_2$ can be formed from 1.5 mol of copper?

59. Balance the equation for the fermentation of glucose and then answer the following questions.
    $$C_6H_{12}O_6(aq) \longrightarrow CH_3CH_2OH(aq) + CO_2(g)$$
    (a) In this reaction, how many moles of ethanol (CH_3CH_2OH) can be prepared from 6 mol of glucose?
    (b) If, during the fermentation process, 10.5 mol of carbon dioxide is found to be produced, how many moles of ethanol will have been produced?

Acid–Base
Reactions

Acid! Most people immediately think of danger when they hear this word. Base! In the United States, most people think baseball when they hear this word. To chemists and a great many other scientists, these two words call to mind chemical compounds that are quite common in our natural surroundings. Chemical reactions between acids and bases dramatically influence our personal health, the health of our environment and every living organism, and the health of our economy.

Some acids and bases can be quite dangerous in the hands of those inexperienced with working with such chemicals. Concentrated sulfuric acid and concentrated sodium hydroxide (a chemical base often referred to as *lye*) can cause serious chemical burns. However, batteries containing sulfuric acid and oven cleaners containing lye, both in potentially dangerous concentrations, can be found in every garage and kitchen.

Carbonated soft drinks owe their fizz to the presence of carbonic acid. We think nothing of drinking this acid. Even so, one of the authors of this book was confronted with the statement "our lawyers won't let us have acids in a shopping mall due to liability problems" when he proposed to do some chemical demonstrations involving carbonic acid as part of an observance of National Chemistry Week activities. There's that association with danger again.

Face and hand soaps contain bases, but no one who has read the warning label on a common oven cleaner would even consider getting *this* base near the eyes. If, in fact, you did get a base in your eye, you might want to be prepared to wash your eyes with boric acid. Confused?

In this chapter, you will learn that the danger associated with acids and bases is a function of their concentration and identity. Most acids and bases are not inherently dangerous. They are all around us. The acid–base balance in our blood is critical to good health, and many biological reactions depend on acids and bases in the body. The acid content of our rainfall (and other forms of precipitation) plays a critical role in determining the health of our rivers and forests and our ability to raise the foodstuffs upon which we depend. Acids and bases are key components of many manufacturing processes. In fact, sulfuric acid is required by so many industrial processes that the amount of it sold each year is taken as a measure of a nation's economy.

SECTIONS

9-1 Acids and Bases: Chemical Opposites

Everyone should know a few practical things about acids and bases, which we usually encounter as solutions in water. They can be harmful. The harm can range from the stinging sensation when you accidentally squirt lemon juice (citric acid) into your eye to the severe and persistent burns or blindness that result if you spill battery acid (sulfuric acid) on your skin or get it in your eyes and do not flush it immediately with lots of water. The effects of drain cleaner and some oven cleaners can be equally devastating because they contain lye (sodium hydroxide, a strong base). Strongly acidic and basic solutions can also "eat" holes in clothing, very quickly. The first lesson, then, is that strongly acidic or basic substances must be handled with care.

The potential for harm from strong acids and bases, however, does not mean all acids and bases are to be avoided. The dilute solutions of acids and bases that are in everyday use would be sorely missed. Orange juice, vinegar, soda pop, household ammonia, milk, and most of our soaps and detergents fall in this category of dilute solutions.

It turns out that all acidic water solutions, whatever their sources or applications might be, have some chemical properties in common. The same is true of all basic water solutions. Acids and bases are closely related classes of chemical compounds that are often highly reactive, both with other substances and with each other.

Both acids and bases are used in cleaning supplies.

Acid

Base

What about these?

Acid–base indicator:
A substance that changes color with changes in acidity or basicity of a solution

Hydronium ion:
A hydrated proton, H_3O^+

$$H - \overset{+}{\underset{|}{O}} - H$$
$$|$$
$$H$$

Lewis structure

The word *acid* comes from the Latin *acidus*, meaning sour or tart, because in water solutions, acids have a sour or tart taste. Lemons, grapefruit, and limes taste sour because they contain citric acid and ascorbic acid (vitamin C). Vinegar is sour because it contains acetic acid. Another common property of acids is their ability to change the color of compounds known as **acid–base indicators** (Figure 9.1).

The properties that acids have in common, listed in Table 9.1, result from the ability of acids to release a hydrogen ion (H^+) in a water solution (an *aqueous solution*, symbolized by *aq*). The ionization of an acid is therefore often represented by an equation like the following for hydrochloric acid:

$$HCl(aq) \longrightarrow H^+(aq) + Cl^-(aq)$$

Hydrogen ions, because of their small size and resulting concentration of positive charge in a small area, however, do not exist as such in water solution. Instead, they bond to water molecules to give **hydronium ions** (H_3O^+).

$$H^+(aq) + H_2O(\ell) \longrightarrow H_3O^+(aq)$$

Therefore, it is more nearly correct and, as you will see, more useful to write an equation for the ionization of hydrochloric acid (and other acids) with the hydronium ion as a product. This is done by adding the first of the preceding two chemical equations to the second, then canceling like terms:

$$HCl(aq) \longrightarrow \cancel{H^+(aq)} + Cl^-(aq)$$

$$\text{plus } \cancel{H^+(aq)} + H_2O(\ell) \longrightarrow H_3O^+(aq)$$

$$\overline{HCl(aq) + H_2O(\ell) \longrightarrow H_3O^+(aq) + Cl^-(aq)}$$

Hydrochloric acid, HCl(*aq*), which is formed when hydrogen chloride gas [HCl(*g*)] dissolves in water, contains hydronium ions and chloride ions (the anions from the acid).

Bases also have several properties in common (Table 9.1). Water solutions of bases are slippery or soapy to the touch. The classic properties by which bases are recognized are caused by the presence in water solution of hydroxide ions (OH^-). The hydroxides of sodium and potassium (NaOH and KOH) and of calcium and magnesium [Ca(OH)$_2$ and Mg(OH)$_2$] are among the most common bases used in industry and chemical laboratories. These are ionic compounds that produce hydroxide ions when they dissolve. For example, sodium hydroxide dissolves as represented by the following equation:

$$NaOH(s) \xrightarrow{\text{Water}} Na^+(aq) + OH^-(aq)$$

FIGURE 9.1 Acidity of some household products.

A few drops of an acid–base indicator have been added to aqueous solutions of each of the products. The indicator color shows that the club soda is more acidic than the vinegar and the cleaning solution is slightly basic.

9-1a Neutralization Reactions

Though we've indicated that bases yield hydroxide ions when dissolved in water, and acids yield hydronium ions, it might surprise you to know that both of these ions are present together in *all* aqueous solutions, no matter whether an acid or base has been used to prepare the solutions. It is the *relative* amounts of these ions that determines whether a solution is acidic or basic.

A solution that contains more H_3O^+ ions than OH^- ions and has the properties common to acids is referred to as an **acidic solution**. A solution that contains a higher amount of OH^- ions than hydronium ions and has the properties common to bases is described as a **basic**, or **alkaline, solution**.

A water solution cannot be acidic and basic simultaneously. An acidic solution contains a higher concentration of H_3O^+ ions than OH^- ions, and a basic solution contains a higher concentration of OH^- ions than H_3O^+ ions. When a solution of an acid is mixed with a solution of a base, the hydronium ions from the acidic solution react with the hydroxide ions from the basic solution to produce water:

$$H_3O^+(aq) + OH^-(aq) \longrightarrow 2\,H_2O(\ell)$$

TABLE 9.1 Properties of Acids and Bases

Acids	Bases
Sour taste	Bitter taste
Provide H^+ ions	Provide OH^- ions
React with active metals to give hydrogen	Slippery feeling
Change colors of indicators (e.g., litmus turns from blue to red)	Change colors of indicators (e.g., litmus turns from red to blue)
Produce CO_2 when added to limestone ($CaCO_3$)	Neutralize acids
Neutralize bases	

Some Acidic Substances	Some Basic Substances
Vinegar	Household ammonia
Tomatoes	Baking soda
Citrus fruits	Soap
Carbonated beverages	Detergents
Black coffee	Milk of magnesia
Gastric fluid	Oven cleaners
Vitamin C	Lye
Aspirin	Drain cleaners
Ant venom	Antacids
Battery acid	

© Cengage Learning

With bases such as potassium hydroxide and acids such as hydrochloric acid, the second product is a salt, which in this case is potassium chloride (KCl). A **salt** is a compound composed of the positive metal ion from a base and the negative ion from an acid.

$$\underset{\text{Base}}{KOH(aq)} + \underset{\text{Acid}}{HCl(aq)} \longrightarrow$$
$$\underset{\text{Salt}}{KCl(aq)} + \underset{\text{Water}}{H_2O(\ell)}$$

In this reaction, KOH, HCl, and KCl are all water-soluble compounds that yield ions in solution, and water is a molecular compound. A more detailed representation of this process is

$$K^+(aq) + {}^-OH(aq) + H_3O^+(aq) + Cl^-(aq) \longrightarrow$$
$$K^+(aq) + Cl^-(aq) + 2\,H_2O(\ell)$$

When *exactly equivalent amounts* of acid and base react so that all of the H_3O^+ and OH^- ions produced by dissolved acids or bases combine to produce water, the result is a **neutralization reaction**. The reaction of sodium hydroxide with sulfuric acid, like the reaction of potassium hydroxide with hydrochloric acid, is a neutralization reaction.

$$\underset{\text{Base}}{2\,NaOH(aq)} + \underset{\text{Acid}}{H_2SO_4(aq)} \longrightarrow$$
$$\underset{\text{Salt}}{Na_2SO_4(aq)} + \underset{\text{Water}}{2\,H_2O(\ell)}$$

In neutralization reactions, acidic and basic properties are eliminated and a neutral solution is formed.

Acidic solution: A solution that contains a higher concentration of H_3O^+ ions than OH^- ions

Basic solution (alkaline solution): A solution that contains a higher concentration of OH^- ions than H_3O^+ ions

Salt: Ionic compound composed of the cation from a base and the anion from an acid

Neutralization reaction: Reaction of equivalent quantities of an acid and a base

TRY IT 9.1

The previous equation shows that two sodium hydroxide ions react with one molecule of sulfuric acid. Write an equation that shows that the reaction of each molecule of sulfuric acid (H_2SO_4) with water produces two hydronium ions and one sulfate anion (SO_4^{2-}). Then, draw an equation that shows that these two hydronium ions react with two hydroxide ions to produce water, as shown in the first equation in this section on neutralization reactions.

9-1b Acid–Base Definitions

Acid: Molecule or ion able to donate a hydrogen ion to a base

Base: Molecule or ion able to accept a hydrogen ion from an acid

Acid–base reaction: Reaction in which a hydrogen ion is exchanged between an acid and a base

Strong acid: Acid that is 100% ionized in aqueous solution

The reaction between an acid and a base can be viewed as the transfer of a hydrogen ion from an acid to a hydroxide ion formed from a base. In the chemical view, the definitions of acids and bases take into account the role of the hydrogen ion in acid–base reactions. An **acid** is a molecule or ion that donates a hydrogen ion to a base. Because H^+ is a hydrogen atom without an electron, this ion is often called a proton. A **base** is a molecule or ion able to accept a hydrogen ion from an acid. The result of these definitions is that every reaction in which a hydrogen ion is exchanged between reactants is an **acid–base reaction**. A neutralization reaction is one example of such a reaction. The following structural formulas show how a hydrogen ion is exchanged between water (written here as $H-O-H$) and hydrogen chloride gas in the formation of a hydrochloric acid solution. This is an acid–base reaction in which water is the base and HCl is the acid.

$$H-\ddot{O}-H + H-\ddot{\underset{..}{Cl}}: \longrightarrow H-\overset{+}{\underset{|}{\ddot{O}}}-H + :\ddot{\underset{..}{Cl}}:^-$$
$$\qquad\qquad\qquad\qquad\qquad\qquad H$$

Base (H⁺ acceptor)   Acid (H⁺ donor)

In the reaction between sodium hydroxide and the hydronium ion in a hydrochloric acid solution, OH^- is the base and H_3O^+ is the acid.

$$:\ddot{O}-H + H-\overset{+}{\underset{|}{\ddot{O}}}-H \longrightarrow H-\ddot{O}-H + H-\ddot{O}-H$$
$$\qquad\qquad\qquad H$$

Base (H⁺ acceptor)   Acid (H⁺ donor)

The picture of acids and bases as hydrogen ion donors and acceptors explains not only the properties of the acids and bases we have described thus far but also the basicity of ammonia (NH_3) and ions such as the carbonate ion (CO_3^{2-}). The neutral ammonia molecule can accept a hydrogen ion from a water molecule to produce a solution that is basic because it contains hydroxide ions. In this reaction, water donates the hydrogen ion and therefore is the acid.

$$NH_3(aq) + H_2O(\ell) \rightleftharpoons NH_4^+(aq) + OH^-(aq)$$

Base (H⁺ acceptor)   Acid (H⁺ donor)   Ammonium ion

Although they contain no hydroxide ions, carbonate salts such as sodium carbonate (Na_2CO_3, washing soda) and potassium carbonate (K_2CO_3) also dissolve in water to give basic solutions. How does a carbonate salt produce hydroxide ions in water? Like ammonia, the carbonate ion accepts a hydrogen ion from a water molecule, which leaves behind a hydroxide ion. Note the omission of the Na^+ ions. They play no role in the acid–base reaction.

$$CO_3^{2-}(aq) + H_2O(\ell) \rightleftharpoons HCO_3^-(aq) + OH^-(aq)$$

Base (H⁺ acceptor)   Acid (H⁺ donor)   Bicarbonate ion

The equations in this section illustrate a very important point. In the reaction with hydrogen chloride, water reacts as a base. In the reactions with ammonia molecules and the carbonate ion, water reacts as an acid. Because all water molecules are the same, *water must be able to react as an acid or a base depending on whether it reacts with a base or an acid.* This property of water plays an important role in our water-based world. Whenever a substance that can accept or donate a hydrogen ion dissolves in water, the result is a solution with basic or acidic properties. In removing stubborn dirt (Section 9-5), this can work to our advantage. In the production of acid rain (Section 11-2), this causes problems.

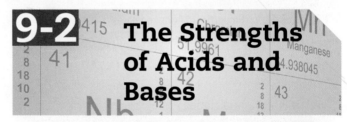

9-2 The Strengths of Acids and Bases

What is the difference between a strong acid such as hydrochloric acid, sold in hardware stores as *muriatic acid* and used to clean brick and concrete, and a weak acid such as the acetic acid in vinegar? *The strength of an acid or base is determined by the extent of ionization in aqueous solution.* The greater the ionization, the stronger the acid or base.

Many common acids such as sulfuric acid, hydrochloric acid, and nitric acid react completely with water to give hydronium ions and anions. Therefore, like ionic compounds (Section 5-2), they are electrolytes; and because they are entirely converted to ions in solution, they are *strong electrolytes*. The solutions of these **strong acids** have high concentrations of hydronium ions and are very acidic (Table 9.2). The equation for the reaction of a strong acid such as nitric acid with water is written with a single arrow, which indicates that the reaction goes to completion.

$$HNO_3(aq) + H_2O(\ell) \longrightarrow H_3O^+(aq) + NO_3^-(aq)$$

TABLE 9.2 Common Acids

Name	Formula	Strength	Use or Occurrence
Sulfuric acid*	H_2SO_4	Strong	Cleaning steel; car batteries; making plastics, dyes, fertilizers
Hydrochloric acid	HCl	Strong	Cleaning metals and brick mortar
Nitric acid	HNO_3	Strong	Making fertilizers, explosives, plastics
Phosphoric acid	H_3PO_4	Moderate	Making fertilizers, detergents, food additives
Acetic acid	CH_3COOH	Weak	Vinegar
Propanoic acid	CH_3CH_2COOH	Weak	Swiss cheese
Citric acid	$HOC(COOH)(CH_2COOH)_2$	Weak	Fruit
Carbonic acid	H_2CO_3	Weak	Carbonated beverages
Boric acid	H_3BO_3	Weak	Eye drops, mild antiseptic

© Cengage Learning

Weak acid: Acid that is only partially ionized in aqueous solution and establishes equilibrium with nonionized acid

Strong base: Base that is 100% ionized in aqueous solution

Weak base: Base that is only partially ionized in aqueous solution and establishes equilibrium with nonionized base

* Sulfuric acid is a polyprotic acid, meaning it has more than one acidic hydrogen. It first ionizes completely to give HSO_4^- (hydrogen sulfate ion, also known as bisulfate ion), which is a weak acid and partially ionizes to give SO_4^{2-} (sulfate ion).

Many acids establish an equilibrium with water at a point where not all the acid molecules have been converted to ions. These are the **weak acids**—they are only slightly ionized in aqueous solution because an equilibrium is established (Section 8-4). Acetic acid is a typical weak acid:

$$CH_3COOH(aq) + H_2O(\ell) \rightleftharpoons$$
Acetic acid

$$H_3O^+(aq) + CH_3COO^-(aq)$$
Acetate ion

The CH_3COOH molecules undergo ionization while H_3O^+ and CH_3COO^- ions simultaneously recombine to give CH_3COOH molecules and water. Since only a few percent of the acetic acid molecules are ionized at any given time,

aqueous solutions of acetic acid contain mostly acetic acid molecules with a few hydronium ions and acetate anions and are only weakly acidic.

As previously explained, many metal hydroxides are ionic compounds and, therefore, **strong bases** because when dissolved in water, they are completely converted to ions (to the extent that they are soluble). Ammonia is a common **weak base**, a base that establishes an equilibrium with water to produce a solution with relatively few ammonium and hydroxide ions.

$$NH_3(g) + H_2O(\ell) \rightleftharpoons NH_4^+(aq) + OH^-(aq)$$

The anions of weak acids are also bases (Table 9.3). Look at the reverse of the reaction of acetic acid with water.

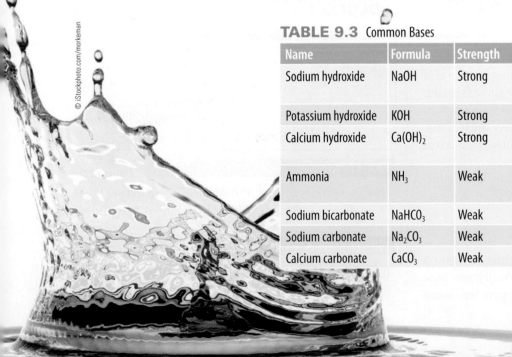

© iStockphoto.com/morkeman

TABLE 9.3 Common Bases

Name	Formula	Strength	Use
Sodium hydroxide	NaOH	Strong	Drain cleaner; producing aluminum, rayon, soaps, detergents
Potassium hydroxide	KOH	Strong	Producing soaps, detergents, fertilizers
Calcium hydroxide	$Ca(OH)_2$	Strong	Producing bleaching powder, paper and pulp; softening water
Ammonia	NH_3	Weak	Producing fertilizer, explosives, plastics, insecticides, detergents
Sodium bicarbonate	$NaHCO_3$	Weak	Antacid
Sodium carbonate	Na_2CO_3	Weak	Detergents, glassmaking
Calcium carbonate	$CaCO_3$	Weak	Chalk

© Cengage Learning

A fire ant from the Amazon Basin. The venom of these and other ants contains the simplest organic acid, formic acid (HCOOH). In the Amazon, fire ant colonies can be so large that streams become polluted by formic acid.

Concentration of a solution: The quantity of a solute dissolved in a specific quantity of a solvent or solution

Molarity: Number of moles of solute per liter of solution

The acetate ion (CH_3COO^-) accepts a hydrogen ion from the acidic hydronium ion and is therefore reacting as a base. You will see that this property of anions plays an important role in solutions, such as blood, in which the acid concentration must remain constant (Section 9-4). In both blood and the many household solutions that use sodium bicarbonate, the basic nature of the bicarbonate ion (HCO_3^-) is put to use in controlling acid concentration.

$$HCO_3^-(aq) + H_2O(\ell) \rightleftharpoons H_2CO_3(aq) + OH^-(aq)$$

Because it is not stable, the carbonic acid (H_2CO_3) formed in this reaction breaks down to form carbon dioxide gas.

$$H_2CO_3(aq) \rightleftharpoons CO_2(g) + H_2O(\ell)$$

9-3 Molarity and the pH Scale

Water, as you have seen, is capable of acting as either a hydrogen ion donor or a hydrogen ion acceptor—that is, as an acid or a base, depending on the properties of the substance with which it reacts. Water can also act as an acid or a base toward itself, although the reaction occurs to only a very small extent. About 1 of every 550,000,000 water molecules is ionized at any given time.

$$H_2O + H_2O \rightleftharpoons H_3O^+ + OH^-$$

Pure water contains equal numbers of hydronium ions and hydroxide ions.

The self-ionization of water provides the basis for a convenient method for expressing numerically just how acidic or basic any water solution is. You may have seen

this quantity in use for a consumer product, for example, on the label of a "pH-balanced" shampoo. (The term refers to a controlled acidity of the solution; thus it is less likely to damage hair.) To understand what pH is requires understanding how to express the **concentration of a solution**, which is the quantity of a solute dissolved in a specific quantity of solvent or solution. We might give the concentration of a solution of sodium hydroxide as 4.0 g/L; that is, 4.0 g of NaOH per liter of solution. In chemistry, however, because of its relation to quantities of chemicals in reactions (Section 8-2), concentration based on the mole is preferred.

9-3a Molarity

Concentrations of solutions in chemistry are usually expressed as **molarity**, which is the number of moles of solute per liter of solution. For example, a 1 molar (1 M) solution contains 1 mol of solute per liter of solution. If the solute is sodium hydroxide, a 1.00 M solution contains 1.00 mol, or 40.0 g, of NaOH per liter of solution. To find the molarity of a solution requires knowing the mass of dissolved solute, the volume of the solution, and the molar mass of the solute. For example, 4.0 g of NaOH is 0.10 mol (4.0 g/40.0 g/mol). Thus, a solution with 4.0 g of NaOH dissolved in 1.0 L of solution is a 0.10 M NaOH solution.

Molarity (M) = moles solute/liters of solution

EXAMPLE 9.1 Molarity

 What is the molarity of a solution that contains 10.0 g of HCl (molar mass, 36.5 g/mol) per liter of solution?

SOLUTION

First, we find the number of moles of HCl equivalent to 10.0 g of HCl.

$$10.0 \text{ g HCl} \times \frac{1 \text{ mol HCl}}{36.5 \text{ g HCl}} = 0.274 \text{ mol HCl}$$

With 0.274 mol dissolved per liter of solution, the molarity is 0.274 M in HCl.

TRY IT 9.2

How many grams of NaOH have been dissolved per liter of solution to make a 4.0 M NaOH solution?

TRY IT 9.3

If 73.0 g of HCl is dissolved in 500 mL of water, what is the molarity of this solution?

9-3b The pH Scale

In pure water (and all neutral solutions), the concentrations of hydronium ions and hydroxide ions are the same. Expressed as molarity, their concentrations are always 1.0×10^{-7} M at 25°C.

The product of the molarity of the hydronium ions and hydroxide ions in pure water is $(1.0 \times 10^{-7})(1.0 \times 10^{-7}) = 1.0 \times 10^{-14}$. *The value 1.0×10^{-14} is important to an understanding of all aqueous solutions because it is a constant that is always the product of the molar concentrations of H_3O^+ and OH^- in the solution.* Using square brackets, as is customary to represent the molarity of a substance, this relationship is written as

$$[H_3O^+][OH^-] = 1.0 \times 10^{-14}$$

If acid is added to pure water, the concentration of H_3O^+ will be greater than 1.0×10^{-7}, and the concentration of OH^- will be less than 1.0×10^{-7}. However, the product of the two must equal 1.0×10^{-14}. This relationship is the basis for calculating the concentration of one of the two ions, hydronium or hydroxide, when the other one is known.

Consider 0.10 M NaOH. Because sodium hydroxide is a strong base that is completely ionized, the molar concentration of hydroxide ions in this solution is also 0.10 M. To compensate, the concentration of hydronium ions must be significantly less than the pure water value of 1.0×10^{-7} M. Using the preceding equation, the concentration of hydronium ions in a 0.10 M NaOH solution is found to be 1.0×10^{-13} M. In the following calculation, the value of 0.10 M is expressed in scientific notation as 1.0×10^{-1} M:

$$[OH^-][H_3O^+] = 1.0 \times 10^{-14}$$

$$(1.0 \times 10^{-1})[H_3O^+] = 1.0 \times 10^{-14}$$

$$[H_3O^+] = \frac{1.0 \times 10^{-14}}{1.0 \times 10^{-1}} = 1.0 \times 10^{-13}\,M$$

An equivalent calculation for a 0.10 M HCl solution, which contains 0.10 M H_3O^+, would show that its concentration of OH^- ions is 1.0×10^{-13} M.

The difference in H_3O^+ concentration between 0.1 M HCl and 0.1 M NaOH is 1 trillion times because each change in the exponent is a power of ten, and the difference between 10^{-1} [H_3O^+] and 10^{-13} [H_3O^+] is 10^{12}, or 1 trillion. The Danish biochemist S. P. L. Sørensen proposed in 1909 that these exponents be used as a measure of acidity. He devised a scale that would be useful in his work of testing the acidity of Danish beer. Sørensen's scale came to be known as the pH scale, from the French *pouvoir hydrogene*, which means hydrogen power. **pH** is defined as *the negative logarithm of the hydronium ion concentration.*

pH: Negative logarithm of the hydronium ion concentration of a solution

$$pH = -\log [H_3O^+]$$

The pH of a solution is pretty easy to calculate if the hydronium ion concentration is given in a form such as 1×10^{-4} M or 1×10^{-11} M. If the hydronium ion concentration is written as 1×10 to some power, take the exponent of 10 and drop the negative sign to get the pH. Thus, we would have pH 4 and pH 11, respectively, for the solutions mentioned in the first sentence of this paragraph. This method depends on the fact that the logarithm of 1 is 0.

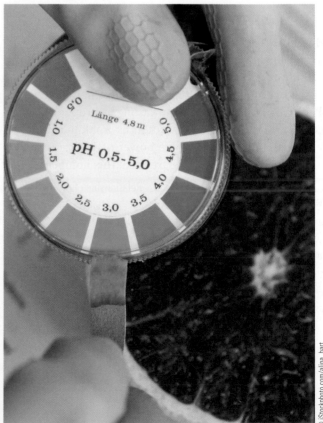

The pH of 3.0 shows that the grapefruit's juice is acidic.

Neutral solution: A solution in which the pH is 7

Since pH = −log [H$_3$O$^+$], we are really asked to do the following math:

pH = −log (1 × 10^{-4}), which equates to
pH = −[log (1) + log(10^{-4})]

Since the log of 1 is 0 (since 10^0 = 1) and the log of 10^{-4} is −4, we see that

pH = −[0 + (−4)], or
pH = 4

Thus, it is easy to determine pH if the concentration is written as 1 times some power of 10. This will not always be the case, however. Let's see how we could estimate the pH of a solution whose hydronium ion concentration is, for example, 6 × 10^{-8} M.

We need to do the following calculation:

pH = −log (6 × 10^{-8}), which equates to
pH = −[log(6) + log(10^{-8})]

Now we have a minor problem because it is unlikely you've memorized the logarithm of the number 6. We could key the number 6 into a calculator and then press the log button to find out that the log of 6 is 0.778. Once we know that, we can determine the following:

pH = −[0.778 +(−8)], or
pH = 7.22

In the absence of a calculator, we may use what we call the *bracketing method* to arrive at a rough estimate of the pH in just a few seconds. Here's how that works. The hydronium ion concentration, in proper scientific notation, will always be written as a number between 1 and 10 multiplied by some power of 10 (for example, 6 × 10^{-8}). It will always be true that this number may be

bracketed as shown here between 1 × 10^{-8} and 10 × 10^{-8}. Thus, if we know that the log of 1 is 0 and that the log of 10 is 1 (since 10^1 = 10), we can quickly follow the logic below.

If [H$_3$O$^+$] = 10 × 10^{-8}, then the pH = −[1 +(−8)], so pH = 7;

and

if [H$_3$O$^+$] = 1 × 10^{-8}, then the pH = −[0 +(−8)], so pH = 8.

Since the hydronium ion concentration in this example was between these two bracketing concentrations, the pH must be between 7 and 8. With a little practice, you can do this bracketing in a matter of a few seconds to arrive at an estimate of the pH. If you want, or need, a more accurate number, you will have to use a calculator.

The larger the concentration of hydronium ion in a solution, the more acidic a solution is. Therefore, as a solution gets more acidic—for example, as [H$_3$O$^+$] goes from 1 × 10^{-3} (or 0.001) M to 1 × 10^{-1} (or 0.1) M—the pH becomes a smaller number. In this case, it goes from pH 3 to pH 1. *The smaller the pH value, the more acidic the solution.*

In a **neutral solution**, the pH is 7 (for a hydronium ion concentration of 1 × 10^{-7}). As the hydronium ion concentration becomes smaller than this, a solution becomes basic. Therefore, in basic solutions the pH is larger than 7. *The larger the pH value, the more basic the solution.*

If pH < 7.0,   solution is acidic   [H$_3$O$^+$] > [$^-$OH].
If pH = 7.0,   solution is neutral   [H$_3$O$^+$] = [$^-$OH].
If pH > 7.0,   solution is basic   [H$_3$O$^+$] < [$^-$OH].

The relationship between pH and the hydronium ion concentration, plus the pH of some common solutions, is illustrated in Figure 9.2.

FIGURE 9.2 Relationship of pH to the concentration of hydrogen ions [H$^+$ or H$_3$O$^+$] and hydroxide ions [OH$^-$] in water at 25°C.

The pH values for some common substances are included in the diagram.

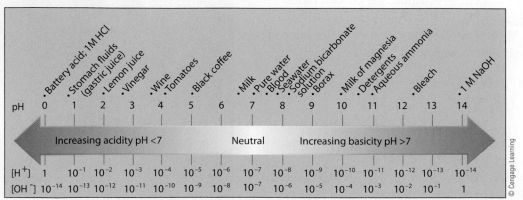

Because weak acids are only slightly ionized, the pH of a weak acid solution is not given by the concentration of the acid but is dependent on its extent of ionization. For example, in a 1.0 M HCl solution, the hydronium ion concentration produced by the strong acid is also 1.0 M and the pH is 0, but in a 1.0 M acetic acid solution, the hydronium ion concentration is only 4.3×10^{-3} M and the pH is 2.4. Because many common substances have hydronium ion concentrations in the range of 0.1 M (1×10^{-1} M) to 10^{-14} M, the pH scale provides a more convenient way of expressing low acid concentrations.

EXAMPLE 9.2 Finding pH

 Many soft drinks contain carbonic acid and phosphoric acid. If the hydronium ion concentration of a cola beverage is 1×10^{-3}, what is the pH?

SOLUTION

pH is defined as the negative of the logarithm of hydronium ion concentration, pH = $-\log [H_3O^+]$. For a concentration of 1×10^{-3}, the exponent -3 shows that the pH is 3.

TRY IT 9.6

Most tomatoes are acidic and have $[H_3O^+] = 1 \times 10^{-4}$ M. What is the pH of these tomatoes? Are they more or less acidic than the cola drink of Example 9.2?

TRY IT 9.7

Use the bracketing method to obtain a pH range for a sample of rainwater that has a hydronium ion concentration of 3×10^{-6} M. Once you have your estimate, determine the exact pH of this solution by using your calculator to find the exact value of the logarithm of 3.

TRY IT 9.8

Use your knowledge of the relationship between hydronium ion concentration and hydroxide ion concentration to determine the pH of a solution of seawater that has a *hydroxide* ion concentration of 2×10^{-6} M. Use your knowledge of the relationship between hydronium ion and hydroxide ion concentrations to determine the hydronium ion concentration. Then use the bracketing method to obtain a pH range. Finally, use the calculator to determine the exact pH.

9-3c The pH Scale Is Nonlinear

By now you may have noticed that the pH scale is not a linear scale. Logarithms do not progress linearly as we proceed from log 1 to log 2 to log 3 to log 4, etc. The logarithm of 1 is 0 and the logarithm of 10 is 1, but the logarithm of 5 is not 0.5. The logarithm of 5 is 0.699, a number that is not exactly halfway between 0 and 1. Thus, the pH scale is not linear.

TRY IT 9.9

Use your calculator to prepare a graph in which you plot the logarithm of the whole numbers 1 to 10 (on the *y*-axis) versus the numbers 1 to 10 (on the *x*-axis) to see what type of relationship emerges. This will show you the nonlinearity of this logarithmic scale and will allow you to confirm for yourself the very important fact that *a tenfold change in hydronium ion concentration results in a pH change of 1 unit!*

TRY IT 9.10

Use your graph to see what the effect on pH would be if the hydronium ion concentration changes from 1×10^{-9} M to 4×10^{-9} M. By what factor would the hydronium ion concentration have to change to cause the pH to increase by three units? By what factor would the hydronium ion concentration have to change to cause the pH to decrease by five units?

Charles D. Winters/Science Source

A pH meter is used to accurately determine the pH of aqueous solutions.

9-4 Acid–Base Buffers

Buffer solution: Solution of an acid plus a base that controls pH by reacting with added base or acid

Buffer: Combination of an acid and a base that can control pH

An acid–base buffer is like a shock absorber—something to prevent a disturbance while retaining the original conditions or structure. The control of pH requires maintaining a steady concentration of H_3O^+ even when sudden "shocks" of acid or base are added. **Buffer solutions** contain a base that can react with an acid and an acid that can react with a base so that the pH of a solution remains close to its original value. An acid–base combination suitable for controlling pH is known as a **buffer.** You see the term on bottles of "buffered" aspirin, meaning that the tablets contain some sort of base to offset the acidity of aspirin. The principal benefit is that such tablets may dissolve faster and thereby go to work faster. The buffering does not, however, actually buffer the highly acidic environment of the stomach, which has a pH of 1.5 to 2.

How does a buffer solution maintain its pH at a nearly constant value? Not only must the acid in the buffer react with added base and the base must react with added acid, but it is also necessary that the acid and base components of a buffer solution not react with each other. To meet these conditions, buffers usually are mixtures of a weak acid and its weakly basic anion (for example, acetic acid and acetate ion, CH_3COOH and CH_3COO^-) or a weak base and its weakly acidic cation (for example, ammonia and ammonium ion, NH_3 and NH_4^+). In a solution that contains an acetic acid–acetate ion buffer, added base will react with the acetic acid:

$$CH_3COOH(aq) + OH^-(aq) \rightleftharpoons$$

$$CH_3COO^-(aq) + H_2O(\ell)$$

and added acid will react with the acetate ion:

$$CH_3COO^-(aq) + H_3O^+(aq) \rightleftharpoons$$

$$CH_3COOH(aq) + H_2O(\ell)$$

Take a closer look at these two reactions, and you'll see that they are the reverse of each other. Reading the first reaction backward produces the second. This balance between two equal but opposite reactions is another example of a chemical equilibrium (Chapter 8). Notice that in these reactions, no new substances are produced. The products are always components of the buffer. Both acid and base are neutralized by different buffer components to maintain the pH at a constant value.

Buffers are very important to many industrial and natural processes. In fact, the control of the pH of your blood is essential to your health. The pH of the blood is about 7.40, and your good health depends on the ability of buffers to maintain the pH within a narrow range. If the pH falls below 7.35, a condition known as *acidosis* occurs; increasing pH above 7.45 leads to *alkalosis*. Both these conditions can be life threatening.

Carbonic acid, which forms when carbon dioxide dissolves in water, and bicarbonate ion form one of several buffer pairs that keep the pH of blood within the necessary safe range.

The blood buffering system depends on two critical equilibria. Carbon dioxide reacts with water to form carbonic acid, and carbonic acid reacts with water to form hydronium ion and bicarbonate ion, as shown here.

$$CO_2(g) + H_2O(\ell) \rightleftharpoons H_2CO_3(aq)$$

$$H_2CO_3(aq) + H_2O(\ell) \rightleftharpoons H_3O^+ + HCO_3^-(aq)$$

If some condition causes the hydronium ion concentration to increase (that is, the pH begins to decrease), the preceding equilibrium involving hydronium ion will shift to the left, according to Le Chatelier's principle, to relieve this stress and maintain the equilibrium. This, in turn, will cause the equilibrium in the first reaction to shift to the left to expel more carbon dioxide. Thus, bicarbonate neutralizes acid in the blood buffering system.

Likewise, if some condition causes the hydronium ion concentration to decrease (that is, the pH begins to increase), the equilibrium involving hydronium ion will shift to the right, according to Le Chatelier's principle, to maintain the equilibrium and produce more hydronium ion. This is another way of saying, in this case, that an increase in hydroxide ion will be neutralized by carbonic acid.

EXAMPLE 9.3 Determining Hydronium Ion Concentration from pH

 What would be the hydronium ion concentration in the blood if the pH is 7.4?

SOLUTION

Since pH = 7.4, we know, by the definition of pH, that

$$-\log [H_3O^+] = 7.4$$

$$\log [H_3O^+] = -7.4$$

We could enter 7.4 into a calculator, hit the $+/-$ button to change sign, and then hit the antilog button to find that the hydronium ion concentration that corresponds to this pH is 4.0×10^{-8} M.

Alternatively, if we don't need an exact answer, the bracketing method can be used to conclude that the hydronium ion concentration would be somewhere between 10^{-7} M and 10^{-8} M, because the pH is between 7 and 8.

TRY IT 9.11

Determine the hydronium ion concentration of a solution of household ammonia whose pH is 10.7.

9-5 Corrosive Cleaners

There are some places in the home where really tough cleaning jobs exist. For these jobs, cleaners are formulated with extremes in pH, which allow the acidity or alkalinity of the cleaner to quickly attack the unwanted dirt, grease, or stain. Toilet-bowl cleaners usually contain hydrochloric acid, which can dissolve most mineral scale (mostly carbonates) and iron stains. Other acids, such as phosphoric acid and oxalic acid, are also used in these products. The pH of toilet-bowl cleaners is usually below 2, and because they contain strong acids, they can be quite harmful to skin and eyes on contact. They should be handled with extreme caution, and rubber gloves should be worn when using them.

On the other end of the pH scale are bleach (Figure 9.3), drain cleaners, and oven cleaners, which have a pH of 12 or higher. These formulations almost always contain the strong base sodium hydroxide (NaOH). Drains usually clog as a result of oils, grease, and hair caught on rough edges inside the drainpipes. As the foreign matter builds up it becomes more tightly packed, until the flow of water from the drain practically stops. The only way to get rid of the material clogging the drain is to either dismantle the drain plumbing (often very difficult), use a plumber's "snake," or dissolve the material. Bases like sodium hydroxide are very good at dissolving drain clogs because they can cause rapid breaking of bonds in oils and greases of animal and vegetable origin. Once the bonds are broken, smaller, more soluble molecules are formed that can be washed down the drain. Hair and other proteins are also broken down by sodium

hydroxide, so a mat of grease and hair in a clogged drain will quickly succumb to the action of the strong base. Numerous products containing sodium hydroxide are available, including solid NaOH pellets, flakes, and concentrated solutions. The solutions offer the easiest way to apply the drain cleaner, but if they become diluted their effectiveness is diminished. Some solid drain cleaners also contain pieces of aluminum metal, which react with aqueous sodium hydroxide to form hydrogen gas (caution—it's flammable) that helps agitate the mixture and hastens the unclogging process.

$$2\,Al(s) + 2\,NaOH(aq) + 6\,H_2O(\ell) \longrightarrow$$
$$2\,Na^+(aq) + 2\,Al(OH)_4^-(aq) + 3\,H_2(g)$$

Oven cleaners also contain strong bases such as sodium hydroxide. Many oven cleaners use aerosol sprays to distribute the cleaner on the inner surface of the oven. Care must be taken not to breathe air containing these aerosols, because strong bases are quite corrosive to nasal tissue, bronchial tubes, and lungs.

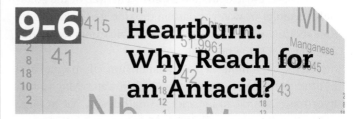

9-6 Heartburn: Why Reach for an Antacid?

The walls of a human stomach contain thousands of cells that secrete hydrochloric acid, the main purposes of which are to kill microorganisms and to aid in digestion. The stomach's inner lining is not harmed by the presence of hydrochloric acid with such a low pH (1.5–2.0), since the mucosa, the inner lining of the stomach, is replaced at the rate of about a half million cells per minute.

The uncomfortable condition known as "heartburn" occurs when the acid contents of the stomach back up into the esophagus and cause a burning sensation in the chest and throat. Among the causes are overeating; spicy, acidic, or fatty foods; and some medications, such as aspirin.

FIGURE 9.3 Corrosive household cleaners. Reading the labels and handling these chemicals with care is essential.

Antacid: Base used to neutralize excess hydrochloric acid in the stomach

Antacids are bases used to neutralize the acid that causes heartburn. The most common antacid ingredients are magnesium and aluminum hydroxides, and bicarbonate or carbonate salts (Table 9.4). Baking soda (sodium bicarbonate) was used to relieve indigestion before many of the other commercial products became available. The bicarbonate ion, a basic anion of a weak acid, reacts with the hydronium ion from hydrochloric acid to form carbonic acid, which decomposes to give carbon dioxide and water. Note that this is the same mechanism by which the blood buffering system neutralizes acid.

$$HCO_3^-(aq) + H_3O^+(aq) \longrightarrow H_2CO_3(aq) + H_2O(\ell)$$

$$H_2CO_3(aq) \longrightarrow CO_2(g) + H_2O(\ell)$$

Alka-Seltzer contains sodium bicarbonate, potassium bicarbonate, citric acid, and aspirin. The fizz of the Alka-Seltzer tablet in water is carbon dioxide gas given off by the reaction of the citric acid with the bicarbonates to give carbonic acid (the preceding reaction).

Milk of magnesia is a suspension of magnesium hydroxide (which is not very soluble) in water. Magnesium hydroxide acts as an antacid in small doses, but in large doses it is a laxative. Calcium carbonate (also known as chalk) is the active ingredient in Tums. The carbonate ion, a base, neutralizes hydronium ion.

$$CO_3^{2-}(aq) + 2\, H_3O^+(aq) \longrightarrow H_2CO_3(aq) + 2\, H_2O(\ell)$$

$$H_2CO_3(aq) \longrightarrow CO_2(g) + H_2O(\ell)$$

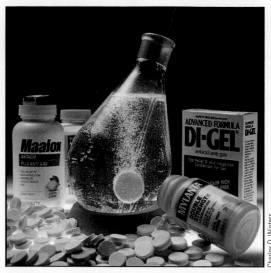

A few common antacids. Some of these neutralize excess stomach acid and produce CO$_2$ gas (center).

Although small amounts of calcium carbonate are safe, regular use can cause constipation. Aluminum hydroxide, the active ingredient in Amphojel, can also cause constipation in large doses. Because calcium carbonate and aluminum hydroxide can cause constipation, antacids such as Maalox and Mylanta contain aluminum hydroxide mixed with magnesium hydroxide to counteract the constipating effects of the former with the laxative action of the latter.

Another class of medications to combat heartburn has recently become more readily available by being reclassified from prescription-only to over-the-counter status. Instead of neutralizing stomach acid, these new drugs (for example, Tagamet and Pepcid) prevent its secretion.

TABLE 9.4 Some Common Antacids

Compound	Formula	Examples of Commercial Products
Magnesium hydroxide	Mg(OH)$_2$	Phillips' Milk of Magnesia
Calcium carbonate	CaCO$_3$	Tums, Titralac
Sodium bicarbonate	NaHCO$_3$	Alka-Seltzer, baking soda
Aluminum hydroxide	Al(OH)$_3$	Amphojel
Aluminum hydroxide and magnesium hydroxide		Maalox, Mylanta, Di-Gel tablets
Aluminum hydroxide, magnesium hydroxide, and magnesium carbonate	MgCO$_3$	Di-Gel liquid
Dihydroxyaluminum sodium carbonate	NaAl(OH)$_2$CO$_3$	Rolaids
Calcium carbonate and magnesium hydroxide		Sodium-free Rolaids

© Cengage Learning

Your Resources

In the back of the textbook:

→ **Review Card on Acid–Base Reactions**

   In OWL for CHEM2 at www.cengagebrain.com

→ **Review Key Terms with Flash Cards (printable or digital)**

→ **Complete Interactive Practice Quizzes to prepare for tests**

→ **Submit Assigned Homework and Exercises**

Applying Your Knowledge

1. What do the following terms mean?
   - (a) Acidic solution
   - (b) Basic solution
   - (c) Neutral solution

2. What is a neutralization reaction?

3. What is a salt?

4. What do the following terms mean?
   - (a) H^+ ion donor
   - (b) H^+ ion acceptor
   - (c) Hydronium ion
   - (d) Hydroxide ion

5. What do the following terms mean?
   - (a) Extent of ionization
   - (b) Strong acid
   - (c) Strong base
   - (d) Weak acid
   - (e) Weak base

6. What is the pH range for acidic solutions?

7. What is the pH range for basic solutions?

8. Label each of the following substances as acidic or basic:
   - (a) Vinegar
   - (b) Citrus fruits
   - (c) Aspirin
   - (d) Black coffee

9. Label each of the following substances as acidic or basic:
   - (a) Gastric fluid
   - (b) Tomatoes
   - (c) Oven cleaners
   - (d) Soap
   - (e) Carbonated beverages
   - (f) Baking soda

10. Indicate which of the following is a property of an acid and which is a property of a base:
    - (a) Sour taste
    - (b) Bitter taste
    - (c) Slippery feeling
    - (d) Changes color of red litmus to blue
    - (e) Changes color of blue litmus to red

11. What do the following terms mean?
    - (a) Molarity
    - (b) Concentration

12. What is the definition of pH?

13. What is a buffer?

14. What is an antacid?

15. Write the balanced equations for the following neutralization reactions.
    - (a) Acetic acid, $CH_3COOH(aq)$, with potassium hydroxide, $KOH(aq)$
    - (b) Sulfuric acid, $H_2SO_4(aq)$, with calcium hydroxide, $Ca(OH)_2(aq)$
    - (c) Sulfuric acid, $H_2SO_4(aq)$, with sodium hydroxide, $NaOH(aq)$

16. Balance the following equations:
    - (a) $HBr(aq) + Ca(OH)_2(aq) \longrightarrow$ _____
    - (b) $HNO_3(aq) + Al(OH)_3(aq) \longrightarrow$ _____

17. What is the equation for the reaction between Tums, $CaCO_3(s)$, and the HCl(aq) in stomach acid?

18. Which is more basic: a solution with pH of 2 or a solution with pH of 10? Explain.

19. Which is more acidic: a solution of black coffee with a pH of 5.0 or milk with a pH of 6.5? Explain.

20. Which of the following are acidic and which are basic?
    (a) Cherries, pH 3.2   (b) Crackers, pH 8.5
    (c) Bananas, pH 4.6   (d) Drinking water, pH 8.0

21. Which is more acidic: a grapefruit, pH 3.2, or a lemon, pH 2.3?

22. Which is more acidic: soap, pH 10, or household ammonia, pH 11?

23. Which component of each of the following buffer solutions would react with added acid and which would react with added base?
    (a) A buffer consisting of equal amounts of ammonium ion (NH_4^+) and NH_3
    (b) A buffer consisting of equal amounts of hydrogen fluoride (HF) and sodium fluoride (NaF)

24. A buffer can be made using a mixture of $NaHCO_3(aq)$ and $H_2CO_3(aq)$. Describe how the pH is maintained when small amounts of acid or base are added to the combination.

25. Two solutions contain 1% acid. Solution A has a pH of 4.6 and solution B has a pH of 1.1. Which solution contains the stronger acid?

26. Moist baking soda is often put on acid burns. Why? Write an equation for the reaction, assuming the acid is hydrochloric acid (HCl).

27. How many grams of NaOH must be dissolved to make 1.0 L of 6.0 M NaOH solution?

28. How many grams of HBr must be dissolved to make 1.0 L of 2.0 M HBr solution?

29. How many moles of HCl are dissolved in 200 mL of 0.100 M HCl?

30. How many moles of KOH are dissolved in 125 mL of 0.500 M KOH?

31. What is the molarity of the solutions formed when the following are dissolved to make 1.0 L of solution?
    (a) 5.0 g HCl   (b) 40.0 g NaOH

32. How many grams of NaCl are dissolved in 1.0 L of a 1.5 M NaCl solution?

33. How many grams of H_2SO_4 are dissolved in 1.0 L of a 0.10 M H_2SO_4 solution?

34. A recipe for preparing dill pickles calls for dissolving 155 g of salt in 3.78 L of water. What is the molarity of the salt solution that is produced?

35. The molarity of a particular glucose solution is 1.8 M. How many moles of glucose are contained in 300 mL of the solution?

36. A researcher added water to 0.500 mol of sodium sulfate to make a 3.8 M solution. How much water did she add?

37. How many grams of HCl are dissolved in a 250.0 mL volume of a 2.00 M aqueous solution of HCl?

38. How many grams of NaOH must be dissolved in 300 mL of water to produce a 1.67 M solution?

39. What is the pH for each of the following solutions?
    (a) 1.0×10^{-2} M HCl   (b) 0.001 M HNO_3

40. What is the pH for each of the following solutions?
    (a) 1.0×10^{-11} M NaOH   (b) 1.0×10^{-3} M HCl

41. The concentration of H_3O^+ in a particular aqueous solution is 1×10^{-3} M. What is the pH of the solution?

42. The concentration of hydronium in an aqueous soulution is found to be 2.3×10^{-8} M. What is the pH of the solution?

43. What is the molarity of H_3O^+ for each of the following solutions?
    (a) Solution with pH 1.0   (b) Solution with pH 0.0
    (c) Solution with pH 5.0   (d) Solution with pH 3.0

44. The pH of tap water can vary widely. The pH of one sample collected in Springfield, Ohio, was found to be 8.2.
    (a) Is this sample of tap water acidic or basic?
    (b) What is the concentration of hydronium in the sample?
    (c) What is the concentration of OH^- in the sample?
    (d) The bedrock in Springfield, Ohio, is composed mainly of limestone $(CaCO_3)$. How might this account for the pH value that was observed?

45. How many milliliters of 2.0 M HCl are needed to neutralize 0.15 mol of NaOH?

46. How many milliliters of 0.50 M HBr are needed to neutralize 0.045 mol of LiOH?

47. The pH of a solution made by dissolving 0.001 mol of NaOH in one liter of water is 11. However, when the amount of NaOH is doubled (to 0.002 mol), the pH is only slightly higher, with a value of 11.3. How can this be?

48. One of the home remedies used to treat bee stings is to make a paste from baking soda $(NaHCO_3)$ and apply it to the site of the sting. What does this tell you about bee venom?

49. When 25 mL of 0.10 M HCl and 35 mL of 0.10 M KOH are mixed, will the resulting 60 mL of solution be neutral? *Hint:* Compare the numbers of moles of dissolved acid and base.

50. When 31 mL of 0.50 M HI and 100 mL of 1.5 M NaOH are mixed, will the resulting 131 mL of solution be neutral? *Hint:* Compare the numbers of moles of dissolved acid and base.

51. A popular caffeinated soda contains 50 mg of caffeine in a 355 mL serving. If the molecular weight of caffeine is 194.19 g/mol, what is the molarity of caffeine in the beverage?

52. If you had two beakers that each contain a clear colorless solution, and one is a buffer, how could you identify which beaker contains the buffer solution?

53. When aluminum hydroxide, $Al(OH)_3$, is added to pure water, the pH does not change, even though this substance is composed of hydroxide ions. How can this be? Think about the definition of a base mentioned in this chapter.

Oxidation–Reduction Reactions

Like the acid–base reactions discussed in Chapter 9, oxidation–reduction (or redox) reactions are common in our surroundings. Oxygen is a reactant in many of these reactions. Although oxidation and reduction reactions always occur simultaneously, most people are probably more familiar with the oxidation component. Some oxidation reactions, such as the burning of a piece of paper or the combustion of gasoline, occur quite readily while others, such as the rusting of an iron tool left outdoors, occur more slowly. In each of these cases, the chemical or object that has been oxidized is quite obvious to a casual observer. It is not so obvious what chemical or object has been reduced, however. We rely on the oxidation of glucose for the biochemical energy we need to think, run, jump, and conduct the activities of our daily lives. Our automobile trips would not be possible without the chemical energy provided by the redox chemistry of the automobile battery. We also depend on batteries to operate flashlights, MP3 players, cell phones, and cameras, to name just a few modern conveniences that we take for granted.

10-1 Oxidation and Reduction

Oxide: A compound of oxygen combined with another element

Combustion: Rapid oxidation that produces heat and (usually) light

"Oxidation" got its name from the chemical changes that occur when oxygen (O_2) combines with other elements or with compounds. Many reactive metals are mined as **oxides** (compounds of oxygen with another element) that were formed by reaction of the metals with oxygen in the air. For example, the major aluminum ore, bauxite, contains aluminum oxide (Al_2O_3), and hematite, a principal iron ore, contains the iron(III) oxide (Fe_2O_3). When materials containing carbon, nitrogen, or sulfur burn in air, the products are carbon oxides (CO_2, CO), nitrogen oxides (NO, NO_2, N_2O, and others), and sulfur oxides (SO_2, SO_3).

A substance that has combined with oxygen is described as having been *oxidized*, and the reaction is classified as *oxidation*. All **combustion** reactions are oxidations. In the combustion of the methane in natural gas, for example, the carbon is oxidized to carbon dioxide and the hydrogen is oxidized to water.

CHAPTER 10

$$CH_4(g) + 2\ O_2(g) \longrightarrow CO_2(g) + 2\ H_2O(g)$$

The gradual rusting away of iron objects begins with the oxidation of iron. This gradual process is not referred to as combustion, however.

$$4\ Fe(s) + 3\ O_2(g) \longrightarrow 2\ Fe_2O_3(s)$$

In the blast furnaces that produce iron from iron ore, one of the important chemical changes is the reaction of the iron oxide in hematite with carbon monoxide.

$$Fe_2O_3(s) + 3\ CO(s) \longrightarrow 2\ Fe(\ell) + 3\ CO_2(g)$$

If the addition of oxygen to iron is oxidation, how is the removal of oxygen from iron described? It is *reduction*, a term sometimes used to mean "to bring something back." To metallurgists hundreds of years ago, reduction meant bringing a metal back from its ore. In today's chemical sense, the iron oxide has been reduced by the removal of oxygen.

The combination of iron with oxygen takes place more slowly than combustion, but the effects are just as dramatic. More than $278 billion was spent in 1998 to repair, remove, or replace rusted transportation structures in the United States.

Notice that in the reaction of iron oxide with carbon monoxide, oxygen has added to the carbon monoxide to produce carbon dioxide—the carbon monoxide has been oxidized. Here is a fundamental concept of chemistry: *Oxidation and reduction always occur together*. If one reactant is oxidized, another must be reduced.

Over time, the definitions of oxidation and reduction have been extended to include processes other than the loss or gain of oxygen atoms. With a more modern definition, you can better see why oxidation and reduction always occur together. An atom or ion is said to be oxidized when it loses electrons. Conversely, an atom or ion is said to be reduced when it gains electrons. Consider the reaction of sodium, a metal, with chlorine, a nonmetal (Figure 10.1):

$$2\,\text{Na}(s) + \text{Cl}_2(\ell) \longrightarrow 2\,\text{NaCl}(s)$$

The product is sodium chloride, a white, crystalline solid that is an ionic compound composed of equal numbers of Na^+ and Cl^- ions.

In the reaction of sodium with chlorine, sodium atoms lose their single valence electron to produce sodium cations (Na^+). **Oxidation** is the loss of electrons, and sodium has been oxidized. When an atom or ion undergoes oxidation, its charge is increased (made more positive). Because electrons have mass, and because mass is conserved in a chemical reaction, we must ask where the electrons lost by atoms during oxidation go. The reaction of sodium with chlorine occurs because as sodium atoms lose electrons, chlorine atoms gain electrons.

Reduction is the gain of electrons, and in the conversion of chlorine to chloride ions (Cl^-) in

© iStockphoto.com/Anest

this reaction, each chlorine atom has been reduced by the gain of a single electron. When an atom or ion undergoes reduction, its charge decreases (becomes more negative).

Reactions in which one reactant is oxidized and another is reduced are known as **oxidation–reduction reactions**, usually referred to as **redox reactions**. For example, every reaction in which a metallic element combines with a nonmetallic element is a redox reaction in which the metal is oxidized and the nonmetal is reduced.

EXAMPLE 10.1 Recognizing Oxidation and Reduction

For the reaction of copper with oxygen to give copper (II) oxide,

$$2\,\text{Cu}(s) + \text{O}_2(g) \longrightarrow 2\,\text{CuO}(s)$$

identify which reactant is oxidized and which is reduced.

SOLUTION

The reaction of Cu with O_2 can be recognized as a redox reaction because it is the addition of oxygen to a reactant and also because it is the combination of a metal and a nonmetal. Since the product is an ionic compound, the electron gain and loss is determined by the charges on the ions. The oxide CuO must be composed of Cu^{2+} and O^{2-} ions. Therefore, the copper metal has been oxidized by the loss of two electrons from each atom, and the oxy-

FIGURE 10.1 Oxidation of sodium by chlorine.
The reaction between sodium metal and chlorine, a nonmetal, releases a large amount of energy. The product is sodium chloride, familiar to everyone as table salt.

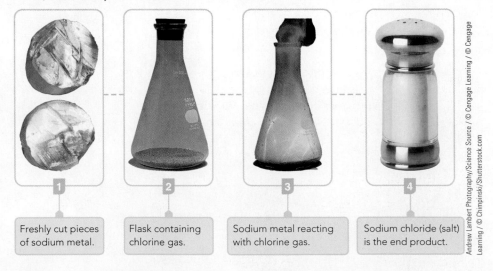

1 Freshly cut pieces of sodium metal.

2 Flask containing chlorine gas.

3 Sodium metal reacting with chlorine gas.

4 Sodium chloride (salt) is the end product.

Andrew Lambert Photography/Science Source / © Cengage Learning / © Cengage Learning / © Chimpinski/Shutterstock.com

gen has been reduced by the gain of two electrons by each atom.

For the reaction of the very active metal lithium with oxygen,

$$4\,Li(s) + O_2(g) \longrightarrow 2\,Li_2O(s)$$

identify what is oxidized and what is reduced.

10-2 Oxidizing Agents: They Bleach and They Disinfect

An agent is something that causes a change to take place. You've seen that whenever oxygen adds to another reactant, that reactant is *oxidized*. The oxygen in such a reaction is the **oxidizing agent**, a reactant that causes an oxidation by gaining electrons. The halogens are also oxidizing agents, for example

$$2\,Al(s) + 3\,Cl_2(g) \longrightarrow 2\,AlCl_3(s)$$
$$Sn(s) + F_2(g) \longrightarrow SnF_2(\ell)$$

In each of these reactions, the neutral halogen molecule has been reduced by the addition of electrons to give -1 ions. Here is another condition common to all redox reactions: *The oxidizing agent is reduced.* Some common oxidizing agents are listed in Table 10.1.

Somewhere around the home, most of us have one or more oxidizing agents. The most common household oxidizing agent is sodium hypochlorite ($NaOCl$), a water-soluble ionic compound. In the laundry, $NaOCl$ is a **bleaching agent**—it removes unwanted color by oxidizing colored

The tablets used to disinfect swimming pools decompose to produce chlorine-containing oxidizing agents.

chemical compounds to produce whiter clean clothes. If you spill an $NaOCl$ solution (for example, Clorox bleach) on your jeans, its effectiveness as a bleach will be immediately obvious, as a white spot will appear. Why does this happen, and how is it related to oxidation?

Substances have color because their molecular structures absorb portions of the visible spectrum of light. Many of the substances that contribute color to textiles or paper are organic compounds with long chains of alternating

Oxidizing agent: A reactant that causes oxidation of another reactant

Bleaching agent: A chemical that removes unwanted color by oxidation of the colored chemical

TABLE 10.1 Some Oxidizing and Reducing Agents

Name	Formula	Uses
Oxidizing Agents		
Lead dioxide	PbO_2	Automobile batteries
Manganese dioxide	MnO_2	Batteries
Sodium hypochlorite solution	$NaOCl(aq)$	Laundry bleach, disinfection
Oxygen	O_2	Metabolism of foods, burning fuels
Ozone	O_3	Water purification, hazardous chemical destruction
Chlorine	Cl_2	Drinking water and wastewater purification, chemical synthesis
Hydrogen peroxide	H_2O_2	Bleaching, antiseptic
Reducing Agents		
Hydrogen	H_2	Fuel, chemical synthesis
Sulfur dioxide	SO_2	Chemical synthesis
Carbon	C	Iron production
Zinc	Zn	Batteries

single and double carbon–carbon bonds (Section 5-3). The color is created by the absorption of some of the light by these bonds. A bleach disrupts the alternating pattern by breaking bonds or converting double bonds to single bonds. The result is loss of the ability to absorb visible light. A very strong bleaching solution can also break bonds in molecules that compose the fabric itself; when this occurs, the fabric develops thin spots and holes.

The ability of NaOCl to act as an oxidizing agent also allows it to serve as a disinfectant. When you use a liquid bleach solution to wash the bathroom floor or the kitchen counter, you are taking advantage of both the disinfecting and bleaching properties of this chemical. By oxidizing molecules in the outer surfaces of bacteria, a disinfectant is able to disrupt the structure of the cells and kill them.

Another familiar oxidizing agent is hydrogen peroxide (H_2O_2), and it too is a bleaching agent. At times, its most popular use has been in bleaching hair to produce "peroxide blonde." A dilute solution (usually 3%) of hydrogen peroxide in water is useful in the medicine cabinet because it is an *antiseptic*, a substance that can be used to cleanse a wound to prevent bacterial infection. An antiseptic acts in the same manner as a disinfectant but, unlike a disinfectant, is mild enough for use on human tissue without causing damage.

Chlorine itself is an oxidizing agent of vital importance in ensuring the purity of drinking water and water that is returned to natural waterways. Treatment of wastewater and drinking water is described in Sections 11.10 and 11.12.

10-3 Reducing Agents: For Metallurgy and Good Health

When a metal reacts with another substance, the metal usually acts as a **reducing agent**, a reactant that causes a reduction by giving up electrons. In reactions with oxygen or the halogens, for example, the metal is the reducing agent and the oxygen or halogen is reduced.

$$2\,Cu(s) + O_2(g) \longrightarrow 2\,CuO(s)$$
$$2\,Al(s) + 3\,Br_2(g) \longrightarrow 2\,AlBr_3(s)$$

In these and all redox reactions, *the reducing agent is oxidized*. Some common reducing agents are included in Table 10.1.

Reducing agents are crucial in the production of metals from their ores because most metals occur in nature only in chemical compounds. Carbon from coal, usually in the form of **coke**, is a valuable metallurgical reducing agent.

$$ZnO(s) + C(s) \longrightarrow Zn(g) + CO(g)$$

In the production of zinc, this reaction is carried out at such a high temperature that the zinc is formed as a gas that is condensed for further purification. In this process, the Zn^{2+} cation in zinc oxide is reduced to zinc metal and $C(s)$ is oxidized to carbon monoxide. It is this process of producing Zn metal that was implicated in the Donora, Pennsylvania, pollution event mentioned in Chapter 4.

Hydrogen gas, like the metals, is a reducing agent in virtually all of its reactions. Hydrogen can, for example, reduce the copper in copper(II) oxide:

$$CuO(s) + H_2(g) \longrightarrow Cu(s) + H_2O(\ell)$$

The addition of hydrogen to another compound is also classified as a reduction. In a reaction that may be of increasing importance in the production of methanol as an

Charles D. Winters

The redox reaction between aluminum and bromine [2Al(s) + 3Br₂(l) ⟶ 2AlBr₃(s)] releases heat and creates a cloud of vaporized bromine as some of the liquid bromine boils away.

Pomegranate juice contains a significant concentration of antioxidants.

© Handypix/Alamy

alternative fuel (Section 12-9), carbon monoxide is reduced to methanol by the addition of hydrogen.

$$CO(g) + 2\,H_2(g) \longrightarrow CH_3OH(\ell)$$

Reducing agents are important to the maintenance of health. Much has appeared in the popular media about the value to human health of consuming fruits and vegetables or vitamin supplements that contain *antioxidants*. Drugstores now offer a variety of "antioxidant vitamins," supplements that usually contain vitamin E, vitamin C, and vitamin A or beta-carotene (which converts to vitamin A in the body).

An antioxidant is, in chemical definition terms, a reducing agent. It prevents oxidation by reducing a potential oxidizing agent. The oxidizing agents of concern in the body are atoms or molecular fragments known as **free radicals**. A free radical contains an unpaired electron and does not stay around for long without grabbing another electron from a nearby molecule (Figure 10.2). If this second molecule happens to have an important function, then that function will be disrupted. For example, if a free radical connects with the part of a DNA molecule that governs cell division (Section 15-11), the result may be abnormal cell division—cancer.

Free radicals arise in the body from normal biochemical reactions and are also produced in the presence of toxic substances from cigarette smoke, polluted air, and other sources. They are quickly deactivated by picking up an electron to form a less-reactive molecule. Thus, the antioxidants are molecules that in one way or another are able to donate an electron to a free radical before it can do any damage.

> **Free radical:** An atom or molecule that contains an unpaired electron

To summarize, we have illustrated three ways to recognize oxidation and reduction:

Oxidation	Reduction
Addition of oxygen	Loss of oxygen
Loss of electrons	Addition of electrons
Loss of hydrogen	Addition of hydrogen

Here's another way to look at it:

Oxidizing agents are REDUCED by:	Reducing agents are OXIDIZED by:
Losing oxygen	Gaining oxygen
Gaining electrons	Losing electrons
Gaining hydrogen	Losing hydrogen

FIGURE 10.2 Action of an antioxidant.
Once a free radical (R·) has formed, it will react as soon as it can with something that will remove or pair with the unpaired electron. By intercepting free radicals, beta-carotene and other antioxidant vitamins prevent them from damaging DNA or other crucial biomolecules.

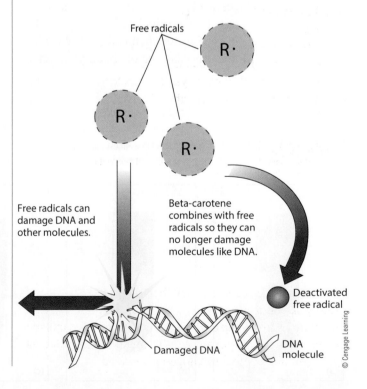

Free radicals

R·

R·

R·

Free radicals can damage DNA and other molecules.

Beta-carotene combines with free radicals so they can no longer damage molecules like DNA.

Deactivated free radical

Damaged DNA

DNA molecule

© Cengage Learning

EXAMPLE 10.2 Oxidizing and Reducing Agents

➡️ *For each of the following redox reactions, identify the oxidizing agent and the reducing agent:*

(a) $2\ CO(g) + O_2(g) \longrightarrow 2\ CO_2(g)$
(b) $CuO(s) + H_2(g) \longrightarrow Cu(s) + H_2O(\ell)$

SOLUTION

(a) Oxygen is the oxidizing agent and is added to carbon monoxide. The carbon monoxide is the reducing agent.
(b) Copper oxide is the oxidizing agent as shown by its loss of oxygen. Hydrogen is the reducing agent and is oxidized by the addition of oxygen.

TRY IT 10.2

Indicate whether the reactant is being reduced or oxidized in the reactions represented by the following incomplete equations, each of which is missing a key reactant.

(a) $Cu(s) \longrightarrow CuO(s)$
(b) $CH_3C \equiv N(g) \longrightarrow CH_3CH_2-NH_2(g)$
(c) $SnO(s) \longrightarrow Sn(s) + H_2O(g)$

10-4 Batteries

A device that produces an electric current from a chemical reaction is called an **electrochemical cell**. Strictly speaking, a **battery** is a series of electrochemical cells. We will, however, stick with the everyday use of "battery" to refer to any device that converts chemical energy to electrical energy.

To function, a battery takes advantage of the relative ease with which metals lose electrons. Consider the reaction that begins to occur as soon as a piece of zinc is placed in a solution of copper ions (Cu^{2+}). The blue color of the copper ions in solution fades as metallic copper is deposited on the surface of the zinc (Figure 10.3). The copper ions are being reduced by gaining electrons and being converted to copper metal.

Since oxidation must accompany reduction, what is being oxidized? That is, what is losing electrons? Careful observation shows that the zinc strip is gradually being consumed in the reaction. The zinc is being oxidized to zinc ions in solution, which are colorless.

The reduction of copper ions by zinc can be thought of as a competition for the available electrons. From observing the reaction in Figure 10.3, it appears that the Zn atoms give up their electrons to the Cu^{2+} ions in the reaction

$$Zn(s) + \underset{\text{Blue}}{Cu^{2+}(aq)} \longrightarrow \underset{\text{Colorless}}{Zn^{2+}(aq)} + Cu(s)$$

If instead Cu atoms more easily gave up their electrons to Zn^{2+} ions, the resulting reaction would be the *reverse* of what is observed. This reverse reaction is unfavorable, however, and does not occur on its own. Apparently, zinc is a stronger reducing agent than copper because it gives up its electrons more easily.

The electrons transferred between a metal that is a good reducing agent and the ion of another metal can provide the electron flow—the *current*—in a battery. This current is produced by physically separating the reducing and oxidizing agents, forcing the transferred electrons to pass through a circuit. A battery is essentially a favorable oxidation–

FIGURE 10.3 Oxidation of zinc by copper ions.

(a) A piece of zinc is immersed in a solution containing Cu^{2+} ions, which give the solution a blue color. (b) After a few minutes, the blue color begins to fade and copper builds up on the remaining zinc. (c) After about an hour, the solution is almost colorless, indicating that most of the Cu^{2+} ions have been reduced to copper atoms, which have formed metallic copper on what is left of the zinc. The Zn^{2+} ions from oxidation of the zinc are colorless. (d) An atomic-level depiction of the process.

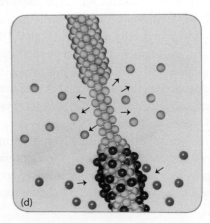

Charles D. Winters

reduction reaction occurring inside a container that has two **electrodes**. At one of these electrodes, oxidation takes place as electrons flow out of the cell (it is marked with a − sign). This electrode is the **anode**. At the other electrode, reduction takes place as electrons flow into the cell (it is marked with a + sign). This electrode is the **cathode**. When all of the reactants inside the battery are used up and it is impossible to convert the reaction products back to their original form, the battery is "dead" and must be discarded. On the other hand, if the reactants can be converted back to their original form, the battery can be used again.

10-4a Throwaway Batteries

The general arrangement of the parts in an electrochemical cell is diagrammed in Figure 10.4(a). The reaction between zinc and copper ions discussed at the beginning of this section provides an example of how such a cell works. The zinc is separated from the copper ion solution and the two are connected so that the reaction proceeds as the electrons are transferred through the connecting wire as shown in Figure 10.4(b). A *salt bridge* allows ions to flow from one electrode chamber to the other. Such a connection is necessary because Zn^{2+} ions are produced at the anode and negative ions must

Electrochemical cell: A device in which a chemical reaction generates an electric current

Battery: A series of electrochemical cells that produces an electric current; commonly refers to any electrochemical, current-producing device

Electrode: A conducting material at which electrons enter and leave an electrochemical cell

Anode: Electrode at which electrons flow out of an electrochemical cell and oxidation takes place

Cathode: Electrode at which electrons flow into an electrochemical cell and reduction takes place

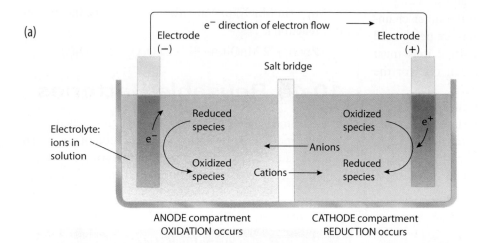

(a)

ANODE compartment
OXIDATION occurs

CATHODE compartment
REDUCTION occurs

FIGURE 10.4 (a) Components of an electrochemical cell.
Oxidation (loss of electrons) occurs in the anode compartment, and electrons flow out of the cell through the external circuit. Electrons re-enter the cell at the cathode, and reduction (gain of electrons) takes place in the cathode compartment. A porous salt bridge (or other connection) allows ions to flow between the two compartments to maintain charge balance.

(b) A simple electrochemical cell.
When the electrodes are connected by a conducting circuit, electrons flow from the zinc electrode, where zinc is oxidized, to the copper electrode, where copper is reduced. The overall reaction in this cell is
$Zn(s) + Cu^{2+}(aq) \longrightarrow Zn^{2+}(aq) + Cu(s)$.

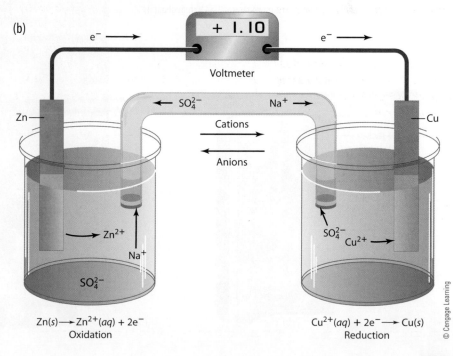

$Zn(s) \longrightarrow Zn^{2+}(aq) + 2e^-$
Oxidation

$Cu^{2+}(aq) + 2e^- \longrightarrow Cu(s)$
Reduction

© Cengage Learning

A collection of disposable batteries.

The products of the reaction in this simple battery cannot easily be converted to their original form. For this to happen, the copper deposited on the copper electrode would somehow have to be dissolved back into solution, and the Zn^{2+} ions would have to be converted back into the zinc metal strip. In other words, because it is not easily reversible, this reaction is best used in a *throwaway* battery, one that cannot be recharged. Batteries of this type are called **primary batteries**.

One of the most common primary batteries today is the *alkaline battery*. Zinc in the presence of potassium hydroxide (KOH, the alkaline substance) serves as the anode. The zinc is separated from the other chemicals (Figure 10.5) by a porous paper, which serves as the salt bridge. The cathode is made of graphite (carbon) combined with manganese dioxide (MnO_2), which is the oxidizing agent. The oxidation at the anode is conversion of Zn to Zn^{2+} (in solid ZnO), and the reduction at the cathode is conversion of Mn^{4+} (in MnO_2) to Mn^{3+} (in Mn_2O_3); thus, the overall cell reaction is

$$Zn(s) + 2\,MnO_2(s) \longrightarrow ZnO(s) + Mn_2O_3(s)$$

10-4b Reusable Batteries

Some batteries allow the oxidation–reduction reactions at the electrodes to be reversed by the addition of energy so that the battery can be recharged. Batteries of this type are called **secondary batteries**. Under favorable conditions,

Primary battery: A battery that cannot be recharged because its reaction is not easily reversible

Secondary battery: A battery that can be recharged by reversing the flow of current, which reverses the current-producing reaction and regenerates the reactants

flow into that chamber to prevent positive charge from building up and stopping the reaction. In the other chamber, Cu^{2+} ions are being used up, and positive ions must move into that one for the same reason.

The electrons flow from the Zn anode through the connecting wire and then flow into the copper cathode, where reduction of Cu^{2+} ions occurs.

FIGURE 10.5 An alkaline battery.
This type of throwaway battery provides a constant voltage of 1.54 V throughout its useful life.

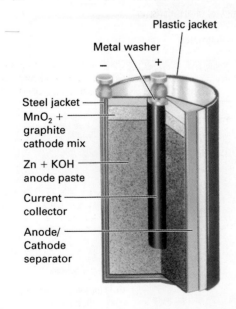

Plastic jacket
Metal washer
− +
Steel jacket
MnO_2 + graphite cathode mix
Zn + KOH anode paste
Current collector
Anode/ Cathode separator

An assortment of batteries. In front of the lead–acid automobile battery are, from left to right, a nonrechargeable zinc-graphite dry cell, several types of nonrechargeable alkaline batteries, and three rechargeable nickel-cadmium batteries.

secondary batteries may be discharged and recharged many times over.

The lead–acid automobile battery is the most familiar secondary battery. As this battery is discharged, metallic lead is oxidized to lead sulfate at the anode, and the Pb^{4+} in lead dioxide (PbO_2) is reduced at the cathode to the Pb^{2+} in lead sulfate ($PbSO_4$). The reaction takes place in the presence of sulfuric acid (battery acid), and the equations for the reactions are

Oxidation (Anode)
$$Pb(s) + SO_4^{-2}(aq) \longrightarrow PbSO_4(s) + 2\ e^-$$

Reduction (Cathode)
$$PbO_2(s) + 4H^+(aq) + SO_4^{-2}(aq) + 2\ e^- \longrightarrow$$
$$PbSO_4(s) + 2\ H_2O(\ell)$$

Net Reaction
$$Pb(s) + PbO_2(s) + 2\ H_2SO_4(aq) \longrightarrow$$
$$2\ PbSO_4(s) + 2H_2O(\ell)$$

The lead–acid battery (Figure 10.6) is reusable because the lead sulfate formed at both electrodes is insoluble and stays on the electrode surface. Then when the battery needs recharging, the lead sulfate is available for the reverse reaction.

Recharging a secondary battery requires reversing the direction of electron flow through the battery, which can be accomplished by a generator or an alternator. When recharging occurs, the reactions at the two electrodes are reversed:

Discharge in a battery when starting car:
 chemical energy $\longrightarrow$ electrical energy

Recharging a battery when car is running:
 electrical energy $\longrightarrow$ chemical energy

Normal recharging of an automobile lead–acid battery occurs during driving. The voltage regulator senses the output from the alternator, and when the alternator voltage exceeds that of the battery, electrical energy is added back into the battery and the battery is recharged. During the recharging cycle in a lead–acid battery, some water is reduced to hydrogen at the cathode, and some water is oxidized to oxygen at the anode. The result is a potentially explosive mixture of hydrogen and oxygen at the top of the battery. Under normal driving conditions, automobile batteries do not explode; however, internal short circuits can produce explosions in older batteries. Batteries must always be protected from sparks.

All in all, the lead–acid battery is relatively inexpensive, reliable, and simple and has an adequate life. Its high weight is its major fault. Newer secondary batteries have found use in some applications such as electronics, but none of these newer batteries can perform like the lead–acid battery does for its cost.

10-4c Fuel Cells

Normal batteries have the advantage of being relatively small, which makes them easily inserted, removed, and transported from place to place. Such batteries are limited in the amount of current they produce by the amount of the reagents inside the battery. When the oxidizable reagent in the battery is consumed, the battery is dead unless it is a rechargeable battery. One way to overcome this problem is to use fuel cells that, like batteries, have an electrode where oxidation takes place and an electrode where reduction takes place. However, fuel cells do not depend on chemicals stored inside the electrode compartments for their energy. Fuel cells produce energy from reactants that continuously flow into their compartments while the chemical reaction products flow out of them.

One type of fuel cell was used in the space program on board the Gemini, Apollo, and space shuttle missions. On April 13, 1970, a dramatic explosion in oxygen tank 1 aboard the Apollo 13 spacecraft triggered the release of oxygen from tank 2. In addition to its use for breathing, the oxygen was necessary for the operation of the fuel cells used to produce electricity and water for the command module and the astronauts. This incident, brought to the movie screen in *Apollo 13*, led to the most dramatic astronaut rescue mission to date.

FIGURE 10.6 Lead–acid battery.
The anodes are lead grids filled with spongy lead. The cathodes are lead grids filled with lead(IV) oxide, PbO_2. Each cell produces a potential of about 2 V. Six cells connected in series produce the desired overall battery voltage. Most lead batteries have a useful life of three years or less. The lead in these batteries can be a health hazard and thus there are stringent environmental requirements concerning their disposal.

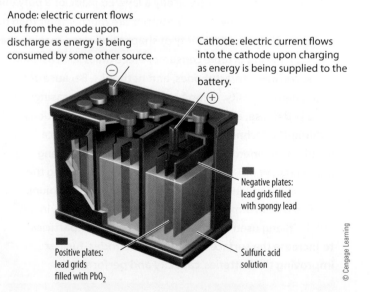

Anode: electric current flows out from the anode upon discharge as energy is being consumed by some other source.

Cathode: electric current flows into the cathode upon charging as energy is being supplied to the battery.

Negative plates: lead grids filled with spongy lead

Positive plates: lead grids filled with PbO_2

Sulfuric acid solution

© Cengage Learning

In these fuel cells, hydrogen gas is pumped onto the anode of the fuel cell, and oxygen gas is pumped onto the cathode, where the following reactions occur:

Oxidation at the anode: $2\,H_2(g) \longrightarrow 4\,H^+ + 4\,e^-$

Reduction at the cathode:
$$O_2(g) + 4\,H^+ + 4\,e^- \longrightarrow 2\,H_2O(g)$$

Net reaction: $2\,H_2(g) + O_2(g) \longrightarrow$
$$2\,H_2O(g) + energy$$

A proton exchange membrane (PEM) separates the two halves of a fuel cell. This membrane allows protons, formed by the oxidation of hydrogen gas, to pass through and react with hydroxide ions, produced by reduction of oxygen gas at the cathode, to form water. The "waste product" of this fuel cell is simply water vapor that can be condensed for use, for instance, by astronauts. Fuel cells deliver as much power as batteries weighing ten times as much. On a typical seven-day mission, the space shuttle fuel cells consumed 1500 pounds of hydrogen and generated 190 gallons of water.

Fuel cell technology applied to cars is currently an area of much interest. The oxygen for such a cell would come from the air, but a car that had to store and transport the hydrogen needed to operate the fuel cell would be problematic because of the explosive nature of hydrogen gas and the technological problems of storing such a light gas. Such cars have been shown to be possible but probably not practical.

A potential solution to the hydrogen storage problem would be to generate the required hydrogen at the point of usage. Much research is currently under way to develop processes capable of converting gasoline to hydrogen and carbon monoxide by the use of appropriate catalysts. The hydrogen could then be used in a fuel cell (with oxygen supplied from the air) to produce electricity to power the car. The carbon monoxide would be converted to carbon dioxide over another catalyst so that the emissions of the car would be water vapor and carbon dioxide. This would create a so-called pollution-free car, although both these emission gases do contribute to global warming. Any carbon-based fuel will ultimately lead to the production of carbon dioxide, so this solution does not totally address the problem it is meant to solve, the generation of carbon dioxide and the consumption of hydrocarbon-based fuels. However, it is a start.

Gasoline–electric hybrid automobiles have been the focus of much industry effort in recent years. These *hybrid cars* rely on a combination of gasoline-generated and battery-generated energy. Honda was the first to bring this type of car to the market. When idling or moving at low speed, this car uses its electric motor. At higher speeds, the gasoline engine and the electric motor both contribute energy. As the car slows down, the wheels drive a generator that returns energy to storage in the battery. These cars need not be plugged in for recharging and can get up to 60 miles per gallon of gasoline. As of 2009, Toyota and Honda were the

Today's Battery: Lithium-ion

Lithium-ion (Li-ion) batteries are a newer type of secondary battery developed throughout the 1980s and first released commercially by Sony in 1991. In these batteries, lithium ions move from an anode (typically graphite) to a cathode (typically a layered oxide or a polyanion) during discharge, and back from the cathode to the anode during charge. Li-ion batteries are much lighter than other energy-equivalent secondary batteries and can be formed into a wide variety of shapes and sizes, making them ideal for use in portable consumer electronics such as handheld game consoles, smartphones, and netbooks. Because of their high energy density, Li-ion batteries are also increasingly being used in defense, automotive, and aerospace applications. Lithium-ion technology has continued to evolve since its initial commercial release. In 2002, Yet-Ming Chiang and a team of MIT professors discovered that doping the batteries' conductive materials with aluminum, niobium, and zirconium improved performance dramatically. In 2004, Chiang used nanoscopic iron-phosphate particles to increase the surface area of the electrodes, further improving the batteries' capacity and performance.

The Tesla Roadster is an all-electric sports car that can travel 244 miles (393 km) on a single charge of its lithium-ion battery pack, and can accelerate from 0–60 mph (0–97 km/h) in 3.7 seconds.

© Drive Images/Alamy

major suppliers of such cars in the United States, although several other manufacturers, including Chevrolet, Ford, Lexus, Mazda, and Nissan have introduced hybrid vehicles, including sedans, SUVs, and pickup trucks.

Fuel cells that do not depend on hydrocarbon-based fuels for the generation of hydrogen are far too costly at present to be practical, and there are still many technological problems to overcome before we see hydrogen-fueled cars on the road.

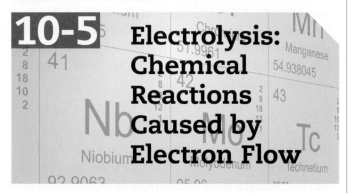

10-5 Electrolysis: Chemical Reactions Caused by Electron Flow

Chemical reactions that are unfavorable can be forced to proceed by the input of energy, as in the recharging of a secondary battery. **Electrolysis reactions** are oxidation–reduction reactions driven by electrical energy from an external power supply. Where electrons flow into the cell, the electrode becomes negatively charged, positive ions in solution migrate toward that electrode, and reduction takes place. In this type of cell (Figure 10.7), the electrodes need not be in separate compartments. The chemical changes associated with electrolysis are essentially the reverse of the changes that take place in batteries. In electrolysis, electrical energy produces chemical change. In batteries, chemical change produces electrical energy.

The metal in some ores is so resistant to reduction that few reducing agents are strong enough to cause the reaction, and electrolysis is a good alternative way to provide enough energy for the reaction. Most importantly, aluminum is a metal of this type. Aluminum, in the form of Al^{3+} ions, is the third most abundant element in Earth's crust (7.4%). The quantity and commercial value of aluminum used in the United States each year is exceeded only by that of iron and steel.

From the time of its discovery in 1825 until near the turn of the century, aluminum was made by reducing $AlCl_3$ with a more active metal (potassium or sodium) but only at a very high cost. Even though there was commercial production of aluminum by 1854, aluminum was considered a precious metal—as gold and platinum are today—and one of its early uses was for jewelry. Napoleon III saw the possibilities of aluminum for military use, however, and commissioned studies on improving its production. The

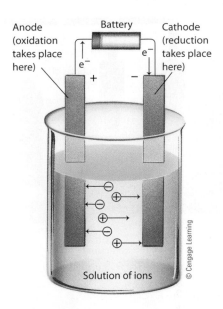

Anode (oxidation takes place here) Battery Cathode (reduction takes place here)

e^- $+$ $-$ e^-

Solution of ions

© Cengage Learning

French had a ready source of aluminum-containing ore, bauxite, named for the French town of Les Baux. In 1886, a 23-year-old Frenchman, Paul Heroult, conceived the electrochemical method that is still in use today. In an interesting coincidence, an American, Charles Hall, who was 22 at the time, announced his invention of the identical process in the same year. Hence, the commercial process is now known as the Hall–Heroult process.

> **Electrolysis reaction:** An oxidation-reduction reaction that requires the input of electrical energy

In the Hall–Heroult process, purified aluminum oxide from bauxite is dissolved in a molten ionic compound (cryolite, Na_3AlF_6). The mixture is electrolyzed in a cell with carbon anodes and a carbon cell lining that serves as the cathode (Figure 10.8).

Because aluminum is such an active metal, producing it from its oxide requires a lot of energy. This, combined with the energy needed to maintain the molten cryolite bath, makes aluminum production highly energy intensive. Indeed, approximately 5% of all electrical energy produced in the United States is devoted to the production of aluminum metal, which accounts for about 40% of the cost of the metal. One reason for the success of aluminum recycling is the large saving in energy cost—making aluminum beverage cans from recycled aluminum requires only 5% of the energy used in making the cans from new aluminum, and the process is about 20% cheaper overall.

FIGURE 10.8 Aluminum production by electrolysis.

At the cathode, aluminum ions are reduced to aluminum metal. The anode reaction is production of oxygen gas, which reacts with the carbon anodes. (The cell reaction is $2 Al_2O_3 + 3 C \longrightarrow 4 Al + 3 CO_2$.) Molten aluminum is denser than the molten salt mixture in the cell and collects at the bottom.

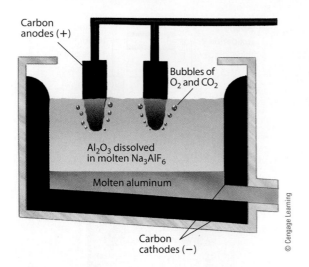

Carbon anodes (+)

Bubbles of O_2 and CO_2

Al_2O_3 dissolved in molten Na_3AlF_6

Molten aluminum

Carbon cathodes (−)

© Cengage Learning

Corrosion: The unwanted oxidation of metals during exposure to the environment

Another practical application of electrolysis is *electroplating*—the coating of one metal, which is made the cathode in an electrolysis cell, with another metal, which is made the anode. For example, dull iron or steel surfaces can be electroplated with silver to produce silver-plated dinnerware.

10-6 Corrosion: Unwanted Oxidation–Reduction

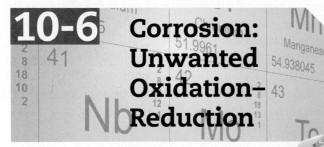

In the United States alone, more than $10 billion is lost each year to **corrosion**—the unwanted oxidation of metals during exposure to the environment. Much of this corrosion is the rusting of iron and steel, although other metals may corrode as well. The oxidizing agent causing all of this unwanted corrosion is usually oxygen. Iron is most severely affected by corrosion because

Drozdowski/Shutterstock.com

rust does not adhere strongly to the metal's surface. The continuing loss of surface iron as rust forms and then flakes off eventually causes structural weakness.

The driving forces behind metal corrosion are the activity of the metal as a reducing agent and the strength of the oxidizing agent. Whenever a strong reducing agent (the metal) and a strong oxidizing agent (such as oxygen) are in contact, a reaction between the two substances is likely. Factors governing the rates of chemical reaction, such as temperature and concentration (Section 8-3), affect the rate of corrosion as well. Consider the corrosion of an iron spike (Figure 10.9). There are tiny microcrystals of loosely bound iron atoms on the surface of the metal that can easily be oxidized.

$$Fe(s) \longrightarrow Fe^{2+}(aq) + 2 e^-$$

Because iron is a good conductor of electricity, the electrons produced by this oxidation can migrate through the metal to a point where they can reduce something. The fact that iron is a conductor of electricity is important because corrosion would come to an abrupt halt as a result of a buildup of excessive negative charge if the electrons were not conducted away. One location on the surface of the iron where electrons can be used is any tiny drop of water containing dissolved oxygen. Here the oxygen gains electrons, forming hydroxide ions.

$$O_2(g) + 2 H_2O(\ell) + 4 e^- \longrightarrow 4 OH^-(aq)$$

The Fe^{2+} ions are further oxidized to Fe^{3+} ions, which react with OH^- ions to form the hydrated iron oxide known as rust.

$$2 Fe^{3+}(aq) + 6 OH^-(aq) \longrightarrow Fe_2O_3 \cdot 3H_2O(s)$$
Rust

The rate of rusting is enhanced by salts, which dissolve in the water on the surface of the iron. The hydroxide ions and iron ions migrate more easily in the ion solutions produced by the presence of the dissolved salts. Automobiles rust more quickly when exposed to road salts in wintry climates. If road salts are used in your driving area, it's a good idea to wash the underside of your automobile after the snowy season ends to remove the accumulated salts.

For rust to form, three reactants are necessary. These are iron, oxygen, and water. Rusting can be prevented by protective coatings such as paint, grease, oil, enamel, or a corrosion-resistant metal such as chromium. Most of these coatings keep out moisture. Some of the metals that are more

FIGURE 10.9 Corrosion of iron.

The site of iron oxidation may be different from the site of oxygen reduction because iron is a conductor of electricity and electrons can move through it from one site to another.

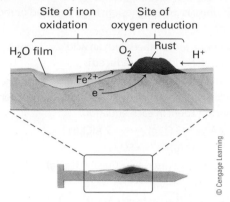

active than iron form adherent oxide coatings when they corrode. Coating iron with these metals provides corrosion protection. One of these active metals is zinc. When the zinc coating of a galvanized object is exposed to air and water, a thin film of zinc oxide forms that protects the zinc from further oxidation. If the zinc coating should get scratched so that iron is exposed to the air, zinc will quickly reduce any Fe^{2+} ions formed because zinc is more active than iron in giving up electrons.

Your Resources

In the back of the textbook:

→ *Review Card on Oxidation–Reduction Reactions*

   In OWL for CHEM2 at www.cengagebrain.com

→ *Review Key Terms with Flash Cards (printable or digital)*

→ *Complete Interactive Practice Quizzes to prepare for tests*

→ *Submit Assigned Homework and Exercises*

Applying Your Knowledge

1. Define the following terms:
   (a) Oxidation in terms of electrons
   (b) Reduction in terms of electrons
   (c) Oxidation in terms of oxygen
   (d) Reduction in terms of hydrogen
   (e) Oxidizing agent
   (f) Reducing agent

2. Give definitions for the following terms:
   (a) Anode   (b) Primary battery
   (c) Secondary battery   (d) Cathode

3. Identify the more highly oxidized substance in each of the following:
   (a) CO or CO_2 (b) NO_2 or NO
   (c) SO_3 or SO_2 (d) N_2 or NH_3

4. Identify two applications of the oxidizing properties of chlorine, Cl_2.

5. What is the *chemical* difference between bleaching and disinfecting?

6. Which is the more common oxide of hydrogen, H_2O or H_2O_2?

7. Describe how combustion and oxidation differ.

8. Which oxide of carbon, CO or CO_2, is produced in incomplete combustion?

9. Which is more highly oxidized, a potassium atom or a K^+ ion?

10. Tell whether the underlined atoms in the following equations are oxidized or reduced. Justify your answers.
    (a) $2\,\underline{H}_2(g) + O_2(g) \longrightarrow 2\,H_2O(\ell)$
    (b) $2\,\underline{Cu}(s) + O_2(g) \longrightarrow 2\,CuO(s)$
    (c) $2\,\underline{Sn}O(s) \longrightarrow 2\,Sn(s) + O_2(g)$
    (d) $\underline{Fe}_2O_3(s) + 3\,C(s) \longrightarrow 2\,Fe(s) + 3\,CO(g)$
    (e) $\underline{N}_2(g) + 3\,H_2(g) \longrightarrow 2\,NH_3(g)$

11. True or false? An oxidizing agent is reduced in a chemical reaction. Explain your response.

12. True or false? A reducing agent is reduced in a chemical reaction. Explain your response.

13. True or false? An oxidizing agent undergoes a decrease in charge during a chemical reaction. Explain your response.

14. True or false? Reactants always undergo a change in their charge during an oxidation-reduction reaction. Explain your response.

15. Why can Earth's atmosphere be described as an oxidizing atmosphere?

16. Would Jupiter's atmosphere, which is rich in hydrogen (H_2) and methane (CH_4), be described as an oxidizing or a reducing atmosphere?

17. The main source of magnesium is seawater, which contains Mg^{2+} ions. Is this magnesium in an oxidized or reduced form?

18. When zinc metal reacts with an acid, Zn^{2+} ions form. Is the zinc metal oxidized or reduced?

19. Tell which substance is reduced and which is oxidized in the following equations. Label the oxidizing agent and reducing agent in each equation.
    (a) $2\,Al(s) + 3\,Cl_2(g) \longrightarrow 2\,AlCl_3(s)$
    (b) $S(s) + O_2(g) \longrightarrow SO_2(g)$
    (c) $CuO(s) + H_2(g) \longrightarrow Cu(s) + H_2O(g)$
    (d) $C_2H_4(g) + H_2(g) \longrightarrow C_2H_6(g)$

20. One of the chemical reactions that takes place when bleach and ammonia are combined is shown here:
    $$NaOCl + NH_3 \longrightarrow NaONH_3 + Cl_2$$
    (a) Which reactant undergoes oxidation?
    (b) Which reactant undergoes reduction?
    (c) Which reactant acts as the oxidizing agent?
    (d) Is this chemical equation balanced? If not, balance it.

21. Why would automobile rusting be less of a problem in southern Arizona than in Chicago?

22. Would you expect a piece of iron to rust on the surface of the moon? Explain.

23. Oxidation always occurs at the anode of an electrochemical cell.
    (a) True   (b) False

24. In an electrochemical cell, oxidation takes place at the _____ and reduction takes place at the _____.
    (a) Cathode, salt bridge
    (b) Anode, cathode
    (c) Cathode, anode
    (d) Electrode, voltmeter

25. Describe a simple battery, naming three essential parts.

26. How is a fuel cell similar to a battery? How is it different?

27. What is the difference between an electrochemical cell and a battery?

28. Besides electricity, what did the fuel cells used on the space shuttle produce?

29. Classify each of the following as a primary or secondary battery.
    (a) The battery in a wristwatch
    (b) A traditional automobile battery
    (c) The batteries in a hybrid vehicle
    (d) The battery in a cellular phone

30. A brand-new battery has a mass of 249.6 g. After it has been fully discharged, its mass is still 249.6 g. Explain this.

31. Think of ways electricity might be distributed to consumers of electric automobiles. Compare all the methods you have thought of with the methods used to distribute hydrocarbon fuels (gasoline). List as many benefits and problems as you can for each.

32. Give an advantage and disadvantage of an electric car that must be recharged by electricity from a fossil-fuel electric power plant.

33. Electric cars were popular in the early 1900s. Why were they replaced by cars that used an internal combustion engine?

34. A cell in a lead storage battery produces 2 volts, but the electrical system of a car operates at 12 volts. How can lead storage battery technology generate the 12 volts?

35. An alkaline battery produces 1.5 volts, but two batteries connected end to end produce 3.0 volts. What is happening?

36. Oxygen in the air reacting with substances in food can cause spoilage. Why are antioxidants added to food?

37. Which term best describes an antioxidant?
    (a) reducing agent   (c) oxidizing agent
    (b) cation   (d) free radical

38. Free radicals are molecules or molecular fragments with unpaired electrons. Which of the following is a free radical: NO, CO, or O_3? Justify your answer.

39. Free radicals are molecules or molecular fragments with unpaired electrons. Which of the following is a free radical: NO_2, CO_2, or N_2O? Justify your answer.

40. Normally, the mass of a new battery is the same as its mass after it is fully discharged. Explain why this is true.

41. The mass of a fuel cell is constant, but the mass of reactant and product changes during operation of the cell. Explain why this is true.

42. Aqueous solutions of Cu^{2+} ion are blue. The higher the concentration, the darker the color. When a strip of zinc metal is immersed in a solution of Cu^{2+}, the blue color fades. What is happening?

43. Aqueous solutions of Cu^{2+} ion are blue—the higher the concentration, the darker the color. When a strip of copper metal is immersed in a solution of Cu^{2+}, the color stays the same. Why does this happen?

44. In addition to iron, what reactants are needed for the formation of rust?

45. Why does iron continue to corrode even after a layer of rust forms on the surface, whereas aluminum stops corroding after a corrosion layer forms?

46. Why is galvanized iron less susceptible to corrosion than bare iron?

47. The equation for the Hall–Heroult process to convert bauxite to pure Al is
$$2\,Al_2O_3 + 3\,C \longrightarrow 4\,Al + 3\,CO_2$$
What is oxidized in the reaction and what is reduced?

Water, Water
Everywhere

A single molecule of water consists of only three atoms. Yet water is so critical to life as we know it that the search for water on other planets has become a major focus of our efforts to determine if life is unique to Earth. How can it be that such a common and simple molecule, two hydrogen atoms covalently bonded to a central oxygen atom, plays such a major role in our world? Are the chemical and physical properties of water so significantly different from molecules of similar size and molecular weight as to make liquid water a key to survival of living organisms? Furthermore, why is there so much concern about the availability of pure water when more than 70% of Earth's surface is covered with water?

Too little water and we die of thirst. Too much water and we drown. Parts of Earth where water is in short supply become arid deserts, and life is difficult. If too much water is present, normal life for land dwellers becomes impossible and aquatic life takes over. Clearly, there is a healthy balance between too little and too much water, but the quality of the water, and even its physical state, are also critical. Although liquid water is abundant in the oceans, the much purer water on which many of life's activities depends is in relatively short supply. There is increasing evidence that the days when cattle ranchers fought over water supplies in the Old West may return to haunt us in ways people would never have imagined. As Earth's population begins to exceed seven billion people, the demand for water for human consumption and for the activities associated with our modern lifestyles has become enormous. There is not enough pure water to satisfy the worldwide demand, and much of the pure water isn't in the areas of the world where it is needed.

Water dissolves all kinds of substances and reacts chemically with many of the atmospheric pollutants we've encountered in earlier chapters. Several nitrogen oxides and

© iStockphoto.com/bratan007

sulfur oxides react readily with water in the atmosphere to produce acidic rainfall that can have a very negative impact on our environment and on many building materials. Dissolved substances in liquid water can include natural minerals and chemicals, man-made fertilizers and pesticides, gasoline and fuel additives, toxic chemicals of all types, and even unused antibiotics that people have flushed down the toilet. To minimize environmental damage and to ensure a safe and adequate water supply, we must limit all types of pollution. We must control both the amount of pollution and the types of pollution to which we subject our water supply because what we've got is all there is. The water must be used, reused, and used again, over and over, in a never-ending cycle. If it becomes polluted, we will have to clean it up.

11-1 The Unique Properties of Water

We explored some of the properties of water in Chapter 5 and are familiar with the shape of the molecule (Section 5-4) and the polar nature of the covalent oxygen–hydrogen bonds in the molecule (Section 5-5). We have seen that water is capable of intermolecular hydrogen bonding (Section 5-7). These features give rise to some unique physical properties of water in comparison to some other small molecules of similar size, number of atoms, and molecular weight.

© iStockphoto.com/setthakan

Water plays a crucial role in weather, food production, the economy, and the environment.

In Table 11.1, we compare water with ammonia, methane, nitrogen, oxygen, and carbon dioxide. All of these molecules contain between two and five atoms and have molecular weights between 16 g/mol and 44 g/mol. An inspection of the data in Table 11.1 shows that is where the similarities end. Water is the only one of these compounds whose physical state at room temperature is a liquid! Its boiling point is hundreds of degrees higher than the others. The melting point of water is also seen to be dramatically different from the others. Of this set of molecules, only ammonia has a boiling point and a melting point that are even remotely close to those of water. A review of the Lewis structures of these molecules reveals that water and ammonia are the only two molecules in this set that are capable of intermolecular hydrogen bonding. These physical properties also reveal that water is the only compound that exists naturally in all three physical states (solid, liquid, and gas) under normal conditions on Earth. Water appears uniquely suited for its role as the liquid essential to life on Earth.

As you read about the properties of water, try to imagine how these properties influence your daily activities and how different your lives would be were any of these properties significantly different.

11-1a Some Properties of Water

1. *Water is a liquid at room temperature as a direct consequence of hydrogen bonding between adjacent water molecules.* Pure water is a liquid between 0°C and 100°C.

2. *The density of solid water (ice) is less than that of liquid water.* Put another way, water expands when it freezes. If ice were a typical solid, it would be denser than liquid water, and lakes would fill with ice from the

Water in all three phases. Fog and mist are suspended water droplets. Water vapor cannot be seen.

Charles D. Winters

bottom up. This would have disastrous consequences for marine life, which could not survive in areas with winter seasons. The application of pressure causes ice to melt. This is a consequence of the structure of ice. The pressure causes ice to change to a form with a smaller volume, and since liquid water occupies a smaller volume than does ice, the ice converts to a liquid.

3. *The heat of fusion (melting) of ice is 80 cal/g.* **Heat of fusion** is the amount of heat required to melt 1 g of ice at 0°C into 1 g of liquid water at 0°C. This amount of energy is much higher than other common materials in our environment. Note here that when ice melts, the temperature remains the same until all of the solid has been converted into liquid.

4. *Water has a relatively high heat capacity.* Water can absorb large quantities of heat without large changes in temperature, because the added heat can break hydrogen bonds instead of increasing the temperature. For comparison, the **specific heat capacity** of water (1.00 cal/g-°C, 4.18 J/g-°C) is about ten times that of copper (0.092 cal/g-°C, 0.385 J/g-°C) or iron (0.106 cal/g-°C, 0.444 J/g-°C). Water's heat capacity accounts

TABLE 11.1 Comparison of Water to Some Similar-Sized Molecules

Compound Property	Water (H_2O)	Ammonia (NH_3)	Methane (CH_4)	Nitrogen (N_2)	Oxygen (O_2)	Carbon Dioxide (CO_2)
Molecular weight (g/mol)	18	17	16	28	32	44
State at room temperature	liquid	gas	gas	gas	gas	gas
Boiling point (°C)	100	−33.4	−161.5	−195.8	−183	sublimes
Melting point (°C)	0	−77	−182.5	−210	−218.8	−56.6 @ 5.2 atm

© Cengage Learning

for the moderating influence of lakes and oceans on the climate. Huge bodies of water absorb heat from the Sun and release the heat at night or in cooler seasons. Earth would have extreme temperature variations if it were not for this property of water. By contrast, the temperatures on the surface of the moon and the planet Mercury vary by hundreds of degrees through the light and dark cycles.

5. *Water has a high heat of vaporization.* For a liquid to vaporize, heat is required. The **heat of vaporization** of a liquid is a measure of the intermolecular attractions holding the molecules together in the liquid. Water has one of the highest heats of vaporization (540 cal/g or 2250 J/g). A consequence of this high heat of vaporization is the cooling effect that occurs when water evaporates from moist skin. Evaporating water molecules take with them a considerable amount of energy, which was needed to overcome the attractions between the leaving molecules and those remaining behind. As was the case for melting, the temperature of the liquid remains the same during vaporization.

6. *Water has a high surface tension.* Unlike gases, liquids have surface properties, and these are extremely important in the overall behavior of many liquids. Molecules beneath the surface of the liquid are completely surrounded by other molecules and experience forces in all directions because of intermolecular attractions. By contrast, molecules at the surface are attracted only by molecules below or beside them (Figure 11.1). This unevenness of attractive forces at the surface of the liquid causes the surface to contract, making it act like a skin. The energy required to overcome the "toughness" of this liquid skin is called the **surface tension,** and it is higher for liquids that have strong intermolecular attractions. Water's surface tension is high compared with those of most other liquids because of the extensive hydrogen bonding that holds water molecules to each other.

7. *Water is an excellent solvent, often referred to as the universal solvent.* Because it is such a good solvent, water from natural sources is not pure water but is instead a solution of substances dissolved by contact with water.

Heat of vaporization: The heat required to vaporize a given quantity of liquid at its boiling point

Surface tension: The amount of energy required to overcome the attraction for one another of molecules at the surface of a liquid

EXAMPLE 11.1 Energy Associated with Phase Changes of Water

 Calculate the total amount of energy that would be required to convert 5.00 g of ice at 0°C to 5.00 g of steam at 100°C.

SOLUTION

The energy applied to solid water is first entirely devoted to melting, and the temperature does not change until all of the ice is melted. Once the water is liquid, additional energy serves to warm the water to 100°C. Once at 100°C, all of the energy applied serves to convert the liquid to gas,

FIGURE 11.1 Surface tension.

(a) The surface is strengthened by intermolecular forces attracting surface molecules. (b) The water strider, a lightweight insect that does not provide enough force per unit area to break through the surface tension. Note that the strider does not walk on the sharp ends of its "toes."

Typical molecule in liquid

Surface molecule

Air   Water surface   H₂O

(a)

(b)

Ingo Arndt/Minden Pictures/Getty Images

and the temperature stays at 100°C. By calculating the energy needed to melt the ice, to warm the water from 0°C to 100°C, and to vaporize the water, we can determine the total amount of energy required.

We need to know the following information to solve this problem:

Heat of fusion of ice = 80.0 cal/g
Heat capacity of water = 1.00 cal/g-°C
Heat of vaporization of water = 540 cal/g

The heat input required to convert the solid ice at 0°C to liquid water at 0°C is found by multiplying the mass of ice (5.00 g) by the heat of fusion of ice (80 cal/g) to give 400 cal. This gives 5.00 g of liquid water at 0°C whose temperature must be increased to 100°C before vaporization can occur. The heat required to increase the temperature is found by multiplying the mass (5.00 g) times the heat capacity of water (1.00 cal/g-°C) times the total temperature change (100°C). The temperature change requires a total of 500 cal. Lastly, the heat required to convert the 5.00 g of water at 100°C to 5.00 g of steam at 100°C is simply the mass (5.00 g) times the heat of vaporization (540 cal/g). The vaporization requires 2700 cal. Thus, the total heat input required (an endothermic process) is 400 cal plus 500 cal plus 2700 cal, or 3600 cal.

By analogy, the reverse of this process, conversion of 5.00 g of steam at 100°C to 5.00 g of ice at 0°C, would release 3600 cal in an exothermic process.

TRY IT 11.1

Determine the amount of heat involved in the following conversions, and indicate whether the process is endothermic (requires energy input in the form of heat) or exothermic.

(a) Conversion of 3.00 g of ice at 0°C to 3.00 g of liquid water at 70°C

(b) Conversion of 6.00 g of water at 50°C to 6.00 g of steam at 100°C

(c) Conversion of 10.0 g of steam at 100°C to 10.0 g of liquid water at 0°C

Acid rain: Rain that has a pH below 5.6

As examples of the importance of water, consider that we need liquid water for drinking, solid water for ice skating and cooling, and water vapor for helping to regulate Earth's temperature in its role as a greenhouse gas. Were the density of solid water (ice) greater than that of liquid water, then lakes would freeze from the bottom up, ice skating would be more challenging, and the *Titanic* would likely have survived her maiden voyage since the iceberg that sank her would have been at the bottom of the ocean. We'd have to more carefully monitor a full glass of iced liquid since it would overflow as the ice melted. Water's high heat capacity and high heat of vaporization are taken advantage of in many chemical and industrial processes in which water is used to regulate temperature. Humans keep cool by sweating because the evaporating water takes up heat from the body. Citrus farmers frequently spray their crops with water on nights the temperature is expected to hover very near freezing, because the heat released as water freezes may be just enough to protect the fragile citrus from damage. The frozen water also serves as an insulating blanket. Water's surface tension explains why water drops and soap bubbles are spherical.

Water's properties as a universal solvent are both a blessing and a curse. Water is incredibly useful as a solvent for so many solutes, and we depend on this in innumerable ways. By the same token, water's ability to dissolve so many substances makes it very easily polluted. Since pure water of relatively neutral pH is critical to many human activities, it will be useful to look at some of the many ways in which water is polluted and the procedures for purification of water.

11-2 Acid Rainfall

The term **acid rain** was first used in 1872 by Robert Angus Smith, an English chemist and climatologist. He used the term to describe the acidic precipitation that fell on Manchester, England, at the start of the Industrial Revolution. Although neutral water has a pH of 7, rainwater becomes naturally acidified from dissolved carbon dioxide, a normal component of the atmosphere. The carbon dioxide, whose solubility in water at room temperature is 3.4 g/L, reacts reversibly with water to form a solution of the weak acid carbonic acid.

$$CO_2(g) + H_2O(\ell) \rightleftharpoons H_2CO_3(aq)$$

$$H_2CO_3(aq) + H_2O(\ell) \rightleftharpoons H_3O^+ + HCO_3^-(aq)$$

At equilibrium, the pH of rainfall saturated with CO_2 from the air is 5.6. Any precipitation with a pH below 5.6 is considered to be acid rain.

As you may have seen in Chapter 4, NO_2, SO_2, and SO_3 can all react with water in the atmosphere to produce acids: NO_2 produces nitric acid (HNO_3) and nitrous acid (HNO_2); SO_3 and SO_2 produce sulfuric acid (H_2SO_4) and sulfurous acid (H_2SO_3). When conditions are favorable, water droplets carrying these acids precipitate as rain or snow with a low pH. Ice core samples taken in Greenland and dating back to 1900 contain sulfate (SO_4^{2-}) and nitrate

(NO_3^-) ions. This indicates that at least from 1900 onward, acid rain has been commonplace.

Acid rain, as well as the precipitation of solid particles that contain acids, is a problem today because of the large amounts of acidic oxides being produced by human activities. When this precipitation falls on natural areas that cannot easily tolerate such acidity, serious environmental problems occur. The average annual pH of precipitation falling on much of the northeastern United States and northeastern Europe is between 4.0 and 4.5. Specific rainstorms in some areas where there are numerous sources of SO_2 and NO_x have had pH values as low as 1.5. Complicating matters further is the fact that acid rain is an international problem—rain and snow don't observe borders (Figure 11.2). Many Canadian residents are offended by the government of the United States because some of the acid rain produced in the United States falls on Canadian cities and forests.

The extent of the problems with acid rain can be seen in "dead" (fishless) ponds and lakes, dying or dead forests, and crumbling buildings. For the most part, dead lakes are still picturesque, but no fish can live in the acidified water. Lake trout and yellow perch die at pH values below 5.0, and smallmouth bass die at pH values below 6.0. Mussels die when the pH is below 6.5.

Acid rain damages trees in several ways. It disturbs the stomata (openings) in tree leaves and causes increased transpiration and a water deficit in the tree. The surface struc-tures of the bark and the leaves can also be destroyed by the acid. Acid rainfall can acidify the soil, damaging fine root hairs and thus diminishing nutrient and water uptake. In addition, acid rain dissolves minerals that are insoluble in groundwater and surface waters of normal pH, and many of these minerals contain metal ions toxic to plant life. For example, acid rain dissolves aluminum hydroxide in the soil, allowing aluminum ions (Al^{3+}) to be taken up by the roots of plants, where they have toxic effects.

$$Al(OH)_3(s) + 3\,H^+(aq) \longrightarrow Al^{3+}(aq) + 3\,H_2O(\ell)$$

Effects of acid rain on a forest in one of the most polluted parts of Europe. The devastation has been caused by emission of sulfur dioxide and nitrogen oxides from factories in the former East Germany and Czechoslovakia.

FIGURE 11.2 A world map of areas affected by acid deposition, which includes both acid rain and the precipitation of solid particles.

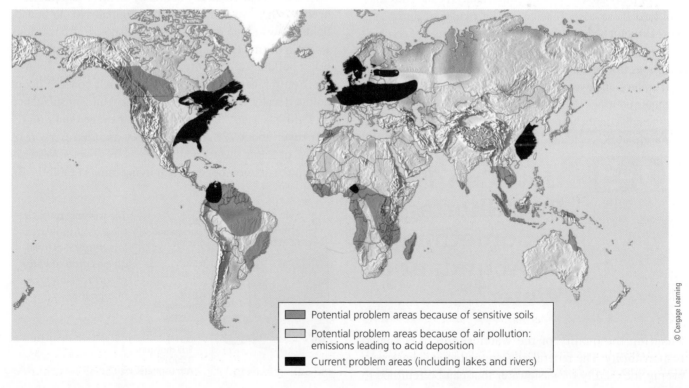

Potential problem areas because of sensitive soils

Potential problem areas because of air pollution: emissions leading to acid deposition

Current problem areas (including lakes and rivers)

The effects of acid rain and other pollution on stone and metal structures are especially devastating because of their irreversibility. By damaging stone buildings in Europe, acid rain is slowly but surely dissolving the continent's historical heritage. The bas-reliefs on the Cologne Cathedral in Germany are barely recognizable today. The Tower of London, St. Paul's Cathedral, and Lincoln Cathedral in Great Britain have suffered the same fate. Other beautifully carved statues and bas-reliefs on buildings throughout Europe and the eastern part of the United States and Canada are slowly passing into oblivion by the action of pollutants—in particular, acid rain (Figure 11.3).

What can be done about acid rain? Obviously, eliminating the emissions of the oxides of nitrogen and sulfur would be the answer. This is not easy, however. Some stopgap measures are being taken, such as spraying hydrated lime, $Ca(OH)_2$, into acidified lakes to neutralize at least some of the acid and raise the pH toward 7.

$$Ca(OH)_2(s) + 2\,H^+(aq) \longrightarrow Ca^{2+}(aq) + 2\,H_2O(\ell)$$

Some lakes in the problem areas have their own safeguard against acid rain. They have limestone-lined bottoms, which supply calcium carbonate ($CaCO_3$) for neutralizing the acid from acid rain (just as an antacid tablet relieves indigestion).

The ultimate answers to acid rain problems lie with those industries that produce the oxides of sulfur and nitrogen and with the regulatory agencies that govern them. Methods exist for the control of SO_2 emissions, although some of these are costly. In the final analysis, the consumer will bear those costs. The control of oxides of nitrogen is more difficult because there are so many sources of combustion exhaust gases. Catalytic converters help control NO_x emissions from automobiles, but most home furnaces, industrial boilers, and even electric generating plants do not have adequate controls. Fortunately for acid rain production, NO_2 is so reactive in the troposphere that it is not the major contributor that SO_2 is to acid rain.

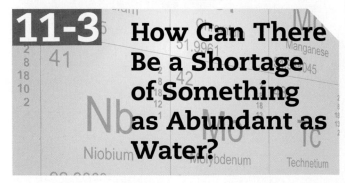

11-3 How Can There Be a Shortage of Something as Abundant as Water?

Ensuring the quality of the water supply is a shared responsibility. The federal government enacts laws governing the quality of wastewater that can be returned to

FIGURE 11.3 Deterioration to Paul Revere's tomb as a result of acid deposition.

the environment and the quality of the water that can be supplied for drinking. The U.S. Environmental Protection Agency (EPA) enforces these laws. State and local governments have their own laws and shoulder the responsibility for enforcing them as well as seeing to it that local industries, sewage treatment plants, and municipal water treatment facilities meet federal standards.

The goal of these regulations is to keep water relatively pure so the next user of the water, whether that is a city wanting a water supply or a trout looking for a nice stream to live in, will find it suitable.

Individual citizens also have a role to play in keeping the water clean. Voluntary action by an informed public is needed to halt water pollution that originates in our households. Laws alone cannot halt the disposal of hazardous waste in the municipal garbage or the pouring of potential water pollutants into our toilets, sinks, and storm drains.

You might ask why there is so much concern about polluted water when water is the most abundant substance on Earth's surface. Oceans (with an average depth of 2.5 mi) cover about 72% of Earth. They are the reservoir of 97.5% of Earth's water. Only 2.5% is freshwater. Water is also the major component of all living things (Table 11.2).

97.5% Oceans

2.5% Fresh water

1.97% Ice caps and glaciers

0.5% Groundwater

0.03% Other water:
0.02% Lakes, rivers
0.01% Soil moisture
0.0001% Atmosphere

The water supply. Of the 2.5% freshwater, less than 1% is available as groundwater or surface water for human use.

© Cengage Learning

TABLE 11.2 Water Content

	Percentage
Marine invertebrates	97
Human fetus (1 month)	93
Adult human	70
Body fluids	95
Nerve tissue	84
Muscle	77
Skin	71
Connective tissue	60
Vegetables	89
Milk	88
Fish	82
Fruit	80
Lean meat	76
Potatoes	75
Cheese	35

© Cengage Learning

For example, the water content of an average adult human is 70%—the same proportion as for Earth's surface.

An average of 4.35 trillion gallons of rain and snow fall on the contiguous United States each day (Figure 11.4). Of this amount, 3.1 trillion gallons return to the atmosphere by evaporation and transpiration. The discharge to the sea and to underground reserves amounts to 0.80 trillion gallons daily, leaving 0.45 trillion gallons of surface water each day for domestic and commercial use. The United States receives enough annual precipitation to cover the entire country to a depth of 30 inches. The 48 contiguous states withdrew 40 billion gallons per day from natural sources in 1900, but that rose to over 450 billion gallons by 2004. Daily water use will increase as the world's population increases. But the total water supply does not increase. Thus, areas where water is pumped out of the ground faster than it is replenished will surely expand. This means that water must be reused on a daily basis. The water molecules you drink from a water fountain may have been in the municipal wastewater of another city just a few days earlier.

The two sources of usable water are *surface water*, such as rivers, lakes, and wetland waters, and *groundwater*, which is beneath Earth's surface. Figure 11.4 shows our water resources and the flow of groundwater. About 90 billion gallons of the total water withdrawn every day is groundwater drawn from wells drilled into **aquifers**, layers of water-bearing porous rock or sediment held in place by impermeable rock. These kinds of wells, which supply water to many cities in the Great Plains and along the East Coast, are called artesian wells.

In the arid western United States, wells used to pump water for irrigation either are going dry or require drilling

> **Aquifer:** A layer of water-bearing porous rock or sediment held in place by impermeable rock

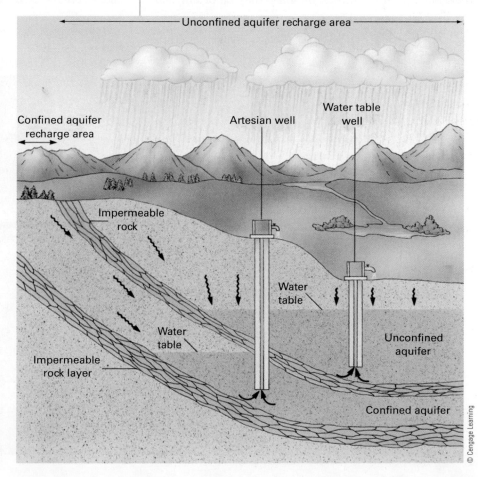

FIGURE 11.4 Water resources.
Surface water includes water that collects in rivers, lakes, and wetlands. Groundwater may be in unconfined aquifers, which are recharged by surface water from directly above, or in confined aquifers, which lie between impermeable rock layers. Where water in a confined aquifer is under pressure, pumping of the water may not be needed.

© Cengage Learning

so deep that irrigation is no longer economically feasible. In some places, water for agriculture has been withdrawn from the huge Ogallala aquifer, which stretches from South Dakota to Texas, 40 times faster than it is replaced by natural processes. Farms have been abandoned as taking water from shallow regions of the aquifer has become impractical.

The depletion of a major aquifer along the eastern seaboard has caused large sinkholes in several areas when the limestone rock strata of the aquifer collapse as the water is withdrawn (Figure 11.5). The entire city of Houston has sunk several feet over the years as the result of extensive use of the underground water sources in that area. Many coastal cities are also experiencing problems with brackish water that comes from aquifers where the freshwater has been drawn off, causing seawater to flow into the depleted aquifer. In Dayton, Ohio, the issue of aquifer use is somewhat different. A decline in industry and related loss of population has led to a rise in the water table, causing basements to flood in the downtown area. Together, these examples indicate the important and sometimes precarious balance that exists between water use and availability.

© iStockphoto.com/Alan Goulet

Who's using the water? Table 11.3 shows the breakdown for water use in the United States. Thermoelectric power and irrigation account for 82% of all water used.

Industrial water usage can be directly related to production of finished products. One gallon of water, for example, is used to produce eight sheets of ordinary typing paper, while 80 gallons of water are needed to produce a gallon of gasoline, and about 25 gallons of water are needed to make a single box of nails. Of course, because none of these products *contains* water, the water used is either recycled into the environment or recycled within the facility.

In many industrial locations, the largest single use of water is in cooling systems. Water can absorb more heat than any other readily available liquid—it has a heat capacity of 1 cal/g, or 4.18 J/g. Recirculating cooling water is an important means of water conservation. A cooling tower allows the warmed water to lose its heat to the surrounding air and helps to reduce thermal pollution of the river or lake where the used water would otherwise be discharged.

FIGURE 11.5 A sinkhole in Richmond, Virginia.
As the result of aquifer depletion, the top of an underground cavern collapsed, leaving five dead and devastating a historic neighborhood.

AP Photo/Richmond Times-Dispatch, Mark Gormus

TABLE 11.3 Water Consumption in the United States as a Percentage of Daily Total of 408 Billion Gallons

	Total (%)
Households	11
Industry	5
Thermoelectric power	48
Agriculture/irrigation	34
Other	2

Source: http://water.usgs.gov

TABLE 11.4 Average Daily per Capita Indoor Domestic Water Use

Domestic Indoor Use	Gallons per Capita in Average Homes	Gallons per Capita in Homes with Water-Efficient Fixtures
Showers	12.6	10.0
Clothes washers	15.1	10.6
Dishwashers	1.0	1.0
Toilets	20.1	9.6
Baths	1.2	1.2
Leaks	10.0	5.0
Faucets*	11.1	10.8
Other domestic uses indoors[†]	1.5	1.5
Total	72.6	49.7

* Americans drink more than 1 billion glasses of tap water per day.
[†] On average, 50% to 70% of all home water is used outdoors for watering lawns and plants.
Source: American Water Works Association website: http://awwa.org

In addition, the high heat capacity of water enables the industrial user to recycle heat energy captured by the cooling water.

Groundwater and surface water sources each provide half of the more than 44 billion gallons of **potable** (drinkable) water that is used each day in the United States. Table 11.4 gives the average amounts for various personal uses. Only a small fraction of municipal water really needs to be potable—that is, of drinking-water quality. The largest portion of a water supply could be disinfected and made bacteriologically safe while avoiding the more costly treatment needed to meet drinking quality standards. Water treated in this way would be suitable for irrigation of parks and golf courses, air-conditioning, industrial cooling, and toilet flushing. Consider the data in Table 11.4, which illustrate the inefficiency of residential water systems. It is obvious that strong arguments could be made for dual water systems—one system that treats water for drinking and a different system for other uses.

Residential water conservation is a way to cut demand for freshwater supplies. Although Table 11.3 shows that residential use is a small part of the total, home water conservation can be an important step in cutting demand for freshwater supplies in large urban areas.

Almost all water molecules have been recycling since they were formed billions of years ago. Have you ever considered that your next glass of water may include some molecules that Aristotle or Abraham Lincoln drank, or some molecules that flooded into the *Titanic* when it sank? In spite of these possibilities, the idea of obtaining potable water from wastewater or even sewage, especially wastewater that was *recently* discharged, is psychologically difficult for many people to accept. Nevertheless, the technology has been developed and was used in NASA's space shuttle flights. What this means is if water (pure or not) is available, it should be considered for reuse. In the southwestern United States, the rate of depletion of aqui-

fers has led to direct recycling of water from sewage, a process called **groundwater recharge**. For example, in El Paso, Texas, 10 million gallons of pure water per day from sewage effluent is pumped into the underground aquifer that is the main source of water for El Paso.

Potable: Describes water suitable for drinking

Groundwater recharge: Return to groundwater of water from treated sewage

Pathogen: A disease-causing microorganism

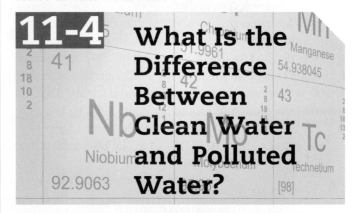

11-4 What Is the Difference Between Clean Water and Polluted Water?

The term *pollution* is used to describe any condition that causes the natural usefulness of air, water, or soil to be diminished. Water that is judged unsuitable for drinking, washing, irrigation, or industrial uses is polluted water. The pollution (Table 11.5) may be heat, radioisotopes, toxic metal cations and anions, organic molecules, acids, alkalies, or organisms that cause disease (**pathogens**). Water suitable for some uses might be considered polluted and

Megan
Claire
Christopher

TABLE 11.5 U.S. Public Health Service Classes of Pollutants

Pollutant	Example
Oxygen-demanding wastes	Plant and animal material
Infectious agents	Bacteria, viruses
Plant nutrients	Fertilizers such as nitrates and phosphates
Organic chemicals	Solvents, pesticides, detergent molecules
Other minerals and chemicals	Acids from mining operations, inorganic chemicals from metal-working operations
Sediment from land erosion	Clay silt from stream beds
Radioactive substances	Mining wastes, used radioisotopes
Heat from industry	Cooling water from electric generating plants

© Cengage Learning

TABLE 11.6 Dissolved Substances Found in "Pure" Water

Name	Formula	Comment
The following come from contact of water with the atmosphere:		
Carbon dioxide	CO_2	Makes water slightly acidic
Dust particles	—	Can be large amounts at times
Nitrogen	N_2	Along with oxygen, causes visible bubbles in hot water
Nitrogen dioxide	NO_2	Formed by lightning
Oxygen	O_2	Supports aquatic life
The following vary considerably depending on the kinds of rock formations the water has contacted:		
Bicarbonate ions	HCO_3^-	Soils and rocks
Calcium ions	Ca^{2+}	From limestone
Chloride ions	Cl^-	Soils, clays, and rocks
Iron(II) ions	Fe^{2+}	Soils, clays, and rocks
Magnesium ions	Mg^{2+}	Soils, clays, and rocks
Potassium ions	K^+	Soils, clays, and rocks
Sodium ions	Na^+	Soils, clays, and rocks
Sulfate ions	SO_4^{2-}	Soils and rocks

© Cengage Learning

therefore unsuitable for other uses—you might go swimming in water you would not consider drinking. Water that is unsuitable for use has often been polluted by human activity, but natural processes can also pollute water. For example, water that contacts organic substances such as decaying leaves and animal wastes will pick up numerous organic compounds, many of which impart odors and color to the water and some of which might be pathogenic. Silt, consisting of small suspended particles of dirt and sand, can also pollute water. Table 11.6 lists some of the substances that can be found in "pure" natural water. By looking at this list, it is clear that absolutely pure (100%) water is not a common commodity.

As human activities have continued to pollute water, governments have passed laws designed to keep our waters clean. The Clean Water Act of 1972 represented a major change in the thinking of Congress concerning who is responsible for keeping water clean. This act shifted the burden of producing water suitable for reuse from the user (a municipality, for example) to the wastewater discharger. Because it is easier to clean wastewater prior to dumping than to clean the water after the untreated waste has been discharged (Figure 11.6), the Clean Water Act was a major step in improving the quality of our natural waters. The act requires the EPA to establish and monitor emission standards—the maximum amounts of water pollutants that can be discharged into natural bodies of water from factories, municipal sewage treatment plants, and other facilities. As it turns out, the wastewater effluent from an industry can now often be clean enough to be used for such purposes as irrigation or industrial cooling.

FIGURE 11.6 The EPA requires that virtually all industrial wastewaters be treated prior to discharge.

This water is reasonably pure, although it is not as pure as drinking water.

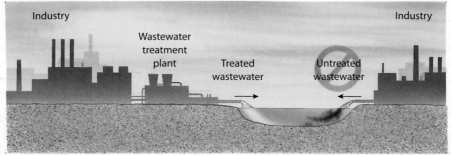

© Cengage Learning

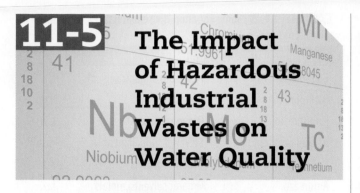

11-5 The Impact of Hazardous Industrial Wastes on Water Quality

Industrial processes, whether making paper, automobiles, or television sets, produce waste materials. Table 11.7 lists some of the industrial pollutants that result from the manufacture of products important to us. For many years, the disposal of solid wastes in *landfills* was considered good engineering practice. Many of the substances present in those wastes were partially dissolved by rainwater and became part of the groundwater, causing serious pollution of water supplies. Many of these older landfills also allowed surface runoff that contained dissolved substances from the wastes to be carried into natural bodies of water.

It was also common practice in the past to place waste and discarded chemicals into metal drums and bury them directly in the ground. After a few years, the drums developed leaks due to corrosion, allowing the drum contents to leak into water that would ultimately become groundwater or surface water. In recognition of this common practice and what it was doing to water quality, the U.S. Congress in 1976 passed the Resource Conservation and Recovery Act (RCRA). In 1980, Congress established the "Superfund," a $1.6 billion program designed to clean up hazardous waste sites that were threatening to contaminate the nation's water supplies. As of 1997, it was known that an average hazardous waste site cleanup took nine years and cost $7.4 million. By 1999, 205 hazardous waste sites had been cleaned to the extent that they could be removed from the Superfund list. The need for cleanup of some remaining sites is of growing urgency. These sites were established in once-rural locations that have now been surrounded by residential development.

> **Hazardous waste:**
> Industrial and household wastes responsible for water pollution

While the Superfund is dealing with existing hazardous waste sites, the RCRA and its regulations govern the disposal of newly generated wastes that have the potential to harm the environment. The EPA defines certain industrial wastes as **hazardous wastes** and closely regulates how they are generated, stored, transported, and disposed of. The RCRA was designed to give "cradle-to-grave" (origin to disposal) responsibility to *generators* of hazardous wastes. Before the RCRA, an industry could hire almost anyone to haul away its waste without regard to where it was taken or how it would be disposed of. Today, a generator of a hazardous waste must know who is transporting it, where it is going, and how it will be disposed of. Each shipment of hazardous waste is accompanied by a *manifest*, a document listing the hazardous waste by name, how much is present, and how it will be disposed of. As the load of hazardous waste is transported, temporarily stored, and finally disposed of, parts of the manifests are signed by each responsible party and returned to the generator at each step. The states also receive copies of the manifests, and annual reports of hazardous waste activities are filed with the EPA by everyone who handles hazardous wastes routinely.

Today, industrial hazardous wastes must be placed in *secure* landfills, incinerated, or treated in some way to render them nonhazardous. No hazardous waste is allowed to be disposed of in a way that could pollute the environment. Some hazardous wastes go to secure landfills with plastic linings that prevent their contents from easily reaching surrounding water supplies. These landfills also have carefully spaced monitoring wells so that any substances escaping from the landfill's contents may be detected, allowing the problem to be corrected. Other hazardous wastes may no longer be placed in secure landfills but must be incinerated or destroyed in some other way. Although incineration seems like a logical best choice to dispose of hazardous wastes, current incinerators are operating

TABLE 11.7 Important Industrial Products and Pollutants Associated with Their Manufacture

Products	Pollutants
Plastics	Solvents, organic chlorine compounds
Pesticides	Organic chlorine compounds, organic phosphate compounds
Medicines	Solvents, metals such as mercury and zinc
Paints	Metals, pigments, solvents, organic residues
Petroleum products	Oils, organic solvents, acids, alkalies
Metals	Metals, fluorides, cyanide, acids, oils
Leather	Chromium, zinc
Textiles	Metals, pigments, organic chlorine compounds, solvents

© Cengage Learning

at near capacity, and it is difficult to get proper permits for new ones because of community opposition. (Incineration of *some* hazardous wastes does produce small quantities of other, even more harmful combustion products.)

States like California that have severe water shortage problems have taken drastic steps to protect their ground- and surface water. In California, Proposition 65, the Safe Drinking Water and Toxic Enforcement Act of 1986, requires the yearly update of a list of chemicals known to the state to cause cancer, birth defects, or other reproductive harm. Discharge of these chemicals into any water that might become a drinking water supply is prohibited, and violation of this prohibition may incur financial penalties.

11-6 Household Wastes That Affect Water Quality

Often we do not think about the things we discard in our garbage, but what we throw away and how we do it can affect the quality of natural waters as much as what industry does. Household wastes that are incinerated can contribute to air pollution, but because the bulk of our household waste goes to landfills, we too can be responsible for causing pollution of groundwater as well as of rivers, streams, and lakes. Table 11.8 lists some common household products and the kinds of chemicals they contain. Because we are the consumers of industrial products, we can put the very same chemicals into the water as industry can. Although the individual amounts of harmful chemicals used in a household are less than those used in a large industry, the total amounts disposed of daily by all households can be very large, even for a medium-sized city.

TABLE 11.8 Some Common Household Hazardous Wastes and Recommended Disposal

Type of Product	Harmful Ingredients	Disposal*
Bug sprays	Pesticides, organic solvents	Haz.
Bathroom cleaners	Bases and acids	Drain
Furniture polish	Organic solvents	Haz.
Aerosol cans (empty)	Solvents, propellants	Trash
Nail polish and remover	Organic solvents	Haz.
Antifreeze	Organic solvents, metals	Haz.
Insecticides	Pesticides, solvents	Haz.
Auto battery	Sulfuric acid, lead	Haz.
Medicine (expired)	Organic compounds	Trash
Paint (latex)	Organic polymers	Trash
Gasoline	Organic solvents	Haz.
Motor oil	Organic compounds, metals	Haz.
Drain cleaners	Bases	Drain
Paints (oil-based)	Organic solvents	Haz.
Mercury and Ni-Cd batteries	Heavy metals	Haz.

* Haz: Professional disposal as a hazardous waste.

Drain: Disposal down the kitchen or bathroom drain with plenty of water.

Trash: Treat as normal trash—will not harm groundwater. In most households, the items marked Haz. are disposed of as normal trash, which results in groundwater pollution.

Source: Water Environment Federation Fact Sheet, 2001.

Households have a greater problem disposing of hazardous wastes than industry does. Even where there is an active recycling project for glass, paper, metals, and plastics, there is often no pickup of chemicals that should be separated from the ordinary trash destined for the landfill. If these chemicals are mixed with ordinary garbage, they go to the city landfill or incinerator. If they are poured out in the sink, driveway, or backyard, they will eventually reach natural waters. The EPA estimates that each year 350 million gallons of waste motor oil are poured on the ground or flushed down the drain by individual citizens. Compare this to the size of the *Deepwater Horizon* oil spill in the Gulf of Mexico (2010), which released approximately 206 million gallons of crude oil.

How can we dispose of hazardous household wastes without danger to the groundwater supply? We can ask our city's municipal waste authorities to provide disposal sites for these wastes or to sponsor periodic household hazardous waste days when these materials can be brought to a central site. In some U.S. cities and some European countries (such as the Netherlands), special trucks routinely pick up hazardous household wastes.

11-7 Toxic Elements Often Found in Water

Heavy metal ions are perhaps the most common of all water pollutants. The heavy metals include such frequently encountered elements as lead and mercury, as well as many less common ones like cadmium, chromium, nickel, and copper. These metals can, at times, be acutely toxic, causing immediate symptoms, but often they are chronically toxic in very small quantities. Chronic toxicity is characterized by nagging symptoms that lessen normal body functions. Inadequate disposal of wastes from mining or industrial activities causes these metals to find their way into water supplies. In addition, some farming activities and the disposal of household wastes can contribute to the presence of heavy metals in our water supplies.

11-7a Mercury

Mercury is a fairly common heavy metal. Its elemental form at room temperature is that of a liquid that is volatile with the characteristic shininess of a metal. Mercury atoms are readily oxidized to Hg_2^{2+} [mercury(I) ion] and Hg^{2+} [mercury(II) ion]. Both of these ionic forms are toxic, and their effects can be cumulative (that is, repeated exposures will increase the toxic effects because the body does not easily rid itself of the element).

Mercury is widely used in industry and can still occasionally be found in homes in mercury thermometers. The disposal of used fluorescent lamps represents a major source of mercury in water. Fluorescent lamps contain small amounts (about 60 mg per lamp) of mercury, which vaporizes inside the lamp and helps carry the electric current between the electrodes. Discarded lamps of this type are ultimately crushed when

they are disposed of in landfills. This releases the mercury to become oxidized and then to dissolve in any water that comes in contact with the waste.

Once released into streams, lakes, or oceans, metallic mercury sinks to the bottom, where it is converted by microorganisms to methylmercury (CH_3Hg). In this stable and persistent form, it accumulates in the flesh of large marine creatures such as tuna, swordfish, and marine mammals. Once there, it becomes a hazard to the health of humans who consume these creatures, and numerous health advisories have been issued about consuming mercury-contaminated seafood.

11-7b Lead

Lead is another widely encountered heavy metal. Like mercury, lead tends to accumulate in the body on repeated exposures. Lead has been used for centuries in plumbing (the Latin name for lead is *plumbum*, the word from which *plumber* is derived). Until recently, lead-based solders were routinely used in almost all plumbing fixtures, including the valves in drinking fountains as well as shower fixtures and faucets found in both the bathroom and kitchen. The current lead level for drinking water allowed by the EPA is 15 mg per liter.

In addition, lead has been widely used in various paint formulations. For the past 30 years, paints containing lead have been banned for interior use, but lead-containing paints are still available for outdoor and industrial use. When these paints or objects coated with them are discarded, lead can find its way into water supplies. The Engelhard Corporation received a 2004 Presidential Green Chemistry Challenge Award for their development of a line of paint pigments that do not contain heavy metals.

11-7c Arsenic

Arsenic occurs naturally in small amounts in many foods. Shrimp, for example, contain about 19 parts per million (ppm) arsenic, and corn may contain 0.4 ppm arsenic. The amount of naturally occurring arsenic in foods depends on the surroundings where they are grown and the metabolism of the plant or animal. While many soils contain arsenic, which causes an accumulation of the element as a plant grows, some insecticides also contain arsenic, which causes an arsenic residue when the insecticide is applied. The U.S. Food and Drug Administration (FDA) has set a limit of 76 ppm for arsenic levels in shellfish. In its ionic forms, arsenic is much more toxic than in its covalently bound compounds.

The typical toxic arsenic compounds contain ions such as arsenate (AsO_4^{3-}) or arsenite (AsO_2^-).

Arsenic and the heavy metal ions are toxic primarily because of their ability to react with sulfhydryl groups (iSH) in enzymes (Section 15-8).

11-8 Measuring Water Pollution

11-8a Biochemical Oxygen Demand

Many organic compounds that find their way into water can easily be oxidized by microorganisms that are also there. This is a natural process that prevents a buildup of organic waste in natural waters. To change this organic material into simple substances (such as CO_2 and H_2O) requires oxygen. The amount of dissolved oxygen required is called the **biochemical oxygen demand (BOD)**, and it is a measure of the quantity of dissolved organic matter. The oxygen is necessary so that the bacteria and other microorganisms can metabolize the organic matter that constitutes their food. Ultimately, given near-normal conditions and enough time, the microorganisms will convert huge quantities of organic matter into the following end products:

$$\text{Organic carbon} \longrightarrow CO_2$$
$$\text{Organic hydrogen} \longrightarrow H_2O$$
$$\text{Organic oxygen} \longrightarrow H_2O$$
$$\text{Organic nitrogen} \longrightarrow NO_3^- \text{ or } N_2$$

At 20°C (68°F), the solubility of O_2 in water under normal air pressure of 1 atmosphere (atm) is only 0.0092 g O_2/L. But a stream containing 10 ppm by weight (just 0.001%) of an organic material with the formula $C_6H_{10}O_5$ has a BOD of 0.012 g O_2/L of water. Clearly, this BOD value exceeds the equilibrium concentration of dissolved O_2 at this temperature. As the bacteria utilize the dissolved oxygen in a stream or lake with this BOD, the oxygen concentration of the water may soon drop too low to sustain any form of fish life. Whether this happens depends on the opportunities for new oxygen to become dissolved in the water. Life-forms can survive in water where the BOD exceeds the dissolved oxygen if the water is flowing vigorously in a shallow stream (this facilitates the absorption of more oxygen from the air via aeration).

BOD values can be greatly reduced by treating industrial wastes and sewage with oxygen or ozone. Numerous commercial cleanup operations now being developed and used employ this type of "burning" of the organic wastes. Another benefit of treating wastewater with oxygen is that some of the nonbiodegradable material becomes biodegradable as a result of partial oxidation.

Highly polluted water often has a high concentration of organic material, with resultant large biochemical oxygen demand. In extreme cases, more oxygen is required than is available from the environment. The result is that fish and other aquatic life can no longer survive. The aerobic bacteria (those that require oxygen for the decomposition process) die. As a result, even more lifeless organic matter results, and the BOD soars. Nature, however, has a backup system for such conditions. A whole new set of microorganisms (anaerobic bacteria) takes over; these organisms take oxygen from oxygen-containing compounds to convert organic matter to CO_2 and water.

Organic nitrogen is converted to elemental nitrogen by these bacteria. Given enough time, enough oxygen may become available, and aerobic oxidation will then return.

11-9 How Water Is Purified Naturally

Water is a natural resource that, within limitations, is continuously renewed. The water cycle offers a number of opportunities for nature to purify its water. The worldwide *distillation* process results in rainwater containing only traces of nonvolatile impurities, along with gases dissolved from the air. *Crystallization* of ice from ocean saltwater results in relatively pure water in the form of icebergs. *Aeration* of groundwater as it trickles over rock surfaces, as in a rapidly running brook, allows volatile impurities to be released to the air and allows oxygen to be

©iStockphoto.com/ranplett

dissolved. *Sedimentation* (or *settling*) of solid particles occurs in slow-moving streams and lakes. *Filtration* of water through sand rids the water of suspended matter such as silt and algae. Of very great importance are the *oxidation* processes carried out by bacteria and other microorganisms. Practically all naturally occurring organic materials—plant and animal tissue as well as their waste materials—can be oxidized in surface waters as long as oxygen is available and their concentration is not too high. Finally, another natural process is *dilution*. Most, if not all, pollutants found in nature are rendered harmless if reduced below certain levels of concentration by dilution.

Before the explosion of the human population and the advent of the Industrial Revolution, natural purification processes were quite adequate to provide ample water of very high purity in all but desert regions. Nature's purification processes can be thought of as massive but somewhat delicate.

Today, the activities of humans often push the natural purification processes beyond their limits, and polluted

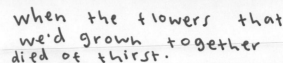

An aeration plant. Water is purified naturally at a treatment facility on the Des Plaines river near Chicago.

© iStockphoto.com/davidtidwell86

water accumulates. A simple example comes from dragging gravel from streambeds. The excavation leaves large amounts of suspended matter in the water, and for miles downstream, aquatic life is destroyed. Eventually, the solid matter settles, and normal life can be found again in the stream.

A more complex example—one that perhaps cannot be solved by relying on natural purification processes—is pollution by organic molecules that cannot be easily oxidized by microorganisms. A **biodegradable** substance is composed of molecules that are broken down to simpler ones by microorganisms. For example, cellulose suspended in water will be converted to carbon dioxide and water. A **nonbiodegradable** substance, on the other hand, cannot be easily converted to simpler molecules by microorganisms. If the conversion process is extremely slow, or if it cannot be done at all by natural microorganisms, nonbiodegradable substances tend to accumulate in the environment.

Some organic compounds, notably some of those produced synthetically, are nonbiodegradable. When these substances are introduced into the environment, they simply stay in the natural waters or are absorbed by life forms and remain intact for a long time. Branched-chain detergent molecules, for example, cannot easily be decomposed by microorganisms.

> **Biodegradable:** Describes a substance that can be broken down into simple molecules by the action of microorganisms
>
> **Nonbiodegradable:** Describes a substance that cannot be broken down by microorganisms and therefore persists in the environment

A branched-chain sodium alkylbenzenesulfonate detergent molecule

The first detergents that contained such molecules accumulated and caused noticeable foaming in rivers and streams. The branched-chain detergents were soon replaced by linear-chain alkylbenzenesulfonate detergents, which are easily decomposed by microorganisms—they are biodegradable.

A linear sodium alkylbenzenesulfonate detergent molecule

Other examples of nonbiodegradable organic pollutants are the chlorinated and polychlorinated hydrocarbons. Many of these compounds are used as insecticides (Section 19-5). The insect-killing ability of dichlorodiphenyltrichloroethane (DDT) was first recognized in 1939. By the end of World War II, its insecticidal properties were legendary because of its ability to kill everything from malaria-causing mosquitoes to lice. This success prompted the introduction of numerous other chlorinated hydrocarbons as insecticides, and by the early 1960s their use was widespread throughout the world. These compounds are broad-spectrum insecticides, killing most insects rather effectively; however, they also biodegrade very slowly, so they tend to accumulate in the environment. This persistence is especially troublesome since their slow biodegradation allows such compounds to accumulate in the food chain. Fish-eating birds, for example, can accumulate large quantities of these insecticides that have accumulated in fish that ate smaller organisms containing these compounds. The populations of falcons, pelicans, bald eagles, ospreys, and other birds have been endangered by persistent pesticides. DDT causes reproductive failure in birds by interfering with the mechanisms that produce strong eggshells. After the use of DDT was banned in the United States in 1972, the numbers of surviving hatchlings increased rather dramatically.

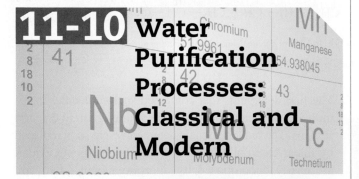

© iStockphoto.com/CarolinaSmith

11-10 Water Purification Processes: Classical and Modern

Sewage, which is 99.9% water, includes everything that flows from the sinks, tubs, washing machines, and toilets in our homes, factories, and public buildings. The outhouses common in rural areas in years past had their counterparts in city cesspools, which were basically holes in the ground into which sewage flowed. In cesspools, organic matter is decomposed by anaerobic bacteria, producing some pretty bad-smelling chemicals such as hydrogen sulfide, which has a characteristic rotten-egg odor. The terrible job of cleaning cesspools inspired the development of cesspools that could be flush-cleaned with water, followed by a connecting series of such pools that could be flushed from time to time. City sewage systems with no holding of the wastes were the next step.

At first, city sewage systems did little but channel sewage water to rivers and streams, where natural purification processes were expected to clean the water for the next users downstream. Today, however, sewage is treated using a combination of methods that can render the treated municipal wastewater almost as clean as the natural waters into which it is being discharged. The simplest treatment method is *primary wastewater treatment*, which copies two of nature's purification methods, settling and filtration. In primary treatment, sewage goes from the primary sedimentation tank shown in Figure 11.7 to the chlorinator.

Primary treatment removes 40% to 60% of the solids present in sewage and about 30% of the organic matter present. Calcium hydroxide and aluminum sulfate are added to produce aluminum hydroxide, which is a sticky, gelatinous precipitate that settles out slowly, carrying suspended dirt particles and bacteria with it.

$$3\ Ca(OH)_2(aq) + Al_2(SO_4)_3(aq) \longrightarrow$$
$$2\ Al(OH)_3(s) + 3\ CaSO_4(s)$$

For many years, municipal sewage treatment plants had only primary treatment, followed by chlorination of the treated wastewater (see Section 11-12) before it was discharged into a suitable river or stream. Chlorination kills any remaining harmful pathogens. Presumably, natural processes would get rid of the remaining solids and dissolved organic matter.

In realizing that this treatment was not sufficient to protect the public from contaminated water, the writers of the 1972 Clean Water Act required that sewage treatment plants also provide *secondary wastewater treatment*, which revives the old cesspool idea but under more controlled conditions. Modern secondary treatment operates in an

FIGURE 11.7 Primary and secondary sewage treatment.

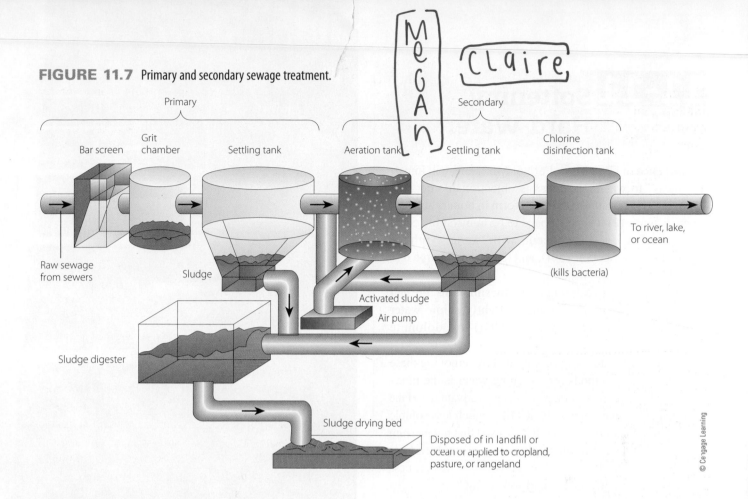

oxygen-rich environment (aerobic), whereas the cesspool operates in an oxygen-poor environment (anaerobic). The results are the same: The organic molecules that will not settle are consumed by microorganisms, and the resulting sludge will settle.

Even a combination of primary and secondary wastewater treatment systems will not remove dissolved inorganic materials such as toxic metal ions, nutrients such as nitrate ions (NO_3^-) or ammonium ions (NH_4^+), or nonbiodegradable organic compounds such as chlorinated hydrocarbons. These materials can be removed by a variety of

Various Definitions of "Pure Water"

Chemist: "Pure H_2O—no other substance."

Parent: "Nothing harmful to my child."

Game and Fish Commission: "Nothing harmful to animals."

Sunday boater: "Pleasing to the eye and nose—no debris."

Ecologist: "Natural mixture containing necessary nutrients."

tertiary wastewater treatment methods that are selectively introduced where the nature of the wastewater requires them. One obstacle to tertiary treatment is the initial expense of modifying sewage treatment plants and the ongoing expense of additional treatment.

Filtration of the water through *carbon black* is a type of tertiary treatment effective for removing soluble organic compounds that are nonbiodegradable and thus remain in the water after secondary treatment. Carbon black consists of finely divided carbon particles with a large surface area on which solutes, including certain potentially toxic substances, can be *adsorbed*.

A different kind of tertiary treatment is needed to remove ammonia or ammonium ion. Because nitrogen is a nutrient for aquatic microorganisms, excessive nitrogen released to natural waters can cause a soaring BOD with the accompanying fish kills and other problems. The water is exposed to **denitrifying bacteria** that convert ammonium ion or ammonia to harmless nitrogen gas.

Denitrifying bacteria:
Bacteria that convert ammonia or ammonium ion to nitrogen

$$NH_4^+(aq) \text{ or } NH_3(aq) \xrightarrow{\text{Denitrifying bacteria}} N_2(g)$$

Softening Hard Water

The presence of Ca^{2+}, Mg^{2+}, Fe^{3+}, or Mn^{2+} ions will impart "hardness" to water. Hardness is objectionable because (1) it causes precipitates (scale) to form in boilers and hot-water systems, (2) it causes soaps to form insoluble curds (this reaction does not occur with some synthetic detergents—see Section 15-4), and (3) it can impart a disagreeable taste to the water.

Hardness due to calcium or magnesium ions, present as carbonates, is produced when slightly acidic water trickles through limestone ($CaCO_3$) or dolomite $CaMg(CO_3)_2$.

Such "hard water" can be softened by removing these ions. One of the methods for softening water is the lime–soda process. The lime–soda process takes advantage of the facts that calcium carbonate ($CaCO_3$) is much less soluble than calcium bicarbonate [$Ca(HCO_3)_2$] and that magnesium hydroxide is much less soluble than magnesium bicarbonate. The raw materials added to the water in this process are hydrated lime [$Ca(OH)_2$] and soda (Na_2CO_3).

The overall result of the lime–soda process is to precipitate almost all of the calcium and magnesium ions in the form of calcium carbonate and magnesium hydroxide and to leave sodium ions as replacements.

Iron present as Fe^{2+} and manganese present as Mn^{2+} can be removed from water by oxidation with air (aeration) to higher oxidation states. If the pH of the water is 7 or above (either naturally or through the addition of lime), the insoluble compounds $Fe(OH)_3$ and $MnO_2(H_2O)_x$ are produced and precipitate from solution.

The desire for and achievement of soft water for domestic use has sparked a rather heated health debate. Soft water is usually acidic and contains Na^+ ions in the place of di- and trivalent metal ions. An increased intake of Na^+ is known to be related to heart disease. The acidic soft water is also more likely to attack metallic pipes, joints, and fixtures, resulting in the dissolution of toxic ions such as Pb^{2+}. One way to avoid sodium ions in drinking water and to use less soap when washing would be to drink only naturally hard water and to do your washing in soft water.

Many commercially available water-softening systems consist of an ion-exchange column containing some type of insoluble, negatively charged resin material that has been loaded up with sodium cations. As water passes through the resin column, the doubly charged cations

© Martin Shields/Science Source

Chlorine contact chamber at a municipal sewage treatment plant in West Caldwell, New Jersey. The water is chlorinated to kill disease-causing organisms before it is returned to natural waterways. The baffles slow the movement of the water to increase potential contact time with the chlorine.

(Ca^{2+}, Mg^{2+}, and Fe^{2+}) common in hard water are attracted more tightly to the column than are the singly charged Na^+ ions. The sodium ions are released into the water and the doubly charged cations are removed. This water-softening process is illustrated in Figure 11.8. A given ion-exchange column has a finite capacity to soften water since it will eventually become loaded with the hard water ions. Many modern refrigerators with drinking water dispensers and ice makers contain such water-softening systems.

Sodium carbonate (Na_2CO_3), sometimes known as washing soda, can also be used to soften water. The addition of this substance to water allows for formation of the more insoluble calcium carbonate and magnesium carbonate, which will settle out as solids.

FIGURE 11.8 Water softening by ion exchange in the home.

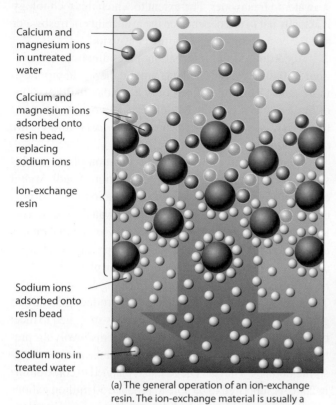

Calcium and magnesium ions in untreated water

Calcium and magnesium ions adsorbed onto resin bead, replacing sodium ions

Ion-exchange resin

Sodium ions adsorbed onto resin bead

Sodium ions in treated water

(a) The general operation of an ion-exchange resin. The ion-exchange material is usually a polymeric material formed into small beads.

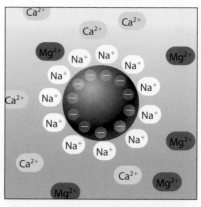

(b) An ion-exchange resin generally has negatively charged groups on the surface of the resin. Sodium ions are adsorbed onto the surface and balance the negative charge.

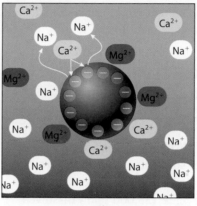

(c) Ion-exchange resin after Ca^{2+} and Mg^{2+} ions have been adsorbed. The more highly charged Ca^{2+} and Mg^{2+} ions are more strongly attracted to the resin beads than the sodium ions. This leads to the replacement of the sodium ions on the resin by the alkaline earth ions.

© Cengage Learning

11-12 Chlorination and Ozone Treatment of Water

With the advent of chlorination of water supplies in the early 1900s, the number of deaths in the United States caused by typhoid and other waterborne diseases dropped from 35 per 100,000 population in 1900 to 3 per 100,000 population in 1930.

Chlorine is introduced into water as the gaseous element Cl_2, and it acts as a powerful oxidizing agent for the purpose of killing bacteria in water. This process is used in treating water that will become both tap water and wastewater before it is released. Chlorination largely prevents the principal waterborne diseases spread by bacteria, which include cholera, typhoid, paratyphoid, and dysentery. Our municipal water supplies, however, remain vulnerable to contamination. In 1993, for example, in the Milwaukee area, a major outbreak of gastrointestinal disease was caused by *Cryptosporidium* bacteria. Lesser outbreaks have been caused by *Salmonella* bacteria. In some cases, the pathogens are found to have entered the water supply via contamination with surface water runoff.

Chlorination of industrial wastewater and city water supplies presents a potential threat because of the reaction of chlorine with residual concentrations of organic compounds to produce **disinfection by-products**.

Water containing organic compounds $\xrightarrow{\text{Chlorine}}$ Chlorinated organic disinfection by-products

These disinfection by-products, which may be present at levels of a few parts per million or less, include dichloromethane, chloroform, trichloroethylene, and chlorobenzene—all suspected carcinogens. The presence of chlorinated hydrocarbons can be prevented by more efficient removal of the organic matter that becomes chlorinated.

One way to eliminate chlorinated hydrocarbons as disinfection by-products is to use ozone (O_3) as the disinfectant.

Disinfection by-product: A compound formed by the action of disinfectants on substances found in water

Permeable: Allows substances to pass through

Semipermeable membrane: A membrane that allows water molecules but not ions or larger molecules to pass through

The ozone is produced on site by passing oxygen or air through an electric discharge. This process normally gives about a 20% ozone–oxygen mixture that is a very strong oxidizer.

Ozonation, like chlorination, is also not without potentially harmful disinfection by-products. Bromide ion (Br⁻), which is found in most natural waters, is oxidized by ozone to bromate ion (BrO_3^-), which is a suspected carcinogen. Generally, this single known harmful disinfection by-product is considered less of a risk factor than the numerous chlorinated hydrocarbons produced by chlorine disinfection.

11-13 Freshwater from the Sea

Because seawater covers 72% of Earth, it is not surprising that it is considered a water source for areas where freshwater supplies aren't sufficient to meet the demand. The oceans contain an average 3.5% (35,000 ppm) dissolved salts by weight, a concentration too high to make ocean water useful for drinking, washing, or agricultural use (Table 11.9). The total of dissolved ions must be reduced to below 500 ppm before the water is suitable for human consumption.

TABLE 11.9 Ions Present in Seawater at Concentrations Greater Than 0.001 g/kg

Ion	Seawater (g/kg)
Cl^-	19.35
Na^+	10.76
SO_4^{2-}	2.71
Mg^{2+}	1.29
Ca^{2+}	0.41
K^+	0.40
HCO_3^-, CO_3^{2-}	0.106
Br^-	0.067
$H_2BO_3^-$	0.027
Sr^{2+}	0.008
F^-	0.001
Total	35.129

© Cengage Learning

Technology has been developed for the conversion of seawater to freshwater. The extent to which this technology is actually put to use depends on the availability of freshwater and the cost of the energy for the conversion.

According to the International Desalination Association, 13,080 desalination plants around the world produced more than 12 billion gallons of water a day in 2009; 60% of these plants are in the Middle East. Two methods used to purify seawater are *reverse osmosis* and *solar distillation*.

Approximately 20% of the desalination plants worldwide depend on thermal (solar) distillation. Saudi Arabia currently has the world's largest desalination plant. It produces 128 million gallons of drinkable water per day; collectively, desalination plants produce 70% of Saudi Arabia's drinkable water. In the United States, Tampa Bay, San Diego, and El Paso already have large desalination plants that utilize reverse osmosis or have such plants nearing completion. The Tampa Bay plant has the capacity to produce 25 million gallons of drinkable water per day, but has struggled to meet this capacity because of technical problems involving pretreatment of the water to prevent fouling of the reverse osmosis membranes. The San Diego plant, planned to come online in 2016, is supposed to be able to produce 50 million gallons per day of drinkable water. A future plant in San Francisco is planned to produce over 119 million gallons per day.

11-13a Reverse Osmosis

An extremely thin piece of material such as a sheet of synthetic polymer or animal tissue can allow some molecules to pass through it. Such a material is called a membrane and is said to be **permeable** to those molecules and ions that can pass through. Permeability is dependent on the presence of tiny passages within the membrane. A membrane permeable to water molecules but not to ions or molecules larger than water molecules is called a **semipermeable membrane**. Many membranes made from synthetic polymers have this characteristic. One such polymer is cellulose acetate. If a semipermeable membrane is placed between seawater (brine) and pure water, the pure

© iStockphoto.com/joannawnuk

water will pass through the membrane to dilute the seawater. This is a process called **osmosis**. The liquid level on the seawater side rises as more water molecules enter than leave, and pressure is exerted on the membrane until the rates of diffusion of water molecules in both directions are equal. **Osmotic pressure** is defined as the external pressure required to *prevent* osmosis. Parts a and b of Figure 11.9 illustrate the concepts of osmosis and osmotic pressure.

Reverse osmosis is the application of pressure to cause water to pass through the membrane from the salty (aqueous solution) side to the pure-water side (Figure 11.9b). The osmotic pressure of normal seawater is 24.8 atm. As a result, pressures greater than 24.8 atm must be applied to cause reverse osmosis. Pressures up to 100 atm are used to provide a reasonable rate of reverse osmosis and to account for the increase in salt concentration that occurs as the process proceeds. Small, portable reverse osmosis devices can be used in emergencies to produce small amounts of pure water for personal use. These hand-operated desalinators can typically produce 4.5 L of clean water per hour from seawater.

11-14 Pure Drinking Water for the Home

In spite of all the efforts taken to purify public water supplies, many consumers are concerned about the quality of the water that comes out of the taps in their homes, schools, and places of business. Parents of small children are especially worried about chemicals such as lead and carcinogenic organic compounds that are chlorine disinfection by-products. Many have turned to bottled water or home water treatment devices.

The FDA regulates the quality of bottled water. The water passes through one or more purification steps (Figure 11.10). The three purification methods (which can also be done in home treatment systems)—distillation, carbon filtration, and reverse osmosis—have already been discussed as methods used for treating municipal water. The maximum levels of contaminants allowed by the EPA after these treatments are listed in Table 11.10. Each of these methods is expensive and results in a high cost per gallon of treated water. In the case of bottled water or home water treatment, the cost of treatment is not the major factor, because only the small amount of water needed for human consumption needs to be specially purified. Figure 11.10 shows which trace pollutants are removed or allowed to remain in the water by each treatment method, color-coded to examples of the types of contaminants in Table 11.10. Thus, the analysis on the label of the bottled

Osmosis: The flow of water molecules through a semipermeable membrane from a less-concentrated to a more-concentrated solution

Osmotic pressure: The external pressure required to prevent osmosis from taking place

Reverse osmosis: The application of pressure on a solution to cause water molecules to flow through a semipermeable membrane from a more-concentrated to a less-concentrated solution

FIGURE 11.9 (a) Purification of water by reverse osmosis. (b) Reverse osmosis at the nanoscopic level.

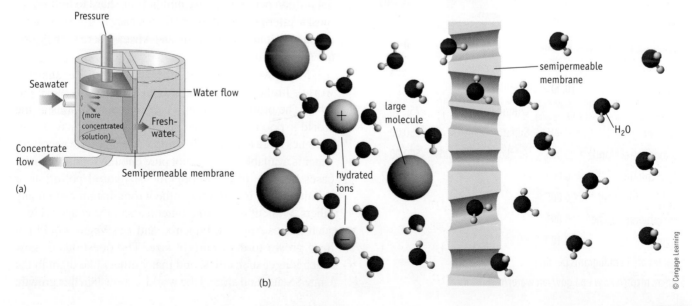

FIGURE 11.10 Final steps that can be used in water purification.   Both bottled water and municipal tap water are purified in these ways. The color code shows which pollutants are removed by each method. Pathogenic bacteria can and do pass through all of these methods. This is why municipal tap water must be treated with chlorine or some other disinfectant before release into the system.

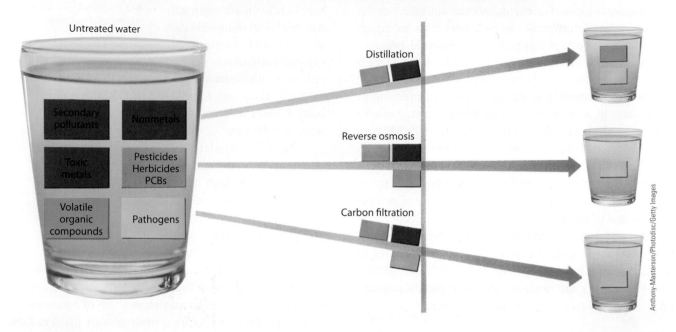

TABLE 11.10 A Partial List of Maximum Contaminant Levels (MCL) for Drinking Water Allowed by the EPA*

Metals		Herbicides, Pesticides, PCBs	
Beryllium	0.004	Chlordane	0.002
Cadmium	0.005	Endrin	0.0002
Chromium	0.100	Heptachlor	0.0004
Copper	1.300	Lindane	0.0002
Lead	0.015	Methoxychlor	0.040
Metalloids		Toxaphene	0.003
Arsenic	0.010	PCBs	0.0005
Antimony	0.006	**Secondary Contaminants**	
Nonmetals		Iron	0.30
Fluoride	4.00	Manganese	0.05
Nitrate	10.00	Zinc	5.00
Volatile Organic Compounds		Chloride	250.00
Benzene	0.005	Sulfate	250.00
Carbon tetrachloride	0.005	Total dissolved solids	500.00
Vinyl chloride	0.002		
Trichloroethylene	0.005		
Hexachlorobenzene	0.001		
Styrene	0.100		

* Values are in milligrams per liter.

Source: http://www.epa.gov/safewater/mcl.html

water is highly important and should be read with care before purchase.

The World Heath Organization (WHO) reports that drinking unclean water causes diarrhea that kills 1.6 million children under the age of five each year. An estimated 1 billion people in the world do not have access to safe, pure drinking water. By the time deaths attributable to diarrhea, cholera, schistosomiasis, and other diseases spread by contaminated water or aggravated by the lack of pure water for personal hygiene are included, the world-wide death toll is estimated at 9300 people per day or over 3.4 million per year! These numbers are hard to fathom for most Westerners, but they are far too easy to grasp for those living in certain areas of Asia and Africa where such deaths are almost routine.

Much work remains to be done to ensure that the world's limited water resources are used wisely and equitably. The problem will only become more serious as the world's population increases and as access to water becomes more and more difficult. Ultimately, the necessary water is available. It is just not pure enough for use in many cases because of anthropogenic or natural pollution or salinity, and it is certainly in the wrong locations in many others. Without water from other parts of the country, cities such as Los Angeles, Phoenix, and Las Vegas would not have grown to their current size. The questions of how much longer such cities, and many others like them in the United States and around the world, can sustain this growth, and at what cost, remain to be answered. *Time* magazine reported in April 2004 that the 2.5 billion gallons of water used to irrigate the world's golf courses each day would be enough to support 4.7 billion people at the UN daily minimum of water. *National Geographic* reported in an excellent look at the world's water resources in September 2002 that bone-dry Dubai had 2.6 million gallons of expensively desalinated seawater coursing through its popular Wild Wadi Water Park. It may be that money is a universal cure for this problem.

Your Resources

In the back of the textbook:

→ *Review Card on Water, Water Everywhere*

   In OWL for CHEM2 at www.cengagebrain.com

→ *Review Key Terms with Flash Cards (printable or digital)*

→ *Complete Interactive Practice Quizzes to prepare for tests*

→ *Submit Assigned Homework and Exercises*

Applying Your Knowledge

1. Define the following terms:
   (a) Surface water   (b) Groundwater
   (c) Brackish water   (d) Pollution
   (e) Groundwater recharge (f) Potable

2. Define the following terms:
   (a) Distillation   (b) Hard water
   (c) Disinfection by-products (d) Semipermeable
   (e) Dilution   (f) Hazardous waste

3. Where does most of the water go that falls on the continental United States every day?

4. Explain how rainwater becomes groundwater.

5. Explain how groundwater can become contaminated with pollutants.

6. What is the origin of brackish water?

7. What activity is the largest single user of water?

8. What causes the level of an aquifer to drop? Cite some examples of the effects of aquifers dropping in level.

9. How much water do you think you use per day? List your uses and include water that might be used "for you," such as in food preparation in a restaurant.

10. What would you expect to find dissolved in "clean" water?

11. What does the term "groundwater recharge" mean? What is the source of water that is used for this purpose?

12. Name five kinds of pollution often found in water. Give a source for each.

13. Prior to the enactment of the Clean Water Act, who was responsible for ensuring that water being used was pure? After passage of this act, who is now responsible?

14. Both surface water and groundwater (natural waters) often contain dissolved ions. Name three positive and three negative ions that are often found in natural waters.

15. Name two methods of disposal for solid wastes from industry and households. Which one of these has the greater possibility to adversely impact water quality?

16. What is the "Superfund"? Explain how it is used to improve water quality.

17. Describe a way by which a landfill can be made more "secure" in terms of water quality protection.

18. Go to the hardware department in a large department store and choose five products. Then list the kinds of hazardous wastes the manufacture of these products might produce. Use Table 11.7 as a guide.

19. Name three common household wastes and the kinds of chemicals they contain that might be harmful to water quality.

20. Describe how measuring the biochemical oxygen demand (BOD) of a sample of water indicates something about its purity.

21. Describe how pure water can become contaminated with lead in the home.

22. Describe how the BOD of wastewater can be lowered.

23. Explain how distillation purifies a sample of water.

24. Explain how aeration purifies a sample of water containing dissolved organic compounds.

25. Explain how settling and filtration purify water samples.

26. What is the difference between "biodegradable" and "nonbiodegradable"? If you had a choice between using a biodegradable and a nonbiodegradable detergent to clean your clothes, which would you choose?

27. Chlorinated hydrocarbons and branched-chain hydrocarbons are nonbiodegradable. What happens to them when they are released into the environment?

28. Name the two methods of primary sewage treatment.

29. Too-high concentrations of nitrogen compounds like ammonia (NH_3) adversely affect water quality. What tertiary sewage treatment method gets rid of these compounds?

30. Name the ions commonly present in hard water. What kinds of problems do they cause?

31. How is chlorination of water similar to aeration of water? How are these different?

32. What are "disinfection by-products," and how are these potentially harmful?

33. Explain how reverse osmosis can be used to purify seawater.

34. Which method of purification of drinking water for the home would most likely get rid of dissolved organic compounds?

35. Determine the amount of heat required to convert 14.0 g of ice at 0°C to 14.0 g of liquid water at 50°C.

36. How many calories of energy are required to raise the temperature of 5000 g of water from a temperature of 25°C to 100°C? The heat capacity of water is 1.00 cal/g-°C.

37. Convert 1000 calories to Joules. 1 cal = 4.18 J.

38. How much energy, in calories, is required to convert 355 g of solid water at 0°C to liquid? The heat of fusion for water is 80 cal/g. If 1 cal = 4.18 J, what is this quantity in Joules?

39. Because energy is conserved, the amount of heat absorbed in one process is the same as the amount of energy released in the reverse of that process. Keeping this in mind, how much energy, in calories, is released when 100 g of liquid water at 0°C freezes?

40. Which requires more energy, heating 1000 g water from 25°C (room temperature) to 100°C, or boiling 1000 g of water at 100°C?

41. Is rainwater the purest form of liquid water?

42. Is it possible to heat ice, but not produce liquid water? Explain.

43. Is it possible to heat water vapor above a temperature of 100°C?

44. The pH of rainwater that is saturated with CO_2 is 5.6.
    (a) Are raindrops of this pH acidic or basic?
    (b) What is the concentration of hydronium in rainwater having a pH of 5.6?
    (c) What is the concentration of OH⁻ in rainwater at a pH of 5.6?

45. Why is rainfall with a pH value of 7.0 not commonly observed?

46. It was stated in Section 11-3 that 4.35 trillion gallons of rain and snow fall on the continental United States each year. Use the following equalities to express this quantity as a number of bottles of purified water.

1 gallon = 4 quarts
1 quart = 2 pints
1 pint = 2 cups
1 cup = 8 fl. oz.
1 fl. oz. = 29.57 mL
1 bottle purified water = 20 fl. oz.

47. Use a search engine on your computer to discover the relationship of the following to water or water pollution.
(a) Love Canal
(b) *A Civil Action*, a 1995 book by Jonathan Harr
(c) *Erin Brockovich*, the movie
(d) Times Beach, Missouri
(e) Minamata disease

48. It may not be immediately obvious, but a steam burn is often more serious than a burn from boiling water. Why do you think this might be?

12

Energy and
Hydrocarbons

The average person in the United States uses far more energy (per capita energy consumption) than the citizens of any other country in the world, and the level of usage continues to rise at an alarming rate. The United States has about 5% of the world's population, but we consume around 25% of the daily global supply of energy, 85% of which comes from fossil fuel combustion. We have already seen (Chapter 6) that carbon dioxide, a product of the combustion of hydrocarbons (organic compounds containing only carbon and hydrogen), may play a major role in global warming. It should come as no surprise that there is much concern over the implications of humans continuing to burn enormous quantities of fossil fuels (coal, petroleum, and natural gas) and the products that can be derived from them (for example, gasoline, methanol, and ethanol). On one hand, there is concern that the supply of these fossil fuels, and, thus, the energy derivable from them, is limited. On the other hand, there is concern that continued exploitation of the energy derivable from the combustion of these fuels may accelerate global warming even if anthropogenic carbon dioxide emissions turn out not to be the principal cause of the phenomenon. All things considered, maybe the best use of fossil fuels is not to burn them to derive the energy contained within their chemical bonds but to exploit them for some higher purpose. We will address this issue in this chapter, consider one alternative fuel source in Chapter 13, and consider other uses of hydrocarbons in Chapter 14.

Hydrocarbons are the principal component of fossil fuels. Natural gas is primarily methane, crude petroleum is a complex mixture of thousands of hydrocarbons, and coal is an even more complex mixture of hydrocarbons. Many of the fuels we use, such as gasoline and jet fuel, are obtained from petroleum.

SECTIONS

Although the International System of Units or Système International (SI) unit of energy is the joule, the more familiar unit of heat is the calorie. A calorie is the amount of heat required to raise the temperature of 1 g of water 1°C. One calorie equals 4.18 joules (J); 1000 calories (cal) equals 1 kilocalorie (kcal). For example, you use 140 kcal/h walking and 80 kcal/h even when you are asleep.

Fossil fuels are also the major source of hydrocarbons that are used to make thousands of consumer products. This chapter describes the chemistry of hydrocarbons and their importance to our energy needs, and Chapter 14 emphasizes the industrial uses of hydrocarbons. Alcohols and ethers are part of the energy discussion because of the need to improve emissions and reduce pollution.

12-1 Energy from Fuels

In our homes and our industries, we obtain most of our energy by burning fossil fuels (Figure 12.1). Why does burning fuels provide energy? Fuels are reduced forms of matter that burn readily in the presence of oxygen, and combustion is an oxidation reaction (Section 10-1) that produces heat. The burning of methane, the principal component of natural gas, can be used to illustrate how the combustion of fuels produces energy.

The use of fossil fuels such as petroleum accounts for over 85% of annual energy production worldwide. The amount of energy produced from petroleum in one year is less than that provided by the Sun in one hour.

FIGURE 12.1 Sources of energy used in the United States in 2008.

Taken together, the fossil fuels (petroleum, natural gas, and coal) accounted for 84% of our fuel. These are not renewable sources of energy. As fossil fuel supplies diminish, the use of renewable energy will become more and more important. Renewable energy comes from plants, wind, flowing water, geothermal vents, and the Sun. (Source: U.S. Department of Energy, Energy Information Administration, Annual Energy Review, 2008)

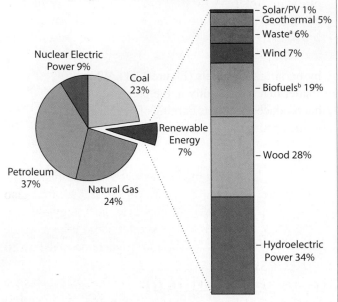

ª Municipal solid waste from biogenic sources, landfill gas, sludge waste, agricultural by-products, and other biomass.
ᵇ Fuel ethanol and biodiesel consumption, plus losses and co-products from the production of fuel ethanol and biodiesel.

Fuels are burned so that we may utilize the energy released from that process for some purpose. Remember from Chapter 8 that processes that release energy, including heat energy, are said to be exergonic. When we focus only on energy released in the form of heat, the process is said to be exothermic. The reverse of these changes are endergonic or endothermic processes. Although the energy changes associated with chemical reactions are most appropriately considered to be exergonic or endergonic, the heat energy often represents the majority of energy change in chemical reactions.

Let's consider what determines whether energy is released or absorbed in a chemical reaction, such as the combustion of a fuel. We know that a chemical reaction involves the conversion of reactants into products. The conversion requires that some chemical bonds in the reactants be broken. Bond cleavage (breaking) requires an energy input. This endothermic process is represented in mathematical terms with a plus (+) sign by chemists, as will be illustrated later. New chemical bonds must be made as product is formed. The formation of chemical bonds releases energy, and chemists represent this exothermic process in mathematical terms by a minus (−) sign. The sum of these is the change in energy associated with a chemical reaction.

Some average bond energies are shown in Table 12.1. This is the heat energy, or enthalpy, that is required to break one mole of the bond in question. Since bond cleavage and bond formation are the reverse of each other, the bond energy is also equal to the heat energy that is released upon formation of one mole of the bond in question. The bond energies may differ slightly from one compound to another, but these average bond energies suffice for our purposes.

The energy changes associated with chemical reactions fall into three different categories:

- **Energy Is Consumed Overall**—If the sum of the energies required to break bonds in the reactants is greater than the energy released by all new bonds formed in the products, energy will be consumed by the chemical reaction. The endothermic, bond-breaking processes (+ sign) require more energy than the exothermic, bond-forming processes (− sign) release, so the sum of these energy changes is endothermic overall (+ sign). This type of chemical reaction requires an overall energy input, and, thus, such a reaction would not be useful as a source of energy.

- **Energy Is Released Overall**—If the sum of the energies required to break bonds in the reactants is less than the energy released by all new bonds formed in the products, energy will be released in the chemical reaction. Thus, the endothermic, bond-breaking processes (+ sign) require less energy than the exothermic, bond-forming processes (− sign) release, so the sum of these energy changes is exothermic overall (− sign). This type of chemical reaction produces energy overall and thus may be a useful source of energy. In other words, the reactants may represent a usable fuel. Factors other than the amount of energy released by combustion of a fuel may also be important in determining its suitability as a fuel, however.

TABLE 12.1 Average Bond Energies

Bond	Energy (kJ/mol)
C — H	416
O = O	498
C = O	803
O — H	467
C — C	356
H — H	436
C — O	336

© Cengage Learning

- **No Net Energy Change Occurs**—If the sum of the energies required to break bonds in the reactants is exactly equal to the energy released by all new bonds formed in the products, there is no net energy change. Thus, the endothermic, bond-breaking processes (+ sign) require exactly the same amount of energy released by the exothermic, bond-forming processes (− sign), so the sum of these energy changes is zero. This is a relatively rare, although not impossible, occurrence for a chemical reaction.

One can, in principle, calculate the heat energy (enthalpy) change in a chemical reaction if the balanced chemical equation is known and the bond energies of all the bonds broken and made are known. For our purposes, we have only included a very abbreviated table (Table 12.1) of the types of bonds that might be involved in the combustion of a hydrocarbon or related fuel.

Let's begin our mathematical analysis of the combustion of some fuels by looking at methane. The complete combustion of any hydrocarbon will produce only carbon dioxide and water, so it is a relatively simple matter to write such an equation for methane. By using the bond energies in Table 12.1, we see that it takes a total of 2660 kJ to break all the bonds in the reactants (one mole of methane and two moles of oxygen). Since those processes are endothermic, we indicate this energy requirement as +2660 kJ. Similarly, we see that the formation of all bonds in the product molecules releases 3474 kJ. We designate this exothermic process as −3474 kJ. The sum of these two processes is −814 kJ (the negative sign indicates that the overall process is exothermic). This is the amount of heat energy released in burning one mole of methane. The heat released when one mole of a fuel is burned is known as the **heat of combustion**.

By repeating this type of calculation, we can compare the energy released for several different types of fuels. However, there is something else we need to consider before comparing the amount of energy released for different fuels. Our calculations would yield the amount of energy released per mole of fuel. This is not a good way to compare fuels because one mole of methane has a mass of 16 g, but one mole of octane has a mass of 114 g. We get a better number for comparison when we divide the heat of combustion per mole by the molecular weight of the compound. This is the number we should use when comparing the heat released by different fuels. For methane, −814 kJ/mol divided by 16 g/mol for methane gives a value of −50.8 kJ/g. This number is in good agreement with the experimental number of −50.1 kJ/g in Table 12.2 that also includes the entropy contribution to the energy change. A generalization is that most hydrocarbons release about 40 to 50 kJ/g of energy upon complete combustion to carbon dioxide and water.

Octane and 2,2,4-trimethylpentane (both of which have the same formula, C_8H_{18}, a molecular weight of 114 g/mol) release similar amounts of energy (roughly −5049 kJ/mol or −44.3 kJ/g) upon combustion. Both reactions involve the breaking and formation of exactly the same types and numbers of bonds, but because it is easier to break branched-chain bonds (nonlinear arrangements of carbon atoms) than straight-chain bonds, the combustion of 2,2,4-trimethylpentane releases slightly more energy. Note in the following analysis that the balanced equation initially gives the amount of energy for combustion of two moles of octane, so we must divide the value obtained by 2 to arrive at the heat of combustion per mole.

> **Heat of combustion:** The quantity of heat released when a fuel is burned

$$\text{H}-\underset{\overset{\displaystyle |}{\text{H}}}{\overset{\overset{\displaystyle \text{H}}{|}}{\text{C}}}-\text{H} \ + \ 2\,\text{O}=\text{O} \ \xrightarrow{\text{Spark}} \ \text{O}=\text{C}=\text{O} \ + \ 2\,\text{H}-\text{O}-\text{H}$$

Bonds Broken Endothermic (+)		Bonds Formed Exothermic (−)	
4 (C—H)	2 (O=O)	2 (C=O)	4 (O—H)
4 mol (416 kJ/mol)=	2 mol (498 kJ/mol)=	2 mol (−803 kJ/mol)=	4 mol (−467 kJ/mol)=
+1664 kJ	+996 kJ	−1606 kJ	−1868 kJ
Sum = +2660 kJ		Sum = −3474 kJ	

Net = (+2660 kJ) + (−3474 kJ)

Net = −814 kJ

$$2\ CH_3(CH_2)_6CH_3 + 25\ O{=}O \xrightarrow{\text{Spark}} 16\ CO_2 + 18\ H_2O$$

or

$$2\ H_3C-\overset{\underset{|}{CH_3}}{\underset{|}{\underset{CH_3}{C}}}-H_2C-\overset{CH_3}{\underset{|}{CH}} + 25\ O{=}O \xrightarrow{\text{Spark}} 16\ CO_2 + 18\ H_2O$$

Bonds Broken
Endothermic (+)

2×7 (C—C)	25 (O=O)
14 mol (356 kJ/mol)=	25 mol (498 kJ/mol)=
+4984 kJ	+12,450 kJ

2×18 (C—H)
36 mol (416 kJ/mol)
+14,976 kJ

Sum = +32,410 kJ for
2 moles C_8H_{18}

Bonds Formed
Exothermic (−)

32 (C=O)	36 (O—H)
32 mol (−803 kJ/mol)=	36 mol (−467 kJ/mol)=
−25,696 kJ	−16,812 kJ

Sum = −42,508 kJ

Net = (+32,410 kJ) + (−42,508 kJ)

Net = −10,098 kJ for 2 mol of C_8H_{18}

Net = −5049 kJ for 1 mol of C_8H_{18}

As we shall see later (Section 12-2), octane is a very poor fuel and 2,2,4-trimethylpentane is a very good fuel even though they release similar amounts of energy upon combustion. Why is that? Clearly the quality of a fuel must be related to more than its energy content.

Let's compare four other fuels (hydrogen, ethane, methanol, and ethanol) by writing out the complete balanced equations for combustion and doing the simple mathematics using the bond energies in Table 12.1.

$$2\ H_2 + O_2 \xrightarrow{\text{Spark}} 2\ H_2O$$

$$2\ CH_3CH_3 + 7\ O_2 \xrightarrow{\text{Spark}} 6\ H_2O + 4\ CO_2$$

$$2\ CH_3OH + 3\ O_2 \xrightarrow{\text{Spark}} 4\ H_2O + 2\ CO_2$$

$$CH_3CH_2OH + 3\ O_2 \xrightarrow{\text{Spark}} 3\ H_2O + 2\ CO_2$$

We can determine that all of these combustion reactions are exothermic; that is, they all release heat. Table 12.2 is a summary of the heats of combustion of the fuels just considered and includes some additional numbers for comparison.

EXAMPLE 12.1 Heat of Combustion of Methanol

Determine the value of the heat of combustion of methanol by using the bond energies in Table 12.1. Compare it to that of methane.

SOLUTION

Using the balanced equation, we see that the combustion of two moles of methanol requires three moles of oxygen and produces four moles of water and two moles of carbon dioxide. This requires breaking six moles of C—H bonds (+416 kJ/mol), two moles of C—O bonds (+336 kJ/mol), two moles of O—H bonds (+467 kJ/mol), and three moles of O=O bonds (+498 kJ/mol) for a total of +5596 kJ. Bond formation in the products involves eight moles of O—H bonds (−467 kJ/mol) and four moles of C=O bonds (−803 kJ/mol) for a total heat release of −6948 kJ. The sum of energy required (+5596 kJ) and energy released (−6948 kJ) is −1352 kJ for two moles of methanol or −676 kJ/mol. Since methanol has a molecular weight of 32 g/mol, this gives a calculated heat of combustion for methanol of -21.1 kJ/g, which compares favorably with the experimental value of -20 kJ/g. Thus, the introduction of an oxygen atom into a methane molecule has reduced the heat of combustion from about -50 kJ/g to around -20 kJ/g.

TRY IT 12.1

Use the procedure in Example 12.1 to calculate the heats of combustion of ethane (C_2H_6) and ethanol (C_2H_5OH) on a per gram basis. Based upon these two examples—methane versus methanol, and ethane versus ethanol—what can you conclude about the introduction of an oxygen atom into a hydrocarbon fuel? Does this result in an increase or a decrease in the heat of combustion?

Three things are apparent from inspection of the data in Table 12.2:

- Hydrogen releases more heat per gram than any of the hydrocarbon or alcohol fuels.

- All hydrocarbons release *about* the same amount of heat per gram.

- Adding oxygen to a hydrocarbon to convert it to an alcohol reduces the amount of heat released upon combustion. This may be seen by comparing methane and methanol or ethane and ethanol.

Three other fuels (glucose, wood, and coal) are included in Table 12.2 for comparison. Note that no simple chemical equations can be written for the combustion of coal and wood because coal and wood are not pure compounds. However, we can weigh a certain mass of coal or wood, burn it, and determine the amount of heat released per gram. Most woods release about 10 to 14 kJ/g, and coal releases about 16 to 30 kJ/g depending on whether we are talking about lignite (soft coal) or anthracite (hard coal). The importance of glucose will be considered later.

As has been shown, the energy content of various fuels may vary quite considerably. Factors other than the actual energy content of the fuel play a big role in determining which of these fuels we use. Availability, ease of transport and storage, pollution-producing potential, cost, and even perceived danger may affect our perception and use of the fuels. Let's look at where some of the petroleum-based fuels come from before addressing the question "Which is the best fuel?"

12-2 Petroleum

Crude petroleum is a complex mixture of thousands of hydrocarbon compounds, and the actual composition of petroleum varies with the location in which it is found. For example, Pennsylvania crude oils are primarily straight-chain hydrocarbons, whereas California crude oil is composed of a larger portion of aromatic hydrocarbons (Section 12-8).

TABLE 12.2 Experimental Heats of Combustion of Some Common Fuels

Fuel	Heat of Combustion (kJ/mol)	Heat of Combustion (kJ/g)
Hydrogen	-242	-121
Methane	-802	-50.1
Ethane	-1428	-47.6
Propane	-2016	-45.8
"Oil"—a hydrocarbon mixture	—	~ -45
Anthracite coal	—	-30.5
Ethanol	-1367	-29.7
Methanol	-640	-20
Lignite coal	—	-16
Glucose	-2800	-15.5 (reverse of photosynthesis)
Wood	—	-10 to -14

© iStockphoto.com/yenwen

© Cengage Learning

Many products, including gasoline, are isolated or produced from petroleum. However, limited petroleum reserves may force us to consider how we use this resource.

Fractional distillation:
Separation of a mixture into fractions that differ in boiling points

Petroleum fraction:
A mixture of hundreds of hydrocarbons with boiling points in a certain range that are obtained by fractional distillation of petroleum

How long will petroleum be viable as a source of energy and starting materials for consumer products? This is difficult to estimate because of continued revisions of recoverable crude oil resources. For example, estimates of global petroleum reserves increased 43% between 1984 and 1994, primarily because of re-evaluation of oil reserves in the Middle East, where more than 65% of the world's oil resources are located. Counterbalancing this increase in oil reserves is the substantial increase in energy demand expected during the next 20 to 30 years from developing countries in Asia and Latin America. Current projections are for oil production to peak between 2010 and 2025. Oil production would continue for several decades after this; however, the increasing cost and lower availability will favor increased use of natural gas, coal, and alternative sources such as wind energy and solar energy, particularly electricity obtained through the use of photovoltaic cells.

12-2a Petroleum Refining

The refining of petroleum begins with the separation of fractions according to boiling point ranges by a process called **fractional distillation**. The difference between simple distillation and fractional distillation is the degree of separation achieved for the mixture being distilled. For example, water that contains dissolved solids or other liquids can be purified by simple distillation. The impure solution is heated to boiling; the water vapor is condensed and collected in a separate container. Since petroleum contains thousands of hydrocarbons, separation of the pure compounds is not feasible or even necessary. The products obtained from distillation of petroleum are still mixtures of hundreds of hydrocarbons, so they are called **petroleum fractions**.

Figure 12.2 illustrates a fractional distillation tower used in the petroleum-refining process. The crude oil is first heated to about 400°C to produce a hot vapor-and-liquid mixture that enters the fractionating tower. The vapor rises and condenses at various points along the tower. The lower-boiling petroleum fractions (those that are more volatile) will remain in the vapor stage longer than the higher-boiling fractions. These differences in boiling point ranges allow the separation of fractions. Some of the gases do not condense and are drawn off the top of the tower, while the unvaporized residual oil is collected at the bottom of the tower. Typical products of the fractional distillation of petroleum are listed in Table 12.3.

12-2b Octane Rating

The burning properties of hydrocarbons depend on their molecular structure. The octane rating is an arbitrary scale for rating the relative knocking properties of gasolines, and it is based on the operation of a standard test engine. Isooctane (2,2,4-trimethylpentane) is the standard used to assign octane ratings. If a fuel does not burn smoothly in an engine, the result is knocking, a sound caused by collision of the piston with the valves or cylinder. Heptane knocks considerably and is assigned an octane rating of 0, whereas 2,2,4-trimethylpentane burns smoothly and is assigned an octane rating of 100. The octane rating of a gasoline is determined by first using the gasoline in a standard engine and recording its knocking properties.

Typical octane ratings for gasoline available at gas stations.

FIGURE 12.2 Petroleum fractionation.

Crude oil is first heated to 400°C in the pipe still. The vapors then enter the fractionation tower. As they rise in the tower, the vapors cool down and condense so that different fractions can be drawn off at different heights. This shows how the rising vapor is repeatedly condensed and collected at the numerous bell caps.

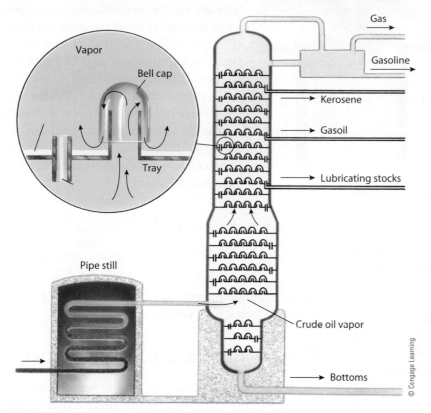

© Cengage Learning

is called the *octane rating* of the gasoline. Thus, if a gasoline has the same knocking characteristics as a mixture of 13% heptane and 87% isooctane, it is assigned an octane rating of 87. This corresponds to regular unleaded gasoline. Other higher grades of gasoline available at gas stations have octane ratings of 89 (regular plus) and 92 (premium).

The "straight-run" gasoline fraction obtained from the fractional distillation of petroleum has an octane rating of only 55 and needs additional refinement because it contains primarily straight-chain hydrocarbons that burn too rapidly to be suitable for use as a fuel in internal combustion engines. Rapid burning causes uncontrolled explosion of the fuel as evidenced by a "knocking" or "pinging" sound in the engine. This reduces engine power and may damage the engine.

The octane rating of a gasoline can be increased either by increasing the percentage of branched-chain and aromatic hydrocarbon fractions or by adding octane enhancers (or a combination of both). Since the octane rating scale was established, fuels superior to isooctane have been developed, so the scale has been extended well above 100. Table 12.4 lists octane ratings for some hydrocarbons and octane enhancers. Note that the octane rating of a fuel is not a

The test results are then compared with the behavior of mixtures of heptane and isooctane, and the percentage of isooctane in the mixture with identical knocking properties

TABLE 12.3 Hydrocarbon Fractions from Petroleum

Fraction	# of C atoms	Boiling Point Range (°C)	Uses
Gas	C_1–C_4	0–30	Gas fuels
Straight-run gasoline	C_5–C_{12}	30–200	Motor fuel
Kerosene	C_{12}–C_{16}	180–300	Jet fuel, diesel oil
Gasoil	C_{16}–C_{18}	Over 300	Diesel fuel, cracking stock
Lubricants	C_{18}–C_{20}	Over 350	Lubricating oil, cracking stock
Paraffin wax	C_{20}–C_{40}	Low-melting solids	Candles, wax paper
Asphalt	Above C_{40}	Gummy residues	Road asphalt, roofing tar

© Cengage Learning

TABLE 12.4 Octane Numbers of Some Hydrocarbons and Gasoline Additives

Name	Octane Number
Octane	−20
Heptane	0
Pentane	62
1-Pentene	91
2,2,4-Trimethylpentane (isooctane)	100
Benzene	106
Methanol	107
Ethanol	108
Tertiary-butyl alcohol	113
Methyl *tertiary*-butyl ether (MTBE)	116
Para-xylene	116
Toluene	118

© Cengage Learning

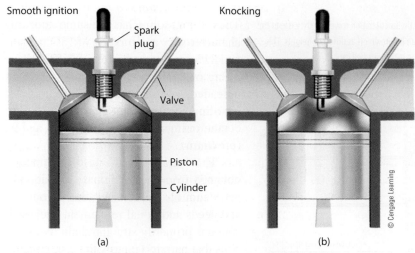

Smooth ignition

Knocking

Spark plug

Valve

Piston

Cylinder

(a)   (b)

© Cengage Learning

Octane rating relates to smoothness of ignition.

measure of the energy content of the fuel. For instance, octane (a poor fuel) and isooctane or 2,2,4-trimethylpentane (a good fuel) release the same amount of energy but burn at different rates. One cannot merely find the fuel with the highest octane rating and decide that is the best fuel. Toluene, for instance, has a very high octane rating (118), but when other properties of toluene, including its energy content, considerable toxicity, combustion properties, and cost are considered, it is not the fuel of choice. Nonetheless, most gasoline mixtures do contain about 5% to 7% toluene.

12-2c Catalytic Re-Forming

The **catalytic re-forming process** is used to increase the octane rating of straight-run gasoline by converting straight-chain hydrocarbons to branched-chain hydrocarbons and aromatics. This is accomplished by using certain catalysts, such as finely divided platinum on a support of alumina (Al_2O_3).

$$CH_3CH_2CH_2CH_2CH_3 \xrightarrow{\text{Catalyst}} \underset{\substack{\text{2-Methylbutane} \\ \text{(octane rating 94)}}}{CH_3\overset{\overset{\displaystyle CH_3}{|}}{C}HCH_2CH_3}$$

Pentane
(octane rating 62)

In this process, straight-chain hydrocarbons with low octane numbers can be re-formed into their branched-chain

isomers, which have higher octane numbers. Catalytic re-forming is also used to produce aromatic hydrocarbons such as benzenes, toluene, and xylenes by using different catalysts and petroleum mixtures. For example, when the vapors of straight-run gasoline, kerosene, and light oil fractions are passed over a copper catalyst at 650°C, a high percentage of the original material is converted into a mixture of aromatic hydrocarbons, from which benzene, toluene, xylenes, and similar compounds may be separated by fractional distillation. This process can be represented by the equation for converting hexane into benzene.

$$CH_3CH_2CH_2CH_2CH_2CH_3 \xrightarrow{\text{Catalyst}} C_6H_6 + 4H_2$$

Hexane
(octane rating 25)

Benzene
(octane rating 106)

The catalytic re-forming process is a major source of hydrogen gas, which is also a potential fuel.

12-2d Catalytic Cracking

Part of the petroleum refinement process involves adjusting the percentage of each hydrocarbon fraction to match commercial demand. For example, the demand for gasoline is higher than that for kerosene. As a result, chemical reactions convert the larger kerosene-fraction molecules into molecules in the gasoline range in a process called *cracking*. The **catalytic cracking process** uses a zeolite catalyst and involves heating saturated hydrocarbons under pressure in the absence of air. The hydrocarbons break into shorter-chain hydrocarbons—both alkanes and alkenes, some of which will be in the gasoline range.

$$C_{16}H_{34} \xrightarrow[\text{Heat}]{\text{Pressure}} C_8H_{18} + C_8H_{16}$$

An alkane

An alkane   An alkene
in the gasoline range

Since alkenes have a higher octane rating than alkanes, the catalytic cracking process also increases the octane rating of the mixture. Catalytic cracking is also important for the production of alkenes used as starting materials in the organic chemical industry.

12-2e Octane Enhancers

The octane number of a given blend of gasoline can also be increased by adding antiknock agents, or octane enhancers.

Prior to 1975, the most widely used antiknock agent was tetraethyllead, $(C_2H_5)_4Pb$. The addition of 3 g of $(C_2H_5)_4Pb$ per gallon increases the octane rating by 10 to 15. Before the Environmental Protection Agency (EPA) required reductions in lead content, both regular and premium gasoline contained an average of 3 g of $(C_2H_5)_4Pb$ or tetramethyllead, $(CH_3)_4Pb$, per gallon. However, the Clean Air Act of 1970 required that 1975-model cars emit no more than 10% of the carbon monoxide and hydrocarbons emitted by 1970 models. The platinum-based catalytic converter chosen to reduce emissions of carbon monoxide and hydrocarbons required lead-free gasolines, since lead deactivates the platinum catalyst by coating its surface. For this reason, and the fact that lead released into the environment is a neurological poison, automobiles manufactured since 1975 have been required to use lead-free gasoline to protect the catalytic converter.

Since tetraethyllead can no longer be used, other octane enhancers must be added to gasoline to increase the octane rating. These have included benzene, toluene, xylenes, 2-methyl-2-propanol, methyl *tertiary*-butyl ether (MTBE), methanol, and ethanol.

12-2f Oxygenated and Reformulated Gasolines

The 1990 amendments to the Clean Air Act require cities with excessive levels of ozone and carbon monoxide pollution to use oxygenated and reformulated gasolines to reduce hydrocarbon and toxic compound emissions (Section 4-9). **Oxygenated gasolines** are blends of gasoline with organic compounds that contain oxygen, such as MTBE, methanol, ethanol, and 2-methyl-2-propanol (*tertiary*-butyl alcohol). The oxygenated gasolines can be produced either by blending in additives at the refinery or by adding ethanol or methanol at the distribution terminals. MTBE has been identified as a pollutant in the water supply in many locations in the United States, and its use has been severely restricted or banned in many states.

Reformulated gasoline is gasoline whose composition has been changed to reduce the percentage of unsaturated hydrocarbons, aromatics, volatile components, and sulfur and to add oxygenated additives. This requires significant changes in the refining process, which makes reformulated gasoline more expensive to produce.

Oxygenated gasolines are required for use during the four winter months in cities that have serious carbon monoxide pollution (Section 4-10). This gasoline must contain enough oxygenated organic compounds to provide an average of 2.7% oxygen by weight. Oxygenated gasolines ignite more easily and burn more cleanly, and this reduces carbon monoxide emissions.

Nine cities with the most serious ozone pollution were required by the 1990 regulations to use reformulated gasolines, and another 87 cities that were not meeting the ozone air-quality standards were given the option of using these gasolines.

The composition of reformulated gasoline later became an issue of political and economic dissension. A ruling issued by the EPA on June 30, 1994, required that 30% of the oxygenated organic compounds (called *oxygenates*) used in reformulated gasoline had to come from renewable sources. Most gasoline producers had been using MTBE, which is made from methanol, to meet the 1990 Clean Air Act regulations. Because methanol is produced from synthesis gas (a product derived from coal, Section 12-4), it is not renewable. The new mandate would have required more use of ethanol, a renewable resource because it can be made from corn, and ethyl *tert*-butyl ether (ETBE), which is made from ethanol. Proponents of the mandate argued that the use of ethanol would reduce reliance on oil imports, cut farmers' reliance on federal farm subsidies, and increase the use of renewable resources.

In response to the EPA ruling, the American Petroleum Institute and the National Petroleum Refiners Association filed suit in the U.S. Court of Appeals for the District of Columbia. The suit accused the EPA of violating the regulatory process by ruling in favor of ethanol producers and farmers who grow corn. On April 28, 1995, the U.S. Court of Appeals approved the petition by prohibiting the EPA from requiring the use of renewable oxygenates in reformulated gasolines.

In subsequent years, a new issue that bears on the composition of reformulated gasoline has surfaced. Concentrations of MTBE, presumably from gasoline leaks and spills that wash into natural waterways, have been found in drinking water. Because MTBE is not easily decomposed by natural processes or water treatment, it tends to remain where it has accumulated. It is reported to cause off tastes and odors in water. Also, it may be a carcinogen. In response to these concerns, a blue-ribbon panel of experts reviewed the evidence. In 2000, at their recommendation, efforts were begun to eliminate use of MTBE and promote further use of renewable oxygenates like ethanol.

> **Oxygenated gasoline:** A blend of gasoline that contains oxygenated organic compounds such as alcohols or ethers to cause the gasoline to burn more cleanly
>
> **Reformulated gasoline:** A gasoline whose composition has been changed to reduce the percentage of olefins, aromatics, and sulfur and to add oxygenated compounds

12-3 Natural Gas

Natural gas is a mixture of gases trapped with petroleum in Earth's crust and is recoverable from oil wells or gas wells where the gases have migrated through the rock. The natural gas found in North America is a mixture of C_1 to C_4 alkanes—methane (60%–90%), ethane (5%–9%), propane (3%–18%), and butane (1%–2%)—with a number of other gases, such as CO_2, N_2, H_2S, and the noble gases, present in varying amounts. In Europe and Japan, the natural gas is essentially all methane.

Natural gas is the fastest-growing energy source in the United States, and U.S. production of natural gas supplies 17% more energy than does U.S.-produced oil. About half of the homes in the United States are heated by natural gas, followed by electricity (18.5%), fuel oil (14.9%), wood (4.8%), and liquefied gas such as butane and propane (4.6%). Coal and kerosene come in at a low 0.5%, and the percentage of homes using solar heating is even lower. However, the United States has only about 5% of the known world reserves of natural gas, which at the present rate of use is enough to last until the year 2050.

Natural gas is also being used as a vehicle fuel, and worldwide there are more than 2.5 million vehicles powered by natural gas. Argentina, with more than 800,000 natural gas-powered vehicles, leads the world in use of these vehicles. California and several other states are encouraging the use of natural gas vehicles to help meet new air-quality regulations. Vehicles powered by natural gas emit minimal amounts of carbon monoxide, hydrocarbons, and particulates, and the price of natural gas is about one-third that of gasoline. The main disadvantages of natural gas vehicles include the need for a cylindrical pressurized gas tank and the lack of service stations that sell compressed natural gas.

Although most natural gas is used as an energy source, it is also an important source of raw materials for the organic chemical industry. (Figure 14.1 shows the uses of alkanes obtained from natural gas and petroleum.)

12-4 Coal

Coal is a complex mixture of high molecular weight hydrocarbons that are about 85% carbon by mass. It usually contains a relatively small but variable amount of sulfur. The small amount of sulfur can be a significant factor when one considers its potential as a contributor to air pollution and acid precipitation, however. By way of contrast with petroleum, coal has more fused rings of carbon atoms, and the organic structure of coal is much more complicated.

About 88% of our annual coal production is burned to produce electricity. Only 1% is used for residential and commercial heating. Although the use of coal is on the rise, its use as a heating fuel has declined because it is a relatively dirty fuel, bulky to handle, and a major cause of air pollution (because of its sulfur content). The dangers of deep coal mining and the environmental disruption caused by strip mining contributed to the decline in the use of coal.

Given our great dependence on coal for the production of electricity and our smaller but still significant dependence on coal for the production of industrial chemicals, just how much coal do we have and how long is it likely to last? World coal reserves are vast relative to supplies of the other fossil fuels. Coal represents about 91% of the world's known

Many kitchen stoves use natural gas to produce the heat necessary to cook food.

© iStockphoto.com/Patricia Hofmeester

fossil fuel reserves with approximately equal percentages of natural gas and oil making up the remaining 9%. Only a small percentage of the world's coal has been mined to date. It is estimated that the world's proven coal reserves (Figure 12.3a) will last at least 200 years at the current rate of usage, but world coal consumption (Figure 12.3b) is predicted to increase significantly as oil and natural gas become more scarce, so this prediction may prove to be an overestimate.

Coal can be converted into a combustible gas (by coal gasification) or a liquid fuel (by coal liquefaction). In each case, environmental problems can be minimized but at additional costs per energy unit obtained from these fuels.

12-4a Coal Gasification

When coal is pulverized and treated with superheated steam, a mixture of CO and H_2 (synthesis gas) is obtained in a process known as coal gasification.

$$C + H_2O + 31 \text{ kcal} \longrightarrow CO + H_2$$

Synthesis gas is used both as a fuel and as a starting material for the production of organic chemicals and gasoline.

In a newer coal gasification process, methane is the end product. Crushed coal is mixed with an aqueous catalyst, the mixture is dried, and CO and H_2 are added. The resulting mixture is then heated to 700°C to produce methane and carbon dioxide. The overall reaction is

$$2 C + 2 H_2O + 2 \text{ kcal} \longrightarrow CH_4 + CO_2$$

Although the overall reaction is slightly endothermic, the subsequent combustion of the methane produced releases 192 kcal/mol; thus, the process is an energy-efficient way to obtain methane, an environmentally clean fuel.

12-4b Coal Liquefaction

Liquid fuels are made from coal by reacting the coal with hydrogen gas under high pressure in the presence of catalysts (hydrogenating the coal). The process produces hydrocarbons like those in petroleum. The resulting crude oil type of material can be fractionally distilled to give fuel oil, gasoline, and certain hydrocarbons used in the manufacture of plastics, medicines, and other commodities. About 5.5 barrels of liquid are produced for each ton of coal. At the present time, the cost of a barrel of liquid from coal liquefaction is about double that of a barrel of crude oil. However, as petroleum supplies diminish and the cost of crude oil increases, coal liquefaction will become economically feasible.

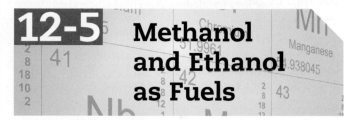

12-5 Methanol and Ethanol as Fuels

Methanol and ethanol are being considered as replacements for gasoline, especially in urban areas that have extremely high levels of air pollution caused by motor vehicles. For example, Southern California has been testing methanol-powered cars since 1981. About half of these cars use 100% methanol (M100). The other half are flexible-fueled vehicles (FFVs) that use either M85, a blend of 85% methanol and 15% gasoline, or gasoline. Although methanol fuels M85

FIGURE 12.3a World Recoverable Coal Reserves.

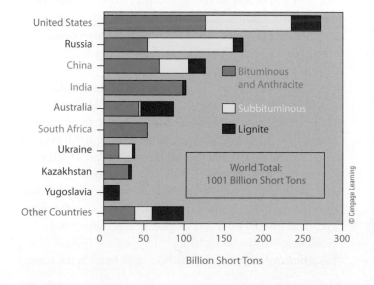

FIGURE 12.3b World Coal Consumption by Region, 1970–2025.

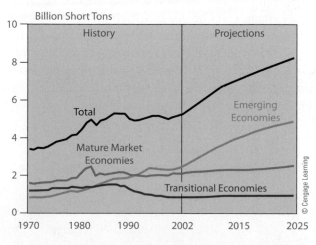

and M100 have in the past received more attention than corresponding ethanol fuels E85 and E100, ethanol-based FFVs have been growing steadily in popularity. By 2009, there were more than 40 E85-compatible vehicles in production. However, at this time, 100% pure ethanol (E100) is not approved as a motor vehicle fuel in the United States.

What are the advantages and disadvantages of switching to methanol- and ethanol-powered vehicles? Both methanol and ethanol burn more cleanly than gasoline, and levels of troublesome pollutants such as carbon monoxide, unreacted hydrocarbons, nitrogen oxides, and ozone are reduced (Chapter 4). However, there is concern about the higher exhaust emissions of carcinogenic formaldehyde from alcohol-powered vehicles. Since the number of these vehicles is relatively limited, it is difficult to assess the extent to which these formaldehyde emissions will contribute to the total aldehyde levels from other sources.

The technology for alcohol-powered vehicles has existed for many years, particularly in racing cars that burn methanol because of its high octane rating of 107. However, methanol and ethanol have only about one-half the energy content of gasoline, which would require fuel tanks to be twice as large to give the same distance per tankful. This is partially compensated for by the fact that alcohol-based fuels cost about half as much as gasoline, so the price per mile would be competitive. While ethanol burns with a blue flame, methanol burns with a colorless flame, so to avoid accidents, something needs to be added (a small amount of gasoline, for example) to methanol so that it can be seen when it burns. Another disadvantage of methanol that is shared by ethanol is the tendency to corrode regular steel, so it will be necessary to use stainless steel for the fuel system or have a resistant coating. Until sufficient numbers of alcohol-powered vehicles are on the road, cars equipped to run on *either* alcohol or gasoline will be necessary because of the lack of service stations selling methanol or ethanol. As the problems of distribution and storage are solved, better-engineered alcohol-fueled engines will be designed and produced, which will lead to more efficient utilization of these substances as fuels.

Another option is to use methanol to make gasoline. Mobil Oil Company has developed a methanol-to-gasoline process, although it is currently not competitive with refined gasoline prices in the United States.

$$2 \; CH_3OH \xrightarrow[\text{Catalyst}]{} \underset{\text{Dimethyl ether}}{(CH_3)_2O} + H_2O$$

$$2 \; (CH_3)_2O \xrightarrow[\text{Catalyst}]{} \underset{\text{Ethylene}}{2 \; C_2H_4} + 2 \; H_2O$$

$$\text{Many } C_2H_4 \xrightarrow[\text{Catalyst}]{} \underset{\text{Gasoline}}{\text{Hydrocarbon mixture in the } C_5\text{–}C_{12} \text{ range}}$$

12-6 Classes of Hydrocarbons

We have demonstrated that hydrocarbons and many related compounds are often used as fuels in our highly industrialized society, but we have not looked closely at their structures and differences. At this point, it is useful to take a closer look at the many types or classes of hydrocarbons and some related compounds, as these have many uses beyond those as fuels. There are four classes of hydrocarbons: the *alkanes*, which contain carbon–carbon single bonds; the *alkenes*, which contain one or more carbon–carbon double bonds; the *alkynes*, which contain carbon–carbon triple bonds; and the *aromatics*, which consist of benzene, benzene derivatives, and fused benzene rings.

12-6a Alkanes: Backbone of Organic Chemistry

The simplest alkane is methane (CH_4), the principal component of natural gas. Alkanes are saturated hydrocarbons (Section 5-3) with the general formula C_nH_{2n+2} (Table 12.5). Notice that all hydrocarbon formulas are traditionally written with the C atom first, followed by the H atom, and that all alkanes have -*ane* as the suffix in their name. When $n = 1$ to 4, the first part of the name is something of historical origin; these are common names that we just have to remember. When $n = 5$ or more, the Greek prefixes (Table 12.5) tell how many carbon atoms are present. For example, the compound with six carbons is called hexane.

The tetrahedral structure of CH_4 has been discussed, but it is important to recognize that *every* carbon in an alkane has a tetrahedral geometry because all carbon atoms in saturated hydrocarbons have four single bonds. The tetrahedral geometry of the carbon atoms in alkanes is difficult to draw in two dimensions.

Chains of carbon atoms are usually represented with a straight line as in Figures 12.4a and 12.5a. However, keep in mind that these drawings are not an accurate representation of the tetrahedral bond angles, which are 109.5 degrees.

Our ability to understand these tetrahedral structures is helped by the use of models. Two types of models are generally used—the "ball-and-stick" model and the "space-filling" model. In a ball-and-stick model, balls represent the atoms and short pieces of wood or plastic represent bonds. For example, the ball-and-stick model for methane (Figure 12.4b) has a blue ball representing carbon, with holes at the correct

TABLE 12.5 The First Eight Straight-Chain Saturated Hydrocarbons

Name	Formula	Boiling Point (°C)*	Structural Formula	Use
Methane	CH_4	−162		Principal component in natural gas
Ethane	C_2H_6	−88.5		Minor component in natural gas
Propane	C_3H_8	−42		Bottled gas for fuel
Butane	C_4H_{10}	0		
Pentane	C_5H_{12}	36		
Hexane	C_6H_{14}	69		Some of the components of gasoline
Heptane	C_7H_{16}	98		
Octane	C_8H_{18}	126		

© Cengage Learning

* Notice the gradual increase in boiling point as the molecular weight increases. Fractional distillation of petroleum is possible because of these differences (Section 12-2).

angles connected by sticks to four white balls representing hydrogen atoms. The space-filling model (Figure 12.4c) is a more realistic representation because it depicts both the relative sizes of the atoms and their spatial orientation in the molecule. This is done by scaling the pieces in the model according to the experimental values of atom sizes. The pieces are held together by links that are not visible when the model is assembled. The wedge-dash projection (Figure 12.4d) is frequently used as a model to represent the geometry of molecules. Solid lines represent bonds in the plane of the paper, dashes represent bonds projecting behind the plane of the paper, and wedges represent bonds projecting out of the plane of the paper.

FIGURE 12.4 The structure of methane, as represented by (a) its structural formula, (b) a ball-and-stick model, (c) a space-filling model, and (d) a wedge-dash projection showing the geometry of the molecule.

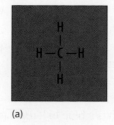
(a)

(b)

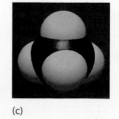
(c)

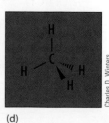
(d)

Charles D. Winters

FIGURE 12.5 The structure of propane, as represented by (a) its structural formula, (b) a ball-and-stick model, (c) a space-filling model, and (d) a wedge-dash projection.

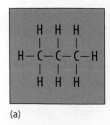
(a)

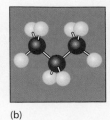
(b)

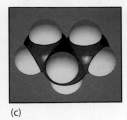
(c)

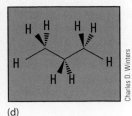
(d)

Charles D. Winters

Isomers: Two or more compounds with the same molecular formula but different arrangements of atoms

Structural isomers: Isomers that differ in the order in which the atoms are bonded together

Branched-chain isomers: Structural isomers of hydrocarbons that have carbon–carbon bonds in side chains

Straight-chain isomers: Structural isomers of hydrocarbons with no side chains

Alkyl groups: Alkanes with a hydrogen atom removed

Illustrations of ball-and-stick models, such as those shown in Figures 12.4(b) and 12.5(b), will be used extensively in the discussion of hydrocarbons and hydrocarbon derivatives to help you visualize the molecular geometry of molecules. For example, notice that the three carbon atoms in the ball-and-stick model of propane in Figure 12.5(b) do not lie in a straight line because of the tetrahedral geometry about each carbon atom. This illustrates why straight-chain drawings, such as the one for propane in Figure 12.5(a), are not accurate representations of the tetrahedral H—C—H and C—C—C bond angles and why projections such as Figure 12.5(d) are used.

12-6b Straight- and Branched-Chain Isomers of Alkanes

The first three alkanes, CH_4, C_2H_6, and C_3H_8, each have only one possible structural arrangement. However, two structural arrangements are possible for C_4H_{10}—a straight-chain arrangement and a branched-chain arrangement.

Structural formulas:

The condensed formulas and properties of the two compounds, butane and methylpropane, are as follows:

Condensed Formulas	$CH_3CH_2CH_2CH_3$ Butane	$\begin{array}{c}CH_3\\ |\\ CH_3CHCH_3\end{array}$ Methylpropane (isobutane)
Melting point	−138.3°C	−160°C
Boiling point (1 atm)	0.5°C	−12°C
Density (at 20°C)	0.579 g/mL	0.557 g/mL

These molecules have different properties even though they have the same number of atoms in the molecule. Ball-and-stick models of the two structures are shown in Figure 12.6.

Two or more compounds with the same molecular formula but different arrangements of atoms are called **isomers**. Isomers differ in one or more physical or chemical properties such as boiling point, color, solubility, reactivity, and density. Several different types of isomerism are possible for organic compounds. **Structural isomers** differ in the order in which the atoms are bonded together. Structural isomerism can be compared to the results you might expect from a child building many different structures with the same collection of building blocks and using all of the blocks in each structure. **Branched-chain** and **straight-chain isomers** are examples of structural isomers that differ in the order in which the atoms are bonded together.

The branched-chain isomer for C_4H_{10}, methylpropane, a common component of bottled gas, has a "methyl" group ($—CH_3$) attached to the central carbon atom. This is the simplest example of the fragments of alkanes known as **alkyl groups**. In this case, removal of H from methane gives a *methyl* group.

$$\text{H}-\underset{\underset{\text{H}}{|}}{\overset{\overset{\text{H}}{|}}{\text{C}}}-\text{H} \xrightarrow{-\text{H}} \text{H}-\underset{\underset{\text{H}}{|}}{\overset{\overset{\text{H}}{|}}{\text{C}}}-$$

or $CH_3—$ $\left(\begin{array}{c}\text{also written}\\ \text{as}—CH_3\end{array}\right)$

FIGURE 12.6 Ball-and-stick models.

Butane, a four-carbon straight-chain hydrocarbon, and methylpropane, a four-carbon branched-chain hydrocarbon, are illustrated.

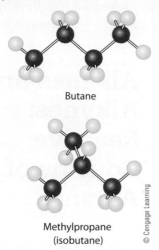

Butane

Methylpropane
(isobutane)

© Cengage Learning

Removal of an H from ethane gives an *ethyl* group.

$$H-\underset{\underset{H}{|}}{\overset{\overset{H}{|}}{C}}-\underset{\underset{H}{|}}{\overset{\overset{H}{|}}{C}}-H \xrightarrow{-H} H-\underset{\underset{H}{|}}{\overset{\overset{H}{|}}{C}}-\underset{\underset{H}{|}}{\overset{\overset{H}{|}}{C}}-$$

$$\text{or}\quad CH_3CH_2- \quad \left(\begin{array}{c}\text{also written}\\ \text{as} -C_2H_5\end{array}\right)$$

Notice that more than one alkyl group is possible when an H atom is removed from C_3H_8.

$$H-\underset{\underset{H}{|}}{\overset{\overset{H}{|}}{C}}-\underset{\underset{H}{|}}{\overset{\overset{H}{|}}{C}}-\underset{\underset{H}{|}}{\overset{\overset{H}{|}}{C}}-H$$

−H from either end carbon

−H from middle carbon

$$H-\underset{\underset{H}{|}}{\overset{\overset{H}{|}}{C}}-\underset{\underset{H}{|}}{\overset{\overset{H}{|}}{C}}-\underset{\underset{H}{|}}{\overset{\overset{H}{|}}{C}}- \quad \text{or}\quad CH_3CH_2CH_2-$$

Propyl

$$H-\underset{\underset{H}{|}}{\overset{\overset{H}{|}}{C}}-\underset{|}{\overset{\overset{H}{|}}{C}}-\underset{\underset{H}{|}}{\overset{\overset{H}{|}}{C}}-H \quad \text{or}\quad (CH_3)_2CH-$$

Isopropyl

Alkyl groups are named by dropping "-ane" from the parent alkane and adding "-yl." Theoretically, any alkane can be converted to an alkyl group. Some of the more common examples of alkyl groups are given in Table 12.6.

TABLE 12.6 Some Common Alkyl Groups

Name	Condensed Structural Representation
Methyl	CH_3-
Ethyl	CH_3CH_2- or C_2H_5-
Propyl	$CH_3CH_2CH_2-$ or C_3H_7-
Isopropyl	CH_3CH- or $(CH_3)_2CH-$ with $\overset{|}{CH_3}$
Butyl	$CH_3CH_2CH_2CH_2-$ or C_4H_9-
t-Butyl*	$CH_3\underset{\underset{CH_3}{|}}{\overset{\overset{CH_3}{|}}{C}}-$ or $(CH_3)_3C-$

© Cengage Learning

* *t* stands for tertiary, sometimes abbreviated *tert*, which means that the central C atom is bonded to three other C atoms.

The number of structural isomers predicted for C_6H_{14}, C_7H_{16}, and C_8H_{18} is 5, 9, and 18, respectively. Every predicted isomer, *and no more*, has been isolated and identified for these hydrocarbons. The large number of structural isomers illustrates the complexity and variety that organic chemistry can have even for simple hydrocarbons.

12-6c Naming Branched-Chain Alkanes

Many alkanes and other organic compounds have both common names and systematic names. Why are both common and systematic names used? Usually, the common name came first and is widely known. Many consumer products are labeled with the common name, and when only a few isomers are possible, the common name adequately identifies the product for the consumer. For example, "isobutane," the common name for methylpropane, is sufficient because there is only one branched-chain isomer possible for C_4H_{10}. However, a system of common names quickly fails when several structural isomers are possible.

We have discussed the "octane rating" of gasoline. Octane (C_8H_{18}) has 18 possible isomers. One of these isomers, 2,2,4-trimethylpentane, known in the fuel industry as "isooctane,"

$$\underset{1}{CH_3}-\underset{2}{\underset{\underset{CH_3}{|}}{\overset{\overset{CH_3}{|}}{C}}}-\underset{3}{CH_2}-\underset{4}{CH}-\underset{5}{CH_3}$$
with CH_3 on C4

2,2,4-Trimethylpentane
(isooctane)

is used as a standard in assigning octane ratings of various gasolines. In this case, a common name such as isooctane would not provide enough information about which isomer was actually being used as the standard. To a chemist, the name isooctane actually refers to an entirely different structure. However, the systematic name provides complete information. The "pentane" part, which means a straight five-carbon chain, identifies the longest chain in the molecule. The numbers "2,2,4-" indicate the locations of the three groups attached to the pentane chain, and "tri" is used as a prefix for "methyl" to indicate that all three groups are methyl groups.

EXAMPLE 12.2 Structural Isomers

 Three isomers are possible for the isomeric pentanes (C_5H_{12}). Draw condensed formulas for these isomers.

SOLUTION

A good plan to follow in drawing all possible isomers—no more and no less—is to start with the straight-chain isomer and then remove one methyl group at a time, placing that methyl group on the remaining chain and checking all possibilities before removing another methyl group. In this case, start with the straight-chain five-carbon pentane as the first isomer.

$$CH_3CH_2CH_2CH_2CH_3$$

Condensed formula: pentane

Removing one methyl and placing it on the second carbon gives a second isomer, 2-methylbutane. Convince yourself that this is the only possible isomer with one methyl attached to a four-carbon chain, since putting the methyl group on the next C gives the same isomer. The third possible isomer is obtained by removing a second methyl and placing it on the second C to give 2,2-dimethylpropane.

$$
\begin{array}{c}
CH_3 \\
| \\
CH_3CHCH_2CH_3
\end{array}
$$

Condensed formula: 2-methylbutane

$$
\begin{array}{c}
CH_3 \\
| \\
CH_3CCH_3 \\
| \\
CH_3
\end{array}
$$

Condensed formula: 2,2-dimethylpropane

TRY IT 12.2

Draw the condensed structural formulas for the following compounds: (a) 2-methylpentane, (b) 3-methylpentane, (c) 2,2-dimethylbutane, and (d) 2,3-dimethylbutane.

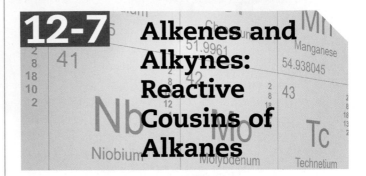

12-7 Alkenes and Alkynes: Reactive Cousins of Alkanes

Alkenes are hydrocarbons that contain a carbon–carbon double bond within their structure. Petroleum contains alkenes, and their presence in gasoline raises the octane rating. Alkenes for use in commercial applications are also obtained from petroleum by a cracking process. Ethene, best known by its common name, ethylene, is the first member of the alkene series of hydrocarbons, compounds that have one or more $C=C$ double bonds.

$$
\begin{array}{ccc}
H & & H \\
\diagdown & & \diagup \\
& C=C & \\
\diagup & & \diagdown \\
H & & H
\end{array}
$$

More ethylene is manufactured each year than any other organic chemical—109 million tons were produced in 2009. Much of the ethylene is used in the production of a plastic known as polyethylene (Section 14-5).

Green tomatoes. On their way to market, these tomatoes may be ripened by exposure to ethylene gas.

© Jan Halaska/Science Source

Ethylene is also found in plants, where it is a hormone that controls seedling growth and regulates fruit ripening. The discovery of this property led to the use of ethylene by food processors for ripening fruits and vegetables after harvest.

The general formula for alkenes with one double bond is C_nH_{2n}. The second member of the alkene series is propene (propylene). Propylene is manufactured in large quantities for use in the production of the plastic known as polypropylene (Section 14-5). Representations for ethene and propene are shown in Figure 12.7. Unlike alkanes, which undergo few chemical reactions easily (except for combustion), alkenes are quite reactive. The site of the chemical change is usually the double bond. This reactivity is essential to making many kinds of plastics, as discussed in Section 14-5. Also, addition of hydrogen to the double-bond carbon atoms converts unsaturated fats (those containing carbon–carbon double bonds) and oils, which are liquids, to the more solid consistency needed for margarine (Section 15-3).

12-7a Structural Isomers of Alkenes

In the alkene series, the possibility of locating the double bond between two different carbon atoms creates additional structural isomers. Ethene and propene have only one possible location for the double bond. However, the next alkene in the series, butene, has two possible locations for the double bond.

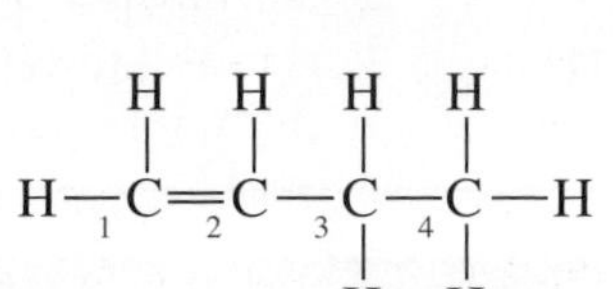

1-Butene

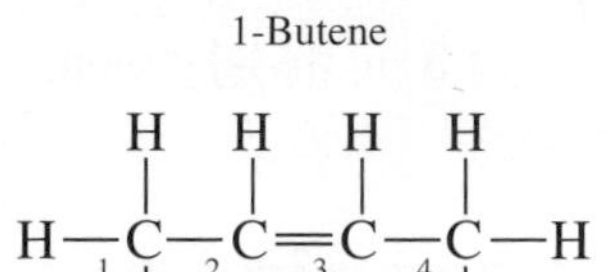

2-Butene

When groups such as methyl or ethyl are attached to carbon atoms in an alkene, the longest hydrocarbon chain is numbered from the end that will give the double bond the lowest number, and then numbers are assigned to the attached group. For example, in the following compound, the longest chain has seven carbons (heptene); the double bond is between C2 and C3 (2-heptene); and the three (tri-) methyl groups are on the third, fourth, and sixth carbons (3,4,6-trimethyl-).

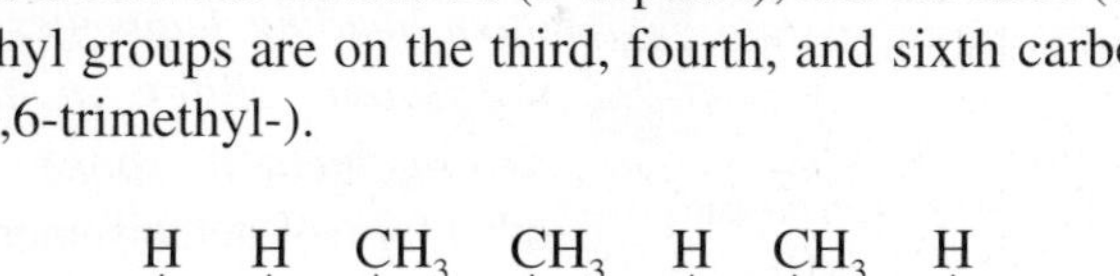

Hence, the name is 3,4,6-trimethyl-2-heptene.

FIGURE 12.7
The two smallest alkenes: (a) ethene, commonly known as ethylene, and (b) propene, commonly known as propylene. Note the planar arrangement of atoms around the carbon–carbon double bond.

Ethene
(a)

Propene
(b)

© Cengage Learning

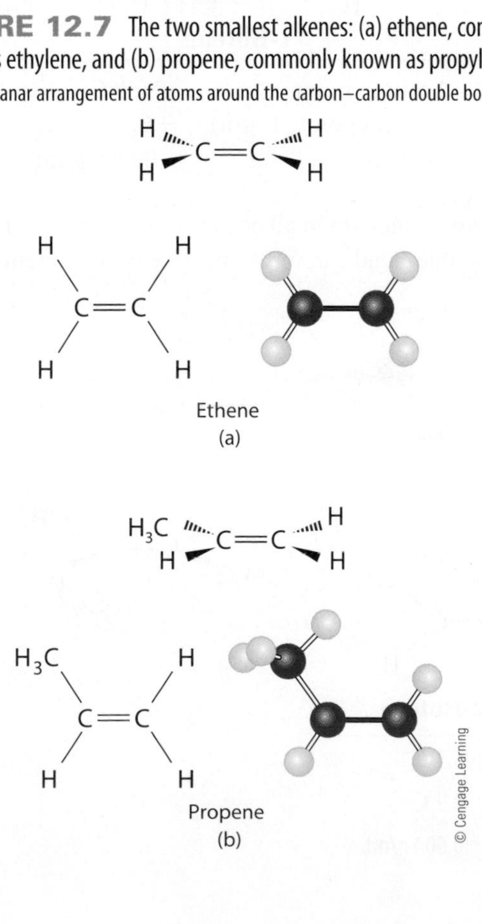

EXAMPLE 12.3 Drawing Alkenes

→ *Draw the structure of 2,3-dimethyl-2-pentene.*

SOLUTION

First, draw the parent alkene and put the double bond in the correct location. In this case, 2-pentene is the parent.

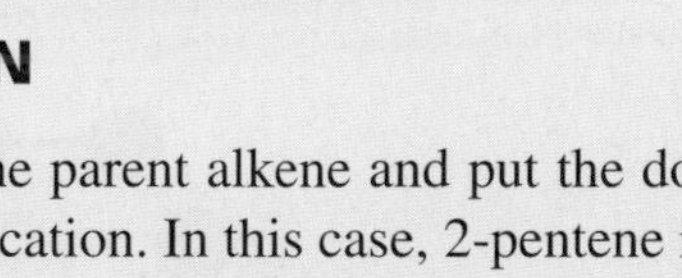

Then place the alkyl groups on the appropriate carbons. In this case, there are two methyl groups, one on C2 and one on C3. Remember that numbering the double bond takes precedence. Also check your drawing to make sure you don't have more than four bonds per carbon.

$$\begin{array}{c} \quad\quad\; H \quad CH_3 \quad CH_3 \quad H \quad\; H \\ \quad\quad | \quad\quad | \quad\quad\quad | \quad\quad | \quad\quad | \\ H-C-C=C-C-C-H \\ \quad\quad | \quad\quad\quad\quad\quad\quad\; | \quad\quad | \\ \quad\quad H \quad\quad\quad\quad\quad\quad H \quad\; H \end{array}$$

TRY IT 12.3

Draw the structure of 3-methyl-1-butene.

12-7b Stereoisomerism: *Cis* and *Trans* Isomers in Alkenes

Cis isomers: Stereoisomers with groups on the same side of a carbon–carbon double bond

Trans isomers: Stereoisomers with groups on opposite sides of a carbon–carbon double bond

Stereoisomerism: Describes isomers with the same molecular formulas and the same atom-to-atom bonding sequence but a different arrangement of the atoms in space

Some alkenes can also have **cis isomers** and **trans isomers,** one of two forms of **stereoisomerism.** *Here the isomers have the same molecular formulas and the same atom-to-atom bonding sequences, but the atoms differ in their arrangement in space.* The other form of stereoisomerism, optical isomerism, is discussed in Section 15-1.

An important difference between alkanes and alkenes is the degree of flexibility of the carbon–carbon bonds in the molecules. Rotation around single carbon–carbon bonds in alkanes occurs readily at room temperature, but the carbon–carbon double bond pre-vents this from taking place in alkenes. Consider ethene (C_2H_4). Its six atoms lie in the same plane, with bond angles of approximately 120 degrees.

If two methyl groups replace two hydrogen atoms, one on each carbon atom of ethene ($H_2C=CH_2$), the result is 2-butene ($CH_3CH=CHCH_3$). Experimental evidence confirms the existence of two compounds with the same set of bonds. The difference in the two compounds is in the location in space of the two methyl groups: the *cis* isomer has two methyl groups on the same side in the plane of the double bond, and the *trans* isomer has two methyl groups on opposite sides of the double bond. The physical properties of the *cis* and *trans* isomers of 2-butene are quite different.

The third possible isomer, 1-butene (a structural isomer of the *cis* and *trans* isomers), does not have *cis* and *trans* structures. Since one carbon atom has two identical groups (H atoms), its properties are different from those of the 2-butene isomers.

1-Butene

Melting point	−185.3°C
Boiling point (1 atm)	−6.3°C
Density (at 20°C)	0.595 g/mL

Cis–trans isomerism in alkenes is possible only when both of the double-bond carbon atoms have two different groups.

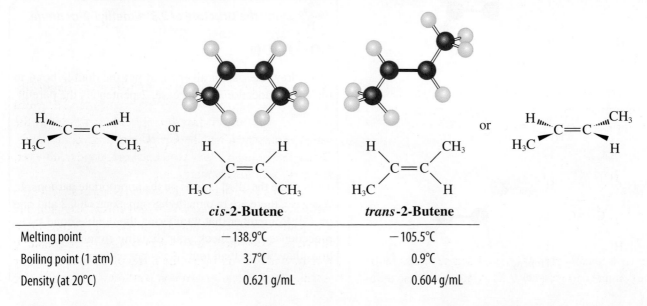

	cis-2-Butene	*trans*-2-Butene
Melting point	−138.9°C	−105.5°C
Boiling point (1 atm)	3.7°C	0.9°C
Density (at 20°C)	0.621 g/mL	0.604 g/mL

The **alkynes** have one or more triple bonds ($-C\equiv C-$) per molecule and have the general formula C_nH_{2n-2}. The simplest one is ethyne, commonly called acetylene (C_2H_2). The naming of alkynes is similar to that of alkenes, with the lowest number possible being used for locating the triple bond.

$$\underset{\underset{H}{|}}{\overset{\overset{H}{|}}{H-\underset{1}{C}}}-\underset{2}{C}\overset{180°\quad180°}{\equiv}\underset{3}{C}-\underset{\underset{H}{|}}{\overset{\overset{CH_3}{|}}{\underset{4}{C}}}-\underset{\underset{H}{|}}{\overset{\overset{H}{|}}{\underset{5}{C}}}-H$$

The name of the preceding compound is 4-methyl-2-pentyne. As with alkenes, changing the location of the multiple bond produces an isomer. For example, 4-methyl-1-pentyne is a different compound from 4-methyl-2-pentyne.

$$\underset{1}{H-C}\equiv\underset{2}{C}-\underset{\underset{H}{|}}{\overset{\overset{CH_3}{|}}{\underset{3}{C}}}-\underset{\underset{H}{|}}{\overset{\overset{H}{|}}{\underset{4}{C}}}-\underset{\underset{H}{|}}{\overset{\overset{H}{|}}{\underset{5}{C}}}-H$$

However, *cis* and *trans* isomers are not possible for alkynes because the geometry around the triple-bond carbon atoms is linear.

12-8 The Cyclic Hydrocarbons

Hydrocarbons can form rings as well as straight chains and branched chains. Two important classes of cyclic hydrocarbons found in petroleum and coal are the cycloalkanes and the aromatics.

12-8a Cycloalkanes

Cycloalkanes are saturated hydrocarbons with ring structures. The simplest cycloalkane is cyclopropane, a highly strained ring compound.

$$\overset{\overset{H\quad\quad H}{\overset{\diagdown\diagup}{C}}}{\underset{\underset{H}{\diagup}\quad\underset{H}{\diagdown}}{\underset{C-C}{}}}\qquad or\qquad \triangle$$
$$C_3H_6$$

The ring is strained because of the 60-degree angles in the ring. Carbon that is bonded to four atoms prefers the tetra-

hedral angle of 109.5 degrees. The deviation from this "ideal" angle creates the strain in some small ring systems. Cyclopropane, a volatile, flammable gas (bp −32.7°C), is a rapidly acting anesthetic. A cyclopropane–oxygen mixture is useful in surgery on babies, small children, and "bad risk" patients because of its rapid action and the rapid recovery of the patient. Helium gas is added to the cyclopropane–oxygen mixture to reduce the danger of explosion in the operating room.

The cycloalkanes are commonly represented by polygons in which each corner represents a carbon atom with two attached hydrogen atoms and the lines represent $C-C$ bonds. The $C-H$ bonds are not shown but are understood. Other common cycloalkanes include cyclobutane (C_4H_8), cyclopentane (C_5H_{10}), and cyclohexane (C_6H_{12}). These cyclic compounds are represented as planar projections even though chemists know that the rings are not planar in reality.

Cyclobutane	Cyclopentane	Cyclohexane

For instance, cyclohexane exists in the so-called chair conformation or shape as shown here because this allows for the preferred 109.5-degree bond angles.

12-8b Aromatic Compounds

Hydrocarbons containing one or more benzene rings (Figure 12.8) are called **aromatic compounds**. The word *aromatic* was derived from *aroma*, which describes the rather strong and often pleasant odor of these compounds. Benzene and many other aromatic compounds, however, are toxic and often carcinogenic.

The main structural feature, which is responsible for the distinctive chemical properties of the benzene-like aromatic compounds, is the six-carbon benzene ring. Figure 12.8a illustrates "smearing" of some of the carbon–carbon bonding electrons above and below the plane of the ring. In other words, all of the carbon–carbon bonds are equivalent, and

FIGURE 12.8 Benzene, the smallest aromatic compound.

(a) The equal distribution of bonding electrons around the ring can be represented by electron clouds above and below the plane of the ring. (b) Another way to represent the bonding electrons of benzene is as alternating double and single bonds, shown here in a ball-and-stick model.

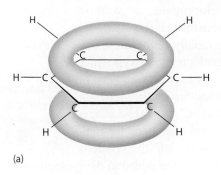

(a)

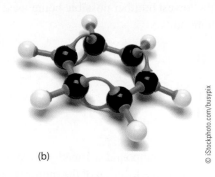

(b)

© iStockphoto.com/busypix

benzene is a planar molecule. The even distribution of electrons around the ring makes benzene and other aromatic compounds less reactive than alkenes. Benzene can be represented equally well by two resonance structures (Chapter 3)

where the circle represents the evenly distributed, smeared electrons. Chemists find the different representations useful in different situations.

When hydrogen and carbon atoms are not shown, benzene is represented by a circle in a hexagon. Each corner in the hexagon represents one carbon atom with one attached hydrogen atom. Remember that this symbol stands for C_6H_6:

Benzene
C_6H_6

Cyclohexane
C_6H_{12}

and a hexagon without a circle stands for cyclohexane (C_6H_{12}).

12-8c Derivatives of Benzene

Benzene and many of its derivatives are among the most widely used chemicals because of their use in the manufacture of plastics, detergents, pesticides, drugs, and other organic chemicals. Several important derivatives are monosubstituted benzenes, with one atom or group replacing one of the hydrogen atoms. For example, substitution of a methyl group for one of the hydrogen atoms in benzene gives methylbenzene, usually called toluene, a common solvent. Ethylbenzene, another common chemical, contains an ethyl group substituted for one of the hydrogen atoms of benzene.

CH_3

CH_2CH_3

Toluene

Ethylbenzene

12-8d Structural Isomers of Aromatic Compounds

Since the benzene molecule has a planar structure, structural isomers are possible when two or more groups are substituted for hydrogen atoms on the benzene ring.

© iStockphoto.com/ TommL

Three isomers are possible if two groups are substituted for two hydrogen atoms on the benzene ring. The prefixes *ortho-*, *meta-*, and *para-* or numbers are used to distinguish among the isomers. When the name of the compound is written, usually only the first letter of one of these terms is given. For example, when the two groups are methyl groups, the three isomers are commonly known as *o*-xylene, *m*-xylene, and *p*-xylene, where *o*, *m*, and *p* refer to *ortho* (1,2), *meta* (1,3), and *para* (1,4) substitution.

1,2-Dimethylbenzene
(*ortho*-xylene)
mp −25°C

1,3-Dimethylbenzene
(*meta*-xylene)
mp −47.9°C

1,4-Dimethylbenzene
(*para*-xylene)
mp 13.3°C

(The xylenes are used in making dyes, insecticides, and drugs.)

Another type of aromatic compound has two or more benzene rings sharing ring edges. Examples include naphthalene, anthracene, benzopyrene, and phenanthrene.

Napthalene
(mothballs)

Anthracene

Benzo(α)pyrene
(found in charcoal smoke and
cigarette smoke)

Phenanthrene

Many organic compounds found in nature are cyclic hydrocarbons that include both aromatic rings and cycloalkane or cycloalkene rings fused together. Steroids (Section 15-3) are good examples. Chemists who isolate organic compounds from plants and develop methods for making them in the laboratory are called *natural product chemists*. For example, Percy Julian was the first chemist to synthesize hydrocortisone, a steroid, and physostigmine, a compound useful in the treatment of glaucoma.

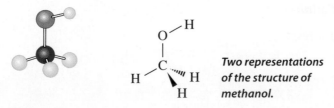

Two representations of the structure of methanol.

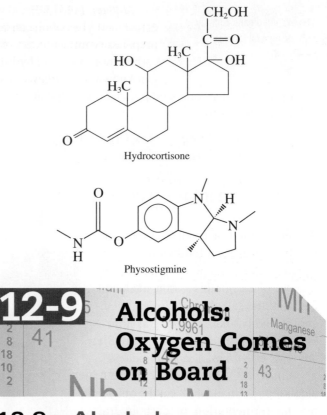

Hydrocortisone

Physostigmine

12-9 Alcohols: Oxygen Comes on Board

12-9a Alcohols

Alcohols, although not strictly hydrocarbons since they also contain oxygen atoms, will be briefly considered here because we included them in our discussion of fuels. Alcohols have the general formula ROH, with R representing an alkyl group. **Alcohols** are essentially alkanes in which one of the hydrogen atoms has been replaced by the hydroxyl (—OH) functional group. A **functional group** is an atom or group of atoms in a molecule that gives the compound its characteristic chemical behavior. Thus, replacement of one of the H-atoms of methane by the —OH group gives an alcohol known as methanol (also known as methyl alcohol). The compound is named by replacing the final -*e* of methane with -*ol*. Likewise, we see that ethanol (also known as ethyl alcohol) and propanol (propyl alcohol) can be formulated from ethane and propane. We can obtain two different alcohol isomers from propane depending on which hydrogen we replace with the —OH group. Thus, we can have isomeric alcohols, propanol and 2-propanol (more commonly referred to as isopropanol or isopropyl alcohol). Isopropyl alcohol is commonly known as "rubbing alcohol" and is found in the medicine cabinet in most homes.

> **Alcohol:** An organic compound containing a hydroxyl (OH) functional group
>
> **Functional group:** An atom or group of atoms that is part of a larger molecule and that has a characteristic chemical reactivity

Ether: An organic compound with the general formula R—O—R′

Methanol (CH_3OH), also called methyl alcohol, can be prepared from a mixture of carbon monoxide and hydrogen known as *synthesis gas*. High pressure, high temperature, and a catalyst are used to increase the yield.

$$C(s) + H_2O(g) \longrightarrow CO(g) + H_2(g)$$

Coal   Steam   Synthesis gas

$$CO(g) + 2\ H_2(g) \xrightarrow[300°C]{ZnO,\ Cr_2O_3} CH_3OH(g)$$

An old method of producing methanol involved heating a hardwood such as beech, hickory, maple, or birch in the absence of air. For this reason, methanol is sometimes called *wood alcohol*. Methanol is highly toxic. Drinking as little as 30 mL can cause death, and smaller amounts (10–15 mL) cause blindness.

Ethanol (C_2H_5OH), also called ethyl alcohol or grain alcohol, can be obtained by the fermentation of carbohydrates (starch, sugars).

$$C_6H_{12}O_6 \xrightarrow{Yeast} 2\ C_2H_5OH + 2\ CO_2$$

Glucose   Ethanol

The yeast contains enzymes that are catalysts for the fermentation process. A mixture of 95% ethanol and 5% water can be recovered from the fermentation products by

Ethanol is the active ingredient of alcoholic beverages.

distillation. Ethanol is the active ingredient of alcoholic beverages. Ethanol is receiving increased attention for use as an alternative fuel and as a fuel additive for oxygenated fuels. By the mid-1980s, 90% of all new auto sales in Brazil were for ethanol-only cars. Today, many people have migrated back to gasoline-powered cars because of an ethanol shortage in 1990. However, all gasoline now sold in Brazil contains 25% ethanol. A fuel mixture of 85% ethanol and 15% gasoline, known as E85 or *gasohol*, is available in at least 22 states in the United States. These are mostly midwestern states.

12-9b Ethers

Ethers have the general formula R—O—R′, where R and R′ stand for alkyl groups (Table 12.6), which may be the same or different. MTBE was once an important commercial ether because of its use in oxygenated and reformulated gasolines.

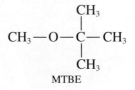

MTBE

Diethyl ether ($C_2H_5OC_2H_5$), an organic solvent, and methyl propyl ether ($CH_3OCH_2CH_2CH_3$), an anesthetic known as neothyl, are also common and well-known ethers.

Your Resources

In the back of the textbook:

→ *Review Card on Energy and Hydrocarbons*

   In OWL for CHEM2 at www.cengagebrain.com

→ *Review Key Terms with Flash Cards (printable or digital)*

→ *Complete Interactive Practice Quizzes to prepare for tests*

→ *Submit Assigned Homework and Exercises*

Applying Your Knowledge

1. What is the definition of a fossil fuel?
2. What are the three major fossil fuels?
3. What is a hydrocarbon?
4. What is the primary component of natural gas?
5. What is the heat of combustion?
6. Give definitions for the following terms:
   (a) Alkane   (b) Alkene
   (c) Alkyne   (d) Aromatic
7. Give definitions for the following terms:
   (a) Isomer
   (b) Straight-chain hydrocarbon
   (c) Branched-chain hydrocarbon
8. Write the formula for a methyl group, an ethyl group, and an alkyl group in general, C_nH_n.
9. Saturated hydrocarbons, the alkanes, have the general formula C_nH_{2n+2} where n is a whole number. Give the names and formulas for the first four alkane compounds in the series.
10. What is the structural formula for 1-pentene?
11. Give the structural formulas for the following:
    (a) 2-Methylpentane
    (b) 4,4-Dimethyl-5-ethyloctane
    (c) 2-Methyl-2-hexene
12. Draw as many different isomers as you can that have the formula C_5H_{12}.
13. Do the two structures in each pair represent identical chemical compounds, isomers or different chemical compounds?
    (a)

    (b)

(c)

(d)

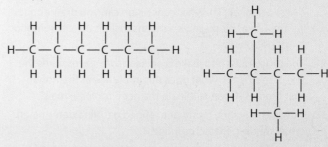

14. Which of the structures in question 13 would you expect to release the most energy in a combustion reaction?
15. Name the isomers in question 12.
16. Draw the condensed structural formulas of all possible structural isomers for C_6H_{14} and name them.
17. Draw the structures of all possible isomers that are dimethylbenzenes.
18. Draw the structure of 2,3,3-trimethyl-1-pentene.
19. Why does 2-butene have *cis* and *trans* isomers but 1-butene doesn't?
20. Draw the *cis* and *trans* isomers of 1,2-dichloroethene.
21. Explain how fractional distillation is used in the refinement of petroleum.
22. What is "straight-run" gasoline?
23. List three gasoline additives that increase the octane rating of gasoline.
24. What is gasohol? Why is gasohol controversial?
25. Explain how synthesis gas and methane can be obtained from coal, and write equations that represent these processes.
26. What do the following terms mean?
    (a) Catalytic re-forming   (b) Catalytic cracking
27. Describe what is meant by the term octane rating.
28. What types of hydrocarbons have high octane ratings?
29. What factors are likely to lead to an increased demand for methanol in the next decade?

30. One suggested alternative to crude oil is the production of biodiesel from soybean oil and other plant oils.
    (a) Use the following information to calculate the (maximum) number of gallons of biodiesel that could have been produced from the U.S. soy crop in 2011:
        42.8 bushels of soybeans are produced per acre
        75.0 million acres of soy were harvested
        11.28 pounds of soybean oil are produced per bushel of soybeans
        7.7 pounds of soybean oil needed per gallon of biodiesel produced
    (b) The total amount of petroleum-based liquid fuel (gasoline, diesel) used by the United States in 2011 was 6.8 billion gallons. Compare this to the amount you calculated in (a). What conclusion can you draw about the use of biodiesel as a means to limit the consumption of petroleum?

31. The process used to make biodiesel from soy requires the use of methanol. What does this imply about the notion that biodiesels can eliminate the need for fossil fuels?

32. How is the octane rating of a refined gasoline determined? What are oxygenated gasolines?

33. 2,2,4-trimethylpentane (isooctane) has an octane rating of 100. Octane, an isomer of isooctane, has a rating of –20. What accounts for this difference?

34. Do oxygenated gasolines cause less pollution than regular gasolines? Explain.

35. How do oxygenated and reformulated gasolines differ?

36. How is coal gasified?

37. What are the advantages and disadvantages of using methanol as an alternative fuel for vehicles?

38. What do the terms M100, E100, M85, and E85 refer to when describing alternate fuels? What is an FFV?

39. Give two reasons for and two against the use of farmland to produce grain for oxygenated fuel instead of food.

40. Label each of the following as an alkane, alkene, aromatic, or alkyne.

    (a) Methane (CH₄)

    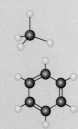

    (b) Benzene (C₆H₆)

(c) 1-Butene (CH₂CHCH₂CH₃)

(d) Acetylene (HCCH)

41. The following show the formulas and ball-and-stick models for various compounds. Label each of the following as an alkane, alkene, aromatic, alkyne, alcohol, or ether.

    (a) Propane (CH₃CH₂CH₃)

    (b) 1-Butyne (CHCCH₂CH₃)

    (c) Diethyl ether (CH₃CH₂OCH₂CH₃)

    (d) Ethanol (CH₃CH₂OH)

    (e) Ethylbenzene (CH₃CH₂C₆H₅)

    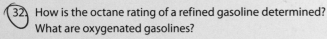

42. An exergonic process is one that releases energy. An exothermic process is one that releases heat energy.
    (a) What additional forms of energy might be associated with an exergonic process that are not associated with an exothermic process?
    (b) Is it possible for a process to be described as both exergonic and exothermic? Explain.
    (c) Can you think of a situation in which a process might be exergonic and endothermic?

43. Using the bond energy data provided in Table 12.1, calculate the theoretical energy change for the production of water from hydrogen and oxygen:
    $$2\,H{-}H + O{=}O \longrightarrow 2\,H{-}O{-}H$$

ONE APPROACH.
70 UNIQUE SOLUTIONS.

13
Nuclear Changes
and Nuclear Power

*R*adioactivity. With what do you associate that word? Hazardous waste? Nuclear power plants? Medical diagnoses? Cancer risks? Nuclear weapons? Cancer cures? All of these are appropriate associations with the kind of change that happens only in atomic nuclei. Nuclear changes are very different from ordinary chemical reactions, as you will see. Nuclear changes are usually accompanied by the emission of radiation and can, in some cases, be accompanied by the release of large amounts of energy. In this chapter, you will learn something about how and when nuclear changes occur.

You will also see that nuclear changes are pictured in much the same way as chemical changes: reactants going to products. Equations can be written for nuclear changes, but although there are similarities between nuclear and chemical changes, nuclear changes are different in some rather significant ways. Certainly the discovery of nuclear changes has affected our lives, some might say for the worse; however, you or one of your friends may be alive today because of an application of what is known about how radioactive isotopes undergo change.

The nuclear age is often considered to have started either in the late 1800s and early 1900s with the discovery of radioactive elements or in 1945 with the first explosion of an atomic bomb. It is true that early atomic theory said nothing about radioactivity, but that was because radiation cannot be detected directly by our five senses. It took the maturing of the sciences—with such diverse discoveries as how to produce a vacuum, photographic film, electricity, and magnetic fields—to lead to the knowledge that some atoms disintegrate spontaneously and, in the process, produce radiation.

13-1 The Discovery of Radioactivity

In February 1896, Henri Becquerel was experimenting in France with the relation between the recently discovered X-rays and the phosphorescence (the emission of visible light after exposure to sunlight) of certain minerals. X-rays had been found to penetrate substances like paper and expose photographic plates. Becquerel had already discovered that phosphorescing uranium minerals also exposed the plates. During several

cloudy days, some samples were left waiting in a closed drawer. Out of curiosity, Becquerel developed the plates. He had quite a surprise—the plates were exposed. Obviously, the emission of radiation that penetrated the paper surrounding the plates had nothing to do with phosphorescence but was a property of the mineral. In fact, all uranium compounds, and even the metal itself, exposed photographic plates. Becquerel also discovered that uranium (U) emitted radiation that was capable of causing air molecules to ionize (that is, to lose electrons and become positively charged particles).

Before long, it was recognized that the radiation from elements like uranium and radium consisted of three types, now known as alpha particles, beta particles, and gamma rays (Chapter 3).

In 1899, Ernest Rutherford found that alpha particles could be stopped by thin pieces of paper and had a range of only about 2.5 cm to 8.5 cm in air, whereas beta particles were capable of penetrating far greater distances in air.

In 1900, Paul Villard identified a third form of natural radiation—gamma (γ) rays. He discovered that these were not streams of particles but instead had the general characteristics of light or X-rays. Gamma rays,

Henri Becquerel

© World History Archive/Alamy

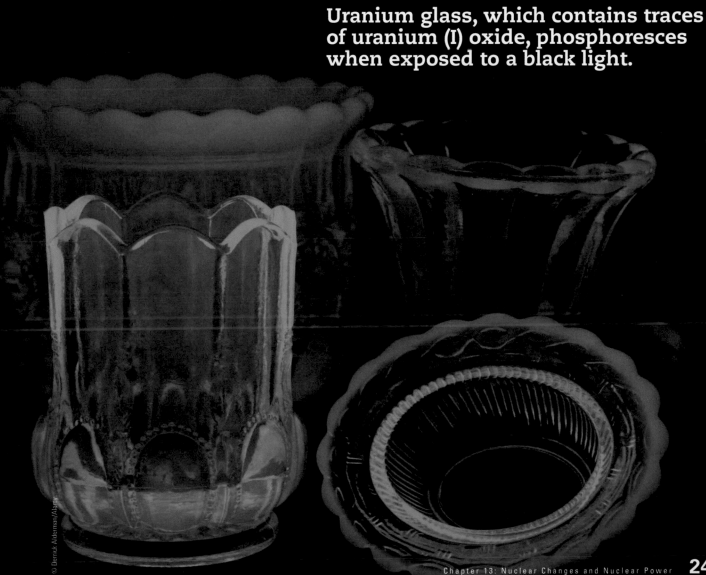

Uranium glass, which contains traces of uranium (I) oxide, phosphoresces when exposed to a black light.

© Derrick Alderman/Alamy

FIGURE 13.1 The relative penetrating abilities of alpha, beta, and gamma radiation.
The heavy, highly charged alpha particles are stopped by a piece of paper (or the skin). The lighter, less highly charged beta particles penetrate paper but are stopped by a 0.5-cm sheet of lead. Because gamma rays have no charge and no mass, they are the most penetrating but can be stopped by several centimeters of lead.

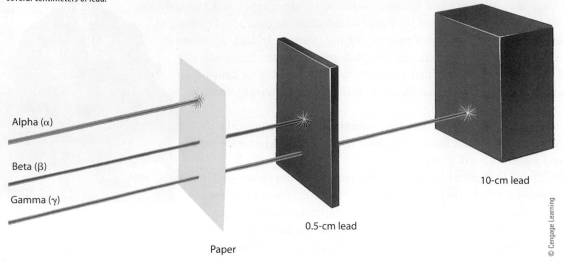

Alpha (α)

Beta (β)

Gamma (γ)

10-cm lead

0.5-cm lead

Paper

© Cengage Learning

Nuclear reaction: A process by which an isotope of one element is transformed into an isotope of another element

Nucleon: A nuclear particle, such as a proton or neutron

a high-energy form of electromagnetic radiation (see Figure 3.7), are extremely penetrating; they are capable of passing through more than 22 cm of steel and about 2.5 cm of lead. Figure 13.1 compares the penetrating ability of the three forms of natural radiation.

13-2 Nuclear Reactions

After discovery of the natural radioactivity of uranium, thorium, and radium, many other elements were found to have radioactive isotopes. All of the elements heavier than bismuth (Bi, atomic number 83) and a few lighter than bismuth have natural radioactivity. While studying radium, Rutherford found that besides emitting alpha particles, radium was also producing radioactive radon gas (Rn). This led Rutherford and one of his students, Frederick Soddy, in 1902 to propose the revolutionary theory that *radioactivity is the result of a natural change of an isotope of one element into an isotope of a different element.* Such a change is a **nuclear reaction**, a process in which an unstable nucleus emits radiation and is converted into a more stable nucleus of a different element. Thus, a nuclear reaction results in a change in atomic number and often a change in mass number as well.

13-2a Equations for Nuclear Reactions

In nuclear reactions, the total number of nuclear particles, called **nucleons** (protons plus neutrons), remains the same, but the identities

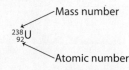

Mass number

$^{238}_{92}U$

Atomic number

of atoms can change. Just as with chemical equations, nuclear equations reflect the fact that matter is conserved. As a result, *the sum of the mass numbers of reacting nuclei must equal the sum of the mass numbers of the product nuclei. There must also be nuclear charge balance—the sum of the atomic numbers of the products must equal the sum of the atomic numbers of the reactants.*

Consider the equation for the nuclear reaction

$$^{226}_{88}Ra \longrightarrow {}^{4}_{2}He^{2+} + {}^{222}_{86}Rn$$

Radium-226   Alpha particle   Radon-222

The mass number on the left equals the sum of the mass numbers on the right. Similarly, the atomic number on the left equals the sum of the atomic numbers on the right.

Mass number: $226 = 4 + 222$

Atomic number: $88 = 2 + 86$

The isotope of uranium with atomic mass 238 is also an alpha emitter. Recall from Chapter 3 that an alpha particle is a helium atom without its electrons, so it has a +2 charge. When the $^{238}_{92}U$ nucleus gives off an alpha particle, made up of two protons and two neutrons, four units of atomic mass and two units of atomic charge are lost. The resulting nucleus has a mass of 234 and a nuclear charge of 90, showing that it is an isotope of thorium, which has 90 protons in its nucleus and an atomic number of 90.

$$^{238}_{92}\text{U} \longrightarrow {}^{4}_{2}\text{He}^{2+} + {}^{234}_{90}\text{Th}$$

Uranium-238   Alpha particle   Thorium-234

Mass number:	238	=	4	+	234

Atomic number:	92	=	2	+	90

As is seen in these two examples, loss of an alpha particle from an atom of a given element results in the formation of an atom whose atomic number is 2 less, and whose mass is 4 less, than that of the original element.

Some unstable nuclei are beta emitters. A beta particle is the same as an electron. For example, uranium-235 emits a beta particle ($_{-1}^{0}\text{e}$) upon decay.

$$^{235}_{92}\text{U} \longrightarrow {}^{0}_{-1}\text{e} + {}^{235}_{93}\text{Np}$$

Uranium-235   Beta particle   Neptunium-235

Mass number:	235	=	0	+	235

Atomic number:	92	=	−1	+	93

Knowing that some nuclei emit beta particles leads to a basic question: How can a nucleus containing protons and neutrons emit a beta particle, which is an electron? It has been established that an electron and a proton can combine outside the nucleus to form a neutron. Therefore, the reverse process is proposed to occur inside the nucleus. When a beta particle is emitted from a decaying nucleus, a neutron decomposes, giving up an electron and changing itself into a proton. The ejected electron is the beta particle.

Beta particle production: $^{1}_{0}\text{n} \longrightarrow {}^{0}_{-1}\text{e} + {}^{1}_{1}\text{H}$

Neutron   Electron   Proton

The resulting proton remains in the nucleus and increases the atomic number by 1. Therefore, loss of a beta particle from an atom of an element results in the formation of an atom of an element whose atomic number has increased by 1 but whose atomic mass is the same.

Gamma (γ) radiation may or may not be given off simultaneously with alpha or beta particles, depending on the particular nuclear reaction involved. Gamma rays are emitted when the product nucleus must lose some additional energy to become stable. Being electromagnetic

radiation, gamma rays have no charge and essentially no mass. The emission of a gamma ray, therefore, cannot alone account for production of a different element.

EXAMPLE 13.1 Writing an Equation for an Alpha Emission

→ *Write an equation for alpha emission from a polonium-218 isotope.*

SOLUTION

First, write the partial equation and set up a table of mass and atomic number changes under it. The atomic number of polonium is 84.

$$^{218}_{84}\text{Po} \longrightarrow {}^{4}_{2}\text{He}^{2+} + ?$$

Mass number:	218	$\longrightarrow$	4	+ ?
Atomic number:	84	$\longrightarrow$	2	+ ?

The mass number of the product must be 214 because its mass number plus that of the alpha particle must equal 218, the mass number of the decaying polonium-218 isotope. The atomic number of the product must be 82 because its atomic number plus that of the alpha particle must equal 84, the atomic number of the decaying isotope. Inasmuch as the element with an atomic number of 82 is lead, the product is $^{214}_{82}\text{Pb}$.

$$^{218}_{84}\text{Po} \longrightarrow {}^{4}_{2}\text{He} + {}^{214}_{82}\text{Pb}$$

TRY IT 13.1

Write an equation showing the emission of an alpha particle by an isotope of neptunium ($^{237}_{93}\text{Np}$).

EXAMPLE 13.2 Writing an Equation for a Beta Emission

→ *Write an equation for beta emission from a lead-210 nucleus.*

SOLUTION

First, write a partial equation that includes what is known: Lead-210 is a reactant, and a beta particle is a product. Then, to aid in determining the mass and atomic number changes, set up a table like those used earlier. The atomic number of lead is 82.

$$^{210}_{82}\text{Pb} \longrightarrow {}^{0}_{-1}\text{e} + ?$$

Mass number:	210	$\longrightarrow$	0 + ?
Atomic number:	82	$\longrightarrow$	−1 + ?

The sum of the mass numbers of the products must equal 210, the mass number of the decaying lead isotope. Since the mass of the beta particle is essentially zero, the mass number of the product nucleus must be 210. The sum of the atomic numbers of the products must also equal the atomic number of lead, 82. So, the atomic number of the product must be 82 [83 + (−1) = 82], which is the atomic number of bismuth (Bi). The product nucleus is $^{210}_{83}\text{Bi}$.

$$^{210}_{82}\text{Pb} \longrightarrow {}^{0}_{-1}\text{e} + {}^{210}_{83}\text{Bi}$$

TRY IT 13.2

Write an equation showing the emission of a beta particle from $^{234}_{91}\text{Pa}$.

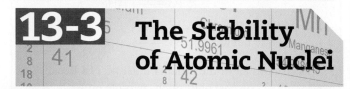

13-3 The Stability of Atomic Nuclei

Positron: A positively charged electron

Why are some nuclei unstable and radioactive while others are stable and not radioactive? The fact that there are strong repulsions among all of those protons packed inside the nucleus of an atom has something to do with nuclear stability. In addition, the relative numbers of neutrons, which are not charged, play some role. Figure 13.2 shows a plot of the number of protons versus the number of neutrons in the known isotopes from hydrogen ($Z = 1$; Z is the symbol for atomic number) to bismuth ($Z = 83$). The nonradioactive (stable) isotopes (blue and green dots) are far fewer in number than the radioactive (unstable) isotopes (red dots).

The stability of nuclei is apparently dependent on the relative numbers of protons and neutrons. The nucleus of the simplest atom, hydrogen, contains only a proton. Its two isotopes, deuterium (^2_1H) and tritium (^3_1H), contain one and two neutrons, respectively. By looking at Figure 13.2, you can see that the band of stable nuclei, the blue and green dots, curves upward toward the neutron axis; this shows that in stable nuclei, the number of neutrons is equal to or greater than the number of protons. From hydrogen to bismuth, except for ^1_1H and ^3_2He, *the mass numbers of stable isotopes are always twice as large as (or even larger than) the atomic number*. It appears that the larger numbers of protons in the nuclei of heavier atoms require extra neutrons to gain stability. Any unstable isotope (a red dot in Figure 13.2) will decay in such a way that its decay product falls closer to the stable band (the blue and green dots).

Beta emission occurs in isotopes that have *too many neutrons* to be stable. These isotopes appear as red dots above the stable band in Figure 13.2. When beta decay occurs, the conversion of a neutron into a proton and an electron increases the atomic number while lowering the number of neutrons, as was illustrated in Example 13.2, and the new isotope moves toward the stable region.

Those isotopes with *too few neutrons* (red dots below the band of stability) decay as well but in a manner that increases the number of neutrons relative to the number of protons. One way this can happen is by emission of a type of subatomic particle discovered in 1932, a **positron**—a positively charged electron, $^0_{+1}\text{e}$. For example, the decay of nitrogen-13, an isotope with too few neutrons, is by positron emission.

$$^{13}_{7}\text{N} \longrightarrow {}^{0}_{+1}\text{e} + {}^{13}_{6}\text{C}$$

The positron results from the decay of a proton.

$$\underset{\text{Proton}}{^{1}_{1}\text{H}} \longrightarrow \underset{\text{Neutron}}{^{1}_{0}\text{n}} + \underset{\text{Positron}}{^{0}_{+1}\text{e}}$$

Because the positron, like the electron, has a mass number of zero, the mass number of the product nucleus is the same as that of the starting nucleus. Therefore, emission of a positron from an atom of an element results in the formation of an atom of an element whose atomic number is reduced by 1 but whose mass is unchanged.

All isotopes of the elements beyond bismuth ($Z = 83$) are unstable. Most of them decay by ejecting an alpha particle. This kind of decay, as illustrated in Example 13.1, decreases the mass number by 4 and the atomic number by 2. The types of radioactive decay we have discussed are summarized in Table 13.1.

© iStockphoto.com/Pixlmaker

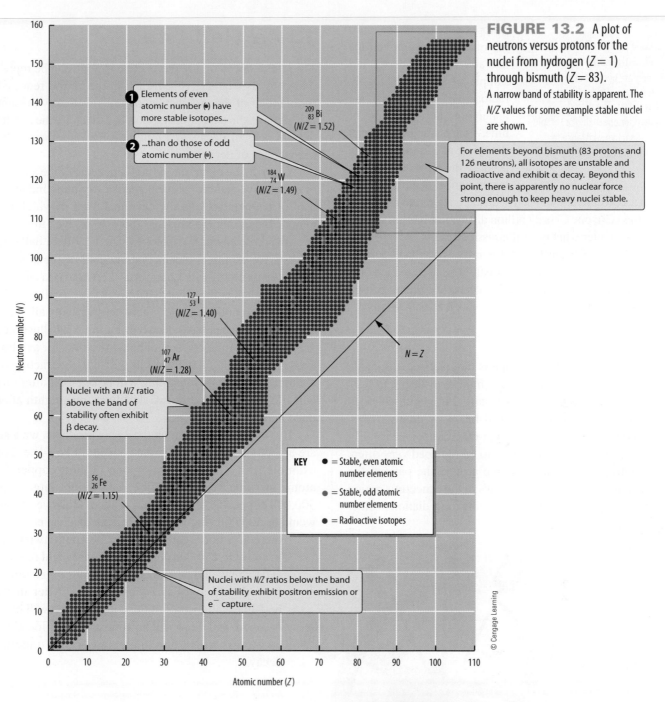

FIGURE 13.2 A plot of neutrons versus protons for the nuclei from hydrogen ($Z = 1$) through bismuth ($Z = 83$). A narrow band of stability is apparent. The N/Z values for some example stable nuclei are shown.

For elements beyond bismuth (83 protons and 126 neutrons), all isotopes are unstable and radioactive and exhibit α decay. Beyond this point, there is apparently no nuclear force strong enough to keep heavy nuclei stable.

❶ Elements of even atomic number (●) have more stable isotopes...

❷ ...than do those of odd atomic number (●).

$^{209}_{83}$Bi
($N/Z = 1.52$)

$^{184}_{74}$W
($N/Z = 1.49$)

$^{127}_{53}$I
($N/Z = 1.40$)

$^{107}_{47}$Ar
($N/Z = 1.28$)

Nuclei with an N/Z ratio above the band of stability often exhibit β decay.

$N = Z$

KEY
● = Stable, even atomic number elements
● = Stable, odd atomic number elements
● = Radioactive isotopes

$^{56}_{26}$Fe
($N/Z = 1.15$)

Nuclei with N/Z ratios below the band of stability exhibit positron emission or e^- capture.

© Cengage Learning

TABLE 13.1 Changes in Atomic Number and Mass Number Accompanying Radioactive Decay

Type of Decay	Symbol	Charge	Mass	Change in Atomic Number	Change in Mass Number
Beta	$^0_{-1}e$	−1	0	+1	None
Positron	$^0_{+1}e$	+1	0	−1	None
Alpha	$^4_{-2}He^{2+}$	+2	4	−2	−4
Gamma	$^0_0\gamma$	0	0	None	None

© Cengage Learning

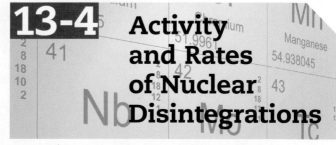

13-4 Activity and Rates of Nuclear Disintegrations

The number of radioactive nuclei that disintegrate in a sample per unit of time is called its **activity** (this is the "activity" in radioactivity). The activity of a sample containing radioactive isotopes depends on the number of nuclei present and the rate at which they decay. If a sample of matter is "highly radioactive," many atoms are undergoing decay per unit of time. A small number of nuclei decaying at a rapid rate can produce the same activity as a larger number of atoms decaying at a slower rate. Radioactive disintegrations are measured in *curies* (Ci); one Ci is 37 billion disintegrations per second (dps), no matter what type of emission results.

To illustrate the differences in rates of decay of radioactive nuclei, consider first how cobalt-60 is used in medicine to treat cancerous tumors in the human body. When cobalt-60 decays, it produces beta particles as well as gamma rays.

$$^{60}_{27}Co \longrightarrow ^{60}_{28}Ni + ^{0}_{-1}e + ^{0}_{0}\gamma$$

Although the cobalt-60 isotope is radioactive, it is stable enough so that only half of a sample will decay in 5.27 years. A cobalt-60 sample is installed in a well-shielded apparatus that emits a focused beam of gamma rays. Because the half-life of the cobalt-60 radioisotope is fairly long, the sample does not have to be replaced very often. By rotating the radiation source around the patient, the physician can concentrate the rays in the cancerous region being treated while somewhat limiting the radiation to other parts of the body (Figure 13.3).

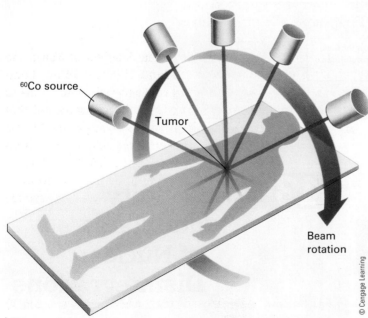

FIGURE 13.3 Treatment for cancer with gamma radiation from cobalt-60.

By adjusting the rotation of the radiation source, the radiation is concentrated where the beams cross at the location of the diseased tissue.

13-4a Half-Life

The rate of decay of any radioactive isotope can be represented by its characteristic **half-life**, the period required for one-half of the radioactive material originally present to undergo radioactive decay. Short half-lives are the results of high rates of decay, and long half-lives are the results of low rates of decay.

The half-life of an isotope is independent of the amount of radioactive material present and is essentially independent of temperature and the chemical form in which the radioactive atoms are present. Table 13.2 gives the half-lives of some radioactive isotopes. The 12.9-hour half-life of $^{64}_{29}Cu$, for example, means that one-half of the amount of copper-64 atoms will remain after 12.9 hours. In another 12.9 hours, half of the original half, $(\frac{1}{2})^2 = (\frac{1}{4})$, will remain. This process continues indefinitely until virtually all of the copper-64 isotopes have decayed. Figure 13.4 illustrates graphically how the concept of half-life works for a radioactive isotope. No matter what its half-life, the fraction of a radioactive isotope remaining will be one-half after one half-life, one-fourth after two half-lives, one-eighth after three half-lives, and so on.

By focusing attention on the decay products, we also are helped in understanding the concept of half-life. For example, if a sample contains one million copper-64 atoms at some beginning time, 12.9 hours later only 500,000 copper-64 atoms would remain. However, there would be 500,000 zinc-64 atoms present that had not been there 12.9 hours earlier. After 25.8 hours (two half-lives), only 250,000 copper-64 atoms would remain, and there would be 750,000 zinc-64 atoms resulting from the decay of the copper atoms. After many half-lives, almost all of the copper atoms will have decayed, and there will be almost one million zinc-64 atoms, which are stable and do not undergo decay.

TABLE 13.2 Half-Lives of Some Radioactive Isotopes

Decay Process	Half-Life
$^{238}_{92}U \longrightarrow ^{234}_{90}Th + ^{4}_{2}He$	4.51×10^9 years
$^{3}_{1}H \longrightarrow ^{3}_{2}He + ^{0}_{-1}e$	12.3 years
$^{14}_{6}C \longrightarrow ^{14}_{7}N + ^{0}_{-1}e$	5730 years
$^{131}_{53}I \longrightarrow ^{131}_{54}Xe + ^{0}_{-1}e$	8.05 days
$^{64}_{29}Cu \longrightarrow ^{64}_{30}Zn + ^{0}_{-1}e$	12.9 hours
$^{69}_{30}Zn \longrightarrow ^{69}_{31}Ga + ^{0}_{-1}e$	55 minutes

FIGURE 13.4 Radioactive decay of 20 mg of oxygen-15, which has a half-life of 2.0 minutes.

The plotted data are given in the following table. After each half-life period, the quantity present at the beginning of the period is reduced by one-half.

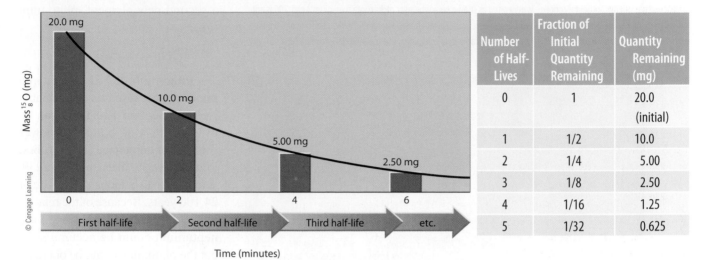

Number of Half-Lives	Fraction of Initial Quantity Remaining	Quantity Remaining (mg)
0	1	20.0 (initial)
1	1/2	10.0
2	1/4	5.00
3	1/8	2.50
4	1/16	1.25
5	1/32	0.625

© Cengage Learning

13-4b Natural Radioactive Decay Series

As one would expect for an element with such a long half-life (Table 13.2), relatively large amounts of uranium-238 can be found in certain rocks and mineral deposits. Uranium-238 decays first to thorium-234, which then itself decays. These first two steps are part of the **uranium series**, which ends with a stable, nonradioactive isotope of lead, lead-206 (Figure 13.5). You would not expect to find much of the short-lived isotopes in a sample of rock, and indeed you have to look carefully for them, but they are there. The longer-lived isotopes such as uranium-234 and thorium-230 are readily detected. Two other similar natural decay series exist, each of which starts out with a different isotope and proceeds through a different set of radioactive decay products. Most of the naturally occurring radioactive isotopes are members of one of the three decay series.

13-5 Artificial Nuclear Reactions

In 1919, Rutherford was successful in producing the first artificial nuclear change by bombarding nitrogen (N_2) with alpha particles. All of the results of the experiment could be explained if one assumed the nuclear reaction to be

$$^{14}_{7}N + {}^{4}_{2}He^{2+} \longrightarrow [{}^{18}_{9}F] \longrightarrow {}^{17}_{8}O + {}^{1}_{1}H$$

where $^{18}_{9}F$ is an unstable nucleus that quickly disintegrates to $^{17}_{8}O$ and $^{1}_{1}H$. Both product nuclei are stable. Rutherford had observed an **artificial transmutation,** the conversion of one element into another, during a laboratory experiment. Following Rutherford's original experiment, there was considerable interest in discovering new nuclear reactions.

In 1934, Irene Curie Joliot, daughter of Marie and Pierre Curie, and her husband, Frédéric Joliot, bombarded aluminum (Al) with alpha particles and observed neutrons and a positron. The Joliots discovered that when the flow of alpha particles striking the Al was stopped, the neutron emissions stopped but the positron emissions continued. They reasoned that the alpha particles reacted with aluminum nuclei to produce phosphorus-30 nuclei, which then decayed to produce positrons.

$$^{27}_{13}Al + {}^{4}_{2}He^{2+} \longrightarrow {}^{30}_{15}P + {}^{1}_{0}n$$

$$^{30}_{15}P \longrightarrow {}^{30}_{14}Si + {}^{0}_{+1}e$$
Positron

The second reaction continued because the phosphorus-30 was decaying more slowly than it was being produced. Phosphorus-30 was the *first radioactive isotope to be produced artificially*. Today, more than 1000 other radioactive isotopes have been produced.

Uranium series: The series of steps in the naturally occurring decay of uranium-238 to lead-206

Artificial transmutation: Experimental conversion of one element into another

FIGURE 13.5 The uranium-238 decay series.

Radium (Ra) and polonium (Po), the two elements discovered by Marie Curie, are part of this series. Radon (Rn), the radioactive gas of environmental concern, is generated as shown here wherever rocks contain uranium. Lead-206 is not radioactive. The half-lives of the isotopes in this decay series vary considerably. Uranium-238 has a half-life of 4.5 billion years, whereas the half-life of lead-210 is 22 years. Some radioactive elements, like thallium-210, have half-lives of only a few minutes.

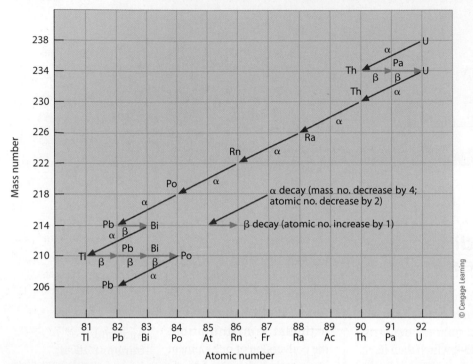

© Cengage Learning

13-6 Transuranium Elements

Transuranium element: An artificial element with an atomic number greater than 92

In 1940, at the University of California, E. M. McMillan and P. H. Abelson prepared element 93, the synthetic element neptunium (Np). Neptunium was the first of the **transuranium elements**, those with atomic numbers greater than 92. The neptunium was made by directing a stream of high-energy deuterons ($_1^2$H) onto a target of uranium-238. The initial reaction was the conversion of uranium-238 to uranium-239:

$$_{92}^{238}\text{U} + {}_1^2\text{H} \longrightarrow {}_{92}^{239}\text{U} + {}_1^1\text{H}$$

Uranium-239 has a half-life of 23.5 minutes and decays spontaneously to the element neptunium by the emission of beta particles:

$$_{92}^{239}\text{U} \longrightarrow {}_{93}^{239}\text{Np} + {}_{-1}^{0}\text{e}$$

Neptunium is also unstable, with a half-life of 2.33 days; it converts into a second new element, plutonium (Pu):

$$_{93}^{239}\text{Np} \longrightarrow {}_{94}^{239}\text{Pu} + {}_{-1}^{0}\text{e}$$

Plutonium-239 is considered important because it is a fissionable isotope (Section 13-9) and can be used for making bombs or power sources for unmanned space probes. Plutonium-239, like neptunium-239, is radioactive, but it has a half-life of 24,100 years. Because of the relative values of the half-lives, very little neptunium could be accumulated, but the plutonium could be obtained in larger quantities. The names neptunium and plutonium were taken from the mythological names Neptune and Pluto in the same atomic number sequence as the order of the planets Uranus (uranium), Neptune, and Pluto (now classified as a dwarf planet).

Although Neptune and Pluto were the last bodies in our solar system classified as planets, their namesakes are not the last in the list of elements. The rush of experiments that followed the synthesis of plutonium produced additional manmade elements. Elements 95 (americium) through 106 (seaborgium) had been created by the end of the 1970s. Up to element 101 (mendelevium), all of the elements can be made by hitting the target nuclei with small particles such as $_2^4$He^{2+} or $_0^1$n. Beyond element 101, special techniques that use heavier bombarding particles are required. The energy of the particles is matched to the energy needed to produce the initial unstable nucleus. For example, lawrencium is made by hitting californium-252 with boron nuclei. In this reaction, five neutrons are produced.

$$_{98}^{252}\text{Cf} + {}_5^{10}\text{B} \longrightarrow {}_{103}^{257}\text{Lr} + 5\,{}_0^1\text{n}$$

By 1996, elements up to 112 had all been created, many with extremely short half-lives. Element 112, for example, has a half-life of 240 microseconds.

Four atoms each of element 113 and element 115 were produced in a fusion reaction between calcium-48 and americium-243 in 2004. These two elements, combined with element 114, produced earlier, extend the periodic table to 115 elements. Element 113 was produced by the alpha decay of element 115. Most recently, elements 116 and 118 have been produced through similar methods. The element to round out the last row of the periodic table as we know it is element 117. Russian and American chemists are currently working to synthesize this element at the Joint Institute of Nuclear Research in Dubna, Russia. In this process, the United States is providing the target berkelium isotope necessary to instigate a reaction.

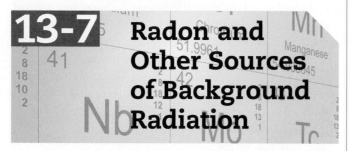

13-7 Radon and Other Sources of Background Radiation

Sources of background radiation, which we receive during normal daily life, fall into two categories: man-made and natural (Figure 13.6). Of these, natural sources produce the bulk of our exposures. When natural sources are coupled with desirable exposures from medical applications, scarcely 3% of the exposures are truly avoidable by society as a whole. To illustrate how unavoidable most exposures to radiation are, consider the element potassium, an element essential to human life that helps regulate water balance in our bodies and is involved in nerve transmissions. A 60-kg (132-lb) person contains about 200 g of potassium. It turns out that just over 0.01% of all potassium atoms are potassium-40 atoms, which are radioactive, with a half-life of 1.25×10^9 years. This means the 60-kg person has roughly 20 mg of radioactive potassium,

which disintegrates at a rate determined by the half-life of that isotope. These disintegrations contribute to the background radiation we all receive. In addition, the beta particle from the decay of a potassium-40 disintegration in your neighbor might pass into you to contribute to your background radiation. Normal background radiation from all sources is 2 or 3 dps.

Another element that contributes to our background radiation, and hence to our risks of radiation-caused damage, is radon. Radon-222, the most common isotope of radon, is radioactive, with a half-life of 3.82 days. It is a product of the uranium decay series (see Figure 13.5) and results from the alpha decay of radium-226.

$$^{226}_{88}\text{Ra} \longrightarrow {}^{222}_{86}\text{Rn} + {}^{4}_{2}\text{He}^{2+}$$

When radon decays, it produces an alpha particle and another short-lived radioisotope, polonium-218.

$$^{222}_{86}\text{Rn} \longrightarrow {}^{218}_{84}\text{Po} + {}^{4}_{2}\text{He}^{2+}$$

Polonium-218 (half-life 3.05 minutes) also decays, producing an alpha particle and lead-214 (half-life 26.8 minutes).

$$^{218}_{84}\text{Po} \longrightarrow {}^{214}_{82}\text{Pb} + {}^{4}_{2}\text{He}^{2+}$$

These alpha particles from the products of radon decay make breathing air containing radon dangerous.

Radon exists as a gas, and all rocks contain *some* uranium, although the amount in most is small (one to three parts per million). Therefore, radon is constantly being formed in amounts that vary with the type of rock or soil. How much of this radon escapes into the outside air or a building on the surface depends on the porosity and moisture content of the soil and how finely divided it is (escape is easier from a small particle than from a large one). The overall geology of the site is also important.

Because radon is chemically unreactive, radon atoms in the air we breathe are inhaled and exhaled without any chemical change, although some may dissolve in lung fluids. If a radon atom happens to decay within the lungs, however, the nongaseous and radioactive *radon daughters* can remain inside the lungs, where they will continue to decay. Those up to lead-210 have short half-lives and include several alpha emitters. Because alpha particles can travel up to 0.7 mm, the approximate thickness of the lining of the lung, they can damage delicate lung tissue and create a higher than normal risk of lung cancer.

Miners in deep mines are exposed to far higher than average radon levels, and as early as 1950 government agencies began monitoring radon exposures and incidences of lung cancer. Today, it is well known that radon

FIGURE 13.6 Sources of average background radiation exposure in the United States.

The sources are expressed as percentages of the total. As seen from the figure, background radiation from natural sources far exceeds that from artificial sources.

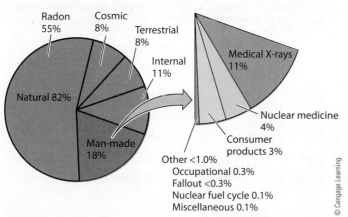

© Cengage Learning

exposure increases one's chance of developing lung cancer. If you smoke, the chances are even greater since there seems to be a synergistic effect between smoking and radon levels in causing lung cancer.

It has been estimated that nearly 1 in 15 homes in the United States is affected by radon contamination. Levels higher than 4 picocuries per liter of air (4 pCi/L) have been designated the "action level" by the Environmental Protection Agency (EPA). Some level of radon could probably be detected in homes in almost every state.

This EPA action level is important for two reasons. First, levels of radon higher than 4 pCi/L should be reduced because levels above this are judged to lead to unacceptable risks of lung cancer. Second, it is difficult to lower the level of radon in most contaminated homes below 4 pCi/L. This last point is an acceptance of the fact that radiation exposure will always be with us.

13-8 Useful Applications of Radioactivity

The damaging aspects of nuclear radiation must always be kept in mind, especially when the possibilities of accidental or unintended exposures are great. However, the radiation from radioisotopes can be put to beneficial use.

13-8a Food Irradiation

In some parts of the world, stored-food spoilage may claim up to 50% of the food crop. In the United States, refrigeration, canning, and chemical additives reduce this figure considerably. Still, there are problems with food spoilage and contamination. Food protection costs amount to a sizable fraction of the final cost of food.

Foods may be irradiated to retard the growth of organisms such as bacteria, molds, and yeasts. This practice, using gamma rays from sources such as cobalt-60 or cesium-137, is common in European countries, Canada, Mexico, and the United States. Such irradiation prolongs shelf life under refrigeration in much the same way that heat pasteurization protects milk. For example, chicken normally has a three-day refrigerated shelf life. After irradiation, chicken may have a three-week refrigerated shelf life. In December 1997, the Food and Drug Administration (FDA) approved irradiation of beef in response to concerns due to food poisoning and product recalls of contaminated hamburger meat. In 2008, iceberg lettuce and spinach were added to the list

after outbreaks of the infectious *Escherichia coli* and *Salmonella* bacteria were traced to bags of fresh spinach.

Foods irradiated at sufficient levels will keep indefinitely when sealed in plastic or aluminum foil packages. At present, the FDA has approved irradiation of many classes of foods (Table 13.3). It is important to note that in no case is there any chance of irradiated food becoming radioactive. The ongoing debate about the safety of irradiated foods revolves around the possibly harmful nature of would-be "radiolytic products"—products of chemical changes in food caused by the high energy of the radiation. For example, could irradiation of food produce a chemical capable of causing genetic damage? To date, no evidence has been found for harmful radiolytic products. Meanwhile, based on extensive review of current data, the FDA-approved radiation limits have been conservatively set at relatively low levels.

13-8b Radiocarbon Dating

Carbon has three isotopes. Two of these, carbon-12 and carbon-13, comprise 98.9% and 1.1%, respectively, of all carbon. A third isotope, carbon-14, is present in very small amounts under natural conditions. Only about 1 in 10^{12} atoms of carbon is carbon-14, a radioactive isotope that emits a beta particle and has a half-life of 5730 years. Atmospheric carbon dioxide contains small amounts of this isotope. Consequently, any living plant will incorporate carbon-14 by photosynthesis since carbon dioxide is the source of carbon for these plants. Upon death, plants cease to incorporate carbon dioxide. These facts may be used to estimate the age of carbon-containing materials.

TABLE 13.3 Irradiated Foodstuffs Approved by the Food and Drug Administration

Food	Purpose
Uncooked beef	Control of food-borne pathogens
Uncooked pork	Control of *Trichinella spiralis*
All fresh foods	Growth and maturation inhibition
All foods	Disinfestation of anthropod pests
Dry enzyme preparations	Microbial disinfection
Dried herbs and spices	Microbial disinfection
Uncooked poultry	Control of food borne pathogens
Eggs	Control of *Salmonella*
Iceberg lettuce and spinach	Control of *Salmonella* and *Escherichia coli*

Source: U.S. FDA regulations.

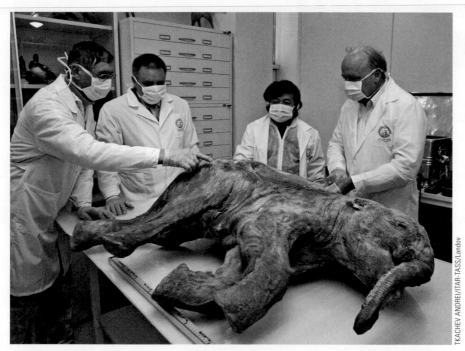

In 2007, the remains of Lyuba, a preserved baby woolly mammoth, were determined to be over 40,000 years old after tissue samples were analyzed using carbon-14 dating.

TKACHEV ANDREI/ITAR-TASS/Landov

of Jesus Christ. However, radiocarbon dating put the age of the cloth at around 650 years, which means the plant material used to make the cloth could not have been grown at the time of Jesus, about 2000 years ago. It was estimated to have been grown between 1260 and 1390 A.D. The cloth and the image on it have remained controversial in spite of this scientific analysis. In contrast to this study, the Dead Sea Scrolls, found in 1947 in a cave in the Qumran area near Jerusalem, have been authenticated as being approximately 2000 years old, an age consistent with the Hebrew writings contained within. Carbon-14 dating has also been used to determine that the remains of the Ice Man found frozen in the Italian Alps in 1991 are about 5000 years old.

A 1-g sample of carbon taken from a living plant would give about 15 decompositions per minute. This level of decomposition will drop by half every 5730 years. Thus, with modern instrumentation, it is possible to estimate the age of dead plants or materials made from these plants. Since most animals, including humans, eat plants, this procedure may be used to calculate when a plant or animal died.

The procedure does have some limitations. It is necessary to know how the atmospheric concentration of carbon-14 has changed over thousands of years. Even with this information in hand, practical considerations dictate that the procedure may only be used to determine the age of objects in a range of about 100 years. The radiocarbon level of an object less than 100 years old will not have declined significantly enough from that of the living organism to allow an accurate determination of age. Likewise, objects older than 40,000 years (more than seven half-lives) will have so little radioactive carbon remaining that an accurate determination will not be possible.

Radiocarbon dating has been used to estimate the age of many artifacts, and its accuracy has been confirmed in tests against articles of known age. Perhaps the most famous, and controversial, example of the use of this technique involved the Shroud of Turin. This piece of cloth, which bears a remarkable human image that appears almost like a photographic negative, has long been purported by many religious groups to be the burial shroud

13-8c Medical Imaging

Radioisotopes are used in medicine in two distinctly different ways: diagnosis and therapy. In the diagnosis of internal disorders, physicians need information on the locations of the disorders. An appropriate radioisotope is introduced into the patient's body, either alone or combined with some other chemical, and it accumulates at the site of the disorder. There the radioisotope disintegrates and emits its characteristic radiation, which can be detected. Modern medical diagnostic instruments not only determine where the radioisotope is located in the patient's body but also construct an image of the area.

The isotopes used for imaging are called medical isotopes and must be prepared artificially. Only four facilities in the world can produce these important radioactive substances. In 2010 there was a shortage of these isotopes when two of the four facilities were shut down for maintenance.

Four diagnostic radioisotopes are listed in Table 13.4. Each produces gamma radiation, which in low doses is less harmful to the tissue than ionizing radiations such as beta or alpha particles. By the use of special carriers, these radioisotopes can be made to accumulate in specific areas of the body. For example, the pyrophosphate ion ($P_4O_7^{4-}$), a simple polyatomic ion, can bond to the technetium-99m radioisotope (the *m* denotes *metastable*, meaning that the isotope decays by emitting a gamma ray to form a more stable isotope with the same

TABLE 13.4 Diagnostic Radioisotopes

Radioisotope	Name	Half-Life (hours)	Uses
^{99m}Tc*	Technetium-99m	6	As TcO_4^- to the thyroid, brain, and kidneys
^{201}Tl	Thallium-201	21.5	To the heart
^{123}I	Iodine-123	13.2	To the thyroid
^{67}Ga	Gallium-67	78.3	To various tumors and abscesses

© Cengage Learning

*The technetium-99m isotope is the one most commonly used for diagnostic purposes. The *m* stands for *metastable*, a term explained on page 251.

Fission: The splitting of a heavy nucleus into smaller nuclei

Fusion: The combination of small nuclei into a heavier nucleus

mass number), and together they accumulate in the skeletal structure where abnormal bone metabolism is taking place. Such investigations often pinpoint bone tumors.

The imaging method is based on the emission of gamma rays from the target organ. As the gamma rays strike a gamma-ray camera, the signal is processed by a computer and displayed as a video image (Figure 13.7).

13-8d Smoke Detectors

Another application of radioactivity is found in ionizing smoke detectors (Figure 13.8). A source of radioactivity, 0.0002 g of americium-241 oxide, is positioned at one electrode of an ionizing chamber so that it emits a mixture of alpha particles and gamma rays, which produce ions from nitrogen and oxygen in the air. The ions help to carry an electrical current to the second electrode. The alarm remains silent as long as this current is maintained. If smoke particles enter the chamber, the current is interrupted and the alarm is activated.

13-9 Energy from Nuclear Reactions

A vast amount of energy is released when heavy atomic nuclei split—the nuclear **fission** process—and when small atomic nuclei combine to make heavier nuclei—the **fusion** process. In 1938, Otto Hahn, Fritz Strassman, Lise Meitner, and Otto Frisch discovered that $^{235}_{92}$U is fissionable by neutrons (Figure 13.9). In less than a decade, this discovery led to two important applications of this energy release accompanying fission—the atomic bomb and nuclear power plants.

There is a huge difference between the amount of energy liberated in an ordinary chemical reaction, such as the burning of methane in air, and the energy liberated in a nuclear fission reaction. If you compared the energy from the burning of only 16 g of methane with that from the fission of an equivalent amount of uranium-235, the fission reaction would produce almost 25 million times as much energy. One kilogram of uranium fuel undergoing fission in a nuclear reactor can produce the same amount of energy as the combustion of 3000 tons of coal or 14,000 barrels of oil.

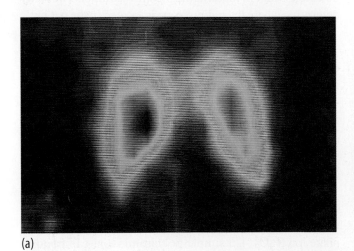

(a)

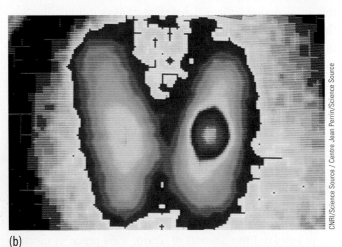

CNRI/Science Source / Centre Jean Perrin/Science Source

(b)

FIGURE 13.7 Thyroid imaging.

(a) Healthy human thyroid gland. (b) Thyroid gland showing effect of hyperthyroidism. Technetium-99m concentrates in sites of high activity. Images of this gland, which is located at the base of the neck, were obtained by recording γ-ray emission after the patient was given radioactive technetium-99m. Current technology creates a computer color-enhanced scan.

FIGURE 13.8 Ionizing Smoke Detectors.

A stream of ions formed by americium-241 alpha and gamma emission maintains a current. Smoke particles interrupt the stream of ions and cause an alarm to be activated.

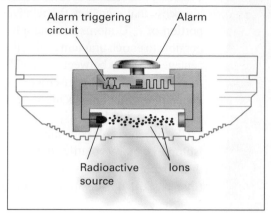

Alarm triggering circuit

Alarm

Radioactive source

Ions

13-9a Fission Reactions

Nuclear fission can occur when a neutron ($_0^1n$) enters a heavy nucleus. Certain heavy nuclei with an odd number of neutrons ($_{92}^{235}U$, $_{92}^{233}U$, $_{94}^{239}Pu$) will undergo fission when struck by slow-moving *thermal* neutrons (neutrons with a kinetic energy about the same as that of a gaseous molecule at ordinary temperatures). The splitting of the heavy nucleus produces two smaller nuclei, two or more neutrons (an average of 2.5 neutrons for $_{92}^{235}U$), and much energy. Uranium-235 atoms, each undergoing fission, will produce a number of different reaction products. One fission reaction that can take place follows:

$$_{92}^{235}U + _0^1n \longrightarrow _{92}^{236}U \longrightarrow$$
Unstable nucleus

$$_{56}^{141}Ba + _{36}^{92}Kr + 3\,_0^1n + 8.3 \times 10^{-15} \text{ kcal}$$
Typical fission products

Note that the same isotope may split in more than one way. The lighter nuclei produced by the fission reaction are the *fission products*. These fission products, such as $_{56}^{141}Ba$ and $_{36}^{92}Kr$, can also be unstable and emit beta particles ($_{-1}^0e$) and gamma rays (γ) and may have long half-lives (see Section 13-4). Eventually, the decay of these fission products leads to stable isotopes.

The neutrons emitted by the fission of one uranium-235 atom can cause the fission of other uranium-235 atoms. For example, the 3 neutrons emitted in the uranium fission (Figure 13.9) could produce fission in three more uranium atoms, the 9 neutrons emitted by those nuclei could produce nine more fissions, the 27 neutrons from these fissions could produce 81 neutrons, the 81 neutrons could produce 243 additional neutrons, those 243 neutrons could produce 729, and so on. This process is called a **chain reaction** (Figure 13.10), and it occurs at a maximum rate when the uranium sample is large enough for most of the neutrons emitted to be captured by other nuclei before passing out of the sample. A sample of fissionable material of sufficient size to self-sustain a chain reaction is termed the **critical mass**. If a critical mass of fissionable material is suddenly brought together, an explosion will occur because the energy released during each fission reaction cannot dissipate rapidly enough.

In a nuclear fission bomb, the critical mass is kept separated into several smaller subcritical masses until detonation, at which time the masses are driven together by an explosive device. During the split second when the chain reaction occurs, the tremendous energy of billions of fission reactions is liberated, and everything in the immediate vicinity is heated to temperatures of 5 to 10 million degrees Celsius and vaporized. The sudden expansion of hot gases literally pushes aside everything nearby and scatters the radioactive fission fragments over a wide area (Figure 13.11).

Of the uranium found in nature, only 0.711% is the easily fissionable $_{92}^{235}U$. The other 99.289% is $_{92}^{238}U$, which is

Chain reaction: A process by which neutrons from one fission reaction can cause multiple fission reactions in nearby nuclei

Critical mass: The mass of fissionable material able to sustain a chain reaction

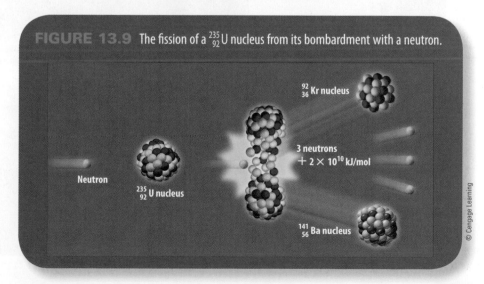

FIGURE 13.9 The fission of a $_{92}^{235}U$ nucleus from its bombardment with a neutron.

$_{36}^{92}Kr$ nucleus

3 neutrons
$+ 2 \times 10^{10}$ kJ/mol

Neutron

$_{92}^{235}U$ nucleus

$_{56}^{141}Ba$ nucleus

© Cengage Learning

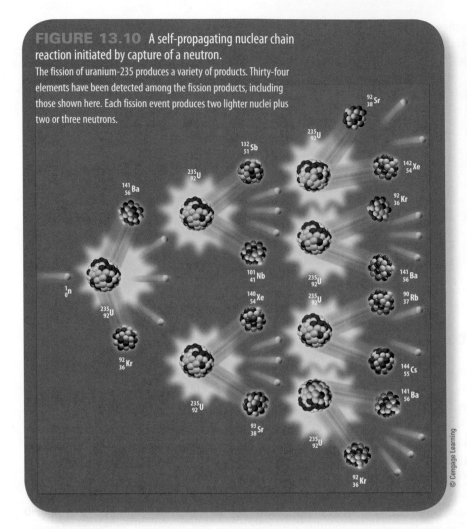

FIGURE 13.10 A self-propagating nuclear chain reaction initiated by capture of a neutron.

The fission of uranium-235 produces a variety of products. Thirty-four elements have been detected among the fission products, including those shown here. Each fission event produces two lighter nuclei plus two or three neutrons.

not fissionable by thermal neutrons. To make nuclear bombs or nuclear fuel for generation of electricity, the naturally occurring uranium must first be *enriched*, a process that increases the relative proportion of $^{235}_{92}U$ atoms in a sample. It is possible to enrich uranium so that the percentage of $^{235}_{92}U$ is between 2% and 5% by making use of slight differences in the volatility of uranium hexafluoride (UF_6). Of the two kinds of UF_6 ($^{238}UF_6$ and $^{235}UF_6$), the molecule containing the lighter isotope is more volatile and will pass through porous barriers faster. After multiple passes, the desired separation is achieved.

13-9b The Mass–Energy Relationship

What is the source of the tremendous energy of the fission process? It ultimately comes from the conversion of mass into energy, according to Einstein's famous equation, $E = mc^2$, where E is energy that results from the loss of an amount of mass m, and c is the speed of

FIGURE 13.11 Implosion.

By carefully shaping a conventional explosive, enough fissionable material can be brought together quickly to produce a massive chain reaction, causing many of the atoms to undergo fission almost simultaneously.

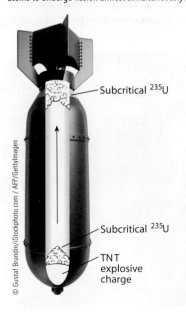

Subcritical ^{235}U

Subcritical ^{235}U

TNT explosive charge

The Doomsday Clock. In 1947, the **Bulletin of Atomic Scientists** *began to use this clock, set at 7 minutes before midnight, as a reminder of the potential for disaster from nuclear weapon use. The clock hands have been set a total of 18 times. In 1991, the clock was reset to a record 17 minutes when the United States and the Soviet Union agreed to begin nuclear disarmament. In 2007, the clock was reset to 5 minutes before midnight, where it currently remains, in response to North Korea's test of a nuclear weapon.*

together. This force is called the nuclear **binding energy** and is directly related to the stability of the nucleus. The binding energy depends on the number of particles in the nucleus. For example, if separate neutrons and protons are combined to form any particular nucleus, the resulting nucleus always has less mass than the starting nucleons. This mass difference, converted into energy units according to Einstein's equation, is the binding energy and can be expressed for each atom as its *binding energy per nucleon* by dividing the total binding energy for the atom by the number of nucleons in the nucleus.

> **Binding energy:** The force holding neutrons and protons together in a nucleus

Figure 13.12 shows that light atoms and heavy atoms have lower binding energies per nucleon than atoms with mass numbers between 50 and 65. This means that energy can be released when extremely light atoms are combined to make heavier atoms and when extremely heavy atoms are split to make lighter atoms. The greatest nuclear stability (greatest binding energy per nucleon) is at iron-56 ($^{56}_{26}$Fe). This is why iron is the most abundant of the heavier elements in the universe.

Because of their relative stabilities, most fission products fall into the intermediate range of mass numbers (Figure 13.12). Therefore, when fission occurs and smaller, more stable nuclei result, these nuclei will contain less mass per nuclear particle. In the process, mass must be changed into energy—the tremendous energy released in a nuclear bomb or, under controlled conditions, in a nuclear power plant.

light (186,000 miles/s, or 3.00×10^8 m/s). If the masses of the products of the fission of a uranium-235 atom by a neutron are compared with the masses of the reactants, it is found that the products have less mass than the reactants. In the case of the fission reaction

$$^{1}_{0}n + ^{235}_{92}U \longrightarrow$$

$$^{93}_{37}Rb + ^{141}_{55}Cs + 2\,^{1}_{0}n + energy$$

the difference in mass is about 0.00023 kg for every 235 g of $^{235}_{92}$U used. This mass difference is equivalent to 5.0×10^{12} cal. The mass "lost" as a result of the fission process is the source of the tremendous energy that is released.

The small nucleus of an atom is crowded with neutrons and protons. The very fact that nuclei hold together indicates that there must be some sort of force binding the particles

FIGURE 13.12 Relative stability of nuclei with different mass numbers.
Both lighter and heavier isotopes are less stable (less binding energy per nuclear particle) than isotopes with masses between 50 atomic mass units (amu) and about 65 amu. The most stable isotope is that of iron-56. Light isotopes can be fused together to form more stable atoms (nuclear fusion), and heavier isotopes can be split into more stable, lighter atoms (nuclear fission).

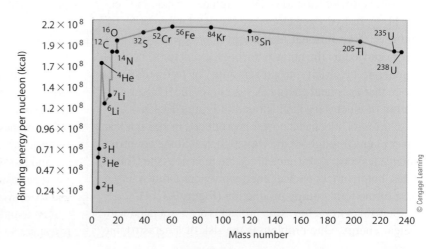

13-10 Useful Nuclear Energy

13-10a Electricity Production

Enrico Fermi (an Italian scientist who had immigrated to the United States in the late 1930s) and others believed that nuclear fission might somehow be made to proceed at a controlled rate. They reasoned that if a way could be found to control the number of thermal neutrons, their concentration could be maintained at a level sufficient to keep the fission process going but not high enough to allow an uncontrolled chain reaction. It would then be possible to drain the heat energy away on a continuing basis to do useful work. In 1942, while working at the University of Chicago, Fermi was successful in building the first atomic reactor.

An atomic reactor has several essential components. The reactor fuel must contain significant concentrations of atoms such as $^{235}_{92}U$, $^{233}_{92}U$, or $^{239}_{94}Pu$ that are fissionable by slow-moving neutrons. Typically, reactor fuel will contain uranium in the form of an oxide, U_3O_8, that has been enriched to contain about 3% or 4% $^{235}_{92}U$. A *moderator* is required to slow the speed of the neutrons produced in the reactions without absorbing them. Graphite and water have been used as moderators. A *neutron absorber*, such as cadmium or boron steel, is present to provide fine control over the neutron concentration. *Shielding* to protect the workers from dangerous radiation is an absolute necessity. Shielding tends to make reactors heavy and bulky installations. Finally, a *heat-transfer fluid* provides a large and even flow of heat energy away from the reaction center. Water is used as a heat-transfer fluid in many nuclear reactor designs. In addition to water's high heat capacity (Section 11-1), its hydrogen atoms are excellent moderators for neutrons. Conventional technology then allows the heat energy carried by the hot water from the reactor to be used to generate electricity, to power ships, or to operate any device that uses heat energy. A system for the nuclear production of electricity is illustrated in Figure 13.13.

Currently, electrical power is produced in 103 nuclear power plants throughout the United States. Some countries produce less energy by nuclear fission than does the United States, but in others nuclear energy provides a much larger share of the total energy production (Figure 13.14).

Several extremely vexing problems are associated with nuclear energy. One problem is the risk of a catastrophic accident at a nuclear power facility. Two such accidents, of different degrees of seriousness, are widely known and create a negative public viewpoint of all nuclear power. In late March 1979, an accident occurred at the Three Mile Island power plant near Harrisburg, Pennsylvania. A water pump failed in the reactor and caused a partial reactor core meltdown. Steam vented inside the reactor vessel, and hydrogen gas was produced by the decomposition of the steam at the very high temperatures there. This hydrogen gas, with the oxygen that was also produced, caused a risk of a chemical explosion that would have blown apart the safety containment, releasing fission products. In fact, a small amount of radioactive gas was vented into the atmosphere, but the containment building remained intact. There were no deaths directly associated with the Three Mile Island accident. Studies of individuals living within a 10-mile radius of the plant have revealed no elevation of cancer rates. Partly as a result of safety concerns uncovered in the Three Mile Island incident, a moratorium was placed on the construction of new nuclear power plants in the United States. In 2012, the first two plants in a generation were approved for construction in Georgia.

Certainly the most catastrophic nuclear accident occurred on April 26, 1986, at the Chernobyl unit 4 reactor near Kiev, Ukraine. The accident resulted in a core meltdown, explosion, and fire.

The Chernobyl reactor, built in 1983, had a design quite different from those elsewhere in the world. Graphite was used exclusively as the moderator. Although this design allows for a higher efficiency than reactors with other types of moderators, the graphite can be ignited if sufficiently high temperatures are reached. In addition, the Chernobyl reactor lacked an adequate containment vessel to withstand an explosion in its reactor core.

While engineers were running an unauthorized test on the electrical generator, the Chernobyl reactor suddenly increased its power output. Before neutron absorbers could be lowered into the core, a meltdown occurred. A vast amount of steam formed, which together with burning blocks of graphite and radioactive fuel caused the entire reactor roof to blow off. The result was the release of a radioactive plume that rose almost 5000 meters into the atmosphere, scattering an estimated 100 million curies of radioisotopes, some with years-long half-lives, throughout countries west of Chernobyl and as far away as Great Britain and Norway.

The results of the released radiation are widespread and extensive. More than 170,000 people had to permanently abandon their homes. Large areas of farmland and forest are so contaminated that they may remain unusable

FIGURE 13.13 A nuclear power plant.

In this reactor design, ordinary water (called light water to differentiate it from D₂O—heavy water) is pressurized and allowed to carry heat energy from the reactor core to a heat exchanger, where high-pressure steam is generated. This steam passes through a turbine, which generates electricity. Although simple in concept, safety considerations make the design, testing, and operation of a nuclear power plant a complex and costly operation.

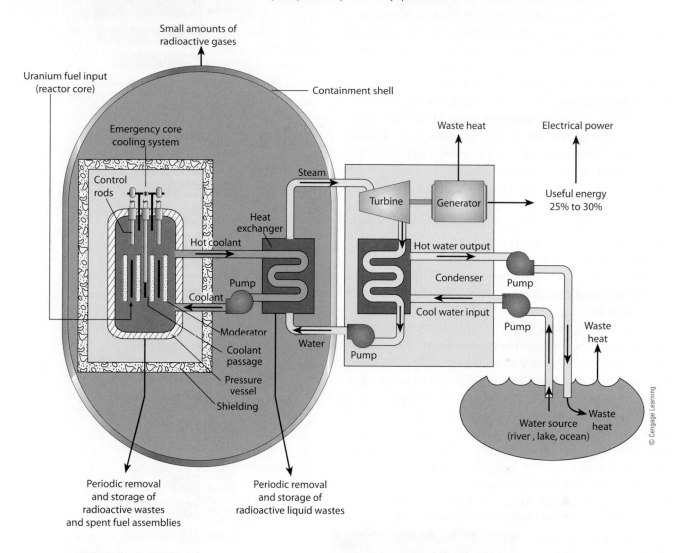

for 100 years after the accident. More than 4300 workers who participated in cleanup at the reactor have died, and cancer rates have risen among those exposed to radiation from the disaster. The Ukrainian government has spent the equivalent of billions of dollars on Chernobyl-related projects and continues to monitor the health of those in the region.

Today, the reactor is entombed in more than 300,000 tons of concrete, including an underground concrete liner to protect the groundwater. Work has been undertaken to prevent leaks in the concrete encasement. Plans are being developed for construction of a new, more stable encasement.

13-10b Radioactive Waste: A Problem to Be Solved

Nuclear waste is classified as *high-level radioactive waste* or *low-level radioactive waste*, depending on the amount of ionizing radiation given off and the time period for which the waste must be safely stored. High-level radioactive waste is mostly spent nuclear fuel and waste associated with the weapons industry. This waste gives off large amounts of ionizing radiation and must be stored safely for thousands of years because of the long half-lives of the radioactive isotopes involved. For instance, plutonium-239 has a half-life of about 24,100 years.

As with many industrial activities that are located in a relatively localized geographic area but generate benefits that may be spread over a much broader geographic area, there are concerns about where and how to store the waste associated with the processes. Citizens are, in many cases, happy to take advantage of the benefits of nuclear power as long as they personally aren't negatively affected by the less desirable activities associated with the industry. This has led to the widespread use of NIMBY as an acronym that stands for "not in my backyard." People may be willing to utilize nuclear power generated far away from their homes and may have little concern if the waste is stored in remote areas unlikely to directly affect their health and property values. However, any proposal to build a power plant or waste storage facility (nuclear or not) just down the road from a group of politically and environmentally active citizens is often met with immediate opposition. Such concerns may multiply as we are forced to use more and more alternative energy sources in the future.

As of 2002, about 43,000 metric tons of radioactive waste from nuclear power plants was stored in the United States, an amount expected to grow by 2000 metric tons per year. The storage facilities at the power plants are considered temporary, and an effort to develop a permanent storage site for high-level radioactive waste has been under way since 1982. The location and development of such a site is fraught with technical, social, and political problems.

A nuclear power plant in Bavaria, Germany.

FIGURE 13.14 The approximate fraction of electricity generated by nuclear power in various countries.

About 20% of the electricity in the United States is produced by nuclear power.

Country	Percent
Lithuania	80%
France	78%
Belgium	60%
Slovak Republic	55%
Bulgaria	47%
Sweden	46%
Ukraine	45%
Armenia	42%
Slovenia	41%
South Korea	39%
Hungary	39%
Switzerland	36%
Japan	34%
Finland	31%
Germany	31%
Spain	26%
Czech Republic	25%
United Kingdom	23%
United States	20%
Russia	16%
Canada	14%

Percent of electric power

A radioactive waste storage site must be located far from populated areas and must be free of the potential for leakage or deterioration for well over thousands of years. Any potential for environmental damage is considered unacceptable. The consensus now is that the safest storage will be in underground rock formations in isolated locations. Since 1987, as directed in a congressional act, Yucca Mountain in Nevada has been designated as the potential permanent underground storage site in the United States.

Many billions of dollars have been spent on construction and feasibility studies at Yucca Mountain, located in

The Yucca Mountain site in Nevada.

USGS

the desert 145 kilometers northwest of Las Vegas. The facility would occupy 3 square miles of tunnels through dense volcanic rock 800 feet beneath the surface of an almost 5000-foot mountain. The location is almost 1000 feet above the water table in a region that gets about 6 inches of rain a year. The waste would be placed inside metal alloy containers that have titanium drip shields (titanium doesn't rust).

Nevertheless, concerns have arisen over the possibility that some rainwater may leach into the storage site through cracks in the rock. There has also been some concern because the site is near a volcano that last erupted 20,000 years ago and may erupt again, although the possibility is considered remote. Another concern is the possibility of an earthquake in the area. Scientific agreement on the stability of the site for thousands of years to come has seemed an impossible goal. There is also significant opposition to the use of Yucca Mountain from the many states through which the waste would have to travel via trains or trucks to reach it.

In 2002, in a move to accelerate progress toward the date when Yucca Mountain would be receiving radioactive waste, President George W. Bush endorsed the Yucca Mountain site and the $58 billion needed for its further construction. Some predicted that the site would be operational by 2010, while others proposed a longer time frame for planning. Opponents argued that the uncertainties about potential hazards were too great and the site should never be used. Both the House and the Senate voted in 2002 to proceed with activation of the Yucca Mountain site. In June 2008, the Department of Energy requested a license from the Nuclear Regulatory Commission (NRC) to construct and operate the nation's first geological repository for high-level nuclear waste at Yucca Mountain. But three months later, the application was docketed by the NRC. Congress gave the NRC a three-year schedule to reach a decision on whether to approve construction. In 2010, the site was ultimately deemed "untenable" by the Obama administration.

Low-level radioactive waste gives off small amounts of ionizing radiation, is usually generated in small quantities, and need only be safely stored for relatively short periods of time because of the half-lives of the radioisotopes involved. Low-level nuclear waste includes such things as contaminated laboratory clothing, cleaning equipment and supplies, medical waste that is radioactive, and discarded radioactive devices such as smoke detectors. It is only necessary to safely store this waste for periods of 100 to 500 years. Prior to about 1979, most waste of this type was sealed in steel drums and dumped into the ocean. Current procedures require that such waste be stored in steel drums and buried in secure sites under several feet of soil (Figure 13.15).

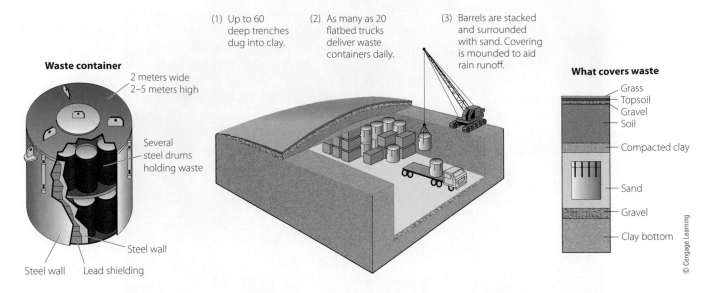

Waste container

2 meters wide
2–5 meters high

Several steel drums holding waste

Steel wall

Steel wall   Lead shielding

(1) Up to 60 deep trenches dug into clay.

(2) As many as 20 flatbed trucks deliver waste containers daily.

(3) Barrels are stacked and surrounded with sand. Covering is mounded to aid rain runoff.

What covers waste

Grass
Topsoil
Gravel
Soil

Compacted clay

Sand

Gravel

Clay bottom

© Cengage Learning

Plasma: A high-temperature state of matter consisting of unbound nuclei and electrons

However, such procedures have met with much opposition, especially from those living in the vicinity of the repositories. Even though such waste is less dangerous than high-level radioactive waste, the NIMBY philosophy comes into play when the reality of the issue of storage becomes personal.

13-10c Fusion Reactions

Nuclear fusion produces about the same amount of energy as fission on a per-mole basis but produces fewer radioactive by-products; the products that are radioactive have short half-lives. Hydrogen ($_1^1H$) has two isotopes: deuterium ($_1^2H$) and tritium ($_1^3H$). The nuclei of deuterium and tritium can be fused together at high temperatures to form a helium-4 atom. The energy released is 4.1×10^8 kcal.

$$_1^2H + {}_1^3H \longrightarrow {}_2^4He^{2+} + {}_0^1n + 4.1 \times 10^8 \text{ kcal}$$

When very light nuclei, such as those of hydrogen, helium, or lithium, are combined, or *fused*, to form an element of higher atomic number, energy equivalent to the difference between the total mass of the reacting atoms and the smaller total mass of the more stable products is released. This energy, which comes from a decrease in mass, is the source of the energy released by the Sun, the stars, and hydrogen bombs. Typical examples of fusion reactions are

$$4\,_1^1H \longrightarrow {}_2^4He^{2+} + 2\,_{+1}^0e + 6.14 \times 10^8 \text{ kcal}$$
$$2\,_1^2H \longrightarrow {}_2^4He^{2+} + {}_0^1n + 7.3 \times 10^7 \text{ kcal}$$
$$2\,_1^2H \longrightarrow {}_1^3H + {}_1^1H + 9.2 \times 10^7 \text{ kcal}$$
$$_1^3H + {}_1^2H \longrightarrow {}_2^4He^{2+} + {}_0^1n + 4.04 \times 10^8 \text{ kcal}$$

If fusion were to be used as a source of energy here on Earth, a suitable source of fusible atoms (fuel) would be needed. Fortunately, the oceans are a potential source of fantastic amounts of deuterium. There are 1.03×10^{22} atoms of deuterium in a *single liter* of seawater. If all of the deuterium atoms in a cubic kilometer of seawater were fused to form heavier atoms, the energy released would be equal to that released from burning 1360 billion barrels of crude oil, and this is approximately the total amount of oil originally present on this planet.

Fusion reactions occur rapidly only when the temperature is of the order of 100 million °C or more. At these high temperatures, atoms do not exist as such; instead, there is a **plasma** consisting of unbound nuclei and electrons. In this plasma, nuclei can combine. The first fusion reactions that scientists were able to create artificially were produced in hydrogen bombs, or thermonuclear bombs. In a thermonuclear bomb, the high temperatures needed to initiate fusion are achieved by using the heat of a fission bomb (atomic bomb).

In one type of hydrogen bomb, lithium deuteride ($_3^6Li\,_1^2H$, a solid salt) is placed around an ordinary $_{92}^{235}U$ or $_{94}^{239}Pu$ fission bomb. The fission reaction is set off in the usual way. A $_3^6Li$ nucleus absorbs one of the neutrons produced and splits into tritium and helium.

$$^{6}_{3}\text{Li} + ^{1}_{0}\text{n} \longrightarrow ^{3}_{1}\text{H} + ^{4}_{2}\text{He}^{2+}$$

The temperature reached by the fission of $^{235}_{92}\text{U}$ or $^{239}_{94}\text{Pu}$ is sufficiently high to bring about the fusion of tritium and deuterium.

Because there is so much available potential fuel in the oceans (as deuterium), controlled fusion seems like a natural candidate for an alternative source of energy. Another attractive feature is the rather limited production of dangerous radioisotopes. Most radioisotopes produced by fusion have short half-lives and therefore are a serious hazard for only a short time.

Three critical requirements must be met for controlled fusion to be a source of energy. First, the temperature must be high enough for fusion to occur, 100 million °C or more. Second, the plasma must be confined long enough to release a net output of energy. Third, the energy must be recoverable in some usable form.

Fusion reactions are extremely difficult to control, principally because of the difficulty of holding the hot plasma together long enough for the particles to react. A further problem is reaching the very high temperature and maintaining it long enough to sustain the reaction. Nevertheless, progress is being made. Only time will tell if fusion can be controlled to the extent that would make it practical. If it does happen, abundant and low-cost energy may truly become available to everyone.

Your Resources

In the back of the textbook:

→ *Review Card on Nuclear Changes and Nuclear Power*

   In OWL for CHEM2 at www.cengagebrain.com

→ *Review Key Terms with Flash Cards (printable or digital)*

→ *Complete Interactive Practice Quizzes to prepare for tests*

→ *Submit Assigned Homework and Exercises*

Applying Your Knowledge

1. Which of the following three types of radiation is the most penetrating?
   (a) Alpha particles
   (b) Beta particles
   (c) Gamma rays

2. Give the type and approximate amount of material required to stop each of the following:
   (a) Alpha particles
   (b) Beta particles
   (c) Gamma rays

3. Which radioisotope sample is more hazardous, 1 g of $^{238}_{92}\text{U}$ with a half-life of 4.5 billion years or 1 g of $^{222}_{86}\text{Rn}$ with a half-life of 3 days? Explain your choice.

4. What is the nuclear reaction for the production of a beta particle?

5. What makes Rn-222 a health hazard? What health problem results from Rn-222 exposure?

6. Why must the sum of the mass numbers of the products of a nuclear reaction always equal the sum of the mass numbers of the reactants?

7. What is wrong with the following nuclear equation?
$$_2^4\text{He} + {}_{13}^{27}\text{Al} \longrightarrow {}_{15}^{49}\text{P} + {}_0^1\text{n}$$

8. What kinds of projectile particles were used for the transmutations to form elements up to atomic number 101?

9. Iodine-131 has a half-life of 8 days. Approximately what fraction of iodine-131 will remain after 1 month (31 days)?

10. A pure sample of a radioactive isotope weighing 156 mg and with a half-life of 128 days is allowed to decay for one year. Approximately how much of the isotope remains?

11. A pure 300-mg sample of a radioactive isotope undergoes decay for 3.5 years. After that time, 37.5 mg of the radioactive isotope remain. What is the half-life of the isotope?

12. A pure 36.7-mg sample of an unidentified radioactive isotope was allowed to decay for 32 days. After that time, it was found that only 2.3 mg of the isotope remained. Use Table 13.2 and your knowledge of half-life to propose an identity for the isotope.

13. What are transuranium elements? How do they differ from the other elements in the periodic table?

14. What is meant by the term *uranium series*?

15. One outcome of drilling for natural gas is that significant amounts of helium are obtained. How do you think that helium came to be present in natural gas deposits?

16. What is the final product of the uranium series?

17. What are the two major applications of radioisotopes in nuclear medicine?

18. What is the reason for irradiating food with gamma rays? What is the effect of the radiation?

19. What are the electrical charges and relative masses for the following?
    (a) Beta particles
    (b) Alpha particles
    (c) Gamma rays
    (d) Positrons
    (e) Neutrons

20. Tell why it is necessary to use high-energy accelerators to produce the transuranium radioisotopes from smaller nuclei.

21. Describe what effect gamma rays from cobalt-60 have on cancer cells and why this makes them useful in radiation therapy for cancer.

22. Why does airplane travel increase a person's annual dose of ionizing radiation?

23. Which source of background radiation contributes more to a person's annual dose, weapons test fallout or natural radiation in food, water, and air?

24. Fusion of two deuterium nuclei to form helium-3 and hydrogen-1 yields 1.53×10^{-16} kcal of energy. A liter of seawater is estimated to contain 1.03×10^{22} atoms of deuterium. How many kilocalories can be liberated if all the 1.03×10^{22} deuterium atoms in a liter of seawater undergo fusion?

25. Carbon-14 dating uses the carbon-14 half-life of 5730 years to estimate the age of carbon-containing samples. How old is a wooden throne if the current carbon-14 activity is one-eighth of the original?

26. As described in this chapter, radiocarbon dating seems rather straightforward. However, this is not the case. It is necessary to provide corrections because of the variability of levels in the environment over time. One cause of this variability is the combustion of fossil fuels. Another cause took place in the 1950s and 1960s. What event(s) might have led to the need for additional correction to radiocarbon dating results?

27. Copper-64 has a half-life of 12.9 hours. How many half-lives and hours are needed for the activity to fall to one-fourth the initial level of 1 microcurie (37,000 disintegrations per second)?

28. What was the contribution made by each of the following scientists in the study of radioactivity?
    (a) Becquerel
    (b) Marie Curie
    (c) Rutherford

29. Radon gas in homes is a common problem in many parts of North America because radon-222 is radioactive. This nuclide has a half-life of 3.82 days. What fraction of a radon-222 sample will remain after 15.28 days have passed?

30. What effect does the emission of a beta particle have on the mass number of a nucleus?

31. Describe nuclear fission:
    (a) What is the starting material?
    (b) What is needed to cause fission?
    (c) What are the products of fission?

32. Name three problems that are associated with nuclear energy from fission.

33. What happened at the Chernobyl unit 4 reactor in 1986?

34. (a) What does the term *reprocessing of nuclear fuels* mean?
    (b) What danger is associated with it?

35. What is nuclear fusion?

36. Compare nuclear fission and nuclear fusion as sources of energy. Name the fuels, benefits, problems, and current status.

37. How might you estimate whether an isotope is likely to be radioactive if you are provided with its identity?

38. What is meant by "too many neutrons" in relation to the radioactivity of atoms?

39. In which way do isotopes having too many neutrons decay?

40. In which way do isotopes having too few neutrons decay?

41. Complete the following nuclear equations:
    (a) $^{1}_{0}n + ^{235}_{92}U \longrightarrow ^{142}_{56}Ba + [\underline{\hspace{1cm}}] + 3\,^{1}_{0}n$
    (b) $^{1}_{0}n + ^{235}_{92}U \longrightarrow [\underline{\hspace{1cm}}] + ^{129}_{50}Sn + 2\,^{1}_{0}n$
    (c) $^{1}_{0}n + ^{239}_{94}Pu \longrightarrow [\underline{\hspace{1cm}}] + ^{123}_{52}Te + 2\,^{1}_{0}n$

42. Balance the following nuclear equations, giving symbols, nuclear charges, and mass numbers:
    (a) $^{64}_{29}Cu \longrightarrow \underline{\hspace{1cm}} + ^{0}_{-1}e$
    (b) $^{69}_{30}Zn \longrightarrow \underline{\hspace{1cm}} + ^{69}_{31}Ga$
    (c) $^{131}_{53}I \longrightarrow ^{131}_{54}Xe + \underline{\hspace{1cm}}$

43. Balance the following beta decay equations, giving symbols, nuclear charges, and mass numbers:
    (a) $^{14}_{6}C \longrightarrow \underline{\hspace{1cm}} + ^{0}_{-1}e$
    (b) $^{210}_{82}Pb \longrightarrow \underline{\hspace{1cm}} + ^{0}_{-1}e$

44. Write a balanced nuclear equation for each of the following:
    (a) decays by positron emission
    (b) decays by alpha emission
    (c) decays by beta emission

45. Complete the following nuclear reactions:
    (a) $^{218}_{84}Po \longrightarrow \underline{\hspace{1cm}} + ^{4}_{2}He$
    (b) $\underline{\hspace{1cm}} \longrightarrow ^{206}_{82}Pb + ^{4}_{2}He$
    (c) $^{99}_{43}Tc \longrightarrow \underline{\hspace{1cm}} + ^{0}_{-1}e$
    (d) $^{226}_{88}Ra \longrightarrow ^{4}_{2}He + \underline{\hspace{1cm}}$
    (e) $^{195}_{79}Au + \underline{\hspace{1cm}} \longrightarrow ^{195}_{78}Pt$

46. (a) What is the half-life of plutonium-239?
    (b) How does this add to the dangers of this isotope?

47. In what important way are the isotopes uranium-235 and plutonium-239 similar?

48. Describe how the "magnetic bottle" is used to contain a nuclear fusion reaction.

49. How many atoms of tritium will remain after four half-lives if there were initially 300,000 atoms?

50. Radioactive decay of ^{238}U can yield ^{234}Th and an alpha particle. An initial sample contained 200,000 atoms of ^{238}U. How many alpha particles will be produced from this sample after one half-life of 4.5 billion years?

Organic Chemicals
and Polymers

Fossil fuels are not only our major source of energy; they are also the major source of the hydrocarbons that we use to make thousands of consumer products. About 3% of the petroleum refined today is the starting material for the synthesis of organic chemicals of commercial importance. (Organic chemicals are substances that contain a C—H bond within their molecular structure.) These chemicals are essential to making plastics, synthetic rubber, synthetic fibers, drugs, dyes, pesticides, fertilizers, and thousands of other consumer products. For this reason, the organic chemical industry is often referred to as the *petrochemical* industry.

A few of the major classes of organic compounds and some of their reactions, especially those used in making polymers, are described in this chapter, with the goal of introducing you to a major segment of chemistry and the chemical industry.

Carbon compounds hold the key to life on Earth. Consider what the world would be like if all carbon compounds were removed; the result would be much like the barren surface of the Moon. If carbon compounds were removed from the human body, there would be nothing left except water and a small residue of minerals. The same would be true for all living things. Carbon compounds are also an integral part of our lifestyle. Fossil fuels, foods, and most drugs are made of carbon compounds. Since we live in an age of plastics and synthetic fibers, our clothes, appliances, and most other consumer goods contain a significant portion of carbon compounds.

More than 85% of the millions of known compounds are carbon compounds, and a separate branch of chemistry, **organic chemistry**, is devoted to the study of them. Why are there so many organic compounds? The discussion of hydrocarbons and their structural and geometric isomers in Chapter 12 indicates two reasons: (1) the ability of many carbon atoms to be linked in sequence with stable carbon–carbon single, double, and triple bonds in a single molecule and (2) the occurrence of isomers. A third reason will be discussed further in this chapter: the variety of functional groups that bond to carbon atoms.

The economic importance of the organic chemical industry can be seen by looking at the list in Table 14.1 of chemicals produced in the United States in very large quantities. Four are organic chemicals.

Organic chemistry: The chemistry of carbon compounds

14-1 Organic Chemicals

Many of the organic chemicals used in the chemical industry are obtained from fossil fuels. For example, ethylene, propylene, butylene, and acetylene are obtained by catalytic cracking of natural gas or petroleum. Figure 14.1 summarizes the uses of these as starting materials. Petroleum and coal tar, obtained by heating coal at high temperatures in the absence of air, are the primary sources of benzene-like aromatic compounds used in the chemical industry (Figure 14.2). Distilling coal tar yields the aromatic compounds listed in Table 14.2.

TABLE 14.1 Major Chemicals Produced in the United States*

Name	How Made	End Uses
Sulfuric acid	Burning sulfur to SO_2, oxidation of SO_2 to SO_3, reaction with water; also recovered from metal smelting	Fertilizers, petroleum refining, manufacture of metals and chemicals
Ethylene	Cracking hydrocarbons from oil and natural gas	Plastics, antifreeze production, fibers, and solvents
Ammonia	Catalytic reaction of nitrogen and hydrogen	Fertilizers, plastics, fibers, and resins
Phosphoric acid	Reaction of sulfuric acid with phosphate rock; burning of elemental phosphorus and dissolution in water	Fertilizers, detergents, and water-treating compounds
Chlorine	Electrolysis of NaCl, recovery from HCl users	Chemical production, plastics, solvents, pulp and paper
Propylene	Cracking oil and oil products	Plastics, fibers, solvents
Sodium hydroxide	Electrolysis of NaCl solution	Chemicals, pulp and paper, aluminum, textiles, oil refining
Ammonium nitrate	Reaction of ammonia and nitric acid	Explosives, fertilizers
Urea	Reaction of NH_3 and CO_2 under pressure	Fertilizers, animal feeds, adhesives, and plastics
Styrene	Dehydrogenation of ethylbenzene	Polymers, rubber, polyesters

© Cengage Learning

* Organic chemicals are highlighted.

FIGURE 14.1 Hydrocarbons from petroleum and natural gas as raw materials.
Catalytic cracking produces ethylene, butylene, acetylene, and propylene, which are converted into other chemical raw materials and many kinds of consumer products.

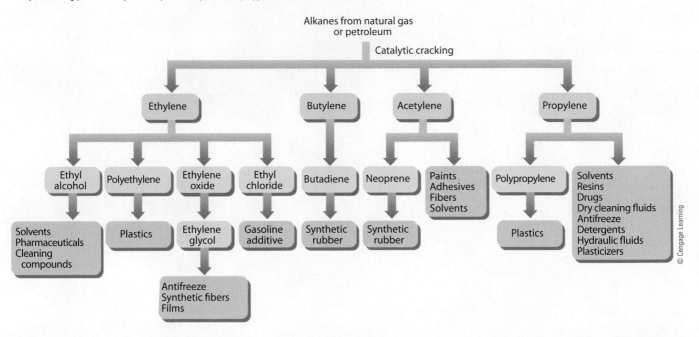

© Cengage Learning

Organic chemicals were once obtained only from plants, animals, and fossil fuels, and these are still direct sources of hydrocarbons (Figures 14.1 and 14.2) and many other important chemicals, such as sucrose from sugarcane and ethanol from fermented grain mash. However, the development of organic chemistry has led to cheaper methods for the synthesis of both naturally occurring substances and new substances.

The millions of organic compounds include classes of compounds that are obtained by replacing hydrogen atoms of hydrocarbons with atoms or groups of atoms known as *functional groups* (page 235). The functional groups for alcohols and ethers were discussed in Section 12-9. Recall that a functional group is an atom or group of atoms that lends a set of properties to a molecule, including physical properties and the ability to undergo specific chemical reactions. Alcohols and their oxidation products—aldehydes, ketones, and carboxylic acids—are among the most useful functional group classes of compounds.

FIGURE 14.2 Aromatic compounds from petroleum and coal and their uses.

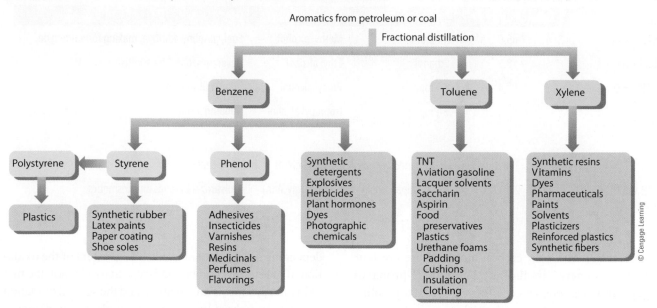

TABLE 14.2 Fractions from Distillation of Coal Tar

Boiling Range (°C)	Name	Tar, Mass (%)	Primary Constituents
Below 200	Light oil	5	Benzene, toluene, xylenes
200–250	Middle oil (carbolic oil)	17	Naphthalene, phenol, pyridine
250–300	Heavy oil (creosote oil)	7	Naphthalenes and methylnaphthalenes, cresols
300–350	Green oil	9	Anthracene, phenanthrene
Residue		62	Pitch or tar

14-2 Alcohols and Their Oxidation Products

Alcohols contain one or more hydroxyl (—OH) groups bonded to carbon atoms and are a major class of organic compounds. The importance of methanol, ethanol, and 2-methyl-2-propanol as fuels and fuel additives was described in Sections 12-2 and 12-5. Additional uses of these and other commercially important alcohols are listed in Table 14.3. Alcohols are classified according to the number of carbon atoms bonded directly to the —C—OH carbon and are termed primary (one additional C atom), secondary (two additional C atoms), or tertiary (three additional C atoms). The reactivities of these classes of alcohols are different. An "R" is used to represent any kind of carbon-based group. The use of R, R′, and R″ indicates that the R groups may be different.

$$R—\overset{\overset{\displaystyle H}{|}}{\underset{\underset{\displaystyle H}{|}}{C}}—OH \qquad R′—\overset{\overset{\displaystyle R}{|}}{\underset{\underset{\displaystyle H}{|}}{C}}—OH \qquad R′—\overset{\overset{\displaystyle R}{|}}{\underset{\underset{\displaystyle R″}{|}}{C}}—OH$$

Primary   Secondary   Tertiary

Ethanol and 1-propanol are primary alcohols. 2-Propanol, or isopropyl alcohol, is familiar to us as rubbing alcohol, which is a 70% solution of 2-propanol in water. 2-Propanol is a secondary alcohol and is one of the two structural isomers of an alcohol with three carbon atoms.

2-Propanol   1-Propanol

TABLE 14.3 Some Important Alcohols

Condensed Formula	Boiling Point (°C)	Systematic Name	Common Name	Use
CH_3OH	65.0	Methanol	Methyl alcohol	Fuel, gasoline additive, making formaldehyde
CH_3CH_2OH	78.5	Ethanol	Ethyl alcohol	Beverages, gasoline additive, solvent
$CH_3CH_2CH_2OH$	97.4	1-Propanol	Propyl alcohol	Industrial solvent

CH_3CHCH_3
   │
  OH
	82.4	2-Propanol	Isopropyl alcohol	Rubbing alcohol

$HOCH_2CH_2OH$	198.0	1,2-Ethanediol	Ethylene glycol	Antifreeze

$HOCH_2CHCH_2OH$
     │
    OH
	290.0	1,2,3-Propanetriol	Glycerol (glycerin)	Moisturizer in foods and cosmetics

© Cengage Learning

For an alcohol with four carbon atoms, there are four structural isomers, including 2-methyl-2-propanol (*tertiary*-butyl alcohol), whose former use as a gasoline additive was described in Section 12-2.

CH₃CH₂CH₂CH₂OH
1-Butanol
A primary alcohol

$$CH_3CHCH_2OH$$
    │
   CH₃
2-Methyl-1-propanol
A primary alcohol

$$CH_3CHCH_2CH_3$$
    │
   OH
2-Butanol
A secondary alcohol

$(CH_3)_3C$—OH
2-Methyl-2-propanol
A tertiary alcohol

Alcohols can serve as the starting substances for the preparation of many other types of organic compounds. One important reaction involves **oxidation**, which for most organic substances refers to a reaction in which the number of covalent bonds between carbon and oxygen atoms is increased. Oxidation of alcohols may yield **aldehydes**, **ketones**, or **carboxylic acids**,

Oxidation: For organic compounds, a process in which the number of bonds between carbon and oxygen is increased

Aldehyde: An organic compound containing a —C=O functional group
             │
             H

Ketone: An organic compound containing a carbonyl (C=O) functional group between two carbon atoms

Carboxylic acid: An organic compound containing a —C=O functional group
              │
              OH

     O
     ‖
R—C—H
Aldehydes

     O
     ‖
R—C—R′
Ketones

     O
     ‖
R—C—OH
Carboxylic acids

depending on the alcohol used and the extent of the oxidation. If the starting compound is a primary alcohol, the first oxidation product is an aldehyde and the second oxidation product is a carboxylic acid. For example, the oxidation of ethanol, a primary alcohol, can be used to make acetaldehyde and acetic acid.

      O
      ‖
H—C—H
Formaldehyde

Formaldehyde (HCHO), the simplest aldehyde, is obtained by oxidation of methanol. It is a gas at room temperature but is often used as a 40% aqueous solution called formalin. Although formaldehyde is an important starting material for polymers, it presents a number of health hazards because of its toxicity and carcinogenicity. Formaldehyde is also an air pollutant, being produced in trace amounts in the incomplete combustion of fossil fuels. One of the concerns about methanol as a fuel (Section 12-5) is the potential for increased levels of formaldehyde in urban areas because of the production of formaldehyde from incomplete combustion of methanol.

Secondary alcohols are oxidized to give ketones. The oxidation of 2-propanol (isopropyl alcohol) gives acetone, a ketone widely used as an organic solvent and in nail polish remover solution.

    OH   O
    │   Oxidation   ‖
CH_3CCH_3 ⟶ CH_3CCH_3
    │
    H
2-Propanol   Acetone

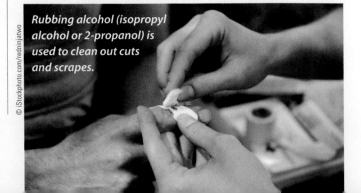

Rubbing alcohol (isopropyl alcohol or 2-propanol) is used to clean out cuts and scrapes.

© iStockphoto.com/redninjatwo

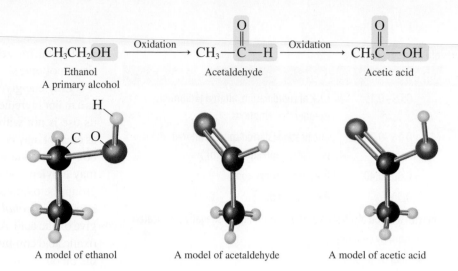

CH_3CH_2OH $\xrightarrow{\text{Oxidation}}$ $CH_3-\overset{\displaystyle O}{\underset{\displaystyle \|}{C}}-H$ $\xrightarrow{\text{Oxidation}}$ $CH_3\overset{\displaystyle O}{\underset{\displaystyle \|}{C}}-OH$

Ethanol   Acetaldehyde   Acetic acid
A primary alcohol

A model of ethanol   A model of acetaldehyde   A model of acetic acid

14-2a Hydrogen Bonding in Alcohols

The physical properties of alcohols provide an example of the effects of hydrogen bonding between molecules in liquids. This is illustrated by the boiling points in Table 14.3. Hydrogen bonding explains why methanol (molar mass 32 g) is a liquid, whereas propane, which has a slightly higher molar mass (44 g/mol) but no hydrogen bonding capability, is a gas at room temperature. The boiling point of methanol is lower than that of water because methanol has only one O — H hydrogen through which it can bond.

The higher boiling point of ethylene glycol can be attributed to the presence of two — OH groups per molecule. Glycerol, with three — OH groups, has an even higher boiling point and allows this substance to be used to remove heat from a car engine without boiling.

The alcohols listed in Table 14.3 are also very soluble in water because of hydrogen bonding between water molecules and the — OH group in alcohol molecules. However, as the length of the hydrocarbon chain increases, the resulting alcohols are less soluble because the nonpolar hydrocarbon portion has greater influence on the solubility than the hydrogen bonding by the — OH group.

14-2b Ethanol

Ethanol is the "alcohol" of alcoholic beverages and is prepared for this purpose by fermentation of carbohydrates (starch, sugars) from a wide variety of plant sources (Table 14.4). The growth of yeast is inhibited at alcohol concentrations higher than about 14%, and fermentation comes to a stop. Beverages with a higher ethanol concentration are prepared either by distillation or by fortification with ethanol that has been obtained by the distillation of another fermentation product. The maximum concentration of ethanol that can be obtained by distillation of ethanol–water mixtures is 95%. The "proof" of an alcoholic beverage is twice the volume percent of ethanol: 80-proof vodka, for example, contains 40% ethanol by volume; 95% ethanol is 190 proof.

Although ethanol is not as toxic as methanol (Section 12-9), 1 pint of pure ethanol, rapidly ingested, would kill most people. Ethanol is a depressant, and the effects of different blood levels of ethanol are shown in Table 14.5. Rapid consumption of two 1-ounce "shots" of 90-proof whiskey or of two 12-ounce beers can cause one's blood alcohol level to reach 0.05%. Ethanol is quickly absorbed into the bloodstream and metabolized by enzymes produced in the liver. The rate

TABLE 14.4 Common Alcoholic Beverages

Name	Source of Fermented Carbohydrate	Amount of Ethyl Alcohol	Proof
Beer	Barley, wheat	5%	10
Wine	Grapes or other fruit	18% maximum, unless fortified	20–28
Brandy	Distilled wine	40%–45%	80–90
Whiskey	Barley, rye, corn, etc.	45%–55%	90–110
Rum	Molasses	~45%	90
Vodka	Potatoes	40%–50%	80–100

TABLE 14.5 Effects of Blood Alcohol Level*

Number of Drinks[†]	Alcohol by Volume (%)	Effect
1 to 4	0.05–0.15	Lack of coordination, altered judgment, exaggerated emotions
5 to 10	0.15–0.20	Intoxication (slurred speech, altered perception and equilibrium)
10 to 15	0.30–0.40	Unconsciousness, coma
Over 15	0.50	Possible death

* In the United States, a person with a blood alcohol level of 0.08% or higher is legally intoxicated.

[†] 1 drink = 12 oz of beer (4.5% alcohol by volume)

4 oz of wine (14% alcohol by volume)

1½ oz of liquor (45% alcohol by volume)

© Cengage Learning

Denatured alcohol:
Ethanol with small added amounts of a toxic substance that cannot be removed easily by chemical or physical means

of detoxification is about 1 ounce of pure ethanol per hour. Ethanol is oxidized to acetaldehyde, which is further oxidized to acetic acid; eventually, CO_2 and H_2O are produced and eliminated through the lungs and kidneys.

Ethanol → (Liver enzymes) → Acetaldehyde

The federal tax on alcoholic beverages is about $26 per gallon. Since the cost of producing ethanol is only about $1 per gallon, ethanol intended for industrial use must be *denatured* to avoid the beverage tax. **Denatured alcohol** contains small amounts of a toxic substance, such as methanol or gasoline, that cannot be removed easily by chemical or physical means.

Apart from being used in the alcoholic beverage industry, ethanol is used widely in solvents, in the preparation of many other organic compounds, and as a gasoline additive (Section 12-9). For many years, industrial ethanol was also made by fermentation. However, in the last several decades, it has become cheaper to make the ethanol from petroleum by-products, specifically by the catalyzed addition of water to ethylene. More than 1 billion pounds of ethanol are produced each year by this process.

14-2c Ethylene Glycol and Glycerol

More than one alcohol group can be present in a single molecule. Ethylene glycol, a di-alcohol (Table 14.3), is

used in antifreeze and in the synthesis of polymers.

Ethylene glycol works very well as a major ingredient in antifreeze, but its use is not without problems. Just as ethanol may be oxidized to the carboxylic acid known as acetic acid, so may ethylene glycol's two hydroxyl groups be oxidized by liver enzymes known as *alcohol dehydrogenases* to give oxalic acid. At physiological pH, oxalic acid combines with calcium in the bloodstream to form an insoluble salt, calcium oxalate.

$HOCH_2CH_2OH$ — (Liver enzymes) →
Ethylene glycol

Oxalic acid — (Under physiological conditions) → Ca^{2+} Calcium oxalate

Calcium oxalate crystallizes in the kidneys and in the brain, leading to serious illness or even death. Most such poisonings occur when radiator fluid is changed. Carelessly stored antifreeze or antifreeze spilled on a driveway is frequently ingested by pets and even small children. Many antifreezes are sold as brightly colored, often green, yellow, or red, mixtures that attract small children. Children and pets also find the sweet taste of ethylene glycol attractive, and these combined effects can lead to deadly consequences. As little as 5 mL can kill a cat, a small child can die from drinking as little as 30 mL, and 60 mL can kill an adult or a large dog. Coma and death often result if treatment is not begun shortly after ingestion.

One treatment for ethylene glycol poisoning involves flooding the system with ethanol. Alcohol dehydrogenases in the liver have a natural preference for metabolizing ethanol over ethylene glycol. If these enzymes can be kept busy metabolizing ethanol, the unmetabolized ethylene glycol is excreted from the kidneys with a half-life of about 14 hours. After five half-lives (70 hours), only about 3% of the initial amount of ethylene glycol remains. Thus, routine treatment for ethylene glycol poisoning involves administering ethanol intravenously under carefully controlled conditions if the victim cannot be induced to regurgitate the ingested antifreeze shortly after ingestion. Careful monitoring by a physician or veterinarian is required because the acetic acid produced by ethanol metabolism

A small structure difference makes a big difference in properties. Glycerol, which is composed of three carbon atoms with an — OH group on each one, is nonpoisonous, sweet, and syrupy and holds moisture to the skin. It is used in soap and also as a food additive (Section 16-9). Antifreeze contains ethylene glycol, which is composed of two carbon atoms with an — OH group on each one, and is extremely poisonous.

HOCH₂CH₂OH

HOCH₂CHCH₂OH
$\quad\quad$ OH

can lead to excessive buildup of acid in the bloodstream, resulting in a condition known as metabolic acidosis, which can be fatal. You may wish to review the blood buffering system in Section 9-4 to see how this would work. Doctors often administer sodium bicarbonate solutions intravenously to combat this buildup of acidity.

Ethylene glycol's use as a de-icer for airplanes preparing to fly in icing conditions has been problematic because hundreds of thousands of gallons of the solvent were used each year at large airports around the world. Much of this solvent often found its way into the local ecosystem around the airports, causing serious environmental problems. In recent years, much more attention has been given to restricting the amount of ethylene glycol and other chemicals used for de-icing, and to minimizing their release into the environment. Recovery and recycling systems have been put in place, and there now is much more stringent monitoring of storm water runoff from airports.

Propylene glycol is gaining popularity as an antifreeze because its structure renders it incapable

$$\text{OH}$$
$$|$$
$$\text{CH}$$
$$\text{H}_3\text{C} \diagup \quad \diagdown \text{CH}_2\text{OH}$$

Propylene glycol

of oxidation to oxalic acid. More environmentally friendly antifreezes, such as Sierra® brand, use this nontoxic alcohol.

These antifreezes are very popular at zoos and other places where large numbers of animals may be exposed to leaking antifreeze. One of the drawbacks of propylene glycol, however, is that it costs more than ethylene glycol.

Glycerol, a tri-alcohol (Table 14.3), is a by-product of the manufacture of soaps. Because of its moisture-holding properties, glycerol has many uses in foods and tobacco as a digestible and nontoxic humectant (a substance that gathers and holds moisture) and in the manufacture of drugs and cosmetics. It is also used in the manufacture of nitroglycerin and numerous other chemicals. Perhaps the most important compounds of glycerol are its natural esters, which are the fats and oils found in plants and animals (Section 15-3).

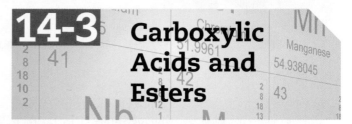

14-3 Carboxylic Acids and Esters

14-3a Carboxylic Acids

Carboxylic acids contain the —COOH functional group and are prepared by the oxidation of alcohols or aldehydes. These reactions occur quite easily, as evidenced by the souring of wine to form vinegar, which results from the oxidation of ethanol to acetic acid in the presence of oxygen from the air.

Carboxylic acids are polar and readily form hydrogen bonds. This hydrogen bonding results in relatively high boiling points for the acids, even higher than those of alcohols of comparable molar mass. For example, formic acid (46 g/mol) has a boiling point of 101°C, whereas ethanol (46 g/mol) has a boiling point of only 78°C.

All carboxylic acids are weak acids (see Section 9-2) and react with bases to form salts:

$$\underset{\text{Acetic acid}}{\text{CH}_3\overset{\displaystyle O}{\overset{\|}{\text{C}}}-\text{OH}(aq)} + \text{NaOH}(aq) \longrightarrow$$

$$\underset{\text{Sodium acetate}}{\text{CH}_3\overset{\displaystyle O}{\overset{\|}{\text{C}}}\text{O}^-\text{Na}^+(aq)} + \text{H}_2\text{O}(\ell)$$

Salts of carboxylic acids where the chain of carbon atoms is much longer have been used as soaps for centuries. A number of carboxylic acids are found in nature and have been known for many years. As a result, some of the familiar

The odor typically associated with manure is produced by a concentration of valeric acid (see Table 14.7).

© iStockphoto.com/GlobalP

TABLE 14.6 Some Simple Carboxylic Acids

Structure	Odor	Common Name	Systematic Name	Boiling Point (°C)
$\overset{O}{\overset{\|}{HCOH}}$	Sharp	Formic acid	Methanoic acid	101
$\overset{O}{\overset{\|}{CH_3COH}}$	Vinegar	Acetic acid	Ethanoic acid	118
$\overset{O}{\overset{\|}{CH_3CH_2COH}}$	Swiss cheese	Propionic acid	Propanoic acid	141
$\overset{O}{\overset{\|}{CH_3(CH_2)_2COH}}$	Rancid butter	Butyric acid	Butanoic acid	163
$\overset{O}{\overset{\|}{CH_3(CH_2)_3COH}}$	Manure	Valeric acid	Pentanoic acid	187

© Cengage Learning

Ester: An organic compound containing a —COOR functional group

carboxylic acids are almost always referred to by their common names (Table 14.6).

Acetic acid, the same acid that is diluted to make white vinegar, is produced in concentrated form and in large quantities for use in making cellulose acetate, a polymer used in the manufacture of photographic film base, synthetic fibers, plastics, and other products (Section 14-5).

Three other acids produced in large quantity have two carboxylic acid groups and are known as dicarboxylic acids.

$$\overset{O}{\overset{\|}{HOC}}(CH_2)_4\overset{O}{\overset{\|}{COH}}$$

Adipic acid

$$HO-\overset{O}{\overset{\|}{C}}-\bigcirc-\overset{O}{\overset{\|}{C}}-OH$$

Terephthalic acid

$$\bigcirc\overset{O}{\underset{\|}{\overset{\|}{\overset{}{C}}}}-OH$$
$$\overset{C}{\underset{\|}{\overset{}{\overset{}{}}}}-OH$$
$$O$$

Phthalic acid

These three acids are used to manufacture polymers (Section 14-5). There are other acids, however, whose sources are familiar to you (Table 14.7) since they occur in nature.

14-3b Esters

Carboxylic acids react with alcohols in the presence of strong acids to produce **esters**, which contain the —COOR functional group. In an ester, the —OH of the carboxylic acid is replaced by the OR group from the alcohol. For example, when ethanol is mixed with acetic acid in the presence of sulfuric acid, ethyl acetate is formed. This reaction is a dehydration (loss of water) in which sulfuric acid acts as a catalyst.

$$\overset{O}{\overset{\|}{CH_3C}}-OH \ + \ H-OCH_2CH_3 \xrightarrow{H_2SO_4}$$

Acetic acid   Ethanol

$$\overset{O}{\overset{\|}{CH_3C}}-OCH_2CH_3 \ + \ H_2O$$

Ethyl acetate

Names of esters are derived from the names of the alcohol and the acid used to prepare the ester. The alkyl group from the alcohol is named first followed by the name of the acid changed to end in *-ate*. For example, ethyl acetate is the name of the ester prepared from the reaction of ethanol and acetic acid. Ethyl acetate is a common solvent for lacquers and plastics and is also used as fingernail polish remover.

Unlike the acids from which they are derived, esters often have pleasant odors (Table 14.8). The characteristic odors and flavors of many flowers and fruits are due to the

TABLE 14.7 Some Naturally Occurring Carboxylic Acids

Name	Structure	Natural Source
Glycolic acid	HO—CH$_2$—COOH	Sugarcane
Citric acid	HOOC—CH$_2$—$\overset{\displaystyle OH}{\underset{\displaystyle COOH}{C}}$—CH$_2$—COOH	Citrus fruits
Lactic acid	CH$_3$—$\underset{\displaystyle OH}{CH}$—COOH	Sour milk
Oleic acid	CH$_3$(CH$_2$)$_7$—CH = CH—(CH$_2$)$_7$—COOH	Vegetable oils
Oxalic acid	HOOC—COOH	Rhubarb, spinach, cabbage, tomatoes
Stearic acid	CH$_3$(CH$_2$)$_{16}$—COOH	Animal fats
Tartaric acid	HOOC—$\underset{\displaystyle OH}{CH}$—$\underset{\displaystyle OH}{CH}$—COOH	Grape juice, wine

Nail polish contains ethyl acetate, a solvent with a unique odor.

© iStockphoto.com/chaoss

presence of natural esters. For example, the odor and flavor of bananas is primarily due to the ester 3-methylbutyl acetate (also known as isoamyl acetate).

Food and beverage manufacturers often use mixtures of esters as food additives. The ingredient label of a brand of imitation banana extract reads "water, alcohol (40%), isoamyl acetate and other esters, orange oil and other essential oils, and FD&C Yellow #5." Except for the water, these are all organic compounds.

of synthetic polymers in consumer products is indicated by the fact that about half of the top 50 chemicals produced in the United States are used in the production of plastics, fibers, and rubbers. The average production of synthetic polymers in the United States exceeds 200 pounds per person annually. Many synthetic organic polymers are *plastics* of one sort

Polymer: A large molecule composed of repeating units derived from smaller molecules

14-4 Synthetic Organic Polymers

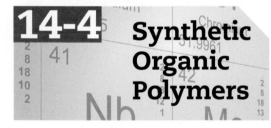

It is impossible for us to get through a day without using a dozen or more synthetic organic **polymers**. The word *polymer* means "many parts" (Greek *poly*, meaning "many," and *meros*, meaning "parts"). Polymers are giant molecules with molar masses ranging from thousands to millions. Our clothes are polymers, our food is packaged in polymers, and our appliances and cars contain a number of polymer components.

Approximately 80% of the organic chemical industry is devoted to the production of synthetic polymers. The prominence

TABLE 14.8 Some Acids and Alcohols and Their Esters

Acid	Alcohol	Ester	Odor of Ester
CH$_3$COOH Acetic acid	$\overset{\displaystyle CH_3}{\underset{}{CH_3CHCH_2CH_2OH}}$ 3-Methyl-1-butanol	$\overset{\displaystyle O \qquad CH_3}{CH_3COCH_2CH_2CHCH_3}$ 3-Methylbutyl acetate	Banana
CH$_3$CH$_2$CH$_2$CH$_2$COOH Pentanoic acid	$\overset{\displaystyle CH_3}{\underset{}{CH_3CHCH_2CH_2OH}}$ 3-Methyl-1-butanol	$\overset{\displaystyle O \qquad CH_3}{CH_3CH_2CH_2CH_2COCH_2CH_2CHCH_3}$ 3-Methylbutyl pentanoate	Apple
CH$_3$CH$_2$CH$_2$COOH Butanoic acid	CH$_3$CH$_2$CH$_2$CH$_2$OH 1-Butanol	$\overset{\displaystyle O}{CH_3CH_2CH_2COCH_2CH_2CH_2CH_3}$ Butyl butanoate	Pineapple
CH$_3$CH$_2$CH$_2$COOH Butanoic acid	⬡—CH$_2$OH Benzyl alcohol	$\overset{\displaystyle O}{CH_3CH_2CH_2COCH_3}$⬡ Benzyl butanoate	Rose

© Cengage Learning

or another. A **plastic** is a polymer that can be molded into a variety of shapes. *All plastics are polymers, but not all polymers are plastics.* Examples of items often made of plastics include dishes and cups, containers, telephones, plastic bags for packaging and wastes, plastic pipes and fittings, automobile steering wheels and seat covers, and cabinets for appliances, radios, and television sets. In fact, such plastic items, along with textile fibers and synthetic rubbers, are so widely used that they are commonly taken for granted.

Some of our most useful polymer chemistry has resulted from copying giant molecules in nature. Rayon is remanufactured cellulose; synthetic rubber is copied from natural latex rubber. As useful as these polymers may be, however, polymer chemistry is not restricted to nature's models. Polystyrene, nylon, and Dacron are a few examples of synthetic molecules that do not have exact duplicates in nature.

Polymers are made by chemically joining together many small molecules into one giant molecule, or macromolecule. The small molecules used to synthesize polymers are called **monomers**. Synthetic polymers can be classified as **addition polymers**, made by monomer units directly joined together, or **condensation polymers**, made by monomer units combining so that a small molecule, usually water, is split out between them.

As you proceed to study polymers in this chapter, you may find it particularly fascinating to couple your reading of this text with an inspection of a website called The Macrogalleria (http://www.pslc.ws/macrog.htm) devoted solely to polymers. This site, developed and maintained by the polymer department at the University of Southern Mississippi, uses the stores of a modern shopping mall to illustrate the variety of polymers and their practical uses before allowing you, should you so choose, to explore the chemistry of these polymers, including structure, properties, and synthesis, in as much detail as you wish.

14-4a Addition Polymers

Polyethylene

The monomer for addition polymers normally contains one or more double or triple bonds. The simplest monomer of this group is ethylene ($CH_2{=}CH_2$). When ethylene is

heated to between 100°C and 250°C at a pressure of 1000 atmospheres (atm) to 3000 atm in the presence of a catalyst, polymers are formed with molar masses of up to several million. A reaction of ethylene begins with breaking of one of the bonds in the carbon–carbon double bond, so an unpaired electron, a reactive site, remains at each end of the molecule. This step, the *initiation* of the polymerization, can be accomplished with initiator chemicals such as organic peroxides that are unstable and easily break apart into free radicals, which have unpaired electrons (Section 10-3). The free radicals readily add to molecules containing carbon–carbon double bonds to produce new free radicals. This destroys the double bond.

The growth of the polyethylene chain then begins as the unpaired electron combines with a double bond electron in an unreacted ethylene molecule. This leaves another unpaired electron to react with yet another ethylene molecule. For example,

Polyethylene
n ranges from 1000
to 50,000

In the process, the unsaturated hydrocarbon monomer, ethylene, is changed to a saturated hydrocarbon polymer, *polyethylene.*

Polyethylene is the world's most widely used polymer, with many millions of tons produced annually in the United States alone. What are some of the reasons for this popularity? The wide range of properties of polyethylene leads to many uses.

Polyethylenes formed under various pressures and catalytic conditions have different molecular structures and hence different physical properties. For example, chromium oxide as a catalyst yields almost exclusively the linear polyethylene shown at the end of this paragraph—a polymer with no branches on the carbon chain. The zigzag structure represents the shape of the chain more closely because of the tetrahedral arrangement of bonds around each carbon in the saturated polyethylene chain. Long, linear chains of polyethylene can pack closely together to give a material with high density (0.97 g/mL) and high molar mass, referred to as high-density polyethylene (HDPE). This material is hard, tough, and rigid. A plastic milk bottle is a good example of an application of HDPE. Other HDPE containers are shown in Figure 14.3.

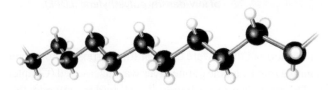

A portion of a polyethylene molecule

or

Model of linear polyethylene

Each corner represents a CH$_2$ group

If ethylene is heated to 230°C at a pressure of 200 atm, free radicals attack the polyethylene chain at random positions, causing irregular branching (Figure 14.4a). Branched chains of polyethylene cannot pack closely together, so the resulting material has a lower density (0.92 g/mL) and is called low-density polyethylene (LDPE). This material is soft and flexible (Figure 14.5). Sandwich bags are made from LDPE. Other conditions can lead to cross-linked polyethylene, in which branches connect long chains to each other (Figure 14.4b). If the linear chains of polyethylene are treated in a way that causes cross-links between chains to form cross-linked polyethylene (CLPE), a very tough form of polyethylene is produced. The plastic caps on soft-drink bottles are made from CLPE.

Polymers of Ethylene Derivatives

Many different kinds of addition polymers are made from mono-mers in which one or more of the

FIGURE 14.3 Some bottles made of HDPE, one of the most commonly recycled kinds of polymeric materials.

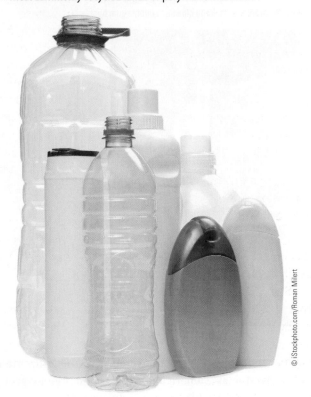

hydrogen atoms in ethylene have been replaced with either halogen atoms or a variety of organic groups. If the formation of polyethylene is represented as

$$n\,H_2C=CH_2 \longrightarrow \left(\!CH_2-CH_2\!\right)_n$$

then the general reaction

$$n\,H_2C=CH \longrightarrow \left(\!CH_2-CH\!\right)_n$$
$$\qquad\quad |\qquad\qquad\qquad\quad |$$
$$\qquad\quad X\qquad\qquad\qquad\quad X$$

FIGURE 14.4 Models of (a) branched and (b) cross-linked polyethylene.

(a)

(b)

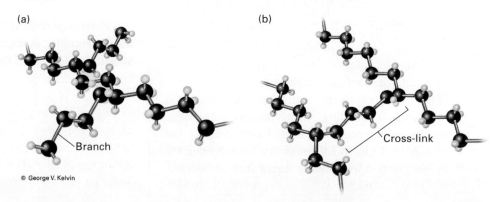

Branch

Cross-link

FIGURE 14.5 Making LDPE film.

(a) The blown tube of polyethylene film. (b) Diagram showing how the tube is blown and the film collected in a roll. The process is similar to blowing a bubble gum bubble.

(a)

© Pascal Goetgheluck/Science Source

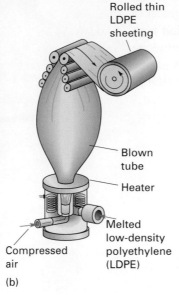

Rolled thin LDPE sheeting

Blown tube

Heater

Melted low-density polyethylene (LDPE)

Compressed air

(b)

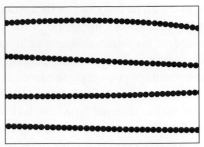

Representation of the relationship between the polymer chains of high-density polyethylene (HDPE).

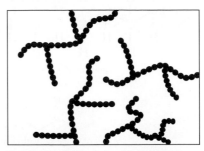

Representation of the relationship between the branched polymer chains of low-density polyethylene (LDPE).

can be used to represent a number of other important addition polymers, where X is Cl, F, or an organic group (Table 14.9).

For example, the monomer for making *polystyrene* is styrene, and *n* is about 5700.

$$n\,H_2C = CH \longrightarrow \left(\!CH_2 - CH\!\right)_n$$

Styrene is a major industrial chemical, primarily because of its use in making polystyrene. Polystyrene is a clear, hard, colorless solid at room temperature that can be molded easily at 250°C, making it a plastic. More than 6 million tons of polystyrene are produced in the United States each year to make food containers, toys, electrical parts, insulating panels, appliance and furniture components, and many other items. The variation in properties shown by polystyrene products is typical of synthetic polymers. For example, a clear polystyrene drinking glass that is brittle and breaks into sharp pieces somewhat like glass is very different from a polystyrene coffee cup that is soft and pliable.

A major use of polystyrene is in the production of rigid foams by "expansion molding." In this process, polystyrene beads are placed in a mold and heated with steam or hot air. The beads, 0.25 mm to 1.5 mm in diameter, contain 4% to 7% by weight of a low-boiling liquid such as pentane.

Chlorofluorocarbons were used for this purpose before their negative impact on the ozone layer was discovered (Chapter 7). The steam causes the low-boiling liquid to vaporize; this expands the beads, and as the foamed particles expand, they are molded in the shape of the mold cavity. Expanded polystyrene is used for egg cartons, meat trays, coffee cups, and packing material.

Polypropylene, used in indoor–outdoor carpeting, bottles, fabrics, and battery cases, is made from propylene.

$$n \quad \begin{matrix} H \\ \end{matrix} C = C \begin{matrix} H \\ CH_3 \end{matrix} \longrightarrow \left(\!\begin{matrix} H & H \\ C - C \\ H & CH_3 \end{matrix}\!\right)_n$$

Propylene   Polypropylene

Poly(vinyl chloride) (PVC), used in floor tiles, garden hoses, plumbing pipes, and trash bags, has a chlorine atom substituted for one of the hydrogen atoms in ethylene.

$$n \quad \begin{matrix} H \\ \end{matrix} C = C \begin{matrix} H \\ Cl \end{matrix} \longrightarrow \left(\!\begin{matrix} H & H \\ C - C \\ H & Cl \end{matrix}\!\right)_n$$

Vinyl chloride   Poly(vinyl chloride)

TABLE 14.9 Ethylene Derivatives That Undergo Addition Polymerization

Formula	Monomer Common Name	Polymer Name (Trade Names)	Uses
$H_2C = CH_2$	Ethylene	Polyethylene (Polythene)	Squeeze bottles, bags, films, toys and molded objects, electrical insulation

$H_2C = C - CH_3$
       $\mid$
       H
	Propylene	Polypropylene (Vectra, Herculon)	Bottles, films, indoor–outdoor carpets

$H_2C = C - Cl$
       $\mid$
       H
	Vinyl chloride	Poly(vinyl chloride) (PVC)	Floor tile, raincoats, pipe

$H_2C = C - CN$
       $\mid$
       H
	Acrylonitrile	Polyacrylonitrile (Orlon, Acrilan)	Rugs, fabrics

$H_2C = C$ — ⬡
       $\mid$
       H
	Styrene	Polystyrene (EPS, Styron)	Food and drink coolers, building material insulation

$H_2C = C - O - C - CH_3$
       $\mid$ $\parallel$
       H   O
	Vinyl acetate	Poly(vinyl acetate) (PVA)	Latex paint, adhesives, textile coatings

$H_2C = C$ — CH₃ , CH₃ , C, O	Methyl methacrylate	Poly(methyl methacrylate) (Plexiglas, Lucite)	High-quality transparent objects, latex paints, contact lenses
$F_2C = CF_2$	Tetrafluoroethylene	Polytetrafluoroethylene (PTFE, Teflon)	Gaskets, insulation, bearings, nonstick coatings

© Cengage Learning

Although the representation

$$\left(\begin{array}{cc} H & H \\ | & | \\ -C - C - \\ | & | \\ H & H \end{array} \right)_n$$

saves space, keep in mind how large the polymer molecules are. Generally, n is 500 to 50,000, and this gives molecules with molecular weights ranging from 10,000 to several million. The molecules that make up a given polymer sample are of different lengths and thus are not all of the same molecular weight. As a result, only the average molecular weight can be determined.

In summary, the numerous variations in substituents, length, branching, and cross-linking make it possible to produce a variety of properties for each type of addition polymer. Chemists and chemical engineers can fine-tune the properties of a polymer to match desired properties. Appropriate selection of monomer and reaction conditions accounts for the widespread use of these giant molecules.

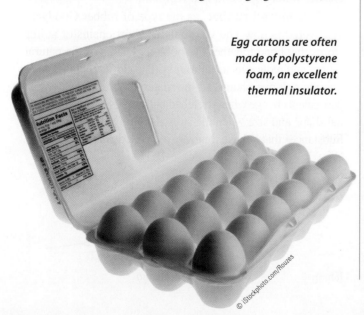

Egg cartons are often made of polystyrene foam, an excellent thermal insulator.

© iStockphoto.com/Rouzes

EXAMPLE 14.1 Addition Polymers

 Draw the structural formula of the repeating unit for the following addition polymers: (a) polypropylene, (b) poly(vinyl acetate), and (c) poly(vinyl alcohol).

SOLUTION

The names show that the monomers for these polymers are propylene ($CH_2 = CHCH_3$), vinyl acetate ($CH_2 = CHOOCCH_3$), and vinyl alcohol ($CH_2 = CHOH$). The repeating units in the polymers therefore have the same structures but without the double bonds.

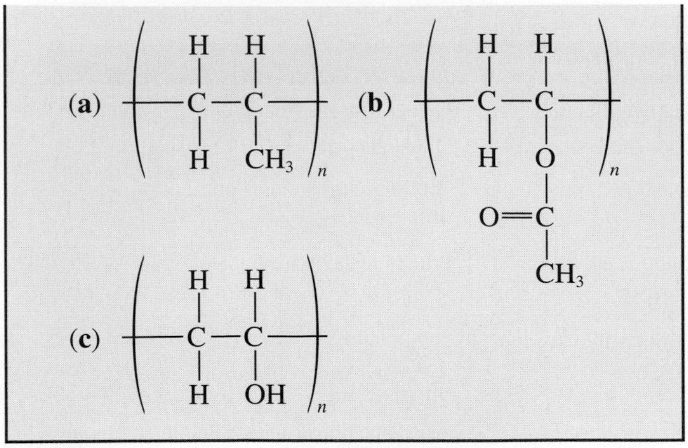

(a) Structure of polypropylene, (b) structure of poly(vinyl acetate), (c) structure of poly(vinyl alcohol)

TRY IT 14.1

Draw the structural formula of the monomers used to prepare the following polymers: (a) polyethylene, (b) poly(vinyl chloride), and (c) polystyrene.

Natural rubber:
Poly-*cis*-isoprene from the *Hevea brasiliensis* tree

Vulcanized rubber:
Rubber with short chains of sulfur atoms that bond together the polymer of natural rubber

Although there are many uses of addition polymers, perhaps one of the most interesting applications is the use of sodium polyacrylate in superabsorbent diapers and some feminine hygiene products. The monomer for this polymer is sodium acrylate.

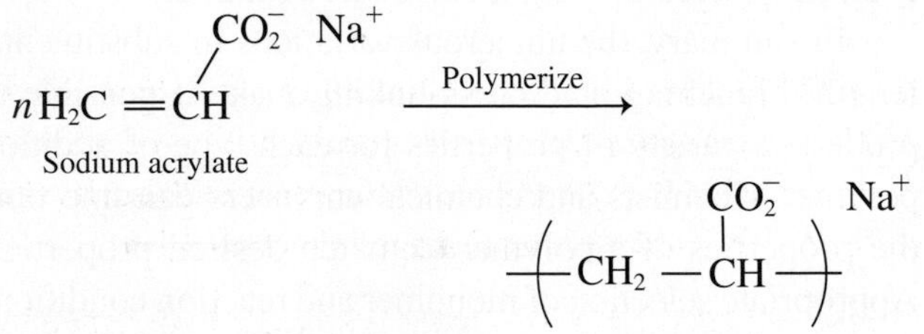

Sodium acrylate → Polymerize → Sodium polyacrylate

This polymer, available in powder form, is found in most modern disposable diapers in a layer consisting of paper fluff and the polymer. The polymer has been marketed under the trade name Waterlock because of its ability to absorb up to 800 times its own weight in water and convert the liquid water to a gel. Since there is actually no chemical reaction of the water with the polymer, the water will, over relatively long periods of time, be released from the polymer. For this reason, a crystalline version of the polymer is sold in garden stores for use in the soil surrounding plants to help cut down on

watering frequency. These crystals can also be placed around the base of live Christmas trees to prevent premature drying. The technology has also been used in the toy industry, where small figures such as animals and fish can be expanded to many times their original size simply by placing them in water for a period of time. Distilled water works best for this since salts dissolved in the water inhibit the swelling. In fact, the gelatinous goo created by placing powdered Waterlock polymer in distilled water can be quickly converted back to liquid just by adding common table salt (NaCl).

Rubber: Natural and Synthetic

Natural rubber, a product of the *Hevea brasiliensis* tree, is a hydrocarbon with the empirical formula C_5H_8. When rubber is decomposed in the absence of oxygen, the monomer isoprene is obtained.

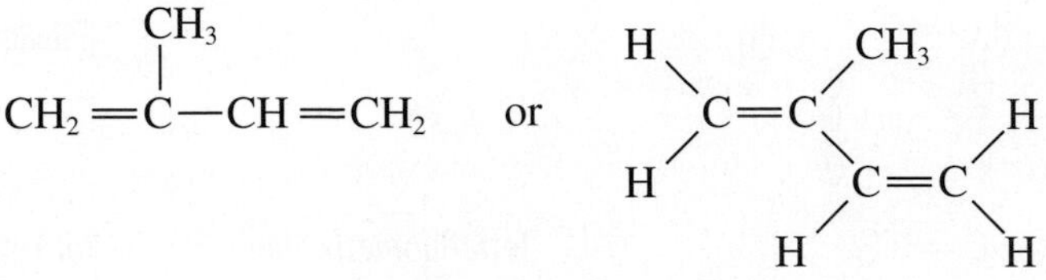

Isoprene (2-methyl-1,3-butadiene)

Natural rubber occurs as *latex* (an emulsion of rubber particles in water), which oozes from rubber trees when they are cut. Precipitation of the rubber particles yields a gummy mass that is not only elastic and water-repellent but also very sticky, especially when warm. In 1839, after five years of work on this material, Charles Goodyear (1800–1860) discovered that heating gum rubber with sulfur produces a material that is no longer sticky but is still elastic, water-repellent, and resilient.

Vulcanized rubber, as the type of rubber Goodyear discovered is now known, contains short chains of sulfur atoms that bond together the polymer chains of the natural rubber and reduce its unsaturation. The sulfur chains help to link the polymer chains, so the material does not undergo a permanent change when stretched but springs back to its original shape and size when the stress is removed (Figure 14.6). Substances that behave this way are called *elastomers*.

FIGURE 14.6 Stretching vulcanized rubber.
After stretching, the cross-links pull it back into the original arrangement.

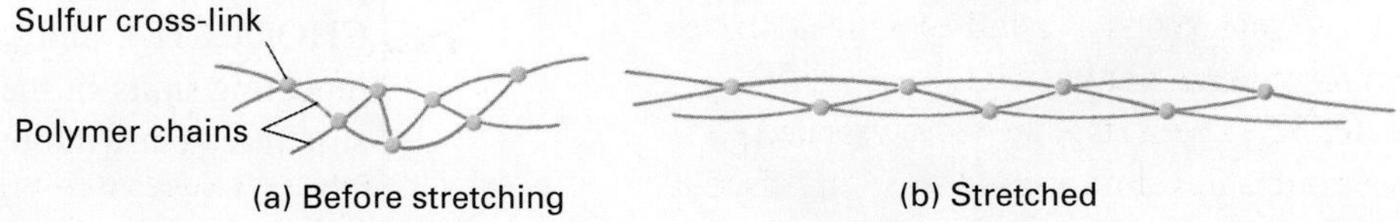

(a) Before stretching   (b) Stretched

1,3-Butadiene

Poly-*cis*-butadiene (Natural rubber; the —CH₂—CH₂— groups are *cis*)

Poly-*trans*-isoprene (Gutta-percha; the —CH₂—CH₂— groups are *trans*)

In later years, chemists searched for ways to make a synthetic rubber so that we would not be completely dependent on imported natural rubber during emergencies, such as during the first years of World War II. In the mid-1920s, German chemists polymerized butadiene (obtained from petroleum and structurally similar to isoprene but without the methyl-group side chain). Polybutadiene is used in the production of tires, hoses, and belts.

The behavior of natural rubber (polyisoprene), it was learned later, is due to the specific molecular geometry within the polymer chain. We can write the formula for polyisoprene with the CH₂ groups on opposite sides of the double bond (the *trans* arrangement) or with the CH₂ groups on the same side of the double bond (the *cis* arrangement).

Natural rubber is poly-*cis*-isoprene. However, the *trans* material also occurs in nature in the leaves and bark of the sapotacea tree and is known as *gutta-percha*. It is brittle and hard and is used for golf ball covers, electrical insulation, and other applications not requiring the stretching properties of rubber. Without an appropriate catalyst, polymerization of isoprene yields a solid that is like neither rubber nor gutta-percha because it is a random mixture of the *cis* and *trans* geometries. Neither the *trans* polymer nor the randomly arranged material is as good as natural rubber (*cis*) for making automobile tires.

In 1955, chemists at the Goodyear and Firestone companies almost simultaneously discov-

ered how to use *stereoregulation* catalysts to prepare synthetic poly-*cis*-isoprene. This material is structurally identical to natural rubber. More than 10 million tons of synthetic rubber are produced worldwide every year.

14-4b Condensation Polymers

A **condensation reaction** is one in which two molecules, each containing at least one reactive functional group, react by forming a larger molecule and eliminating a small molecule, frequently water. The reaction of alcohols with carboxylic acids to give esters and water (Section 14-3) is an example of condensation. This important type of chemical reaction does not depend on the presence of a carbon–carbon double bond in the reacting molecules. Instead, it requires the presence of two different kinds of functional groups on two different molecules. If each reacting molecule has two functional groups, both of which can react, it is possible for condensation reactions to produce long-chain polymers.

Polyesters

A molecule with two carboxylic acid groups, such as terephthalic acid, and another molecule with two alcohol groups, such as ethylene glycol, can react with each other at both ends to form ester linkages (Section 14-3).

If *n* molecules of acid and alcohol can react in this manner, the process will continue until a large polymer molecule, known as a *polyester*, is produced.

Poly(ethylene terephthalate)

Poly-*cis*-isoprene (the —CH₂—CH₂— groups are *cis*)

More than 2 million tons of poly(ethylene terephthalate), commonly referred to as PET or PETE, are produced in the United States each year for use in beverage bottles, apparel, tire cord, film for photography, food packaging, coatings for microwave and conventional ovens, and home furnishings. Polyester textile fibers are marketed under such names as Dacron and Terylene. Films of the same polyester, when magnetically coated, are used to make audiotapes and videotapes. This film, Mylar, has unusual strength and can be rolled into sheets 1/30 the thickness of a human hair.

The inert, nontoxic, non-allergenic, noninflammatory, and non-blood-clotting properties of Dacron polymers make Dacron tubing an excellent substitute for human blood vessels (Figure 14.7) in heart bypass operations. Dacron sheets are used as a skin substitute for burn victims.

Polyamides (Nylons)

Another useful and important type of condensation reaction is that between a carboxylic acid and a **primary amine**, which is an organic compound containing an — NH₂ functional group. Amines can be considered derivatives of ammonia (NH_3), and most of them are weak bases, similar in strength to ammonia. An amine reacts with a carboxylic acid to split out a water molecule and form an **amide**, for example,

$$\underset{\text{Carboxylic acid}}{R-\overset{\overset{\displaystyle O}{\|}}{C}-OH} + \underset{\text{Amine}}{H-\underset{\underset{\displaystyle H}{|}}{N}-R} \xrightarrow{\text{Heat}} \underset{\text{Amide}}{R-\overset{\overset{\displaystyle O}{\|}}{C}-\underset{\underset{\displaystyle H}{|}}{N}-R} + H_2O$$

Nylon polymers are produced when diamines react with dicarboxylic acids. Reactions of this type yield a group of polymers that perhaps have had a greater impact on society than any other type. These are the *polyamides*, or nylons.

In 1928, the DuPont Company embarked on a program of basic research headed by Dr. Wallace Carothers (1896–1937), who came to DuPont from the Harvard University faculty. His research interests were high-molecular-weight compounds, such as rubber, proteins, and resins, and the reaction mechanisms that produced these compounds. In February 1935, his research yielded a product known as nylon-6,6 (Figure 14.8), prepared from butane-1,4-dicarboxylic acid (adipic acid, a diacid) and 1,6-diaminohexane (hexamethylenediamine, a diamine). The name nylon-6,6 is based on the presence of six carbons in each of the monomers used to make the polymer.

Golf balls with gutta-percha covers.

© Corbis

$$2\ HO-\overset{\overset{\displaystyle O}{\|}}{C}-\underset{\text{Terephthalic acid}}{\bigcirc}-\overset{\overset{\displaystyle O}{\|}}{C}-OH + 2\ HO-\underset{\text{Ethylene glycol}}{CH_2-CH_2}-OH \longrightarrow$$

$$HO-\overset{\overset{\displaystyle O}{\|}}{C}-\bigcirc-\overset{\overset{\displaystyle O}{\|}}{C}-O-CH_2-CH_2-O-\overset{\overset{\displaystyle O}{\|}}{C}-\bigcirc-\overset{\overset{\displaystyle O}{\|}}{C}-O-CH_2-CH_2-OH + 2\ H_2O$$

Ester links — Ester link

FIGURE 14.7 Medical uses of Dacron.
A Dacron patch is used to close an atrial septal defect in a heart patient.

MICHAEL ENGLISH, M.D./Custom Medical Stock Photo, Inc.

image. All four had to be pressed after cleaning. As women's hemlines rose in the mid-1930s, silk stockings were in great demand, but they were very expensive and short-lived. Nylon changed all that almost overnight. Nylon could be knitted into the sheer hosiery women wanted, and it was much more durable than silk. The first public sale of nylon

FIGURE 14.8 Making nylon.

The diacid is dissolved in hexane, and the diamine is dissolved in water. They do not mix with each other. The layers form because hexane and water are immiscible. Water, which is more dense than hexane, forms the bottom layer. The nylon polymer forms at the interface between the two reactants and continues to form there as the nylon "rope" is pulled away.

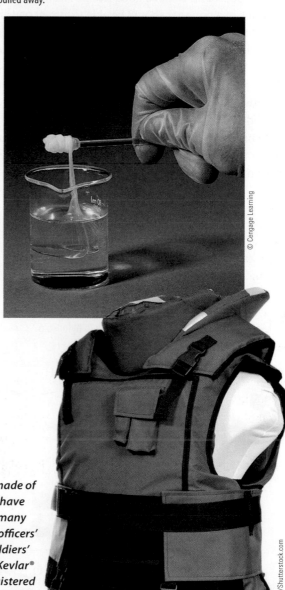

© Cengage Learning

$$n\ \mathrm{HO-\overset{\overset{O}{\|}}{C}-(CH_2)_4-\overset{\overset{O}{\|}}{C}-OH} + n\ \mathrm{H_2N-(CH_2)_6-NH_2} \longrightarrow$$

Adipic acid   Hexamethylenediamine

$$\left(\mathrm{-\underset{H}{\overset{}{N}}-(CH_2)_6-\underset{H}{\overset{}{N}}-\overset{\overset{O}{\|}}{C}-(CH_2)_4-\overset{\overset{O}{\|}}{C}-}\right)_n + n\ \mathrm{H_2O}$$

Nylon-6,6

This material could easily be extruded (pushed through an opening to make a long strand) into fibers that were stronger than natural fibers and chemically more inert. The discovery of nylon jolted the American textile industry at almost precisely the right time. Natural fibers were not meeting the needs of 20th-century Americans. Silk was not durable and was very expensive, wool was scratchy, linen crushed easily, and cotton did not have a high-fashion

Vests made of Kevlar have saved many police officers' and soldiers' lives. (Kevlar® is a registered trademark of DuPont for its aramid fiber.)

© risteski goce/Shutterstock.com

hose took place in Wilmington, Delaware (the location of DuPont's main office), on October 24, 1939. World War II caused all commercial use of nylon to be abandoned until 1945, as the industry turned to making parachutes and other war materials. Not until 1952 was the nylon industry able to meet the demands of the hosiery industry and to release nylon for other uses.

Figure 14.9 illustrates another facet of the structure of hydrogen bonding in nylon that explains why nylons make such good fibers. To have good tensile strength, the chains of atoms in a polymer should be able to attract one another, but not so strongly that the plastic cannot be initially extended to form the fibers. Ordinary covalent chemical bonds linking the chains together would be too strong. Hydrogen bonds, with a strength about one-tenth that of an ordinary covalent bond, link the nylon chains in the desired manner. Kevlar, another polyamide, is used to make bulletproof vests, vests worn by bull riders, military helmets, and fireproof garments. Kevlar is made from *p*-phenylenediamine and terephthalic acid.

Kevlar

FIGURE 14.9 Hydrogen bonding in nylon-6,6.
Hydrogen bonds are highlighted in blue shading.

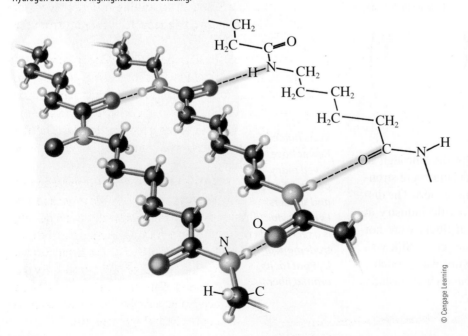

EXAMPLE 14.2 Condensation Polymers

→ *Write the repeating unit of the condensation polymer obtained by combining HOOCCH₂CH₂COOH and H₂NCH₂CH₂NH₂. Identify the amide linkage.*

SOLUTION

The dicarboxylic acid reacts with the diamine to split out water molecules and form a polyamide. The repeating unit is

$$\left(\begin{matrix} O & & O & \text{Amide linkage} \\ \| & & \| & \\ -CCH_2CH_2CNCH_2CH_2N- & \\ & & | & | \\ & & H & H \end{matrix}\right)_n$$

TRY IT 14.2

Draw the structure of the repeating unit of the condensation polymer obtained from reacting terephthalic acid with ethylene glycol.

14-5 New Polymer Materials

Few plastics produced today find end uses without some kind of modification. For example, body panels for the GM Saturn and Corvette automobiles are made of **reinforced plastics**, which contain fibers embedded in a matrix of a polymer. These are often referred to as **composites**. The strongest geometry for a solid is a wire or a fiber, and the use of a polymer matrix prevents the fiber from bending or buckling. As a result, reinforced plastics are stronger than steel on a weight basis. In addition, the composites have a low density—from 1.5 g/cm³ to 2.25 g/cm³ compared with 2.7 g/cm³ for aluminum, 7.9 g/cm³ for steel, and 2.5 g/cm³ for concrete. The only structural material with a lower density is wood, which has an average value of 0.5 g/cm³. In addition, polymers do not corrode. The low density, high strength, and high chemical resistance of composites are the basis for their use in the automobile, airplane, construction, and sporting goods industries.

© Cengage Learning

Glass fibers currently account for more than 90% of the fibrous material used in reinforced plastics because glass is inexpensive and glass fibers possess high strength, low density, good chemical resistance, and good insulating properties. In principle, any polymer can be used for the matrix material. Polyesters are the number one polymer matrix at the present time. Glass-reinforced polyester composites have been used in structural applications such as boat hulls, airplanes, missile casings, and automobile body panels.

Other fibers and polymers have also been used, and the trend is toward increased utilization of composites in automobiles and aircraft. For example, a composite of graphite fibers in a polymer matrix is used in the construction of the F-117 Stealth fighter and other military aircraft. Graphite–polymer composites are used in a number of sporting goods such as golf club shafts, tennis racquets, fishing rods, and skis. The F-16 military aircraft was the first to contain graphite–polymer composite material, and the technology has advanced to the point where many aircraft, such as the F-22, use graphite composites.

Although few automobiles contain exterior body panels made of plastics, a number of components are plastic. Examples include bumpers, trim, light lenses, grilles, dashboards, seat covers, and steering wheels—enough plastics to account for an average of 250 pounds per car. The increased emphasis on improving fuel efficiency will likely

© iStockphoto.com/Sascha Hahn

lead to the use of greater amounts of plastics in the construction of automobiles, in both interior components and exterior body panels.

14-6 Recycling Plastics

Recycling metals such as aluminum, iron, and lead has been occurring for years (Section 8-6). However, programs for recycling plastics developed much more slowly because of the costs associated with separating different types of plastics and producing usable recycled products from the used plastics.

Disposal of plastics has been the subject of considerable debate as municipalities face increasing problems in locating sufficient landfill space. The number one waste is paper products, which make up about 40% of the volume in landfills. (Newspaper alone accounts for 16% of the volume.) Next is plastics, which make up about 20% of the volume in landfills.

Four phases are needed for a successful recycling of any waste material: *collection*, *sorting*, *reclamation*, and *end use*. Public enthusiasm for recycling and state laws requiring recycling have resulted in a dramatic increase in the collection of recycled items. This led to an annual average growth in recycling plastics of 32% per year for the 1988–1993 period. This trend of growth slowed for nearly ten years but has since resumed its uphill climb (Figure 14.10).

Composites combine great strength with impressive flexibility. Advanced fighter planes and high-performance skis are two of many products that make use of composite technology.

© iStockphoto.com/Sportstock

Codes are stamped on plastic containers to help consumers identify and sort their recyclable plastics (Figure 14.11). PET and high-density HDPE are both widely used for making soft-drink bottles and, for that reason, are the two most commonly recycled plastics. In 2003, just under 20% of all PET bottles and just under 25% of all HDPE bottles were recycled, according to the American Plastics Council. Bottles made from PVC, LDPE, and polypropylene were recycled at far lower rates, with a 3.4% rate of recycling for polypropylene bottles being by far the largest for these three types of bottles. Overall, about 21% of plastic bottles were recycled in 2003. Figure 14.10 also

FIGURE 14.10 Growth in post-consumer plastic bottle recycling.

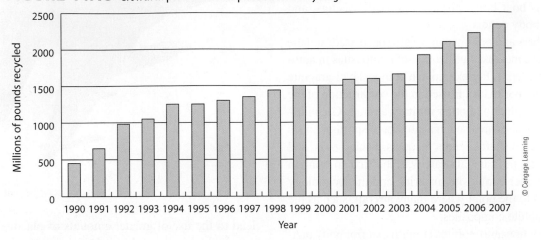

FIGURE 14.11 Plastic container codes.

These symbols are used to identify types of plastic so that containers can be sorted for recycling.

Code	Material	Total bottles (recycled %)
1 PETE	Poly(ethylene terephthalate) (PET or PETE)*	20–30
2 HDPE	High-density polyethylene	50–60
3 V	Poly(vinyl chloride) (PVC)*	5–10
4 LDPE	Low-density polyethylene	5–10
5 PP	Polypropylene	5–10
6 PS	Polystyrene	5–10
7 OTHER	All other resins and layered multi-material	5–10

*Bottle codes are different from standard industrial identification to avoid confusion with registered trademarks.

shows the recent trend in recycling of plastic bottles, and Figure 14.12 shows that recycled PET bottles are used largely for fibers for carpets, clothing, and furniture. Their use for new food and beverage containers and industrial strapping is also quite common. HDPE is recycled for similar uses as well as for trash containers, drainage pipe, garbage bags, and plastic lumber.

FIGURE 14.12 Domestic recycled PET bottle end use.

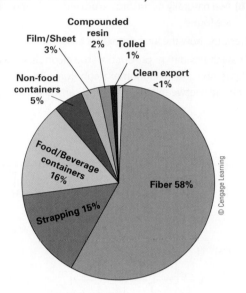

© Cengage Learning

Patagonia has recycled more than 86 million plastic soda bottles in making post-consumer recycled (PCR®) clothing such as this women's Synchilla® jacket. The company notes that no "synchillas" are harmed in the production of its clothing line.

© Charles Gullung/The Image Bank/Getty Images

Your Resources

In the back of the textbook:

➜ *Review Card on Organic Chemicals and Polymers*

  In OWL for CHEM2 at www.cengagebrain.com

➜ *Review Key Terms with Flash Cards (printable or digital)*

➜ *Complete Interactive Practice Quizzes to prepare for tests*

➜ *Submit Assigned Homework and Exercises*

Applying Your Knowledge

1. Why are there more than 11 million organic compounds?

2. Draw the structures of four alcohols that have the formula $C_4H_{10}O$.

3. Identify the class of each of the following compounds:
   (a) $CH_3CH_2CH_2CH_2COOH$
   (b) $CH_3CH = CHCH_2CH_3$
   (c) $C_2H_5OC_2H_5$
   (d) $CH_3CH_2CHOHCH_2CH_3$
   (e) $(CH_3)_3COH$
   (f) $CH_3\overset{\displaystyle O}{\overset{\|}{C}}H$

4. Name the compounds in question 3.

5. Draw the structures, classify, and name the following:
   (a) CH_3COOH (b) CH_3OCH_3
   (c) CH_3CH_2OH (d) CH_3COCH_3
   (e) $HCHO$ (f) CH_3COOCH_3

6. What is meant by the following terms?
   (a) Proof rating of an alcohol
   (b) Denatured alcohol

7. What is the volume percent ethanol in 90 proof gin?

8. What is the proof of pure ethanol?

9. What is a functional group, and what examples are included in this chapter?

10. How do the structures for ethanol, ethylene glycol, and glycerol differ? Give one use for each.

11. Many naturally occurring carboxylic acids have more than one acid group (Table 14.7). What other functional group is often present?

12. Identify each of the following molecules as an alcohol, an ether, an aldehyde, a ketone, a carboxylic acid, or an ester:
   (a) $H\overset{\displaystyle O}{\overset{\|}{C}}CH_2CH_3$
   (b) CH_3CH_2OH
   (c) $CH_3\overset{\displaystyle O}{\overset{\|}{C}}CH_2CH_3$
   (d) $CH_3CH_2CH_2\overset{\displaystyle O}{\overset{\|}{C}}-OH$
   (e) $CH_3\overset{\displaystyle O}{\overset{\|}{C}}-OCH_2CH_3$
   (f) CH_3OCH_3
   (g) $CH_3\overset{\displaystyle O}{\overset{\|}{C}}H$

13. Write the equation for the formation of an ester from acetic acid (CH_3COOH) and ethanol (C_2H_5OH), and give its name.

14. How do the structures of primary, secondary, and tertiary alcohols differ?

15. Give examples of
   (a) two naturally occurring esters and where they are found.
   (b) two naturally occurring carboxylic acids and where they are found.

16. Describe how the liver detoxifies ethanol.

17. Draw the oxidation products of the alcohols shown. Remember, the oxidation of alcohols results in an increase of bonds between already bound carbon and oxygen atoms.

(a)
```
      H   H   OH  H
      |   |   |   |
  H—C—C—C—C—H
      |   |   |   |
      H   H   H   H
```

(b)
```
      H   OH  H
      |   |   |
  H—C—C—C—H
      |   |   |
      H   H   H
```

(c)
```
      OH  H   H   H   H   OH
      |   |   |   |   |   |
  H—C—C—C—C—C—C—H
      |   |   |   |   |   |
      H   H   H   H   H   H
```

18. Ethylene glycol is oxidized to form oxalic acid, in which each carbon–hydrogen bond has been replaced by a carbon–oxygen bond. The safer alternative to ethylene glycol, propylene glycol, is oxidized to pyruvic acid. What is the structure of pyruvic acid?

```
      OH  OH                        O   O
      |   |                         ||  ||
  H—C—C—H   ──Oxidation──→   HO—C—C—OH
      |   |
      H   H
```

```
      OH  OH
      |   |
  H_3C—C—C—H   ──Oxidation──→   Pyruvic acid
      |   |
      H   H
```

19. In what way is an alcohol functional group different from that of a carboxylic acid?

20. In what way is an aldehyde functional group different from that of a ketone?

21. Which of the following carboxylic acids would react with sodium hydroxide to give the best soap?

(a)

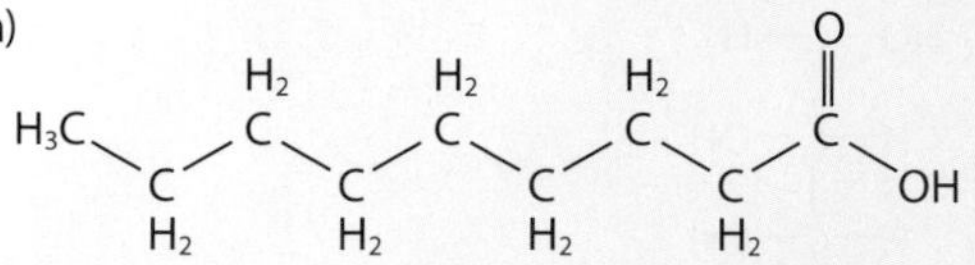

H_3C—$\underset{H_2}{C}$—$\underset{H_2}{C}$—$\underset{H_2}{C}$—$\underset{H_2}{C}$—$\underset{H_2}{C}$—$\underset{H_2}{C}$—$\underset{O}{\overset{O}{C}}$—OH

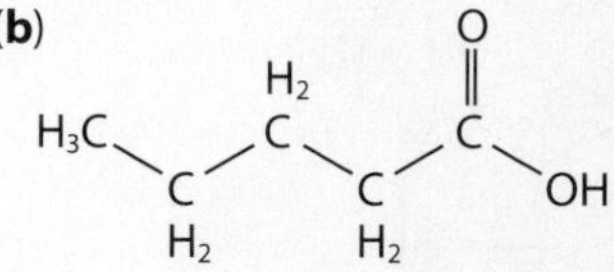

(b)

H_3C—$\underset{H_2}{C}$—$\underset{H_2}{C}$—$\overset{O}{\underset{}{C}}$—OH

(c)

H_3C—$\underset{H_2}{C}$—$\underset{H_2}{C}$—$\underset{H_2}{C}$—$\underset{H_2}{C}$—$\underset{H_2}{C}$—$\underset{H_2}{C}$—$\underset{H_2}{C}$—$\overset{O}{\underset{}{C}}$—OH

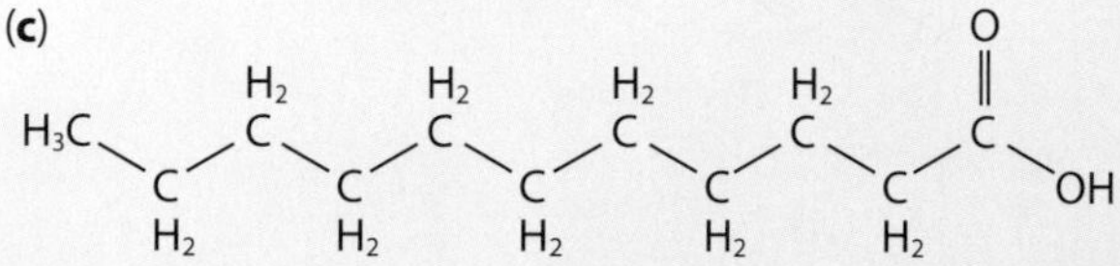

(d)

H_3C—$\underset{}{C}$—$\underset{H_2}{C}$—$\underset{H_2}{C}$—$\overset{O}{\underset{}{C}}$—OH

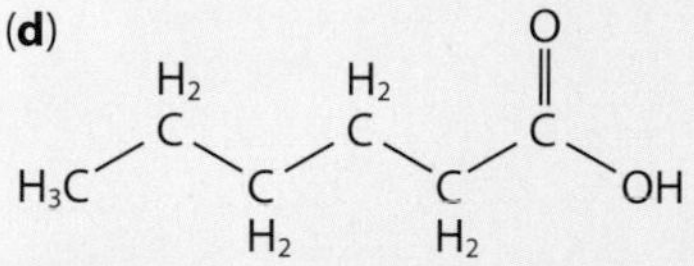

22. Sketch the structure for the monomer of polystyrene.
23. In what ways is a railroad train like polystyrene?
24. What is the origin of the word *polymer*?
25. What is meant by the term *macromolecule*?
26. What do the following terms mean?
    (a) Monomer   (b) Polymer
27. What structural features must a molecule have in order to undergo addition polymerization?
28. What do the following mean?
    (a) Polyethylene   (b) HDPE
    (c) LDPE   (d) CLPE
29. What is the repeating unit of natural rubber?
30. What is the difference in the structures of poly-*trans*-isoprene and poly-*cis*-isoprene? Sketch the arrangements for the two polymers. How do the two differ in their physical properties?

31. What is the role of sulfur in the vulcanization process?
32. What is the effect of vulcanization on the physical properties of natural rubber?
33. What property does a polymer have when it is extensively cross-linked?
34. By using polyethylene as an example, draw a portion of
    (a) a linear polymer.
    (b) a branched polymer.
    (c) a cross-linked polymer.
35. Indicate which of the following can undergo an addition reaction and which cannot. Explain your choices.

    (a) Styrene (C_6H_5CH=CH_2)

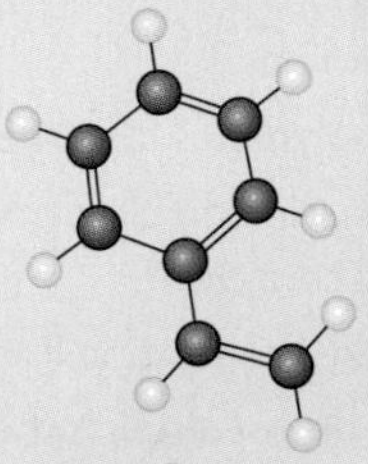

    (b) Propene (CH_2=$CHCH_3$)

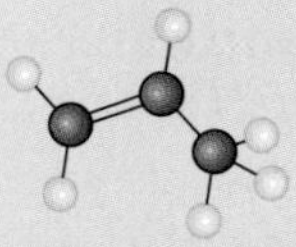

    (c) Ethane (CH_3CH_3)

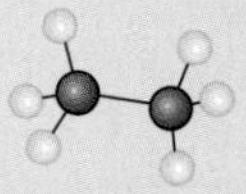

36. Explain how polymers could be prepared from each of the following compounds (other substances may be used):
    (a) CH_3CH=$CHCH_3$
    (b) $HOOCCH_2CH_2COOH$
    (c) $H_2NCH_2CH_2CH_2CH_2NH_2$
37. Draw the structural formula of the monomer used to prepare the following polymers:
    (a) Poly(vinyl chloride)   (b) Polystyrene
    (c) Polypropylene
38. Orlon has the polymeric chain structure shown here:

    —CH_2—$\underset{\underset{CN}{|}}{CH}$—$CH_2$—$\underset{\underset{CN}{|}}{CH}$—$CH_2$—$\underset{\underset{CN}{|}}{CH}$—

    What is the monomer from which this structure can be made?

39. Poly(lactic acid), also known as PLA, is a biodegradable polymer formed from lactic acid in a condensation polymerization process that produces water as a by-product. The structure of PLA is shown below. What is the structure of lactic acid?

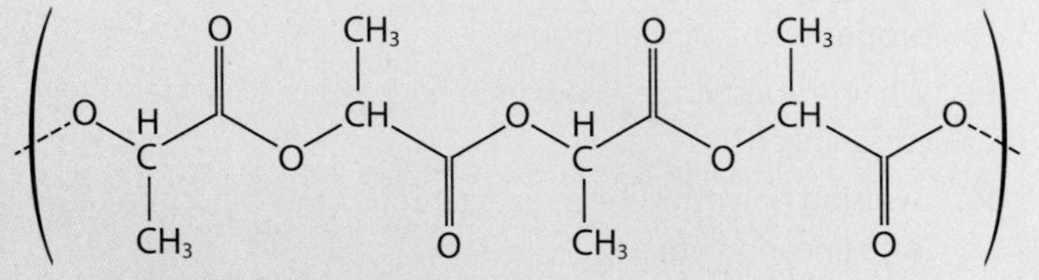

40. Give definitions for the following terms:
    (a) Polyester
    (b) Polyamide
    (c) Nylon-6,6
    (d) Diacid
    (e) Diamine
    (f) Peptide linkage

41. What polymer is identified as PET?

42. What feature do all condensation polymerization reactions have in common?

43. What are the starting materials for nylon-6,6?

44. Which do you think is the source of most polymers used today, green plants or petroleum? Do you think this will ever change? Explain.

45. In Section 14-4, the statement "All plastics are polymers, but not all polymers are plastics" was made. Can you think of an example of a polymer that is not a plastic?

46. Do you think that plastic production will increase in the future? What advantages, if any, do you see when plastics are used in place of other materials?

47. What are the four phases involved in a successful recycling program? What happens if the public is cooperative in turning in recyclable materials but there isn't an adequate infrastructure to process the collected materials?

48. Chemical waste is formed when a reaction produces a product that is not used after its formation. Which process, addition or condensation polymerization, produces more chemical waste. Why?

49. Composite materials are typically reinforced plastics. What properties do the embedded glass fibers or graphite fibers give to composites? Do you foresee any recycling problems with these materials?

50. Many organic compounds have more than one type of functional group. Identify the functional groups in the following naturally occurring molecules:

(a)

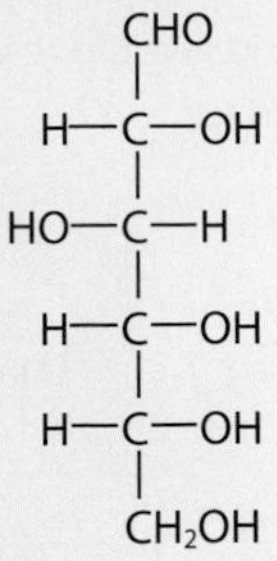

(b)

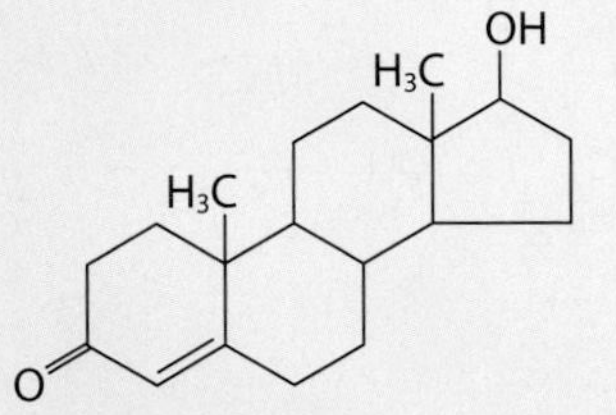

Testosterone

(c)

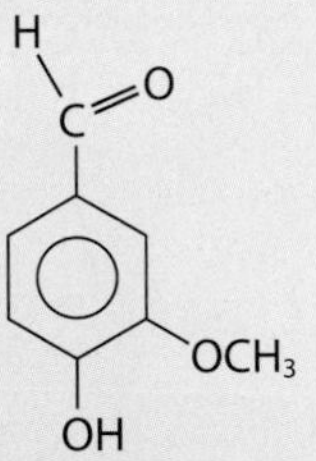

Vanillin
(vanilla bean)

(d)

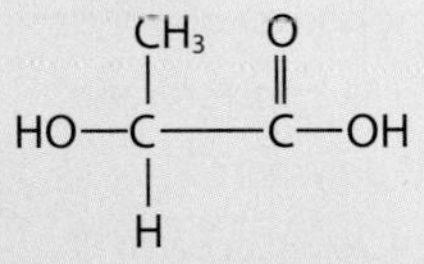

Lactic acid

51. Discuss what plastics are currently being recycled, and give examples of some products being made from these recycled plastics.

52. What are the two main sources of ethanol, and which is likely to be the main source due to its lower cost?

53. What are polymer composite materials? Give two examples that illustrate the importance of polymer composites in the manufacture of consumer products.

54. What is the polymer formed by the reaction of the following monomers?

(a) Ethylene glycol (HOCH$_2$CH$_2$OH) and terephthalic acid,

HOOC—⬡—COOH

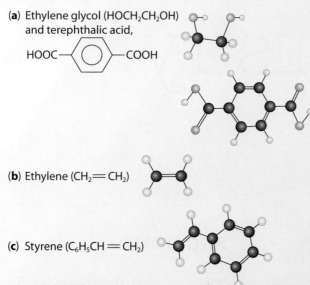

(b) Ethylene (CH$_2$=CH$_2$)

(c) Styrene (C$_6$H$_5$CH=CH$_2$)

55. How many ethylene monomer units, CH$_2$=CH$_2$, are linked together in a polyethylene molecule with a molecular weight of 280,000?

56. How many propylene monomer units of propene, (CH$_2$=CHCH$_3$), are linked together in a polymer molecule with molecular weight of 84,000?

57. What is the predicted molecular weight for a polymer molecule made up of 400,000 styrene units (C$_6$H$_5$CH=CH$_2$)?

58. What is the predicted molecular weight for a polymer molecule made up of 200,000 1,3-butadiene units (CH$_2$=CHCH=CH$_2$)?

59. Use the information in this chapter to determine the approximate percentage of refined petroleum that is used to prepare polymers.

60. If the U.S. per capita production of polymers is 200 pounds, what is the per capita production of refined petroleum? Use your answer from question 59.

15

The Chemistry
of Life

Cooking, nutrition, personal health, drugs, medicine and dentistry, agriculture, our natural environment—biochemistry is fundamental to them all. Sometimes biochemistry is applied in a practical fashion—for example, when a cook scrambles an egg or someone grabs an aspirin tablet to quiet a headache. Sometimes it is applied by a practicing professional or research scientist—for example, when an agriculture expert recommends the appropriate pesticide to a farmer or a pharmaceutical chemist designs a molecule to combat disease. In this chapter, you will be introduced to the major classes of biochemicals and their functions in the body. Some of the applied aspects of biochemistry are also included.

Biochemistry is the study of chemistry in living systems. It could also be correctly called the chemistry of life, and it primarily deals with carbon-based molecules. These organic chemicals found in living organisms are *biochemicals*. A petrochemical manufacturing plant and the cells in your body are doing the same thing—taking in raw materials, using energy and carefully controlled conditions to perform chemical reactions, and putting out valuable products and waste materials. Many biochemicals are polymers. Starches are condensation polymers of simple sugars; proteins are condensation polymers of amino acids; and nucleic acids are condensation polymers of simple sugars, nitrogenous bases, and phosphoric acid groups. All of these very large biomolecules and other essential, smaller ones must be assembled from the raw materials in food. At the same time, burning food must provide energy. To carry out these functions simultaneously, living things have evolved an exquisite system for breaking down food molecules and putting them back together again.

The term *biochemistry* was coined in 1903 by Carl Neuberg, a German chemist. Previously, this area of study would have been referred to as *physiological chemistry*. The overview of biochemistry in this chapter focuses on the molecular structure of the most important kinds of biomolecules. Many of them exhibit "handedness," another type of isomerism, so that's where we begin.

Biochemistry: The study of chemistry in living systems, including plants and animals

15-1 Handedness and Optical Isomerism

Are you right-handed or left-handed? Regardless of our preference, we learn at a very early age that a right-handed glove doesn't fit the left hand, and vice versa. Our hands are not identical; they are mirror images of one another and are not superimposable (Figure 15.1). *An object that cannot be superimposed on its mirror image is called* **chiral**. Objects that are superimposable on their mirror images are **achiral**. Stop and think about the extent to which chirality is a part of our everyday life. We've already discussed the chirality of your hands. Helical seashells are chiral, and most spiral to the right like a right-handed screw. Ignoring any writing or logos, golf clubs are chiral, but baseball bats are achiral. Any object, including a molecule, that is symmetric is achiral and has an identical mirror image.

What is not as well known is that a large number of the molecules in plants and animals are chiral, and usually only one form of the chiral molecule (left-handed or right-handed) is found in nature. For example, all but 1 of the 20 naturally occurring amino acids in our proteins are chiral, and they have the same "handedness." Most of the natural sugars exhibit the opposite "handedness" when compared to amino acids.

A chiral molecule and its nonsuperimposable mirror image are called **enantiomers (optical isomers)**. Recall that isomers are molecules that have the same composition, but a different arrangement of atoms. Straight- and branched-chain alkane isomers were discussed in Section 12-3, and *cis-trans* isomers of alkenes

Chiral: Cannot be superimposed on its mirror image

Achiral: Can be superimposed on its mirror image

Enantiomers (optical isomers): A chiral molecule and its nonsuperimposable mirror-image molecule

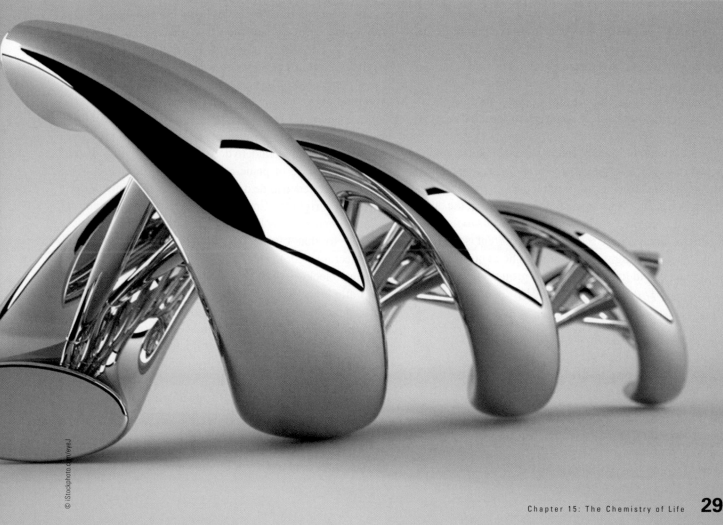

FIGURE 15.1 Mirror images.

When you hold your left hand up to a mirror, it looks like your right hand. But if you place one hand over the other with both hands palms up, they are not identical. This shows that your hands are nonsuperimposable mirror images.

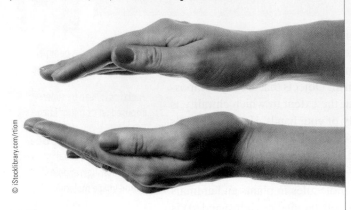

Stereogenic center: A carbon atom with four different attached atoms or groups of atoms that gives rise to enantiomers

were discussed in Section 12-7. Enantiomers are two different molecules, just as your left hand and right hand are different. To have enantiomers, a molecular structure must be asymmetric (without symmetry). The simplest case is a tetrahedral carbon atom bonded to four *different* atoms or groups of atoms. Such a carbon atom is asymmetric and is said to be a **stereogenic center** (from the Greek *stereo*, meaning "handed," and *genesis*, meaning "source"). A molecule that contains a chiral atom is likely to be a chiral molecule, although there are some exceptions that we will not focus on here.

Some compounds are found in nature in both enantiomeric forms. For example, both forms of lactic acid are found in nature. During the contraction of muscles, the body produces only one enantiomer of lactic acid; the other enantiomer is produced when milk sours. Let's look at the structure of lactic acid to see why two different isomers or chiral molecules might be possible. The central carbon atom of lactic acid has four different groups bonded to it: $-CH_3$, $-OH$, $-H$, and $-COOH$.

$$
\begin{array}{c}
H \\
| \\
HO \diagup\overset{\displaystyle C}{\underset{\displaystyle CH_3}{|}}\diagdown COOH
\end{array}
$$

Lactic acid

As a result of the tetrahedral arrangement around the central carbon atom, it is possible to have two different arrangements of the four groups. If a lactic acid molecule is placed so the $C-H$ bond is vertical, as illustrated in

Figure 15.2, one possible arrangement of the remaining groups would be that $-OH$, $-CH_3$, and $-COOH$ are attached in a clockwise manner (isomer I, looking from below). Alternatively, these groups can be attached in a counterclockwise fashion (isomer II). To see further that the arrangements are different, we place isomer I in front of a mirror (Figure 15.2b). Now you see that isomer II is the mirror image of isomer I. What is important, however, is that these mirror-image molecules *cannot be superimposed* on one another, no matter how the molecules are rotated. *These two nonsuperimposable, mirror-image chiral molecules are enantiomers.*

Enantiomers of a chiral compound have the same melting point, the same boiling point, the same density, and other identical physical and chemical properties. They differ in only two ways. They rotate a beam of *plane-polarized light* in opposite directions, as shown in Figure 15.3. For this reason, chiral molecules are sometimes referred to as optical isomers and are said to be *optically active*. Enantiomers also differ in the way in which they interact with other chiral molecules. The difference can sometimes be quite significant, as will be explained later in this section.

One must be able to specify the "handedness" of chiral centers in order to distinguish between them. Enantiomers are mirror-image molecules that have all the same groups attached but are arranged differently in space. It has been necessary to develop an artificial system to describe the arrangement of these atoms so that scientists may be certain they are talking about the same molecule. Historically, enantiomers were referred to as D-enantiomers if they had a spatial arrangement of groups similar to that of the enantiomer of glyceraldehyde (D-glyceraldehyde), which rotated the plane of plane-polarized light in a clockwise fashion, referred to as dextrorotatory (after the Latin *dexter*, meaning "right"). Enantiomers were designated as L-enantiomers if they had a spatial arrangement of groups similar to that of the enantiomer of glyceraldehyde (L-glyceraldehyde), which rotated plane-polarized light in a counterclockwise fashion, referred to as levorotatory (after the Latin *laves*, meaning "left"). It was soon apparent that this simple system would be difficult to apply to more complex molecules with multiple chiral centers or to molecules bearing little similarity to glyceraldehyde. This system is still used in referring to naturally occurring amino acids and naturally occurring sugars, however. Natural amino acids are L-amino acids and natural sugars are D-sugars because of their structural relationship to the glyceraldehyde enantiomers, but it is not true that all D-sugars are dextrorotatory, nor are all L-amino acids levorotatory. Quite confusing! This confusion can be illustrated with lactic acid.

FIGURE 15.2 The enantiomers of lactic acid.

(a) In isomer I, the —OH, —CH₃, and —COOH groups are attached in clockwise order. In isomer II, the same groups are attached in counterclockwise order, seen from below. (b) Isomers I and II are mirror images, a result of their having four different groups on the same carbon atom. (c) The two isomers cannot be turned in any way to make them superimposable.

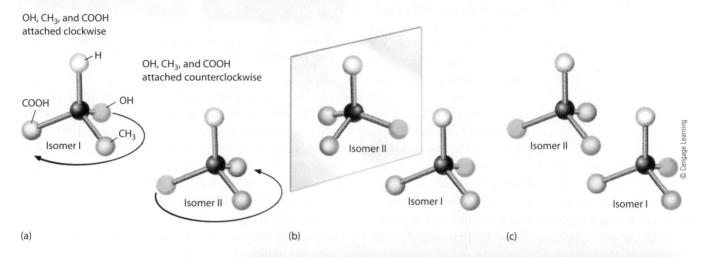

OH, CH₃, and COOH attached clockwise

OH, CH₃, and COOH attached counterclockwise

Isomer I

Isomer II

(a)

Isomer II   Isomer I

(b)

Isomer II   Isomer I

(c)

FIGURE 15.3 Rotation of plane-polarized light by an optical isomer.

(*Top*) Monochromatic light (light of only one wavelength) is produced by a sodium lamp. After it passes through a polarizing filter, the light vibrates in only one direction—it is polarized. Polarized light will pass through a second polarizing filter if this filter is positioned parallel to the first filter but not if the second filter is perpendicular. (*Bottom*) A solution of an optical isomer placed between the first and second polarizing filters causes rotation of the plane of polarized light. The angle of rotation can be determined by rotating the second filter until maximum light transmission occurs. The magnitude and direction of rotation are unique physical properties of the optical isomer being tested.

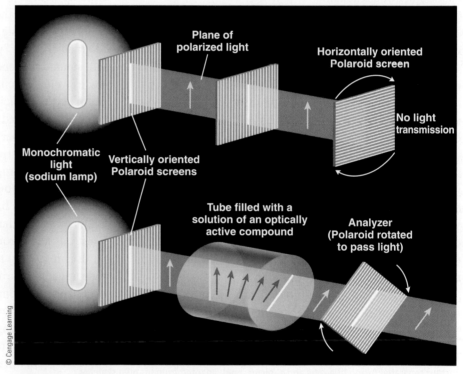

Plane of polarized light

Horizontally oriented Polaroid screen

No light transmission

Monochromatic light (sodium lamp)

Vertically oriented Polaroid screens

Tube filled with a solution of an optically active compound

Analyzer (Polaroid rotated to pass light)

D-Lactic acid is found in souring milk and is levorotatory. L-Lactic acid, which accumulates in muscle tissue during exercise and may cause cramps, is dextrorotatory.

To remove this confusion, chemists now use a **configuration** to refer to the specific arrangement of groups around a stereogenic center, much the way *cis* and *trans* are used to describe the arrangement of groups surrounding an alkene. In this system, one enantiomer is said to have the *R* configuration (Latin *rectus*, right-handed or clockwise) and its mirror image the *S* configuration (Latin *sinister*, meaning left or counterclockwise). Neither of these configurations, however, determines whether the compound is dextro- or levorotatory.

CHO

CHO

H—C····CH₂OH

OH

HOH₂C····C—H

HO

D-Glyceraldehyde

L-Glyceraldehyde

Enantiomers may also differ with respect to their biological properties. They react at different rates in a *chiral*

Configuration: A description of the specific arrangement of atoms in a chemical compound

Racemic mixture: A mixture of equal amounts of enantiomers

Teratogen: A chemical or factor that causes malformation of an embryo

 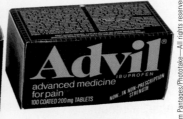

Naproxen (Aleve) is sold as a single enantiomer, and ibuprofen (Advil) is sold as a racemic mixture.

environment, and many of the molecules in plants and animals are chiral. To understand why this difference in activity might exist, think of the hand-in-glove analogy. Although you can put a right-handed glove on your left hand, it takes longer to put it on and doesn't fit very well. Since nature has a preference for L-isomers of amino acids and since enzymes, the catalysts for biochemical reactions, are proteins made from L-amino acids, enzymes are chiral molecules. The catalytic activity of enzymes is dependent on their three-dimensional structure, which in turn depends on their L-amino acid sequence. As a result, enzymes have a binding preference for one enantiomer of a reactant.

Even though nature usually has a preference for one enantiomer, laboratory synthesis of a chiral compound normally gives a mixture of equal amounts of the enantiomers, which is called a **racemic mixture**. The separation and purification of enantiomers are difficult because their physical properties are identical. The first separation of enantiomers by Louis Pasteur in 1848 was based on his observation of mirror-image crystals in a racemic mixture of sodium ammonium tartrate (Figure 15.4). This method is rarely an option, since racemic chemical compounds seldom form mixtures of crystals recognizable as mirror images. Instead, a common method of separating enantiomers is to react them with optically active reagents that have a greater affinity for one enantiomer than the other. Pasteur's discovery of a mold that would selectively destroy one enantiomer of tartaric acid was the first example of this approach.

The chirality of sugars, proteins, and DNA makes the human body highly sensitive to enantiomers. For example,

one enantiomer of a drug may be more active (or more toxic) than the other. A tragic example that called attention to the need for testing both enantiomers occurred in 1963, when horrible birth defects were induced by thalidomide use by pregnant women. After this was discovered, it was determined that one enantiomer was useful for treating morning sickness and the other enantiomer was a **teratogen**. The teratogenic enantiomer caused serious birth defects in the children of women who took thalidomide during the first trimester of their pregnancy.

As a result of these incidents, thalidomide became the most notorious drug ever marketed, and the so-called "thalidomide children" were, in most cases, forced to face life with stunted flippers for arms or legs. They and their families, as well as scientists around the world, vowed never to let this type of incident happen again. Remarkably, the tragedy was limited primarily to Great Britain and Europe thanks, in large part, to some stubborn skepticism about the safety of the drug voiced by Dr. Frances Kelsey at the U.S. Food and Drug Administration (FDA). Assigned to review the thalidomide drug approval application, she fought a dogged battle against the drug company that sought to market thalidomide in the United States until the disastrous news from Europe was indisputable.

It is important to consider how the European thalidomide tragedy occurred. Scientists knew before this particular series of incidents that enantiomers could, in principle, exhibit quite different biological properties, but nothing of this magnitude had ever occurred before. The safety testing done on thalidomide before it was marketed did not reveal its teratogenic effect, although more extensive testing would have done so. As a result of this event, pharmaceutical companies around the world initiated much more stringent testing of both enantiomers of drugs, and such testing continues today.

It may surprise you to learn that there are currently many drugs marketed as racemic mixtures in the United States. In these cases, scientists have been able to show that one enantiomer elicits a desired positive therapeutic response while the other enantiomer shows no biological activity or only a modest undesired

FIGURE 15.4 A depiction of the crystals of sodium ammonium tartrate that Pasteur separated (using tweezers and a microscope), adapted from his own sketches. Note how one crystalline form is the mirror image of the other.

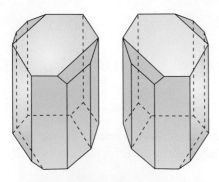

$$
\begin{array}{c}
CO_2^-\ Na^+ \\
| \\
H{-}C{-}OH \\
| \\
HO{-}C{-}H \\
| \\
CO_2^-\ NH_4^+
\end{array}
$$

Sodium ammonium tartrate

biological activity. In such cases, the drugs may be marketed as racemic mixtures. In other cases, where one enantiomer elicits an intolerable biological response, it is necessary for scientists to use available technology either to separate the good enantiomer from the bad or to develop methods that allow the synthesis of only the good enantiomer. Pharmaceutical companies continue to invest hundreds of millions of dollars in developing these technologies to market enantiomerically pure drugs. A commercial produced by Eli Lilly, a major pharmaceutical company, that aired during the Super Bowl in 2012 stated that the cost of developing a drug and bringing it to market was about $1.3 billion. Because many proposed pharmaceutical compounds do not ultimately obtain FDA approval, others have estimated that the total cost per approval is between $4 billion and $11 billion.

It is interesting to note that many drugs currently in use are isolated from natural sources where they are almost always found as single enantiomers. This is because nature generally produces single enantiomers to the exclusion of other possible **stereoisomers**.

One might think the story of thalidomide is over. However, like many drugs, thalidomide has recently been shown to have other medical uses. It is currently marketed in the United States under the trade name Thalomid for use in treating people with HIV, certain neuroblastoma cancers, and leprosy, among other things. However, the teratogenic effect of thalidomide still exists, and the situation became even more complicated when it was shown that the "good" enantiomer of thalidomide is actually converted into the teratogenic enantiomer in the body. Thus, even if the thalidomide is administered only as the pure enantiomer, the teratogenic effect will still be manifested. Fortunately, this interconversion of enantiomers in a biological system is not universal with all drugs.

For this reason, thalidomide is said to be the most regulated drug ever marketed, and women who are given the drug must agree in writing to use two separate forms of birth control so that they will avoid the consequences of a thalidomide-influenced pregnancy. In spite of the potential for good that can come from thalidomide treatment in some cases, there are still people and organizations who take the approach that because it is known that there is great potential for harm from this drug, it should forever be banned from use. The benefits, according to these people, who include many thalidomide victims and their families, do not outweigh the risks under any circumstances.

How do you feel about this issue and its broader applicability to other drugs and technologies?

Often, the biological difference between enantiomers is one of activity or effectiveness, with no difference in toxicity. Aspartame (NutraSweet), widely used as an artificial sweetener, has two enantiomers. However, one enantiomer has a sweet taste, while the other enantiomer is bitter. This indicates that the receptor sites on our taste buds must be chiral, since they respond differently to the "handedness" of aspartame enantiomers. This becomes clearer when looking at the properties of the simple sugars. D-Glucose is sweet and nutritious, whereas L-glucose is tasteless and cannot be metabolized by the body. Another natural substance, carvone, exists as one enantiomer that has a minty odor (found in mint) and another that has the odor of rye bread (found in caraway seeds).

> **Stereoisomers:** Molecules with the same molecular formula but different three-dimensional structures

EXAMPLE 15.1 Chiral Molecules

For each of the following molecules, decide whether the underlined carbon atom is or is not a stereogenic center: (a) $\underline{C}H_2Cl_2$, (b) $H_2N—\underline{C}H(CH_3)—COOH$, and (c) $Cl—\underline{C}H(OH)—CH_2Cl$.

SOLUTION

To be a stereogenic center, an atom must be bonded to four different groups. The underlined carbon atoms in molecules **(b)** and **(c)** meet this condition and are stereogenic centers. The underlined carbon in **(a)** is bonded to a pair of H atoms and a pair of Cl atoms and is therefore not a stereogenic center.

TRY IT 15.1

Which of the following molecules is chiral? Draw the enantiomers for any chiral molecule.

(a)
$$\begin{array}{c} OH \\ | \\ Cl\cdots C \\ \diagup \diagdown \\ Cl \quad CH_3 \end{array}$$

(b)
$$\begin{array}{c} OH \\ | \\ H\cdots C \\ \diagup \diagdown \\ Cl \quad CH_3 \end{array}$$

(c)
$$\begin{array}{c} H \\ | \\ H_2N\cdots C \\ \diagup \diagdown \\ H \quad COOH \end{array}$$

(d)

Ibuprofen

15-2 Carbohydrates

Carbohydrate:
A biomolecule composed of simple sugars

The word *carbohydrate* literally means "hydrate of carbon." Carbohydrates have the general formula $C_x(H_2O)_y$, in which x and y are whole numbers. However, even though the reaction of the carbohydrate sucrose with sulfuric acid produces carbon (Figure 15.5), this does not mean that sucrose is a simple combination of carbon and water. The carbon, hydrogen, and oxygen in carbohydrates are arranged primarily into three organic functional groups

—O—H	—C—H (=O)	—C— (=O)
Alcohol hydroxy group	Aldehyde group	Ketone group

Carbohydrates are divided into groups depending on how many monomers are combined by condensation polymerization: *monosaccharides* (Latin *saccharum*, "sugar"), *disaccharides*, *trisaccharides* (etc.), and *polysaccharides*. Monosaccharides are simple sugars that cannot be broken down into smaller carbohydrate units by acid hydrolysis. In contrast, hydrolysis of a disaccharide or trisaccharide yields two or three monosaccharides (either the same or different), while complete hydrolysis of a polysaccharide produces many monosaccharides (sometimes thousands of them).

Carbohydrates, which make up about half the average human diet, form an essential part of the energy cycle for living things (Section 15-10). Besides energy storage in plants and energy production in animals, carbohydrates serve many other biological purposes. Cellulose is the main structural component of plants. The nucleic acids (Section 15-11) incorporate carbohydrate units in their repeating structure.

15-2a Monosaccharides

The most common simple sugar is D-glucose (also known as dextrose, grape sugar, and blood sugar), which is found in fruit, blood, and living cells. A solution of glucose is often given intravenously when a source of quick energy is needed to sustain life. D-Glucose, along with D-galactose and D-fructose (fruit sugar), are the three common monosaccharides found in the body. They have different struc-

FIGURE 15.5 Reaction of sulfuric acid with sugar.
When sulfuric acid is poured onto the sugar, carbon can be seen forming as water is lost from the sugar.

Charles D. Winters

tures but the same molecular formula, $C_6H_{12}O_6$. D-Galactose and D-glucose are stereoisomers of each other and differ at only a single chiral center.

D-Glucose

D-Galactose

D-Fructose

Planar representations of the monosaccharides, as shown above, are frequently used for convenience, but it should be understood that the tetrahedral bonding of the carbon atoms in these molecules results in the three-dimensional structures shown below. This should be kept in mind when viewing planar representations of monosaccharides and similar or more complex molecules.

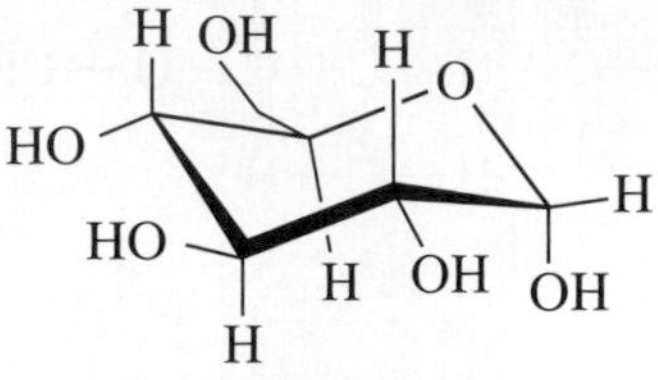

D-Glucose

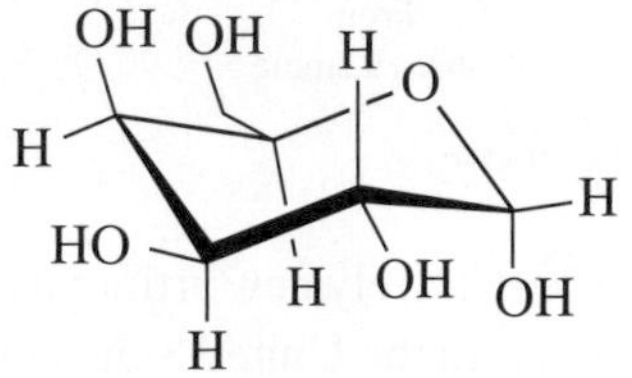

D-Galactose

drugs and infant foods, and in baking. Lactose also serves as a nutrient for bacteria as raw cheeses ripen and develop their distinctive flavors.

The formula for these disaccharides ($C_{12}H_{22}O_{11}$) is not simply the sum of two monosaccharides, $C_6H_{12}O_6$ + $C_6H_{12}O_6$. A water molecule is eliminated as two monosaccharides are united to form the disaccharide. In the body, enzymes catalyze the breakdown (hydrolysis) of disaccharides to their monosaccharides.

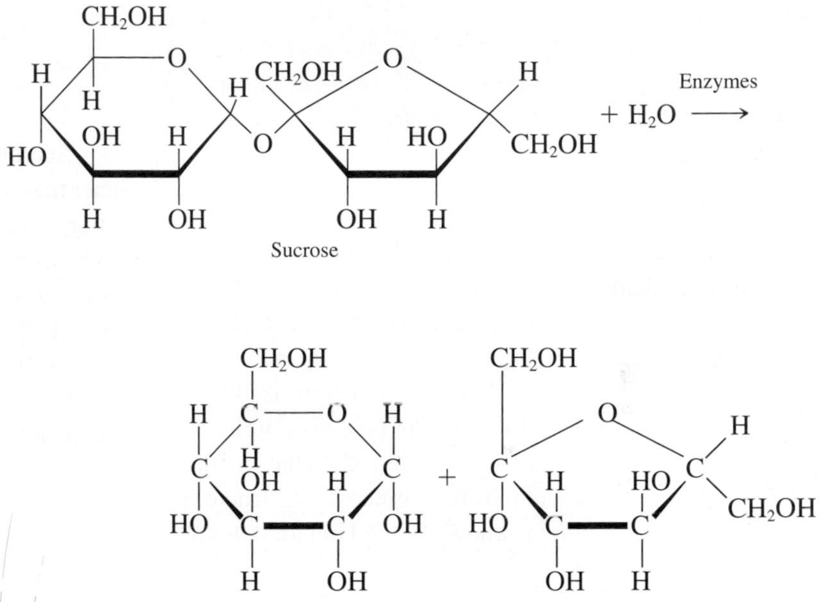

Sucrose

15-2b Disaccharides

Disaccharides result when two monosaccharides are joined together with the elimination of water to give compounds with the general formula $C_{12}H_{22}O_{11}$. The three most common disaccharides are

- **Sucrose** (from sugarcane or sugar beets), composed of a D-glucose monomer and a D-fructose monomer—table sugar;

- **Maltose** (from starch), composed of two D-glucose monomers—used as a sweetener in prepared foods; and

- **Lactose** (from milk), composed of a D-glucose monomer and a D-galactose monomer—used in formulating

D-Glucose   D-Fructose

Sucrose is produced in a high state of purity on an enormous scale—more than 80 million tons per year. About 40% of the world's sucrose production comes from sugar beets and 60% comes from sugarcane. A comparison of the sweetness of common sugars and artificial sweeteners relative to sucrose is given in Table 15.1. Honey, which is a mixture of the monosaccharides glucose and fructose,

Sugarcane is a common source of the disaccharide sucrose.

TABLE 15.1 Sweetness of Common Sugars and Artificial Sweeteners Relative to Sucrose

Substance	Sweetness Relative to Sucrose as 1.0
Lactose	0.16
Galactose	0.32
Maltose	0.33
Glucose	0.74
Sucrose	1.00
Fructose	1.17
Aspartame*	180
Saccharin*	300
Splenda*	600

© Cengage Learning

* Artificial sweeteners.

has been used for centuries as a natural sweetener for foods and is sweeter than sucrose, or cane sugar (Table 15.1). To convert cane sugar into glucose and fructose requires treatment with acid or with a natural enzyme called "invertase." The product, known as *invert sugar*, is often used as a sweetener in commercial food products. Cornstarch can be processed to provide pure glucose, and when the glucose is treated with enzymes a mixture of glucose and fructose is produced. This mixture is known in the United States as high-fructose corn syrup.

Artificial Sweeteners

Saccharin

Saccharin was the first common artificial sweetener. Saccharin passes through the body undigested and consequently has no caloric value. It has a somewhat bitter aftertaste that is offset in commercial products by the addition of small amounts of naturally occurring sweeteners. Such products do have a small caloric value because of the natural sweeteners added.

Aspartame (NutraSweet), which has replaced saccharin as the principal artificial sweetener, is used in more than 3000 products and accounts for 75% of the $1 billion worldwide artificial sweetener market. It is a dipeptide derivative made from the amino acid aspartic acid and the methyl ester of another amino acid, phenylalanine. Aspartame can be digested, and its caloric value is approximately equal to that of proteins. However, since much smaller amounts of aspartame than of table sugar are needed for sweetness, many fewer calories are consumed in the sweetened food. Aspartame is not stable at cooking temperatures,

which limits its use as a sugar substitute to cold foods and soft drinks.

From aspartic acid

From phenylalanine

Aspartame

Sucralose (Splenda) is a relatively new artificial sweetener that was approved for use in the United States in 1998 after several years of use in other countries. It is purported to be as much as 600 times sweeter than sucrose. A comparison of its structure to that of sucrose (previous page) reveals that sucralose is prepared by the replacement of three OH groups in sucrose by three Cl atoms. Like that of other artificial sweeteners, the safety of sucralose has been publicly questioned by some consumer advocate groups.

Sucralose (Splenda)

15-2c Polysaccharides

Nature's most abundant polysaccharides are the starches, glycogen, and cellulose. Some polysaccharides have more than 5000 monosaccharide monomers and more than 1 million molar masses. The monosaccharide most commonly used to build polysaccharides is D-glucose.

Starches and Glycogen

Plant starch is found in protein-covered granules. If these granules are ruptured by heat, they yield a starch, *amylose*, that is soluble in hot water. Amylose constitutes about 25% of most natural starches.

Structurally, amylose is a straight-chain condensation polymer (Chapter 14) with an average of about 200 glucose monomers per molecule. Each monomer is bonded to the next with the loss of a water molecule, just as the two units are bonded in maltose. A representative portion of the structure of amylose is shown in Figure 15.6a.

Glycogen is an energy reservoir in animals, just as starch is in plants. Glycogen is stored in the liver and muscle tissues and is used for "instant" energy until the process of fat metabolism can take over and serve as the energy source.

Cellulose

Cellulose is the most abundant organic compound on Earth, and its purest natural form is cotton. The woody part of trees, the supporting material in plants and leaves, and the paper we make from them are also mainly cellulose. Like amylose, it is composed of D-glucose units. The difference between cellulose and amylose lies in the bonding between the D-glucose units (Figure 15.6b). This subtle structural difference in the connectivity between the glucose units in amylose and cellulose is the reason we cannot digest, or metabolize, cellulose. *Metabolism* is a general term for the sum of all the chemical and physical processes in a living organism. When something is metabolized, it is changed by these processes. Human beings do not have the necessary enzymes to metabolize cellulose, but they do have enzymes that can recognize and break the linkage between glucose units in amylose. On the other hand, termites, a few species of cockroaches, and ruminant mammals (such as

cows, sheep, goats, and camels) are able to digest cellulose because bacteria living in their guts provide the necessary enzyme. D-Glucose can be obtained from cellulose by heating a suspension of the polysaccharide in the presence of a strong acid.

FIGURE 15.6 (a) Amylose structure.
From 60 to 300 glucose units bond together in the manner shown.
(b) Cellulose structure.
From 900 to 6000 glucose units bond together in the pattern shown here.

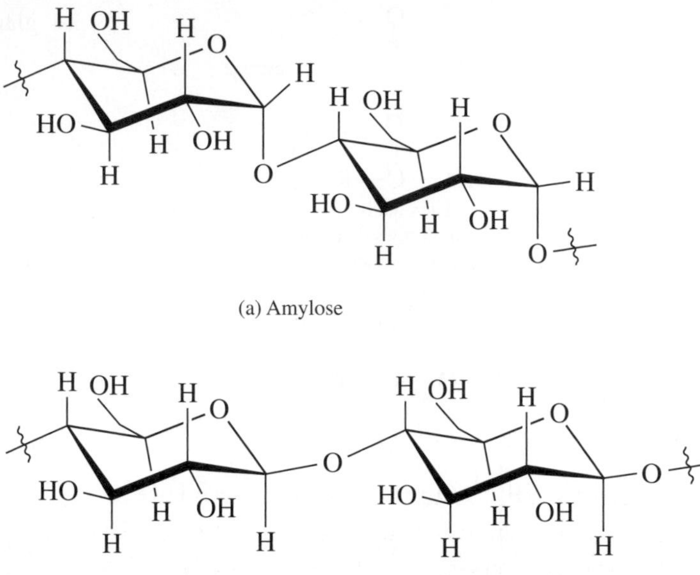

(a) Amylose

(b) Cellulose

© Cengage Learning

Lipid: A class of biomolecules not soluble in water but soluble in organic solvents

Triglyceride: A triester of glycerol and fatty acids

15-3 Lipids

A **lipid** is an organic substance found in living systems that has limited solubility in water but is soluble in organic solvents. Because their classification is based on insolubility in water rather than on a structural feature such as a functional group, lipids vary widely in their structure and, unlike proteins and polysaccharides, are not polymers. Lipids include fats, oils, steroids, and waxes. The predominant lipids are fats and oils, which make up 95% of the lipids in our diet. The other 5% are steroids and several other lipids that are important to cell function.

15-3a Fats and Oils

Fats and oils are **triglycerides**—triesters of glycerol (glycerin) and fatty acids. The general equation for the formation of a triester of glycerol is

Bread is particularly high in starch.

© iStockphoto.com/baytunc

The three R groups can be the same or different groups within the same fat or oil, and they can be saturated or unsaturated. The most common fatty acids in fats and oils are listed in Table 15.2.

Fatty acids such as oleic acid, which contain only one double bond, are referred to as *monounsaturated acids*. One of the unsaturated acids, linoleic acid, is an *essential fatty acid*. The human body cannot produce this acid, but it is required for the synthesis of the *prostaglandins*, an important group of more than a dozen related compounds.

The prostaglandins have potent effects on physiological activities such as blood pressure, relaxation and contraction of smooth muscle, gastric acid secretion, body temperature, food intake, and blood platelet aggregation. Many prostaglandins cause inflammation and fever. The fever-reducing effect of aspirin results from the inhibition of cyclooxygenase, the enzyme that catalyzes the synthesis of prostaglandins.

The term *fat* is usually reserved for *solid* triglycerides (such as butter and lard), and *oil* is the term for *liquid* triglycerides (olive, soybean, and corn oils, for example). The R groups in the fatty acid portions of fats are generally saturated, with only C—C single bonds. The R groups in oils are usually either monounsaturated (one C=C double bond) or polyunsaturated (two or more C=C double bonds). Since the C=C bonds interrupt the zigzag pattern of tetrahedral angles with 120 degree angles, the molecules are irregular in shape and do not pack together efficiently enough to form a solid. Table 15.3 illustrates the percentages of saturated and unsaturated fat found in common dietary oils and fats.

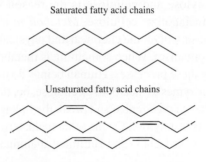

Hydrogen can be added catalytically to the double bonds of an oil to convert it into a semisolid fat. For

TABLE 15.2 Common Fatty Acids in Fats and Oils

Acids		Melting Point (°C)	Source
Saturated (All Solids at Room Temperature)			
Lauric	$CH_3(CH_2)_{10}COOH$	44	Coconut oil
Palmitic	$CH_3(CH_2)_{14}COOH$	63	Animal and vegetable fats
Stearic	$CH_3(CH_2)_{16}COOH$	69	Animal and vegetable fats
Unsaturated (All Liquids at Room Temperature)			
Oleic	$CH_3(CH_2)_7CH=CH(CH_2)_7COOH$	4	Animal and vegetable fats
Linoleic*	$CH_3(CH_2)_4CH=CHCH_2CH=CH(CH_2)_7COOH$	−5	Linseed oil, cottonseed oil
Linolenic	$CH_3CH_2CH=CHCH_2CH=CHCH_2CH=CH(CH_2)_7COOH$	−11	Linseed oil

* An essential fatty acid that must be part of the human diet.

TABLE 15.3 Amounts of Saturated and Unsaturated Fatty Acids in Fats and Oils

Dietary Oil/Fat	Saturated Fat	Polyunsaturated Fat	Monounsaturated Fat
Canola oil	6%	36%	58%
Safflower oil	9%	78%	13%
Sunflower oil	11%	69%	20%
Corn oil	13%	62%	25%
Olive oil	14%	9%	77%
Soybean oil	15%	61%	24%
Peanut oil	18%	34%	48%
Lard	41%	12%	47%
Palm oil	51%	10%	39%
Butterfat	66%	4%	30%
Coconut oil	92%	2%	6%

© Cengage Learning

example, liquid soybean and other vegetable oils are *hydrogenated* to produce cooking fats and margarine.

$$H_2C-O-\underset{\displaystyle O}{\overset{\displaystyle \|}{C}}-(CH_2)_7CH=CH(CH_2)_7CH_3$$

$$HC-O-\underset{\displaystyle O}{\overset{\displaystyle \|}{C}}-(CH_2)_7CH=CH(CH_2)_7CH_3 \quad \xrightarrow[200°C]{H_2,\,Ni}$$

$$H_2C-O-\underset{\displaystyle O}{\overset{\displaystyle \|}{C}}-(CH_2)_7CH=CH(CH_2)_7CH3$$

Triolein (a liquid oil)

$$H_2C-O-\underset{\displaystyle O}{\overset{\displaystyle \|}{C}}-(CH_2)_7CH_2CH_2(CH_2)_7CH_3$$

$$HC-O-\underset{\displaystyle O}{\overset{\displaystyle \|}{C}}-(CH_2)_7CH_2CH_2(CH_2)_7CH_3$$

$$H_2C-O-\underset{\displaystyle O}{\overset{\displaystyle \|}{C}}-(CH_2)_7CH_2CH_2(CH_2)_7CH_3$$

Tristearin (a solid fat)

If it is better to consume unsaturated fats instead of saturated ones, why do food companies hydrogenate oils to reduce their unsaturation? There are several answers to this question. First, the double bonds in the fatty acid are reactive, and oxygen can attack the fat at these bonds. When the oil is oxidized, unpleasant odors and flavors develop. Hydrogenating an oil reduces the likelihood that the food will oxidize and become rancid. Second, hydrogenating an oil makes it less liquid. There are many times when a

food processor needs a solid fat in a food to improve its texture and consistency. For example, if liquid vegetable oil were used in a cake icing, the icing would slide off the cake. Instead of using animal fat, which also contains cholesterol, the manufacturer turns to a hydrogenated or partially hydrogenated oil.

Trans-*fatty acid, a by-product of the hydrogenation process, is not easily metabolized in the human system.*

© iStockphoto.com/lisafx

301

15-3b Steroids

Steroids are found in all plants and animals, and all contain the following four-ring skeleton:

Steroid: A lipid with a four-ring structure

Hormone: A chemical messenger secreted in the bloodstream by endocrine glands

The skeletal four-ring structural drawing on the left is chemical shorthand similar to that described for cyclic hydrocarbons in Section 12-8. There is a carbon at each corner, and the lines represent $C-C$ bonds. Since every carbon atom forms four bonds, additional bonds between carbon atoms and hydrogen atoms are understood to be present whenever the skeletal structure shows fewer than four bonds. The structure on the right shows the hydrogen atoms understood to be present in the four-ring structure shown on the left. Although all of the rings in the skeletal drawing are shown as saturated rings, steroids often have one ring that is unsaturated or aromatic (benzene-like). For example, cholesterol has one double bond in the second ring. Note that its structure also includes alkyl groups and an alcohol group. These groups replace hydrogen atoms in the skeletal representation.

Saturated ring C_6H_{12} Unsaturated ring C_6H_{10} Aromatic ring C_6H_6

dant animal steroid—it is not found in plants. The human body synthesizes cholesterol and readily absorbs dietary cholesterol through the intestinal wall. Therefore, a diet with absolutely no cholesterol does not preclude the presence of cholesterol in the body. An adult human contains about 250 g of cholesterol. Cholesterol receives a lot of attention because high blood cholesterol levels are associated with heart disease (Section 17-11). Proper amounts of cholesterol are essential to our health, because cholesterol undergoes biochemical modification to give milligram amounts of many important **hormones**, such as vitamin D, cortisone, and sex hormones. Cholesterol is also an essential component of cell membranes, and without this important molecule many of our cells would be incapable of maintaining their structure at normal body temperatures.

Sex Hormones

Cholesterol is the starting material (the precursor molecule) for the synthesis of steroid sex hormones. One female sex hormone, *progesterone*, differs only slightly in structure from the male hormone *testosterone*. In fact, relatively minor structural differences between many female and male sex hormones account for significantly different biological activities.

So important is this molecule that Heinrich Otto Wieland (1877–1957) and Adolf Wintaus (1876–1959) received Nobel Prizes in chemistry in 1927 and 1928, respectively, for their work leading to the determination of the structure of cholesterol. Cholesterol is the most abun-

Cholesterol

Progesterone

Testosterone

Other female hormones are estradiol and estrone, together called *estrogens*. The estrogens contain an aromatic ring, which differentiates them from the steroids progesterone and testosterone.

Estradiol

Estrone

The estrogens and progesterone are produced by the ovaries. Estrogens are important to the development of the egg in the ovary, whereas progesterone causes changes in the wall of the uterus and, after fertilization, prevents release of a new egg from the ovary (ovulation). Birth control drugs use derivatives of estrogens and progesterone to simulate the hormonal process resulting from pregnancy and thereby prevent ovulation (Section 17-4).

Testosterone and *androsterone* are the two most important male sex hormones, known as *androgens*. They are synthesized in the testes from cholesterol and are responsible for the development of male secondary sex characteristics and for promoting muscle and tissue growth.

Androstenedione, another male sex hormone, also has muscle-building properties that have been exploited, both legally and illegally. In fact, these *anabolic* (muscle-building) properties have led to the development of a wide variety of synthetic (not naturally occurring) steroids to mimic the tissue-building effects of the natural male sex hormones.

Androstenedione

Methandrostenolone
(Dianabol)

Synthetic steroids such as methandrostenolone (Dianabol) have been abused by professional athletes and, increasingly, amateur athletes as young as junior high school age to gain a competitive advantage. The variety of such steroids readily available on the open market and the Internet, without a prescription in many cases, is frightening when one considers the possible consequences of abuse of these powerful chemical agents. Among the side effects of steroid abuse in men are shrinking testes, enlarged breasts and feminization, increased balding, high blood pressure, unpredictable and rapid mood changes (sometimes known as "roid rage"), and even death. Women who

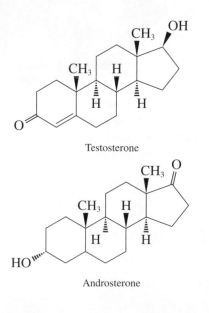

Testosterone

Androsterone

abuse these steroids tend to become very masculine, often to the point of being visually almost unrecognizable as women. The competitive advantages to be gained often seem to outweigh the risks of steroid abuse in the minds of many athletes. *Sports Illustrated* magazine did a survey of world-class athletes regarding the use of banned steroids in competition. More than 90% of these athletes said they would use the banned drugs if they knew they would win all of their competitions and were guaranteed to go uncaught. A follow-up question revealed that 50% would still do so if they knew they would not be caught and would win all of their competitions for a five-year period, even if told they would die from the side effects of using the banned drugs after only five years.

15-3c Waxes

Waxes are esters formed from long-chain (16 or more carbon atoms) fatty acids and long-chain alcohols. The general formula of a wax is the same as that of a simple ester, RCOOR′, with the qualification that R and R′ are limited to alkyl groups with a large number of carbon atoms. Natural waxes are usually mixtures of several esters. Wax coatings on leaves help to protect the leaves from disease and also help the plant to conserve water. The feathers of birds are also coated with wax. Our ears are protected by wax. Several natural waxes have been used in consumer products. These include carnauba wax (from a Brazilian palm tree), which is used in floor waxes, automobile waxes, and shoe polishes, and lanolin (from lamb's wool), which is used in cosmetics and ointments. Lanolin also contains cholesterol.

15-4 Soaps, Detergents, and Shampoos

In strongly basic solutions, fats and oils undergo hydrolysis to produce glycerol and salts of fatty acids. Such reactions are called *saponification* reactions, and the sodium or potassium salts of the fatty acids formed are *soaps*.

$$CH_3-(CH_2)_{16}-COO-CH_2$$
$$CH_3-(CH_2)_{16}-COO-CH_2 + 3\ NaOH \longrightarrow$$
$$CH_3-(CH_2)_{16}-COO\quad CH_2$$

Tristearin (an animal fat)   Sodium hydroxide

$$3\ CH_3-(CH_2)_{16}-COO^-Na^+ + \begin{array}{l} HO-CH_2 \\ HO-CH \\ HO-CH_2 \end{array}$$

Sodium stearate (a soap)   Glycerol (glycerin)

Pioneers prepared their soap by boiling animal fat with an alkaline solution obtained from the ashes of hardwood. The resulting "lye" soap could be "salted out" by adding sodium chloride, because soap is less soluble in a salt solution than in water. The crude soap made this way contained considerable caustic material (sodium hydroxide or potassium hydroxide) in addition to the soap molecules, but it did its job of cleaning quite well.

Substances that are water soluble can readily be removed from the skin or a surface by simply washing with an excess of water. To remove a sticky sugar syrup from your hands, you can dissolve the sugar in water and rinse it away. Many times, the material to be removed is oily, and water will merely run over the surface of the oil. Since the skin has natural oils, even substances such as ordinary dirt that are not oily themselves can adhere quite strongly to the skin and to clothing containing these oils. The hydrogen bonding holding water molecules together is too large to allow the oil and water to intermingle (Figure 15.7), so something like soap is needed to loosen the dirt and wash it away.

The cleaning action of soap is explained by its molecular structure. When present in an oil–water system, the fatty acid anion in the soap moves to the interface between the oil and the water.

© iStockphoto.com/ starfotograf

$$CH_3CH_2CH_2CH_2CH_2CH_2CH_2CH_2CH_2CH_2CH_2CH_2CH_2CH_2CH_2CH_2CH_2 \overset{O}{\underset{\|}{C}}O^-Na^+$$

Sodium stearate molecule

The molecular structure of a soap or synthetic detergent molecule consists of a long oil-soluble (**hydrophobic**, meaning not liking or fearing water) group and a water-soluble (**hydrophilic**, meaning liking or loving water) group. When dissolved in water, soaps and detergents cluster together in a way that buries the hydrophobic "tails" within a spherical particle called a *micelle*. The hydrophobic, nonpolar hydrocarbon interior readily encapsulates the nonpolar oil or grease molecules and makes it possible to remove the oil from the surface by rinsing with water (Figure 15.7).

One undesirable property of soaps is their tendency to form precipitates with Ca^{2+}, Mg^{2+}, and Fe^{2+} ions found in "hard" water. The resulting fatty acid salts of these doubly positive ions are not as soluble in water as the Na^+ ion salts. These less-soluble molecules appear as a scum that sticks to laundry and bathtubs, often containing trapped dirt, which makes it appear even worse.

Detergents are artificial compounds derived from organic molecules and are designed to have even better cleaning action than soaps, but less reaction with the doubly positive ions found in hard water. As a consequence, detergents are often more economical to use and are more effective in hard water than soap. There are many different detergents on the market. An inventory of cleaning materials in a typical household might include half a dozen or more formulated products designed to be the most suitable for a spe-cific job—skin, hair, clothes, floors, or the family car.

Detergents generally have a benzene-like ring at one end. This allows greater variability in the types of head groups.

Typical hydrophilic head groups include negatively charged sulfate ($-OSO_3^-$), sulfonate ($-SO_3^-$), and phosphate ($-OPO_3^{2-}$) groups. Compounds with these groups are *anionic surfactants*. The hydrophobic tail groups are typically unbranched, which tends to render the detergents biodegradable.

Cationic (positively charged) *surfactants* are almost all quaternary ammonium halides (four groups attached to the central nitrogen atom) with the general formula

$$R_1 - \overset{\overset{\displaystyle R_2}{|}}{\underset{\underset{\displaystyle R_3}{|}}{N^+}} - R_4 \; X^-$$

$$CH_3CH_2CH_2CH_2CH_2CH_2CH_2CH_2CH_2CH_2CH_2CH_2CH_2CH_2 - \bigcirc - SO_3^-Na^+$$

Oil-soluble tail
(unbranched, hydrophobic; nonpolar)
A typical synthetic detergent molecule

Water-soluble head
(hydrophilic; polar)

FIGURE 15.7 How soaps and detergents work.

The hydrophilic, ionic ends of the molecules interact strongly with water. The hydrophobic, hydrocarbon ends of the molecules avoid water and are drawn to the oily portion of the dirt. As greasy dirt is broken up by agitation, the particles are surrounded and isolated from one another. This prevents them from coming together again and allows them to be carried away in the wash water.

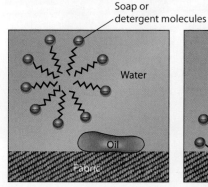
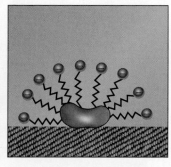
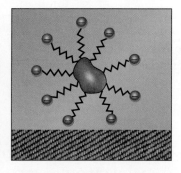
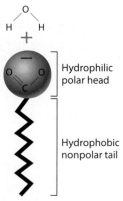

Soap or detergent molecules

Water

Oil

Fabric

Hydrophilic polar head

Hydrophobic nonpolar tail

© Cengage Learning

where one of the R groups is a long hydrocarbon chain and another frequently includes an —OH group. The X^- in the formula represents a halide ion such as chloride (Cl^-) or bromide (Br^-) ion.

Shampoos are generally more complex formulations than simple soap solutions, with a number of ingredients to satisfy different requirements for maintaining clean and healthy-looking hair. If you use soap to wash your hair, incomplete rinsing will result in a "soap film" on the hair, which makes it appear dull. If soap is used to wash hair in hard water, a very noticeable film can usually be seen and is very difficult to remove. In addition, soaps tend to produce solutions with basic pH values. These harsh conditions are damaging to the hair.

Most shampoos contain anionic detergents. These give the product good foaming characteristics because anionic detergents generally foam more than cationic and nonionic detergents. Nonionic surfactants, such as the products obtained by reacting diethanolamine and lauric acid, are often also present in shampoos. While not as good surfactants as anionic surfactants, the nonionics are useful as thickeners and foam stabilizers. They make a shampoo pour from the bottle more slowly and cause the lather to remain thick for a longer period.

A little knowledge of chemistry goes a long way when it comes to shampooing your hair, doing the dishes, or doing the laundry. Anyone who has ever tried to wash oily, dirty hair using only water quickly reaffirms the old adage that oil and water don't mix because the hair remains quite oily upon drying. However, the addition of a little shampoo solves the problem. Anionic detergents and soaps form roughly spherical clusters when placed in water. These spherical clusters (Figure 15.7) may contain hundreds of individual detergent or soap molecules that cluster together, with the hydrophobic tails pointing toward the interior of the cluster and the hydrophilic head groups on the surface of the cluster where they are solubilized by hydrogen bonding to the many water molecules with which they are in contact. Manual agitation of the shampoo through the hair allows the hydrophobic oils and dirt on the hair to become solubilized in the hydrophobic, oily interior of the micelles. These oils and dirt are then flushed away when the hair is rinsed in running water.

$$HN(CH_2CH_2OH)_2 + CH_3(CH_2)_{10}COOH \longrightarrow$$

Diethanolamine   Lauric acid

$$CH_3(CH_2)_{10}\overset{\overset{\displaystyle O}{\|}}{-C}-N(CH_2CH_2OH)_2 + H_2O$$

Lauric diethanolamide
(an amide detergent)

Similar processes take place when detergents are used to remove grease and oils from dishes or clothing. Some clothing stains do not lend themselves to this type of chemical removal, and other methods of cleaning must be tried.

Hair is more manageable, has a better sheen, and has less tendency to attract static charges (causing "flyaway" hair) if all of the shampoo is removed after washing. An anionic detergent can be removed from the hair by using a *rinse*, or conditioner, containing a dilute solution of a cationic detergent, usually a quaternary ammonium compound, which electrically attracts the anions and facilitates their removal. The positive charged end of the quaternary ammonium surfactant also neutralizes negative charges on damaged hair (—COO^- groups from disrupted protein chains), while the alkyl chain attaches to the hair and gives it a smooth feel.

An after-shampoo conditioner also attempts to replace some of the oils that were removed by the detergent back into the hair. A typical conditioner has a water–alcohol dispersing medium as well as skin softeners, oils, waxes, resins, and even short amino acid polymers that can adhere to the hair to produce a more pliable and elastic fiber that is not as likely to become dry or be affected by atmospheric conditions. Holding the correct amount of moisture is the key to hair control, because too much water causes the hair to be limp and too little causes the individual hairs to attract static charge. Although many hair preparations make direct or indirect claims that various proteins and other beneficial ingredients can penetrate the hair and "repair" and even strengthen it, there is no scientific evidence for these claims. Protein molecules, for example, are simply too

large to pass through the surface of the hair. Only much smaller molecules can do this. In fact, *if* hair preparations did function in this way, they would have to be classified as drugs by the FDA.

Lanolin and mineral oil (or their substitutes) are often added to shampoos to replace the natural oils in the scalp, thus preventing it from drying out and scaling. The presence of oil additives and stabilizers sometimes gives the shampoo a pearlescent appearance. These ingredients also make the shampoo less foamy, a quality that is popular in European countries.

15-5 Creams and Lotions

If dry skin is treated with an oily substance after washing, the skin will be protected until enough natural oils have been regenerated. The oily substance used can be derived from animal oils, vegetable oils, or even oils from petroleum (the mineral oils). It is not uncommon to see some kind of rare animal oil, such as mink oil, used in a skin preparation, yet the oil from a mink is not any more effective at holding the skin's moisture than is a less expensive vegetable or mineral oil. When choosing a skin product containing an oil, perhaps a more important factor than the kind of oil is whether it will be soothing or irritating to the skin.

Any substance that holds moisture in the skin can be called a *moisturizer*. Since these substances are all oily in nature, they can lubricate the skin (restore the feeling of normal oiliness) and have the noticeable effect of softening and soothing the skin. These substances are also called *emollients*. All of the oils listed in the previous paragraph that help to hold moisture in the skin would be called emollients.

Getting an emollient distributed evenly on your skin is not as easy as it may seem. Instead of just pouring oil over your body after a shower (this would leave your skin too oily), it would be better to use a mixture containing the emollient. Two kinds of mixtures are commonly used: *Creams* and *lotions* are made by mixing an oily component with water and other ingredients in the right proportions to form a stable mixture that can be more like a solid (a cream) or more like a liquid (a lotion). Mixtures like these are called **emulsions**. Emulsions consist of two substances that would normally not mix, such as an oil and water, but that are made to mix by an **emulsifying agent**, a compound whose molecules have a part that is soluble in water and a part that is soluble in oil. With its dual solubility, the emulsifying agent stabilizes the mixture. Thus, a cream can contain a rather high percentage of an oil and not feel oily or greasy because of the presence of the water. When the cream is applied to the skin, some of the water is absorbed by the skin, and the oil (emollient) remains on the surface to hold in the moisture.

Emulsions are examples of **colloids**, which are quite common. Fog, foams, foods such as milk, and aerosol sprays are all colloids (Table 15.4). Colloidal mixtures differ from solutions. The colloid particles (called the *dispersed phase*) distributed in the solvent-like medium (called the *continuous phase*) are much larger than the molecules or ions that are the solutes in true solutions. Two kinds of emulsions can be formed between oil and water: oil droplets of colloid size dispersed in water and water droplets of colloid size dispersed in oil. As you might expect, oil-in-water emulsions have more of the properties of an aqueous solution, while the water-in-oil emulsions have more oil-like properties—for example, they tend to feel more greasy. An oil-in-water emulsion has tiny droplets of an oily or waxy substance dispersed throughout a water medium; homogenized milk is an example. A water-in-oil

Emulsion: A stable mixture of water and an oily component

Emulsifying agent: A compound that has a water-soluble part as well as an oil-soluble part and that stabilizes an emulsion

Colloid: A particle larger than most molecules or ions dispersed in a solvent-like medium

TABLE 15.4 Types of Colloids

Continuous Phase	Dispersed Phase	Type	Examples
Gas	Liquid	Aerosol	Fog, clouds, aerosol sprays
Gas	Solid	Aerosol	Smoke, airborne viruses, automotive exhaust
Liquid	Gas	Foam	Shaving cream, whipped cream
Liquid	Liquid	Emulsion	Mayonnaise, milk, face creams
Liquid	Solid	Solution	Milk of magnesia, mud
Solid	Liquid	Gel	Jelly, cheese, butter
Solid	Solid	Solid solution	Milk glass, some gemstones, many alloys such as steel

© Cengage Learning

Note: Most colloids will separate unless stabilized. Food and cosmetic emulsions, foams, and aerosols are usually stabilized with emulsifying agents to give them a long shelf life and consistent properties, such as color and texture, throughout their life.

Amino acid: A biomolecule containing an alpha-amino group and a carboxyl group

Essential amino acid: An amino acid that is not synthesized by the body and, therefore, must be obtained from the diet

emulsion has tiny droplets of a water solution dispersed throughout an oil; examples are natural petroleum and butter. An oil-in-water emulsion can be washed off the skin surface with tap water, whereas a water-in-oil emulsion gives skin a greasy, water-repellent surface that resists being washed off by running water. With careful formulation of the emulsion, a *barrier cream* can be made that will effectively resist aqueous solutions that might contain harmful ingredients.

The ingredient listed first on a cream or lotion label is an indication of the kind of emulsion present. If water is listed first, the emulsion is probably an oil-in-water type. If an oil is listed first, the emulsion is probably the water-in-oil type (Figure 15.8). A lotion might contain the same emollient as a cream but contain a different emulsifier and a different ratio of water to oil. Whether an emollient is distributed on the skin in a cream or a lotion depends more on how the product is perceived by the user than on which kind of product is

more effective. All creams and lotions have "shelf lives." Creams and lotions, like all colloids, can tend to "settle out" over a period of time, a property not observed in solutions.

15-6 Amino Acids

Amino acids are the building blocks of all proteins, including enzymes, connective tissue (collagen), and hair (keratin). A large number of proteins exist in nature. For example, the human body is estimated to have 100,000 different proteins. It is amazing that all of these proteins are derived from only 20 different **amino acids** (Table 15.5). All proteins are condensation polymers of amino acids. Even more amazing is nature's preference for only the L-enantiomer of these amino acids. All but one of the 20 amino acids found in nature have the general formula

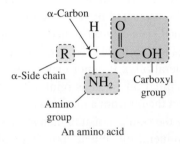

An amino acid

with an amino ($-NH_2$) group, a carboxylic acid group ($-COOH$), and an R group attached to the *alpha*-carbon (the first carbon next to the $-COOH$). R is a unique group for each amino acid (see Table 15.5), and the α-carbon is a stereogenic center, with one exception—glycine. R is a hydrogen atom in glycine, the simplest amino acid. Therefore, glycine is achiral (not chiral) and is the only naturally occurring amino acid that does not have enantiomers. The polarity of the R groups in amino acids affects the structure and function of proteins. The amino acids are grouped in Table 15.5 according to whether the R group is nonpolar, polar, acidic, or basic.

The **essential amino acids** must be ingested from food; they are indicated by asterisks in Table 15.5. A diet that includes meat, milk, eggs, or cheese provides all the essential amino acids. The other amino acids can be synthesized by the human body.

FIGURE 15.8 Two Pond's moisturizing creams.
The cold cream (*left*) has a water-in-oil emulsion, and the dry skin cream (*right*) has an oil-in-water emulsion.

Water-in-oil emulsion

Oil-in-water emulsion

Courtesy of Chapel House Photography / Cengage Learning

TABLE 15.5 Common L-Amino Acids Found in Proteins (R Groups Highlighted)

Amino Acid	Abbreviation	Structure		Amino Acid	Abbreviation	Structure
		Nonpolar R Groups				*Polar but Neutral R Groups*
Glycine	Gly	H—CH—COOH, NH₂		Threonine*	Thr	CH₃—CH—CH—COOH, OH, NH₂
Alanine	Ala	CH₃—CH—COOH, NH₂		Cysteine	Cys	HS—CH₂—CH—COOH, NH₂
Valine*	Val	CH₃—CH—CH—COOH, CH₃, NH₂		Asparagine	Asn	H₂N—C(=O)—CH₂—CH—COOH, NH₂
Leucine*	Leu	CH₃—CH—CH₂—CH—COOH, CH₃, NH₂		Glutamine	Gln	H₂N—C(=O)—CH₂CH₂—CH—COOH, NH₂
Isoleucine*	Ile	CH₃—CH₂—CH—CH—COOH, CH₃, NH₂		Tyrosine	Tyr	HO—C₆H₄—CH₂—CH—COOH, NH₂
Proline	Pro	H₂C–CH₂ / H₂C–CHCOOH / N–H (ring)				*Acidic R Groups*
Phenylalanine*	Phe	C₆H₅—CH₂—CH—COOH, NH₂		Glutamic acid	Glu	HO—C(=O)—CH₂CH₂—CH—COOH, NH₂
Methionine*	Met	CH₃—S—CH₂CH₂—CH—COOH, NH₂		Aspartic acid	Asp	HO—C(=O)—CH₂—CH—COOH, NH₂
Tryptophan*	Trp	indole—CH₂—CH—COOH, NH₂				*Basic R Groups*
		Polar but Neutral R Groups		Lysine*	Lys	H₂N—CH₂CH₂CH₂CH₂—CH—COOH, NH₂
Serine	Ser	HO—CH₂—CH—COOH, NH₂		Arginine†	Arg	H₂N—C(=NH)—NH—CH₂CH₂CH₂—CH—COOH, NH₂
				Histidine	His	imidazole—CH₂—CH—COOH, NH₂

* Essential amino acids that must be part of the human diet. The other amino acids can be synthesized by the body.

† Growing children also require arginine in their diet.

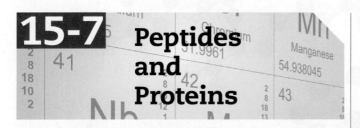

15-7 Peptides and Proteins

How are amino acids polymerized to give proteins? The formation of an amide from the reaction of an amine and a carboxylic acid is described in Section 14-5 in the discussion of polyamides such as nylons.

$$R-\overset{\displaystyle O}{\overset{\|}{C}}-OH + H_2NR' \longrightarrow R-\overset{\displaystyle O}{\overset{\|}{C}}-\overset{\displaystyle H}{\overset{|}{N}}-R' + H_2O$$

Because amino acids have both an amine group and a carboxylic acid group, the —COOH of one amino acid can combine with the —NH₂ of a second amino acid.

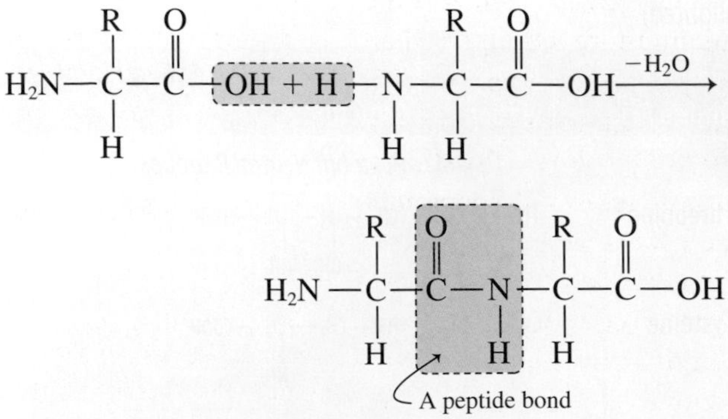

In the condensation reaction, one molecule of water is eliminated between the carboxylic acid of one amino acid and the amine group of another. The result is a **peptide bond**

(called an *amide group* in simpler molecules), and the molecule is a *dipeptide*. Two different orders of attachment are possible between two amino acids, depending on which amine reacts with which acid group. For example, when glycine and alanine react, both glycylalanine and alanylglycine can be formed. Either end of the dipeptide can then react with another amino acid.

Since each dipeptide has a —COOH and an —NH$_2$ group, a tripeptide can be formed from each dipeptide by reaction at either end, and the polymerization process can continue until a large **polypeptide** chain is formed (Figure 15.9). Names of peptides are written from left to right starting with the amino- or N-terminal end. The *-ine* ending of all amino acid residues (except the carboxyl or C-terminal

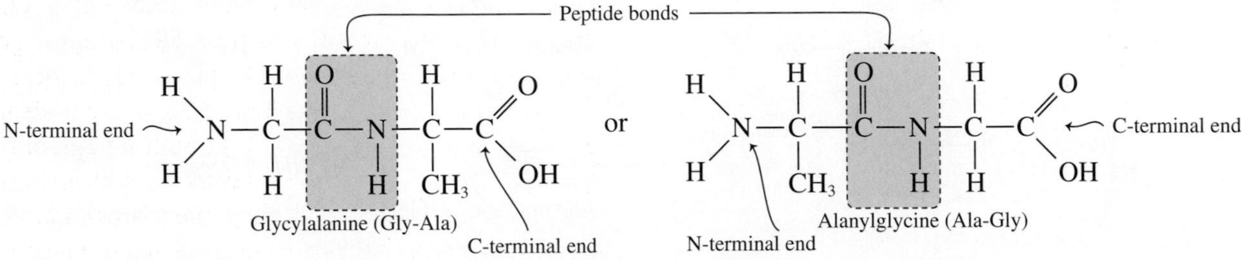

FIGURE 15.9 Two different representations of the same polypeptide chain.
The color coding of atoms in the model is carbon (black), nitrogen (blue), oxygen (red), and hydrogen (white). Identify each peptide bond, the R groups of each amino acid, and the N-terminus and C-terminus in this fragment.

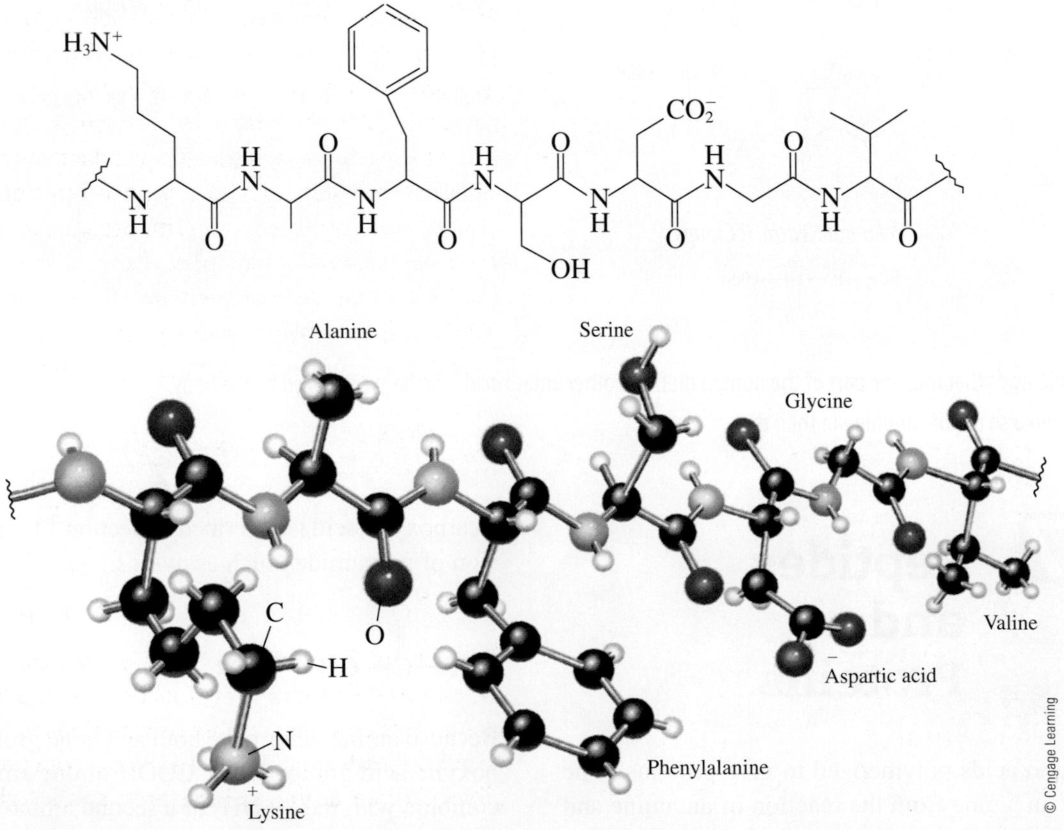

end) is changed to *-yl*. For example, Gly-Ala-Ser is the tripeptide glycylalanylserine.

Proteins are polypeptides containing from 50 to thousands of amino acid residues, and they vary greatly in structure and composition. They can be divided into two classes: simple and conjugated. *Simple proteins* consist only of amino acids. Two examples of simple proteins are insulin and chymotrypsin. Insulin, a hormone that is essential to controlling the concentration of glucose in the blood, has 51 amino acid residues in two linked chains (Figure 15.10). Chymotrypsin, an enzyme that aids in the digestion of proteins in our diet, contains 245 amino acid residues.

FIGURE 15.10 The amino acid sequence of insulin.

Sequence contains 51 amino acids in two chains linked by disulfide bonds (Section 15-9) between the side chains of cysteine residues. The sequence of amino acids of any protein is also called the primary structure.

Conjugated proteins are much more common than simple proteins and contain nonprotein parts called *prosthetic groups*. Prosthetic groups are small nonprotein molecules covalently bonded to the protein. Both myoglobin (Figure 15.11a) and hemoglobin (Figure 15.11b) contain a prosthetic group known as a *heme group* (Figure 15.11c). Myoglobin, which binds oxygen in muscles, consists of a single protein chain (Figure 15.11a) with one heme group (Figure 15.11c). Hemoglobin binds oxygen in blood. Human hemoglobin contains four protein chains (Figure 15.11b), each of which contains a heme group (Figure 15.11c). Two of these four chains are identical with 141 amino acid residues, and the other two are identical with 146 amino acid residues. In both myoglobin and hemoglobin, the oxygen binds reversibly with an iron ion (Fe^{2+}) in the center of the heme group. Carbon monoxide binds to the iron in the heme of hemoglobin much more strongly than does oxygen. Hemoglobin complexed with carbon monoxide cannot carry the oxygen necessary for metabolism, and death may result in severe cases of carbon monoxide poisoning.

Proteins known as enzymes (Section 11-8) often have small nonprotein parts called *cofactors* that are necessary for biological activity. These cofactors are frequently inorganic cations but may be small organic molecules that are referred to as *coenzymes*. The distinction between a prosthetic group and a coenzyme is basically associated with the type of protein involved; conjugated proteins contain prosthetic groups and enzymes contain coenzymes.

To summarize, then, *proteins are polypeptides that are condensation polymers of amino acids*. They vary greatly in size, and some proteins include non–amino acid groups.

> **Peptide bond:** Amide group that connects two amino acids
>
> **Polypeptide:** A condensation polymer of amino acids
>
> **Protein:** A class of polypeptides that may also include non–amino acid parts

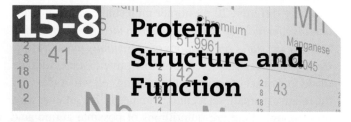

15-8 Protein Structure and Function

The order of the amino acid residues in a peptide or protein molecule is essential to its function. As the length of the chain increases, the number of variations in the sequence of amino acids quickly increases. Six tripeptides are possible if three amino acids (for example, glycine, Gly;

© Cengage Learning

Only the polypeptide backbones are shown in these "ribbon" diagrams; side chains are omitted. Myoglobin has a single protein chain (yellow) and one heme group (Fe ion shown in white). Hemoglobin has four protein chains of two different kinds (yellow and purple), with one heme group in each protein chain. In this type of representation of proteins, the alpha-helix parts of the protein chain are shown as ribbon-like spirals.

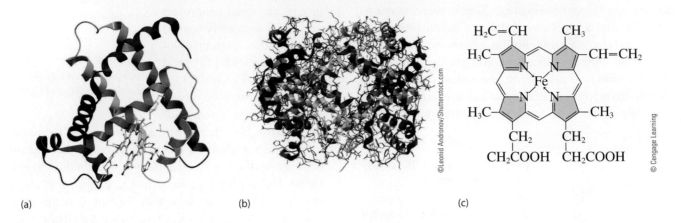

(a)   (b)   ©Leonid Andronov/Shutterstock.com   (c)   © Cengage Learning

alanine, Ala; and serine, Ser) are linked in all possible combinations. They are:

- Gly-Ala-Ser

- Ser-Gly-Ala

- Ala-Gly-Ser

- Gly-Ser-Ala

- Ser-Ala-Gly

- Ala-Ser-Gly

If *n* amino acids are all different, the number of arrangements is *n*!. The notation *n*!, called *n factorial*, is mathematical shorthand for the result of multiplying *n* by all of the whole numbers between it and zero. For four different amino acids, the number of different arrangements is 4!, or $4 \times 3 \times 2 \times 1 = 24$. For five different amino acids, the number of different arrangements is 5!, or 120. If all 20 different naturally occurring amino acids are bonded, the sequences alone make 2.43×10^{18} (2.43 quintillion) uniquely different 20-monomer molecules. *Since proteins can also include more than one molecule of a given amino acid, the possible combinations are essentially infinite.* However, of the many different possible proteins that could be made from a set of amino acids, a living cell will make only the relatively small, select number it needs. Put another way, even though there are quintillions of possible amino acid combinations, only a small fraction of these has biological activity, and even fewer are produced by living systems.

Many short-chain peptides are important biochemicals. For example, enkephalins and endorphins (Figure 15.12) are referred to as "natural opiates" because they moderate

pain in the same manner as opium derivatives. Our bodies synthesize enkephalins and endorphins to moderate pain, and our pain threshold is related to levels of these neuropeptides in our central nervous system. Individuals with a high tolerance for pain produce more of these neuropeptides and consequently tie up more receptor sites than normal; hence, they feel less pain. A dose of heroin temporarily bonds to a high percentage of the sites, resulting in little or no pain. Continued use of heroin causes the body to reduce or cease its production of enkephalins and endorphins. If use of the narcotic is stopped, the receptor sites become empty, and withdrawal symptoms occur.

EXAMPLE 15.2 Peptides

 Use the structures of amino acids in Table 15.5 to draw the structure of the tripeptide represented by Ala-Ser-Gly, and give its name.

SOLUTION

The amino acid sequence in the abbreviated name shows that alanine should be written at the left with a free H_2N— group, glycine should be written at the right with a free —COOH group, and both should be connected to serine by peptide bonds:

$$H_2N-\underset{\underset{CH_3}{|}}{\overset{\overset{H}{|}}{C}}-\overset{\overset{O}{\|}}{C}-\underset{\underset{H}{|}}{N}-\underset{\underset{CH_2OH}{|}}{\overset{\overset{H}{|}}{C}}-\overset{\overset{O}{\|}}{C}-\underset{\underset{H}{|}}{N}-\underset{\underset{H}{|}}{\overset{\overset{H}{|}}{C}}-\overset{\overset{O}{\|}}{C}-OH$$

The name is alanylserylglycine.

FIGURE 15.12 Enkephalins and endorphins, polypeptides that are natural opiates.

Note that all three polypeptides have one identical amino acid sequence.

Peptide	Amino acid sequence
Met-enkephalin	Tyr —Gly —Gly —Phe — Met
Leu-enkephalin	Tyr —Gly —Gly —Phe —Leu
β-endorphin	Lys—Arg—Tyr —Gly —Gly —Phe— Met —Thr — Ser — Glu —Lys — Ser — Glu —
	Thr —Pro—Leu—Val —Thr —Leu—Phe—Lys—Asn—Ala—Ile —Ile —Lys —
	Asp—Ala—Tyr —Lys—Lys —Gly —Glu

TRY IT 15.2

Use the structures of amino acids in Table 15.5 to draw the structure of the tetrapeptide Cys-Phe-Ser-Ala.

Proteins are important in a wider variety of ways than other kinds of biomolecules. As *enzymes*, they serve as catalysts in biological synthesis and degradation reactions. As *hormones*, they serve a regulatory role; as *antibodies*, they protect us against disease. They make up the muscle fibers that contract so that we can move. And proteins are the major constituents of cellular and intracellular membranes, skin, hair, muscle, and tendons. Each individual protein has its own group of amino acids arranged in a definite molecular structure that is specific to the function of that protein.

The sequence of amino acids bonded to one another in a protein by peptide bonds is the protein's **primary protein structure**. Changing the sequence alters the properties of a protein, and just one change may produce a new protein unable to function like the original one. For example, *sickle cell anemia* manifests itself when a mutant variation of normal hemoglobin known as sickle cell hemoglobin is present in the body. Sickle cell hemoglobin results when a single acidic amino acid (glutamic acid) is replaced by a single nonpolar amino acid (valine) in two of the four protein chains of normal hemoglobin. This change of only two amino acids, out of a total of 574 amino acids in the four protein chains, leads to the formation of fragile, sickle-shaped cells. The inability of these cells to flow easily through capillaries causes pain and inflammation and can lead to severe anemia, organ damage, and even an early death.

In some parts of some proteins and in the entirety of others, the shape of the backbone of the molecule (the chain containing peptide bonds) has a regular, repetitive pattern that is referred to as its **secondary protein structure**. The two most common secondary structures are the α-helix and the β-pleated sheet. The α-helix is held together by *intramolecular* (within the molecule) hydrogen bonding between backbone peptide bonds. An N—H group of one amino acid forms a hydrogen bond with the oxygen atom in the third amino acid down the chain (Figure 15.13).

The α-helix is the basic structural unit of the α-keratins in wool, hair, skin, beaks, nails, and claws.

Silk has the β-sheet structure (Figure 15.14) in which several chains of amino acids are joined side-to-side by *intermolecular* (between protein chains) hydrogen bonds. The resulting structure is not elastic, because stretching the fibers would break either covalent bonds or the many

Primary protein structure: The amino acid sequence in a protein

Secondary protein structure: The shape of the backbone structure of a protein

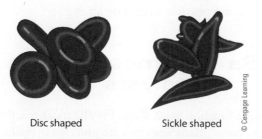

Disc shaped   Sickle shaped

© Cengage Learning

Red blood cells (erythrocytes). Normally they are disc shaped, but with a single amino acid defect in two of the protein chains, they adopt the pointed sickle shape. Sickle cell anemia is an inherited (genetic) disease.

FIGURE 15.13
Helical protein structure.

An illustration of how hydrogen bonding connects a peptide bond nitrogen atom to an oxygen atom in the third amino acid unit down the chain, resulting in a coiled structure.

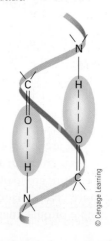

© Cengage Learning

FIGURE 15.14 β-Pleated sheet protein structure.
Hydrogen bonds are shown as dotted lines.

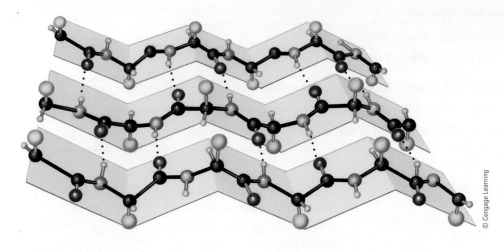

© Cengage Learning

FIGURE 15.15
Collagen, a fibrous protein.
Collagen is composed of three polypeptide chains twisted together into a rope-like fiber.

© Cengage Learning

Tertiary protein structure: The shape of an entire folded protein chain

Quaternary protein structure: The structure of a protein with more than one polypeptide

Denaturation: Destruction of proper functioning of a protein by changing its structure

Enzyme: A biomolecule that functions as a catalyst for biochemical reactions

Active site: The region of an enzyme where the catalytic chemistry occurs

hydrogen bonds holding the individual protein strands in the sheet. However, just as you can bend the stack of pages in this book, so too can the stack of protein sheets be bent.

Tertiary protein structure refers to how helices and sheets come together. One kind of tertiary structure is found in collagen, a fibrous protein: three amino acid chains (primary structure) each twist into left-handed helices (secondary structure), which in turn are twisted into a right-handed superhelix (tertiary structure) to form an extremely strong fibril (Figure 15.15). Bundles of fibrils make up the tough collagen. A visual analogy for the first three levels of protein structure appears in Figure 15.16.

A second kind of tertiary structure is found in globular proteins, which contain regions where the polypeptide chain forms α-helices, other regions where parallel parts of the chain are organized into β-pleated sheets, and some parts that are just coiled randomly.

Quaternary protein structure is the shape assumed by the entire group of chains in a protein composed of two or more chains. Hemoglobin (Figure 15.11) is a prime example of quaternary structure—its four protein chains,

together with the nonprotein heme molecule carried by each, can function only when combined in precisely the right shape.

All of the structural features—primary, secondary, tertiary, and quaternary—are critical to the proper functioning of a protein and give a protein its "native" or "natural" structure. Any physical or chemical process that changes the native protein structure and makes it incapable of performing its normal function is called a **denaturation** process. For example, heating an aqueous solution of a protein breaks hydrogen bonds that help to maintain tertiary structure and causes the protein molecule to unfold and precipitate. This is what happens when an egg is cooked. Another way to cause denaturation is through the use of chemical compounds. Denaturing chemicals include reducing agents, which break disulfide link ages, and acids and bases, which affect the hydrogen bonds and ionic interactions between polypeptide chains. Whether denaturation is reversible depends on the protein and the extent of denaturation. Certainly, it is not possible to uncook an egg.

15-8a Enzymes: Protein Catalysts

Enzymes function as catalysts for chemical reactions in living systems. Many enzymes are globular proteins. One outcome of the highly organized, though seemingly random, structure of globular proteins is creation of the region (the **active site**) that allows an enzyme to function as a catalyst. Like all catalysts, enzymes increase the rate of a

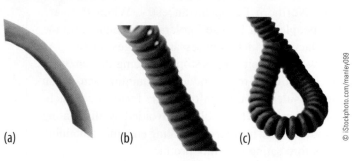

FIGURE 15.16 Structure of a telephone cord.
(a) The straight cord is like the primary structure of a protein. (b) The curling of the cord is like the secondary structure of a protein. (c) When the curled cord is twisted up, it is like the tertiary structure of a protein.

(a)   (b)   (c)

© iStockphoto.com/manley099

reaction by lowering the energy of activation (Figure 8.3). This lowering occurs because the reaction path or process is changed. Enzymes are very effective catalysts and typically increase reaction rates by anywhere from 10^6 to 10^{16} times.

Many biomolecules are broken down during digestion by hydrolysis reactions, which are essentially the reverse of condensation reactions. In hydrolysis, a larger molecule is split into smaller molecules with the addition of H— and —OH of water where a bond was broken. The enzyme maltase catalyzes the hydrolysis of the sugar maltose into two molecules of D-glucose. This is the only function of maltase, and no other enzyme can substitute for it. Sucrase, another enzyme, hydrolyzes only sucrose. Lysozyme catalyzes hydrolysis of sugar polymers (polysaccharides) found

in bacterial cell walls (Figure 15.17). Some enzymes are less specific. The digestive enzyme trypsin, for example, primarily hydrolyzes peptide bonds in proteins. However, the structure and polarity of trypsin are such that it can also catalyze the hydrolysis of some esters.

Since the structure of the active site of an enzyme is important, the same factors that cause denaturation will also destroy the activity of the enzyme. For example, most enzymes are effective only over a narrow temperature range and a narrow pH range. Enzymes are denatured irreversibly at high temperatures or at pH values outside their effective range.

Many inherited diseases or defects affect how enzymes function. An example is found in *lactose intolerance*, which is an inability to digest lactose, the sugar present in milk

FIGURE 15.17 Lysozyme with substrate in the active site.
The structure of lysozyme is shown at (1) as a space-filling model. The active site is a cleft in the surface of the lysozyme that stretches horizontally across the middle of the enzyme. The active site is occupied by a portion of a polysaccharide molecule, the substrate (green atoms). (The part of the polysaccharide not bound to the active site has been omitted so that you can see the enzyme better.) The diagram at (4) shows noncovalent interactions (red dotted lines) that hold the substrate to the enzyme. The bond that will be broken when the substrate is hydrolyzed is marked by an arrow (5).

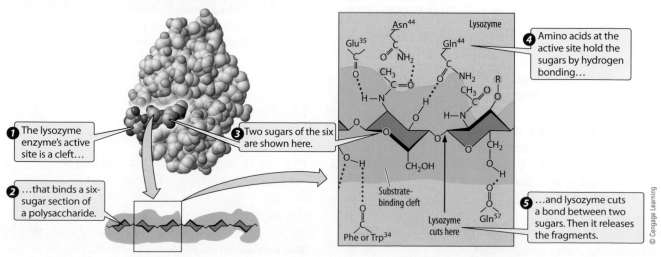

© Cengage Learning

from all mammals. Well over half the world's population has this problem to some degree. The pattern of inheritance is noticed largely in those with Asian or African ancestry, Native Americans and Latinos, and to a lesser extent in those of Northern European ancestry. When they are infants, people with lactose intolerance manufacture the enzyme lactase, which is necessary to digest lactose. As they grow older, their bodies stop producing this enzyme, and the ingestion of milk products containing lactose can lead to considerable discomfort in the form of stomachaches and diarrhea. The problem is avoided by eliminating milk products from the diet or by taking enzyme-containing tablets before eating any milk product.

15-9 Hair Protein and Permanent Waves

Hair, fingernails, and toenails are composed principally of keratin. An important difference between hair keratin and other proteins is its high content of the amino acid cysteine, which contains the —SH (sulfhydryl) group in its side chain.

Cysteine plays an important role in the structure of hair by forming disulfide bonds (—S—S—) between protein chains. The disulfide bonds, or "bridges," function the same way that cross links do in other polymers (Chapter 14). In keratin, these protein chains are twisted together into spirals that group together to constitute individual strands of hair (Figure 15.18).

In addition to the disulfide bonds between hair protein structures, ionic bonds and hydrogen bonds between protein side chains also affect the behavior of hair. In fact, these bonds explain "bad hair days." Consider, for example, the interaction between a lysine —NH$_2$

A freshly created permanent wave.

group and a —COOH of glutamic acid on a neighboring protein chain. The acidic —COOH groups lose their protons, forming negatively charged —COO$^-$ groups, while the basic NH$_2$ groups gain protons to form positively charged —NH$_3^+$ groups. When the —NH$_3^+$ and —COO$^-$ groups on adjacent chains approach each other, an ionic bond is formed that helps hold the two protein chains together.

Ionic bond between two protein chains

Or consider how a hydrogen bond might form between adjacent protein chains.

$$O = C \qquad \qquad N - H$$

$$H - C - CH_2 - OH \cdots O = C - CH_2 - C - H$$

$$H - N \qquad \qquad OH \qquad \qquad C = O$$

←————— Protein chains —————→

Hydrogen bond

Hydrogen bond between two protein chains

Moisture affects hydrogen bonds between protein chains, and when their number and arrangement changes, hair

FIGURE 15.18

Permanent waving, a chemical oxidation–reduction process.

The disulfide bonds in hair protein are broken by a reducing agent (Step 1). The protein strands are twisted into a more curly shape (Step 2). Then the new shape is fixed by an oxidizing agent that re-forms the disulfide bonds (Step 3).

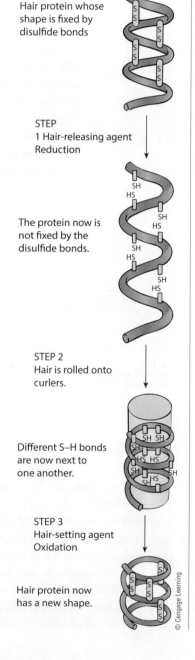

Hair protein whose shape is fixed by disulfide bonds

STEP 1 Hair-releasing agent Reduction

The protein now is not fixed by the disulfide bonds.

STEP 2 Hair is rolled onto curlers.

Different S–H bonds are now next to one another.

STEP 3 Hair-setting agent Oxidation

Hair protein now has a new shape.

© Cengage Learning

changes. When hair is wet, it can be stretched to one and one-half times its dry length. On a very humid day, enough bonds will break simultaneously for hair to lose its curl. On a very dry day, enough static electrical charge can build up for individual hairs to repel each other. Different people have "bad hair" on different days because the bonding patterns in their hair vary.

For more permanent curling, a more permanent chemical change in the hair structure is needed. The disulfide bonds between protein chains hold hair in its natural shape. In "permanent waving," these cross links are broken by a reducing agent (Step 1, Figure 15.18), which relaxes the tension. An oxidizing agent generates new cross links, and the hair retains the shape of the roller, so it appears curly.

The most commonly used hair-waving reducing agent is the ammonium salt of thioglycolic acid ($HSCH_2COO^-NH_4^+$). A typical waving solution contains 5.7% thioglycolic acid to break disulfide bonds, 2.0% ammonia to disrupt ionic bonds, and 92.3% water. The usual oxidizing agent used to re-form the disulfide bonds in the second step of the reaction is hydrogen peroxide (H_2O_2) or a perborate compound that releases hydrogen peroxide in solution ($NaBO_3 \cdot 4\,H_2O$).

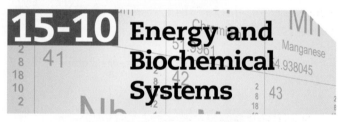

15-10 Energy and Biochemical Systems

The energy to sustain all but a few forms of life comes from the Sun. During photosynthesis, green plants absorb energy from the Sun to make glucose and oxygen from carbon dioxide and water. Glucose is a major energy source for all living organisms. During metabolism, much of the protein, fat, and carbohydrate in our diets is converted to glucose (Section 16-2). The energy stored in glucose is eventually transferred to the bonds in molecules such as adenosine triphosphate (ATP, Figure 15.19). When living organisms need energy, phosphate bonds in the ATP molecules are hydrolyzed to give adenosine diphosphate (ADP, Figure 15.20) and energy for other biochemical reactions.

In the complex process of photosynthesis, carbon dioxide is reduced to make sugar, and water is oxidized to oxygen:

$$6\,CO_2 + 6\,H_2O + 688\ kcal \longrightarrow$$

Carbon   Water   Energy
dioxide   (sunlight)

$$C_6H_{12}O_6 + 6\,O_2$$

Glucose   Oxygen

FIGURE 15.19 Stylized structure (*left*) and molecular structure (*right*) of adenosine triphosphate (ATP).

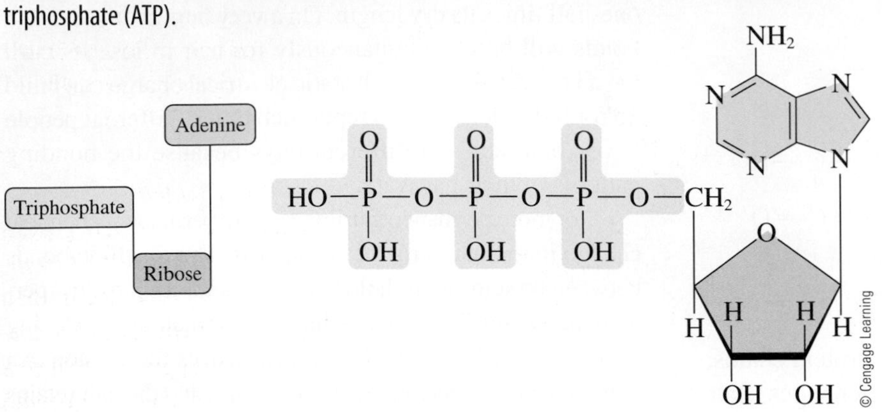

FIGURE 15.20 Hydrolysis of ATP to produce ADP (adenosine diphosphate).

Adenosine diphosphate
(ADP)

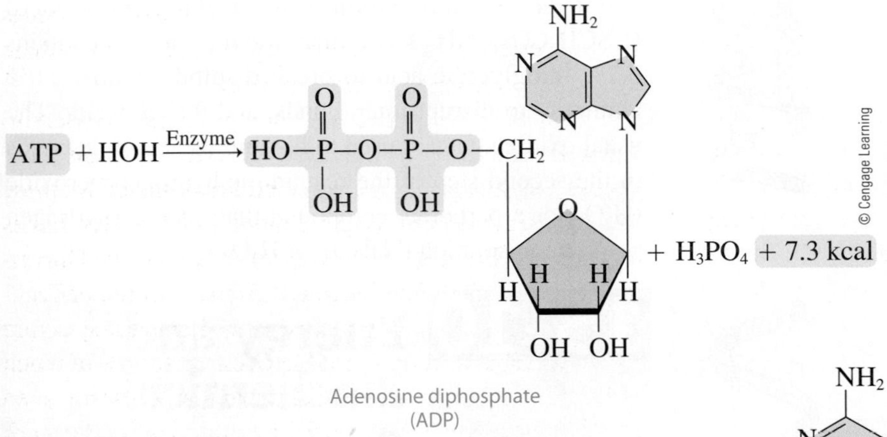

Nucleic acids: Polymers of nucleotides that contain deoxyribose or ribose, nitrogen bases, and phosphate groups

Deoxyribonucleic acid (DNA): Nucleic acid that functions as a genetic information storage molecule

Ribonucleic acid (RNA): Nucleic acid that transmits genetic information and directs protein synthesis

---

specialized enzymes that catalyze their synthesis and decomposition, the nucleic acids constitute a remarkable system that accurately copies millions of pieces of data with very few mistakes.

Like polysaccharides and polypeptides, nucleic acids are condensation polymers. Each monomer in these polymers includes one of two simple sugars, one phosphoric acid group, and one of a group of heterocyclic nitrogen compounds that behave chemically as bases. A particular nucleic acid is a **deoxyribonucleic acid (DNA)** if it contains the sugar 2-deoxy-D-ribose, and it is a **ribonucleic acid (RNA)** if it contains the sugar D-ribose.

The five organic bases that play a key role in the mechanism for information storage are *adenine* (A), *guanine* (G), *thymine* (T), *cytosine* (C), and *uracil* (U). These bases are mentioned so often in any discussion of nucleic acid chemistry that to save space they are usually referred to only by the first letter of each name.

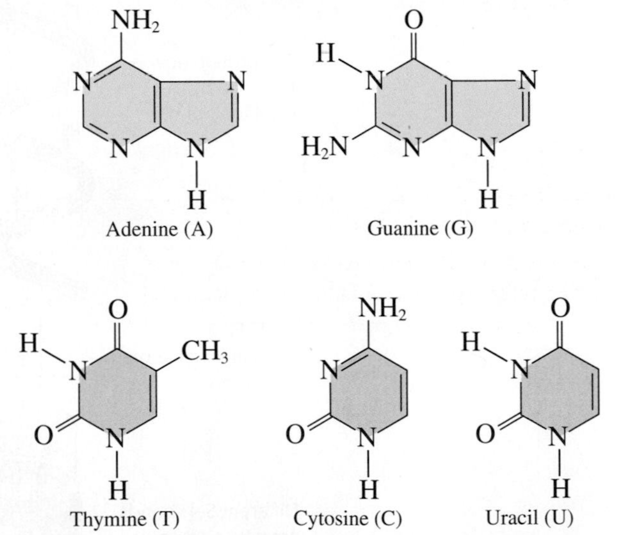

Adenine (A)   Guanine (G)

Thymine (T)   Cytosine (C)   Uracil (U)

Nucleic acids are found in all living cells, with the exception of the red blood cells of mammals. DNA occurs primarily in the nucleus of the cell, and RNA is found mainly in the cytoplasm, outside the nucleus. There are three major types of RNA, each with its own characteristic size, base composition, and function in protein synthesis (as described later in this section): messenger RNA (mRNA), transfer RNA (tRNA), and ribosomal RNA (rRNA).

The oxygen produced in photosynthesis is the continuing source of all the oxygen in our atmosphere. We are dependent on the plant life of our planet, and we must live in balance with the oxygen output of that plant life as well as with the food output of the same plant life. Photosynthesis is absolutely vital to life on Earth, and this endergonic reaction requires energy from the Sun.

15-11 Nucleic Acids

The genetic information that makes each organism's offspring look and behave like its parents is encoded in molecules called **nucleic acids**. Together with a set of

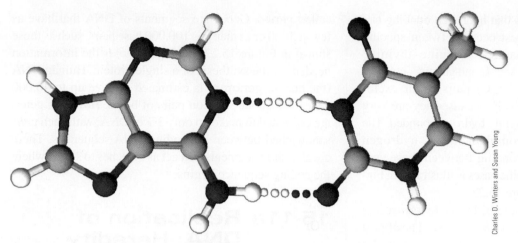

Hydrogen bonding between two of the nucleic acid bases, thymine and adenine.

Charles D. Winters and Susan Young

The monomers that polymerize to make both DNA and RNA are known as **nucleotides**. They have structures like the one shown in Figure 15.21a. Each nucleotide contains a phosphoric acid unit, a ribose sugar unit, and one of the five bases. The nucleotides of DNA and RNA have two structural differences: (1) they contain different sugars (deoxyribose and ribose); and (2) the base uracil occurs only in RNA, whereas the base thymine occurs only in DNA. The other bases—adenine, guanine, and cytosine—are found in both DNA and RNA.

Polynucleotides have molar masses ranging from about 25,000 for tRNA molecules to billions for human DNA. The sequence of nucleotides in the polymer chain (as shown by the base sequence) is its primary structure. Polynucleotides are formed by the polymerization of nucleotides to make esters. As an example, Figure 15.21b shows three monomers condensed to a trinucleotide.

To understand how DNA functions, it is necessary to know its structure. In the early 1950s three teams were focused on this goal: Drs. Rosalind Franklin and Maurice Wilkins, separately, at King's College, London; Drs. James Watson and Francis Crick together at Cambridge University; and Dr. Linus Pauling at California Institute of Technology. In 1953, Pauling published an incorrect structure of DNA. What occurred next has been the source of much controversy. It is known that Watson visited King's College in an effort to unify the work of his, Wilkins's, and Franklin's laboratories. During his visit, but without Franklin's knowledge, Watson was shown data that she had acquired. These data, now known as Photo 51, provided Watson and Crick with some of the information needed to propose an accurate structure for DNA. In 1953, Watson and Crick published their hypothesis for the structure of DNA and revolutionized our understanding of heredity and genetic diseases. Watson, Crick, and Wilkins were jointly awarded the Nobel Prize in Physiology or Medicine in 1962. Rosalind Franklin was not a recipient of the prize, as she died in 1958. Nobel Prizes are not awarded posthumously. It was not until the early 2000s that Dr. Franklin's contribution was widely acknowledged. The published structure of DNA included pairs of polynucleotides arranged in a double helix, stabilized by hydrogen bonding between the base groups lying opposite each other in the two chains. The critical point of the Watson–Crick model

FIGURE 15.21

(a) A nucleotide.

Because its sugar unit is ribose, this is a ribonucleotide.

(b) Bonding in a trinucleotide.

The sugars could also be deoxyribose units, and the bases can be any of the five bases shown in the text. The primary structures of both DNA and RNA are extensions of this structure.

Adenine unit— a nitrogenous base

Phosphoric acid unit

Ribose unit—a simple sugar

(a)

Base 1

Base 2

Base 3

(b)

© Cengage Learning

is that hydrogen bonding can best occur between specific bases. Adenine–thymine (A — T) and guanine–cytosine (G — C) pairs occur exclusively because they are very tightly hydrogen bonded. The complementary hydrogen bonding between these specific bases is illustrated in Figure 15.22.

The function of polynucleotides is to transcribe hereditary information so that *like begets like*. The almost infinite variety of primary structures of polynucleotides likewise allows an almost infinite variety of information to be recorded in the molecular structures of the strands of nucleic acids. The different arrangements of just a few bases produce a wide diversity of structures. In a somewhat similar fashion, multiple arrangements of just a few language symbols convey the many ideas in this book. The coded information in the polynucleotide controls the inherited characteristics of the next generation as well as most of the continuous life processes of the organism.

Twenty-three double-stranded DNA pairs form the 46 human chromosomes, which have specialty heredity areas

FIGURE 15.22 Complementary hydrogen bonding in the DNA double helix.

Hydrogen bonds in the thymine–adenine (T — A) and cytosine–guanine (C — G) pairs stabilize the double helix.

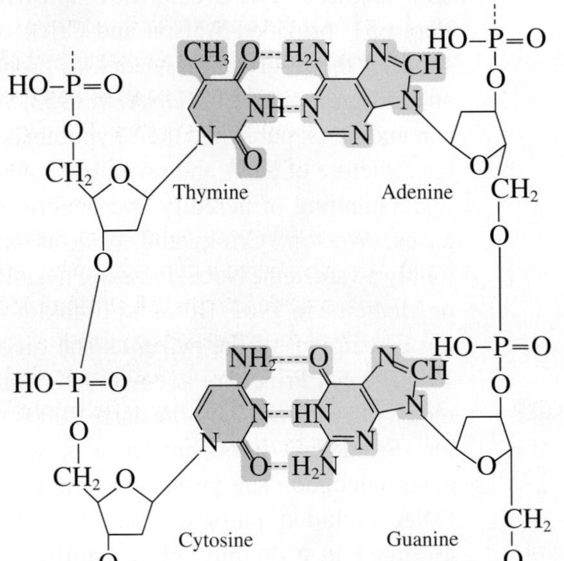

called **genes**. Genes are segments of DNA that have as few as 1000 or as many as 100,000 base pairs, such as those shown in Figure 15.22. Each gene holds the information needed for the synthesis of a single protein. Human DNA (the human **genome**) is estimated to have up to 40,000 genes and about 3 billion pairs of bases. However, genes are estimated to make up only 3% of DNA, with each gene sandwiched between noncoding DNA sequences. There are also short segments that act as switches to signal where the coding sequence begins.

15-11a Replication of DNA: Heredity

The transfer of coded information begins with the replication of DNA and continues with natural protein synthesis. Almost all nuclei in an organism's cells contain the same chromosomal composition. This composition remains constant regardless of whether the cell is starving or has an ample supply of food materials. Each organism begins life as a single cell with this same chromosomal composition; in sexual reproduction, half of a chromosome comes from each parent. The DNA structure is faithfully copied during normal cell division (mitosis—both strands); only half is copied in cell division that produces reproductive cells (meiosis—one strand).

In **replication**, the double helix of the DNA structure unwinds, and each half of the structure serves as a template, or pattern, from which the other complementary half can be reproduced from the molecules in the cell environment (Figure 15.23). Replication of DNA occurs in the nucleus of the cell before the cell divides.

15-11b Natural Protein Synthesis

The proteins of the body are continually being replaced and resynthesized from the amino acids available to the body. The use of isotopically labeled amino acids has made possible studies of the average lifetimes of amino acids as constituents in proteins—that is, the time it takes the body to replace a protein in a tissue. Considering that this process must be extremely complex, replacement is very rapid. Only minutes after radioactive amino acids are injected into animals, radioactive protein can be found. Although all of the proteins in the body are continually being replaced, the rates of replacement vary. Half the proteins in the liver and plasma are replaced in 6 days; the time needed for replacement of muscle proteins is about 180

© Cengage Learning

FIGURE 15.23 Replication of DNA.

When the double helix of DNA (blue) unwinds, each half serves as a template on which to assemble nucleotides to form new DNA strands.

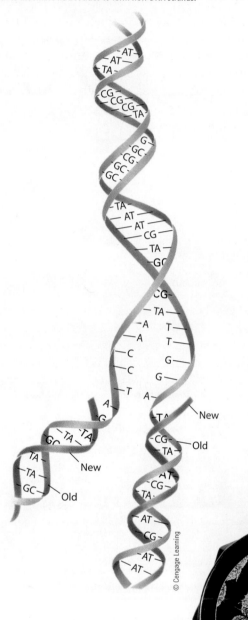

New

Old

New

Old

© Cengage Learning

15.24. First, messenger RNA, like all forms of RNA, is synthesized in the cell nucleus. The sequence of bases in one strand of the chromosomal DNA serves as the template from which a single strand of a messenger ribonucleotide (mRNA) is made in a process known as **transcription**. The bases of the mRNA strand complement those of the DNA strand. Two bases are complementary when each one fits the other and forms one or more hydrogen bonds. Messenger RNA contains only the four bases: adenine (A), guanine (G), cytosine (C), and uracil (U). DNA contains the four bases adenine (A), guanine (G), cytosine (C), and thymine (T). The bases in DNA are transcribed into bases in mRNA as follows:

DNA	mRNA
A	U
G	C
C	G
T	A

This means that, provided the necessary enzymes and energy are present, wherever a DNA has an adenine base (A), the mRNA will transcribe a uracil base (U), and so on. After transcription, mRNA passes from the nucleus of the cell to a ribosome (which contains the rRNA), where mRNA serves as the template for the sequential ordering of amino acids during protein synthesis by the process known as **translation**. As its name implies, messenger RNA contains the sequence message, in the form of a three-base code (called a **codon**), for ordering amino acids into proteins (Table 15.6). Each of the thousands of different proteins synthesized by cells is coded by a specific mRNA or segment of an mRNA molecule.

Transcription: The process by which the information in DNA is read and used to synthesize RNA

Translation: The process for sequential ordering of amino acids that is directed by RNA during protein synthesis

Codon: A three-base sequence carried by mRNA that codes a specific amino acid in protein synthesis

days; and replacement of protein in other tissues, such as bone collagen, takes even longer.

Recall that each organism has its own kinds of proteins. The number of possible unique arrangements of 20 amino acid units is 2.43×10^{18}, yet proteins characteristic of a given organism can be synthesized by the organism in a matter of a few minutes.

The DNA in the cell nucleus holds the code for protein synthesis, which is carried out in a series of steps summarized in Figure

© iStockphoto.com/ claylib / Biophoto Associates/Science Source/Colorization by: Mary Martin

Human chromosomes (magnified about 8000 times). DNA assembles into chromosomes when a cell is about to divide.

FIGURE 15.24 Protein synthesis.

A section of DNA unwinds and transcription results in production of messenger RNA. Outside the cell nucleus, at a ribosome, the code carried by messenger RNA is translated into the amino acid sequence of a newly synthesized protein. Transfer RNA brings the amino acids into position one at a time.

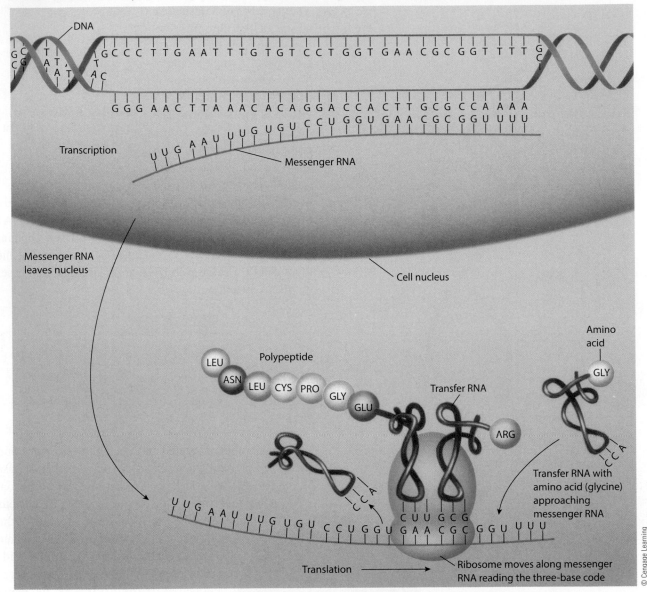

Transfer RNAs carry the amino acids to the mRNA one by one. Each of the 20 amino acids found in proteins has at least one corresponding tRNA, and some have multiple tRNAs (Table 15.6). Table 15.6 lists the RNA codes and shows in the first line that, for example, UUU is code for phenylalanine and UCU is code for serine.

In the schematic illustration in Figure 15.24, which summarizes DNA⟶RNA⟶protein, pick a three-base sequence on the bottom DNA strand of the two DNA strands at the top of the figure and follow it through to the bottom, where a tRNA attaches to a three-base codon on mRNA. Assume the DNA strand you have selected serves as the template for the synthesis of the single strand of

mRNA shown in the figure. Do you agree with the changes that occur in the letters representing the three-base sequence?

EXAMPLE 15.3 Nucleic Acids

→ If the base sequence in a DNA segment is ...GCT-GTA..., what is the base sequence in the complementary mRNA? What is the order in tRNA?

SOLUTION

The base pairs between DNA and mRNA are G...C, A...U, and T...A. Therefore, the complementary mRNA segment for a DNA segment of GCTGTA is determined by using

TABLE 15.6 Messenger RNA Codes for Amino Acids*

First Letter of Code	Second Letter of Code				Third Letter of Code
	U	C	A	G	
U	Phenylalanine	Serine	Tyrosine	Cysteine	U
	Phenylalanine	Serine	Tyrosine	Cysteine	C
	Leucine	Serine	STOP	STOP	A
	Leucine	Serine	STOP	Tryptophan	G
C	Leucine	Proline	Histidine	Arginine	U
	Leucine	Proline	Histidine	Arginine	C
	Leucine	Proline	Glutamine	Arginine	A
	Leucine	Proline	Glutamine	Arginine	G
A	Isoleucine	Threonine	Asparagine	Serine	U
	Isoleucine	Threonine	Asparagine	Serine	C
	Isoleucine	Threonine	Lysine	Arginine	A
	START or methionine	Threonine	Lysine	Arginine	G
G	Valine	Alanine	Aspartic acid	Glycine	U
	Valine	Alanine	Aspartic acid	Glycine	C
	Valine	Alanine	Glutamic acid	Glycine	A
	Valine	Alanine	Glutamic acid	Glycine	G

© Cengage Learning

* In groups of three (called *codons*), bases of mRNA code the order of amino acids in a polypeptide chain. A, C, G, and U represent adenine, cytosine, guanine, and uracil, respectively. Some amino acids have more than one codon, and hence more than one tRNA can bring the amino acid to mRNA.

these allowed base pairs. Start with GCTGTA, and place the correct base below it. The resulting mRNA segment is CGACAU. The order in tRNA is the complement of the mRNA segment. The allowed base pairs are G...C and A...U, so the order in tRNA is GCUGUA.

Your Resources

In the back of the textbook:

→ **Review Card on The Chemistry of Life**

   In OWL for CHEM2 at www.cengagebrain.com

→ **Review Key Terms with Flash Cards (printable or digital)**

→ **Complete Interactive Practice Quizzes to prepare for tests**

→ **Submit Assigned Homework and Exercises**

1. What is meant by the word *chiral*?
2. Which of these objects is chiral?
   (a) A baseball
   (b) The human body
   (c) A sphere
   (d) A baseball glove
   (e) Scissors
   (f) The Statue of Liberty
   (g) The Washington Monument
   (h) A double helix
   (i) A nucleotide
   (j) Glycine
   (k) Proline
3. What do the following terms mean?
   (a) L-Isomers   (b) D-Isomers
   (c) Optically active   (d) Polarimeter
4. Identify the following statements about chiral molecules as true or false. If false, explain.
   (a) All chiral substances rotate plane-polarized light.
   (b) Enantiomers have the same physical properties.
   (c) The direction in which plane-polarized light is rotated by a chiral molecule allows us to determine the arrangement of groups surrounding a stereogenic center.
   (d) All D-sugars are dextrorotatory.
5. Do the structures in each pair represent molecules that are the same, different, or enantiomers?
   (a) 
   (b) 
   (c) 
6. How many stereogenic centers are contained within the structure of each molecule shown here?

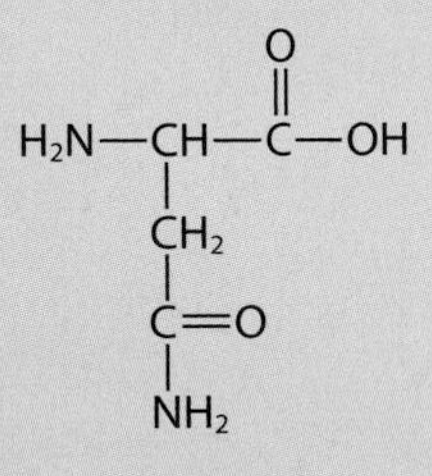

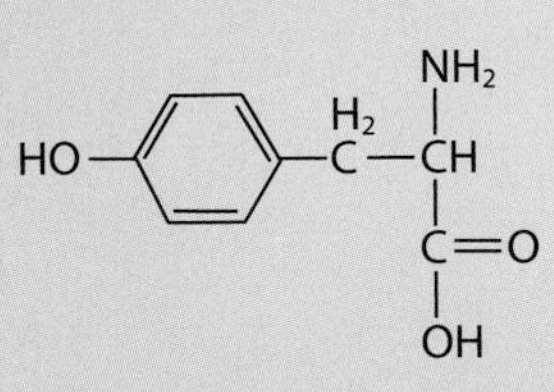

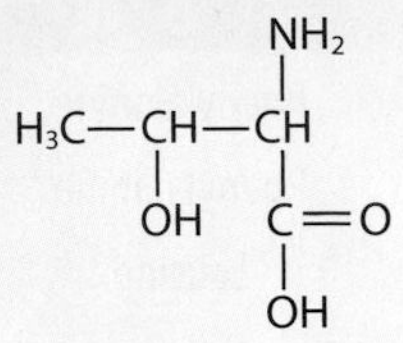

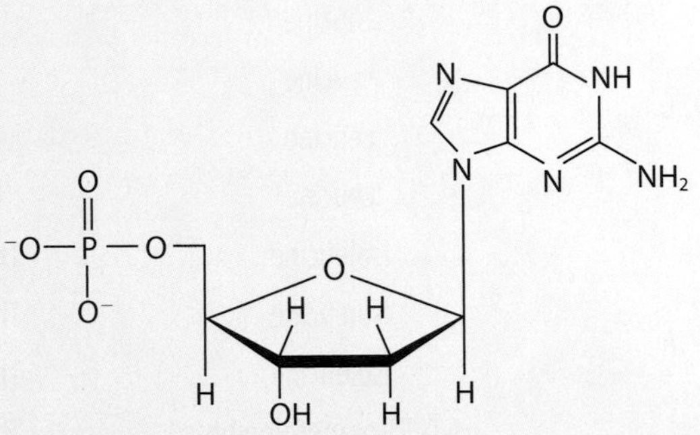

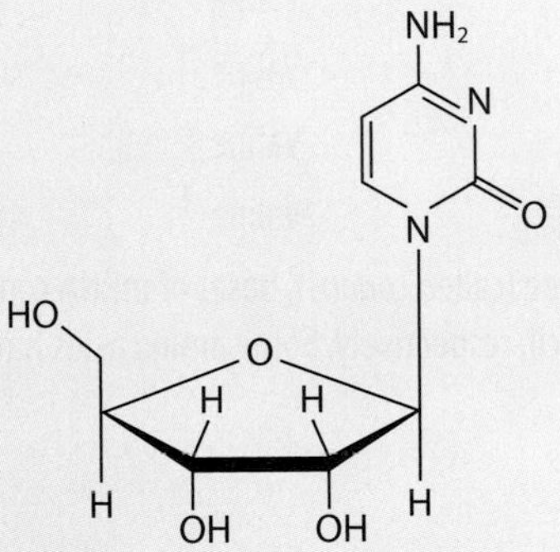

7. Which of the following molecules is chiral? Draw the enantiomer for any chiral molecule.
   (a) 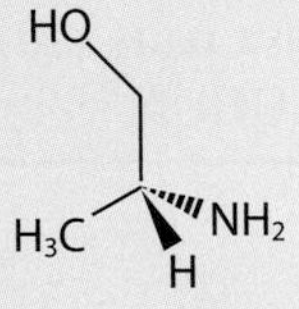
   (b) 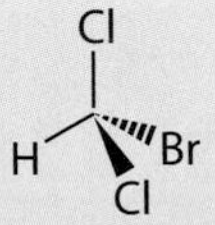
   (c)

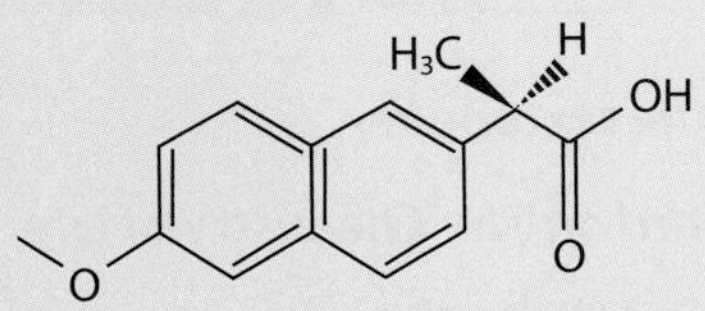

8. What is a racemic mixture?
9. Give definitions for the following:
   (a) Carbohydrate   (b) Monosaccharide
   (c) Disaccharide   (d) Polysaccharide

10. What is an artificial sweetener?

11. What are the following substances?
    (a) Starches   (b) Glycogen
    (c) Cellulose

12. What is a lipid?

13. How do the structures of unsaturated fatty acids and saturated fatty acids differ?

14. What is a partially hydrogenated fatty acid?

15. What is a *trans*-fatty acid?

16. Describe how a surfactant such as a soap or detergent molecule works.

17. What is the difference between a soap and a detergent?

18. Describe a micelle. Why do they form?

19. It is possible to mix oil and water and get a stable emulsion using a soap. It is also possible to make a stable oil-in-water emulsion using the yolk of an egg (mayonnaise). Explain how these two emulsifiers work.

20. Explain why synthetic detergents are more effective than soaps in hard water.

21. A certain cream imparts a somewhat greasy feel to the skin after its application. What kind of emulsion do you suspect this cream to be based on?
    (a) Water-in-oil   (b) Oil-in-water

22. A skin cream has its ingredients listed on the back of the jar. Water is listed *after* several other ingredients, some of which have the word *oil* in their names. On what kind of emulsion do you think this cream is based?

23. What is an amino acid? Draw and label a generalized structure for an amino acid.

24. What is a peptide bond?

25. What is an essential amino acid?

26. What functional groups are always present in each molecule of an amino acid?

27. What is meant by the following terms?
    (a) Dipeptide   (b) Polypeptide
    (c) Amino acid residue (d) Protein

28. What is meant by the term "amino acid sequence"?

29. Define "neuropeptides" and give two examples.

30. What are the meanings of the terms "primary," "secondary," "tertiary," and "quaternary" in reference to the structures of proteins?

31. Which of the following biochemicals are polymers?
    (a) Starch   (b) Cellulose
    (c) Glucose   (d) Fats
    (e) Glycylalanylcysteine (f) Proteins
    (g) DNA   (h) RNA

32. Classify each molecule as a fat, oil, steroid, wax, carbohydrate, amino acid, peptide, deoxyribonucleic acid, or ribonucleic acid.

(a)

(b)

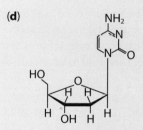

(c)

(d)

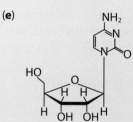

(e)

(f)

(g)

(h)

33. What is meant by the following terms?
    (a) α-Helix   (b) β-Pleated sheet
34. What does it mean to "denature" a protein?
35. All enzymes are proteins, but not all proteins are enzymes. What is the difference between these two amino acid–based classes of molecules?
36. What is the active site of an enzyme?
37. Which amino acids contain the element sulfur?
38. Which amino acids contain a hydroxyl functional group in their side chain?
39. Which amino acids contain a benzene-like ring in their side chain?
40. Which amino acids have a carboxylic acid functional group in their side chain?
41. Which amino acids have a hydrocarbon side chain?
42. Which amino acids contain an amide functional group in their side chain?
43. What term is used to describe the amide bond that links two amino acids together in a peptide?
44. Glutathione is an important tripeptide found in all living tissues. It is also named glutamylcystylglycine. Draw the structure of glutathione. Which enantiomeric forms of the amino acids would you predict are used to synthesize this peptide?
45. (a) How many tetrapeptides are possible if four amino acids are linked in different combinations that contain all four amino acids?
    (b) Write these combinations for tetrapeptides made from glycine, alanine, serine, and cystine. Use three-letter abbreviations for the amino acids. For example, one combination is Gly-Ala-Ser-Cys.
46. Use three-letter abbreviations to write the possible tripeptides that can be formed from phenylalanine, serine, and valine if each tripeptide contains all three amino acids.
47. How many peptides can be made from five different amino acids?
48. Name the following tripeptide:

49. Name the following peptide.

50. Draw the structure of alanylglycylphenylalanine.
51. Explain why the human body can metabolize D-glucose but not L-glucose.
52. How is it possible to identify a molecule as being a carbohydrate based on its molecular formula alone?
53. What is the structural difference between amylose and cellulose?
54. What is lactose intolerance?
55. Describe the two kinds of bonding that hold protein strands together in hair.
56. What is the effect of water on the natural protein bonding in hair?
57. What is the disulfide linkage, and what role does it play in protein structure?
58. What three molecular units are found in nucleotides?
59. What are the differences in structure between DNA and RNA?
60. What is the purpose of ATP?
61. What are the fundamental components of nucleic acids?
62. Explain the word *codon*.
63. What is meant by the term "complementary bases"?
64. How many hydrogen bonds are possible between the following complementary base pairs?
    (a) Cytosine and guanine   (b) Thymine and adenine
65. A segment of a DNA strand has the base sequence ...GCTGTAACCGAT...
    (a) What is the base sequence in the complementary mRNA?
    (b) What is the order in tRNA?
    (c) Consult Table 15.6 and give the amino acid sequence in the portion of the peptide being synthesized.
66. A segment of a DNA strand has the base sequence ...TGTCAGTGGGCCGCT...
    (a) What is the base sequence in the complementary mRNA?
    (b) What is the order in tRNA?
    (c) Consult Table 15.6 and give the amino acid sequence in the portion of the peptide being synthesized.

67. Translate the DNA sequence GGGAACTTA into:
    (a) The mRNA sequence
    (b) The amino acid sequence

68. Write the overall equation for photosynthesis.

69. What stabilizing forces hold the double helix together in the secondary structure of DNA proposed by Watson and Crick?

70. (a) What are the three major types of RNA?
    (b) What are their functions?

71. A base sequence in a DNA segment is ...GTAGC... What is the base sequence in the complementary mRNA?

72. (a) What are the three bases present in both messenger RNA and DNA?
    (b) What is the fourth base found in mRNA?
    (c) What is the fourth base found in DNA but not in mRNA?

73. Give the mRNA complementary base sequence that matches the following DNA segment base sequence:

DNA sequence	mRNA sequence
G...	
A...	
C...	
A...	

74. Many of the chapters in this text include a mention of Nobel Prizes. Which are mentioned in this chapter, and for what were they awarded?

75. This problem has you translating backward. Starting with the amino acid sequence proline-glutamic acid-threonine-glutamic acid, derive the mRNA code and the DNA.

MEGAN
CLAIRE

16

Nutrition:
The Basis of Healthy Living

We are living in a health-conscious society. Websites, magazines, and television programs constantly offer us advice on what to eat. The advice ranges from responsible journalism to scare tactics and sensationalism. Fads about what to eat and what not to eat are rapidly translated into marketable products. And advertisements bombard us from every direction with information about food. The advertisements and food stories in the media are bound to attract our attention. After all, many times every day, we make choices about food.

How can we sort out the information so that these choices are good ones? The task is impossible without knowing something about the basic principles of nutrition, which requires starting with the fundamentals of biochemistry presented in Chapter 15.

Nutrition is the science that deals with diet and health. The old saying "we are what we eat" is true. The skin that covers us now is not the same skin that covered us seven years ago. The fat beneath our skin is not the same fat that was there just a year ago. Our oldest red blood cells are 120 days old. The entire lining of our digestive tract is renewed every three days. Many chemical reactions are required to replace these tissues, and the energy and raw materials for these reactions are supplied by what we eat.

Nutrition, then, is concerned with the chemical requirements of the body—the nutrients and the chemical energy we get from them. **Nutrients** are chemical substances in foods that provide the energy and raw materials required by biochemical reactions. *The five classes of nutrients are carbohydrates, fats, proteins, vitamins, and minerals.* In addition, an average adult needs about 2.5 L of water each day.

Nutrition: The science of diet and health

Nutrient: Chemical substance in food that provides energy and raw materials required by chemical reactions

16-1 Digestion: It's Just Chemical Decomposition

Even before you swallow your lunch, chemical reactions begin the process of **digestion**. Like all biochemical processes, digestion is under the control of enzymes (biochemical catalysts, Section 15-8). Amylase, an enzyme in saliva, begins the digestion of the carbohydrates in starch. Carbohydrates are polymers of simple sugars (Section 15-2), and their digestion is the reverse of their formation—bonds are broken to set free the simple sugars. The same is true

Digestion: The breakdown of food into substances small enough to be absorbed from the digestive tract

Foods like these can provide all five classes of nutrients.

for digestion of proteins (polymers of amino acids) and the triglycerides in fats and oils (composed of glycerol and fatty acids).

$$\text{Carbohydrates} \xrightarrow{\text{Digestion}} \text{Simple sugars (monosaccharides)}$$

$$\text{Proteins} \xrightarrow{\text{Digestion}} \text{Amino acids}$$

$$\text{Fats (triglycerides)} \xrightarrow{\text{Digestion}} \text{Glycerol + fatty acids}$$

Digestion begins in the mouth, continues in the stomach, and is completed in the small intestine. By the time a meal has been digested, all of the nutrient molecules too large to cross cell membranes and enter the bloodstream have been converted into smaller molecules and absorbed.

Suppose you have a peanut butter and jelly sandwich for lunch. Digestion of starch from the bread begins as you chew and continues in your stomach. It is completed in the small intestine, where sugars from the jelly are also converted to monosaccharides. The indigestible carbohydrates (dietary fiber) from the bread and peanut butter pass unchanged through the small *and* large intestines.

Digestion of protein from the bread and peanut butter starts in your stomach, where large polypeptides are broken down into smaller ones. The job is finished in the small intestine, where the individual amino acids enter the bloodstream. Lipids, mainly fats and oils from the peanut butter, are digested mainly in the small intestine.

Once digestion is finished, what happens to the small molecules that have been produced? They may undergo (1) complete breakdown to produce energy, carbon dioxide, and water; (2) recycling into new biomolecules; or (3) placement into storage as triglycerides (fat) for future use (Figure 16.1). The energy yield from foods is described later in this chapter. The following sections examine the roles of the nutrients in our diets.

16-2 Sugar and Polysaccharides: Digestible and Indigestible

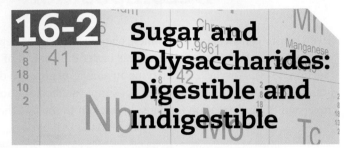

The major kinds of digestible carbohydrates in foods are the simple sugars (glucose and fructose), disaccharides (sucrose, maltose, and lactose), and polysaccharides (amylose and amylopectin in starch from plants, and glycogen from meat). The indigestible carbohydrates include cellulose and its derivatives, pectin (the substance that makes jam and jelly gel), and plant gums.

Glucose is our major fuel molecule. It is the end product of carbohydrate digestion, and if more is needed, there are biochemical pathways for making it from stored fat and even protein. These pathways are essential because glucose is the only fuel that the brain and red blood cells can use. To provide energy, glucose molecules are "burned" throughout our bodies by a tightly integrated series of biochemical reactions whose final products are carbon dioxide and chemical energy stored in adenosine triphosphate (ATP), whose structure was given in Figure 15.19. ATP is often called the body's energy currency. It carries energy to be spent wherever in the body energy is needed.

FIGURE 16.1 The release and use of chemical energy in the breakdown and synthesis of biomolecules.

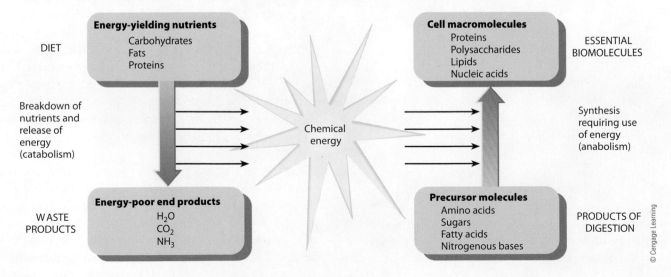

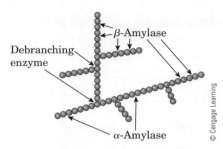

FIGURE 16.2 Digestion of glycogen.

Polysaccharides like glycogen, shown in Figure 16.2 as a series of circles representing individual glucose monomers, is digested beginning in the mouth. There, the enzyme alpha-amylase begins to "prune" the branches, breaking glycogen into smaller polysaccharides. Further digestion is accomplished in the stomach, aided by the highly acidic environment. Last, two enzymes in the small intestine, a debranching enzyme and beta-amylase, break down any remaining polysaccharides to glucose. The glucose is then transported across the intestinal lining and into the bloodstream.

The indigestible polysaccharides are collectively referred to as **dietary fiber**. All dietary fiber comes from plants. There is insoluble fiber, mainly from the structural cellulose parts of plants, and soluble fiber—the gums and pectins. Barley, legumes, apples, and citrus fruits are foods with a high content of gums and pectins.

It has long been known that dietary fiber prevents constipation. Evidence has been accumulating that dietary fiber has significant other benefits. Adding soluble fiber to the diet decreases blood cholesterol levels, thereby decreasing the risk of heart disease. There is also evidence of a connection between insufficient dietary fiber and colorectal cancer. In our increasingly health-conscious society, such information provides a marketing advantage. We have all seen evidence of this in the bran muffins now found on every breakfast counter.

Dietary fiber: Food polysaccharides that are not broken down by digestion

16-3 Lipids: Mostly Fats and Oils

Most of the lipids in the diet, which we usually refer to as "fats," are triglycerides. We get them from meats and fish, vegetables and vegetable oils, and dairy products. They may be solid fats or oils, and they may incorporate saturated or unsaturated fatty acids (Section 12-3). The fat content of some common foods is shown in Figure 16.3.

Fatty tissue is composed mainly of specialized cells, each featuring a large globule of triglycerides. When their energy is needed, the triglycerides in fat cells are hydrolyzed to give glycerol and free fatty acids, which leave the cells and are transported to the liver. There, the fatty acids are broken down to two-carbon molecular fragments that can enter the main energy-producing pathway.

A fat content of 20% to 35% is strongly recommended in the 2005 dietary guidelines for Americans released by the U.S. Department of Health and Human Services. The fat in today's diet is about 40% saturated, 40% monounsaturated, and 20% polyunsaturated. Lowering the saturated and monounsaturated fat and raising the polyunsaturated fat content of the diet is also strongly recommended. What is the basis for these recommendations? Heart disease is the primary cause of death in the United States (Section 17-1), and *atherosclerosis*, the buildup of fatty deposits called *plaque* on the inner walls of arteries, reduces the flow of blood to the heart. If a coronary artery is blocked by plaque, a heart attack occurs as a result of the reduced blood flow carrying oxygen to the heart. About 98% of all heart attack victims have atherosclerosis, and the major components of atherosclerotic plaque are saturated fatty acids and cholesterol.

The relation between blood levels of cholesterol and heart disease is well established. The more saturated fats and cholesterol in the diet, the higher the blood cholesterol is likely to be. Cholesterol (a lipid) has a waxy consistency, so to be transported in the bloodstream it must bond to a more water-soluble substance. Cholesterol combines with

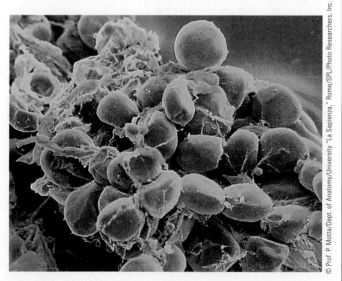

Where the fat goes. The yellow cells are adipocytes, the fat-storing cells of our bodies. Each adipocyte is almost entirely composed of a single droplet of triglycerides.

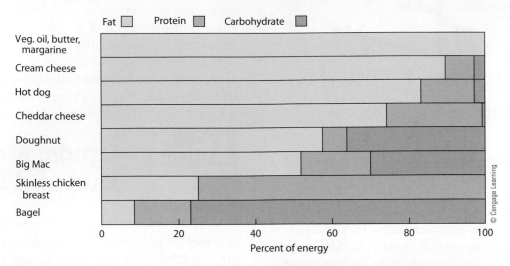

Fat ☐ Protein ☐ Carbohydrate ☐

Veg. oil, butter, margarine
Cream cheese
Hot dog
Cheddar cheese
Doughnut
Big Mac
Skinless chicken breast
Bagel

Percent of energy

© Cengage Learning

Lipoprotein: An assemblage of lipids, cholesterol, and water-soluble proteins

Low-density lipoprotein (LDL): A lipoprotein that transports cholesterol from the liver to tissues throughout the body

High-density lipoprotein (HDL): A lipoprotein that transports cholesterol from body tissues to the liver

proteins to form **lipoproteins** (Figure 16.4), which are water soluble because of their many $-NH_3^+$ and $-COO^-$ ions. About 65% of the cholesterol in the blood is carried by **low-density lipoproteins (LDLs)**, whereas about 25% is carried by **high-density lipoproteins (HDLs)**. (The density difference is caused by the ratios of lipid to protein.)

LDLs are the "bad" cholesterol and HDLs are the "good" cholesterol referred to in discussions of heart disease. LDLs transport cholesterol away from the liver and throughout the body; they are therefore "bad" because they distribute cholesterol to arteries, where it can form the deposits of atherosclerosis. HDLs are "good" because they transport excess cholesterol from body tissues to the liver, where it is converted to bile acids that are needed in digestion.

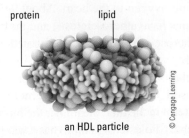

protein   lipid

an HDL particle

© Cengage Learning

FIGURE 16.4 Structure of an HDL particle, showing the combination of protein and lipid.

16-4 Proteins in the Diet

Meat, fish, eggs, cheese, and beans are high-protein foods. The major function of dietary protein is to provide amino acids for new protein synthesis. Proteins are also the major dietary source of nitrogen for the synthesis of other kinds of nitrogen-containing biomolecules. Although the air we breathe is 78% nitrogen gas, this nitrogen cannot be used directly for biosynthesis by humans.

The beginning stage of the protein digestion process takes place outside the body, if the food is cooked. Heating causes proteins to denature (Section 15-8) and makes them somewhat more easy to digest. Once in the stomach, the acidic environment plus the action of the enzyme pepsin begin to break the polypeptide chains into smaller fragments. In the small intestine a set of enzymes further degrades the peptides to individual amino acids, and these pass into the intestinal lining and are absorbed into the bloodstream (Figure 16.5).

Excess amino acids from dietary proteins are not stored in the body. The nitrogen is converted to ammonia and then excreted as urea, while the carbon atoms are cycled to glucose and energy generation or stored as fat (Figure 16.6). Therefore, it is necessary to eat some protein every day.

The average amount of protein in the American diet is, however, well beyond what is necessary, and a protein-deficient diet is rare in the United States. *Protein–energy malnutrition*, a group of disorders due to various combinations of deficient protein and energy intake, is most common in children in underdeveloped countries.

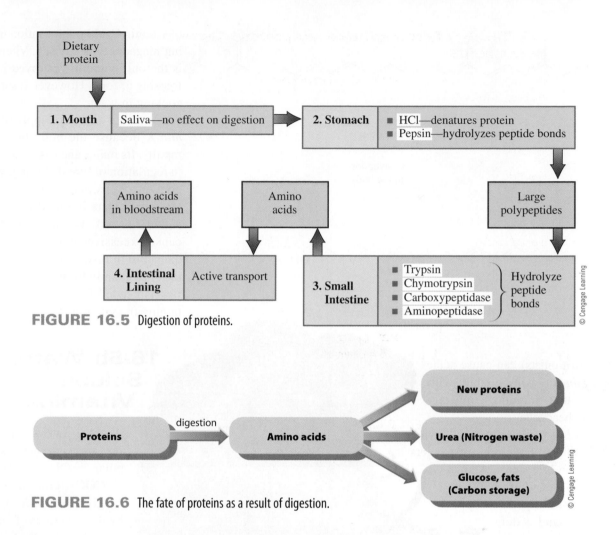

FIGURE 16.5 Digestion of proteins.

FIGURE 16.6 The fate of proteins as a result of digestion.

16-5 Vitamins in the Diet

A **vitamin** is an organic compound essential to health that must be supplied in small amounts by the diet. Vitamins provide no energy and are unchanged by digestion. The body does not synthesize vitamins but needs them to function as **coenzymes**, nonprotein molecules that make enzyme function possible. There are two kinds of vitamins: fat soluble and water soluble.

The fat-soluble vitamins—A, D, E, and K—can be stored in the fatty tissues of the body (especially the liver). They have nonpolar hydrocarbon chains and rings that are compatible with nonpolar oils and fats. It is important to store enough fat-soluble vitamins but not too much, as excesses are not excreted and can be toxic.

The water-soluble vitamins are the eight B vitamins and vitamin C. Because excesses are excreted rather than stored, it had previously been assumed that overdosing on water-soluble vitamins has no toxic effects. With the rise in popularity of megadoses of vitamins, some toxic reactions have been observed. It does, however, take much greater quantities of water-soluble vitamins than fat-soluble vitamins to create harmful effects.

> **Vitamin:** Organic compound essential to health that must be supplied in small amounts by the diet
>
> **Coenzyme:** A nonprotein molecule that makes enzyme function possible

16-5a Fat-Soluble Vitamins

Vitamin A, also known as retinol, is essential to vision because it is a component of vision receptors in the eye. Carrots, long associated with good vision, provide beta-carotene, not vitamin A itself. As it passes through the intestinal wall, beta-carotene is converted into vitamin A. Vitamin A also aids in the prevention of infection by barring bacteria from entering and passing through cell membranes. A single dose of vitamin A greater than 200 mg (660,000 international units [IU], a unit for vitamin doses

Beta-carotene
(orange pigment in carrots)

Conversion
in the body

Vitamin A
(retinol)

of vitamin E, although caution in taking megadoses is advised. Vitamin E is the only vitamin destroyed by the freezing of food. However, it survives cooking in water.

We need a steady supply of *vitamin K* because the body uses it up rapidly. Its major and essential role is in regulation of blood clotting, which goes on to repair not only cut skin but also small tears in blood vessels (on a daily basis). Because overdoses cause excessive blood clotting and danger of brain damage, vitamin K is the only vitamin for which a prescription is needed for supplements containing the vitamin alone.

used by nutritionists) can cause acute *hypervitaminosis A* in an adult, with symptoms that include vomiting, fatigue, and headache.

Vitamin D is produced when ultraviolet light (UV) shines on the skin and triggers the conversion of a steroid known as *ergosterol* to vitamin D. Its major role is to help the body use calcium, and a deficiency causes rickets in children, the same condition caused by calcium deficiency. Vitamin D supplements are rarely needed, except by those who are almost never exposed to the Sun. Both vitamin A and vitamin D are essential to normal growth and development. Overdosage of vitamin D can have serious consequences, however. Calcium deposits can form in the kidney, lungs, or tympanic membrane of the ear (leading to deafness). Infants and small children are especially susceptible to vitamin D toxicity.

Vitamin E is now well established as an *antioxidant* (Section 10-3). It is particularly effective in preventing the oxidation of polyunsaturated fatty acids to form peroxides (which contain —O—O— groups). Perhaps this is why vitamin E is always found distributed among fats in nature. The fatty acid peroxides are particularly damaging because they can cause runaway oxidation in the cells. Vitamin E protects the integrity of cell membranes. No toxicity has been specifically associated with large doses

Broccoli is a good source of vitamin K.

© iStockphoto.com/LizV

16-5b Water-Soluble Vitamins

All of the water-soluble vitamins are coenzymes that have polar —OH, —NH₂, and —COOH groups, as illustrated on the following page by the structures of vitamin C and several of the B vitamins.

Once thought to be a single substance, the eight *B vitamins* are often found together and interact so that a deficiency of one can cause deficiencies of others. The B vitamins are B_6, B_{12}, folic acid, niacin, thiamin, riboflavin, biotin, and pantothenic acid. Vitamin B_6, considered the master vitamin, is known to be involved in 60 enzymatic reactions, many in the metabolism and synthesis of proteins. It is also needed for the synthesis of hemoglobin and the white blood cells of the immune system. Consuming more than 60 times the recommended daily quantity of vitamin B_6 causes nerve damage, and in huge doses (2–6 g per day) it can cause paralysis.

Large doses of niacin (50 mg per day or more) can also be toxic, with reactions ranging from skin rash and nausea to abnormal liver function and

Vitamin B_6

Niacin

changes in heart rhythm. Niacin deficiency is common in countries with a corn-based diet.

When the hulls of wheat, rice, and other grains are removed and the kernels ground to produce flour, a large proportion of riboflavin and certain other vitamins and minerals is lost, as is the soluble fiber. To counteract this loss, since the 1940s flour sold in the United States has been *enriched* by adding back some of the lost vitamins and minerals. Customarily, enriched wheat flour contains added riboflavin, thiamin, niacin, and iron. When enriched flour came into use in the United States, the incidence of pellagra, the niacin deficiency disease, almost disappeared.

Thiamine

Riboflavin

In 1996, based on growing scientific evidence, the Food and Drug Administration (FDA) ruled that folic acid should also be added to enriched flour. Folic acid can reduce risks of certain birth defects, anemia, and heart disease.

Vitamin C, also known as *ascorbic acid*, helps destroy invading bacteria; aids the synthesis and activity of interferon, which prevents the entry of viruses into cells; and combats the ill effects of toxic substances, including drugs and pollutants. It also aids in the synthesis of collagen, which is present in cartilage, bone, tendons, and connective tissue, and is therefore important in the healing of wounds and for infants and pregnant women. In addition, vitamin C is an antioxidant.

Vitamin C

The role of vitamin C in preventing the common cold has long been debated, and numerous studies have found it to be either effective or ineffective. Currently, there is more evidence in favor of its ability to decrease the severity of cold symptoms than for any ability to prevent colds. It is reported, however, that one-third of the U.S. population takes vitamin C supplements to ward off colds. Daily doses of 1 g or more can cause diarrhea, nausea, and abdominal cramps in some people. Vitamin C is a strong enough acid to damage tooth enamel if vitamin capsules are chewed.

Diets low in vitamin C result in a condition called *scurvy*, characterized in its early stages by lack of energy, and which can ultimately result in death. This disease was known to affect sailors (including pirates) during long voyages and was also responsible for the failure of more than one expedition to the South Pole. Diets rich in protein and grain but low in fresh fruits and vegetables were to blame. It was not until 1932 that the link between vitamin C deficiency and scurvy was made.

Mineral: As a dietary nutrient, an element other than carbon, hydrogen, nitrogen, and oxygen that is needed for good health

16-6 Minerals in the Diet

As nutrients, **minerals** are elements, other than carbon, hydrogen, nitrogen, and oxygen, that are needed for good health. Although referred to on labels as "potassium" or "iodine" and so on, most minerals are present in foods, food supplements, and our bodies in ionic form and are not present in their elemental state. Not only does the human body need minerals, but these minerals must be maintained in balanced amounts, with no deficiencies and no excesses. Many of the body's minerals are excreted daily and must therefore be replenished each day.

The *seven macronutrient minerals* make up about 4% of body weight. They are *calcium, phosphorus, magnesium, sodium, potassium, chlorine,* and *sulfur*; their sources and functions are listed in Table 16.1. Except for the role of long-term calcium deficiency in osteoporosis, deficiencies of these minerals are rare because of their abundance in a variety of foods.

Sodium, potassium, and *chloride*, as ions (Na^+, K^+, and Cl^-), are essential to *electrolyte balance* in body fluids. Electrolyte balance, in turn, is essential for fluid balance, acid–base balance, and transmission of nerve impulses. Table salt is the principal source of sodium and

TABLE 16.1 The Macrominerals

Name	Sources	Major Functions	Deficiency Symptoms	Groups at Risk of Deficiency
Sodium (Na^+)	Table salt, processed foods	Major extracellular ion, nerve transmission, regulates fluid balance	Muscle cramps	Those consuming a severely sodium-restricted diet
Potassium (K^+)	Fruits, vegetables, grains	Major intracellular ion, nerve transmission	Irregular heartbeat, fatigue, muscle cramps	Those consuming diets high in processed foods, those taking high blood pressure medication
Chloride (Cl^-)	Table salt	Major extracellular ion	Unlikely	No one
Calcium (Ca^{2+})	Milk, cheese, bony fish, leafy green vegetables	Bone and tooth structure, nerve transmission, muscle contraction, blood clotting	Increased risk of osteoporosis	Postmenopausal women, teenage girls, those with kidney disease
Phosphorus*	Meat, dairy, cereals, and baked goods	Bone and tooth structure, buffers, membranes, ATP, DNA	Bone loss, weakness, lack of appetite	Premature infants, elderly
Magnesium (Mg^{2+})	Nuts, greens, whole grains	Reactions involving ATP, nerve and muscle function	Nausea, vomiting, weakness	Alcoholics, those with kidney and gastrointestinal disease
Sulfur†	Protein foods, preservatives	Part of amino acids and vitamins, glutathione, acid–base balance	None when protein needs are met	No one

* Phosphorus is covalently bonded in organic compounds and is also present in phosphate ions (PO_4^{3-}).

† Sulfur is covalently bonded in organic compounds and is also present in sulfate ions (SO_4^{2-}).

Source: Adapted from L. A. Smolin and M. B. Grosvenor: *Nutrition: Science & Applications*. Philadelphia: Saunders College Publishing, 1997.

Osteoporosis: Reduction of bone mass associated with calcium loss

chloride ions, and dietary deficiencies are unlikely. When there is extreme fluid loss through vomiting, diarrhea, or traumatic injury, electrolytes must be supplied to restore their concentration in body fluids.

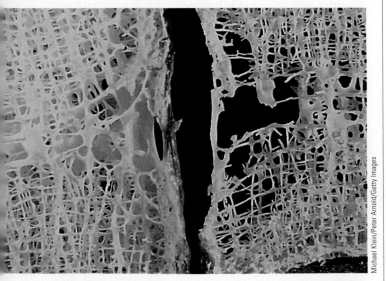

Michael Klein/Peter Arnold/Getty Images

Osteoporosis. The effect of calcium loss on bone is shown by the comparison between a normal bone (left) and a bone afflicted with osteoporosis (right).

About 99% of the *calcium ions* in the body are in bones and teeth. Together with sodium and potassium ions, calcium ions also participate in transmission of nerve impulses and regulation of the heartbeat. To be absorbed, calcium must be present in solution as Ca^{2+}. Absorption is enhanced in the presence of lactose (milk sugar) and also by a fatty meal, which passes through the intestine more slowly, allowing more time for absorption. A deficiency of vitamin D decreases calcium ion absorption and contributes to the bone deformities that accompany rickets.

A gradual loss of bone mass and density during adulthood is a normal process, and one of every three people older than 65 years has some degree of **osteoporosis**. The role of a calcium-deficient diet over a lifetime in hastening osteoporosis is sufficiently well established that a food label health claim on this basis is allowed. Postmenopausal white or Asian women of slight body build with inactive lifestyles are at greatest risk for osteoporosis. After menopause, the body produces less estrogen, which has the ability to slow bone dissolution. A preventive measure against later osteoporosis is consumption of adequate calcium during the adolescent years (ages 12 to 18), when bone is forming. Estrogen replacement for postmenopausal women is effective in slowing bone loss (Section 17-4).

Several minerals are essential in small amounts (micronutrient minerals). Of these, four—iron, copper, zinc, and iodine—are allowed to be listed on Nutrition Facts labels, iron being mandatory and the others optional. Other minerals known to be essential are selenium, manganese, fluorine, chromium, and molybdenum. Still other minerals (ten or more) are known to be essential in other mammals and are present and probably essential in humans. The list of trace minerals and our knowledge of their functions are constantly evolving.

Iodide ion has long been known to be essential to the thyroid gland. Hormones produced by the thyroid gland contain iodine and are responsible for growth, development, and maintenance of all body tissues. If there is a deficiency of iodine, the thyroid glands can become extremely enlarged, a condition called *goiter*. The routine use of iodized salt (0.1% potassium iodide) is the best way to ensure adequate iodine in the diet. More than half the table salt sold in the United States is iodized, and the use of such salt is mandatory in some countries.

Iron deficiency is associated with *anemia*—a decrease in the oxygen-carrying capacity of the blood, as indicated by low red blood cell or heme concentrations. Iron-deficiency anemia is the most common nutritional deficiency disease in both developed and underdeveloped nations. Early symptoms include muscle fatigue and lethargy. The ability to fight off invading bacteria is also diminished. Continued anemia creates defects in the structure and function of skin, fingernails, mouth, and stomach.

Zinc deficiency is known to cause poor growth and development, decreased immune function, and poor wound healing. Because meat is a better source of zinc than plant-based foods, vegetarians are among those at greatest risk of zinc deficiencies (others are the elderly and children with poor diets). Zinc supplements are often touted as able to improve immune function, enhance sexual performance, and increase fertility. It's a good idea to bring a degree of skepticism to such claims for zinc and other minerals. In the case of zinc, adding it to the diet of a person with a mild deficiency might improve wound healing, immune function, and appetite. Currently, however, there is little evidence for such changes in a healthy person with no existing zinc deficiency.

16-7 Food Additives

Many chemicals with little or no nutritive value are added to commercially processed food (Figure 16.7). Some of these food additives serve to protect the food from being spoiled by oxidation, bacterial attack, or aging. Others add and enhance flavor or color. Still others control pH; prevent caking; or stabilize, thicken, emulsify, sweeten, leaven, or tenderize the food.

CHIP: IOD VEGETABLE OIL (SA-MALT), BEAN AND/OR COTTONSEED OIL), RAISIN UGAR, JUICE CONCENTRATE, SORBITOL, SALT, TTER, NATURAL AND ARTIFICIAL FLAVORS, BHA ANILLA (A PRESERVATIVE), CITRIC ACID (A STABI-T BUT- LIZING AGENT).

CHOCOLATE CHIP: SEMISWEET CHOCO-LATE CHIPS [SUGAR, CHOCOLATE LIQUOR, COCOA BUTTER, LECITHIN (AN EMULSIFIER), VANILLA FLAVOR], CORN SYRUP, CRISP RICE (RICE, SUGAR, SALT, MALT), INVERT SUGAR, BROWN SUGAR, CORN SYRUP SOLIDS, GLYCERIN, PAR-TIALLY HYDROGENATED VEGETABLE OIL), SORBITOL, SALT, NATURAL AND ARTIFI-CIAL FLAVORS, BHA (A PERSERVATIVE), CITRIC ACID (A STABIILIZING AGENT).

STRAWBERRY: DEXTROSE, SUGAR, FLA-VORED FRUIT PIECES [DEHYDRATED APPLES (TREATED WITH SULFUR DIOX-IDE, SODIUM SULFITE, AND

APPLE PIECES SODIUM FITE (TO SUGAR, A PUREE, SU FLAVOR WIT SODIUM BEN CORN SYR RICE (RIC ERIN, I GENA AND SU

Note the variety of food additives described on the label of a package of cookies.

16-7a The GRAS List

The FDA lists hundreds of substances "generally recognized as safe" (GRAS) for their intended use by the FDA. The GRAS list was published in several installments in 1959 and 1960. It was compiled from the results of a questionnaire asking experts in nutrition, toxicology, and related fields to give their opinions about the safety of various seasonings, artificial flavorings, and other substances customarily added to foods. Since its publication, some substances have been removed, mostly due to suspicion that they are cancer-causing agents. One of these was the artificial sweetener *cyclamate*, which was correlated with an increase in bladder cancer in 1969. *New* food additives—substances not on the original GRAS list—must receive FDA approval for use in foods based on test results submitted by the manufacturers.

> **Hypertonic solution:**
> A solution with a concentration of solute higher than that found in its surroundings

16-7b Food Preservation

Oxidation and microorganisms (bacteria, fungi, and others) are the major enemies in the decomposition of food. Any process that prevents the growth of microorganisms or retards oxidation is generally an effective preservation process. Drying grains, fruits, and meat is one of the oldest preservation techniques. Drying is effective because water is necessary for both the growth of microorganisms and the chemical reactions of oxidation.

There are also chemical additives that can preserve food. Salted meat and fruit in a concentrated sugar solution are protected from microorganisms. The dissolved sodium

Types of Food Additives:

- Acidity regulators
- Anticaking agents
- Antifoaming agents
- Antioxidants
- Color fixatives
- Color retention agents
- Emulsifiers
- Firming agents
- Flavor enhancer
- Flour treatment agents
- Food acids
- Food coloring
- Gelling agents

- Glazing agents
- Humectants
- Improving agents
- Mineral salts
- Preservatives
- Propellants
- Seasonings
- Sequestrants
- Stabilizers
- Sweeteners
- Thickeners
- Vegetable gums

Additives can kill bacteria, prevent oxidation, increase vitamin content, and add flavor, texture, and color to many types of fresh and processed food. Milk, meat, grains, and greens all typically go through an additive process. Organic foods are no exception, though they are usually treated with natural alternatives to artificial additives.

chloride or sucrose creates a **hypertonic solution** in which water flows by osmosis (page 208) from the micro-organism to its environment. Thus, salt and sucrose have the same effect on microorganisms as dryness; both dehydrate them.

A preservative is effective if it prevents multiplication of microorganisms during the shelf life of the product. In general, food is spoiled by toxic substances secreted by the microorganisms. Sterilization by heat or radiation, or inactivation by freezing, is often undesirable, since they may affect the quality of the food. Chemical agents seldom achieve sterile conditions but can preserve foods for considerable lengths of time. Two common chemical preservatives in packaged foods are sodium benzoate (which is permitted in nonalcoholic beverages and in some fruit juices, fountain syrups, margarines, pickles, relishes, olives, salads, pie fillings, jams, jellies, and preserves) and sodium propionate (which can be used in bread, chocolate products, cheese, pie crust, and fillings).

16-7c Antioxidants

The direct action of oxygen in the air is the chief cause of the destruction of the fats in food. Carbon–carbon double bonds in polyunsaturated fatty acids are particularly susceptible. Oxidation produces a complex mixture of volatile aldehydes, ketones, and acids that causes a rancid odor and taste. Foods kept wrapped, cold, and dry are relatively protected from air oxidation. The most common antioxidant food additives are butylated hydroxyanisole (BHA) and butylated hydroxytoluene (BHT), which act by releasing a hydrogen atom from their —OH groups as a free radical (H·).

OH
C(CH₃)₃   and

OH
C(CH₃)₃

OCH₃

OCH₃

BHA (two isomers)

OH
(CH₃)₃C   C(CH₃)₃

CH₃

BHT

16-7d Sequestrants

Trace amounts of metals get into food from the soil and from machinery during harvesting and processing. Copper, iron, and nickel, as well as their ions, catalyze the oxidation of fats. The class of food additives known as **sequestrants** react with trace metals in foods, tying them up in complexes so the metals will not catalyze decomposition of the food. With the competitor metal ions tied up (sequestered), antioxidants such as BHA and BHT can accomplish their task much more effectively.

> **Sequestrant:** A chemical that ties up metal atoms or ions by forming bonds with them; used as a food additive

The sodium and calcium salts of ethylenediaminetetraacetic acid (EDTA) are common sequestrants in many kinds of foods and beverages. The structural formulas of tetrasodium EDTA (NaEDTA) and EDTA bonded to a metal ion are shown in Figure 16.8.

16-7e Food Flavors

Much of the sensation of taste in food is from our sense of smell. For example, the flavor of coffee is determined largely by its aroma, which in turn is due to a mixture of more than 500 compounds, mostly volatile oils. Most flavor additives, like many perfume ingredients, were originally derived from plants. The plants were crushed, and the compounds were extracted with various solvents such as ethanol or carbon tetrachloride. Sometimes a single compound was extracted; more often, the residue contained a mixture of several compounds. By repeated efforts, relatively pure oils were obtained. Oils of wintergreen, peppermint, orange, lemon, and ginger, among others, are still obtained in this way. These oils, alone or in combination, are then added to foods to produce the desired flavor. Today, synthetic preparations of flavors are also common food additives.

16-7f Flavor Enhancers

Flavor enhancers have little or no taste of their own but amplify the flavors of other substances. They exert synergistic and potentiation effects. *Synergism* is the cooperative action of two different substances (or people, or anything) such that the total effect is greater than the sum of the effects of each used alone. *Potentiators* do not have an effect themselves but exaggerate the effects of other chemicals. Some nucleotides (Section 15-11), for example, have no taste but enhance the flavor of meat and the effectiveness of salt. Flavor enhancers were first used in meat and fish but now are also used to

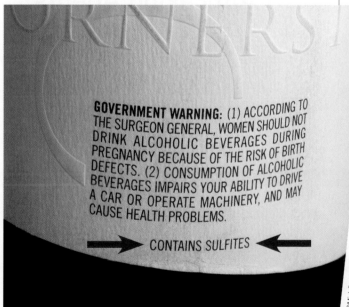

GOVERNMENT WARNING: (1) ACCORDING TO THE SURGEON GENERAL, WOMEN SHOULD NOT DRINK ALCOHOLIC BEVERAGES DURING PREGNANCY BECAUSE OF THE RISK OF BIRTH DEFECTS. (2) CONSUMPTION OF ALCOHOLIC BEVERAGES IMPAIRS YOUR ABILITY TO DRIVE A CAR OR OPERATE MACHINERY, AND MAY CAUSE HEALTH PROBLEMS.

→ CONTAINS SULFITES ←

Sulfites as additives. Because many individuals are allergic to sulfites, foods that contain more than 10 parts per million (ppm) of sulfites must have them listed on the label. The sulfites include chemicals such as sulfur dioxide (SO₂), sodium sulfide (Na₂S), and sodium bisulfite (NaHSO₃).

FIGURE 16.8 The sodium salt of ethylenediaminetetraacetic acid (NaEDTA).

The EDTA molecule has the remarkable ability to "sequester," or isolate, a metal ion by forming six bonds to it—two from nitrogen atoms in amino groups and four from oxygen atoms in ionized carboxyl groups (—COO⁻).

Na EDTA

© Cengage Learning

EDTA with metal ion

intensify flavors or mask unwanted flavors in vegetables, bread, cakes, fruits, and beverages. Three common flavor enhancers are monosodium glutamate (MSG), 5'-nucleotides, and maltol. While MSG can cause headaches and profuse sweating (the so-called Chinese restaurant syndrome) in some people, it is a natural constituent of many foods.

16-7g Food Colors

Some 30 chemicals are used as food colors. About half of them are laboratory synthesized and half are extracted from natural materials. Most food colors are large organic molecules with several double bonds and aromatic rings. Such structures have electrons that can absorb certain wavelengths of light and pass the rest; the wavelengths passed give the substances their characteristic colors. Beta-carotene (page 334), the orange-red substance

in carrots and a variety of plants (and also an antioxidant), is an example of a natural food color that may be noted on the ingredient list as "artificial color." Synthetic food colors must be "certified food colors," tested by the FDA to ensure safety, quality, consistency, and strength of color; these must be listed by name.

16-7h pH Control in Foods

Weak organic acids are added to such foods as cheese, beverages, and salad dressings to give a mildly acidic taste. They often mask undesirable aftertastes, but in some cases, such as in fruit-flavored sodas, the acidic taste is expected. Weak acids and acid salts react with bicarbonate to form CO_2 in the baking process. Buffers are also added to adjust and maintain a desired pH. Potassium acid tartrate, for example, is a buffer because it is a salt of an organic acid that can act as either an acid or a base.

They also function as preservatives to prevent the growth of microorganisms, as antioxidants to prevent rancidity and browning, as viscosity modifiers in dough, and as melting point modifiers in such food products as cheese spreads and hard candy. Citric acid is the most common and well-known acidulant.

16-7i Anticaking Agents

Anticaking agents are added to *hygroscopic* foods (in amounts of 1% or less) to prevent caking in humid weather. Table salt is particularly prone to caking unless an anticaking agent is present. The additive (magnesium silicate, $MgSiO_3$, for example) incorporates water into its structure as water of hydration and does not appear wet like sodium

One use of FDA-approved food dyes.

©iStockphoto.com/Devonyu

chloride does when it absorbs water physically on the surface of its crystals. As a result, the anticaking agent keeps the surface of sodium chloride crystals dry and prevents crystal surfaces from co-dissolving and joining together.

16-7j Stabilizers and Thickeners

Stabilizers and thickeners improve the texture and blends of foods. Like carrageenan, most stabilizers and thickeners are polysaccharides, which have numerous hydroxyl groups as a part of their structure. The hydroxyl groups form hydrogen bonds with water and help to provide a more even blend of the water and oils throughout the food. Stabilizers and thickeners are particularly effective in icings, frozen desserts, salad dressings, whipped cream, confections, and cheeses. In reduced-fat foods, they replace some of the "fatty" texture that would otherwise be missing. The plant gum thickeners (for example, gum arabic, guar gum, locust bean gum, gum tragacanth) are also good sources of soluble dietary fiber (Section 16-2).

16-8 Energy: Use It or Store It

16-8a How We Use Energy

A certain amount of energy is needed just to stay alive. While you are sitting completely still, your heart beats, your chest rises and falls as you breathe, your body temperature is maintained, chemical reactions proceed in cells, and messages that control these activities flow through your nervous system. The energy needed for these activities is the **basal metabolic rate (BMR)**. It is measured when a person is at rest at a comfortable temperature but not asleep, has not eaten for 12 hours, and has not engaged in vigorous activity for several hours.

The BMR is affected by many factors, including body weight and activity level. An increased BMR can come from anxiety, stress, lack of sleep, low food intake, congestive heart failure, fever, increased heart activity, and the ingestion of drugs (including caffeine, amphetamine, and epinephrine). A decreased BMR can result from malnutrition, inactive tissue due to obesity, and low-functioning adrenal glands. Infants and children have higher BMRs than adults, and after early adulthood the BMR decreases about 2% to 3% per decade.

As soon as voluntary activity begins, the metabolic rate speeds up. Some examples are given in Table 16.2.

> **Basal metabolic rate (BMR):** The minimum energy required to sustain an awake but resting body

TABLE 16.2 Calories/Hour Expended in Common Physical Activities

Some examples of physical activities commonly engaged in and the average amount of food Calories a 154-pound individual will expend by engaging in each activity for 1 hour.	
Moderate Physical Activity	**Food Calories/Hour Expended in Common Physical Activity***
Hiking	370
Light gardening/yard work	330
Dancing	330
Golf (walking and carrying clubs)	330
Bicycling (<10 mph)	290
Walking (3.5 mph)	280
Weight lifting (general light workout)	220
Stretching	180
Vigorous Physical Activity	
Running/jogging (5 mph)	590
Bicycling (>10 mph)	590
Swimming (slow freestyle laps)	510
Aerobics	480
Walking (4.5 mph)	460
Heavy yard work (chopping wood)	440
Weight lifting (vigorous effort)	440
Basketball (vigorous)	440

* Food Calories burned per hour will be higher for persons who weigh more than 154 lb (70 kg) and lower for persons who weigh less.

Source: Adapted from the 2005 Dietary Guidelines for Americans: www.healthierus.gov

© Crystal Cartier/StockFood Creative/Getty Images

16-8b Energy from Foods

Fats in the diet provide more than twice as many calories per gram as do carbohydrates and proteins.

Fats: 9 kcal/g; 9 Cal/g

Carbohydrates: 4 kcal/g; 4 Cal/g

Proteins: 4 kcal/g; 4 Cal/g

For example, if a steak is 49% water, 15% protein, 0% carbohydrate, 36% fat, and 0.7% minerals, then 3.5 ounces of steak (about 100 g) would provide about 384 kcal, or 384 food Cal. Note that a food Calorie (Cal) is indicated with a capital C and is actually equivalent to 1000 calories (cal) or 1 kilocalorie (kcal). Some representative caloric values are given in Table 16.3.

TABLE 16.3 The Approximate Percentages of Carbohydrates, Fats, Proteins, and Water in Some Whole Foods

Food	Water	Protein	Fat	Carbo-hydrates	kcal/100 g	Food	Water	Protein	Fat	Carbo-hydrates	kcal/100 g
Vegetables						*Meats and Fish (cont'd)*					
Spinach, raw	90.7	3.2	0.3	4.3	26	Oysters, raw	84.6	8.4	1.8	3.4	66
Collard greens, cooked	89.6	3.6	0.7	5.1	33	*Grains and Grain Products*					
Lettuce, Boston, raw	91.1	2.4	0.3	4.6	25	Wheat grain, hard	13.0	14.0	2.2	69.1	330
Cabbage, cooked	93.9	1.1	0.2	4.3	20	Brown rice, dry	12.0	7.5	1.9	77.4	360
Potatoes, cooked	75.1	2.6	0.1	21.1	93	Brown rice, cooked	70.3	2.5	0.6	25.5	119
Turnips, cooked	93.6	0.8	0.2	4.9	23	Whole-wheat bread	36.4	10.5	3.0	47.7	243
Carrots, raw	88.2	1.1	0.2	9.7	42	White bread	35.8	8.7	3.2	50.4	269
Squash, summer, raw	94.0	1.1	0.1	4.2	19	Whole-wheat flour	12.0	14.1	2.5	78.0	361
Tomatoes, raw	93.5	1.1	0.2	4.7	22	White cake flour	12.0	7.5	0.8	79.4	364
Corn kernels, cooked on cob	74.1	3.3	1.0	21.0	91	*Dairy Products and Eggs*					
Snap beans, cooked	92.4	1.6	0.2	5.4	25	Milk, whole	87.4	3.5	3.5	4.9	65
Green peas, cooked	81.5	5.4	0.4	12.1	71	Yogurt, whole milk	89.0	3.4	1.7	5.2	50
Lima beans, cooked	70.1	7.6	0.5	21.1	111	Ice cream	62.1	4.0	12.5	20.6	207
Red kidney beans, cooked	69.0	7.8	0.5	21.4	118	Cottage cheese	79.0	17.0	0.3	2.7	86
Soybeans, cooked	73.8	9.8	5.1	10.1	118	Cheddar cheese	37.0	25.0	32.2	2.1	398
Meats and Fish						Eggs	73.7	12.9	11.5	0.9	163
Lean beef, broiled	61.6	31.7	5.3	0	183	*Fruits, Berries, and Nuts*					
Beef fat, raw	14.4	5.5	79.9	0	744	Apples, raw	84.4	0.2	0.6	14.5	58
Lean lamb chops, broiled	61.3	28.0	8.6	0	197	Pears, raw	83.2	0.7	0.4	15.3	61
Lean pork chops, broiled	69.3	17.8	10.5	0	171	Oranges, raw	86.0	1.0	0.2	12.2	49
Lard, rendered	0	0	100.0	0	902	Cherries, sweet	80.4	1.3	0.3	17.4	70
Calf's liver, cooked	51.4	29.5	13.2	4.0	261	Bananas, raw	75.7	1.1	0.2	22.2	85
Beef heart, cooked	61.3	31.3	5.7	0.7	188	Blueberries, raw	83.2	0.7	0.5	15.3	62
Brains	78.9	10.4	8.6	0.8	125	Red raspberries, raw	84.2	1.2	0.5	13.6	57
Chicken, whole, broiled	71.0	23.8	3.8	0	136	Strawberries, raw	89.9	0.7	0.5	8.4	37
Cod, raw	81.2	17.6	0.3	0	78	Almonds	4.7	18.6	54.2	19.5	598
Salmon, broiled	63.4	27.0	7.4	0	182	Pecans	3.4	9.2	71.2	14.6	689
Freshwater perch, raw	79.2	19.5	0.9	0	91	Walnuts	3.5	14.8	64.0	15.8	651

© Cengage Learning

The human body, like everything else, is subject to the law of conservation of energy. In our case, it translates into the following equation:

$$\begin{matrix} \text{Energy taken} \\ \text{in (food)} \end{matrix} = \begin{matrix} \text{Energy} \\ \text{used} \end{matrix} + \begin{matrix} \text{Energy} \\ \text{stored (fat)} \end{matrix}$$

For most people, the secret to dieting is little more than applying this equation. When energy taken in exceeds energy used, the excess enters storage, mostly as triglyceride molecules in fat cells. When energy taken in remains the same but more energy is used, less is stored as fat. When energy used is more than energy taken in, some must be removed from storage. To be genuinely successful at providing weight loss, a program must integrate management of both diet and exercise.

To estimate your BMR, multiply your weight in pounds by 10 kcal/lb. Then, to get a rough estimate of your daily caloric needs, choose your general level of activity according to Table 16.4 and multiply your BMR by the appropriate factor given in the table.

EXAMPLE 16.1 Daily Energy Needs

Julie weighs 110 lb and is a college freshman—she plays tennis or skates several times a week and keeps in shape by running every day. Using Table 16.4, estimate her daily caloric needs.

These two individuals have different levels of activity and different caloric needs.

© iStockphoto.com/Gerville /
© iStockphoto.com/FotoShoot

SOLUTION

Based on her weight, Julie's estimated BMR is

110 lb × 10 kcal/lb = 1100 kcal = 1100 food Calories

From what we know about Julie, her activity level is moderate. Therefore, using the factor of 1.6 from Table 16.4, her estimated daily caloric needs are

1100 kcal × 1.6 = 1760 kcal = 1760 food Calories

TRY IT 16.1

Jack weighs 160 lb. As a graduate student, he spends most of his time reading in the library or working at the computer. Estimate his daily caloric needs.

16-9 Our Daily Diet

Following the regulation of a comprehensive food-labeling law established in 1994 and revised in 2008, the most readily available information about the nutritive value of food is on food packages. An example of the Nutrition Facts label is given in Figure 16.9. The total food Calories and food Calories from fat are listed at the top. Next comes a section that lists the weight in grams or milligrams per serving, plus the % Daily Value for the *macronutrients* (nutrients needed by the body in large amounts) judged most important to health. It is mandatory that this section

TABLE 16.4 Daily Energy Needs According to Physical Activity

Activity Level	Factor	
Very light		To estimate your daily energy needs, multiply your estimated BMR (10 kcal/lb × your weight in pounds) by the factor listed in the table that best represents your general level of daily activity.
Men	1.3	
Women	1.3	
Light		
Men	1.6	
Women	1.5	**Very light:** mostly sitting and standing activities
Moderate		
Men	1.7	**Light:** mostly walking activities
Women	1.6	**Moderate:** cycling, tennis, dancing
Heavy		**Heavy:** heavy manual digging, climbing, basketball, soccer
Men	2.1	
Women	1.9	

© Cengage Learning

FIGURE 16.9 The Nutrition Facts label.

An example of the label required by the law since 1994. Variations are allowed for small packages or foods that do not contain significant amounts of certain nutrients (e.g., vitamins and minerals need not be listed on canned soda).

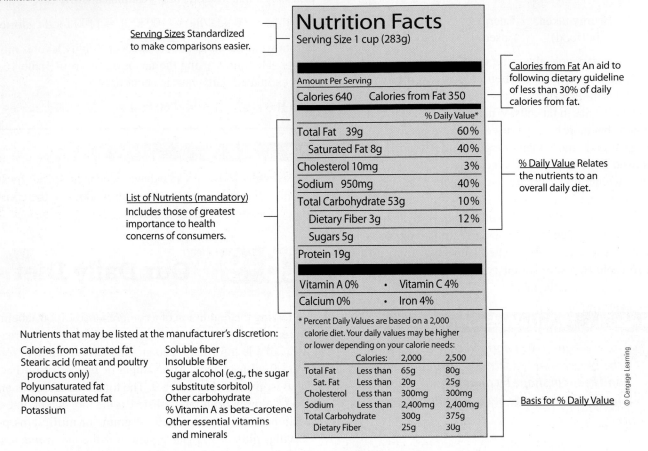

Serving Sizes Standardized to make comparisons easier.

List of Nutrients (mandatory) Includes those of greatest importance to health concerns of consumers.

Calories from Fat An aid to following dietary guideline of less than 30% of daily calories from fat.

% Daily Value Relates the nutrients to an overall daily diet.

Basis for % Daily Value

© Cengage Learning

Nutrition Facts

Serving Size 1 cup (283g)

Amount Per Serving

Calories 640	Calories from Fat 350

	% Daily Value*
Total Fat 39g	60%
Saturated Fat 8g	40%
Cholesterol 10mg	3%
Sodium 950mg	40%
Total Carbohydrate 53g	10%
Dietary Fiber 3g	12%
Sugars 5g	
Protein 19g	

Vitamin A 0%	•	Vitamin C 4%
Calcium 0%	•	Iron 4%

* Percent Daily Values are based on a 2,000 calorie diet. Your daily values may be higher or lower depending on your calorie needs:

		Calories:	2,000	2,500
Total Fat	Less than		65g	80g
Sat. Fat	Less than		20g	25g
Cholesterol	Less than		300mg	300mg
Sodium	Less than		2,400mg	2,400mg
Total Carbohydrate			300g	375g
Dietary Fiber			25g	30g

Nutrients that may be listed at the manufacturer's discretion:

Calories from saturated fat
Stearic acid (meat and poultry products only)
Polyunsaturated fat
Monounsaturated fat
Potassium

Soluble fiber
Insoluble fiber
Sugar alcohol (e.g., the sugar substitute sorbitol)
Other carbohydrate
% Vitamin A as beta-carotene
Other essential vitamins and minerals

include total fat, saturated fat, cholesterol, sodium, total carbohydrate, dietary fiber, sugars, and protein, as illustrated. Except for dietary fiber and protein, these are nutrients that should be limited.

The *reference daily values* are established as listed at the bottom of the food label and are further explained in Table 16.5. The percentage of daily values on a label (the *DVs*) are calculated from these values, as represented in a 2000-Cal daily diet. This many calories is about right for moderately active women, teenage girls, and sedentary men. Teenage boys, active men, and very active or pregnant women have higher caloric requirements.

Beneath the heavy line on every nutrition facts label are mandatory listings for vitamin A, vitamin C, calcium, and iron. Here also, these required listings are for those vitamins and minerals deemed to be of greatest public health concern. The major deficiency diseases of today are anemia, associated with iron deficiency, and osteoporosis, associated with calcium deficiency. The two vitamins in the mandatory list are those that decrease cancer risk.

TABLE 16.5 Basis for the Reference Daily Values on Nutrition Labels

Total Fat	30% of daily food Calories—an upper limit (65 g/2000-Cal daily diet)
Saturated Fat	10% of daily food Calories—an upper limit (20 g/2000-Cal)
Cholesterol	300 mg—a daily maximum for all diets
Sodium	2400 mg—a daily maximum for all diets
Total Carbohydrate	60% of daily food Calories (300 g/2000-Cal)
Dietary Fiber	23 g (based on 11.5 g/1000-Cal)
Sugars	No specified daily value
Protein	Listing the percentage daily value is not mandatory, but daily value is 10% of dietary food Calories from protein (50 g/2000-Cal) for adults (excluding pregnant women and nursing mothers) and children over 4 years old

© Cengage Learning

We have our choice of a wide variety of vitamin supplements.

© pulp/Photodisc/Getty Images

16-10 Some Daily Diet Arithmetic

Sometimes doing a little arithmetic can help you interpret food labels or identify your own food needs. A few examples of very practical kinds of diet and food calculations follow.

EXAMPLE 16.2 Daily Grams of Fat

A good way to keep track of food Calories from fat is to monitor the total grams of fat in your diet every day. A moderately active 100 lb woman would get 30% of her daily 1750 food Calories by consuming 53 g of fat per day. She is deciding how many Yummie Bite cookies she can eat. Their nutrition label says that there are two cookies per serving and 45 Cal from fat per serving. How many cookies can she eat and leave 33 g of fat for other foods to be eaten that day? What percentage of her 53 g of fat will the cookies provide?

SOLUTION

To leave 33 g of fat for other foods means that she can eat enough cookies to contain

$$53 \text{ g fat} - 33 \text{ g fat} = 20 \text{ g fat}$$

With 9 Cal per gram of fat, the grams of fat per "serving" of two cookies is

$$\frac{45 \text{ Cal}}{\text{Serving of 2 cookies}} \times \frac{1 \text{ g fat}}{9 \text{ Cal}} = \frac{5 \text{ g fat per serving}}{\text{of 2 cookies}}$$

With two cookies and 5 g of fat per serving, she can enjoy eight Yummie Bite cookies, which contain a total of 20 g of fat. Other foods eaten that day will have to be pretty low in fat content. The 20 g of fat in the eight cookies provides 38% of her daily fat allowance.

$$\frac{20 \text{ g fat (cookies)}}{53 \text{ g fat (daily fat)}} \times 100 = 38\%$$

TRY IT 16.2

Comparison of the labels gives the following information per serving of two kinds of soup:

	Calories	Calories from fat
Healthy Chicken:	90	15
Cream of Mushroom:	140	80

©iStockphoto.com/og-vision

How many grams of fat are there in one serving of each soup? For persons who wish to limit their daily fat consumption to 65 g, what percentage of this 65 g is in one serving of each?

EXAMPLE 16.3 Caloric Value of Food and Exercise

 A slice of pepperoni pizza contains 10 g of protein, 20 g of carbohydrate, and 7 g of fat. (a) How many Calories does the pizza slice provide? (b) What percentage of the total food Calories is from fat? (c) How long would a person have to run at 5 miles per hour to burn off the food Calories in the pizza slice (consult Table 16.5)?

SOLUTION

(a) The total caloric value of the pizza is calculated from the quantities of protein, carbohydrate, and fat:

$$10 \text{ g protein} \times \frac{4 \text{ Cal}}{\text{g protein}} = 40 \text{ Cal}$$

$$20 \text{ g carbohydrate} \times \frac{4 \text{ Cal}}{\text{g carbohydrate}} = 80 \text{ Cal}$$

$$7 \text{ g fat} \times \frac{9 \text{ Cal}}{\text{g fat}} = 63 \text{ Cal}$$

$$\text{Total} \quad \overline{183 \text{ Cal}}$$

(b) Percentage of food Calories from fat:

$$\frac{63 \text{ Cal (fat)}}{183 \text{ Cal (total)}} \times 100 = 34\%$$

(c) Running at 5 miles per hour burns ~10 Cal (total) per minute (590 Cal/hr). Therefore, to burn off this slice of pizza would require running for 18.3 minutes.

$$183 \text{ Cal} \times \frac{1 \text{ minute}}{10 \text{ Cal burned}} = 18.3 \text{ minutes}$$

TRY IT 16.3

A slice of devil's food cake with chocolate frosting contains 3 g of protein, 8 g of fat, and 40 g of carbohydrate.

(a) How many Calories does the slice of cake provide?

(b) What percentage of the Calories is from fat?

(c) How long could a person listen to a lecture fueled by the energy from this piece of cake?

Your Resources

In the back of the textbook:

→ *Review card on Nutrition: The Basis of Healthy Living*

   In OWL for CHEM2 at www.cengagebrain.com

→ *Review Key Terms with Flash Cards (printable or digital)*

→ *Complete Interactive Practice Quizzes to prepare for tests*

→ *Submit Assigned Homework and Exercises*

Applying Your Knowledge

1. What is meant by the word *digestion*?

2. What is amylase? To what class of biochemical molecules does it belong?

3. What are the relationships between mass and energy for the following substances?
   (a) Fats   (b) Proteins
   (c) Carbohydrates

4. What are the products formed by digestion of the following nutrients?
   (a) Carbohydrates   (b) Fats
   (c) Proteins

5. What is meant by *BMR* (or basal metabolic rate)?

6. What is meant by the following words and terms?
   (a) Simple sugars
   (b) Digestible carbohydrates
   (c) Indigestible carbohydrates
   (d) Disaccharides
   (e) Polysaccharides
   (f) Dietary fiber

7. What are triglycerides?

8. Give definitions for the following words and terms as applied to fats and fatty acids:
   (a) Saturated   (b) Monounsaturated
   (c) Polyunsaturated   (d) Partially hydrogenated

9. What do the following terms mean?
   (a) Atherosclerosis   (b) Cholesterol
   (c) Plaque   (d) Lipoproteins
   (e) Low-density lipoproteins
   (f) High-density lipoproteins

10. What are the following?
    (a) FDA   (b) Reference daily values

11. Give definitions for the following words:
    (a) Macronutrients   (b) Micronutrients

12. What are the major categories of information required on a Nutrition Facts label?

13. Where do the following substances occur in nature?
    (a) Starches
    (b) Glycogen
    (c) Cellulose

14. What are micronutrient minerals? Which one must be listed on a Nutrition Facts label?

15. What is a vitamin? What is the function of a vitamin?

16. Which of the vitamins listed is *not* one of the B vitamins?
    (a) Niacin
    (b) Folic acid
    (c) Riboflavin
    (d) Ascorbic acid

17. Which of the vitamins listed is fat soluble?
    (a) Vitamin C
    (b) Vitamin B_6
    (c) Vitamin K
    (d) Vitamin B_{12}

18. Refer to the structure of Vitamin A on page 334. What functional groups are contained within the structure?

19. Why is it advised to avoid taking large doses of fat-soluble vitamins?

20. Overdoses of which of the following vitamins should definitely be avoided? Why?
    (a) Riboflavin   (b) Vitamin C
    (c) Vitamin A   (d) Vitamin B
    (e) Vitamin D

21. Match the symptoms associated with vitamin deficiency or overdose to the name of the vitamin.
    Vitamin C deficiency ____   (a) Rickets
    Vitamin K overdose ____   (b) Excessive blood clotting
    Vitamin D deficiency ____   (c) Nerve damage
    Niacin deficiency ____   (d) Pellagra
    Vitamin B_6 overdose ____   (e) Scurvy

22. Another vitamin-related problem experienced by early Arctic explorers was due to the consumption of polar bear liver. Consuming just 10 g led to an acute overdose of vitamin A. Using information in the text, how many international units of vitamin A are found in one gram of the liver?

23. What is an electrolyte? What is meant by the *electrolyte balance* of the body?

24. What is the GRAS list?

25. What are the seven most abundant macronutrient minerals in the body?

26. What is the function of each of the following food additives?
    (a) Sodium benzoate
    (b) Sodium propionate
    (c) BHA (butylated hydroxyanisole)
    (d) BHT (butylated hydroxytoluene)
    (e) Sodium EDTA

27. Why are sequestrants needed as food additives?

28. What is the function of each of the following food additives?
    (a) Monosodium glutamate (MSG)
    (b) Yellow no. 5
    (c) Acetic acid
    (d) Calcium silicate
    (e) Carrageenan
    (f) Lecithin

29. What percentage of fat is currently recommended for today's diet by most public health authorities?

30. Lactose is sometimes referred to as "milk sugar" because it is the principal carbohydrate in milk. The lactose structure follows. Is this a monosaccharide, disaccharide, or polysaccharide? Explain.

31. Lactose intolerance occurs in people who are unable to hydrolyze milk sugar (lactose). This happens because they do not produce enough of the enzyme lactase. Normally infants and children have adequate amounts of lactase, but many adolescents and adults produce less. Should milk be promoted as a healthy food for the general population? Give one argument for and one against.

32. The body stores excess energy in small globules of fat in specialized cells of adipose tissue. The energy available from a gram of fat is 9 Cal. What would happen to the volume of these cells if the body stored energy by collecting carbohydrates instead of fats? Assume that the densities of carbohydrate and fat are the same. (Recall that 1 g carbohydrate = 4 Cal.)

33. Cholesterol has the following structure. Why is cholesterol insoluble in water? Why is it recommended that people monitor their blood serum cholesterol levels? How many carbon atoms does a cholesterol molecule contain?

34. Folic acid is found in citrus fruits and leafy green vegetables. The FDA is now requiring that folic acid be added to cereals, breads, and pastas. The *American Journal of Public Health* says that such grain fortification would prevent 300 to 700 birth defects per year. At the level recommended for addition, approximately 3.25 million adults over 50 would receive too much folic acid. What question would you want answered before you would support such a nationwide program?

35. Why is osteoporosis a more significant problem for women after menopause? What preventive measures can be taken to reduce the chances of experiencing osteoporosis?

36. Should the government embark on a national program to add more calcium to our diets to reduce our chances of calcium deficiency and osteoporosis? Give one reason for such a program and one against.

37. What happens to the human thyroid gland if a person has an iodine deficiency? What is this condition called? What is the link between this problem and iodized salt?

38. When fats and fatty portions of food become rancid, what kind of reactions have probably occurred? What food additives are used to minimize these reactions?

39. Choose a label from a food item and try to identify the purpose of each food additive.

40. What foods have you eaten during the past week that did not contain any food additives?

41. What is the basal metabolic rate (BMR) for a 110-lb woman who has a moderate activity level?

42. What is the BMR for a 145-lb man who is sedentary, with very light activity? What would this person's BMR be if he started to exercise regularly so his activity increased to the moderate level?

43. How much time bicycling at 5 mph is needed to consume the 100 Cal available from a serving of tomato soup? (*Note:* Energy expended bicycling at 5 mph is 5 Cal/min.)

44. A 100-g serving of ice cream contains 207 Cal. How long would you have to run at 5 mph to consume the energy available from the ice cream?

45. Walking on a road at 3.5 mph consumes 3.5 Cal/min. How long would you have to walk to consume the energy available from 100 g of white bread? (*Note:* 100 g white bread = 269 Cal.)

46. A blood serum cholesterol level of 200 mg/100 mL is supposed to give a low heart disease risk level. How much total cholesterol would be carried by the blood serum of an adult with a blood volume of 12 pints? (1 pint = 473 mL.)

47. A blood serum cholesterol level of 240 mg/100 mL is supposed to give a high heart disease risk level. How much total cholesterol would be carried by the blood serum of an adult with a blood volume of 13 pints? (1 pint = 473 mL.)

48. The quantity of fat in a food is important to people who need to restrict fat intake. Which of the following would be the better low-fat choice: a piece of pecan pie weighing 113 g providing 459 Cal with 180 Cal from fat, or a piece of apple pie weighing 125 g providing 305 Cal with 108 Cal from fat?

49. How many Calories are in a fast-food chicken sandwich that contains 27 g of protein, 46 g of carbohydrate, and 34 g of fat?

50. How many kcal are in a fast-food bacon cheeseburger supreme if it contains 34 g of protein, 44 g of carbohydrate, and 46 g of fat?

51. An exceptionally active female volleyball player who weighs 145 lb wants to maintain a diet of 30% calories from fat, 60% from carbohydrates, and 10% from protein. How many Cal does the player need per day? How many grams of fat, carbohydrate, and protein does she need?

52. By using the following Nutrition Facts label, determine what percentage of the Calories in the product come from fat. What percentage of your daily Calorie need (1800 Cal) is provided by a serving? Would this product be a good food item for a person on a low-sodium diet? Justify your answers.

Nutrition Facts		
Serving Size 1/2 cup (120mL) condensed soup		
Servings Per Container About 6		
Amount Per Serving		
Calories 100	Calories from Fat 20	
		% Daily Value*
Total Fat 2g		3%
Saturated Fat 0g		0%
Cholesterol 0mg		0%
Sodium 730mg		30%
Total Carbohydrate 18g		6%
Dietary Fiber 2g		8%
Sugars 10g		
Protein 2g		
Vitamin A 10%	• Vitamin C 30%	
Calcium 2%	• Iron 4%	

53. Why isn't the vitamin B_6 content of a food not always listed on its nutrition label?

Chemistry
and Medicine

Everyone is interested in developments that influence public health and, most particularly, their own personal health. Half of all Americans take at least one prescription medication and the demand for nonprescription, over-the-counter (OTC) medications is enormous. For even a general understanding of the major public health issues, the amount to be learned is large. Knowledge, however, is our personal first line of defense. With chemistry as the framework, in this chapter we take a look at the most important classes of diseases and the medications available to treat them.

The average life expectancy for men in the United States rose from 53.6 years in 1920 to 74.6 years in 2008. During this same period, the life expectancy for women rose from 54.6 years to 82.67 years.

What roles do chemistry and chemical technology play in this ongoing increase in life expectancy and the accompanying improvements in the quality of life?

Most importantly, remarkable progress has been made in recent years in understanding the chemical reactions that regulate biological processes. As this understanding grows, the old trial-and-error method of screening chemicals for use as drugs is being replaced. Instead, drug molecules are being designed to have exactly the molecular shape and chemical reactivity needed to counteract disease. Another positive outcome is a growing understanding of what makes for a healthy lifestyle.

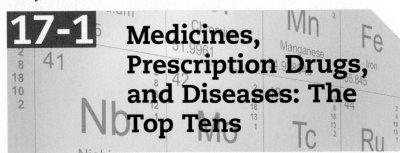

17-1 Medicines, Prescription Drugs, and Diseases: The Top Tens

Americans spent more than $307 billion on legal drugs in 2010, more than seven times more than the amount spent in 1990. More than $15 billion of this is for **over-the-counter drugs**—those that can be bought in any supermarket or drugstore. The rest is for **prescription drugs**, which cannot be purchased without instructions from a physician. A drug is classified as one or the other by the Food and Drug Administration (FDA). In general, a substance is available only by prescription if it has potentially dangerous side effects, if it should be used only by people with specific medical problems, or if it treats a condition so serious that a person with that condition should be under a doctor's care.

The top ten branded prescription drugs, based on the number of prescriptions written during 2008 (Table 17.1), show an interesting cross section of medicinal uses. In first place is a drug that lowers blood cholesterol and stabilizes plaque while acting as an anti-inflammatory. There are two proton pump inhibitors (Nexium and Prevacid), which treat conditions such as indigestion and acid reflux disease by reducing gastric acid production in the stomach. Three of the drugs act against vascular problems (Lipitor, Plavix, and Diovan), two act against respiratory problems (Singulair and Advair Diskus), one addresses hyperthyroidism (Synthroid), and two act as antidepressants (Lexapro and Effexor). Effexor, the ninth most prescribed brand, belongs to the same class of drugs as Schedule I substances 2C-B, mescaline, and MDMA

Over-the-counter drug: A medication that can be purchased without a prescription

Prescription drug: A medication that can be purchased only at the direction of a physician

Pharmacy links chemistry with the medical sciences. The *British Pharmacopoeia* of 1862 was the first attempt to standardize the use of chemistry in the treatment of disease.

(3,4-methylenedioxymeth-amphetamine).

As illustrated in Table 17.1, all drugs have a **trade name** and a **generic name**. The trade name (brand name) is the name used by the drug manufacturer. For example, the antibiotic amoxicillin is sold under trade names such as Amoxil, Amoxidal, Amoxibiotic, Infectomycin, Moxaline, Trimox, Utimox, and Wymox. *Amoxicillin* is a generic name, which is a drug's generally accepted common chemical name. Once the patent protection on a drug has expired, it can be manufactured and marketed competitively by many companies and prescribed by its generic name. Often, prescriptions written by the generic name are cheaper to fill.

In 1900, five of the ten leading causes of death for individuals of all ages in the United States were infectious diseases (pneumonia and influenza, tuberculosis, gastrointestinal infections, kidney infections, diphtheria). Currently, pneumonia and influenza (tabulated together) are the only infectious diseases from this group remaining in this top ten, causing 2.7% of all deaths of individuals of all ages (Table 17.2).

TABLE 17.1 Top Ten Branded Drugs Prescribed in 2008

Rank	Brand Name	Generic Name	Drug Class
1	Lipitor	Atorvastatin	Statin
2	Nexium	Esomeprazole	Proton pump inhibitor
3	Lexapro	Escitalopram	Selective serotonin reuptake inhibitor
4	Singulair	Montelukast	Leukotriene receptor antagonist
5	Plavix	Clopidogrel	Antiplatelet agent
6	Synthroid	Levothyroxine	Thyroid hormone
7	Prevacid	Lansoprazole	Proton pump inhibitor
8	Advair Diskus	Fluticasone/salmeterol	Corticosteroid/beta-2 adrenergic receptor agonist
9	Effexor	Venlafaxine	Phenylethylamine
10	Diovan	Valsartan	Angiotensin II receptor antagonist

Source: SDI/Verispan, VONA, Full year 2008.

TABLE 17.2 Leading Causes of Death in 2006

	All Ages		Ages 15–24		Ages 25–34	
	Total	Percent (%)	Total	Percent (%)	Total	Percent (%)
Total deaths	2,426,264		34,887		42,952	
Heart disease	631,636	26.03	1,076	3.08	3,307	7.70
Cancer	559,888	23.08	1,644	4.71	3,656	8.51
Cerebrovascular diseases (stroke)	137,119	5.65	210	0.60	527	1.23
Chronic lower respiratory diseases	124,583	5.13	151	0.43	255	0.59
Accidents	121,599	5.01	16,229	46.52	14,954	34.82
Diabetes mellitus	72,449	2.99	165	0.47	673	1.57
Alzheimer's disease	72,432	2.99	*	*	1	0.00
Influenza and pneumonia	56,326	2.32	184	0.53	335	0.78
Kidney disease	45,344	1.87	99	0.28	276	0.64
Blood poisoning	34,234	1.41	139	0.40	291	0.68
Suicide	33,300	1.37	4,189	12.01	4,985	11.61
Chronic liver disease and cirrhosis	27,555	1.14	26	0.07	316	0.74
Essential hypertension and hypertensive renal disease	23,855	0.98	21	0.06	110	0.26
Parkinson's disease	19,566	0.81	2	0.01	4	0.01
Assault (homicide)	18,573	0.77	5,717	16.39	4,725	11.00

* Figure does not meet standards of reliability or precision, or category not applicable.

Source: *National Vital Statistics Report*, Vol. 57, No. 14. www.cdc.gov/nchs/data/nvsr/nvsr57/nvsr57_14.pdf

This dramatic decrease in deaths from infectious diseases can be attributed in large measure to the development of antibiotics.

Meanwhile, acquired immunodeficiency syndrome (AIDS), an infectious disease that is caused by the human immunodeficiency virus (HIV), has come onto the scene. In 1993, HIV infection hit the news by moving ahead of all other causes of death for those in the 25- to 44-year-old age group. Only three years earlier, in 1990, HIV infection placed third as the cause of death in this age group, with accidents and cancer coming first and second. Although AIDS continues to be a very serious health issue, AIDS deaths in the United States have dropped dramatically thanks to intensive educational programs regarding AIDS prevention and the development of drugs that, although unable to cure AIDS, can prolong life expectancy dramatically. AIDS deaths now rank as the sixth most common cause of death in the 25- to 44-year-old age group in the United States. Unfortunately, the cost of these drugs has, at least in part, prevented their use on a worldwide basis. AIDS continues to be a devastating disease, especially in sub-Saharan Africa.

A sad reflection of the social condition in the United States is the fact that accidents, homicide, and suicide rank as the first, second, and third most common causes of death, respectively, in the 15- to 24-year-old age group. In the 25- to 44-year-old age group, accidents still rank as the primary cause of death, with suicide now occupying the fourth position, followed by homicide in fifth place.

17-2 Drugs for Infectious Diseases

Modern chemotherapy—the treatment of disease with chemical agents—began with the work of Paul Ehrlich. In 1904, he concluded that **infectious diseases** could be conquered if chemicals could be found that attack disease-causing microorganisms without harming the host. After observing that dyes used to stain bacteria also killed the bacteria, he developed arsenic compounds similar to the dyes. One of these compounds (*arsphenamine*) was the first effective drug for an infectious disease. It revolutionized the treatment of syphilis at the time it was introduced. Syphilis is now treated with penicillin, the oldest of the antibiotics that are now our principal weapons against infectious diseases.

Strictly speaking, an **antibiotic** is a substance produced by a microorganism that inhibits the growth of other microorganisms. Any compound, whether natural or synthetic, that acts in this manner is now referred to as an "antibiotic," however. A person falls victim to an infectious disease when invading microorganisms multiply faster than the body's immune system can destroy them. Antibiotics help the immune system by either destroying invaders or preventing their multiplication.

17-2a Penicillins

The penicillins were discovered by Sir Alexander Fleming, a bacteriologist at the University of London. He was working in 1928 with cultures of *Staphylococcus aureus*, a bacterium that causes boils and some other infections. One day he noticed that one culture was contaminated by a blue-green mold. For some distance around the mold growth, the bacterial colonies had been destroyed. On further investigation, Fleming found that the broth in which this mold had grown had a similar lethal effect on many **pathogenic** (disease-causing) bacteria. The mold was later identified as *Penicillium notatum* (the spores sprout and branch out in pencil shapes, hence the name). Although Fleming showed that the

> **Infectious disease:** A disease caused by microorganisms
>
> **Antibiotic:** A substance that inhibits the growth of microorganisms
>
> **Pathogenic:** Capable of causing disease

*The enemies: bacteria and viruses. (a) The blue rods are **Haemophilus influenza** bacteria lying on human nasal tissue. This bacterium causes bronchitis, pneumonia, and a wide range of diseases in children, including meningitis and blood poisoning. (b) The dots are **Rubella** viruses emerging from the surface of an infected cell, where the viruses are being reproduced. **Rubella** is the virus that causes German measles.*

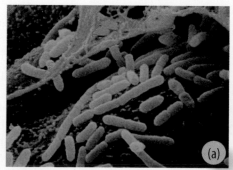

mold contained an antibacterial agent, which he called *penicillin*, he was not able to purify the active substance.

In 1940, the active ingredient, penicillin G, was identified, and by 1943 it was available for clinical use. As World War II drew to a close, penicillin G was saving many lives threatened by pneumonia, bone infections, gonorrhea, gangrene, and other infectious conditions. Since then, numerous penicillins have been developed. Amoxicillin, the most-prescribed antibiotic, is a penicillin.

General penicillin structure

Penicillin G, the first penicillin

Amoxicillin, a broad-spectrum antibiotic—active against a variety of bacteria

All penicillins kill growing bacteria by preventing normal development of their cell walls. Bacteria are members of the plant kingdom and their cells, unlike those of mammals, rely on a rigid cell wall. The rigidity is maintained by cross-linking bonds between peptide chains. Penicillins inhibit the enzyme that forms these cross links and prevent the reaction from occurring, causing the cell to burst. For their work in discovering and determining how penicillin functions as an antibiotic, Howard Florey, Ernest Chain, and Sir Alexander Fleming shared the Nobel Prize in medicine and physiology in 1945.

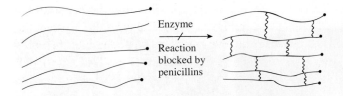

Enzyme

Reaction blocked by penicillins

17-2b Other Classes of Antibiotics

The *cephalosporins* are similar in structure to the penicillins and also act by disrupting cell wall synthesis. They

are widely used in hospitals because of their low toxicity and broad range of antibacterial activity. All cephalosporins have the same general structure in which R_1 is often a hydrogen and R_2 and R_3 are substituents with several functional groups.

General cephalosporin structure

$R_1 = $ —H

$R_2 = $ —CH_3

$R_3 = $ —NH—C—CH—

R groups in cephalexin, used to treat streptococcal infections

Other kinds of antibiotics attack bacteria in different ways from the penicillins and cephalosporins. Many interfere with the synthesis or functioning of bacterial DNA. The *tetracyclines* (such as Achromycin and Terramycin) and *doxycycline*, for example, inhibit bacterial protein synthesis, and *rifampin* (important in treating tuberculosis) inhibits RNA synthesis from DNA.

doxycycline

The defeat of infectious diseases by antibiotics was once thought to be complete. That was a mistake. After more than a half-century of use, many antibiotics are not nearly as effective as they once were. Strains of malaria, typhoid fever, gonorrhea, and tuberculosis have emerged that are resistant to the antibiotics that once defeated them. Researchers have begun to hunt for new antibiotics, in some cases resuming a search that had been abandoned because it was believed

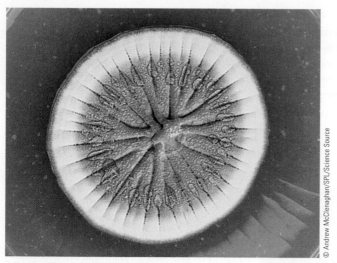

*A **Penicillium notatum** culture.*

unnecessary. The urgency of the search was heightened with the discovery in 1997 of a strain of methicillin-resistant staph (*S. aureus*), the bacteria responsible for deadly infections among the seriously ill in hospitals. Vancomycin has been the only antibiotic able to cure such staph infections. In fact, vancomycin has been designated as the antibiotic of last resort, only to be used when all others fail.

According to the Centers for Disease Control and Prevention (CDC), almost 2 million patients in the United States pick up an infection *in the hospital* each year. In 2009, an estimated 99,000 of those patients died as a result of their infections. This number was about 13,300 in 1992. More than 70% of the bacteria that cause hospital-acquired infections are resistant to at least one of the drugs commonly used to treat them.

Strains of *S. aureus* resistant to methicillin are common in hospitals, and these bacteria are increasingly being found in nonhospital settings such as locker rooms. Minor scrapes incurred by high school football players and wrestlers have become infected and have led to amputation of limbs and even death in very short periods of time. The problem has become so serious that it was the focus of an article in *Sports Illustrated* in the spring of 2005.

While the race for new antibiotics is on, the medical community endeavors to make everyone aware of some basic facts:

- Doctors should take care not to over-prescribe antibiotics.

- People should not demand antibiotic treatment in cases where it is not necessary (for example, antibiotics have no effect on viral infections).

- Antibiotics must always be taken for the full period of the prescription.

- Everyone should wash hands frequently and thoroughly.

17-3 AIDS: A Viral Disease

Since its first appearance in the 1980s, HIV infection has spread rapidly and worldwide. In the United States, the first 100,000 cases of HIV infection were reported over the course of the eight years from 1981 to 1989. The second 100,000 cases were detected in the following two years. An estimated 1,106,400 persons in the United States were living with HIV in 2006. About 25% to 35% of these people did not know they were infected. Worldwide, as estimated by the World Health Organization, about 33 million individuals were HIV infected as of 2007.

From a chemical perspective, it is important to understand that there are very few known ways to combat viral diseases (other than vaccination). The reason for this is the method of attack used by viruses—they are essentially chemical parasites. They take over the DNA of human cells and put it to work for their own reproduction. It is difficult to attack the cells reproducing the virus without also attacking the host's own cells.

HIV is a *retrovirus*. It consists of an outer double lipid layer surrounding a matrix containing proteins, an enzyme called *reverse transcriptase*, and RNA. The word *retrovirus* is used because the virus enzyme carries out RNA-directed synthesis of DNA rather than the usual DNA-directed synthesis of RNA (see Figure 15.23).

The AIDS retrovirus penetrates the T cells of the immune system (1 in Figure 17.1). Once inside, the virus releases its contents, and the *reverse transcriptase* of the AIDS virus translates the RNA code of the virus into double-stranded DNA (2 in Figure 17.1). The virus DNA enters the T-cell nucleus and is incorporated into the cell's own DNA (3 in Figure 17.1). Then, the T cell makes RNA from viral DNA, the proteins needed for a new virus are made from this RNA (4 in Figure 17.1), and the new virus is released (5 in Figure 17.1). Eventually, the T cell swells and dies, releasing more AIDS viruses to attack other T cells. As their T cells are destroyed, individuals are attacked by

Red ribbons promote awareness of HIV and AIDS.

FIGURE 17.1 Attack of HIV on a T cell (a lymphocyte).

The virus enters the lymphocyte (1) and produces its own DNA (2). Antiviral compounds such as efiravenz can interrupt this step. The DNA then enters the lymphocyte nucleus (3), where it combines with the lymphocyte's DNA. Then, through the cell's normal processing, DNA is transcribed to messenger RNA and, back outside the nucleus, the viral RNA provides the template for the production of proteins (4) that assemble into a new virus particle (5). AZT can interfere with the production of viral proteins in (4), and protease inhibitors can lessen the ability of the released virus to infect new cells.

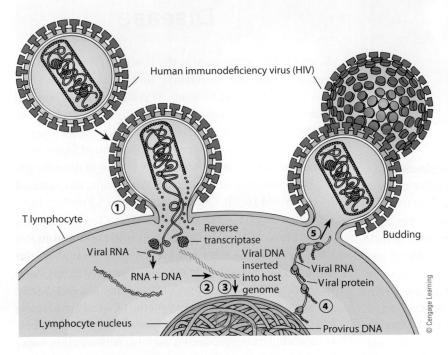

© Cengage Learning

diseases that are normally defeated by the body's immune system and thus are rare in healthy individuals.

The search for weapons against HIV reached a turning point in 1996. Earlier efforts had yielded drugs that somewhat slow the progress of AIDS but in no way approach a cure of the disease. In early 1996, three members of a new class of drugs directed at HIV, the *protease inhibitors*, were approved in rapid succession. Convincing clinical trials had shown that a protease inhibitor taken in *combination* with older AIDS drugs dramatically lowers the concentration of the virus in patients' blood, often to below detectable levels. The drug combination diminishes the number of serious complications and decreases the death rate among seriously ill individuals. Apparently, a combination of drugs overcomes the major problem with single-drug therapy—if even a few virus particles are left alive by a drug treatment, the virus can very quickly adapt by converting to a drug-resistant form.

As their name implies, protease inhibitors act by inhibiting a protease

enzyme. By fitting into the active site of the enzyme (Figure 17.2), the inhibitors prevent the enzyme from functioning. Because this enzyme cuts a long protein chain into smaller proteins essential to survival of the HIV virus, halting its action kills the virus.

The most successful older anti-AIDS drug, *azidothymidine* (AZT, also called *zidovudine*), is one of the drugs used in combination with a protease inhibitor. AZT is a derivative of deoxythymidine, a nucleoside. (Nucleosides are nucleotides without the phosphate group; see Section 15-11.) AZT differs from the natural nucleoside by having an $—N_3$ group instead of an $—OH$ group. Still, because AZT resembles a natural nucleoside, it is accepted by the HIV reverse transcriptase enzyme and placed into the viral DNA. Once there, its structure prevents additional nucleosides being added to the chain, and cell division is halted. AZT has been found to reduce the mortality of babies born to HIV-positive mothers to close to zero.

Despite the new air of optimism, it is important to keep the situation in perspective relative to several facts. The new drug combination is not yet a cure. It is extremely expensive, it is unavailable in many developing countries, and it can have unpleasant side effects. An ominous further problem is that one must take up to 20 pills a day on a precisely timed schedule; wavering from the schedule may provide opportunities for new drug-resistant variations of the virus to develop. Three recently developed AIDS drugs (Enfuvirtide, Atazanavir, and Fosamprenavir) were approved by the FDA in 2003 and are currently available for use where more traditional treatments fail. However, they are inconvenient to administer and are very expensive. Therefore, they are usually not used until three or four other treatments have proven unsuccessful. A 2009 publication in the *New England Journal of Medicine* described the anecdotal case of an HIV-positive patient whose HIV levels dropped below detectable levels after he received a bone marrow transplant from a donor with mutated HIV-resistant genes. While researchers suggest that it's premature to consider such treatment a possible cure for AIDS, the case certainly shows hope for future developments.

17-4 Steroid Hormones

In considering hormones, we turn our attention from drugs that fight invading organisms to drugs that make up deficiencies in natural biochemicals or mimic their action.

Hormones are produced by glands and secreted directly into the blood. They serve as chemical messengers, regulating biological processes by interacting with **receptors** that are sometimes distant from where they are secreted. For example, the adrenal gland just above the kidney releases a group of hormones that act throughout the body to regulate the availability of glucose in the blood. Hormones are chemically diverse but are mostly proteins or steroids. Cortisol, also called *hydrocortisone*, is a steroid hormone released by the adrenal gland in response to stress. Synthetic cortisol has many medical uses, such as the treatment of rheumatoid arthritis. The ring structure common to all steroids is evident in cortisol:

Receptor: A molecule or portion of a molecule that interacts with another molecule to cause a change in biochemical activity

Cortisol

One of the most revolutionary medical developments of the 1950s was the worldwide introduction and use of "the pill." The development of synthetic analogs of the female sex hormones made reliable birth control available to women for the first time. The ongoing search for an equivalent birth control pill for use by men has not yet been equally successful.

Most oral contraceptive pills used by women today contain a combination of two synthetic hormones. One,

FIGURE 17.2 Ritonavir, a protease inhibitor, in the active site of the protease enzyme.

The ritonavir molecule is shown inside the purple box. By occupying the active site of the enzyme, ritonavir prevents the enzyme from carrying out its function. The ribbon-like structure represents the protein chain of the enzyme.

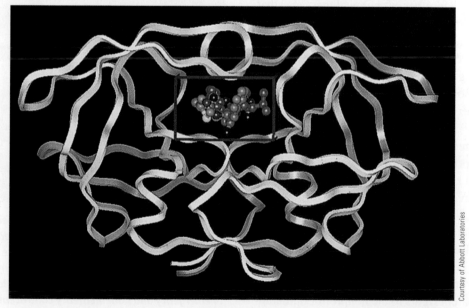

Courtesy of Abbott Laboratories

most commonly ethinyl estradiol, has estrogen-like activity in regulating the menstrual cycle. The other, most commonly norethindrone, has progesterone-like activity and establishes a state of false pregnancy that prevents ovulation. In a theme that is common in drug development, successful contraceptives are very similar in molecular structure to those natural molecules with similar activity. (Compare the following structures with those in Section 15-3.)

Ethinyl estradiol
(synthetic estrogen)

Norethindrone
(synthetic progesterone)

On a monthly cycle in nonpregnant women, release of a hormone by the pituitary gland triggers ovulation, which results in the release of an egg with the potential for fertilization. Once a woman is pregnant, a hormone known as *progesterone* (the "pregnancy hormone") is released. This hormone carries a number of chemical messages that help prepare the woman's uterus for implantation of the fetus and block further ovulation during pregnancy. Synthetic steroid-based contraceptives act by occupying the progesterone-binding site. This occupation effectively signals the woman's body into "thinking" she is pregnant. Thus, ovulation does not occur and the woman does not become pregnant.

Even though there is some controversy surrounding the use of contraceptives that prohibit ovulation, another use of this chemical knowledge has resulted in even more controversy: the development of post-fertilization birth control chemicals. A drug commonly known as RU-486 (mifepristone), developed by the French pharmaceutical company Roussel-Uclaf and used in Europe since 1988 to induce abortion, became available in the United States in 2000 under the brand name Mifeprex after Roussell-Uclaf donated the patent rights to the nonprofit Population

Council to avoid boycotts and lawsuits. Mifeprex is actually a mixture of mifepristone and misoprostol. Mifepristone is a progesterone antagonist, a mimic that occupies the progesterone-binding site but is able to send the appropriate chemical message to implant the fetus, and misoprotol is an abortifacient, so the fetus is spontaneously aborted. Patients must agree that they will have a surgical abortion if this procedure fails.

An alternative to drug-induced abortion involves the use of methotrexate in combination with misoprostol. Methotrexate inhibits normal cell growth and division to inhibit the development of the embryo and placenta. It is worth noting that these drugs have uses other than those discussed here, but they remain controversial because of their potential use as abortion agents.

So-called morning-after pills or emergency contraceptive pills actually consist of large doses of ordinary oral contraceptives to be taken shortly after sexual intercourse. These were initially used for rape victims but are now approved for emergency contraception when taken within 72 hours of unprotected intercourse. The FDA approved emergency contraceptives for over-the-counter use in 2013.

Synthetic steroids with estrogen-like activity are also used medically to replace natural hormones where there is a disease-related deficiency, after a hysterectomy, or after menopause. A major benefit of these drugs is a slowing of the normal bone loss (osteoporosis) that occurs with age.

Former Secretary of State Hillary Clinton speaks at a Planned Parenthood event. Founded in 1916, Planned Parenthood is a reproductive health care provider with more than 850 locations in the United States. Though it provides many other services, Planned Parenthood is one of the country's leading distributors of emergency and nonemergency contraception.

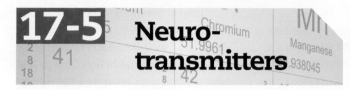

17-5 Neurotransmitters

Hormones and neurotransmitters have in common their roles as chemical messengers within the body. Neurotransmitters, as illustrated in Figure 17.3, carry nerve impulses from one nerve to the next or to the location where a response to the message will occur.

The body has a variety of neurotransmitters, each with its own distinctive molecular receptors and functions. New information about neurotransmitters is being reported frequently and is of great interest because of its usefulness to medicine. Most drugs that affect the brain or the nervous system interact with neurotransmitters or their receptors.

17-5a Norepinephrine, Serotonin, and Antidepressive Drugs

Norepinephrine and serotonin are neurotransmitters with receptors throughout the brain. Norepinephrine helps to control the fine coordination of body movement and balance, alertness, and emotion; it also affects mood, dreaming, and the sense of satisfaction. Serotonin is involved in temperature and blood pressure regulation, pain perception, and mood. Serotonin and norepinephrine appear to work together to control the sleeping and waking cycle.

Norepinephrine

Serotonin

The normal cycle of neurotransmitter action at nerve *synapses*, the gaps between nerve endings (Figure 17.3), is as follows:

1. The neurotransmitter is released from the neuron.

2. The neurotransmitter crosses the synapse to interact with the receptor.

FIGURE 17.3 Transmission of a nerve impulse.
The neurotransmitter is stored in the vesicles until it is needed. When a nerve impulse arrives, the vesicles move to the cell membrane and join with it so that the neurotransmitter is released. By crossing the synapse and binding to receptors on the surface of the adjacent nerve cell, the neurotransmitter transmits the impulse.

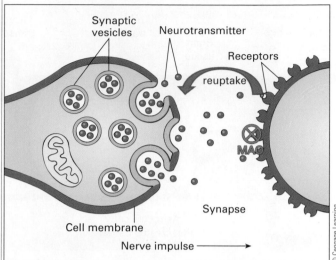

© Cengage Learning

3. The neurotransmitter is inactivated, either by reuptake by the neuron it came from or by conversion to an inactive form by an enzyme, often a monoamine oxidase (MAO).

The biochemistry of mental depression is not fully understood, but a deficiency of norepinephrine and serotonin (possibly also dopamine; see the next section) almost certainly plays a role. Evidence is provided by the manner in which three classes of drugs, illustrated by the following three compounds, influence the action of these neurotransmitters.

Amitriptyline, a tricyclic antidepressant
(Elavil)

Phenelzine, an MAO inhibitor
(Nardil)

Fluoxetine, an SSRI
(Prozac)

FIGURE 17.4 The blood–brain barrier.

Openings in brain capillary membranes are small enough to keep out large molecules, and because the membrane contains mostly lipid molecules, ions cannot cross the membrane either. Only lipid (fat)-soluble molecules can cross unaided. These limitations provide the barrier that protects the brain from toxic compounds but sometimes present a problem by also keeping out beneficial drugs.

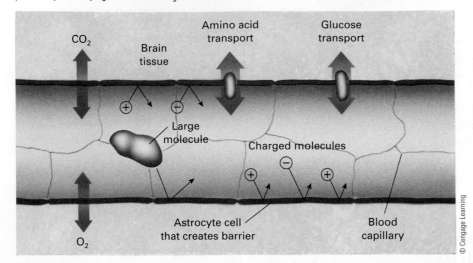

The tricyclic antidepressants such as Elavil (the trade name) prevent inactivation of neurotransmitters by preventing their reuptake by neurons, which increases the concentration of the neurotransmitter in the synapse. The monoamine oxidase (MAO) inhibitors such as Nardil diminish the action of MAO, which is the enzyme that inactivates norepinephrine and serotonin. The third type of drug action is represented by Prozac, which prevents the recapture of serotonin by neurons that release it. Drugs of this type (Zoloft is another), known as *selective serotonin reuptake inhibitors (SSRIs)*, have become the drugs of choice for treating serious clinical depression. Both types of drugs increase the levels of neurotransmitters in the synapse. For each of these three classes of drugs, the major mechanism of action is to increase the concentration of neurotransmitters at synapses.

17-5b Dopamine

Dopamine is produced in several areas of the brain, where it helps to integrate fine muscular movement as well as to control memory and emotion. An understanding of the brain chemistry of dopamine led to development of an effective treatment for Parkinson's disease. Patients with this disease experience trembling and muscular rigidity, among other symptoms, because of a deficiency of dopamine. Dopamine does not cross the blood–brain barrier (Figure 17.4) and thus cannot be administered as a drug. L-Dopa, it was found, can cross the barrier and then be converted to dopamine in the brain. While it does not cure Parkinson's disease, L-dopa can completely alleviate its symptoms for several years.

Dopamine has also been identified as the neurotransmitter that produces the feelings of well-being and reward associated with drug addiction. Drugs that block dopamine receptors have been used to treat schizophrenia, which is, however, a complex condition not attributable solely to dopamine activity.

17-5c Epinephrine and the Fight-or-Flight Response

Epinephrine, also known as adrenaline, is both a neurotransmitter in the brain and a hormone released from the adrenal gland. Its sudden discharge when we are frightened produces the *fight-or-flight response*, which includes increased blood pressure, dilation of blood vessels, widening of the pupils, and erection of the hair. Because of its widespread and rapid effects, epinephrine has a number of medical uses, notably in crisis situations. It is administered to counteract cardiac arrest (by stimulating the heart rate), to elevate dangerously low blood pressure (by constricting blood vessels), to halt acute asthma attacks (by dilating bronchial tubes), and to treat the extreme allergic reaction known as *anaphylactic shock*.

17-6 The Dose Makes the Poison

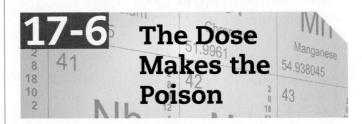

The amount of a chemical substance that enters the body is known as a *dose*. The substance might be a lifesaving medication or might be a poison. What accounts for the difference? *The dose makes the poison.* A German physician and chemist who referred to himself as Paracelsus recognized this in the 16th century, and it is still true. Whether a dose of a given substance is poisonous or not depends on the size of the dose. Most substances can be poisonous in a sufficiently large dose. For a given individual, age, gender, weight, and general state of health also play a role in the effect of a given dose. The term *risk-to-benefit ratio* is commonly used to refer to the balance that exists between the helpful (curative) and harmful (side-effect) properties of a pharmaceutical compound.

Doses of medications and poisons are customarily expressed as milligrams per kilogram of body weight (mg/kg). Aspirin, for example, is used to treat rheumatoid arthritis at a dosage of about 110 mg/kg per day. (A typical aspirin tablet contains just 325 mg; for an average 70-kg adult, two aspirin tablets amount to a dose of 9 mg/kg.)

A quantitative measure of toxicity is obtained by administering various doses of substances to be tested to laboratory animals (such as rats). The dose found to be lethal in 50% of a large number of test animals under controlled conditions is called the LD_{50} (lethal dose—50%, Chapter 2) and is usually reported in milligrams of the substance per kilogram of body weight. Thus, if a statistical analysis of data on a large population of rats showed that a dose of 1 mg/kg were lethal to 50% of the population tested, the LD_{50} for this poison would be 1 mg/kg. Obviously, species differences can produce different LD_{50} values for a given poison. For this reason, defining risk to human beings based on animal data is difficult. It is, however, generally safe to assume that a chemical with a low LD_{50} value for several species will also be quite toxic to human beings. In Table 17.3, compare the LD_{50} values for substances you have probably ingested at one time or another with those for substances you would want to avoid.

TABLE 17.3 LD_{50} Values for Several Chemicals

Chemical	LD_{50} (mg/kg administered orally to rat)
Sodium chloride	3750
Aspirin	1750
Ethanol	1000
Morphine	500
Caffeine	200
Heroin	150
Lead	20
Cocaine	17.5
Parathion	13
Aflatoxin	10
Sodium cyanide	10
Nicotine	2
Strychnine	0.8
Sarin	0.4
Batrachotoxin	0.002*
Tetanus toxin	5×10^{-6}
Botulinum toxin	3×10^{-8}

*From a poisonous frog. LD_{50} in mice.

Every chemical substance has an LD_{50} associated with it. It's just that we normally think of LD_{50} values when we are talking about medicines or toxic substances. However, one should always be aware of the potential toxicity associated with substances we normally would not consider dangerous. Relatively small doses of these substances may elicit the same negative responses as smaller doses of more toxic substances. Some chemicals commonly used as pesticides, insecticides, and nerve agents have incredibly small LD_{50} values, as shown in Table 17.3. These substances should always be used with caution.

Analgesic: A drug that relieves pain

17-7 Painkillers of All Kinds

Drugs that relieve pain are known as **analgesics**. They range from aspirin to cocaine. Some are illegal drugs, some are prescription drugs, and some are OTC drugs.

17-7a Opium and Its Relatives

Opium, obtained from the unripened seed pods of opium poppies, contains at least 20 different compounds. Chemically, they are *alkaloids*—organic compounds that contain nitrogen, are bases, and are produced by plants. About 10% of crude opium is *morphine*, which is primarily responsible for the effects of opium. Named for the Greek god of dreams, Morpheus, by the German pharmacist who first isolated it from opium in 1803, morphine is medically valuable as a strong painkiller able to produce sedation and loss of consciousness. The term *opioid* is now applied to all compounds with morphine-like activity.

Heroin, the diacetate ester of morphine, does not occur in nature but can be synthesized from morphine. As shown in Figure 17.5, their structures differ in only one kind of functional group. Heroin is much more addictive than morphine and for that reason has no legal use in the United States. *Codeine*, a methyl ether of morphine, is one of the alkaloids in opium and is used in cough syrup and for relief of moderate pain. Codeine is less addictive than morphine, but its analgesic activity is only about one-fifth that of morphine. One of the most effective substitutes for morphine is *meperidine*, whose structure was identified in 1931, and is now sold as Demerol. It is less addictive than morphine.

17-7b Mild Analgesics

When milder general analgesics are required, few compounds work as well for many people as *aspirin*. In ancient times, salicylate extracts from willow bark were used as painkillers. Aspirin, first synthesized in the late 1800s and manufactured by the Bayer chemical company, is a derivative of the natural willow salicylates but with milder side effects. Each year, about 40 million pounds of aspirin are manufactured in the United States. Aspirin is also an **antipyretic** (fever reducer) and an **anti-inflammatory** agent. Aspirin inhibits cyclooxygenase, the enzyme that catalyzes the reaction of oxygen with polyunsaturated fatty acids to produce prostaglandins. Excessive prostaglandin production causes fever, pain, and inflammation—just the symptoms aspirin relieves.

Aspirin is not without its problems, however. It is known to reduce the ability of blood to clot, and frequent use of aspirin is a known factor in the incidence of peptic ulcers. These ulcers are crater-like sores that develop when the digestive juices produced by the stomach eat away the lining of the digestive tract. The combination of these two effects suggests that caution should be used by people with a history of gastrointestinal bleeding or clotting disorders, among other things, since aspirin may exacerbate bleeding problems. Health professionals often recommend that patients cease using aspirin at least five days before surgery or dental work that could lead to bleeding. Aspirin has also been implicated as a factor in the development of a condition known as Reye's syndrome. This is an unusual reaction to a viral infection that results in brain swelling and fatty disease of the liver and kidneys and can result in death. It tends to occur primarily in children during the recovery from flu or chicken pox. The exact cause of Reye's syndrome is unknown, but studies have shown that using aspirin or chemically similar salicylate medications to treat symptoms of viral illnesses increases the risk of developing the syndrome.

Despite these concerns, aspirin remains a highly recommended pain reliever.

A bottle of aspirin tablets that has developed the vinegar-like odor of acetic acid should be discarded. Acetic acid forms as aspirin ages and breaks down by reacting with moisture in the air.

Acetylsalicylic acid (aspirin)

Salicylic acid   Acetic acid

Flowering opium poppies.

©iStockphoto.com/AtWaG

FIGURE 17.5 Opioids.

Morphine and codeine are natural alkaloids in the opium poppy. Heroin is a synthetic derivative with similar activity as a drug, but it is dangerously addictive.

	R_1	R_2
Morphine	—O—H	—O—H
Codeine	—O—CH$_3$	—O—H
Heroin	—O—$\overset{O}{\overset{\|}{C}}$—CH$_3$	—O—$\overset{O}{\overset{\|}{C}}$—CH$_3$

Opioid structure

Several OTC aspirin alternatives are now available for pain sufferers. The three principal ones are *acetaminophen (Tylenol)*, *ibuprofen (Advil, Nuprin, Motrin)*, and *naproxen (Aleve)*. All except acetaminophen contain a carboxylic acid group, as does aspirin.

Acetaminophen
(Tylenol)

Ibuprofen
(Advil, Nuprin, Motrin)

Naproxen
(Aleve)

Acetaminophen is the only one that does not promote some bleeding in the stomach. Acetaminophen is an effective analgesic and antipyretic, but it is not an anti-inflammatory agent. Aspirin, ibuprofen, and naproxen are *nonsteroidal anti-inflammatory drugs (NSAIDs)*, as distinguished from anti-inflammatory drugs that are steroids, such as cortisone. Ibuprofen, originally available only by prescription (Motrin), is similar to aspirin in its effectiveness but causes less bleeding. Naproxen, also originally a prescription drug (Anaprox or Naprosyn), has as its principal advantage a long period of activity, making twice-a-day administration possible for round-the-clock pain relief.

A group of NSAIDs known as COX-2 inhibitors recently came under considerable scrutiny because of potential side effects. These drugs block an enzyme known as *cyclooxygenase-2*, and this action inhibits the production of chemical messengers known as prostaglandins that cause the pain and swelling associated with arthritis inflammation. COX-2 inhibitors included celecoxib (Celebrex), rofecoxib (Vioxx), and valdecoxib (Bextra). In late 2004, these drugs were shown to significantly increase the risk of major fatal and nonfatal heart attacks in clinical trial participants taking these drugs. While the absolute risk of a COX-2–induced heart attack remained small, the 50% increase in risk caused considerable concern. The initial response by the FDA was to recommend their removal from the market, but significant demands by arthritis sufferers have resulted in a reconsideration of this decision. Doctors may still prescribe some COX-2 inhibitors, but Merck, the manufacturer of Vioxx, was dealt a major blow in 2005 when a Texas jury awarded more than $250 million to a Texas woman whose husband died of a heart attack. He had been taking the arthritis painkiller. Since then, Merck has been brought to court a number of times on similar charges, with mixed results. Merck still faces thousands of lawsuits filed by some of the more than 20 million people who took the drug, and in 2007 the company set aside $4.85 billion for future legal claims from U.S. citizens. Ultimately, the increased risk of heart attack will be balanced against the benefits in pain relief available to users of these drugs.

17-8 Mood-Altering Drugs: Legal and Illegal

Everyone who drinks coffee or alcoholic beverages has personal experience with the effects of drugs classified as stimulants (the caffeine in coffee) or depressants (ethyl alcohol). Stimulants and depressants with stronger activity are among the drugs that are often *abused*—used in ways that are socially unacceptable or harmful. Although moderate alcohol consumption is an accepted social custom in the United States, alcohol is an abused drug for many individuals. Other kinds of drugs that are subject to abuse are opioids and hallucinogens.

Although the list of drugs of abuse is wide ranging, a central theme is their relationship to brain chemistry and the effect of the drug on neurotransmitters or their receptor sites. Before discussing the mood-altering drugs, it's important to understand the legal control of abused drugs. Table 17.4 lists the various classifications of the U.S. Drug Enforcement Administration. Schedule I drugs are not legally available in any manner. Schedule II drugs can be dispensed only once from a written prescription. A refill requires a new written prescription.

17-8a Depressants

The effect of central nervous system depressants depends more on the dose than on the particular drug. The sequence proceeds from **sedation**, or relaxation, to sleep to general anesthesia to coma and death. Medically, depressants are used to treat anxiety and insomnia.

The best-known depressants are the *barbiturates*. Variations in chemical structure produce a range of barbiturates from very short acting to very long acting. Barbiturates are potential drugs of abuse, and overdoses cause many deaths. They are especially dangerous when ingested

General barbiturate structure

with ethyl alcohol, another depressant, because the two together give a *synergistic* effect. Together, alcohol and barbiturates produce an effect greater than that of their combined individual effects.

The second major class of depressants is the *benzodiazepines*. The familiar *tranquilizers* Librium (chlordiazepoxide) and Valium (diazepam) are members of this group. Both the barbiturates and the benzodiazepines act at receptors for the neurotransmitter gamma-aminobutyric acid (GABA, $HOOCCH_2CH_2CH_2NH_2$), which normally *inhibits* rather than excites transmission of nerve impulses. Because of differences in their molecular mechanism of action, benzodiazepines are safer and less subject to abuse than barbiturates.

Ethyl alcohol (CH_3CH_2OH), like barbiturates, enhances the action of the neurotransmitter GABA. Unlike barbiturates, however, the effects are dose dependent. At low doses, higher brain functions are affected, producing decreased inhibitions, altered judgment, and impaired

control of motion. As dosage increases, reflexes diminish, and consciousness diminishes to the point of coma and death.

17-8b Stimulants

Stimulants are drugs that excite the central nervous system. They stimulate production of the neurotransmitters norepinephrine, serotonin, and dopamine in the brain, heart (which is stimulated to beat faster), and veins (which become constricted). Note the similarity in the structures of amphetamine and methamphetamine to epinephrine and norepinephrine (Section 17-5), which helps to explain their action.

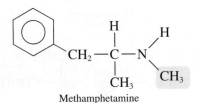

Amphetamine

Methamphetamine

Amphetamines were once available in OTC preparations used to stay awake. Now, the only approved ingredient for "stay-awake" pills is caffeine. Amphetamines have become controlled substances because of their great potential for abuse. The only generally accepted medical use of amphetamines is to treat *narcolepsy*, a condition of uncontrollable attacks of sleep.

Abuse of *methamphetamine* has become a major problem in the United States in recent years. Hardly a week goes by without a story about methamphetamine use making the front page of the newspaper or the evening news. Often these stories focus on relatively rural areas where the manufacture of methamphetamine and its distribution have become a

© iStockphoto.com/dr3amer

TABLE 17.4 Classification of Drugs

Designation	Description	Examples
Over-the-counter (OTC)	Available to anyone	Antacids, aspirin, most cough medicines
Prescription drugs	Available only by prescription	Antibiotics
Unregulated nonmedical drugs	Available in beverages, foods, or tobacco	Ethanol, caffeine, nicotine
Controlled substances*		
Schedule I	Abused drugs with a lack of accepted safety, and there is no accepted medical use.	Heroin, LSD, marijuana, methaqualone
Schedule II	Abuse may lead to severe physical and psychological dependence, but there is a currently accepted medical use with severe restrictions.	Morphine, PCP, cocaine, methadone, methamphetamine
Schedule III	Abuse may lead to moderate or low physical dependence or high psychological dependence, but there is a currently accepted medical use.	Anabolic steroids, codeine, hydrocodone with aspirin or Tylenol, and some barbiturates
Schedule IV	Abuse may lead to limited physical or psychological dependence relative to Schedule III drugs, but there is a currently accepted medical use.	Darvon, Talwin, Equanil, Valium, and Xanax
Schedule V	Abuse may lead to limited physical or psychological dependence relative to Schedule IV drugs, but there is a currently accepted medical use.	Cough medicines with codeine

© Cengage Learning

* The sale, distribution, and possession of these drugs or substances classified as controlled substances are controlled by the Drug Enforcement Administration of the U.S. Department of Justice. Detailed information about scheduling and abused drugs may be found at the following websites: http://www.usdoj.gov/dea/pubs/csa/812.htm#c and http://www.usdoj.gov/dea/concern/concern.htm

cottage industry of the very worst kind. The ready availability of a chemical that can easily be made into methamphetamine using information found on the Internet has created a nightmare drug problem of epidemic proportions. That chemical is *pseudoephedrine*, the major ingredient in many OTC cold medications such as Sudafed and Claritin-D.

A similar chemical known as *phenylpropanolamine* (simply replace the CH_3 group on the nitrogen atom of pseudoephedrine with a hydrogen atom) was, until recently, the active ingredient in Dexatrim, a common OTC diuretic.

Pseudoephedrine is a decongestant that shrinks blood vessels in the nose, lungs, and other mucous membranes. Drug dealers have learned how to isolate this ingredient from cold pills. Then, using common chemicals such as anhydrous ammonia, which is often stolen from farm storage tanks, they easily convert it into methamphetamine in houses, garages, and even traveling meth labs in cars, vans, and pickup trucks. The problem has become so serious that in 2004, Oklahoma became the first state in the nation to classify these common cold remedies as Schedule V narcotics. Oklahoma restricts their sale to pharmacies, requires the pills to be placed behind glass enclosures, limits the amount sold per customer, and requires purchasers to show photo identification and sign a computer register whose data is available statewide. In 2005 the U.S. Congress passed the Combat Methamphetamine Epidemic act, which put in place restrictions on the amount of pseudoephedrine-containing products that can be purchased and mandated the ways in which these products can be stored by pharmacies. One pharmaceutical company has begun to use a different active ingredient in its abused decongestant to help address the problem. This ingredient cannot easily be converted into methamphetamine. Despite these efforts at remediation, in all probability, the problem will continue for many years to come.

Cocaine

Cocaine, derived from the leaves of the coca plant of South America, is a stimulant and a Schedule II drug (used medically as a local anesthetic). By preventing the removal of norepinephrine from nerve endings, it causes uncontrolled

While more than a dozen states have decriminalized personal use or possession of marijuana, the drug remains illegal under the federal Controlled Substances Act of 1970. In October 2009, the Obama administration announced that it would not seek to arrest medical marijuana users or suppliers on federal charges as long as they conform to state laws. In 2007, more than 775,000 arrests were made for possession of marijuana, which amounts to 42.1% of all drug arrests that year.

Hallucinogen: A drug that causes perceptions with no basis in the real world

firing of the nerves. *Crack* is a form of cocaine obtained by heating a mixture of cocaine and sodium bicarbonate. The reaction is an acid–base reaction since the base, sodium bicarbonate, neutralizes cocaine hydrochloride, the usual form in which cocaine is isolated. The term *crack* refers to the crackling sound made by the heated mixture during the release of carbon dioxide as bicarbonate reacts with acid. The appearance of crack cocaine on the illegal drug market has caused an increase in the number of cocaine addicts because crack is much more addictive than cocaine. The "high" lasts less than ten minutes, creating the need to use crack repeatedly over a short period. Many users become addicted after only a single use, and there is a high risk of taking a lethal dose.

17-8c Hallucinogens

Hallucinogens are chemicals that cause vivid illusions, fantasies, and hallucinations. Many hallucinogens have been found in plants, including *mescaline*, which comes from the fruit of the peyote cactus, and *lysergic acid diethylamide (LSD)*, which is made from lysergic acid derived from either the morning glory or ergot, a fungus that grows on grasses. Others, such as MDMA and 2C-E (2,5-dimethoxy-4-ethylphenethylamine), can only be synthesized in laboratories. These drugs are not addic-

tive in the same manner as cocaine but sometimes cause destructive behavior and lingering psychological problems. The medicinal value of hallucinogens, which are believed to activate serotonin receptors, is a hotly contested issue.

Marijuana is a mild hallucinogen and sedative made from the hemp plant, *Cannabis sativa*. Although the millions of marijuana users regard it as a "safe" drug, this is a controversial conclusion. At high doses, marijuana may be moderately addictive and can create paranoia and intense anxiety. Long-term use at moderate doses may cause a general state of disinterest in personal achievements. In addition, tetrahydrocannabinol (THC), the active ingredient in marijuana smoke, can damage the lungs, impede brain function, and hamper the immune system. Animal studies also suggest that it may produce birth defects in offspring.

Phencyclidine (PCP, known as "angel dust") is an especially dangerous drug with a unique pattern of effects. At low doses, its effects resemble those of alcohol. With higher doses, hallucinations set in and behavior can become hostile and self-destructive, promoting psychoses that can last for weeks. Physical effects include seizures, coma, and death from cardiac arrest.

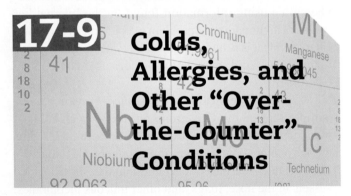

17-9 Colds, Allergies, and Other "Over-the-Counter" Conditions

What is the first thing we do when we feel sick? According to estimates, 75% of all illnesses are treated with products from the drugstore shelves. There are more than 300,000 OTC products on the market. But note that there are about only 800 active ingredients packaged in all of these different combinations. The top three categories of nonprescription medications in terms of sales dollars are used to treat allergies and colds, pain, and gastrointestinal problems.

About one person in ten suffers from some form of allergy. Allergic symptoms occur when special cells in the nose and breathing passages release *histamine*. Histamine is a neurotransmitter that accounts for most of the symptoms of hay fever, bronchial asthma, and other allergies. In a familiar theme in drug design, *antihistamines* are medications that block histamine receptors. Chlorpheniramine and compounds with very similar chemical structures (for example, brompheniramine and diphenhydramine) are ingredients in many OTC preparations for treating hay fever and the "stuffy" noses associated with allergies. A frequent side effect of many antihistamines is drowsiness. In fact, many OTC sleep aids contain antihistamines.

N=CH—CH₂CH₂NH₂ structure

Histamine
(neurotransmitter that
causes allergic reaction)

Chlorpheniramine structure with CHCH₂CH₂N(CH₃)₂ and Cl

Chlorpheniramine
(antihistamine, e.g., in
Chlor-Trimeton)

Diphenhydramine structure CHOCH₂CH₂N(CH₃)₂

Diphenhydramine
(antihistamine used in sleep aids,
e.g., Sleep-Eze)

The common cold is caused by a virus, and like other, more serious viral diseases, it cannot truly be cured by any known method. The best we can do is treat the symptoms. Most OTC preparations for colds contain a variety of ingredients, usually two or more of those listed in Table 17.5. We have already discussed antihistamines and analgesics. There is some doubt about how useful an antihistamine is in treating a cold.

Decongestants shrink nasal passages to relieve the stuffiness that accompanies colds and allergies. The decongestants activate receptors for epinephrine (an ingredient in some cold medications) and similar neurotransmitters. *Antitussives* suppress coughing. Opioids all have antitussive activity, and codeine is often used for this purpose. Dextromethorphan (Table 17.5) is an opioid-related compound that has antitussive activity without the analgesic and other

TABLE 17.5 Typical Ingredients in a Combination OTC Cold Medication

Type of Ingredient (Purpose)	Common Example	
	Name	*Structure*
Decongestant (shrink nasal tissues)	Pseudoephedrine	structure with OH, CH—CHNHCH₃, CH₃
Antitussive (prevent cough)	Dextromethorphan	structure with N—CH₃, CH₃O
Expectorant (loosen fluids in cough)	Guaifenesin	structure with OCH₃, OCH₂CHCH₂OH, OH
Analgesic (diminish pain, fever)	Acetaminophen	structure HO—NH—C(=O)—CH₃
Antihistamine (counteract allergic reaction)	Chlorpheniramine	structure with N, CHCH₂CH₂N(CH₃)₂, Cl

© Cengage Learning

Poison ivy. Contact with the urushial oil found in poison ivy can cause a severe allergic reaction. Calamine lotion, which contains a mixture of zinc oxide (ZnO) and 0.5% iron(III) oxide (Fe_2O_3), acts as a soothing **antipruritic** *(anti-itching agent).*

Andy Crawford and Steve Gorton/Dorling Kindersley/Getty Images

effects of opioids. Dextromethorphan is a nonaddictive opiate. Its name derives from the fact that it is the optically active stereoisomer of methorphan that rotates plane-polarized light in a clockwise (dextrorotatory) direction (Section 15-1). Its mirror-image stereoisomer, levomethorphan, is an addictive opiate that rotates plane-polarized light in a counterclockwise direction. *Expectorants* are meant to stimulate secretions in the respiratory tract so that mucus is dislodged in coughing. Most likely, guaifenesin is the only effective ingredient of this type.

A few guidelines for selecting and using OTC products are recommended by numerous groups concerned with public health:

- Choose single-ingredient products specific to the condition you have.

- Cut down on unnecessary expense by choosing generic products. (The chemical ingredients are the same.)

- *Read* labels, and *follow instructions* for dosage.

- *Pay attention* to cautions with respect to drowsiness or interactions with alcohol or other medications.

17-10 Preventive Maintenance: Sunscreens and Toothpaste

Many drugstore products are directed toward preventing bodily harm rather than curing an existing condition. Among the most important are sunscreens, which protect against skin cancer, and toothpaste, which prevents not only cavity formation but also deterioration of the gums and eventual tooth loss.

Our world is bathed in ultraviolet (UV) radiation that is sufficiently energetic to harm living things exposed to it. The natural protective mechanism against UV radiation is an increase in the skin of the pigment known as melanin, producing what we call a "tan." The melanin molecules absorb some of the UV energy and convert it to heat, thus diminishing damage to the molecular structure of the skin. However, even with melanin's protection, exposure to sunlight causes trouble. Especially in fair-skinned people whose skin contains smaller amounts of melanin, the result is a visible reddening— *erythema* or, in everyday language, sunburn. Although it is less visibly noticeable, dark-skinned people can also experience sunburn. Evidence has been mounting that the risk of

developing skin cancer rises with the amount of lifetime exposure to the Sun's rays (Section 7-5).

In 2004, the UV index developed by the National Weather Service and the U.S. Environmental Protection Agency (EPA) in 1994 was replaced by the Global Solar UV Index. The new scale rates the level of UV exposure expected on a given day as measured at noon, although the actual UV rises and falls as the day progresses. The color-coded scale is shown in Table 17.6 on the next page.

If you are going to be exposed to direct sunlight, it is a good idea to protect yourself from as much UV exposure as you can. Besides physical barriers such as long-sleeve shirts and wide-brimmed hats, a variety of chemicals will selectively absorb UV light. These *sunscreens* can be applied as oils, creams, or lotions. The sunscreen must function in a manner similar to melanin by absorbing UV light and converting it into heat energy. Many sunscreens contain *p*-aminobenzoic acid (PABA) and other chemicals with similar structures as the active ingredients. These compounds absorb an appreciable amount of the most harmful UV radiation.

Sunscreens have sun protection factors listed prominently on their labels. The *sun protection factor (SPF)* is defined as the ratio of time required to produce a perceptible erythema on a site protected by a specified dose of the sunscreen to the time required for minimal erythema development on unprotected skin. An SPF of 4, for example, would provide four times the skin's natural sunburn protection. The time required for sunburn depends on an individual's skin type and on the intensity of the UV radiation. The EPA recommends against any communication of the "time it takes to sunburn" since that implies a safe period in which no protection is required. *There is no safe period since any change in your skin's natural color is a sign of damage to the skin.* Tanning does not protect you from skin cancer. Always use a sunscreen with an SPF of at least 15.

© Lourens Smak/Alamy

TABLE 17.6 The Global Solar UV Index Established in 2004*

Exposure Category	Index Number	Sun Protection Messages
LOW	1–2	• Wear sunglasses on bright days. In winter, reflection off snow can nearly double UV strength. • If you burn easily, cover up and use sunscreen.
MODERATE	3–5	• Take precautions, such as covering up and using sunscreen, if you will be outside. • Stay in shade near midday when the Sun is strongest.
HIGH	6–7	• Protection against sunburn is needed. • Reduce time in the Sun between 11 a.m. and 4 p.m. • Cover up, wear a hat and sunglasses, and use sunscreen.
VERY HIGH	8–10	• Take extra precautions. Unprotected skin will be damaged and can burn quickly. • Try to avoid the Sun between 11 a.m. and 4 p.m. Otherwise, seek shade, cover up, wear a hat and sunglasses, and use sunscreen.
EXTREME	11+	• Take all precautions. Unprotected skin can burn in minutes. Beachgoers should know that white sand and other bright surfaces reflect UV and will increase UV exposure. • Avoid the Sun between 11 a.m. and 4 p.m. • Seek shade, cover up, wear a hat and sunglasses, and use sunscreen.

* UV rays are only about half as intense 3 hours before and after the peak. Physical surroundings such as snow, sand, and water reflect more UV and intensify exposure. Latitude and altitude also play a role; exposure increases with proximity to the equator and with altitude.

Source: http://www.epa.gov/sunwise/doc/uviguide.pdf

Keeping teeth clean requires *toothpaste*, a mixture of detergents and *abrasives*, which are hard substances that help remove unwanted materials on the tooth surface. The structure of tooth enamel is essentially that of a stone composed of calcium carbonate ($CaCO_3$) and calcium hydroxy phosphate [apatite—$Ca_{10}(PO_4)_6(OH)_2$]. Despite being the hardest substance in the human body, tooth enamel is readily attacked by acids. Because the decay of some food particles produces acids, it is important to keep teeth clean.

Within moments after you clean your teeth, a transparent film composed of proteins from saliva begins to coat the teeth and gums. This coating offers a place for food debris to collect and for oral bacteria to multiply. These bacteria convert dextrins, from the breakdown of sugars, into acids. At the same time, a tenacious film composed of these bacteria, food particles, and their breakdown products begins to form. As the film hardens, it becomes dental *plaque*. If this plaque is not removed regularly and completely from the surface of the teeth and beneath the gum line by brushing and flossing, the generation of acids and other harmful substances continues, destroying the tooth or the gum and eventually the bone that holds it in place.

Plaque that is not removed from the teeth becomes calcified from minerals in the saliva. The calcified plaque is known as *tartar*. It is possible to control tartar buildup by using toothpastes containing sodium pyrophosphate ($Na_4P_2O_7$), which interferes with the mineral crystallization that causes tartar buildup. Beneath the gum line, tartar is a special problem because its presence makes it easier for plaque to grow, which irritates gum tissue and allows the gum to become diseased. Only a dentist or oral hygienist can remove tartar from beneath the gum line. By keeping teeth free from plaque and from prolonged contact with the acids produced by plaque bacteria, we can preserve the hard, stone-like enamel of the tooth.

The abrasive material in toothpaste serves to cut into the surface deposits, and the detergent assists in suspending the particles in the rinse water. Abrasives commonly used in toothpastes include hydrated silica (a form of sand, $SiO_2 \cdot n\, H_2O$), hydrated alumina ($Al_2O_3 \cdot n\, H_2O$), and calcium carbonate ($CaCO_3$). It is difficult to select an abrasive that is hard enough to cut the surface contamination yet not so hard as to cut the tooth enamel. The choice of detergent is easier; any good detergent will do quite well. Because the necessary ingredients in toothpaste are not very palatable, it is not surprising to see various flavorings, sweeteners, thickeners, and colors included to appeal to our senses.

Tooth decay occurs when bacteria eat food particles that remain in tiny fissures and crevices in the teeth and produce acids that attack the tooth enamel. Fortunately, it is possible to modify the crystalline structure of tooth enamel and make it more resistant to decay by the addition of fluoride ion (F^-) to toothpaste. When regularly applied, some of the fluoride ions actually replace the hydroxide ions in the hydroxyapatite structure to form fluorapatite [$Ca_{10}(PO_4)_6F_2$]. The fluoride ion forms a stronger ionic bond in the crystalline structure than the hydroxide ion because of its high concentration of negative charge; as a result, the fluoroapatite is harder and less subject to acid attack than the hydroxyapatite. Hence, there is less tooth decay. Fluoride ion is introduced into essentially all public water supplies in the United States for this purpose. Concentrations of fluoride ion of one part per million (ppm) have proved safe and efficient for reducing tooth decay. About 80% of all toothpaste sold in this country contains fluoride ions in some form. Compounds such as stannous fluoride (SnF_2) and sodium monofluorophosphate (Na_2FPO_3) provide a low fluoride ion concentration in toothpastes.

In the United States, more teeth are now lost as a result of gum disease than from decay. Gum disease results from the lack of proper massage, from irritating deposits below the gum line, from bacterial infection, and from poor nutrition. More attention is being given to toothpastes containing disinfectants such as peroxides in addition to the other ingredients.

©iStockphoto.com/deliormanli

17-11 Heart Disease

Many drugs and surgical techniques now in use are able to decrease the death rate and improve the quality of life for persons suffering from heart disease. However, heart disease remains the number one killer of Americans. Known medically as **cardiovascular disease**, heart disease results from any condition that decreases the flow of blood, and consequently oxygen, to the heart or diminishes the ability of the heart to beat regularly and function in a normal manner.

The most common cause of heart disease is the plaque buildup on artery walls known as *atherosclerosis*, which was discussed in Section 16-3 in conjunction with the role of diet in plaque formation. If changes in lifestyle do not successfully combat this condition, the next step is cholesterol-lowering drugs, which include *lovastatin* and *cholestyramine*. Lovastatin acts by interfering with cholesterol synthesis in the liver. Cholestyramine acts by binding to bile acids in the intestines and accelerating their excretion. This causes the liver to convert more cholesterol into bile acids, leaving less to enter the circulatory system.

A result of plaque buildup in blood vessels can be *angina*, or chest pain during exertion, which occurs because of insufficient oxygen delivery to the heart muscle. The attacks are brought on when the heart must work harder and thus increase its oxygen demand. To treat angina, *vasodilators*, drugs that *dilate* veins (make them open wider), are used. When the veins are dilated, the blood pressure against which the heart must work is reduced. The classic vasodilators are organic nitrogen compounds such as nitroglycerin or amyl nitrite.

High blood pressure also contributes to heart disease. For this condition the next step after lifestyle changes is use of a **diuretic**, most commonly a *thiazide* (for example, Diazide), which stimulates the production of urine and excretion of Na^+. With increased urine output, blood volume and consequently blood pressure are decreased.

The development of beta-blockers, drugs used to treat angina and other aspects of heart disease, illustrates how understanding the biochemistry of disease can lead to design of drugs.

In the 1960s, two types of receptors that are part of the natural regulatory system for heart rate were discovered and named *beta-receptors*. The beta-1 receptors are located primarily in the heart—stimulation of these sites speeds up the rate at which the heart beats. The beta-2 receptors are located in the peripheral blood vessels and the bronchial tubes. Stimulation of the beta-2 receptors relaxes muscle fibers, opening up the blood vessels and bronchial tubes so that blood flows more easily, making it easier to breathe deeply and quickly. These receptors are stimulated by the natural hormones epinephrine and norepinephrine during the fight-or-flight response (Section 17-5).

Armed with this information, chemists began to search for compounds that would compete with epinephrine and norepinephrine at the beta-receptor sites. If these sites could be blocked, stimulation of the heart muscle could be prevented. For a heart already overworked from the buildup of plaque in the arteries, this might produce enough relaxation to avoid an impending heart attack. In addition, these drugs might be able to relieve high blood pressure.

The first successful drug of this type, a *beta-blocker*, was *propranolol (Inderal)*, now used to treat cardiac arrhythmias, angina, and hypertension. Look back at Section 17-5 to see its similarity in structure to the compounds whose action it blocks. Propranolol and the other beta-blockers have become widely prescribed drugs.

OH
|
OCH₂CHCH₂NHCH
CH₃
CH₃

Propranolol (Inderal)

A heart attack (a *myocardial infarction*) results from a reduction in blood flow to the heart muscle. About 98% of all heart attack victims have atherosclerosis, and the reduction in blood flow can usually be traced to a clot in the plaque of a coronary artery. If prolonged, the blockage causes part of the heart muscle to die from lack of oxygen; if the damage is sufficient, the heart attack can be fatal. It is important to realize that although death from heart attack is most common among older people, the condition of atherosclerosis begins many years earlier.

New, clot-dissolving drugs given to heart attack victims in the emergency room or the ambulance show promise in reducing the death rate. These drugs are enzymes that act on *plasminogen*, a natural factor in the blood, by converting it to *plasmin*. Once this happens, plasmin proceeds to dissolve blood clots by the body's own natural mechanism. Three enzymes have been developed as drugs that catalyze the plasminogen ⟶ plasmin conversion: (1) *urokinase*, a natural enzyme isolated from human urine; (2) *streptokinase*, isolated from a *Streptococcus* bacterium, and (3) *tissue plasminogen activator (TPA)*, one of the first drugs produced by recombinant DNA technology to receive government approval. TPA is made in genetically altered hamster ovary cells and is identical to the natural human enzyme that activates plasminogen.

Carcinogen: An agent that causes cancer

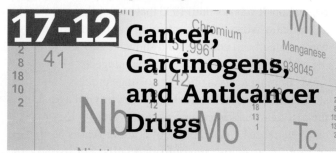

17-12 Cancer, Carcinogens, and Anticancer Drugs

Cancer is not just one but perhaps 100 different diseases. A cancer begins when a cell in the body starts to multiply without restraint and produces descendants that invade other tissues. It seems reasonable then that drugs might be able either to stop this undesirable spreading of cancer cells or to prevent cancer from happening at all. A major obstacle to successful drug treatment for cancer is that its biochemistry is not well understood. There is, however, general agreement that cancer is initiated by damage to DNA, which may be done by physical agents (such as ionizing radiation), biological agents (such as viruses), or chemical agents (such as compounds in cigarette smoke).

Every cancer comes from a single cell—one that is a modification of a normal cell. A normal cell functions according to directions stored in its genetic data bank, the DNA, and when a cell divides, each new cell gets its own exact copy of the parent DNA. If anything disrupts this DNA replication process, the genetic code in one of the descendant cells may cause that cell to grow and function differently from a normal cell.

Carcinogens are chemicals that cause cancer, which manifests itself in at least three ways. First, the *rate of cell growth* (that is, the rate of cellular multiplication) in cancerous tissue differs from the rate in normal tissue. Cancerous cells may

Pink ribbons promote awareness of breast cancer.

©iStockphoto.com/VikaValter

divide more rapidly or more slowly than normal cells. Second, cancerous cells *spread to other tissues* in a process called metastasis; they know no bounds. Normal liver cells divide and remain a part of the liver. Cancerous liver cells may leave the liver and be found, for example, in the lung. Third, most cancer cells show *partial or complete loss of specialized functions*. Although located in the liver, cancerous cells no longer perform the functions of the liver.

Attempts to determine the chemical causes of cancer have evolved from early studies in which the disease was linked to a person's occupation. We now know that a person's lifestyle plays a role as well. In 1775, Percivall Pott, an English physician, first noticed that people employed as chimney sweeps had a higher rate of skin cancer than the general population. It was not until 1933 that benzo[a]pyrene ($C_{20}H_{12}$, an aromatic hydrocarbon containing five fused carbon rings) was isolated from coal dust and shown to be metabolized in the body to produce one or more carcinogens.

Carcinogenesis is often a two-stage process. In the first stage, *initiation*, a chemical, physical, or viral agent alters the cell's DNA. Sometimes a single exposure to some carcinogen causes a rapid onset of a tumor that is composed of rapidly growing, uncontrolled cells, but usually the abnormal cells continue to reproduce in about the same way as normal cells around them. Then, a *promotion* occurs. This is the second stage and may occur days, months, or years after the initiation. This promotion may be a physical irritation or exposure to some toxic chemical that is itself not a carcinogen. In either case, the promotion results in the killing of a large number of cells. The destruction of cells is almost always compensated for by a sudden growth of new cells, and the abnormal cells begin to grow in ways the original DNA coding never intended. The cancer has started.

Cigarette smoke contains over 20 aggressive carcinogens, such as benzo[a]pyrene, a polycyclic aromatic hydrocarbon.

To illustrate the initiation and promotion aspects of carcinogenesis, consider some experiments performed in 1947 at Oxford University in England. First, very small doses of dimethylbenzanthracene (DMBA), a known carcinogenic component of coal tar, were applied to the skin of a group of mice. These mice were then separated into two groups. In one group the exposed skin was daubed with croton oil, a strongly irritating natural oil. The other group of mice had their skin daubed with croton oil four months later. Almost every mouse in both groups developed a tumor where the DMBA had been applied. In other groups of mice tested, neither croton oil nor DMBA alone produced any tumors, and if croton oil was applied first to the skin, followed by the DMBA, tumors failed to appear. Apparently DMBA had an initiation effect, whereas croton oil had a promotion effect.

Cancers are treated by (1) surgical removal of cancerous growths and surrounding tissue; (2) irradiation to kill the cancer cells; and (3) chemicals that kill the cancer cells, a process referred to as cancer *chemotherapy*. Cancer patients are considered cured if, after their treatment, they die at about the same rate as the general population. Another definition of success in cancer therapy is given by the number of patients who survive for five years after the treatment. In the 1930s fewer than 20 cancer patients in 100 were alive five years after treatment; in the 1940s, it was 25 in 100; in the 1960s, it was 33 in 100; and today it is close to 60 in 100.

Men and women show some significant variations in the types of cancers to which they are susceptible. Figure 17.6 shows the estimated number of cancers by type predicted for men and for women in the United States in 2009 along with the estimated numbers of deaths expected from each of these cancer types. Prostate cancer accounts for one-fourth of the cancer cases in men, but lung and bronchial cancers cause the most deaths, with prostate cancer deaths accounting for only 9% of total male cancer deaths. In women, breast cancer accounts for just over one-fourth of the cancer cases but only about 15% of the deaths. Once again, lung and bronchial cancers account for the largest percentage (26%) of cancer deaths in women. Figures 17.7 and 17.8 show the trend in the number of cancer deaths by type for men and women from 1930–2005 and 1930–2001, respectively. It is encouraging that in the past few years, deaths from all types of cancers have declined for both men and women except in one instance. The death rate due to lung and bronchial cancers in women had been increasing for many years but now appears to be, at best, leveling off.

FIGURE 17.6 Ten leading cancer types for the estimated new cancer cases and deaths by gender, U.S. 2009. (Source: Cancer Statistics 2009: A Presentation from the American Cancer Society.)

Cancer Cases*

		Men 766,130	Women 713,220		
Prostate	25%			27%	Breast
Lung & bronchus	15%			14%	Lung & bronchus
Colon & rectum	10%			10%	Colon & rectum
Urinary bladder	7%			6%	Uterine corpus
Melanoma of skin	5%			4%	Non-Hodgkin lymphoma
Non-Hodgkin lymphoma	5%			4%	Melanoma of skin
Kidney & renal pelvis	5%			4%	Thyroid
Leukemia	3%			3%	Kidney & renal pelvis
Oral cavity	3%			3%	Ovary
Pancreas	3%			3%	Pancreas
All other sites	19%			22%	All other sites

Cancer Deaths*

		Men 292,540	Women 269,800		
Lung & bronchus	30%			26%	Lung & bronchus
Prostate	9%			15%	Breast
Colon & rectum	9%			9%	Colon & rectum
Pancreas	6%			6%	Pancreas
Leukemia	4%			5%	Ovary
Liver & intrahepatic bile duct	4%			4%	Non-Hodgkin lymphoma
Esophagus	4%			3%	Leukemia
Urinary bladder	3%			3%	Uterine corpus
Non-Hodgkin lymphoma	3%			2%	Liver & intrahepatic bile duct
Kidney & renal pelvis	3%			2%	Brain/ONS
All other sites	25%			25%	All other sites

* Excludes basal and squamous cell skin cancers and in situ carcinomas except urinary bladder.
ONS = Other nervous system.

FIGURE 17.7 Annual age-adjusted cancer death rates* among males for selected cancer types, U.S. 1930 to 2005. (Source: Cancer Statistics 2009: A Presentation from the American Cancer Society.)

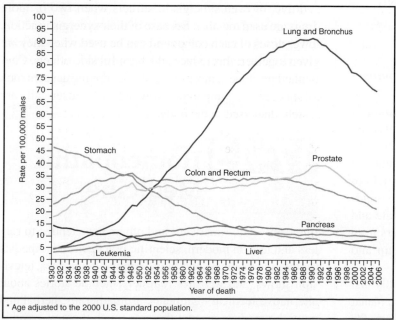

* Age adjusted to the 2000 U.S. standard population.

During World War I, the toxic effects of the mustard gases were found to include damage to bone marrow and changes in DNA (mutations) that created abnormal offspring. In addition, the so-called *nitrogen mustards* caused cancers in some animals.

Mustard gas

Nitrogen mustard (general formula)

A nitrogen mustard

When the wartime-imposed secrecy surrounding these chemicals ended, it occurred to some researchers that cancers might be treated with similar compounds. The result might be to alter the DNA in cancer cells to the extent that these cells could be destroyed selectively.

One of the most widely used anticancer drugs is now cyclophosphamide, a compound that contains the nitrogen mustard group.

Cyclophosphamide

Cyclophosphamide, and other anticancer drugs that act in the same manner, are *alkylating agents*—reactive organic compounds that transfer alkyl groups in chemical reactions. Their anticancer activity results from the transfer of alkyl groups (for example CH_3CH_2—) to the nitrogen bases in DNA, often to guanine. The alkyl group physically gets in the way of base pairing and prevents DNA replication, which stops cell division. Although alkylating agents attack both normal cells and cancer cells, the effect is greater for rapidly dividing cancer cells, a criterion that applies to all cancer chemotherapy agents.

FIGURE 17.8 **Annual age-adjusted cancer death rates* among females for selected cancer types, U.S. 1930 to 2005.** (Source: Cancer Statistics 2009: A Presentation from the American Cancer Society.)

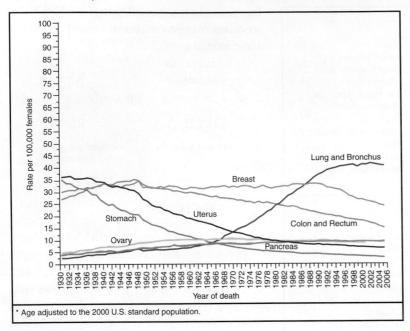

* Age adjusted to the 2000 U.S. standard population.

agent of this type because it is active against an unusually large number of kinds of cancer.

Doxorubicin

All cancer chemotherapy is tedious and has risks and unpleasant side effects. The problem is that no agent has yet been found that kills only cancer cells, so there is always a balance to be struck between killing healthy cells and killing cancer cells. Because chemotherapy drugs kill actively dividing cells, the goal is to kill many more cancer cells than normal cells (ideally, of course, 100% of the cancer cells). In addition to being highly toxic, most of the cancer chemotherapy drugs are themselves carcinogenic and very high doses are usually necessary. As a result, single-agent chemotherapy has largely given way to combination chemotherapy because of positive additive, or even synergistic, effects when two or more drugs are used together. Because of their synergistic action, lower doses of each compound can be used when they are given together; this reduces the harmful side effects. Chemotherapy alone is most successful in treating cancers such as leukemia or lymphoma, in which the cancer cells are widely dispersed in the body.

17-13 Homeopathic Medicine

In recent years, Americans and others have tended to rely less and less on modern, scientific practices and remedies for treating medical conditions. Popular magazines, television shows, and websites routinely include stories about alternative medical treatments, aromatherapy, herbal and natural remedies, magnet therapy, touch therapy, psychic healing, acupuncture, and homeopathy. The debate over the efficacy of such treatments continues. It is safe to say,

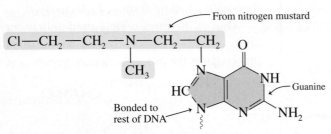

An alkylated guanine in DNA. You can see how the group alters the size of the guanine residue.

Another class of chemotherapy drug, the *antimetabolites*, interferes with DNA synthesis because these drugs are similar in molecular structure to compounds required for DNA synthesis (*metabolites*). Methotrexate, for example, is an antimetabolite similar in structure to folic acid (a B vitamin), used in the synthesis of nucleic acids. Methotrexate prevents reduction of folic acid in the first step of nucleic acid synthesis by strongly binding to the enzyme for that reaction. (Methotrexate and folic acid are large molecules. Methotrexate differs from folic acid by the addition of one $-CH_3$ group and replacement of a $C=O$ by a $C-NH_2$.)

Yet another way to attack DNA is by physically disrupting its shape. Drugs accomplish this by fitting a planar ring portion of their molecular structure in between the base pairs that form the ladder structure of DNA (see Figure 15.22). Doxorubicin is a most valuable chemotherapeutic

however, that most, if not all, of these treatment methods have not been subjected to the same rigorous scientific standards as most modern medical practices and treatments.

A discussion of all types of alternative medical treatments is beyond the scope of this book. We will, instead, focus on one area where it can be easily demonstrated that the science does not always support the claims. That area is **homeopathy**, a system of medical treatment based on the use of minute quantities of remedies that in larger doses produces effects similar to those of the disease or condition being treated.

Homeopathy is often confused by consumers with "natural remedies" or "herbal remedies," but homeopathy is a very particular approach to treating disease that has little or no basis in natural medicine. Modern medicine depends on the *dose-response relationship*, in which there is a direct relationship between the dosage of medicine taken and the effect it elicits. Most scientists believe that a chemical agent must be physically present in the biological system for it to cause an observable effect. The science of **pharmacology** relies on this accepted scientific principle, as does the science of chemistry. Even seemingly mundane activities such as adding sweetener to soft drinks or sugar to iced tea reveal that there is a relationship between concentration and effect.

Homeopathy's founder, Samuel Hahnemann, M.D. (1775–1843), is said to have based his theory on an experience in which he ingested cinchona bark, the source of quinine used to treat malaria. After taking it, he experienced thirst, throbbing in the head, and fever—symptoms common to malaria. He hypothesized that the drug's power to cure the disease arose from its ability to produce symptoms similar to the disease itself. He and his early followers then conducted "provings" in which they administered herbs, minerals, and other substances to healthy people, including themselves, and kept detailed records of what they observed.

Hahnemann was troubled by the toxic side effects of many of the substances he used, so he began experimenting with dilution of the medication. He claims to have discovered that although dilution eliminated the side effects, it did not eliminate the positive effects of the substances. This proposition became known as "the law of infinitesimals." He also developed ritualistic shaking (*succession*) of the medications after each successive serial dilution and claimed that the method of shaking could affect the outcome of the treatment. Hahnemann lived at roughly the same time as Amedeo Avogadro and was thus unaware of Avogadro's ideas concerning the number of molecules (6.02×10^{23}) in one mole of a substance. This may account for the fact that the excessive dilution factors that should worry modern scientists and advocates of homeopathy did not concern Hahnemann. Simple calculations show that many true homeopathic remedies do not contain a single molecule of the purported active ingredient. For example, a highly advertised homeopathic cold and flu cure known as Oscillococcinum, made from an extract of duck liver, is sold at a dilution that would require the consumption of 10^{377} L of the medicine to ensure the ingestion of only six molecules of the active ingredient!

True homeopathy has no scientific basis, and many scientists and consumer protection groups have called for increased control and regulation of homeopathic and herbal cures. Consumers spend billions of dollars yearly on natural supplements and well over $200 million on true homeopathic medicines. As the distinction between natural supplements and homeopathic remedies becomes increasingly blurred, one must urge caution in this area. Many natural herbal supplements and treatments are not homeopathic in nature, even though many have added the word *homeopathic* to their advertising to cash in on the cachet of the word. Even though it is well known that many proven medicines have their origin in indigenous natural remedies, few of the natural supplements currently on the market have been subjected to the rigorous standards required

Homeopathy: A system of medical treatment based on the use of minute quantities of natural substances that, in a healthy person, would produce symptoms of the same disease

Pharmacology: The branch of medicine concerned with the uses, effects, and modes of action of drugs

Samuel Hahnemann

by the FDA. When it is employed, such testing reveals that some supplements contain significant impurities and highly variable quantities of the purported active ingredient. In many cases, the natural supplements contain none of the purported active ingredient. These issues should be of con-cern to those taking their health seriously. Finally, patients are often reluctant to tell physicians they are seeing that they are taking such supplements. This is a serious issue because supplements may interact with physician-prescribed medications.

Your Resources

In the back of the textbook:

→ *Review card on Chemistry and Medicine*

  In OWL for CHEM2 at www.cengagebrain.com

→ *Review Key Terms with Flash Cards (printable or digital)*

→ *Complete Interactive Practice Quizzes to prepare for tests*

→ *Submit Assigned Homework and Exercises*

Applying Your Knowledge

1. Give examples of a bacterial disease and a viral disease.
2. Give definitions of the following terms:
   (a) Over-the-counter drug
   (b) Prescription drug
   (c) Generic name for a drug
   (d) Trade name for a drug
3. Name the agency responsible for classifying drugs in the United States.
4. What are antibiotics?
5. Name three major classes of antibiotics.
6. What do the following three acronyms represent?
   (a) AIDS   (b) HIV
   (c) AZT
7. What is chemotherapy?
8. Describe how penicillin kills bacteria.
9. What is a retrovirus?
10. For what disease or condition is each of the following classes of drugs used?
    (a) Analgesics   (b) Antipyretics
    (c) Antibiotics   (d) Antihistamines

11. For what disease or condition is each of the following classes of drugs used?
    (a) Vasodilators   (b) Alkylating agents
    (c) Beta-blockers   (d) Antimetabolites
12. Describe the role of a receptor in biochemistry.
13. What two classes of natural biomolecules require receptors for their action?
14. To what classes of drugs do the following compounds belong?
    (a) Barbiturates   (b) Amphetamines
15. To what classes of drugs do the following compounds belong?
    (a) Methotrexate   (b) Chlorphenirimine
16. Describe the following and give an example of each:
    (a) Schedule I drugs   (b) Schedule II drugs
17. Which of the following terms apply to codeine?
    (a) Analgesic   (b) Antibiotic
    (c) Opioid
    (d) A scheduled drug with a potential for abuse
    (e) An antitussive

18. How do antihistamines work in the body? What, if any, side effects do antihistamines have?

19. What are the symptoms experienced in angina?

20. Name a drug that might be used to treat angina.

21. What is a barbiturate? What are the physiological effects of barbiturates?

22. Nitrogen mustards are alkylating agents, drugs that interfere with DNA replication. Explain what this means.

23. What happens when beta-receptor sites in heart muscle are stimulated?

24. Name the class of biomolecule that includes dopamine, norepinephrine, and serotonin.

25. The estrogen estradiol has the following structure:

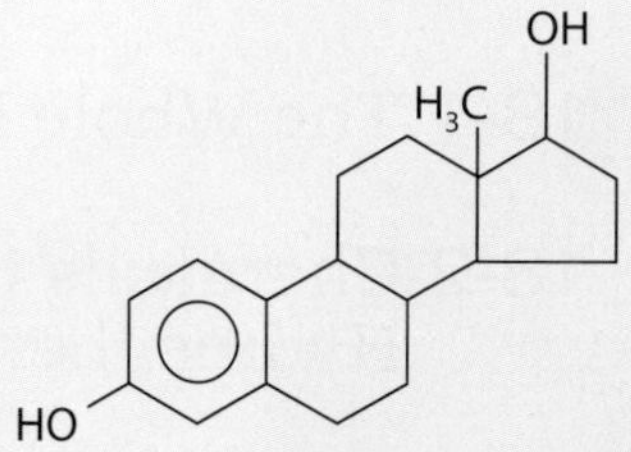

What functional groups are different in ethinyl estradiol?

26. What is the function of the hormone progesterone?

27. Give the functions for the following:
    (a) Dopamine   (b) Epinephrine

28. What are the applications for the following drugs?
    (a) Tylenol   (b) Ibuprofen

29. What are the four classifications of drugs in terms of Drug Enforcement Administration regulations?

30. How are cocaine and crack related?

31. Classify each of the following substances as either a hallucinogen, an antidepressant, or a depressant:
    (a) Mescaline
    (b) Lysergic acid diethylamide (LSD)
    (c) *Cannabis sativa*   (d) Phencyclidine (PCP)
    (e) Barbiturates   (f) Amphetamines

32. What are the functions of the following over-the-counter drugs?
    (a) Antihistamines   (b) Analgesics
    (c) Decongestants   (d) Antitussives

33. What disease is treated with each of the following drugs? Tell the function of each drug.
    (a) Vasodilators   (b) Diuretics

34. What are nitrogen mustards? What was their original purpose? What is their current medical use?

35. What are the three modes of treatment for cancers?

36. Penicillins have the general formula shown as follows. What is the R group in penicillin G? Why is it necessary to have a number of different penicillins?

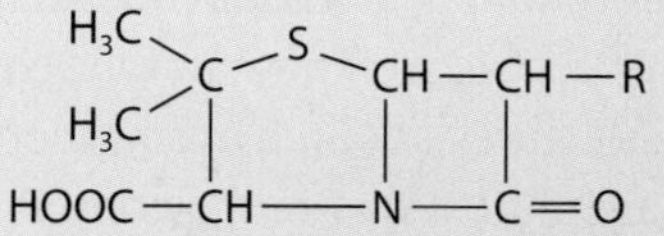

37. What is the effect of a hallucinogen? Name two examples of hallucinogens.

38. Describe the normal steps in the action of a neurotransmitter at a nerve synapse.

39. Of the three compounds heroin, morphine, and codeine,
    (a) which is the most effective pain killer?
    (b) which is not a natural alkaloid?
    (c) which is so addictive that its sale and use are illegal in the United States?

The Chemistry
of Useful Materials

The long view from space has dramatized what we already knew—the crust of Earth is a very unusual environment, uniquely suited, at least in this solar system, for the production and support of life as we know it. Our environment is also quite heterogeneous. Mixtures abound; everywhere we look, the elements and compounds are almost lost in the complicated array of mixtures produced by natural forces acting over very long periods.

Throughout most of history, we had not developed the power to alter our environment significantly. Early everyday objects, such as stone hammers or wooden plows, were only physically changed from the natural material. Then came the chemical reduction of copper from natural minerals, followed by iron, and now the flood of new materials produced each year. We have developed, beyond question, the power to change Earth's natural chemical mixtures in almost any way we choose.

In this chapter, our focus is on inorganic substances—the elements other than carbon, and their compounds. We look at the origins of the raw materials for our pots and pans, homes and office buildings, automobiles and airplanes, and a multitude of other manufactured items.

18-1   The Whole Earth

Although we rely on the Earth as a source of useful materials, we can access only a small fraction of the planet to obtain these materials. The **hydrosphere**, which includes salt water and freshwater above and below Earth's surface, is a commercial source of magnesium, bromine, and sodium chloride, which is not only table salt but also an essential chemical raw material.

The solid portion of the Earth available to us, the crust, is a very small part of the whole—less than 1% by mass. Geologists define Earth's crust as a region between the surface and a depth of about 5 km to 35 km that lies over regions of greater density (Figure 18.1).

Three major types of rocks are found in Earth's crust: *igneous rocks* (such as basalt), formed by solidification of molten rock; *sedimentary rocks* (such as sandstone, which is cemented sand), formed by deposition of dissolved or suspended substances from oceans and rivers; and *metamorphic rocks* (such as marble), formed by the action of heat and pressure on existing rocks. Figure 18.2 gives the average composition of Earth's crust. The most

Hydrosphere: All freshwater and salt water that is part of planet Earth abundant substances in rocks are silicates, which are composed of silicon, oxygen, and positive metal ions (Section 18-5). The more than 2000 kinds of known **minerals** fall into a few major classes (Table 18.1).

Mineral: A naturally occurring, solid inorganic compound

Fortunately for the mining industry, the composition of the crust is not uniform. The deepest mine, the TauTona mine in South Africa, extends only 3.94 km beneath the surface and follows a vein of gold-rich ore. Natural forces have concentrated different minerals in different places. For example, as molten rock gradually cools, the minerals that solidify first (those with the higher melting points) can sink in the remaining liquid and become concentrated. Or minerals can be redistributed according to variations in their solubility in natural waters.

Metals obtained from the Earth's crust are important materials in our modern society.

©iStockphoto.com/ricardoazoury

FIGURE 18.1 A cross section of Earth.

Geologists customarily list the composition of Earth in terms of oxides, as shown here.

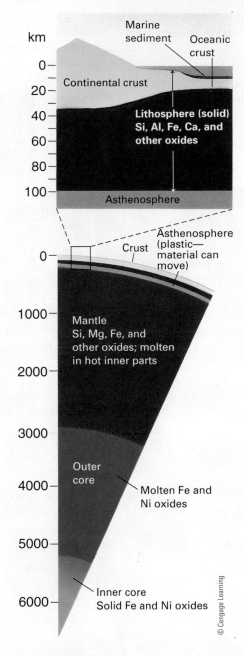

FIGURE 18.2 Relative abundance (by mass) of elements in Earth's crust compared to abundance in the whole Earth.

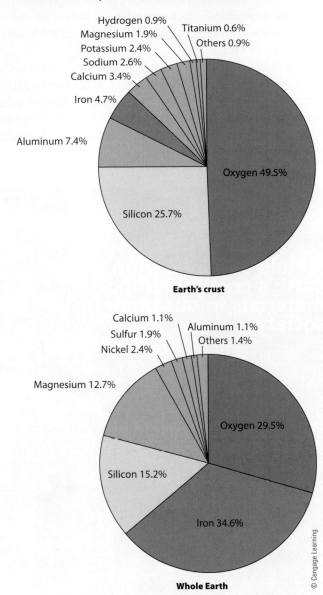

The western United States was once covered by a large, landlocked sea. The water evaporated, leaving huge deposits of sodium carbonate, a soluble salt that is a valuable chemical raw material. While most nations must manufacture sodium carbonate from other chemicals, the United States meets a large proportion of its needs by mining. Other minerals are concentrated elsewhere, of course. Most of the known deposits of nickel are in New Caledonia and Zimbabwe. Most of the chromium is in Botswana and Turkey. Each country must rely on imports of one kind or another.

TABLE 18.1 Major Mineral Groups in Earth's Crust

Mineral Group	Example	Formula	Uses
Silicates	Quartz	SiO_2	Glass, ceramics, alloys
	Feldspar	$KAlSi_3O_8$	Ceramics
Oxides	Hematite	Fe_2O_3	Iron ore, paint pigment
Carbonates	Calcite	$CaCO_3$	Optical instruments (pure crystals), industrial chemicals
Sulfides	Galena	PbS	Lead ore, semiconductors
Sulfates	Gypsum	$CaSO_4 \cdot 2\,H_2O$	Cement, plaster of paris, wallboard, paper sizing
Halides	Fluorite	CaF_2	Lasers and electronics (pure crystals), source of fluorine (F_2)

18-2 Chemicals from the Hydrosphere

Separation of salt from seawater by evaporation in San Francisco Bay.

© Shorelark/Alamy

A single mouthful of seawater is enough to convince anyone that it is salty. Indeed, sodium chloride is the major mineral in seawater. But consider that other dissolved minerals are constantly being deposited into the oceans from rivers, undersea volcanoes, and thermal vents. In addition to sodium and chlorine, the major elements present as dissolved ions are magnesium, sulfur, calcium, potassium, bromine, carbon, nitrogen, and strontium. Lower concentrations of other elements in seawater could also provide huge quantities of economically important metals such as uranium, copper, manganese, and gold.

The lower the concentration of a metal ion in seawater, of course, the higher the cost of isolating the metal is likely to be. Nevertheless, as high-quality mineral deposits on the land are depleted, the economics of mining the sea may become more attractive. Marine organisms may help to solve the problem. For example, one family of such organisms (the *tunicates*) accumulates vanadium to more than 280,000 times its concentration in seawater. Perhaps aquatic farming of such creatures could be put to work extracting metals.

18-2a Salt and the Chloralkali Industry

Perhaps the single most important material extracted from the hydrosphere is salt, which serves as the source of several very important chemical compounds. In coastal regions, salt is separated from seawater by evaporation of the water in large lagoons open to the Sun. The natural *brines*, or salty waters, found in wells and lakes such as the Great Salt Lake in Utah are other sources of salt. After isolation, the salt is purified to the degree required for its use. Much of it is destined for the *chloralkali industry*, which produces chlorine gas and sodium hydroxide.

Chlorine gas is used to disinfect drinking water and sewage and in the production of organic chemicals such as pesticides and vinyl chloride, the building block of plastics called polyvinyl chlorides (PVCs, Section

14-5). Chlorine gas is commonly among the top chemicals produced each year in the United States. Almost all chlorine gas is made by electrolysis of aqueous sodium chloride. The other product of sodium chloride electrolysis, sodium hydroxide, is equally valuable because it is the most commonly used base in industrial processes. The reaction in electrolysis of aqueous NaCl is

$$2\,NaCl(aq) + 2\,H_2O(\ell) \xrightarrow{\text{Electrical energy}} 2\,NaOH(aq) + H_2(g) + Cl_2(g)$$

A complicating factor in designing electrochemical cells for this reaction is that the chlorine and sodium hydroxide react with each other if they remain in contact.

A modern electrochemical cell for producing chlorine and sodium hydroxide is illustrated in Figure 18.3. The

FIGURE 18.3 A chloralkali cell.

Large banks of these cells produce gaseous chlorine and aqueous sodium hydroxide solution, both important industrial chemicals. Because of the need for cheap electricity, chloralkali plants are located near hydroelectric plants at, for example, Niagara Falls.

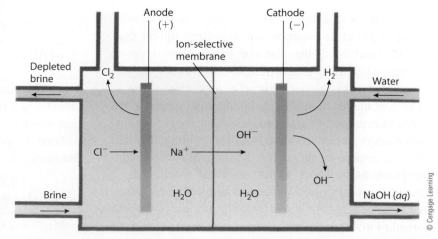

© Cengage Learning

Magnesium-aluminum alloys are used as a lightweight alternative to iron alloys in wheels.

© Riccardo Sergnese/Alamy

Alloy: A mixture of two or more metals

Precipitation: Formation of a solid product during a chemical reaction between reactants in a solution

two electrode compartments are separated by a synthetic membrane that allows only sodium ions that enter as part of the brine to pass through it. The reactions are

$$2\ Cl^-(aq) \longrightarrow Cl_2(g) + 2\ e^- \qquad \text{Oxidation}$$

$$2\ H_2O(\ell) + 2\ e^- \longrightarrow$$
$$H_2(g) + 2\ OH^-(aq) \qquad \text{Reduction}$$

$$2\ Cl^-(aq) + 2\ H_2O(\ell) \longrightarrow$$
$$H_2(g) + Cl_2(g) + 2\ OH^-(aq) \quad \text{Overall reaction}$$

Brine is introduced into the anode compartment, and chloride ion oxidation occurs there. To maintain charge balance within the cell, as Cl^- ions are oxidized, Na^+ ions must pass from the anode to the cathode compartment. Since OH^- ions are produced in the cathode compartment, the product there is aqueous NaOH, with a concentration of 20% to 35% by weight. Most sodium hydroxide is used in industrial processes, although it is also present in some oven, drain, and sewer-pipe cleaners.

18-2b Magnesium from the Sea

Magnesium, with a density of 1.74 g/cm³, is the lightest structural metal in common use. It is 36% lighter than aluminum and 78% lighter than iron. Many **alloys** designed for light weight and great strength contain magnesium. Most manufactured aluminum objects, for example, contain about 5% magnesium, which is added to improve the mechanical properties and corrosion resistance of the aluminum under alkaline conditions. There are also alloys that have the reverse formulation—that is, more magnesium than aluminum. These alloys are used where a high strength-to-weight ratio is needed and where corrosion resistance is especially important. The amount of magnesium used in American cars is expected to rise from about

4 kg per vehicle in 2002 to as much as 150 kg in some models by 2020. This usage is expected to increase as manufacturers produce lighter-weight cars to meet new federal fuel-economy standards.

Because there are 6 million tons of magnesium present as Mg^{2+} in every cubic mile of seawater, which is about 0.14% magnesium, the sea can furnish an almost limitless amount of this element. The recovery of magnesium from seawater begins with the **precipitation** of insoluble magnesium hydroxide [$Mg(OH)_2$] (Figure 18.4). The only thing needed for this step is a ready supply of an inexpensive base (a source of OH^- ions), a need fulfilled nicely by seashells, which contain calcium carbonate ($CaCO_3$). Heating the calcium carbonate converts it to lime, which then reacts with water to give calcium hydroxide, the base used in the precipitation.

$$\underset{\text{Seashells}}{CaCO_3(s)} \xrightarrow{\text{Heat}} \underset{\text{Lime}}{CaO(s)} + CO_2(g)$$

$$CaO(s) + H_2O(\ell) \longrightarrow Ca(OH)_2(aq)$$

$$Mg^{2+}(aq) + Ca(OH)_2(aq) \longrightarrow Mg(OH)_2(s) + Ca^{2+}(aq)$$

The solid magnesium hydroxide is isolated by filtration and then neutralized by another inexpensive chemical, hydrochloric acid.

$$Mg(OH)_2(s) + 2\ HCl(aq) \longrightarrow MgCl_2(aq) + 2\ H_2O(\ell)$$

When the water is evaporated, solid hydrated magnesium chloride is left. After drying, it is melted (at 708°C) and then electrolyzed in a huge steel pot that serves as the cathode. Graphite bars serve as the anodes. The electrode reactions are

$$2\ Cl^- \longrightarrow Cl_2(g) + 2\ e^- \qquad \text{Oxidation}$$

$$Mg^{2+} + 2\ e^- \longrightarrow Mg(\ell) \qquad \text{Reduction}$$

$$Mg^{2+} + 2\ Cl^- \longrightarrow Mg(\ell) + Cl_2(g) \quad \text{Overall reaction}$$

FIGURE 18.4 An overview of the Dow process for magnesium production.

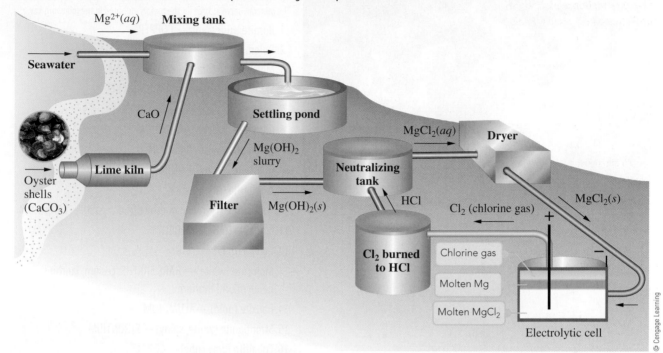

As the molten magnesium forms, it is removed. The chlorine produced in the electrolysis is recycled into the process by reacting it with hydrogen to generate hydrogen chloride, which combines with water to produce hydrochloric acid.

18-3 Metals and Their Ores

According to the U.S. Geological Survey, the average American will require the quantities of minerals listed in Table 18.2 on an annual basis. In this section, we focus on the metals. Some of the chemistry of familiar silicon-based materials—glass, ceramics, cement—is discussed in Section 18-5.

Less reactive metals such as copper, silver, and gold can be found as free elements. The somewhat more reactive metals are present as sulfides formed early in Earth's existence (such as CuS, PbS, and ZnS). Because of their extremely low water solubility, sulfides resist oxidation and reactions with water and other ions. The still more reactive metals have been converted over millennia into oxides (such as MnO_2, Al_2O_3, and TiO_2) and are mined in that form.

The most reactive metals, such as sodium and potassium, are present in nature as soluble salts in the ocean and mineral springs; in solid deposits of these salts; or in insoluble, stable aluminosilicates, such as albite ($NaAlSi_3O_8$)

and orthoclase ($KAlSi_3O_8$). Such silicates are found in all parts of the world, but because of their great stability they are not currently used as sources of the metals they contain.

TABLE 18.2 Approximate U.S. per Capita Yearly Requirement of New Raw Materials from Earth's Crust

Element or Mineral	Quantity (pounds)	Major Uses
Stone, sand, gravel	19,000	Roads and buildings
Coal	7650	Generating electricity; iron, steel, and chemical manufacturing
Iron ore	600	Automobiles and ships, structural support
Clays	280	Bricks, paper, paint, glass, pottery
Salt	420	De-icing, detergents, cooking, chemical manufacturing
Aluminum	75	Food and beverage cans, household items, vehicles
Copper	25	Electrical motors, wiring
Zinc	15	Brass, galvanized iron, steel
Lead	15	Auto batteries, solder, electronics parts
Manganese	6	Iron and steel production; glass decolorizer; brick and ceramic colorant

Source: http://minerals.usgs.gov/west/

© 1988 Paul Silverman/Fundamental Photographs

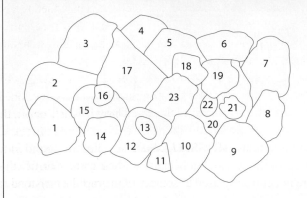

FIGURE 18.5 A collection of native metals and minerals.
The minerals serve as ores for the metals indicated in the following key.

1. Bornite (iridescent)—COPPER
2. Dolomite (pink)— MAGNESIUM
3. Molybdenite (gray)—MOLYBDENUM
4. Skutterudite (gray)—COBALT, NICKEL
5. Zincite (mottled red)—ZINC
6. Chromite (gray)—CHROMIUM
7. Stibnite (gray)—ANTIMONY
8. Gummite (yellow)—URANIUM
9. Cassiterite (rust)—TIN
10. Vanadinite crystal on goethite (red crystal)—VANADIUM
11. Cinnabar (red)—MERCURY
12. Galena (gray)—LEAD
13. Monazite (white)—RARE EARTHS: cesium, lanthium, neodymium, thorium
14. Bauxite (gold)—ALUMINUM
15. Strontianite (white, spiny)—STRONTIUM
16. Cobaltite (gray cube)—COBALT
17. Pyrite (gold)—IRON
18. Columbinite (tan, gray stripe)—NIOBIUM, TANTALUM
19. Bismuth (shiny)
20. Rhodochrosite (pink)—MAGNESIUM
21. Rutile (shiny twin crystal)—TITANIUM
22. Filigree on quartz (gray)—NATIVE SILVER
23. Pyrolusite (black, powdery)—MANGANESE

Slag: Mixture of nonmetallic waste products separated from a metal during its refining

These minerals, like the ocean, represent a resource that may have to be tapped when richer and more easily processed ores are depleted.

A selection of beautiful minerals that are also ores is pictured in Figure 18.5. (A mineral is an *ore* if separating the metal from it is possible and economically practical.) The preparation of metals from minerals requires chemical reduction—the conversion of positive metal ions to free metals. To reduce a metal ion requires a source of electrons, which can be either an electrical current (as in the case of magnesium production mentioned earlier) or a chemical reducing agent.

18-3a Iron

Iron is the fourth most abundant element in Earth's crust and the second most abundant metal. Our economy depends on iron and its alloys, particularly steel. Most of the world's iron is located in large deposits of iron oxides in the United States (Minnesota), Sweden, France, Venezuela, Russia, Australia, and the United Kingdom.

Iron ore is reduced in a blast furnace (Figure 18.6a). The solid material fed into the top of the furnace is a mixture of iron oxide (Fe_2O_3), coke (C), and limestone ($CaCO_3$). A blast of heated air is forced into the furnace near the bottom. The major reactions that occur within the blast furnace result in reduction of iron oxide

$$2\ C(s) + O_2(g) \longrightarrow 2\ CO(g) + heat$$

$$Fe_2O_3(s) + 3\ CO(g) \longrightarrow 2\ Fe(s) + 3\ CO_2(g) + heat$$

and conversion of silica present in the ore to molten calcium silicate ($CaSiO_3$).

$$CaCO_3(s) \xrightarrow{\text{Heat}} CaO(s) + CO_2(g)$$

$$CaO(s) + SiO_2(s) \xrightarrow{\text{Heat}} \underset{\text{Slag}}{CaSiO_3(\ell)}$$

Consequently, as the blast furnace operates, two molten layers collect in the bottom. The lower, denser layer is mostly liquid iron. The upper, less dense layer is the **slag**. From time to time, the furnace is tapped at the bottom, and the molten iron is drawn off. Another outlet somewhat higher in the blast furnace can be opened to remove the liquid slag.

The iron that comes from the blast furnace, known as *pig iron*, contains many impurities (up to 4.5% carbon, 1.7% manganese, 0.3% phosphorus, 0.04% sulfur, and 1% silicon). Iron reacts with the carbon impurity at the temperatures of the blast furnace to form cementite, an iron carbide (Fe_3C), which causes pig iron to be brittle.

$$3\ Fe(s) + C(s) \longrightarrow Fe_3C(s)$$

When molten pig iron is poured into molds of a desired shape (engine blocks, brake drums, transmission housings), it is called **cast iron**. However, pig iron and cast iron contain too much carbon and other impurities for most uses. The structurally stronger material known as **steel** is obtained by removing the phosphorus, sulfur, and silicon impurities and decreasing the carbon content.

18-3b Steels

Many kinds of iron alloys are collectively known as *steels*. The most common is *carbon steel*, an alloy of iron with about 1.3% carbon. To convert pig iron into carbon steel,

the excess carbon is burned out with oxygen.

In the *basic oxygen process* for steel production (Figure 18.6b), pure oxygen is blown into molten iron through a refractory tube, which is pushed below the surface of the iron. A *refractory* is a material that withstands high temperatures without melting. At elevated temperatures, the dissolved carbon reacts very rapidly with the oxygen to produce gaseous carbon monoxide and carbon dioxide, which escape.

During steelmaking, silicon or transition metals such as chromium, manganese, and nickel can be added to give alloys with specific physical, chemical, and mechanical properties. For example, the addition of chromium results in the production of stainless steel.

The physical properties of steel can be adjusted by the temperature and rate of cooling used in its production. If the steel is cooled rapidly by quenching in water or oil, the carbon in the steel remains in the form of cementite (Fe_3C) and the steel will be hard, brittle, and light colored. Slow cooling favors the formation of crystals of carbon (graphite) instead of cementite. The resulting steel is more *ductile* (easily drawn into shape).

All of the processes in steelmaking, from the blast furnace to the final heat treatment, use tremendous quantities of energy, mostly in the form of heat. In the production of one ton of steel, approximately one ton of coal or its energy equivalent is consumed.

Cast iron: Molded pig iron or other carbon–iron alloy

Steel: Malleable iron-based alloy with a relatively low percentage of carbon

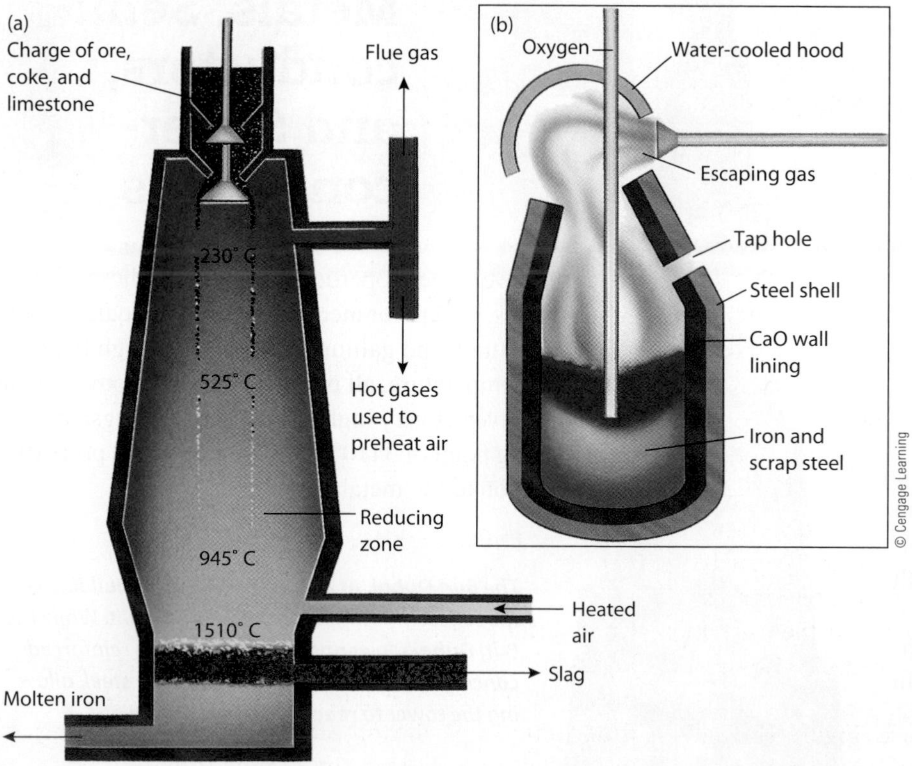

18-3c Copper

Although copper metal occurs in the free (elemental) state in some parts of the world, the supply available from such sources is quite insufficient for the world's needs. The majority of the copper used today is obtained from various copper sulfide ores, such as chalcopyrite ($CuFeS_2$), chalcocite (Cu_2S), and covellite (CuS). Because the copper content of these ores is about 1% to 2%, the powdered ore is first concentrated by the flotation process.

FIGURE 18.6 Iron and steel production.

In the blast furnace (a), the descending materials in the charge are hit by a blast of hot air from coke burning in the heated air. The basic oxygen furnace for steel production (b) is charged with a mixture of molten pig iron from the blast furnace, steel scrap to be recycled, and other metals according to the type of steel being made. After oxygen is blown in for about 20 minutes, the finished steel is poured off through the tap hole, and the furnace is ready for another charge.

(a)
Charge of ore, coke, and limestone
Flue gas
230° C
525° C
945° C
1510° C
Hot gases used to preheat air
Reducing zone
Heated air
Slag
Molten iron

(b)
Oxygen
Water-cooled hood
Escaping gas
Tap hole
Steel shell
CaO wall lining
Iron and scrap steel

© Cengage Learning

In the flotation process, the powdered ore is mixed with water and a frothing agent such as pine oil. A stream of air is blown through the mixture to produce froth. The **gangue** in the ore, which is composed of sand, rock, and clay, is easily wetted by the water and sinks to the bottom of the container. In contrast, a copper sulfide particle is hydrophobic—it is not wetted by the water. The copper sulfide particle becomes coated with oil and is carried to the top of the container in the froth. The froth is removed continuously, and the floating copper sulfide minerals are recovered from it.

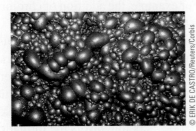

The separation of CuS from gangue by flotation.

© ERIK DE CASTRO/Reuters/Corbis

The preparation of copper metal from copper sulfide ore involves **roasting** the ore in air to convert some of the copper sulfide and any iron sulfide present to the oxides.

$$2\,Cu_2S(s) + 3\,O_2(g) \longrightarrow$$
$$2\,Cu_2O(s) + 2\,SO_2(g)$$

$$2\,FeS(s) + 3\,O_2(g) \longrightarrow$$
$$2\,FeO(s) + 2\,SO_2(g)$$

Subsequently, the mixture is heated to a higher temperature, and some copper is produced by the reaction.

$$Cu_2S(s) + 2\,Cu_2O(s) \xrightarrow{\text{Heat}}$$
$$6\,Cu(s) + SO_2(g)$$

The product of this operation is a mixture of copper metal, sulfides of copper, iron and other ore constituents, and slag. The molten mixture is heated in a converter with silica materials. When air is blown through the molten material in the converter, two reaction sequences occur. In one, the iron is converted to slag.

$$2\,FeS(s) + 3\,O_2(g) \longrightarrow 2\,FeO(s) + 2\,SO_2(g)$$
$$FeO(s) + SiO_2(s) \longrightarrow FeSiO_3(\ell)$$
Molten slag

In the other, the remaining copper sulfide is converted to copper metal by the preceding two reactions shown for Cu_2S. The copper produced in this manner is crude or "blister" copper (96%–99.5% Cu), the blistered surface resulting from the escaping gas. The blister copper is later purified electrolytically.

In the electrolytic purification of copper, the anodes are blister copper bars and the cathodes are made of pure copper. As electrolysis proceeds, copper is oxidized at the anode, moves through the solution as Cu_2^+ ions, and is deposited on the cathode. The voltage of the cell is regulated so that more active impurities (such as iron) are left in the solution and less active ones are not oxidized at all. The less active impurities include gold and silver, which collect as "anode slime," an insoluble residue beneath the anode. The anode slime is subsequently treated to recover the valuable metals.

The copper produced by the electrolytic cell is 99.95% pure and is suitable for use as an electrical conductor. Copper for this purpose must be pure because very small amounts of impurities, such as arsenic, considerably reduce the electrical conductivity of copper.

18-4 Properties of Metals, Semiconductors, and Superconductors

Metals have some properties totally unlike those of other substances. Except for mercury, which is a liquid at room temperature, and gallium, which melts slightly above room temperature, all metals are solids. Some remain solids even at very high temperatures. Tungsten has a melting point of 3410°C. There are several properties common to the metals:

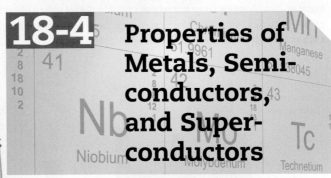

The Burj Dubai. At 818 m (2684 ft), this building is the tallest man-made structure ever built. While the Burj Dubai's lower portion is framed in reinforced concrete, its upper levels are built with steel, allowing the tower to reach its impressive height.

© iStockphoto.com/Philip Lange

- *High electrical conductivity*. Metal wires easily carry electrical currents. The electrons in metals are quite mobile.

- *High thermal conductivity*. Metals are much better conductors of heat than other materials. A plastic spoon used to stir hot tea will not heat up nearly as much as a metal spoon. Some metals conduct heat more than other metals. Silver has twice the thermal conductivity of steel.

- *Ductility and malleability*. Most metals can be drawn into wire (ductility) or hammered into thin sheets (malleability); gold is the most malleable metal. Extremely thin sheets of gold are used for decoration.

- *Luster*. Polished metal surfaces reflect light; most metals have a silvery white color because they reflect all wavelengths of light equally well.

- *Insolubility in water and other common solvents*. No metal dissolves in water, but a few, such as the Group 1 and 2 metals, react with water to form hydrogen gas and solutions of metal hydroxides.

Any theory of the bonding of metal atoms must be consistent with these properties. Structural investigations of metals have led to the conclusion that solid metals are composed of regular arrays, or *lattices*, of metal ions in which the bonding electrons are loosely held. Figure 18.7 illustrates one model for metallic bonding in which the regular array, or lattice, of positively charged metal ions is embedded in a "sea" of mobile electrons. These mobile valence electrons are delocalized over the entire metal crystal, and the freedom of these electrons to move throughout the solid is responsible for the properties associated with metals. In contrast to those in metals, the valence electrons in nonmetals are fixed in bonds between like atoms. This means that nonmetals are nonconductors of electricity.

18-4a Semiconductors: The Basis of Our Modern World

You should not be surprised that somewhere between the excellent electrical conductivity of most metals and the nonconductivity of nonmetals, there are some elements that are **semiconductors**.

> **Semiconductor:**
> A material with electrical conductivity intermediate between those of metals and insulators

FIGURE 18.7 "Electron sea" model of bonding in metals. The positively charged metal atom nuclei are surrounded by a "sea" of negatively charged electrons.

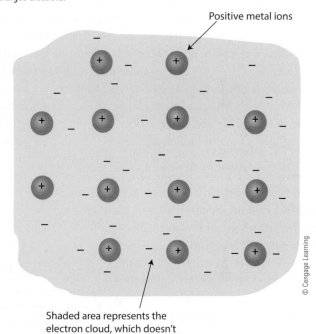

Positive metal ions

Shaded area represents the electron cloud, which doesn't belong to any one ion.

© Cengage Learning

 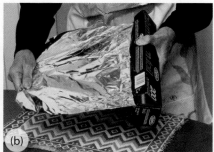 

Putting the properties of metals to use. (a) Joining two wires with molten solder, a fusible alloy. (b) Preserving food by covering it with malleable aluminum foil. (c) Building large, unyielding structures out of iron and steel.

Doping: In semiconduction, the addition of atoms of an element with extra or fewer valence electrons than the element of the semiconductor material

Transistor: A semiconductor device that controls electron flow in circuits

This means they will conduct electricity under certain conditions. Silicon, when it is in a highly purified state, is a semiconductor. Silicon acts like a nonmetal and fails to conduct a current until a certain voltage is applied; then it begins to conduct moderately well. This behavior interested electrical engineers, who recognized that silicon might act like a "gate" for electron flow in electrical circuits. This electron gate–like activity was not realized until a process known as **doping** was discovered. One common dopant is boron, a Group 13 element just to the upper left of silicon in the periodic table. When boron is added to pure silicon, the boron atoms, with one fewer valence electron than silicon, introduce *positive holes* in the lattice arrangement of silicon atoms. The presence of these positive holes (which can move about in the solid just like electrons but in the opposite direction) makes the doped silicon somewhat more conductive—in effect, it becomes a better electron gate. Silicon doped with an element that creates positive holes is called a *p-type* semiconductor. Another dopant is arsenic, from Group 15, which contains one *more* valence electron than silicon. Silicon doped with arsenic is called an *n-type* semiconductor because there are extra negative electrons present in the solid. These, of course, enhance the conductivity of the doped silicon. Through careful control of the amount and type of dopant, the conductivity of the silicon can be adjusted to a fine degree (Figure 18.8).

In 1947, a device consisting of a layer of *p*-type silicon sandwiched between two *n*-type layers was constructed by John Bardeen, Walter Brattain, and William Shockley at Bell Laboratories. This device, called the **transistor**, has revolutionized our world (Figure 18.9). Bardeen, Brattain, and Shockley shared the 1956 Nobel Prize in physics for their discovery. The importance of the transistor was recognized as soon as it was discovered. Although it was first demonstrated at Bell Laboratories in December 1947, the transistor wasn't announced until July 1, 1948, after patent applications had been filed. Because the transistor can control electron flow in circuits with such accuracy, yet is so small and requires so little power to operate, it is now possible to design electronic circuits to fit into extremely small volumes. Such objects as camcorders as small as peas, radios small enough to strap to the back of ants, and other amazing devices can be made using transistors based on doped silicon. Of course, more mundane items such as cell phones, microwave ovens, and video game consoles also owe their existence to the transistor. The central processing unit (CPU) of computers consists of millions of transistors and other circuit elements fabricated on wafers of pure silicon (Figure 18.9). These devices are called integrated circuits.

18-4b Solar Energy to Electricity

The *solar cell* or photovoltaic cell converts energy from the Sun into electron flow. Solar cells are now used in calculators, watches, spaceflight applications, communication satellites, and signals for automobiles and trains as well as the source of electricity in undeveloped regions throughout the world, where electrical power grids are virtually nonexistent. It has been estimated that all of the electricity used in the United States could be made by solar cells having only 10% efficiency and covering about 13,000 km^2, which is 0.13% of the land area in the United States.

FIGURE 18.8 Doping of silicon in semiconducting devices.
Adding atoms with five valence electrons (e.g., arsenic) introduces extra electrons that can move through the crystal. Adding atoms with three valence electrons introduces holes that can also move through the crystal. Note that each silicon atom has four valence electrons, so a perfect crystal has each silicon atom surrounded by an octet of electrons.

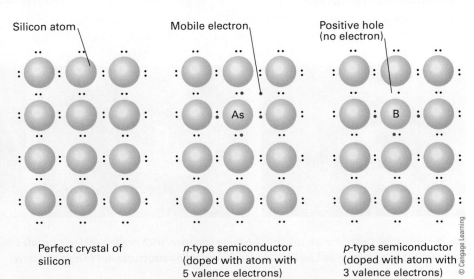

Perfect crystal of silicon

n-type semiconductor (doped with atom with 5 valence electrons)

p-type semiconductor (doped with atom with 3 valence electrons)

© Cengage Learning

FIGURE 18.9 The first transistor, developed at Bell Laboratories, and a modern computer chip (inset) that contains millions of transistors.

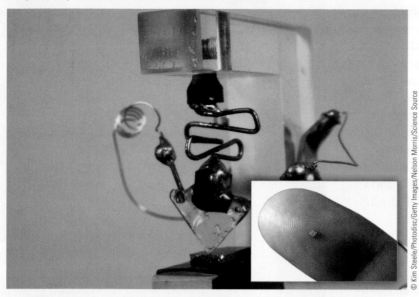

© Kim Steele/Photodisc/Getty Images/Nelson Morris/Science Source

One type of solar cell consists of two layers of almost pure silicon. The lower, thicker layer contains a trace of boron (B); the upper, thinner layer has a trace of arsenic (As). As pointed out earlier, the As-enriched layer is an *n*-type semiconductor, with mobile electrons; the B-enriched layer is a *p*-type semiconductor, with the electron deficiencies known as holes (see Figure 18.8). There is a strong tendency for the mobile electrons in the *n*-type layer to pair with the unpaired electrons in the holes in the *p*-type layer. If the two layers are connected by an external circuit (Figure 18.10) and light of sufficient energy strikes the surface and is absorbed, excited electrons can leave the *n*-type layer and flow through the external circuit to the *p*-type layer. As this layer becomes more negative because of added electrons,

electrons are repelled internally back into the *n*-type layer, which is now positive and attracts the electrons. The process can continue indefinitely as long as the cell is exposed to sunlight.

18-4c Can Anything Conduct Better Than a Metal?

When metals are heated, their electrical conductivity decreases. Lower conductivity at higher temperatures can be explained if the movement of the valence electrons is considered to be limited by rapidly vibrating atoms in the metal lattice. The kinetic molecular theory says that higher temperatures mean more motion, and this applies even when the atoms are in fixed positions, as they are in a metal. When the metal atoms are relatively stationary, as they are when the metal is cool, the electrons can move through the lattice much like a person moving through a room filled with a large number of other people quietly chatting with one another. When the temperature is elevated, the metal atoms begin to vibrate wildly, and the electrons have more trouble getting through the lattice, in much the same way that a person would have trouble moving through the room if all of the other occupants suddenly became agitated.

From this picture of electrical conductivity of metals, you might assume that if a sufficiently low temperature were reached, conductivity might be quite high (almost zero resistance, in other words). In fact, the conductivity of a pure metal crystal does approach infinity (zero resistance) as a temperature of absolute zero (0 Kelvin, or –273°C) is approached. In many metals, however, a more interesting thing happens before absolute zero is reached. At a certain low temperature, the conductivity suddenly increases, as though absolute zero had already been reached. At this temperature, called the *superconducting transition temperature*, the metal becomes a **superconductor** of electricity. The superconductor offers no resistance

FIGURE 18.10 A solar cell.

Beneath the outer glass is a metal grid that allows as much light as possible to strike the *n*-type semiconductor layer while serving as the electrode at which electrons leave the cell. The *n*-type semiconductor layer is almost transparent. Beneath it is the *p*-type semiconductor layer and the electrode at which electrons re-enter the cell.

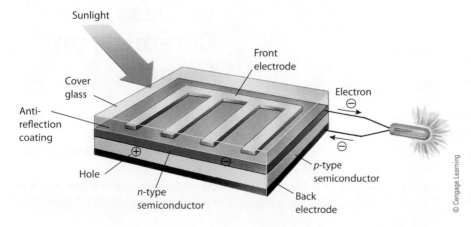

Sunlight

Cover glass

Anti-reflection coating

Front electrode

Electron ⊖

⊖

Hole

n-type semiconductor

p-type semiconductor

Back electrode

© Cengage Learning

whatever to electrical flow. This phenomenon means that electric motors made of superconducting wires would be 100% efficient, and electrical transmission lines could be made 100% efficient. Resistance to electron flow causes energy loss in motors, transmission lines, and other electrical devices. In superconductors, the hindrance of electron flow by vibrating atoms in the metal lattice has been replaced by some kind of cooperative action that allows electron movement. Table 18.3 lists some of the metals that have superconducting transition temperatures. Not all metals display superconducting properties.

The relatively low transition temperatures of the metals shown in Table 18.3 mean that it would be impractical to make superconducting motors or transmission lines from them. Shortly after the superconductivity of metals was discovered, certain alloys (mixtures of metals) were prepared that had much higher transition temperatures than the metals themselves. Niobium alloys showed the most promise, but they still had to be cooled to below 23 K ($-250°C$) to exhibit superconductivity. To maintain such a low temperature would require liquid helium, which costs about $6 per liter—an expensive proposition for all but the most exotic applications.

In January 1986, K. Alex Müller and J. Georg Bednorz, scientists at an IBM laboratory in Switzerland, discovered that a barium–lanthanum–copper oxide became superconducting at 35 K, a temperature that must be maintained with liquid helium. This discovery provoked a flurry of activity that quickly resulted in a substance that became superconducting at 90 K. $LaBa_2Cu_3O_x$ was then found to superconduct at temperatures above the boiling point of nitrogen (77 K, $-196°C$). At less than 50 cents per liter, liquid nitrogen is a much cheaper refrigerant than liquid helium. As of 2010, the record was held by $HgBa_2Ca_2Cu_3O_9$, which is a superconductor at 135 K.

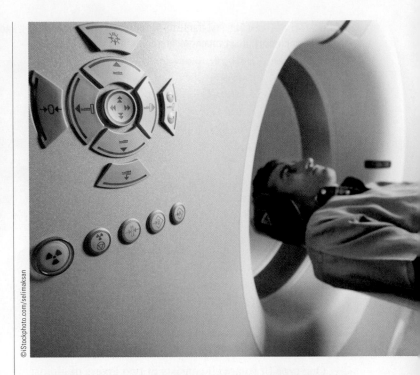

©iStockphoto.com/selimaksan

Why the great excitement over the potentialities of superconductivity? Superconducting materials are being used to build more powerful electromagnets, such as those used in nuclear particle accelerators (Section 13-6) and in magnetic resonance imaging (MRI) machines, which are used in medical diagnosis. One of the main factors affecting the efficiency of MRI machines is the heating of the electromagnet due to electrical resistance. Many scientists are saying that the discovery of high-temperature superconductors may prove to be more important than the discovery of the transistor because of its potential effect on electrical and electronic technology. For example, the use of superconducting materials for transmission of electric power could save as much as 30% of the energy now lost because of the resistance of the wire.

18-5 From Rocks to Glass, Ceramics, and Cement

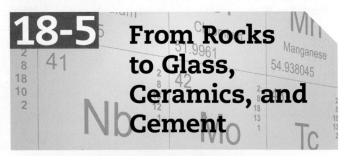

Earth's crust is largely held together by chemical bonds between silicon and oxygen, in either pure **silica** (SiO_2) or **silicate** minerals in which silicon–oxygen anions are combined with metal cations.

The most common of the many crystalline forms of pure silica is quartz. It is a major component of granite and

TABLE 18.3 Superconducting Transition Temperatures of Some Metals

Metal	Transition Temperature (K)
Aluminum	1.183
Gallium	1.087
Lanthanum	4.8
Lead	7.23
Niobium	9.17

© Cengage Learning

sandstone and also occurs as pure crystals. The basic structural unit of quartz and most silicates is the tetrahedron. In quartz, every silicon atom is bonded to four oxygen atoms, and every oxygen atom is bonded to two silicon atoms. The result is an infinite array of tetrahedra sharing corners.

Most silicates consist of networks of silicon–oxygen tetrahedra linked together in ways that range from chains, rings, and sheets to three-dimensional networks.

The simplest network silicates are the *pyroxenes*, which contain extended chains of linked SiO_4 tetrahedra (Figure 18.11). If two such chains are laid side by side, they may link up by sharing oxygen atoms in adjoining chains. The result is an *amphibole*. Because of their double-stranded chain structure, the amphiboles are fibrous materials. Examples of this type of silicate include jade and asbestos.

If the linking of silicate chains continues in two dimensions, sheets of SiO_4 tetrahedral units result (Table 18.4). Various clays and mica have this sheet-like structure. Clays, which are essential components of soils, are *aluminosilicates*—some Si^{4+} ions are replaced by Al^{3+} ions plus other cations that take up the additional positive charge. Feldspar, a component of many rocks and a network silicate, is weathered in the following reaction to form clay.

$$2 \text{ KAlSi}_3O_8(s) + CO_2(g) + 2 H_2O(\ell) \longrightarrow$$

Feldspar

$$\text{Al}_2(\text{OH})_4\text{Si}_2\text{O}_5(s) + 4 \text{ SiO}_2(s) + K_2CO_3(aq)$$

Kaolinite (a clay)

Some medications sold in the United States (for example, Kaopectate) contain highly purified clay that absorbs excess stomach acid and possibly harmful bacteria that cause stomach upset.

TABLE 18.4 Silicates

Si — O Tetrahedra Present as	Class Name	Structure
Individual anions	Orthosilicates	
Chains	Pyroxenes (linear chains)	
	Amphiboles (double chains)	
Sheets	Mica, talc, clays	
Three-dimensional networks	Silica, feldspars, zeolites	—

© Cengage Learning

FIGURE 18.11 Silicate structures with oxygen atoms shown in red and silicon atoms shown in yellow.

(a) Pyroxene; tetrahedral SiO_4 units are joined in chains by silicon–oxygen–silicon bonds. (b) Amphibole, which is asbestos. Chains of SiO_4 units are joined side-by-side by silicon–oxygen–silicon bonds.

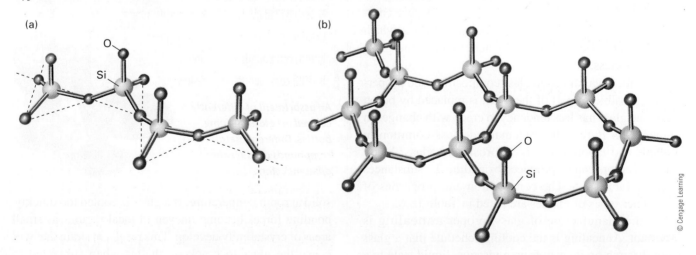

(a)

(b)

© Cengage Learning

18-5a Glass

Glass: A hard, noncrystalline, transparent substance made by melting silicates with other substances

Annealing: Heating and cooling a metal, glass, or alloy in a manner that makes it less brittle

When silica is melted, some of the bonds are broken and the tetrahedral SiO_4 units move with respect to each other. On cooling, reorganization into the same orderly arrangement in crystalline silica is hard to achieve because of the difficulty the groups experience in moving past one another. Instead, cooling produces glass—a hard, noncrystalline, transparent substance with an internal structure like that of a liquid. The random liquid-like molecular arrangement accounts for one of the typical properties of a glass: It breaks irregularly instead of splitting along a plane like a crystal. Glass and other solid substances that lack regular atomic arrangement and properties of crystals are known as amorphous substances.

Common window glass is made by melting a mixture of silica with sodium and calcium carbonates. Bubbles of carbon dioxide are evolved, and the cooled mixture is a glass composed of sodium and calcium silicates. The n in these formulas represents a large number and is necessary to show that glass contains SiO_3 groups linked together rather than individual SiO_3^{2-} ions.

$$n\,Na_2CO_3(\ell) + n\,SiO_2(s) \xrightarrow{\text{Heat}} Na_{2n}(SiO_3)_n + n\,CO_2(g)$$

$$n\,CaCO_3(\ell) + n\,SiO_2(s) \xrightarrow{\text{Heat}} Ca_n(SiO_3)_n + n\,CO_2(g)$$

White sand is the source of the silica for glassmaking. Even the best grade of sand contains a small proportion of iron(III) compounds that gives it a brown or yellow color. When this sand is made into glass, the iron is converted to a mixture of light green iron(II) silicates, explaining the green tint of some old bottles. Adding a manganese compound to the melt produces pink manganese silicates, which offset the green of the iron silicates, making the glass appear colorless.

Countless variations in glass composition and properties are possible. If part of the silica is replaced by boron oxide, the glass has less tendency to crack with changes in temperature. Pyrex, the trademarked glass common in kitchens and laboratories, is a borosilicate glass. Many beautiful colors can be produced by adding the substances listed in Table 18.5. The composition and properties of some other types of glasses are listed in Table 18.6.

In the manufacture of glass, proper **annealing** is important. Annealing is the cooling schedule that a glass is put through on its way from a viscous, liquid state to a

Natural ingredients for making glass: sand (SiO_2), seashells for lime ($CaCO_3$), and seaweed for soda ash (Na_2CO_3).

TABLE 18.5 Substances Used to Color Glass

Substance	Color
Copper(I) oxide	Red, green, blue
Tin(IV) oxide	Opaque
Calcium fluoride	Milky white
Manganese(IV) oxide	Violet
Cobalt(II) oxide	Blue
Finely divided gold	Red, purple, blue
Uranium compounds	Yellow, green
Iron(II) compounds	Green
Iron(III) compounds	Yellow

An assortment of differently colored art glass perfume bottles. Different metal compounds produce the different colors.

solid at room temperature. If a glass is cooled too quickly, bonding forces become uneven in local regions as small areas of crystallinity develop. This results in strain that will cause the glass to crack or shatter when subjected to

TABLE 18.6 Some Special Glasses

Special Addition or Composition	Desired Property
Large amounts of PbO with SiO_2 and Na_2CO_3	Brilliance, clarity, suitable for optical structures; crystal or flint glass
SiO_2, B_2O_3, and small amounts of Al_2O_3	Small coefficient of thermal expansion; borosilicate glass: Pyrex, Kimax, and others
One part SiO_2 and four parts PbO	Ability to stop (absorb) large amounts of X-rays and gamma rays: lead glass
Large concentrations of CdO	Ability to absorb neutrons
Large concentrations of As_2O_3	Transparency to infrared radiation

© Cengage Learning

mechanical shocks or sudden temperature changes. High-quality glass, such as that used in optics, must be annealed very carefully. The huge Mt. Palomar, California, observatory mirror was annealed from 500°C to 300°C over a period of nine months.

18-5b Ceramics

What do you know about *ceramics*? Perhaps you associate the term with pottery vases that you see at a crafts show, bathroom tile, or components of electrical equipment. **Ceramics** are a large and diverse class of materials with the properties of nonmetals. What they have in common is that all are made by baking or firing minerals or other substances, often including silicates and metal oxides.

Ceramic materials have been made since well before the dawn of recorded history. They are generally fashioned from clay or other natural earths at room temperature and then permanently hardened by heat. Silicate ceramics include objects made from clays, such as pottery, bricks, and table china. The three major ingredients of common pottery are clay (from weathering of feldspar as described previously), sand (silica), and feldspar (aluminosilicates). Clays mixed with water form a moldable paste because they consist of many tiny silicate sheets that can easily slide past one another. When the clay–water mixture is heated, the water is driven off, and new Si — O — Si bonds are formed so that the mass of platelets becomes permanently rigid. Like glass, ceramics are amorphous.

A new type of glass-like ceramics with unusual properties has become widely available for home and industrial use. Ordinary glass breaks because once a crack starts, there is nothing to stop the crack from spreading. It was discovered that if glass objects produced in the usual manner are heated until many tiny crystals develop, the resulting material, when cooled, is much more resistant to breaking than normal glass. In molecular terms, the randomness of the glass structure has been partially replaced by the order of a crystalline silicate. The materials produced in this way (an example is Pyroceram) are generally opaque and are used for kitchenware and in other applications in which the material is subjected to stress or high temperatures.

Ceramic materials are attractive for several reasons. The starting materials for making them are readily available and cheap. Ceramics are lightweight in comparison with metals and retain their strength at temperatures above 1000°C, where metal parts tend to fail. They also have electrical, optical, and magnetic properties of value in the computer and electronic industries.

The one severely limiting problem in utilizing ceramics is their brittle nature. Ceramics deform very little before they fail catastrophically, the failure resulting from a weak

Ceramics: Materials and products made by firing mixtures of nonmetallic minerals

Many ceramics, like this pot fragment from A.D. 1200, are made from fire-hardened clays.

Album/Art Resource, NY

point in the bonding within the ceramic matrix. However, such weak points are not consistent from object to object, so the predictability of failure is poor. Since the stress failure of ceramic materials is due to molecular abnormalities resulting from impurities or disorder in the basic atomic arrangements, much attention is now being given to purer starting materials and the control of the processing steps. In addition, *ceramic composite materials*, mixtures of ceramic materials or of ceramic fibers with plastics, can overcome the tendency of plain ceramics to crack.

18-5c Cement and Concrete

A **cement** is a material that can bond mineral fragments into a solid mass. The most common cement, known as portland cement, is made by roasting a powdered mixture of calcium carbonate (limestone or chalk), silica (sand), aluminosilicate mineral (kaolin, clay, or shale), and iron oxide at a high temperature in a rotating kiln. As the materials pass through the kiln, they lose water and carbon dioxide and ultimately form "clinker," in which the materials are partially melted together. The clinker is ground to a very fine powder after the addition of a small amount of calcium sulfate (gypsum). A typical composition of a portland cement, expressed in terms of oxides, is 60% to 67% CaO; 17% to 25% SiO_2; 3% to 8% Al_2O_3; up to 6% Fe_2O_3; and small amounts of magnesium oxide, magnesium sulfate, and potassium and sodium oxides. As in glass, the oxides are not isolated into molecules or ionic crystals, and the submicroscopic structure is quite complex.

Many different reactions occur during the setting of cement. Initially, the calcium silicates react with water to give a sticky gel. The gel has very large surface area and is responsible for the strength of concrete. Reactions with carbon dioxide in the air also occur at the surface. After the initial solidification, small, densely interlocked crystals begin to form, a process that continues for a long time and increases the compressive strength of the cement.

More than 800 million tons of cement are manufactured each year, most of which is used to make concrete. Concrete, like many other materials containing $Si - O$ bonds, is virtually noncompressible but lacks tensile strength. If concrete is to be used where it will be subject to tension, it must be reinforced with steel.

Your Resources

In the back of the textbook:

→ *Review card on The Chemistry of Useful Materials*

    In OWL for CHEM2 at www.cengagebrain.com

→ *Review Key Terms with Flash Cards (printable or digital)*

→ *Complete Interactive Practice Quizzes to prepare for tests*

→ *Submit Assigned Homework and Exercises*

Applying Your Knowledge

1. Define the following terms:
   (a) Sedimentary rock
   (b) Igneous rock

2. Define the following terms:
   (a) Slag
   (b) Ductile
   (c) Brine
   (d) Alloy

3. Define the following terms:
   (a) Annealing
   (b) Amorphous
   (c) Ceramic
   (d) Cement

4. What are the two products of the "chloralkali" process? Name two common uses for each of the products.

5. Name a common metal taken from seawater. Give several uses for this metal.

6. Describe three sources of gold described in this chapter.

7. Complete and balance (if needed) the following equations:
   (a) $2\,H_2O(\ell) + 2\,e^- \longrightarrow$ _____ $+\ OH^-(aq)$
   (b) $Mg^{2+}(aq) + 2\,OH^-(aq) \longrightarrow$ _____
   (c) $Cl^-(aq) \longrightarrow Cl_2(g) +$ _____
   (d) $CaCO_3(s) \longrightarrow$ _____ $+ CO_2(g)$

8. Name two elements that occur in the free metallic state in nature.

9. What are the three ingredients used in a blast furnace to make iron? Which one of these is the reducing agent?

10. Explain why the molten mixture in a blast furnace separates naturally into a layer of slag and a layer of molten iron.

11. What are the differences between pig iron, cast iron, and steel? Mention composition as well as properties and uses.

12. What is the principal impurity in pig iron?

13. What element is used to convert pig iron into steel?

14. What possible atmospheric pollutants are produced by the refining of iron and copper?

15. Complete and balance (if needed) the following equations:
    (a) $Cu_2S(s) + O_2(g) \longrightarrow$ _____
    (b) $FeS(s) + O_2(g) \longrightarrow$ _____

16. How is pure copper produced?

17. A study in 1981 found that there are 7.2×10^{-9} grams of gold per gallon of seawater off the coast of Nova Scotia, Canada.
    (a) How many gallons of this seawater would be required to obtain one gram of gold?
    (b) Amounts of gold are commonly measured using the *troy ounce*. If there are 31.1 grams per troy ounce, how many gallons of this seawater would it take to produce one troy ounce of gold?

18. Name four properties of metals and give an example of the use of a metal taking advantage of each property.

19. Describe the electron structure of metals in terms of their conductivity of electricity.

20. Describe a *p*-type semiconductor made from silicon. Name two elements that might be used as dopants.

21. Use your knowledge of the periodic table to identify elements other than arsenic that could be used with silicon to make an *n*-type semiconductor.

22. Use your knowledge about the periodic table to identify elements other than boron that could be used with silicon to make a *p*-type semiconductor.

23. Use your knowledge of the periodic table to identify another element other than silicon that might be used as the primary component of a semiconductor.

24. Name an important application of *p*- and *n*-type semiconductors in addition to their application in the manufacture of transistors and integrated circuits.

25. Draw the structural unit SiO_4, found in many silicate minerals. What is the name for the shape of this structural unit?

26. Draw a structure consisting of SiO_4 units arranged in a chain with each unit sharing two of its oxygen atoms with neighboring SiO_4 units. What is this silicate structure named?

27. Why is quartz (a silicate) an electrical insulator, but pure silicon a semiconductor of electricity?

28. Some of the superconducting materials mentioned in this chapter are ceramics. What implications does this have on our ability to use these semiconductors to replace wires?

Feeding the World

When hunting and gathering were the primary means of obtaining food, only catastrophic events led to widespread hunger. Early agricultural practices led to the formation of towns and, later, cities. As time passed, population increases and the concentration of people in larger and larger towns and cities forced the development of a more advanced agriculture, which has mostly succeeded but has sometimes dramatically failed. To help grow the enormous amount of food needed to feed today's world population, currently surpassing 7 billion and growing at the rate of 1.2% per year, the chemical industry supplies modern scientific agriculture with a large assortment of chemicals—the *agrichemicals*. These include fertilizers, medicine and food supplements for livestock, and chemicals to destroy unwanted pests and plant diseases. However, even with the use of new agrichemicals, sound agricultural practices, and biotechnology advances, the world's food supply cannot keep pace with the rate of growth of the world's human population. More than 1 billion people in the world depend on agricultural lands that are not productive enough to support them adequately.

Through the 8000 to 10,000 years of recorded human history, food production techniques have developed enormously. It is generally estimated that 90% of the U.S. population worked to provide food and fiber during most of the 19th century. Although we now have more efficient agricultural methods that can provide more food per acre than at any time in the past, the world's farmers are increasingly unable to feed the growing world population. The goal of this chapter is to describe the demands placed on modern agriculture, and to present the ways in which chemistry has enabled agriculture to meet these demands.

SECTIONS

19-1 World Population Growth

Chemistry is essential to modern agriculture, and society is dependent on modern agriculture. So, it is just as important to understand our global dependence on agriculture as it is to understand the chemistry behind it. Important issues to consider are: How close are we to reaching Earth's capacity to support its human population? What is the maximum sustainable population level? What factors determine Earth's capacity? These questions have been studied by organizations such as the Worldwatch Institute and world conferences such as the 2002 World Summit on Sustainable Development held in Johannesburg, South Africa. All such studies recognize that controlling the growth rate of the world's population is essential to maintaining adequate food supplies to feed the world.

The world's population rose to more than 6.67 billion in 2007 (Figure 19.1). This is more than 3.5 times the size of Earth's population at the beginning of the 20th century and double the size in 1960. However, according to the *Global Population Profile: 2002* issued by the U.S. Census Bureau in 2004, the population increased by only 74 million people in 2002–2003 compared with the high annual increment of 87 million people in 1989–1990. Furthermore, the annual growth rate in 2002 was only 1.19%

Growing the quantities of food needed to sustain the world's population would not be possible without the application of chemistry.

compared with the high of 2.2% in 1963–1964. Census experts believe that this slowdown in population growth will continue into the foreseeable future. This slowdown is largely attributable to a decline in the fertility of women around the world. In 1990, women around the world, on average, gave birth to 3.3 children during their lifetime. This number had dropped to an average of 2.58 by 2008. This number varies dramatically from country to country, however, with 2008 data showing a high of 7.34 children per woman in Mali and a low of 1.08 children per woman in Singapore, with the United States coming in at 2.10 children per woman. These encouraging numbers have led to Census Bureau projections that the level of fertility will drop below the replacement level of slightly more than 2.1 children per woman by 2050. Anything higher than this will result in an exponential population growth. Even so, the world's population is expected to exceed 9 billion people by 2050. The assumptions in such projections include birthrates and death rates, which, of course, change as time passes. What is not considered in the projections is the quality of life for those alive at any period.

In 2002, noted biologist Edward O. Wilson provided the following statement in his book *The Future of Life.*

The ecological footprint—the average amount of productive land and shallow sea appropriated by each person in bits and pieces around the world for food, water, housing, energy, transportation, commerce and waste absorption—is about 2.5 acres in developing nations but about 24 acres in the U.S. The footprint for the total human population is 5.2 acres. For every person in the world to reach present U.S. levels of consumption with existing technology would require four more planet Earths.

The ultimate question becomes, what is the *carrying capacity* of Earth—that is, what is the maximum population our planet can support? Here is where the quality of life and human values become important. If everyone were to live as we do in highly developed countries, the high level of resource consumption would quickly reduce the carrying capacity of Earth. Currently, the consumption of resources is vastly different in developed and developing countries, with many in developing countries trapped in poverty and malnourishment. It will be difficult to raise their standard of living without significant resource depletion.

A factor that affects Earth's carrying capacity is food-use efficiency. Up to half of the world's grain is fed to animals each year. A kilogram of feedlot-produced beef requires 7 kg of grain compared to 4 kg for 1 kg of pork, and 2 kg for 1 kg of poultry or 1 kg of fish. Just reducing the world's consumption of beef would free up grain either for direct consumption or as feed for pork, poultry, and fish, which are more efficient meat producers than beef. Consider an average American who uses 800 kg of grain

FIGURE 19.1 The *J*-shaped curve of past exponential world population growth, with projections to 2100. Notice that exponential growth starts off slowly, but as time passes the curve becomes increasingly steep. The current world population of 6.8 billion (2010 data) people is projected to reach 8 to 12 billion people at some time in this century. (This figure is not to scale.) (Source: World Bank and United Nations)

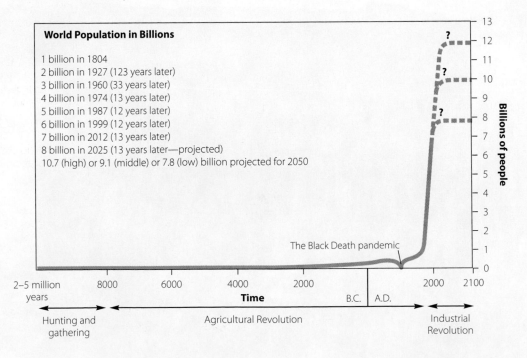

per year, the great bulk of it consumed indirectly in the form of meat, eggs, milk, cheese, yogurt, and ice cream. If Americans cut their annual grain intake in half, 105 million tons of grain would be saved, which is enough to feed two-thirds of the population of India for one year. The good news is that the greater awareness among Americans for lower-fat diets (Section 16-9) has resulted in a decrease in beef consumption and a growing preference for pork, poultry, and fish. But this development is offset by the fact that as developing countries strive to advance, their diets move toward increased meat consumption and decreased grain consumption.

19-2 What Is Soil?

The majority of agriculture depends directly on soil. Soil is a mixture of four components—mineral particles, organic matter, water, and air (Figure 19.2). Weathering processes in nature over thousands of years break rock into small mineral particles found in soil. Organic matter in soil is a mixture that includes leaves, twigs, plant and animal parts in various stages of decomposition, and microorganisms. **Humus**, the dark-colored decomposed organic material, is important to a good soil structure. As a source of nutrients for plants, humus is almost like a time-release capsule, slowly releasing its contents.

© iStockphoto.com/DonNichols

A handful of humus. Humus is partially decomposed organic material from plants and animals. Where soil contains ample humus, it is spongy, holds water well, and is a healthy environment for plants and organisms that live in the soil.

Maintaining humus in the soil is of major concern to the agriculturist. Humus such as peat moss or organic fertilizer can be added. However, there is no real substitute for natural plant growth that is returned to the ground for humus formation. Clover is often grown for this purpose and

> **Humus:** Decomposed organic matter in soil

FIGURE 19.2 Soil formation.

Weather, plants and their litter, animals and other organisms, and topography interact to produce soil.

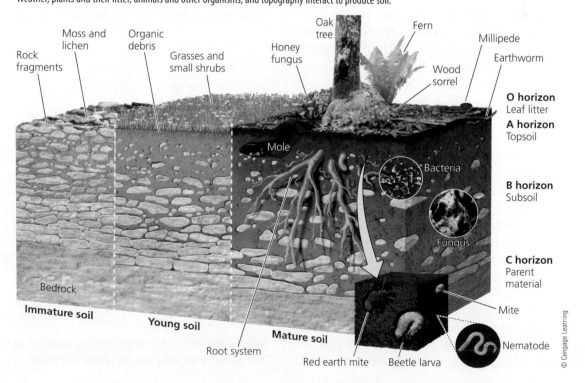

© Cengage Learning

Horizons: Layers within the soil

Topsoil: The layer of soil exposed to the air

Subsoil: The layer of soil just beneath the topsoil

Loam: Soil consisting of a friable mixture of varying proportions of clay, sand, and organic matter

Percolation: Movement of water through subsoil and rock formations

plowed under at the point of its maximum growth. The compost pile of the gardener is another effort to maintain humus for a productive soil.

In addition to being a source of plant nutrients, humus is important in maintaining good soil structure, often keeping it *friable*. Friable material crumbles easily under slight pressure. Soil rich in humus may contain as much as 5% organic matter. Soils in the grasslands of North America have humus to a considerable depth in contrast to rainforest regions, where there is only a thin film of humus on the ground surface.

19-2a Soil Profile

Layers within the soil are called **horizons**. The **topsoil** contains most of the living matter and the humus from dead organisms. Topsoil is usually several inches thick, and in some locations more than three feet of topsoil can be found. The **subsoil**, up to several feet in thickness, contains the inorganic materials from the parent rocks as well as organic matter, salts, and clay particles washed out of the topsoil.

Because healthy topsoil has abundant life forms, it must also contain an abundant supply of oxygen. Soil that supports vegetative growth and serves as a host for insects, worms, and microbes is typically full of pores; such soil is likely to have as much as 25% of its volume occupied by air. The ability of soil to hold air depends on soil particle size and how well the particles pack and cling together to form a solid mass. The particle size groups in soils vary from clays (the finest) through silt and sand to gravel (the coarsest). The particle size of a clay is 0.005 mm or less. The small particles in a clay deposit pack closely together to eliminate essentially all air and thus support little or no life. A typical soil horizon is composed of several particle size groups. A **loam**, for example, is a soil consisting of a friable mixture of varying proportions of clay, sand, and organic matter; a loam has a high air content.

Air in soil has a different composition from the air we breathe. Normal dry air at sea level contains about 21% oxygen (O_2) and 0.04% carbon dioxide (CO_2). In soil, the percentage of oxygen may drop to as low as 15%, and the percentage of carbon dioxide may rise above 5%. This

results from the partial oxidation of organic matter in the closed space (Figure 19.3). The carbon in the organic material combines with oxygen to form carbon dioxide. This increased concentration of carbon dioxide tends to cause groundwater to become acidic; acidic soils are described as *sour* soils because of the presence of aqueous acids.

$$CO_2(g) + 2\,H_2O(\ell) \longrightarrow H_3O^+(aq) + HCO_3^-(aq)$$

Crushed limestone ($CaCO_3$) applied to soil combines with hydrogen ions to form bicarbonate ions, thus raising the pH.

$$H_3O^+(aq) + CO_3^{2-}(aq) \longrightarrow HCO_3^-(aq) + H_2O(\ell)$$

If enough limestone is added to neutralize the acid in the soil and leave an excess of limestone, the pH of the soil becomes alkaline (basic).

19-2b Water in the Soil: Too Much, Too Little, or Just Right

Water can be held in soil in three ways: It can be *absorbed* into the structure of the particulate material, it can be *adsorbed* onto the surface of the soil particles, and it can occupy the pores ordinarily filled with air.

Water is removed from soil in four ways: Plants transpire water while carrying on life processes, soil surfaces evaporate water, water is carried away in plant products, and water moves through the subsoil and rock formations below in a process called **percolation**. Soils with good percolation naturally drain water from all but the small pores.

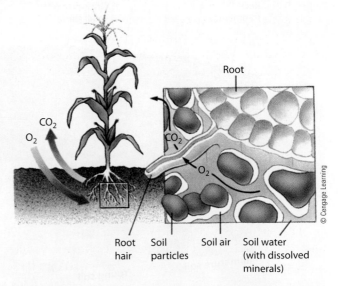

FIGURE 19.3 Exchanges between water and air in soil pores. Plant roots absorb oxygen and water and release carbon dioxide.

The percolation of a soil depends on the soil particle size and its chemical composition. Because of their small particle sizes, clays and, to a lesser degree, silts tend to pack together in an impervious mass with little or no percolation. Of course, sand, gravel, and rock pass water readily. Waterlogged soils that do not percolate support few crops because of their lack of air and oxygen. Rice, which requires waterlogged soil, is an important exception. A negative aspect of the massive flow of water through soil is the **leaching effect**. Water, known as the "universal solvent" because of its ability to dissolve so many different materials, dissolves away, or leaches, many of the chemicals needed to make a soil productive. If the leached material is not replaced, the soil becomes increasingly unproductive.

Soils become acidic, or sour, not only because of the oxidation of organic matter but also because of *selective leaching* by the passing groundwater. Salts of Group 1 and 2 metals are more soluble than salts of the Group 13 and transition metals. For example, a soil containing calcium, magnesium, iron, and aluminum ions is likely to be slightly alkaline, or sweet, before leaching with water. After the selective removal of calcium and magnesium salts, the soil becomes acidic because the iron and aluminum ions each tie up hydroxide ions from water and release hydrogen ions.

$$Fe^{3+} + H_2O \longleftarrow FeOH^{2+} + H^+$$
$$Al^{3+} + H_2O \longleftarrow AlOH^{2+} + H^+$$

A report released by the United Nations Environment Program (UNEP) in 1992 stated that 4.84 billion acres of soil—an area the size of China and India combined—had been degraded to the point at which it will be difficult or impossible to reclaim them. By 2002, the UNEP was describing soil as a "threatened natural resource" and reporting that 17% of the land surface worldwide is strongly degraded. As populations grow, prolonged cultivation prevents cropland from maintaining its productivity.

19-3 Nutrients

At least 18 known elemental nutrients are required for normal green plant growth. Three of these, the **nonmineral nutrients**—carbon, hydrogen, and oxygen—are obtained from air and water (Table 19.1). The **mineral nutrients** must be absorbed through the plant root system as solutes in water. The 15 known mineral nutrients fall into three groups: **primary nutrients**, **secondary nutrients**, and **micronutrients**, depending on the amounts necessary for healthy plant growth.

19-3a Primary Nutrients

The primary nutrients are nitrogen, phosphorus, and potassium. Although bathed in an atmosphere of nitrogen, most plants are unable to use the air as a supply of this vital element. They require nitrogen fixation, the process of changing atmospheric nitrogen into water-soluble compounds that can be absorbed through the plant's roots and assimilated by the plant.

Nature fixes nitrogen in two ways. In the first method, bolts of lightning provide the high energy needed to oxidize nitrogen to nitric oxide (NO). The NO is further oxidized to NO_2 (Section 4-6), which reacts with water to form nitric acid (HNO_3). This method is estimated to provide less than 10% of the nitrogen fixed by nature. Nitric acid is readily soluble in rain, clouds, or ground moisture and thus increases nitrate concentration in soil.

In the most important method, nitrogen-fixing bacteria that live in the roots of legumes, such as soybeans and alfalfa, use an enzyme, **nitrogenase**, to catalyze a complex series of reactions that convert atmospheric nitrogen into ammonia under normal atmospheric conditions. Legume nitrogen fixation can add more than 100 lb of nitrogen per acre of soil in one growing season. Another major source of nitrogen replenishment in soil is dead organisms and animal wastes. Even in the absence of legumes, this can be an adequate source of nitrogen.

Leaching effect: The dissolving of soil material as water moves through the soil

Nonmineral nutrients: Carbon, hydrogen, and oxygen in soil

Mineral nutrients: Plant nutrients absorbed through roots as solutes in water

Primary nutrients: Major elements required by plants

Secondary nutrients: Elements required by plants in moderate amounts

Micronutrients: Elements required by plants in relatively small amounts

Nitrogenase: Bacterial enzyme that catalyzes nitrogen fixation

TABLE 19.1 Essential Plant Nutrients

Nonmineral	Primary	Secondary	Micronutrients
Carbon	Nitrogen	Calcium	Boron
Hydrogen	Phosphorus	Magnesium	Chlorine
Oxygen	Potassium	Sulfur	Copper
			Iron
			Manganese
			Molybdenum
			Sodium
			Vanadium
			Zinc

© Cengage Learning

Like nitrogen, phosphorus must be in a soluble mineral or inorganic form before it can be used by plants. Unlike nitrogen, phosphorus is derived completely from the mineral content of the soil. Salts of the dihydrogen phosphate ion ($H_2PO_4^-$) and monohydrogen phosphate ion (HPO_4^{2-}) are the dominant phosphate ions in soils of normal pH (Figure 19.4). Because of the great concentration of electric charge associated with the phosphate ion (PO_4^{3-}), phosphates are more tightly held to positive ions such as Ca^{2+} and Fe^{3+} and are not as easily leached by groundwater as are nitrate salts, which are generally soluble in water.

Potassium in the form of K^+ ion is a key element in the enzymatic control of the interchange of sugars, starches, and cellulose. Although potassium is the seventh most abundant element in Earth's crust, soil used heavily in crop production can be depleted of this important metabolic element, especially if the soil is regularly fertilized with nitrate, with no regard to potassium content.

19-3b Secondary Nutrients

Calcium and magnesium are available in small amounts as Ca^{2+} and Mg^{2+} ions as well as in complex ions and crystalline formations. These abundant elements are bound tightly enough by soil so they are not readily leached yet are held loosely enough to be available to plants. When held in the soil as sulfate salts (SO_4^{2-}), sulfur is readily available to plants.

© Nigel Cattlin/Alamy

A plant suffering from chlorosis. Chlorophyll, the green plant pigment, requires nitrogen, magnesium, and iron from the soil. Deficiencies of any of these nutrients cause chlorosis, a condition of low chlorophyll content, indicated by yellowing of the leaves.

19-3c Micronutrients

Only very small amounts of micronutrients are required by plants; therefore, unless extensive cropping or other factors deplete the soil of these nutrients, sufficient quantities are usually available.

Iron is also an essential component of the enzyme involved in the formation of chlorophyll. When the soil is iron deficient or when too much lime is present in the soil, iron availability decreases. Often a gardener or lawn worker will apply phosphate and lime to adjust soil acidity, only to see green plants turn yellow because of **chlorosis**. What

FIGURE 19.4 Availability of phosphate in the soil as a function of pH.

The ions present vary with the pH. At pH 5 to pH 8, $H_2PO_4^-$ and HPO_4^{2-} predominate. At very low pH values, phosphorus is in the form of the nonionized acid H_3PO_4. At a very high pH, all three protons are removed and the phosphorus is in the form of the phosphate ion (PO_4^{3-}).

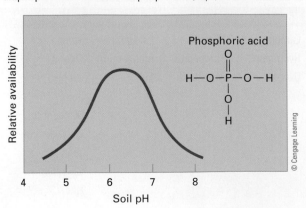
© Cengage Learning

© fabio fersa/Shutterstock.com

A natural method of nitrogen fixation. The energy in a bolt of lightning is sufficient to disrupt the very stable triple bond in a nitrogen molecule (N_2). The result is the reaction with oxygen in the air to produce NO. This is the same reaction that takes place in an internal combustion engine.

$$N_2 + O_2 \xrightarrow{\text{High temp}} 2\,NO$$

happens in such cases is that both phosphate and the hydroxide from the lime tie up the iron and make iron unavailable to the plants.

$$Fe^{3+}(aq) + 2\ PO_4^{3-}(aq) \longrightarrow Fe(PO_4)_2^{3-}(aq)$$
Phosphate   Tightly bound complex

$$Fe^{3+}(aq) + 3\ OH^-(aq) \longrightarrow Fe(OH)_3(s)$$
Insoluble hydroxide

19-4 Fertilizers Supplement Natural Soils

Primitive peoples raised crops on a cultivated plot until the land lost its fertility, then they moved to a virgin piece of ground where they cut down ("slashed") natural vegetation and burned off the stubble to clear the land. In many cases, the slash–burn–cultivate cycle was no more than a year in length, and few found a piece of ground anywhere that could support successful cropping for more than five years without fertilization. In farming villages, developed in ancient times and prevalent throughout the Middle Ages, innovation in fertilization was demanded because the same land had to be used for many years. With the use of legumes in crop rotations, manures, dead fish, or almost any organic matter available, the land was kept in production.

The acreage used worldwide in the cultivation of food crops would probably be sufficient if modern synthetic chemical fertilization were employed on all of it. For example, chemical fertilization accounted for crop yield explosions in the following: (1) U.S. corn—25 bushels per acre in 1800, 110 bushels per acre in the 1980s, and 153.8 bushels per acre in 2008; (2) English wheat—less than 10 bushels per acre from A.D. 800 to 1600, and more than 75 bushels per acre in the 1980s. However, the cost of synthetic chemical fertilization is beyond what developing countries can bear, and the environmental impact of such a large dispersion of fertilizer chemicals would probably be massive. For example, aquifer contamination with nitrates due to corn crop fertilization renders well water from these natural underground basins unfit for drinking in large areas of the U.S. corn belt.

Fertilizers that contain only one nutrient are called **straight fertilizers**. Potassium chloride for potassium is an example of a straight fertilizer. Those containing a mixture of the three primary nutrients are called **complete**, or **mixed, fertilizers**. The primary nutrients are absorbed by plant roots as simple inorganic ions: nitrogen in the form of nitrates (NO_3^-), phosphorus as phosphates ($H_2PO_4^-$ or HPO_4^{2-}), and potassium as the K^+ ion. Organic fertilizers, which have very little mineral content, can supply these ions but only when used in large quantities over a long time. For example, a manure might be a 0.5–0.24–0.5 fertilizer, in contrast to a typical chemical fertilizer, which might carry the numbers 6–12–12. These numbers indicate the analysis in order of the percentage of nitrogen as N, phosphorus as P_2O_5, and potassium as K_2O (commonly known as potash) in the fertilizer (Figure 19.5). In addition to containing the desired ions, chemical fertilizers place the ions in the soil in a form that can be absorbed directly by plants. The problem is that these inorganic ions are relatively easily leached from the soil and may pose pollution problems if not contained. The much slower organic fertilizer tends to stay put. **Quick-release fertilizers** are water soluble as opposed to **slow-release fertilizers**, which require days

> **Straight fertilizer:** A fertilizer that contains only one nutrient
>
> **Complete (mixed) fertilizer:** A fertilizer that contains the three primary nutrients (nitrogen, phosphorus, and potassium)
>
> **Quick-release fertilizer:** A fertilizer that is water soluble
>
> **Slow-release fertilizer:** A combination of plant nutrients that are slow to dissolve in water

FIGURE 19.5 How to read a fertilizer label.

The three numbers, in order, refer to the percentage by weight of N (nitrogen), P_2O_5 (phosphate), and K_2O (potash). Following the lead of J. von Liebig, his German mentor, who was the first scientist to suggest adding nutrients to soils, Samuel William Johnson, an American, burned plants and analyzed their ashes. He expressed the nutrient concentrations as the oxides present in the ashes, a practice that has continued to this day.

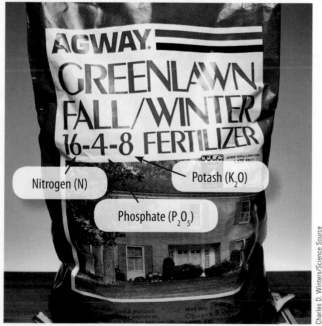

or weeks for the material to dissolve completely. Table 19.2 lists the necessary plant nutrients and suitable chemical sources of each.

19-4a Nitrogen Fertilizers

The cheapest source of nitrogen is the air, but it must be combined with relatively expensive hydrogen, obtained from petroleum, to form ammonia in the **Haber process**.

The complex interactions among social, economic, and political necessities are well illustrated by the development of the industrial synthesis of ammonia. The need for an industrial process for nitrogen fixation was recognized as early as 1890. Scientists in England noted that the world's future food supply would be determined by the amount of nitrogen compounds available for fertilizers. At the time, the sources of such compounds were limited to sodium nitrate from Chile and rapidly depleting supplies of guano (bird droppings) from Peru.

Shortly after 1900, many researchers in England and Germany began investigating methods of preparing ammonia, and in 1908 the German chemist Fritz Haber developed a feasible process for the direct synthesis of ammonia from its elements.

$$N_2(g) + 3 H_2(g) \longrightarrow 2 NH_3(g)$$

The first industrial plant based on the Haber process (see Figure 19.6) for ammonia synthesis began operation in Germany in 1911; in the United States, industrial production was begun in 1921.

Subsequently, the Haber process has provided ammonia for fertilizers on a huge scale and is now largely responsible for our ability to feed a world population of more

TABLE 19.2 Some Chemical Sources of Plant Nutrients

Element	Source Compound(s)
Nonmineral Nutrients	
C	CO_2 (carbon dioxide)
H	H_2O (water)
O	H_2O (water)
Primary Nutrients	
N	NH_3 (ammonia), NH_4NO_3 (ammonium nitrate), H_2NCONH_2 (urea)
P	$Ca(H_2PO_4)_2$ (calcium dihydrogen phosphate)
K	KCl (potassium chloride)
Secondary Nutrients	
Ca	$Ca(OH)_2$ (calcium hydroxide, slaked lime), $CaCO_3$ (calcium carbonate, limestone), $CaSO_4$ (calcium sulfate, gypsum)
Mg	$MgCO_3$ (magnesium carbonate), $MgSO_4$ (magnesium sulfate, Epsom salts)
S	Elemental sulfur, metallic sulfates
Micronutrients	
B	$Na_2B_4O_7 \cdot 10 H_2O$ (borax)
Cl	KCl (potassium chloride)
Cu	$CuSO_4 \cdot 5 H_2O$ (copper sulfate pentahydrate)
Fe	$FeSO_4$ (iron(II) sulfate, iron chelates)
Mn	$MnSO_4$ (manganese(II) sulfate, manganese chelates)
Mo	$(NH_4)_2MoO_4$ (ammonium molybdate)
Na	NaCl (sodium chloride)
V	V_2O_5, VO_2 (vanadium oxides)
Zn	$ZnSO_4$ (zinc sulfate, zinc chelates)

FIGURE 19.6 The Haber process for ammonia production.

A mixture of N_2 and H_2 in the proper proportions for reaction is heated and passed under pressure over the catalyst. The ammonia is collected as a liquid, and unreacted gases are recycled.

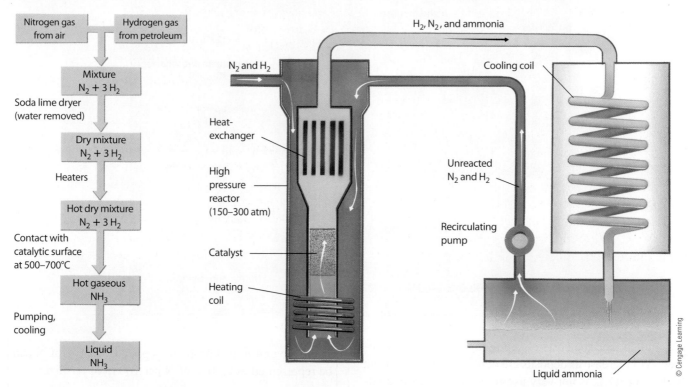

than 7 billion. The Haber process has been so well developed that ammonia is very inexpensive (about $400 per ton) and is usually among the top ten industrial chemicals in terms of quantity produced each year.

FIGURE 19.7 Nitrogen fertilizers produced from anhydrous ammonia.

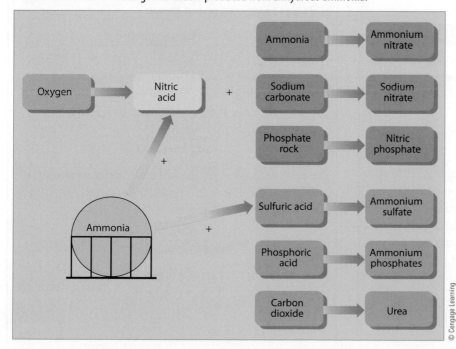

Ammonia can be applied directly to the soil or converted into numerous agrichemicals (Figure 19.7). Two common straight fertilizers for nitrogen are ammonium nitrate (NH_4NO_3) and urea. A slurry of water, urea, and ammonium nitrate is often applied to crops under the name "liquid nitrogen." Such a solution can contain up to 30% nitrogen and is easy to store and apply.

When applied to the surface of the ground around plants, urea is subject to considerable nitrogen loss unless it is washed into the soil by rain or irrigation. When urea decomposes, ammonia is formed, some of which is lost to the air and some that is absorbed by moist soil particles. As much as half of the nitrogen applied to the soil can be lost in this way.

19-4b Phosphate Fertilizers

Phosphorus is readily available in the form of phosphate rock, which can be transformed into the needed fertilizers (Figure 19.8). World deposits of phosphate rock

FIGURE 19.8 Fertilizers produced from phosphate rock.

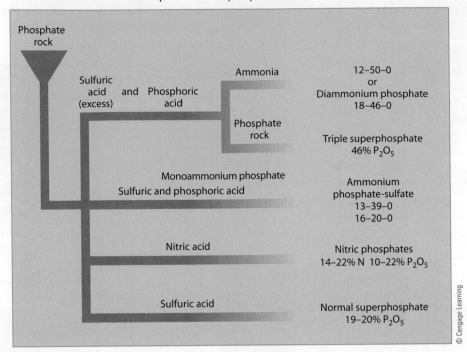

are limited, and the costs of supplying phosphorus fertilizers will increase as deposits are depleted. The phosphate rock, $Ca_3(PO_4)_2$, is not very useful because of its exceedingly low solubility. When treated with sulfuric acid, however, phosphate rock becomes more soluble; the resulting mixture of phosphate and sulfate salts of calcium is called *superphosphate*.

$$Ca_3(PO_4)_2 + 2\ H_2SO_4 \longleftarrow Ca(H_2PO_4)_2 + 2\ CaSO_4$$
Phosphate rock Superphosphate

EXAMPLE 19.1 Fertilizer Labels

 The label on a fertilizer bag lists the numbers 22–6–8. What do these numbers mean? Calculate the pounds of each nutrient present in a 50.0 lb bag of fertilizer.

SOLUTION

The numbers mean that the fertilizer contains 22.0% nitrogen (N) in the form of some nitrogen-containing compound, 6.00% of a phosphorus-containing compound (calculated as a percentage of P_2O_5), and 8.00% of a potassium-containing compound (calculated as a percentage of K_2O). The pounds of each nutrient present are calculated as follows:

Nitrogen: The pounds of nitrogen present are calculated by multiplying the percentage of nitrogen times 50.0 lb.

Since the unit in this problem is pounds, 22.0% of N can be represented as 22.0 lb of N per 100 lb of fertilizer.

$$\frac{22\ lb\ N}{100.0\ \text{lb fertilizer}} \times 50\ \text{lb fertilizer} = 11\ lb\ N$$

Phosphorus: Since the amount of phosphorus present is represented as 6.00% P_2O_5, the actual percentage present as elemental phosphorus is smaller. To calculate the fraction of phosphorus in P_2O_5 requires using the atomic masses of P and O in a conversion factor:

$$\frac{2 \times \text{Atomic mass P}}{(2 \times \text{Atomic mass P}) + (5 \times \text{Atomic mass O})}$$

$$2\ P = 2 \times 31\ amu = 62\ amu$$
$$5\ O = 5 \times 16\ amu = 80\ amu$$

Total for P_2O_3 = 142 amu, so factor is

$$\frac{62\ amu\ P}{142\ amu\ P_2O_5}$$

The relative masses are the same no matter what units are used. Since the unit in this problem is pounds, the conversion factor can be written as

$$\frac{62.0\ lb\ P}{142.0\ lb\ P_2O_5}$$

$$\frac{6.0\ \text{lb } P_2O_5}{100.0\ \text{lb fertilizer}} \times \frac{62.0\ lb\ P}{142.0\ \text{lb } P_2O_5} \times 50.0\ \text{lb fertilizer}$$

$$= 1.3\ lb\ P$$

Potassium: Since the amount of potassium present is represented as 8.00% K_2O, calculating the pounds of potassium present requires using the atomic masses of K and O in a conversion factor:

$$\frac{2 \times \text{Atomic mass K}}{(2 \times \text{Atomic mass K}) + (\text{Atomic mass O})}$$

$2 \text{ K} = 2 \times 39 \text{ amu} = 78 \text{ amu}$, $1 \text{ O} = 16 \text{ amu}$, and the total of 2 K and 1 O = 94 amu, so the conversion factor showing the ratios in pounds is

$$\frac{78.0 \text{ lb K}}{94.0 \text{ lb } K_2O}$$

$$\frac{8.0 \text{ lb } K_2O}{100.0 \text{ lb fertilizer}} \times \frac{78.0 \text{ lb K}}{94.0 \text{ lb } K_2O} \times 50.0 \text{ lb fertilizer}$$

$$= 3.3 \text{ lb K}$$

TRY IT 19.1

Ammonium nitrate (NH_4NO_3) is widely used as a fertilizer. A bag of ammonium nitrate sold as fertilizer lists the numbers 35–0–0. Use the formula weight for NH_4NO_3 to verify that this is the correct number.

19-5 Protecting Food Crops

The natural enemies of crops include more than 80,000 diseases brought on by viruses, bacteria, fungi, algae, and similar organisms; 30,000 species of weeds; 3000 species of nematodes; and about 10,000 species of plant-eating insects. About one-third of the food crops in the world are lost to pests each year, with the loss rising above 40% in some developing countries. A "pest" is any organism that in some way reduces crop yields or human health. **Pesticides**, chemicals used to control pests, are classified according to the pests they control: **insecticides** kill insects, **herbicides** kill weeds, and **fungicides** kill fungi.

19-5a Pesticides

Pesticides are the chemical answer to pest control. Eighteen common classes of pesticides are fortified with more than 2600 active ingredients to fight the battle with pests. More than 5 billion pounds of pesticides are produced worldwide each year. In 1993, pesticide sales in the United States totaled $6.8 billion. Although the dollar cost was up to $11

billion in 2001, according to the Environmental Protection Agency (EPA), the actual poundage use began a decline in 1987. Worldwide expenditures were around $32 billion in 2001. There are three reasons for a decrease in the demand for pesticides in the United States: (1) Cropland planted was less than 330 million acres, down from a high of 383 million acres in 1982. (2) Farming is becoming more cost-effective as farmers learn to use a minimum of pesticide for the desired effect. (3) Farmers are becoming more concerned with environmental and health issues related to the use of pesticides.

Pesticide: A chemical used to kill pests

Insecticide: A pesticide that kills insects

Herbicide: A pesticide that kills weeds

Fungicide: A pesticide that kills fungi

Insecticides

Before World War II, the list of insecticides included only a few compounds of arsenic, petroleum oils, nicotine, pyrethrum (obtained from dried chrysanthemum flowers), rotenone (obtained from roots of derris vines), sulfur, hydrogen cyanide gas, and cryolite (the mineral Na_3AlF_6). Dichlorodiphenyltrichloroethane (DDT), the first of the chlorinated organic insecticides, was originally prepared in 1873, but it was not until 1939 that Paul Müller of Geigy Pharmaceutical in Switzerland discovered the effectiveness of DDT as an insecticide.

The use of DDT increased enormously on a worldwide basis after World War II, primarily because of its effectiveness against mosquitoes that spread malaria and lice that

Some pesticides available for a home garden.

© Tony Freeman/PhotoEdit

carry typhus. The World Health Organization estimates that approximately 25 million lives have been saved through the use of DDT. DDT seemed to be the ideal insecticide—it is cheap and has relatively low toxicity to mammals (oral LD_{50} is 300–500 mg/kg). However, problems related to extensive use of DDT began to appear in the late 1940s. Many species of insects developed resistance to DDT, and DDT was also discovered to have a high toxicity toward fish.

DDT is not metabolized very rapidly by animals; instead, it is deposited and stored in the fatty tissues. The biological half-life of DDT is about eight years; only half the amount of DDT an animal assimilates today will be metabolized in eight years. If ingestion continues at a steady rate, DDT builds up within the animal over time.

The use of DDT was banned in the United States in 1973, although it is still in use in some other parts of the world. The buildup of DDT in natural waters is a reversible process; the EPA reported a 90% reduction of DDT in Lake Michigan fish by 1978 as a result of the ban on the use of the insecticide in the United States.

The most important insecticides can be grouped in three structure classes: chlorinated hydrocarbons, organophosphorus compounds, and carbamates. Chlorinated hydrocarbons such as DDT are referred to as *persistent pesticides* because they persist in the environment for years after their use.

Malathion is a widely used member of a class of pesticides known as *organophospate derivatives*. It is biodegradable and less toxic (LD_{50} for female rats is 1000 mg/kg) than DDT. Several other organophosphorus compounds are modifications of this basic structure. They include dimethoate ($LD_{50} = 387$ mg/kg), chlorpyrifos ($LD_{50} = 95$–270 mg/kg), and methyl parathion ($LD_{50} = 18$–50 mg/kg). These organophosphates often are used in the treatment of many pests that are found on vegetables or, in the case of methyl parathion, on cotton plants.

Chlorpyrifos

Methyl parathion

The carbamate insecticides are derivatives of carbamic acid.

Carbamic acid

Carbaryl

Pyrimicarb

Malathion

Dimethoate

A corn earworm at work.

They are also nonpersistent insecticides. The most widely used carbamate is *carbaryl*, a general-purpose insecticide with a relatively low mammalian toxicity (oral LD_{50} for rats is 250 mg/kg). A serious drawback to the use of carbaryl for spraying crops is its high toxicity toward honeybees. Another carbamate of interest is *pirimicarb* (oral LD_{50} for female rats is 147 mg/kg), which is a selective insecticide for aphids and has the advantage of controlling strains that have developed resistance to organic phosphates. Pirimicarb has a low toxicity to predators such as bees. It is rapidly metabolized and leaves no lasting residues in plant materials. Pirimicarb is expensive, however; its preparation is more complex than that of other pesticides, and special handling procedures must be used.

The choice of solutions to the problems of insecticide use is not an easy one. Their use introduces trace amounts of them into our environment and our water supplies. If we fail to use them, we must tolerate malaria, plague, sleeping sickness, and the consumption of a large part of our food supply by insects. The goal of the insecticide quest is a selectively toxic chemical that is quickly biodegradable. Continuing research in the development of more effective and safer insecticides is intense, and new products are introduced each year.

Herbicides

Herbicides kill plants. They may be **selective herbicides** and kill only a particular group of plants, such as the broad-leaved plants or the grasses, or they may be **nonselective herbicides**, making the ground barren of all plant life.

Selective herbicides act like hormones, very selective biochemicals that control a particular chemical change in a particular type of organism at a particular stage in its development. Most selective herbicides in use today are growth hormones; they cause cells to swell, so leaves become too thick for chemicals to be transported through them and roots become too thick to absorb needed water and nutrients. Nonselective herbicides usually interfere with photosynthesis and thereby starve the plant to death. On application, the plant quickly loses its green color, withers, and dies.

The most widely used herbicide is 2,4-dichlorophenoxyacetic acid (2,4-D). The corresponding trichloro- compound (common name: 2,4,5-T) has also been shown to be highly effective, but it was banned by the EPA because of a number of health problems associated with its use.

2,4-D
(2,4-Dichlorophenoxyacetic acid)

2,4,5-T
(2,4,5-Trichlorophenoxyacetic acid)

The only structural difference between 2,4,5-T and 2,4-D is the additional chlorine atom on the benzene ring in the fifth position. Agent Orange, widely used as a defoliant during the Vietnam War, was a mixture of these two compounds, contaminated with a

> **Selective herbicide:** An herbicide that kills only a particular group of plants
>
> **Nonselective herbicide:** An herbicide that kills all plant life

toxic impurity, tetrachlorodibenzodioxin (TCDD). Many veterans of the Vietnam War have attributed their chronic illnesses to exposure to Agent Orange.

Both 2,4-D and 2,4,5-T result in an abnormally high level of RNA (Section 15-11) in the cells of the affected plants, causing the plants to grow themselves to death.

Triazines have been found to be effective as herbicides; the most important one is atrazine.

1,3,5-Triazine

Atrazine

Atrazine is widely used in no-till corn production and for weed control in minimum tillage. Atrazine is a poison to any green plant if it is not quickly metabolized. Corn and certain other crops have the ability to render atrazine harmless, which weeds cannot do. Hence, the weeds die, and the corn shows no ill effect.

Several herbicides work by inhibiting plant enzymes that catalyze the synthesis of amino acids in plants that are essential amino acids in animals (Section 15-6). Unlike animals, plants can synthesize all 20 amino acids. Since animals do not synthesize essential amino acids, developing inhibitors for amino acids that are synthesized by plants but not by animals would produce safer herbicides. Glyphosate and sulfonylureas are herbicides with this mechanism of action. Glyphosate, the active ingredient in Roundup, is a phosphate derivative of glycine that inhibits the synthesis of the essential amino acids tyrosine and phenylalanine and is used to control perennial grasses.

$$HO-\overset{\overset{\displaystyle O}{\|}}{\underset{\underset{\displaystyle H}{|}}{P}}-CH_2-\overset{\overset{\displaystyle H}{|}}{N}-CH_2-\overset{\overset{\displaystyle O}{\|}}{C}-OH$$

Glyphosate

Sulfometuron methyl, the active ingredient of Oust, is a sulfonylurea that inhibits the synthesis of the essential amino acids valine, leucine, and isoleucine.

Sulfometuron methyl

Paraquat is a herbicide and can be used to kill weeds before the crop sprouts. Such herbicides are called *pre-emergent* herbicides. When applied directly to susceptible plants, paraquat quickly causes a frostbitten appearance and death. Paraquat has a nitrogen atom in each aromatic ring of the two-ring system.

Paraquat
(1,1′-Dimethyl-4,4′-bipyridinium dichloride)

Paraquat has received considerable attention because it was used to spray illegal poppy and marijuana fields in Mexico and elsewhere, which caused drug users to suffer lung damage from residual paraquat.

The amount of energy saved by herbicides used in no-till farming is enormous. The saving of topsoil is also considerable because the cover from the previous crop holds the soil against wind and water runoff. However, agriculturists who use herbicides are highly dependent on agricultural research institutions for the selection of herbicides that will do the desired job without harmful side effects. Such selections depend on considerable research, much of which is carried out on a trial-and-error basis on test plots. A procedure that is recommended today may be outdated by the next growing season.

Fungicides

About 200 of the 100,000 classified fungal species are known to cause serious plant disease. Agricultural fungicide application accounts for about 20% of all pesticide use. Most fungicides are applied to the seed or foliage of the growing plant or to harvested produce to prevent storage losses. The earliest fungicides were inorganic substances such as elemental sulfur and compounds of copper and mercury. In the 19th century, sulfur was used to control mildew on fruit and grapes.

Methyl bromide (CH_3Br) was widely used as a soil fumigant in strawberry and tomato production and in some fruit tree nurseries until it was shown that it was contributing to the destruction of the ozone layer (Section 7-4). Methyl bromide production and use began to be severely regulated and restricted after this discovery with planned total phaseout by 2005. However, in 2005 the parties to the Montreal Protocol granted the United States critical-use exemptions of 37.5% of the historic baseline for methyl bromide use. An additional critical-use exemption was granted for new

©iStockphoto.com/VIDOK

410   CHEM | Chapter 19

production of methyl bromide in 2006 for 27% of the historic baseline production of this compound. It is unclear how long these special critical-use exemptions will be granted. The California Department of Pesticide Regulation began an investigation of the risk assessment of methyl iodide (CH_3I) after it received an application to register products containing this compound as a fumigant in 2004.

In recent times, different types of organic compounds have been used. An important class of fungicides is the dithiocarbamates and their derivatives, which are widely used on many crops such as fruits and field vegetables.

$$(CH_3)_2N-\overset{\overset{S}{\|}}{C}-S-S-\overset{\overset{S}{\|}}{C}-N(CH_3)_2$$

Thiram, a dithiocarbamate derivative

19-6 Sustainable Agriculture

Extensive use of fertilizers and pesticides, and development of higher yielding varieties of crops, have made it possible to increase crop yields dramatically. However, poor farming practices have resulted in soil erosion and loss of soil fertility. The annual amount of topsoil lost to wind and water erosion is from 2 to 4 tons per acre in Africa, Europe, and Australia; 4 to 8 tons per acre in North, Central, and South America; and nearly 12 tons in Asia. These amounts are about 20 times the rate of replenishment of the topsoil. Compaction of soil, which decreases soil fertility, is caused by (1) growing shallow-rooted crops year after year; (2) not incorporating enough organic matter into the soil; (3) practices that alter soil microbial and earthworm populations; and (4) using heavy machinery, particularly on wet soils. Compacted soil, when dry, looks like brick. It restricts root growth and has soil oxygen levels below those necessary for optimum uptake of nutrients.

Another major problem is the increasing resistance of pests to pesticides. About 500 different insect and mite species, 80 fungus species, and 80 weed species are now resistant to commonly used pesticides.

These problems have led to considerable variations of traditional farming methods sometimes referred to as *organic farming*, *alternative farming*, and *sustainable agriculture*. Even the U.S. Department of Agriculture concedes that it is difficult to define these terms, so it is little wonder that there is confusion over their meaning. However, most agree that one goal of agriculture would be to sustain a viable system of food production with minimal negative impact on the environment. This will, in all probability, require a combination of efforts that might fall under one definition or another.

Organic farming is farming without synthetic fertilizers and pesticides. Organic farming uses only about 40% of the energy required for modern inorganic farming and produces about 90% of the yield. The costs of energy saved in organic farming are offset by the costs of human labor required by the use of natural fertilizers. Many claims are made that organic farming produces a better product for human consumption. However, there is no real evidence that these claims are generally true. Organic farming does have one clear advantage, however: it is definitely less of a threat to the environment than regular farming if agrichemicals are not very carefully controlled.

Sustainable agriculture, as defined by the U.S. Department of Agriculture, means an integrated system of plant and animal production practices having a site-specific application that will, over the long term:

- Satisfy human food and fiber needs

- Enhance environmental quality and the natural resource base upon which the agricultural economy depends

- Make the most efficient use of nonrenewable resources and on-farm resources and integrate, where appropriate, natural biological cycles and controls

Organic farming: Farming without synthetic chemical fertilizers and pesticides

Sustainable agriculture: An approach to agriculture whose broad goal is to provide food for the world's population while minimizing environmental impact

Tim McCabe/USDA Natural Resources Conservation Service

Soil erosion caused by water runoff, which carries soil, fertilizer, and pesticides with it.

Integrated pest management (IPM): Limiting the use of pesticides by relying on a combination of disease-resistant crop varieties, natural predators or parasites, and selected use of pesticides

- Sustain the economic viability of farm operations
- Enhance the quality of life for farmers and society as a whole.

One goal of sustainable agriculture would be to limit the use of agrichemicals to that which is absolutely necessary by increasing the use of environmentally friendly procedures, where possible, to fight pests and to produce food and fiber. Other examples of sustainable practices include (1) expanding crop rotation because the same pests do not attack every crop, (2) using multiple crops in alternate plantings with a given planting field (Figure 19.9), (3) using as much natural fertilizer as possible before resorting to agrichemicals, (4) increasing the use of biological pest controls, (5) employing renewed efforts at soil and water conservation, and (6) having a diversification in livestock as well as field crops on the same farm. Some would define some of these efforts as *alternative agriculture*, but the semantics are irrelevant when compared to the goals.

The main sources of plant nutrients in alternative farming systems are animal and green manures. A green manure crop is a grass or legume that is plowed into the soil or surface-mulched at the end of a growing season to enhance soil productivity.

One broad approach to limiting use of pesticides is **integrated pest management (IPM)**, which relies more on disease-resistant crop varieties and biological controls such as natural predators or parasites that control pest populations than on agrichemicals. Farmers can select tillage methods, planting times, crop rotations, and plant-residue management practices to optimize the environment for beneficial insects that control pest species. If pesticides are used as a last resort, they are applied when pests are more vulnerable or when any beneficial species and natural predators are least likely to be harmed.

IPM programs have been most effective for cotton, sorghum, peanuts, and fruit orchards but less effective for corn and soybeans. Some of the biological controls include release of sterilized insect pests, use of insect pheromones to disrupt mating, release of natural predator pests, and use of natural insecticides.

19-6a Natural Insecticides

Many plant species contain natural protection against insects. For example, nicotine protects tobacco plants from sucking insects. However, its commercial use is limited by its high mammalian toxicity (oral LD_{50} for rats is 50 mg/kg).

One of the oldest and best-known natural insecticides is pyrethrum, a contact insecticide obtained from dried chrysanthemum flowers by extraction with hydrocarbon solvents. Pyrethrum is very effective in killing flying insects and is relatively non-toxic toward mammals (oral LD_{50} for rats is 129 mg/kg). Pyrethrum aerosol sprays are excellent home insecticides because of their safety for humans and rapid action on pests. However, pyrethrum lacks persistence against agricultural insects because of its instability to air and light, so it must be mixed with small amounts of other insecticides to kill insects that might recover from sublethal doses of pyrethrum.

FIGURE 19.9 Strip farming, a sustainable practice in which different crops are grown side by side.
Often, some strips contain legumes that produce nitrogen and reduce the need for chemical fertilizers. Soil is preserved by contouring the strips according to the topography of the land.

The insecticidal properties of pyrethrum result from six esters that are collectively called *pyrethrins*. After the isolation and identification of these compounds, a number of derivatives have been synthesized that are more effective than the natural pyrethrins. For example, dimethrin is effective against mosquito larvae and is safe to use (oral LD_{50} for rats is greater than 10,000 mg/kg). The basic pyrethrin structure is shown in black, and the groups that vary in different pyrethrins are shown in yellow.

Dimethrin

Another source of effective natural pesticides is the neem tree, which grows widely in Africa and Asia. For centuries, people of India have known of the insect-fighting ability of the neem tree. The oil extracted from neem tree seeds has been found to be effective against more than 200 species of insects, including locusts, gypsy moths, cockroaches, California medflies, and aphids. *Azadirachtin*, an active ingredient of oil from neem tree seeds, interferes with insect molting, reproduction, and digestion. Tests have shown that it is specific to insects without affecting pest predators. For example, use of azadirachtin on an aphid-infested field killed the aphids without harming ladybugs and lacewings, which are aphid predators.

Transgenic crop: A crop that has been altered by the techniques of genetic engineering

19-7 Agricultural Genetic Engineering

Armed with the ability to insert genes into organisms (Section 15-11), it follows that introduction of genes into food plants and animals to fight pests, control diseases, and improve food quality should be of value.

Natural breeding methods require up to ten years to produce plants suitable for field testing, but genetic engineering requires as little as one year. One of the first examples of a genetically engineered plant was produced by the insertion of DNA segments from the tobacco mosaic virus into the genetic code of tomato plants (Figure 19.10). The DNA segments caused the tomato plants to be strongly resistant to attack by this virus.

The first **transgenic crop** brought to market was the Flavr Savr tomato. Ordinarily, tomatoes are picked green

FIGURE 19.10 Steps necessary to insert a gene from tobacco mosaic virus into a tomato plant to make the plant resistant to a viral disease.

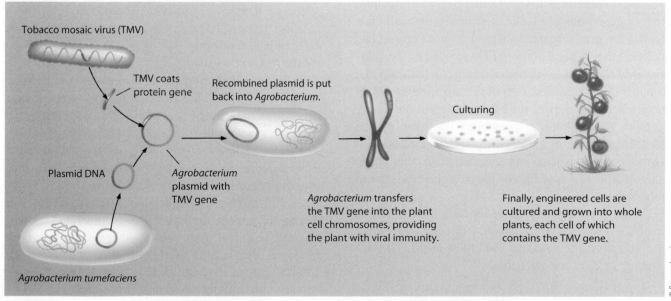

Tobacco mosaic virus (TMV)

TMV coats protein gene

Recombined plasmid is put back into *Agrobacterium*.

Culturing

Plasmid DNA

Agrobacterium plasmid with TMV gene

Agrobacterium transfers the TMV gene into the plant cell chromosomes, providing the plant with viral immunity.

Finally, engineered cells are cultured and grown into whole plants, each cell of which contains the TMV gene.

Agrobacterium tumefaciens

© Cengage Learning

The Bt and Non-Bt corn photos were taken as part of field trial conducted on the main campus of Tennessee State University at the Institute of Agricultural and Environmental Research. The work was supported by a competitive grant from the CSREES, USDA titled Southern Agricultural Biotechnology Consortium for Underserved Communities, (2000–2005). Dr. Fisseha Tegegne and Dr. Ahmad Aziz served as Principal and Co-principal Investigators respectively to conduct the portion of the study in the State of Tennessee.

*The corn on the left has been genetically modified to produce the **Bt** gene. The corn on the right has not.*

and later ripened with ethylene gas to avoid excess softening and spoilage before the tomatoes reach the consumer. The Flavr Savr tomato was genetically modified to ripen more slowly so it could be picked at a ripened stage, enhancing its flavor and increasing its shelf life. The Flavr Savr was ultimately a financial failure. It was not always up to market standards and also was aggressively criticized by groups objecting to genetically modified foods.

The chances that today there are food products containing a genetically modified ingredient on supermarket shelves are high, however. In 2003, there were 87 transgenic crop plants approved for commercial production in the United States. The list included beets, chicory, corn, cotton, flax, melons, papaya, potatoes, rapeseed, rice, soybeans, squash, tobacco, tomatoes, and wheat, among others. Each can be assumed to have passed the scrutiny of three government agencies. The U.S. Department of Agriculture monitors experimental plantings of new crops and their potential for gene transfer to weeds or other crops. The Food and Drug Administration (FDA) evaluates the composition of the food for potential allergens, toxic agents, and nutritional equivalence to the nonmodified food. If there are health concerns, the FDA can forbid marketing of a food. The EPA has responsibility for the effects of pesticides produced by transgenic plants. They deal with the impact of the pesticide on human health or other plants and must approve use of a herbicide on a plant resistant to that herbicide.

The most widely planted genetically modified food crops are soybeans, corn, rapeseed (the source of canola oil), and cotton. The percentages of these crops that were transgenic in 2002 were as follows:

- 74% of the U.S. soybean crop was transgenic. Soybeans are used in oil, flour, lecithin, and protein extracts.

- 32% of the U.S. corn crop was transgenic. The corn provides oil and a variety of other food ingredients, but very little transgenic corn was marketed as fresh corn, canned corn, or popcorn.

- 60% of the Canadian rapeseed crop was transgenic. Canola oil is incorporated into salad dressing, margarine, cheese, fried foods, and many kinds of desserts.

- 71% of the U.S. cotton crop was transgenic. Beyond its use in textiles, cotton is the source of cottonseed oil that reaches the market in cooking oils, salad dressings, and various dessert products.

By genetic engineering, a soil bacterium (*Bacillus thuringiensis* or *Bt*) has been modified to produce pesticides that are more toxic than natural ones. The modified genes have then been inserted into crop plants to create crops that synthesize their own pesticides. There is *Bt* corn with a toxin that destroys the European corn borer and other pests, and there is *Bt* cotton protected against the cotton bollworm and the budworm.

Genetic engineering has also been used to develop transgenic crops that are resistant to certain herbicides. This allows the control of weeds without damaging the plant. For example, cotton, corn, and soybeans have been genetically engineered with resistance to glyphosate, the active ingredient in the herbicide Roundup, which can thus be used to destroy competing weeds.

A different type of application is to use genetic engineering to increase the amount of commercially useful substance in a plant. Transgenic rapeseed plants have been developed using a gene from the California bay tree that shuts off fatty acid synthesis at 12 carbons instead of 18 carbons. The resulting transgenic plant contains up to 40% lauric acid, a saturated fatty acid used to make soaps, detergents, and shampoos (Section 15-4).

The value of transgenic crops is, however, a topic that remains controversial. The potential risks and benefits to humans and the environment are not fully understood. There are different realms of uncertainty for each type of modification and each type of plant. Will overall ecological balance be disrupted? Will there be gene transfer from crops to weeds or other crops? Are beneficial insects at risk? Will insects or plant diseases become so tolerant to toxins that they cannot be stopped at all? These and other questions await further study.

Your Resources

In the back of the textbook:

→ *Review card on Feeding the World*

   In OWL for CHEM2 at www.cengagebrain.com

→ *Review Key Terms with Flash Cards (printable or digital)*

→ *Complete Interactive Practice Quizzes to prepare for tests*

→ *Submit Assigned Homework and Exercises*

Applying Your Knowledge

1. What factors affect Earth's carrying capacity?

2. What are the likely consequences if the present rate of human population growth continues?

3. What is an operational definition of "soil"?

4. Give definitions and descriptions for the following:
   (a) Topsoil   (b) Subsoil
   (c) Loam

5. What is the structure of a typical soil?

6. Name what causes soil to be:
   (a) Sour   (b) Sweet

7. If crushed limestone is spread on soil, will it raise or lower the pH of the soil? Explain.

8. Give definitions for the following:
   (a) Leaching   (b) Selective leaching

9. What are the three nonmineral nutrients obtained from water and air required for normal plant growth?

10. Which is more easily leached from soils, nitrates or phosphates? Why?

11. (a) Which groups of elements are first leached from soils, the Group 1 and 2 metals or the Group 13 and transition metals?
    (b) What is the effect of this selective leaching on soil pH?

12. (a) What are two important roles of humus in the soil?
    (b) Do leaves turned into the soil to produce humus raise or lower the soil pH?

13. Give definitions and/or descriptions for the following:
    (a) Nutrients   (b) Nonmineral nutrients
    (c) Mineral nutrients

14. What are the three primary mineral plant nutrients that are considered in fertilizer formulations?

15. Which is more likely to be a problem in farming, a soil shortage of N, P, and K, or a shortage of Ca, Mg, and S? Give a reason for your answer.

16. (a) What does the term "nitrogen fixation" mean?
    (b) Give two natural ways that this process occurs.

17. Write the balanced equation for the Haber process. Why is this process important in the production of fertilizers?

18. What are legumes? What is nitrogenase?

19. What is chlorosis? What causes it? What are the symptoms of chlorosis?

20. An inorganic fertilizer has a grade of 10–0–0. Is this a complete fertilizer? Explain.

21. A lawn fertilizer is labeled as 32–2–8. What does this mean?

22. Inorganic fertilizers are graded on their content of nitrogen, phosphorus, and potassium. Are any of these actually present as the element? Are any of the reference substances actually in a bag of fertilizer?

23. A fertilizer consisting of Peruvian seabird droppings is labeled as 12–12–2.5. How many pounds of each nutrient are present in a 50-lb bag of this fertilizer? Phosphorus is expressed as a percentage of P_2O_5 and potassium is expressed as a percentage of K_2O.

24. Why is the percentage of potassium in fertilizer less than the amount expressed as % K_2O?

25. Phosphate rock is treated with sulfuric acid to make superphosphate. Why isn't the phosphate rock simply used as a phosphorus-containing fertilizer?

26. Urea (NH_2CONH_2) is a common straight fertilizer supplying nitrogen. What happens to urea when it is applied to soil? What nitrogen compound is formed when urea decomposes?

27. What is a straight fertilizer? What is a complete fertilizer?

28. Give definitions or descriptions for the following:
    (a) Quick-release fertilizers   (b) Slow-release fertilizers

29. Give definitions for the following:
    (a) Pesticide   (b) Insecticide
    (c) Herbicide   (d) Fungicide

30. What are the raw materials used to prepare ammonia via the Haber process?

31. In what way is the production of ammonia for fertilizer production via the Haber process dependent on petroleum resources?

32. Why do you think that the reaction of $N_2(g)$ and $H_2(g)$ in the Haber process must take place at high temperatures?

33. Ammonium nitrate [$NH_4NO_3(s)$] is a common straight fertilizer. It decomposes at high temperatures to produce N_2, H_2O, and O_2. Handling ammonium nitrate has led to accidental explosions. Give one reason for continued use of ammonium nitrate as a fertilizer and one reason why it should be under greater control.

34. What is the reason the use of DDT is banned in the United States? Why do you suppose that DDT is still used in other parts of the world?

35. Why is DDT fat soluble and not water soluble? What two properties make DDT a problem?

36. Trace the rise and fall of the use of DDT in agriculture. Debate whether it has been more good than bad for the human race.

37. What is the approximate percentage of food crops that is lost to pests every year? What is the estimated dollar value of these lost crops?

38. What is a persistent pesticide? Give an example.

39. What is a biodegradable pesticide? Give an example.

40. (a) What two herbicides were formulated to produce Agent Orange?
    (b) Which of these herbicides is currently banned in the United States for agricultural use?

41. (a) Discuss the benefits and possible harms in using pesticides in agriculture.
    (b) What conclusions can you draw on this controversial issue?

42. Organic farming saves energy in one area but loses it in another. Explain.

43. What is sustainable agriculture?

44. If you grew a garden, would you use synthetic fertilizers and pesticides? Why or why not?

45. What is integrated pest management?

46. Two effective natural insecticides are *Bt* toxins and neem tree seeds. Explain what these are and why they are becoming so popular.

47. (a) What are transgenic crops?
    (b) Give some examples.

48. What mechanism of action is followed by glyphosate and sulfonylureas? What is the relationship between "essential" amino acids and these herbicides?

49. Why are pheromones interesting compounds for insect control? How are they used?

50. Assume it is possible to use genetic engineering to improve the storage life of a ripe harvested food like oranges. Give two questions you would want answered before you would buy and eat these oranges.

51. What do you think about the connection between population growth and limitations on the amount of cultivable land? Do you see any conflict? Explain.

52. A 1992 United Nations study indicated that 4.84 billion acres of farmland soil were so degraded that it would be impossible to reclaim them. What are the main causes of this soil degradation?

53. Order the following insecticides in increasing order of toxicity based on the LD_{50} values for rats given in parentheses:
    DDT ($LD_{50} = 100$ mg/kg)
    Malathion ($LD_{50} = 1000$ mg/kg)
    Carbaryl ($LD_{50} = 250$ mg/kg)
    Pirimicarb ($LD_{50} = 150$ mg/kg)
    Dimethrin ($LD_{50} = 10,000$ mg/kg)

Chapter 2

2.1.   (a) Pb   (b) P   (c) HCl   (d) $AlBr_3$ (e) F_2

2.2A. $2H_2(g) + O_2(g) \longrightarrow 2 H_2O(\ell)$

2.2B.   (a) Hydrogen gas reacts with chlorine gas to form hydrogen chloride gas.

(b) *Reactants*: 2 H atoms in 1 H_2 molecule, 2 Cl atoms in 1 Cl_2 molecule; *products*: 2 HCl molecules, which contain 2 H atoms and 2 Cl atoms

2.3.   *mega* = 1 million, 20 megabucks = 20 million dollars ($20,000,000)

Chapter 3

3.1A. 28 protons, 28 electrons, and 31 neutrons

3.1B. The sum of protons and neutrons is equal to 165. Because there are 88 neutrons, the number of protons is equal to $165 - 88 = 77$. The element is Iridium.

3.2. $^{107}_{47}Ag$, $^{109}_{47}Ag$

3.3. (a) Group 2 is the only main group that has all metals. (Group 1 is sometimes regarded as having only metals since hydrogen is a nonmetal. Figure 3.9 shows hydrogen separated from the rest of the members of Group 1 for this reason.)

   (b) Groups 17 and 18 have only nonmetals.

   (c) Groups 13, 14, 15, and 16 include metalloids, nonmetals, and metals. The number of valence electrons are: (a) Group 1, 1; Group 2, 2. (b) Group 17, 7; Group 18, 8. (c) Group 13, 3; Group 14, 4; Group 15, 5; Group 16, 6.

3.4A. (a) Ba (b) Se (c) Si (d) Ga

3.4B. (a) Sr (b) Cl (c) Cs

Chapter 5

5.1A. CaF_2

5.1B. Two. $CoCl_2$ and $CoCl_3$

5.2. **(a)** CsI **(b)** $SrCl_2$ **(c)** BaS

5.3. **(a)** Rubidium chloride **(b)** Gallium oxide

    (c) Calcium bromide **(d)** Iron(II) nitride

5.4. **(a)** CoS **(b)** MgF_2 **(c)** KI

5.5. **(a)** $MgCO_3$ **(b)** NaH_2PO_4

5.6. **(a)** H $:\ddot{I}:$ **(b)** $:\ddot{B}r:\ddot{B}r:$

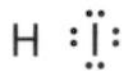

    (c)   H   **(d)** $H:\ddot{O}:\ddot{C}l:$
           H$:\ddot{C}:$H
              H

5.7. **(a)**   H H
      H$:\ddot{C}:\ddot{C}:$H
        H H

    (b)   H H H H   H H H
        H$:\ddot{C}:\ddot{C}:\ddot{C}:\ddot{C}:$H and H$:\ddot{C}:\ddot{C}:\ddot{C}:$H
        H H H H   H | H
                                 H$:\ddot{C}:$H
                                  H

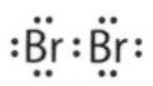

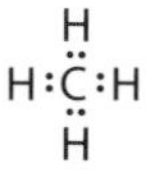

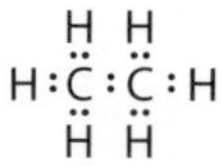

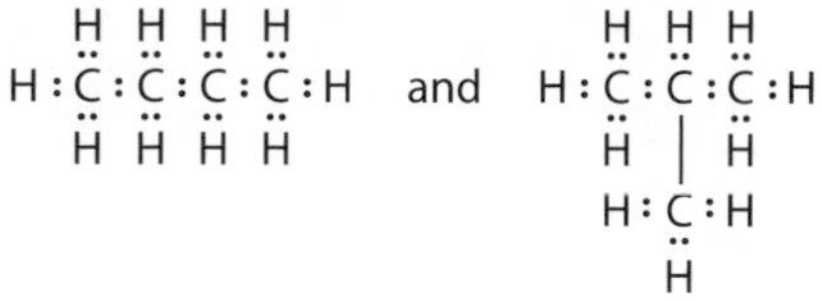

5.8. Tetraphosphorus trisulfide

Chapter 6

6.1. 300−400 ppm; 300,000−400,000 ppb

6.2. 2.5×10^{-5} m $= 2.5 \times 10^4$ nm $= 25,000$ nm

   2.5×10^{-6} m $= 2.5 \times 10^3$ nm $= 2500$ nm

   The IR has longer wavelengths than the UV radiation. UV radiation is more energetic.

6.3. $(5/9)(59° − 32°) = 15°$; $(9/5)450° + 32° = 842°$

6.4. $15° + 273.15° = 288.15°$

6.5.
CO_2	370 ppm	370,000 ppb
CH_4	1.8 ppm	1800 ppb
N_2O	0.31 ppm	310 ppb
O_3	0.04 ppm	40 ppb
CCl_3F	0.00026 ppm	0.26 ppb
CCl_2F_2	0.00052 ppm	0.52 ppb

6.6. Since carbon dioxide is only 27.3% carbon, we need to solve the following equation:

   6 metric tons C $= 0.273$ (X metric tons CO_2)

   Thus, 6/0.273 = 22 metric tons of CO_2

6.7. ~132 gallons

6.8. The high standard of living enjoyed by citizens of the United States has its price, and some feel this standard of living is achieved at the expense of less fortunate countries, which struggle to attain the same standard of living.

Chapter 7

7.1. Molecular oxygen absorbs high-energy ultraviolet radiation (UVC) and is converted to two oxygen atoms. An oxygen atom then combines with an oxygen molecule to produce an ozone molecule.

7.2. Oxygen absorbs high-energy UVC (wavelengths less than 240 nm) radiation, and ozone absorbs UVB (wavelengths less than 320 nm) radiation. The bond in molecular oxygen is stronger than the bond in ozone and requires more energetic radiation to break in the protection process.

7.3. The bathtub analogy uses the level of water in the tub to illustrate the ozone concentration in the stratosphere. Naturally occurring reactions in the atmosphere produce and consume ozone at equal rates to maintain a steady-state concentration, just as the rate of water added to a bathtub could be adjusted to exactly match the rate of water draining from the tub and maintain a constant level in the tub. Additional chemical reactions of anthropogenic origin that consume ozone would correspond to the addition of extra drains to the tub so that the level of water in the tub would drop.

7.4. The substances emitted into the atmosphere elsewhere accumulate at the pole. There, the weather conditions, extreme cold and little wind, coupled with long periods of darkness during the Antarctic winter, are critical to the chemical reactions leading to the production of the reactive species involved in the destruction.

7.5. The C—Cl bond is the critical bond.

7.6. The CFCs are not soluble in or reactive with water.

7.7. Decreased crop yields, increased skin cancer rates, increased biological mutations in organisms.

7.8. Probably not. The increased time required to get a tan in a booth that uses longer-wavelength lamps appears to offset any benefit realized from using such lamps.

Answers to Try It

Chapter 8

8.1A. $2 Al(s) + 3 Cl_2(g) \longrightarrow 2 AlCl_3(s)$

8.1B. **(a)** yes **(b)** no; $SiO_2(s) + 2 C(s) \longrightarrow Si(s) + 2 CO(g)$

8.2A. Because the mass of one mole of any element is equal to the atomic weighted average of the element, in grams, the mass of one mole of selenium atoms is 78.96 g.

8.2B. Consulting the periodic table, we can see that the mass of 1 mole of Mg atoms is 24.3 g. Since this is the same as the mass of the first magnesium sample, that sample must contain 6.022×10^{23} Mg atoms. For the second sample, which weighs half as much as the first, there must be half as many atoms. So, there are $6.022 \times 10^{23}/2 = 3.011 \times 10^{23}$ Mg atoms.

8.3A. Na = 22.99 g/mol
Cl = 35.45 g/mol
NaCl = 58.44 g/mol

8.3B. 6 mol of C = (6 mol)(12.01 g/mol) = 72.06 g
12 mol of H = (12 mol)(1.10 g/mol) = 12.12 g
6 mol of O = (6 mol)(16.00 g/mol) = 96.00 g
So $C_6H_{12}O_6$ = 180.18 g/mol

8.4. $50 \text{ mol Ba(NO}_3)_2 \times \dfrac{261 \text{ g Ba(NO}_3)_2}{1 \text{ mol Ba(NO}_3)_2} = 13{,}000 \text{ g Ba(NO}_3)_2$

8.5.
$CO(g)$	+	$2 H_2(g)$	$\longrightarrow$	$CH_3OH(\ell)$
1 mol CO		2 mol H_2		1 mol CH_3OH
28 g CO		2×2 g/mol = 4 g H_2		32 g CH_3OH

Chapter 9

9.1. $H_2SO_4 + 2 H_2O \longrightarrow 2 H_3O^+ + SO_4^{2-}$
   $2 H_3O^+(aq) + 2 OH^-(aq) \longrightarrow 2 H_2O\ (l)$

9.2. $\dfrac{4.0\ \text{mol NaOH}}{1\ \text{L}} \times \dfrac{40\ \text{g NaOH}}{1\ \text{mol NaOH}} = \dfrac{160\ \text{g NaOH}}{1\ \text{L}}$

9.3. $(73\ \text{g})/(36.5\ \text{g/mol}) = 2.0\ \text{mol}$
   and
   $(2.0\ \text{mol})/(0.5\ \text{L}) = 4\ \text{mol/L} = 4\ \text{M}$

9.4. $(20\ \text{g})/(40\ \text{g/mol}) = 0.50\ \text{mol}$
   and
   $(0.50\ \text{mol})/(2\ \text{L}) = 0.25\ \text{mol/L} = 0.25\ \text{M}$

9.5. The way to do this would be to evaporate some of the solvent as a way of increasing the solute concentration. This is not practical in most cases.

9.6. For $(H_3O^+) = 1 \times 10^{-4}\ \text{M}$, pH = 4. A pH of 4 is less acidic than a pH of 3; the tomatoes are less acidic than the cola.

9.7. $3 \times 10^{-6}\ \text{M}$ is between $1 \times 10^{-6}\ \text{M}$ (pH = 6) and $10 \times 10^{-6}\ \text{M}$ (pH = 5). The exact pH is 5.523, since pH = $-\log (3 \times 10^{-6}) = -\log (3) - \log (10^{-6}) = -0.4471 + 6 = 5.5229$.

9.8. Since the product of the hydroxide ion concentration and the hydronium ion concentration is always 10^{-14}, we can calculate that the hydronium ion concentration is $(1 \times 10^{-14})/(2 \times 10^{-6}) = 5 \times 10^{-9}$ and, by bracketing, the pH must be between 8 and 9. The exact pH is 8.301.

9.9. The graph would be logarithmic in nature and would curve upward in a nonlinear fashion.

9.10. If the hydronium ion concentration changes by a factor of 4 (i.e., from 1×10^{-9} to 4×10^{-9}), the pH only changes by 0.6 pH units. An inspection shows that a tenfold change in hydronium ion concentration causes a 1-unit change in pH, so an increase of 3 pH units would require a thousandfold change in hydronium ion concentration. To decrease by 5 pH units, the hydronium ion concentration would have to decrease by 100,000 (a factor of 10^5), which is to say 100,000 fold.

9.11. If pH is 10.7, this means that $-\log (H_3O^+) = 10.7$. Manipulating this, we see that
   $\log (H_3O^+) = -10.7$
   $(H_3O^+) = $ antilog (-10.7)
   $(H_3O^+) = 2 \times 10^{-11}$

Chapter 10

10.1. Li is oxidized (converted to Li^+); O_2 is reduced (converted to O^{2-}).

10.2. (a) Cu is oxidized (addition of oxygen).

   (b) $CH_3C \equiv N$ is reduced (addition of hydrogen).

   (c) SnO is reduced (removal of oxygen).

Chapter 11

11.1. **(a)** (3 g)(80 cal/g) + (3 g) (1 cal/g-°C)(70°C) = 550 cal; endothermic

(b) (6 g)(1 cal/g-°C)(50°C) + (6 g)(540 cal/g) = 3540 cal; endothermic

(c) (10 g)(−540 cal/g) + (10 g)(1 cal/g-°C)(−100°C) = −6400 cal; exothermic

Answers to Try It

CHEM

Chapter 12

12.1. **(a)** The balanced equation for the combustion of ethane is

$$2\ CH_3CH_3 + 7\ O_2 \longrightarrow 4\ CO_2 + 6\ H_2O$$

This requires the input of heat required to break the following bonds:

12 mol of C—H bonds = (12 mol)(416 kJ/mol) = 4992 kJ

2 mol of C—C bonds = (2 mol)(356 kJ/mol) = 712 kJ

7 mol of O=O bonds = (7 mol)(498 kJ/mol) = 3486 kJ

for an endothermic total of 9190 kJ

Formation of the products releases the heat associated with formation of the following bonds:

8 mol of C=O bonds = (8 mol)(−803 kJ/mol) = −6424 kJ

12 mol of O—H bonds = (12 mol)(−467 kJ/mol) = −5604 kJ

for an exothermic total of −12,028 kJ

The sum of these two is −2838 kJ for 2 moles of ethane. That is equal to −1419 kJ/mol. Since the molecular weight of ethane is equal to 30 g/mol, that equates to −47.3 kJ/g.

(b) The balanced equation for the combustion of ethanol is $CH_3CH_2OH + 3\ O_2 \longrightarrow 2\ CO_2 + 3\ H_2O$

This requires the input of heat required to break the following bonds:

5 mol of C—H bonds = (5 mol)(416 kJ/mol) = 2080 kJ

1 mol of C—C bonds = (1 mol)(356 kJ/mol) = 356 kJ

1 mol of C—O bonds = (1 mol)(336 kJ/mol) = 336 kJ

1 mol of O—H bonds = (1 mol)(467 kJ/mol) = 467 kJ

3 mol of O=O bonds = (3 mol)(498 kJ/mol) = 1494 kJ

for an endothermic total of 4733 kJ.

Formation of the products releases the heat associated with formation of the following bonds:

4 mol of C=O bonds = (4 mol)(−803 kJ/mol) = −3212 kJ

6 mol of O—H bonds = (6 mol)(−467 kJ/mol) = −2802 kJ

for an exothermic total of −6014 kJ.

The sum of these two is −1281 kJ for 1 mole of ethanol. Since the molecular weight of ethanol is equal to 46 g/mol, that equates to −27.8 kJ/g.

The introduction of oxygen into a hydrocarbon reduces the amount of heat released upon combustion.

12.2. **(a)**

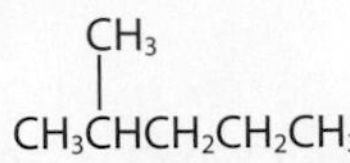

CH₃
|
CH₃CHCH₂CH₂CH₃
2-Methylpentane

(b)

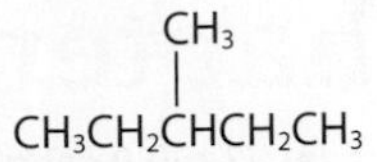

CH₃
|
CH₃CH₂CHCH₂CH₃
3-Methylpentane

(c)

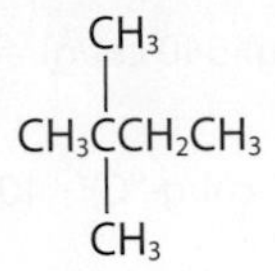

CH₃
|
CH₃CCH₂CH₃
|
CH₃
2,2-Dimethylbutane

(d)

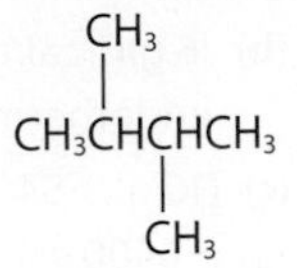

CH₃
|
CH₃CHCHCH₃
|
CH₃
2,3-Dimethylbutane

12.3.

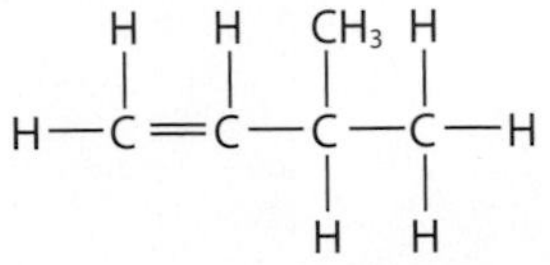

H   H   CH₃ H
| | | |
H—C=C—C—C—H
| |
H   H
3-Methyl-1-butene

Chapter 13

13.1. $^{237}_{93}\text{Np} \longrightarrow \, ^{4}_{2}\text{He} + \, ^{233}_{91}\text{Pa}$

13.2. $^{234}_{91}\text{Pa} \longrightarrow \, ^{0}_{-1}\text{e} + \, ^{234}_{92}\text{U}$

Answers to Try It

Chapter 14

14.1. **(a)**

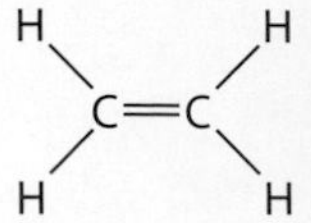

H—C=C—H with H, H substituents (ethylene)

(b)

H—C=C—H with H, Cl substituents (vinyl chloride)

(c)

H—C=C—H with H and phenyl substituents (styrene)

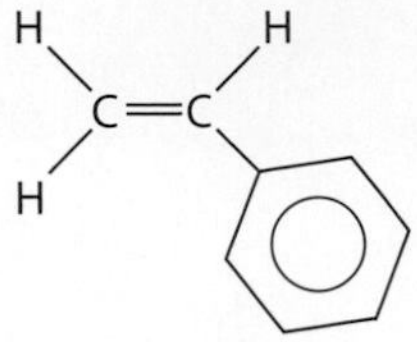

14.2.

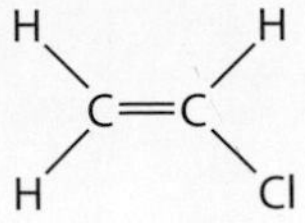

$$\left(\!\begin{array}{c}O\\ \parallel\\ C\end{array}\!-\!\bigcirc\!-\!\begin{array}{c}O\\ \parallel\\ C\end{array}\!-O-CH_2-CH_2-O\right)_n$$

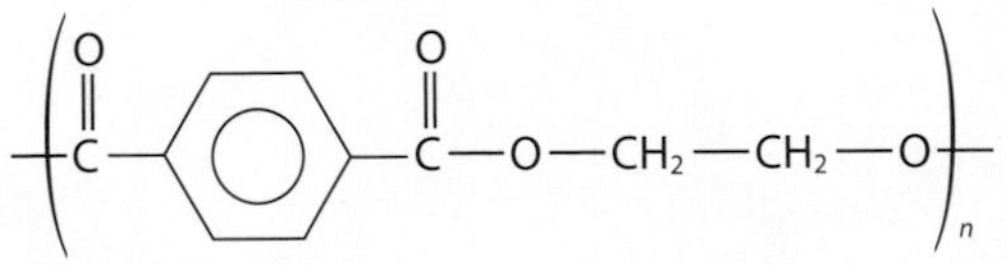

Chapter 15

15.1. **(b)** is chiral.

(d) is chiral.

15.2.

15.3. AGGCTA

Answers to Try It

Chapter 16

16.1.   160. lb $\times$ 10 kcal/lb $=$ 1600 kcal

1600 kcal $\times$ 1.3 $=$ 2080 kcal

16.2.   Healthy Chicken:

$$15 \text{ Cal fat} \times \frac{1 \text{ g}}{9 \text{ Cal}} = 2 \text{ g fat}$$

$$\frac{2 \text{ g fat}}{65 \text{ g fat}} \times 100\% = 3\% \text{ of daily fat allowance}$$

Cream of Mushroom:

$$80 \text{ Cal fat} \times \frac{1 \text{ g}}{9 \text{ Cal}} = 9 \text{ g fat}$$

$$\frac{9 \text{ g fat}}{65 \text{ g fat}} \times 100\% = 14\% \text{ of daily fat allowance}$$

16.3.

$$3 \text{ g protein} \times \frac{4 \text{ kcal}}{\text{g protein}} = 12 \text{ kcal}$$

$$8 \text{ g fat} \times \frac{9 \text{ kcal}}{\text{g fat}} = 72 \text{ kcal}$$

$$40 \text{ g carbohydrate} \times \frac{4 \text{ kcal}}{\text{g carbohydrate}} = 160 \text{ kcal}$$

$$\text{Total} = 244 \text{ kcal per slice}$$

Chapter 19

19.1. Formula weight: $2\,N \times 14 = 28$, $4\,H \times 1 = 4$, $3\,O \times 16 = 48$. Total $= 80$.

Percentage $N = \dfrac{28}{80} \times 100\% = 35\%$

There is no K or P in ammonium nitrate, so their values are zero.

Answers to Applying Your Knowledge

Chapter 1

1. zoology

3. amount of exposure

5. biochemistry or genetics

7. Air conditioning certainly contributed to the growth of these cities in a very hot climate. This growth has led to increased pollution, much of it due to greater automobile traffic in these sprawling cities. Overall, most would say the benefits have outweighed the negative effects.

9. Individual answers may vary but should be based on facts and not on ignorance, unsubstantiated assertions, or fear of chemicals.

Chapter 2

1. Answer will depend on each person's experience.

3. No, they would be the same substance.

5. This fact illustrates that one chemical can have many different uses depending on its quantity. The benefits of using a small amount of nitroglycerin in treating angina outweigh the risks, while when used in larger quantities, the risks can be reversed, such as when it is used as an explosive.

7. This answer can vary between students, but one example is the combustion of gasoline to propel vehicles. This reaction is a chemical change because the liquid gasoline is converted to heat and gases. Another constructive example is the burning of coal to heat water into steam, which is then used to turn a turbine and produce electricity. The combustion of coal results in a flame and in other gases. The above two examples are examples of chemical changes because the products have a different chemical formula than reactants; another way to look at this reaction is that it is not reversible. On the other hand, a destructive reaction is the use of ammonia nitrate to construct bombs for devastation. This compound is a solid but, when mixed with the correct reactants, will produce a flame and a rapid expansion of gases that makes the explosion. The products formed in this reaction are definitely different than the starting materials.

9. (a) tin: solid
   (b) bromine: liquid
   (c) dysprosium: solid
   (d) xenon: gas
   (e) samarium: solid
   (f) lithium: solid
   (g) mercury: liquid
   (h) iodine: solid

11. (a) **S-I-Ne**: sulfur-iodine-neon
    (b) **Cr-Y**: chromium-yttrium
    (c) **V-Ir-U-S**: vanadium-iridium-uranium-sulfur OR **V-I-Ru-S**: vanadium-iodine-ruthenium-sulfur
    (d) **Re-Si-S-Ta-N-Ce**: rhenium-silicon-sulfur-tantalum-nitrogen-cerium OR **Re-S-I-S-Ta-N-Ce**: rhenium-sulfur-iodine-sulfur-tantalum-nitrogen-cerium
    (e) **Cr-Os-Sb-O-W**: chromium-osmium-antimony-oxygen-tungsten OR **Cr-Os-S-B-O-W**: chromium-osmium-sulfur-boron-oxygen-tungsten OR **Cr-O-S-Sb-O-W**: chromium-oxygen-sulfur-antimony-oxygen-tungsten OR **Cr-O-S-S-B-O-W**: chromium-oxygen-sulfur-sulfur-boron-oxygen-tungsten
    (f) **Fe-Nd-Er**: iron-neodymium-erbium

(g) **Ac-Cu-Se**: actinium-copper-selenium OR **Ac-C-U-Se**: actinium-carbon-uranium-selenium

13. The answers depend on the student's choice of word.

15. (a) Cu, Cn
    (b) Cu, Cr, Ce
    (c) W, Ti, Sn
    (d) Tl, Th
    (e) N, Ni
    (f) Carbon, calcium
    (g) Fe, F
    (h) N, Ni, Ne

17. False, a molecule is the smallest part of a compound.

19. False, a molecule is the smallest part of a compound.

21. A mixture of sand and salt can be separated by adding water to the mixture. The salt will dissolve in the water while the sand will not and will settle to the bottom. Filtration of the solution will result in the capture of sand on the filter paper while the dissolved salt will pass through the funnel (filtrate). The sand can be identified because it is not soluble in water and can be recovered by filtering. The salt can be recovered from the filtrate by evaporating the water.

23. Atrazine, $C_8H_{14}N_5Cl$, contains the elements carbon, hydrogen, nitrogen, and chlorine.

25. (a) BonAmi kitchen and bath cleanser
    (b) Coca-Cola
    (c) Gatorade
    (d) Coca-Cola
    (e) Skippy Peanut Butter
    (f) Kraft Grated Parmesan Cheese
    (g) Morton's Iodized Salt
    (h) Oil of Olay
    (i) Mylanta
    (j) Kellogg's Frosted Mini-Wheats

27. (c) The identity of the atoms in the reactants has to be the same as the products. The number of atoms on each side of the equation must also be equal.

29. $N_2 + 3H_2 \longrightarrow 2NH_3$

31. (a) Electrical energy can be generated by chemical reactions used in alkaline batteries and car batteries.
    (b) Heat energy can be generated by chemical reactions involving combustion or burning of gasoline.
    (c) Light energy can be generated by chemical reactions used by fireflies or in glow sticks.
    (d) Mechanical energy can by generated by using water behind a dam to turn a turbine or the alternator on a car using the belt of the engine.

33. Balance

  (a) On the left side of the arrow, "2 Na" means 2 Na atoms; one Cl_2 molecule contains 2 Cl atoms. On the right side 2 NaCl units contain 2 Na atoms and 2 Cl atoms.

  (b) On the left one N_2-molecule contains 2 N atoms and 3 Cl_2 molecules contain 6 Cl atoms. On the right 2 NCl_3 molecules contain a total of 2 N atoms and 6 Cl atoms.

  (c) On the left there are 1 C atom, 2 H atoms, and $2 + 1 = 3$ O atoms. On the right there are 1 C atom, 2 H atoms, and 3 O atoms.

  (d) On the left there are 4 H atoms and 4 O atoms in 2 molecules of H_2O_2. On the right there are 4 H atoms in the 2 molecules of water; there are also 2 O atoms in the 2 water molecules and 2 more O atoms in the O_2 molecule for a total of 4 O atoms.

35. (a) No. The reactant side contains 1 silver atom, 1 nitrogen atom, 1 sulfur atom, 2 sodium atoms, and 7 oxygen atoms while the product side contains 2 silver atoms, 1 nitrogen atom, 1 sulfur atom, 1 sodium atom, and 7 oxygen atoms.

  (b) Yes. The reactant side contains 1 silver atom, 1 nitrogen atom, 1 hydrogen atom, 1 chlorine atom, and 3 oxygen atoms while the product side contains 1 silver atom, 1 nitrogen atom, 1 hydrogen atom, 1 chlorine atom, and 3 oxygen atoms.

37. The tea in tea bags is a mixture. It can be partially separated by dissolving some water-soluble substances with hot water. Instant tea is a mixture of the water-soluble substances in tea.

39. (a) 1 gram = 1000 milligrams
  (b) 1 kilometer = 1000 meters
  (c) 1 gram = 100 centigrams

41. (a) 9 cal/g; no
  (b) 100 cm/m; no.
  (c) 1.5 g/mL; yes, grams/milliliter is mass/volume.
  (d) 454 g/lb.; no.

43. 5.5 acres/55 cows

45. 10 km $\times$ (1000 m/1 km) = 10,000 m

47. (a) 0.04 m
  (b) 43 mg
  (c) 15500 mm
  (d) 0.328 L
  (e) 980 g

49. 70 kg $\times$ (1000 g/1 kg) $\times$ 100 mg/1 g = 70,000,000 mg

51. (a) 8.0×10^7
  (b) 3.0×10^5
  (c) 1.6×10^{-5}
  (d) 9.7×10^1

53. (a) 450,000,000 watts
  (b) 4,500,000 bulbs

55. 22,420 g

57. The mass of the Al object is 0.34 as much as the mass of the Fe object.

59. 128 fluid ounces

61. 7.73 grains

Chapter 3

1. Matter is neither created nor destroyed. Examples: flash bulb before and after use, hard boiling an egg, yarn knitted into clothing, melting ice cubes in a glass. There is no change in mass during the chemical changes (first two) or the physical changes.

3. John Dalton

5. Experiments indicated that matter was conserved. Elements had been identified. Compounds had been shown to be composed of definite amounts of specific elements. The composition of a compound had been shown to always be the same regardless of its source. The composition of a combination of elements could be predicted.

7. The experiments of Rutherford and Curie each were able to detect the presence of charged particles like protons and electrons but not neutral particles like the neutron.

9. Particles smaller than the atom, such as protons, neutrons, electrons.

11. Atoms exist with 1, 2, ..., up to 118 protons; each number of protons corresponds to a different element. Most elements exist as more than one naturally occurring isotopes, i.e., atoms with various numbers of neutrons for a given number of protons. This accounts for the large number of different masses for actual atoms. For example carbon has three different isotopes; ^{12}C, ^{13}C, and ^{14}C.

13. See Tables 3.2 and 3.3.

15. Atoms of the same element all have the same atomic number. This means they have the same number of protons in the nucleus. For example these two atoms both have 10 protons: $^{21}_{10}Ne$, $^{22}_{10}Ne$.

17. (a) an isotope with two fewer neutrons (18 versus 20); the isotope is the same element and has the same number of protons (17) and electrons (17).

19. Iodine is represented by the symbol $^{127}_{53}I$; it has 53 protons and 74 neutrons. The neutron count equals the mass number minus the atomic number. $127 - 53 = 74$ neutrons.

21. (e) $_{34}Se$

23. The atomic number has the symbol Z. The number of protons equals the atomic number.
    (a) Germanium, Z = 32 or 32 protons
    (b) Silicon, Z = 14 or 14 protons
    (c) Nickel, Z = 28 or 28 protons
    (d) Cadmium, Z = 48 or 48 protons
    (e) Iridium, Z = 77 or 77 protons

25. Yes, the statement is correct. The atomic mass of 24.305 is a weighted average based on the masses of individual Mg isotopes and their abundances.

27. $(0.5069 \times 78.91833) + (? \times 80.91629) = 79.904$
    ? = 0.4931, or 49.31%

29. $v = 4 \times 10^{14}$ cycles/sec

31. Wavelength and frequency are inversely proportional, so shorter wavelengths correspond to higher frequency. Light of $\lambda = 600$ nm has a higher frequency.

33. (d) energy increases

35. (c) IR radiation has a lower frequency than UV radiation.

37. Bohr assignments:

Bohr Electron Arrangement	Number of Electrons	Element
2-4	6	Carbon
2-8	10	Neon
2-8-3	13	Aluminum
2-8-8-2	20	Calcium

39. Na, 1; Mg, 2; Al, 3; Si, 4; P, 5; S, 6; Cl, 7; Ar, 8

41. When elements are arranged in order of their atomic numbers, their chemical and physical properties show repeatable trends.

43. The known elements were arranged in order, and gaps between elements occurred. These gaps suggested the existence of elements not yet isolated.

45. Metals are good conductors of heat and electricity. Metals are malleable and ductile. Nonmetals are the opposite: they are not ductile, are not malleable, and are poor conductors of heat and electricity.

47. (a) There are 7 periods.
    (b) There are 8 representative groups.
    (c) There are 2 representative groups that are all metals if you ignore H in 1A.
    (d) The elements in Group 7A and 8A are all nonmetals.
    (e) The elements of period 7 are all metals.

49. 10^5 or 100,000

51. (a) B, Al, Ga, In, and Tl all have 3 valence electrons.
    (b) C, Si, Ge, Sn, and Pb all have 4 valence electrons.
    (c) Cl, Br, I, F, and At all have 7 valence electrons.
    (d) H, Li, Na, K, Rb, Cs, and Fr all have 1 valence electron

53. (a) Be is more metallic than B.
    (b) Ca is more metallic than Be.
    (c) Ge is more metallic than As.
    (d) Bi is more metallic than As.

55. a completed outer shell or set of eight electrons in the outermost energy level

57. Composition of atoms

Atomic Number	Name of Element	Number of Valence Electrons	Period	Metal or Nonmetal
6	C	4	2	NM
12	Mg	2	3	M
17	Cl	7	3	NM
37	Rb	1	5	M
42	Mo	6	5	M
54	Xe	0 or 8	5	NM

59. The size of an atom is determined by the distance of the outermost electrons from the nucleus. The outer or last electron in Cs is in a higher energy level, $n = 6$. Li has outer electrons in $n = 2$. The higher the level, the greater the average distance from the nucleus.

61. The smaller the radius of the nonmetal atom, the more reactive it is. The larger of two atoms is typically more reactive for metals and the smaller is more reactive for nonmetals. Metal atoms at the top of a group are smaller and less reactive. Metals lose electrons. Removing an electron is more difficult for small atoms. Nonmetal atoms at the top of a group are smaller than those at the bottom and are more reactive than atoms at the bottom. The nonmetals gain electrons. The smaller the nonmetal, the stronger the attraction for an additional electron. Noble gases are generally less reactive than other atoms. Note below that in "e" xenon is more like a metal than helium is.

(a) Rb more reactive than Li
(b) Ba more reactive than Mg
(c) Na more reactive than Ar
(d) O more reactive than Ne
(e) Xe more reactive than He
(f) F more reactive than Br

Chapter 4

1. (a) air that contains unwanted and harmful substances
   (b) mixtures containing water droplets and/or tiny particulates suspended in a gas
   (c) smog-containing compounds formed in reactions initiated by sunlight
   (d) substances formed by reactions of emitted compounds with components in the air

3. nitric oxide from internal combustion engines; sulfur oxides from coal combustion; carbon monoxide from inefficient burning of hydrocarbons

5. The abbreviation CAA refers to the Clean Air Act. The first act was passed in 1970 and originally controlled air pollution from cars and industry. Amendments in 1977 imposed stricter auto emission standards. The 1990 CAA extends to manufacturing and commercial activity. This version of the CAA regulates particulates, ozone, carbon monoxide, oxides of nitrogen and sulfur, carbon dioxide, and substances that would deplete stratospheric ozone. A newer version was passed in 1996.

7. Industrial smog is chemically reducing, whereas photochemical smog is chemically oxidizing. Industrial smog contains sulfur dioxide mixed with soot, fly ash, smoke, and partially oxidized organic compounds. Photochemical smog is essentially free of SO_2, but contains ozone, ozonated hydrocarbons, organic peroxide compounds, nitrogen oxides, and unreacted hydrocarbons.

9. b, c, a, and d in order

11. Of the nitrogen oxides (NO_x) in the atmosphere, 97% come from natural sources. Lightning strikes during electrical storms produce NO. Nitric oxide, NO, is so reactive that it combines with O_2 in the atmosphere to produce NO_2. Some bacteria produce N_2O. About 3% of the atmospheric nitrogen oxides come from human activity such as combustion in automobile engines.

13. Hydrocarbons are released into the atmosphere by natural sources and human sources. Hydrocarbons are put into the atmosphere by living plants such as deciduous trees, by the decay of dead plants and animals, and by excrement from insects and animals. Human activity introduces hydrocarbons when organic solvents are used (for example, in the handling of petroleum products, etc.). Generally only human activities are within our control.

15. Nitrogen dioxide plays a role in the formation of ozone in the troposphere.

17. Nitrogen dioxide (NO_2) will dissociate to form an oxygen atom and nitric oxide, if it is excited by a sufficiently energetic photon of UV light. The O atom forms O_3 by reaction with O_2.

$$NO_2(g) + h\nu \longrightarrow NO(g) + O(g)$$
$$O + O_2 \longrightarrow O_3$$

19. All three are oxidizing agents. Ozone, sulfur dioxide, and nitrogen dioxide cause lung damage.

21. Air is precooled below 0°C or 273 K to remove water vapor as ice. The temperature is decreased to less than −78°C or 194 K to remove CO_2. The dry air, free of CO_2, is compressed to more than 100 atmospheres. This compression heats the air because energy is added to compress it. The air is cooled to room temperature. The room temperature air is allowed to expand and cool. This compression-expansion cycle is repeated until all of the air is liquefied. The liquid air is allowed to warm, and each gas in the mixture can be collected when it "boils" off at its own boiling point.

23. lung diseases, cancer, and mutagenic effects

25. Oil-burning electric fuel plants generate SO_2. The amount of SO_2 emitted can be reduced either by using low sulfur fuel oil or by passing plant exhaust gas through molten sodium carbonate to form sodium sulfite.

27. (c) 400 ppm is equal to 0.04%.

Chapter 5

1. (a) A cation is an ion with a positive charge.
   (b) An anion is an ion with a negative charge.
   (c) Atoms react to acquire an electron configuration with 8 electrons in the outermost shell to match the configuration of the nearest noble gas.
   (d) The formula unit is the simplest element ratio for an ionic compound like NaCl instead of Na_2Cl_2.

3. (a) a pair of electrons shared between two atoms
   (b) four electrons (2 pairs) shared by two atoms
   (c) six electrons (3 pairs) shared between two atoms
   (d) a pair of valence electrons on an atom that is not shared with another atom
   (e) a pair of electrons shared between two atoms
   (f) either a double or a triple bond

5. (a) bond between two atoms that share electrons equally
   (b) bond between two atoms that do not attract shared electrons equally

7. (a) a compound consisting only of carbon and hydrogen
   (b) a compound consisting only of carbon and hydrogen with only single bonds
   (c) a compound consisting only of carbon and hydrogen with one or more multiple bonds between carbon atoms
   (d) Alkenes are hydrocarbons with one or more carbon–carbon double bonds.

9. (a) Br^{1-}
   (b) Al^{3+}
   (c) Na^{1+}
   (d) Ba^{2+}
   (e) Ca^{2+}
   (f) Ga^{3+}
   (g) I^{1-}
   (h) S^{2-}
   (i) Group 1 atoms lose one valence electron to form a +1 ion, see 9c above.
   (j) Group 7 atoms have seven valence electrons and gain one to form a 1− ion, see 9g.

11. (a) AlI_3 — aluminum iodide
    (b) $SrCl_2$ — strontium chloride
    (c) Ca_3N_2 — calcium nitride
    (d) K_2S — potassium sulfide
    (e) Al_2S_3 — aluminum sulfide
    (f) Li_3N — lithium nitride

13. (a) calcium sulfate
    (b) sodium phosphate
    (c) sodium bicarbonate
    (d) potassium hydrogen phosphate
    (e) sodium nitrite
    (f) copper(II) nitrate

15. (a) Lithium is in Group 1, so it forms the Li^{1+} ion. Tellurium is in Group 6 and forms the Te^{2-} ion. The formula for the neutral ionic combination is Li_2Te.

    (b) $MgBr_2$
    (c) Ga_2S_3

17. (a) With a ratio of three telluride anions for every two bismuth cations, the most likely formula is Bi_2Te_3.
    (b) Bismuth cations have a +3 charge, so for bismuth telluride to be electrically neutral, the telluride anion must have a charge of −2.

19. (a) $(NH_4)_3PO_4$
    (b) Na_2SO_4
    (c) $CuCl_2$
    (d) $Cr(NO_3)_3$
    (e) KBr
    (f) $CaCO_3$
    (g) $NaClO$

21. (a) ionic
    (b) covalent
    (c) ionic
    (d) covalent
    (e) covalent

23. (a) NO, nitrogen monoxide
    (b) SO_3, sulfur trioxide
    (c) N_2O, dinitrogen oxide
    (d) NO_2, nitrogen dioxide

25. Lewis structures
    (a) :C≡O:
    (b)

$$:\ddot{F}:$$
$$|$$
$$:\ddot{F}-Si-\ddot{F}:$$
$$|$$
$$:\ddot{F}:$$

    (c)

$$\begin{array}{ccc} H & & H \\ \ \backslash & & / \\ & C=C & \\ / & & \backslash \\ H & & H \end{array}$$

(d) H—S̈—H

(e) H—C≡C—H

(f)
```
    H        H
    |        |
H—C—C—H
    |        |
    H        H
```

(g) [:Ö—H]⁻

(h)
```
:F̈—N—F̈:
     |
    :F̈:
```

27. Shapes

(a) Tetrahedral
```
        :F̈:
         |
:C̈l—C—C̈l:
         |
        :F̈:
```

(b) Pyramidal
```
:C̈l—N̈—C̈l:
      |
     :C̈l:
```

(c) Bent
```
:C̈l—Ö—C̈l:
```

29. The octet rule states that in order for an atom to be stable, it must have a total of 8 electrons or 4 pairs (either lone or bonding pairs) around it. A molecule with an odd number of electrons will have one electron that is not paired and will not satisfy the octet rule. For example, NO has 11 valence electrons with only oxygen satisfying the octet rule with 8 valence electrons while nitrogen only has 7 electrons and will never satisfy the octet rule unless nitrogen gains an electron.

31. H-F because the difference in electronegativity is the greatest. Fluorine draws electrons away from hydrogen more than chlorine and bromine because it has the highest electronegativity.

33. (a) Polar; oxygen draws electrons and makes that end of the molecule negative.

(b) Nonpolar; there are no polar bonds in the butane molecule.

(c) Polar; nitrogen draws electrons so the nitrogen end of the molecule is negative; the molecule is a trigonal pyramid.

35. (a) The total valence electron count is done by adding all contributions from each atom.

Valence electrons from hydrogen	Valence electrons from oxygen	Valence electrons from carbon	Total valence electrons
6 × 1	1 × 6	2 × 4	20

(b) 8 single bonds

(c) 8 bonding pairs

There are 2 nonbonding pairs of electrons on the oxygen (see figure below).

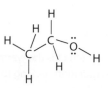

37. (b) solid

39. The number of gas molecules in a given volume of gas decreases as you move away from the Earth's surface. The fewer number of molecules collide with any object less often and exert less push or pressure. Additionally, the force of Earth's gravity decreases with altitude. This results in less weight or force per unit area at greater altitudes.

41. The gas molecules of the perfume vapor diffuse away from the person wearing the perfume.

43. Water has an unusually high normal boiling point and surface tension. Additionally, it expands on freezing.

45. Evaporation of water from one's skin draws energy from the skin. This produces a resulting "chilling" sensation. We are cooled when the energy flows into the evaporating liquid.

47. The solubility of an ionic substance is dependent on the strength of the attraction between the ions and the strength with which the solvent (water, here) can hold onto dissolved ions. The insolubility must be a result of an attraction between barium and sulfate ions that is stronger than the attraction between water and barium or sulfate ions.

49. There are great distances between particles in the gas state while particles in the liquid and solid states are in direct contact. The gas molecules are separated by distances equal to about 1000 molecular diameters. This distance can be decreased by the application of pressure.

51. The increased pressure creates more collisions of gas molecules with the solvent surface. These more frequent collisions create more opportunities for the gas to dissolve in the solvent. Champagne, sparkling waters, Coca-Cola, and Pepsi are examples.

53. one half as much

55. solid because temperature is lower than the melting point

57. below 646°C, at approximately 500°C

Answers to Applying Your Knowledge

Chapter 6

1. **(a)** The greenhouse effect is the warming of Earth by the trapping of IR radiation seeking to escape Earth by certain so-called greenhouse gases such as carbon dioxide. This keeps Earth warmer than it otherwise would be and allows life to flourish.
   (b) Global warming is the term used to describe warming of Earth above and beyond what is normal and desirable.
   (c) CFC is the abbreviation for chlorofluorocarbons, a class of compounds implicated in the destruction of the ozone layer and also contributing to global warming.
   (d) Kyoto Protocol was a global warming policy developed in Kyoto, Japan in 1997 as a first response to this environmental problem.
   (e) Infrared radiation is the type of electromagnetic radiation that is sufficiently energetic to stretch and bend chemical bonds.
3. True
5. **(b)**
7. False
9. False
11. True
13. False
15. True
17. **(d)**
19. False
21. **(e)**
23. Each student's answer may vary, depending on his or her social and political views.
25. Each student's answer may vary, depending on his or her social and political views.
27. Each student's answer may vary, depending on his or her social and political views.
29. 2.5×10^{-3} cm $= 2.5 \times 10^{-5}$ m. This allows us to calculate a frequency of $1.2 \times 10^{13}\ s^{-1}$. This corresponds to an $E_{photon} = 1.9 \times 10^{-21}$ cal.
    2.5×10^{-4} cm $= 2.5 \times 10^{-6}$ m. This corresponds to a frequency of $1.2 \times 10^{14}\ s^{-1}$ and an $E_{photon} = 1.9 \times 10^{-20}$ cal.

Chapter 7

1. (a) an area of decreased stratospheric ozone largely centered above the Antarctic
    (b) chlorofluorocarbons
    (c) hydrochlorofluorocarbons
    (d) a first attempt at an agreement to regulate the production and use of ozone-destroying chemicals
    (e) two or more Lewis dot structures that together represent the actual bonding in a real molecule
    (f) ultraviolet radiation with a wavelength between 320–400 nm
    (g) ultraviolet radiation with a wavelength between 280–320 nm
    (h) ultraviolet radiation with a wavelength between 200–280 nm
    (i) a set of chemical reactions that explains how oxygen and ozone protect us from UVB and UVC
    (j) hydrofluorocarbons
    (k) the region of the atmosphere (from 10–50 km) in which the ozone layer lies
    (l) a state in which the concentration of a chemical remains essentially constant even though the chemical may be undergoing numerous chemical reactions
    (m) a measure of ozone concentration in the atmosphere equal to 1 ppb of ozone

3. (a) true   (b) false   (c) true   (d) true   (e) true
    (f) false   (g) false   (h) true   (i) false   (j) false

5. Once in the stratosphere, CFCs react with sunlight to release chlorine atoms that react with ozone to produce chlorine monoxide and oxygen. The C—Cl bond is the critical bond.

7. About the thickness (3 mm) of two nickels stacked on top of each other

9. UVB ($<$320 nm)

11. Because the chemical bonds in oxygen are stronger than the chemical bonds in ozone, oxygen absorbs higher-energy UV radiation than does ozone.

13. CH_3Br

15. CCl_2F_2 and CCl_3F

17. Weather conditions and sunlight are critical for many of the chemical reactions leading to ozone destruction.

19. Any molecule that is capable of absorbing IR radiation may act as a greenhouse gas in the atmosphere.

21. (b) oxygen

23. (d)

25. (a)

27. UV–C

29. Ground-level ozone is very reactive toward many molecules it encounters and is destroyed before rising into the stratosphere.

31. Global Warming
    (a) troposphere
    (b) IR
    (c) CO_2, H_2O, CH_4, etc.
    (d) Kyoto Protocol
    (e) Mixed acceptance
    (f) Little impact at present

    Ozone Hole
    stratosphere
    UV
    CFCs primarily
    Montreal Protocol
    Widely accepted
    Major impact in reducing problem

Answers to Applying Your Knowledge

Chapter 8

1. (a) 6 and 6

 (b) 7 and 7

 (c) 1, 3, 2, 3

 (d) The number of atoms of each element is the same on both reactant and product side.

3. The number of moles of molecules is not conserved in a chemical reaction, but the number of atoms is conserved.

5. (a) $2\ CH_3OH(l) + 3\ O_2(g) \longrightarrow 2\ CO_2(g) + 4\ H_2O(g)$
 (b) $C_3H_8(l) + 5\ O_2(g) \longrightarrow 3\ CO_2(g) + 4\ H_2O(g)$
 (c) $2\ C_6H_6(l) + 15\ O_2(g) \longrightarrow 12\ CO_2(g) + 6\ H_2O(g)$
 (d) $2\ C_3H_8O(l) + 9\ O_2(g) \longrightarrow 6\ CO_2(g) + 8\ H_2O(g)$

7. (a) $HBr(aq) + KOH(aq) \longrightarrow KBr(aq) + H_2O(l)$
 (b) $H_2S(g) + 2\ NaOH(aq) \longrightarrow Na_2S(aq) + 2\ H_2O(l)$
 (c) $2\ HNO_3(aq) + Ca(OH)_2(aq) \longrightarrow Ca(NO_3)_2(aq) + 2\ H_2O(l)$
 (d) $3\ HCl(aq) + Al(OH)_3(aq) \longrightarrow AlCl_3(aq) + 3\ H_2O(l)$

9. (a) 10 molecules N_2
 (b) 45 molecules N_2
 (c) 200 molecules NH_3

11. (a) $0.5\ \text{mol } C_6H_{12}O_6 \text{ reacting} \times \dfrac{6\ \text{mol water produced}}{1\ \text{mol } C_6H_{12}O_6 \text{ reacting}}$
 $= 3\ \text{mol water produced}$

 (b) $180\ \text{g } C_6H_{12}O_6 \times \dfrac{1\ \text{mol } C_6H_{12}O_6}{180\ \text{g } C_6H_{12}O_6} = 1\ \text{mol } C_6H_{12}O_6$

 $1\ \text{mol } C_6H_{12}O_6 \text{ reacting} \times \dfrac{6\ \text{mol water produced}}{1\ \text{mol } C_6H_{12}O_6 \text{ reacting}}$
 $= 6\ \text{mol water produced}$

 (c) $90\ \text{g water} \times \dfrac{1\ \text{mol water}}{18.01\ \text{g water}} \times \dfrac{1\ \text{mol } C_6H_{12}O_6 \text{ reacting}}{6\ \text{mol water produced}}$
 $= 0.8\ \text{mol } C_6H_{12}O_6$

 (d) $72\ \text{g water} \times \dfrac{1\ \text{mol water}}{18.01\ \text{g water}} \times \dfrac{1\ \text{mol } C_6H_{12}O_6 \text{ reacting}}{6\ \text{mol water produced}}$
 $\times \dfrac{180\ \text{g } C_6H_{12}O_6}{1\ \text{mol } C_6H_{12}O_6} = 120\ \text{g } C_6H_{12}O_6$

13. (a) $4\ \text{mol } SiO_2 \text{ reacting} \times \dfrac{1\ \text{mol Si produced}}{1\ \text{mol } SiO_2 \text{ reacting}} = 4\ \text{mol Si}$

 (b) $4\ \text{mol } SiO_2 \text{ reacting} \times \dfrac{2\ \text{mol CO produced}}{1\ \text{mol } SiO_2 \text{ reacting}}$
 $\times \dfrac{28\ \text{g CO}}{1\ \text{mol CO}} = 224\ \text{g CO}$

 (c) $1\ \text{mol Si produced} \times \dfrac{2\ \text{mol C reacting}}{1\ \text{mol Si produced}} = 2\ \text{mol C}$

 (d) $150\ \text{g } SiO_2 \times \dfrac{1\ \text{mol } SiO_2}{60.1\ \text{g } SiO_2} \times \dfrac{1\ \text{mol Si produced}}{1\ \text{mol } SiO_2 \text{ reacting}}$
 $\times \dfrac{28.09\ \text{g Si}}{1\ \text{mol Si}} = 70.1\ \text{g Si}$

15. Each is the mass of 1 mole of that element. A mole of Cl atoms weighs 35.5 g and a mole of C atoms weighs 12.011 g. Care must be taken not to confuse the molar mass for the naturally occurring form of chlorine, Cl_2, with the value for the atomic version.

17. Each is 0.500 mol of that element.

19. Avogadro's number is 6.022×10^{23} to 4 significant digits or 6.02×10^{23} to 3 s.d.

21. (a) 180.1 g/mol
 (b) 18.01 g/mol
 (c) 114 g/mol
 (d) 149 g/mol
 (e) 404.5 g/mol
 (f) 324 g/mol

23. For the *complete* combustion,

 $2\ C_8H_{18}(l) + 25\ O_2(g) \longrightarrow 16\ CO_2(g) + 18\ H_2O(l)$

 2 moles C_8H_{18} react with 25 moles O_2 to form 16 moles CO_2 and 18 moles of H_2O

 $2 \times 114\ \text{g } C_8H_{18}$ reacts with $25 \times 32\ \text{g } O_2$ to form $16 \times 44\ \text{g } CO_2$ and $18 \times 18\ \text{g } H_2O$

 228 g C_8H_{18} react with 800 g O_2 to form 704 g CO_2 and 324 g H_2O

25. Some reactions are fast because reactants have weak bonds. These reactions have low activation energies; the reactants have a "low hill to climb" to form products. Other reactions are slow because reactants have strong bonds and a high activation energy; these reactants have a high "energy hill to climb" to form products.

27. Increasing the temperature of a reaction usually increases reaction rate. Conversely, decreasing the temperature usually decreases reaction rate. Kinetic theory predicts that average velocities for molecules will increase with temperature. The molecules will collide more often and therefore increase the chances for reaction. At high temperature a higher percentage of molecules will have the required activation energy for reaction so reaction rates usually increase.

29. The low temperature slows the rate at which chemical reactions lead to the decomposition of DNA, preserving it.

31. Reactions with high or large activation energies are slower than reactions with low or small activation energies.

33. The activation energy is high so there is no reaction at room temperature. Once started by the spark (source of activation energy), the reaction continues with production of energy because the products have lower potential energy than the reactants.

35. Faster because contact with oxygen molecules would be more frequent.

37. Coating the eggs with vegetable oil prevents contact between oxygen in the air and the interior of the egg. In the absence of oxygen, the rate of the reactions leading to spoiling is reduced.

39. Reversible reactions occur in both forward and reverse directions.

41. (a) shift to form $CaCO_3$
    (b) shift to form CaO and CO_2

43. If a stress is applied to a system at equilibrium, the system will adjust to relieve the stress.

45. A reaction that shifts to favor reactants is one where the relative amount of products decreases and the amount of reactants increases.

47. When $N_2(g) + 3 H_2(g) \rightleftharpoons 2 NH_3(g)$ reaches equilibrium, all three substances will be present in the reaction mixture.

49. Entire books have been written to discuss the meaning of the first law. A simple statement says: The energy of the universe is a constant. Energy can be converted from one form to another but cannot be destroyed or created. Any energy changes in a reaction occur because there is a difference between potential energy of reactants and potential energy of products.

51. (a) Decrease. The number of gas molecules decreases and organization increases.
    (b) Decrease. The number of gas molecules decreases and organization increases.

53. The molar mass is figured by multiplying the atomic mass of each atom type by the count for the atoms and adding the contributions from each element. This number gives the number of grams for a mole of the substance.

    Example: $C_6H_{12}O_6$

From carbon	6×12.01 amu
From hydrogen	12×1.01 amu
From oxygen	6×16.00 amu
Total	180.18 amu or 180.1 or 180 depending on number of significant digits desired

    (a) 1 mole $C_6H_{12}O_6$ = 180.18 g $C_6H_{12}O_6$
    (b) 1 mole H_2SO_4 = 98.08 g H_2SO_4
    (c) 1 mole Na_2HPO_4 = 141.96 g Na_2HPO_4
    (d) 1 mole $Ca(NO_3)_2$ = 164.10 g $Ca(NO_3)_2$

55. The mol ratios for the equation are read from the coefficients in the balanced equation.
    (a) The coefficients in the equation give the mol ratios. 1 mol CH_3CH_2OH/3 mols O_2.
    (b) (1 mol CH_3CH_2OH) (3 mols O_2/1 mol CH_3CH_2OH) (32 g O_2/1 mol O_2) = 96 g O_2

(c) 500 g ethanol (1 mol C_2H_5OH/46.08 g C_2H_5OH) = 10.85 = 10.9 mols C_2H_5OH

$$10.9 \text{ mol } C_2H_5OH \times \frac{3 \text{ mol } O_2}{1 \text{ mol } C_2H_5OH} \times \frac{32.0 \text{ g } O_2}{1 \text{ mol } O_2}$$
$$= 1046 = 1050 \text{ g } O_2$$

Starting from 10.85 mol CH_3CH_2OH gives 1042 = 1040 mol O_2 to 3 s.d.

57. The formation of water from the elements occurs according to the equation written here. The masses for reactants and products can be decided from the balanced equation. This is done using the numbers of mols required and molar masses.

Use the balanced equation.

$$2 H_2(g) + O_2(g) \longrightarrow 2 H_2O(l)$$

The mol amounts are read from the coefficients in the equation.

$$2 \text{ mol } H_2 + 1 \text{ mol } O_2 \longrightarrow 2 \text{ mol } H_2O$$

Convert each #mols to mass using molar mass.

$$2 \text{ mol } H_2 + 1 \text{ mol } O_2 \longrightarrow 2 \text{ mol } H_2O$$
$$4.0 \text{ g } H_2 + 32.0 \text{ g } O_2 \longrightarrow 36.0 \text{ g } H_2O$$

(a) 32 g O_2 (1 mol O_2/32.0 g O_2) (2 mol H_2/1 mol O_2) (2.0 g H_2/1 mol H_2) = 4.0 g H

(b) The mass ratios are the same for any size unit. The mass unit of grams can be replaced by the units of pounds. This means 32 pounds of oxygen will require 4.0 pounds H_2.

$$2 H_2(g) + O_2(g) \longrightarrow 2 H_2O(l)$$
$$4.0 \text{ g } H_2(g) + 32.0 \text{ g } O_2(g) \longrightarrow 36.0 \text{ g } H_2O$$

4 pounds $H_2(g)$ + 32 pounds $O_2(g) \longrightarrow$ 36 pounds H_2O

The relative amounts are all scaled up by the same factor.

(c) The mass units can all be replaced by the unit "tons" and the ratios will stay the same because all are multiplied by the same factor. This means 32 tons of oxygen will require 4.0 tons of H_2.

$$2 H_2(g) + O_2(g) \longrightarrow 2 H_2O(l)$$
$$4.0 \text{ g } H_2 + 32.0 \text{ g } O_2 \longrightarrow 36.0 \text{ g } H_2O$$
$$4.0 \text{ tons } H_2 + 32 \text{ tons } O_2 \longrightarrow 36 \text{ tons } H_2O$$

$$32 \text{ g } O_2 \times \frac{1 \text{ mol } O_2}{32.0 \text{ g } O_2} \times \frac{2 \text{ mol } H_2}{1 \text{ mol } O_2} \times \frac{2.0 \text{ g } H_2}{1 \text{ mol } H_2} = 4.0 \text{ g } H_2$$

Each problem also can be done using one set-up.

59. The first step is to balance the equation:

$$C_6H_{12}O_6(aq) \longrightarrow 2 CH_3CH_2OH(aq) + 2 CO_2(g)$$

1 mole $C_6H_{12}O_6$ 2 moles CH_3CH_2OH 2 moles CO_2

(a) The short method to solve this problem is to determine the mole-ratio between glucose and ethanol from the balanced equation and use this mole-ratio as a conversion factor.

$$\frac{1 \text{ mol } C_6H_{12}O_6}{2 \text{ mol } CH_3CH_2OH} \quad \text{or} \quad \frac{2 \text{ mol } CH_3CH_2OH}{1 \text{ mol } C_6H_{12}O_6}$$

$$6 \text{ mols } C_6H_{12}O_6 \times \frac{2 \text{ mol } CH_3CH_2OH}{1 \text{ mol } C_6H_{12}O_6} = 12 \text{ mols } CH_3CH_2OH$$

(b) The moles of ethanol can be determined from moles of carbon dioxide using the mole-ratio that comes from the balanced equation: 2 mols ethanol/2 mols carbon dioxide.

$$10.5 \text{ mols } CO_2 \times \frac{2 \text{ mol } CH_3CH_2OH}{2 \text{ mol } CO_2} = 10.5 \text{ mols } CH_3CH_2OH$$

CHEM

Answers to Applying
Your Knowledge

Chapter 9

1. (a) Excess hydronium ion, $[H_3O^{1+}] > [OH^{1-}]$, pH < 7
    (b) Excess hydroxide ion, $[H_3O^{1+}] < [OH^{1-}]$, pH > 7
    (c) Equal amounts of hydronium and hydroxide, pH $= 7$ and $[H_3O^{1+}] = [OH^{1-}]$

3. A salt is the substance formed between the anion of an acid and the cation of a base.

5. (a) measures the fraction of molecules that ionize
    (b) ionizes 100%
    (c) ionizes 100%
    (d) only partially ionized, the majority of acid molecules are still intact
    (e) partially ionized, the majority of base molecules or ions are still intact

7. pH above 7

9. (a) acidic (b) acidic (c) basic
    (d) basic (e) acidic (f) basic

11. (a) Molarity is a concentration measure equal to the moles of solute dissolved in one liter of solution.
    (b) Concentration is a measure of the relative amount of solute in a specific amount of solution.

13. A buffer is an aqueous mixture of both a weak acid and its anion that will maintain a stable pH when either acid or base is added. A basic buffer is a mixture of a weak base and its cation that will maintain a stable pH.

15. (a) $CH_3COOH(aq) + KOH(aq) \longrightarrow H_2O(l) + CH_3COOK(aq)$
    (b) $H_2SO_4(aq) + Ca(OH)_2(aq) \longrightarrow CaSO_4(aq) + 2 H_2O(l)$
    (c) $H_2SO_4(aq) + 2 NaOH(aq) \longrightarrow Na_2SO_4(aq) + 2 H_2O(l)$

17. $2 HCl(aq) + CaCO_3(s) \longrightarrow CaCl_2(s) + CO_2(g) + H_2O(l)$

19. black coffee with the pH $= 5.0$
    The lower the pH, the more acidic the solution.

21. lemon, pH $= 2.3$

23. (a) Ammonium ion reacts with base and ammonia reacts with acid.
    (b) Hydrogen fluoride reacts with base and sodium fluoride reacts with acid.

25. B with pH $= 1.1$ contains the stronger acid.

27. 240 g NaOH

29. 0.0200 mol HCl

31. (a) 0.137 M HCl (b) 1.00 M NaOH

33. 9.8 g H_2SO_4

35. Because solutions are homogeneous, the molarity of a portion of a solution is the same as the molarity of the entire solution. Use the provided information in the formula used to calculate molarity and solve for the number of moles dissolved.

$$1.8\ M = \frac{\#\ \text{moles dissolved}}{0.300\ L}$$

0.54 = number of moles dissolved (in 300 mL of the solution)

37. To determine the number of grams dissolved, it will first be necessary to calculate the number of moles dissolved. Then, using the formula weight of HCl, the number of grams can be determined.

$$2.00\ M = \frac{\#\ \text{moles dissolved}}{0.250\ L}$$

\# moles dissolved $= 0.500$ mol

$$0.500\ \text{moles HCl} \times \frac{36.45\ g\ HCl}{1\ mol\ HCl} = 18.2\ g\ HCl$$

39. (a) The pH for a 1×10^{-2} M HCl solution can be figured two ways. pH $= -\log[H_3O^+]$; pH $= -\log 1 \times 10^{-2}$; pH $= 2$.
    (b) The pH for a 1×10^{-3} M HNO_3 solution can be figured two ways. The $[H_3O^{1+}] = 0.001$; pH $= -\log[H_3O^{1+}]$; pH $= -\log 1 \times 10^{-3}$; pH $= 3$.

41. Using the relationship pH $= -\log[H_3O^+]$, the pH of the solution is 3.

43. The molarity of the hydronium ion $= [H_3O^{1+}] = 10^{-pH}$
    (a) pH $= 1$; $[H_3O^{1+}] = 10^{-pH}$; $[H_3O^{1+}] = 10^{-1} = 1 \times 10^{-1}$ or 0.1
    (b) pH $= 0$; $[H_3O^{1+}] = 10^{-pH}$; $[H_3O^{1+}] = 10^{-0} = 1 \times 10^{-0}$ or 1
    (c) pH $= 5$; $[H_3O^{1+}] = 10^{-pH}$; $[H_3O^{1+}] = 10^{-5} = 1 \times 10^{-5}$ or 0.00001
    (d) pH $= 3$; $[H_3O^{1+}] = 10^{-pH}$; $[H_3O^{1+}] = 10^{-3} = 1 \times 10^{-3}$ or 0.001

45. 75 mL

47. The pH scale is a logarithmic scale and is not linear. Doubling the concentration of base therefore does not double the pH. (In fact, it increases the pH by $\log(2) = 0.3$.)

49. No. Mixture is basic. 0.0035 mol of KOH remains after all HCl has been neutralized.

51. To calculate the molarity of the caffeine, the number of moles of caffeine needs to be determined:

$$50 \text{ mg} \times \frac{1 \text{g}}{1000 \text{ mg}} \times \frac{1 \text{ mol}}{194.19 \text{ g}} = 2.6 \times 10^{-4} \text{ mols}$$

Then, the molarity can be calculated:

$$\text{Molarity } (M) = \frac{2.6 \times 10^{-4} \text{ mol}}{0.355 \text{ L}} = 7.3 \times 10^{-4} M$$

53. Aluminum hydroxide does not alter the pH of pure water, so it must not dissolve to produce hydroxide ions, which would be necessary for a change in pH to occur. In fact, this is exactly the case. $Al(OH)_3$ is very insoluble in water.

Answers to Applying Your Knowledge

Chapter 10

1. **(a)** Oxidation is the loss of electrons.
   (b) Reduction is the gain of electrons.
   (c) Oxidation is the gain of oxygen.
   (d) Reduction is the gain of hydrogen.
   (e) An oxidizing agent is a substance that accepts electrons, causes oxidation to occur.
   (f) A reducing agent is a substance that gives up electrons, causes reduction to occur.

3. **(a)** CO_2
   (b) NO_2
   (c) SO_3
   (d) N_2

5. The difference is in the identity of the reactants. In bleaching, an oxidizing agent is used in a reaction with pigment molecules. In disinfection, an oxidizing agent is used to react with the cell walls of bacteria and viruses.

7. Combustion is a reaction in which oxygen combines with another element; there is rapid production of heat. Combustion is limited to the reaction with oxygen. Oxidation is not restricted to a reaction with oxygen. Oxidation refers to a reaction in which electrons are lost, hydrogen is lost, or oxygen is gained by another element.

9. The potassium ion is more oxidized than the potassium atom. The atom loses an electron when it forms the K^+ ion. The oxidation number goes up from zero to $+1$.

11. True. An oxidizing agent causes oxidation to occur and takes up the electrons lost by the chemical species that is undergoing oxidation. An oxidizing agent therefore gains electrons, which is a reduction.

13. True. An oxidizing agent causes oxidation to take place and gains electrons in an oxidation-reduction reaction. The gain in electrons is a reduction, and a reduction always results in a decrease in charge (more negative charge)

15. The earth's atmosphere is an oxidizing atmosphere because of its oxygen content. The oxygen in the atmosphere is reduced when it reacts with other substances to gain electrons and form O^{2-} ions. The materials reacting with oxygen are oxidized because they lose electrons to the O atoms.

17. Magnesium is oxidized in the Mg^{2+} ion because Mg metal has a zero oxidation number. The element loses electrons when it forms the Mg^{2+} ion.

19. Agents and effects

	oxidized	reduced	oxidizing agent	reducing agent
(a)	Al	Cl_2	Cl_2	Al
(b)	S	O_2	O_2	S
(c)	H_2	CuO	CuO	H_2
(d)	H_2	C_2H_4	C_2H_4	H_2

21. Rusting is less of a problem in Arizona because there is less water in the air.

23. True

25. Oxidation occurs at the anode where electrons are produced. The electrons travel through an external circuit to the cathode where they are picked up by the substance that is reduced. Electricity flows internally through the battery because ions can flow through the salt bridge that connects the anode and cathode compartments.

27. An electrochemical cell is a single cell in which a reduction-oxidation reaction takes place. A battery is a series of electrochemical cells.

29. **(a)** Primary
   (b) Secondary
   (c) Secondary
   (d) Secondary

31. Cars could use battery packs for electric power. Electricity could be routed to homes, recharging hook-ups in parking lots, etc. to recharge the packs. Long trips could be made possible if depleted packs could be exchanged for fully charged ones at recharging stations along interstate highways. A benefit would be that this system would eliminate underground gasoline storage tanks, fuel trucks, and the pollution they produce. Air pollution from car exhaust would decrease. Disposal problems might be connected with the battery pack design. Increases in pollution from electric power plants may offset the decrease from the reduction in auto exhaust. The inefficient production of electricity from fossil fuels would increase fossil fuel consumption by as much as a factor of ten times. Electric cars could be powered from cables embedded in the roads. This would eliminate batteries and concerns about running out of fuel. The costs of building this grid would be high. Maintenance would be expensive.

33. Electric cars had shorter range and endurance than the gasoline internal combustion engine cars. Electricity was not as widely available in rural areas so the internal combustion engine car that carried its own fuel was more practical in the countryside. Electricity was and still is "mysterious" so the need to recharge the electric car was less well understood. The mechanics of the internal combustion engine were more understandable and therefore more accepted.

35. The 3.0 volts results from the addition of the voltage from two 1.5 volt cells. When the cells are connected "in series," the voltages add.

37. The answer is **(a)**. An antioxidant prevents free radicals from carrying out oxidation reactions by donating an electron to the radical. In the process, the antioxidant loses an electron, so antioxidants are reducing agents.

39. NO_2 has an odd number (17) of electrons so one of its electrons is unpaired.

41. The fuel cell is a reaction vessel. Reactants are converted to products, but total mass stays the same.

43. The concentration of Cu^{2+} ions does not increase.

45. Iron oxide forms a loose layer so corrosion of Fe continues. Al_2O_3 bonds tightly to Al, protecting Al from additional oxidation. Galvanized iron does not rust as readily as bare iron because the zinc metal in the galvanizing coating oxidizes more readily than the iron does.

47. C is oxidized; Al^{3+} is reduced.

Chapter 11

1. (a) water available in rivers, lakes, and streams
   (b) water beneath Earth's surface
   (c) fresh water contaminated by salty seawater
   (d) any condition that causes the natural usefulness of water, air, or soil to be diminished
   (e) pumping water into aquifers to maintain water volume
   (f) drinkable water; water that is safe to drink

3. Most of the rainwater that falls on the United States each day returns to the atmosphere by evaporation or transpiration from plants.

5. Groundwater can be contaminated with pollutants when rainwater runs over or filters through materials, dissolves the pollutants, then percolates into the ground water.

7. Agriculture is the largest single user of water in the United States.

9. Water use per day differs from person to person.

11. Groundwater recharge describes the process of pumping water into aquifers to maintain water volume. Purified recycled water from sewage effluent is used as recharge water.

13. The user was responsible for water quality prior to the 1977 Clean Water Act. The discharger is now responsible for maintaining water quality.

15. Two methods for disposal of solid wastes from industry and households are landfills and incineration. Landfills have the greater potential for water contamination.

17. A landfill can be made more secure with respect to water quality protection by controlling the materials placed in the landfill. The materials can be immobilized.

19. batteries, heavy metals
    furniture polish, organic solvents
    bathroom cleaners, acids or caustics
    paint, organic polymers
    oven cleaners, caustics

21. Some of the solder used in joints in older copper pipe plumbing contains lead. This lead will dissolve in acidic water passing through these pipes, so the lead can be picked up by the water.

23. Water is vaporized from one container and condensed in another while dissolved, nonvolatile substances remain behind.

25. Settling separates high-density suspended solids such as sand, while filtration removes suspended low-density matter such as algae.

27. The concentration of these chlorinated hydrocarbons and branched chain hydrocarbons rises because they accumulate in the environment. They are fat soluble and stored by living organisms. They accumulate in organisms at the top of the food chain.

29. Ammonia, $NH_3(aq)$, and ammonium ion, $NH_4^+(aq)$, can be removed from wastewater by using denitrifying bacteria. These bacteria convert the ammonia and ammonium ion to nitrogen, $N_2(g)$. The unbalanced reaction is:

$$NH_3(aq) \text{ or } NH_4^+(aq) \xrightarrow{\substack{\text{denitrifying} \\ \text{bacteria}}} N_2(g)$$

31. Both kill bacteria by oxidizing organic compounds. Aeration depends on O_2 as the oxidizing agent, whereas chlorination uses Cl_2.

33. Mechanical pressure is used to force water molecules through a semi-permeable membrane from the salty aqueous side to the pure water side of the membrane.

35. 1820 cal endothermic

37. $1000 \text{ cal} \times \dfrac{4.18 \text{ J}}{1 \text{ cal}} = 4180 \text{ J}$

39. Because of the conservation of energy, we can calculate how much energy is required to melt the same quantity of ice and express this quantity as the amount of heat released by the same quantity of water when freezing.

$100 \text{ g} \times \dfrac{80 \text{ cal}}{1 \text{ g}} = 8,000 \text{ cal}$

41. No, it is not. Rainwater contains particulate matter (dust) in addition to dissolved substances (acids and pollutants). The processes of distillation and filtration could be used to purify rainwater.

43. Yes. Once formed at a temperature of 100°C, water vapor can be increased in temperature.

45. Natural rainwater is mildly acidic due to the presence of carbon dioxide in the atmosphere, and common pollutants also serve to make rainwater more acidic.

   Because of the presence of CO_2 in the atmosphere, there is always some dissolved CO_2 in rainwater. When CO_2 dissolves in water, carbonic acid is produced and the resulting solution is acidic with a pH below 7.0.

47. Answers will vary.

Chapter 12

1. fuel formed by decomposition of plant and animal matter

3. compound containing only carbon and hydrogen

5. the energy released when a compound reacts completely with oxygen

7. (a) molecules with the same general formula but different molecular structures
   (b) compounds with a linear chain of CH_2 units and terminal CH_3 units
   (c) compounds with a backbone like a straight-chain hydrocarbon, but with some C_nH_{2n+1}, alkyl groups replacing Hs along the chain

9. $n = 1$ methane CH_4
   $n = 2$ ethane C_2H_6
   $n = 3$ propane C_3H_8
   $n = 4$ butane C_4H_{10}

11. (a)

    $$CH_3CHCH_2CH_2CH_3$$
    with CH_3 branch

    (b)

    $$CH_3CH_2CH_2CCHCH_2CH_2CH_3$$
    with CH_3 branch and H_3C CH_2CH_3 branches

    (c)

    $$CH_3C = CHCH_2CH_2CH_3$$
    with CH_3 branch

13. (a) identical
    (b) different
    (c) different
    (d) isomers

15. $CH_3CH_2CH_2CH_2CH_3$
    Pentane

    $$CH_3CHCH_2CH_3$$
    with CH_3 branch
    2-Methylbutane

    $$CH_3CCH_3$$
    with CH_3 branches top and bottom
    2,2-Dimethylpropane

17.

Ortho-dimethylbenzene   Meta-dimethylbenzene   Para-dimethylbenzene

19. There are *cis*- and *trans*- isomers for 2-butene because there are carbon chains on both double-bonded carbon atoms. The 1-butene has only hydrogens on the #1 carbon, so there is no group that can be *cis* or *trans* to the groups on the #2 carbon.

21. Hot petroleum at about 400°C is distilled in a fractionation tower. There are trays at various levels in the tower, each at a unique temperature. Each volatile component will vaporize. These will condense at a temperature range of a specific tray. The compounds that are most volatile are collected near the top. The compounds that are least volatile with the highest boiling point, such as tars, are collected at the bottom.

23. Aromatic hydrocarbons, branched-chain hydrocarbons, 2-methyl-2-propanol, methanol, ethanol, or methyl-*tertiary*-butyl ether

25. Synthesis gas, a mixture of CO and H_2, is produced by passing super-heated steam over pulverized coal.

    $$C + H_2O + 31\ kcal \longrightarrow CO + H_2$$

    Coal gasification using a catalyst, crushed coal, and synthesis gas will produce methane.

    $$2\ C + 2\ H_2O + 2\ kcal \longrightarrow CH_4 + CO_2$$

27. the percentage of isooctane in an isooctane/heptane mixture with the same knocking properties

29. Legal requirements for enhanced oxygenated gasoline; a need for cleaner burning gasoline; higher prices for gasoline from crude oil, petroleum

31. Because methanol is produced from synthesis gas, and because the source of synthesis gas is coal, the production of biodiesel is partly dependent on the use of fossil fuels.

33. Branched hydrocarbons have higher octane ratings than unbranched hydrocarbons. Because 2,2,4-trimethyloctane is a branched hydrocarbon, its rating is higher than that of octane.

35. Oxygenated gasolines are produced by adding oxygenated compounds (ethanol, methanol, MTBE) to reformulated gasoline. Reformulated gasoline is directly produced by the refining process and contains lower percentages of unsaturated hydrocarbons, aromatics, and volatile compounds.

37. Methanol burns more cleanly than gasoline and emits fewer pollutants. However, one of the pollutants produced is carcinogenic formaldehyde.

39. Answers to this question will vary. Possible advantages are: Lesser dependence on nonrenewable petroleum resources; greater diversity of crops being planted. Possible disadvantages are: The use of cropland to grow fuel could reduce the amount dedicated to the growth of foods; the amount of land available cannot satisfy current energy needs.

41. (a) alkane
(b) alkyne
(c) ether
(d) alcohol
(e) aromatic

43. $$\left[\left(2 \times 436\frac{kJ}{mol}\right) + 498\frac{kJ}{mol}\right] - \left[\left(4 \times 467\frac{kJ}{mol}\right)\right]$$
$$= -498\frac{kJ}{mol}$$

Chapter 13

1. gamma rays

3. The more hazardous radioisotope is $^{222}_{86}$Rn , since more ionizing radiation is emitted over a shorter time period.

5. It is a gas and can be inhaled. The high density makes it difficult to exhale, and emitted alpha particles produce cell damage. Radioactive daughters can cause more damage.

7. The mass number on the phosphorus atom should be 31, not 49.

9. very slightly more than 1/16 of original

11. There are two ways to arrive at the correct answer. The first would be to determine how many half lives are required to reduce the amount of the isotope from 300 to 37.5 mg. So, half of 300 is 150 mg (one half life), half of 150 mg is 75 mg (two half lives), and half of 75 mg is 37.5 mg (that makes three half lives). So, three half lives equal 3.5 years, and one half life is equal to 1.16 years.

    Another way would be to calculate the number of half lives mathematically:

    $$37.5 \text{ mg} = 300 \text{ mg} \times \left(\frac{1}{2}\right)^n$$

    where n = the number of half lives

    $$\log (37.5) = \log (300) + n \log \left(\frac{1}{2}\right)$$

    $$1.57 = 2.48 + n \, (-0.3)$$

    $$n = 3$$

    $$3.5 \, \frac{\text{years}}{3} = 1.16 \text{ years}$$

13. Transuranium elements have atomic numbers greater than 92 and are all artificial. None of the transuranium elements are found in nature.

15. The natural decay of several isotopes in the uranium series releases an alpha-particle, $^{4}_{2}$He^{+2}. Once two electrons are obtained by the alpha particle, elemental helium is formed.

17. diagnosis and treatment of diseases

19. (a) beta particles: −1 and 0
    (b) alpha particles: +2 and 4
    (c) Gamma rays (photons): 0 and 0
    (d) Positrons: +1 and 0
    (e) Neutrons: 0 and 1

21. Gamma rays can damage the DNA sequence needed for duplication of the genetic code in the rapidly multiplying cancer cells that duplicate more often than normal cells.

23. natural radiation in food, water, and air

25. 17,200 years (three half lives)

27. The activity of a radioactive material in disintegrations per second decreases by 1/2 when the sample goes through one half life. The activity will decrease to 1/4 after a second half life passes. The copper-64 sample will need to go through two half lives to decrease to 1/4 of the original activity. This period is 2 × 12.9 hours = 25.8 hours.

29. The first thing to do is to determine the number of half lives for the radon-222. Number of half lives = 15.28 days × (1 half life/3.82 days) = 4.00 half lives. The fraction of radon remaining is actually = (½)$^{4.00}$ = 0.0625. At this level, only 1/16 of the original amount is remaining.

31. (a) A starting (fissionable) material has to have a large atomic mass and an odd mass number with the "right" proportion of neutrons to protons. Examples are $^{235}_{92}$U, $^{239}_{94}$Pu.

    (b) Thermal neutrons are typically needed to excite a fissionable nucleus to make it unstable; this unstable nucleus then can undergo fission.

    (c) Nuclei with lower atomic weights and lower atomic numbers, neutrons, and energy are produced in a fission reaction.

33. During an unauthorized test, the Chernobyl Unit 4 reactor had a power surge. The core overheated and a meltdown occurred. A steam explosion and graphite moderator fire blew the roof off. This released an estimated 100 million curies of radioisotopes into the environment. The major cause of the release was operator error and a design that used a core built of flammable graphite with a weak containment structure.

35. Nuclear fusion occurs when small nuclei combine to form a heavier nucleus.

37. If the element has an atomic number equal to or less than that of Bismuth (Z = 83), then any isotope with a number of neutrons that is less than the atomic number is likely to be radioactive. For these elements, those with even atomic numbers tend to have more stable isotopes than those with odd atomic numbers. For atomic numbers greater than 83, all isotopes are unstable and exhibit alpha-decay.

39. If the number of neutrons is far greater than the number of protons, the isotope will undergo beta-decay.

41. (a) $^{1}_{0}$n + $^{235}_{92}$U $\longrightarrow$ $^{142}_{56}$Ba + $^{91}_{36}$Kr + 3^{1}_{0}n

    (b) $^{1}_{0}$n + $^{235}_{92}$U $\longrightarrow$ $^{105}_{42}$Mo + $^{129}_{50}$Sn + 2^{1}_{0}n

    (c) $^{1}_{0}$n + $^{239}_{94}$Pu $\longrightarrow$ $^{115}_{42}$Mo + $^{123}_{52}$Te + 2^{1}_{0}n

43. (a) $^{14}_{6}$C $\longrightarrow$ $^{14}_{7}$N + $^{0}_{-1}$e

    (b) $^{210}_{82}$Pb $\longrightarrow$ $^{210}_{83}$Bi + $^{0}_{-1}$e

45. (a) $^{218}_{84}\text{Po} \longrightarrow {}^{214}_{82}\text{Pb} + {}^{4}_{2}\text{He}$

(b) $^{218}_{90}\text{Po} \longrightarrow {}^{206}_{82}\text{Pb} + {}^{4}_{2}\text{He}$

(c) $^{99}_{43}\text{Tc} \longrightarrow {}^{99}_{44}\text{Ru} + {}^{0}_{-1}e$

(d) $^{226}_{88}\text{Ra} \longrightarrow {}^{4}_{2}\text{He} + {}^{222}_{86}\text{Rn}$

(e) $^{195}_{79}\text{Au} + {}^{0}_{-1}e \longrightarrow {}^{195}_{78}\text{Pt}$

47. Both uranium-235 and plutonium-239 are fissionable.

49. If initially there were 300,000 atoms of ^{238}U, there will be only 18,750 atoms left after four half lives.

Chapter 14

1. Carbon atoms can bond to other carbon atoms in almost unlimited numbers and in a variety of ways. Introducing other elements allows different molecules due to different atom sequences (functional groups and isomers).

3. (a) Carboxylic acid, pentaonic acid

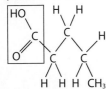

(b) Alkene, 2-pentene

(c) Symmetric ether, diethyl ether

(d) Secondary alcohol, 3-pentanol

(e) Tertiary alcohol, 2 methyl-2-propanol

(f) Aldehyde, ethanal

5. (a) Acid, acetic acid

(b) Ether, dimethyl ether

(c) Alcohol, ethanol

(d) Ketone, acetone

(e) Aldehyde, formaldehyde

(f) Ester, methyl acetate

7. $90 \text{ proof} = \dfrac{90}{2}\% = 45\%$

9. A functional group is an atom or group of atoms that impart a set of physical and chemical properties on the molecule containing the group. Examples included in this chapter are alcohols, aldehydes, ketones, carboxylic acids, esters, amines, and amides.

11. A hydroxyl functional group, –OH, is often present in naturally occurring carboxylic acids. An example is lactic acid which is present in milk.

13. Acetic acid with ethanol combine to form the ester ethyl acetate.

$$C_2H_5 - OH + H_3C - COOH \longrightarrow$$

15. (a) Two naturally occurring esters are 3-Methylbutyl acetate in bananas and butyl butanoate in pineapple. See Table 14.8 for other naturally occurring esters.
    (b) Two naturally occurring carboxylic acids are formic acid in ants and lactic acid in milk. See Tables 14.6, 14.7, and 14.8 for simple carboxylic acids and naturally occurring acids.

17. **(a)**

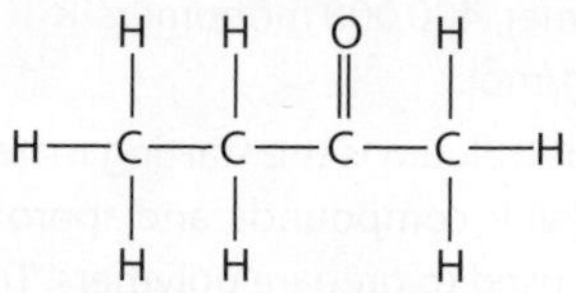

(b)

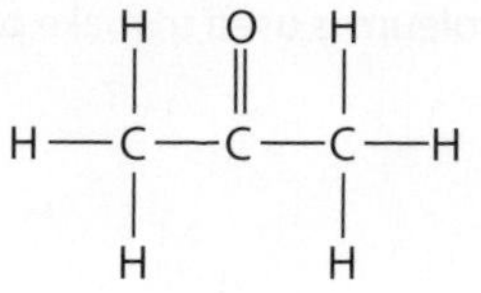

(c)

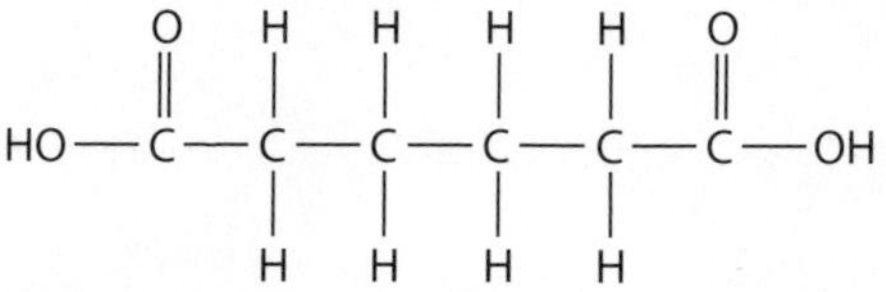

19. Though both groups have a bond between a C atom and a hydroxyl (-OH) group, the C atom in a carboxylic acid also forms a double bond to an oxygen atom.

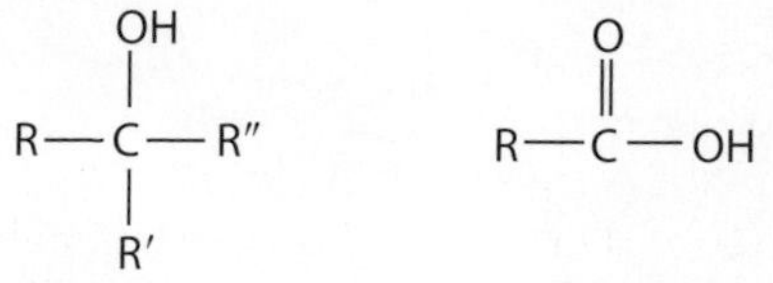

an alcohol   a carboxylic acid

21. Compound **c** would make the best soap. Carboxylic acids with long hydrocarbon chains act as soaps to a greater extent than those with short hydrocarbon chains.

23. A railroad train has a series of repeating, identical, linked units, the railroad cars. Polystyrene is a long chain of repeating, identical, linked units, the styrene molecules.

25. A macromolecule is a molecule with very high molecular mass.

27. Monomers need to contain a multiple bond in order to form addition polymers.

29. Natural rubber is poly-*cis*-isoprene. The individual monomer is isoprene.

isoprene

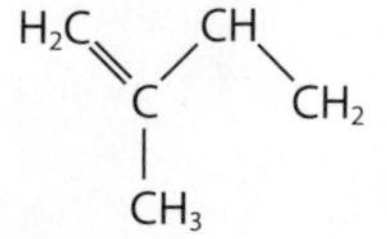

Poly-*cis*-isoprene

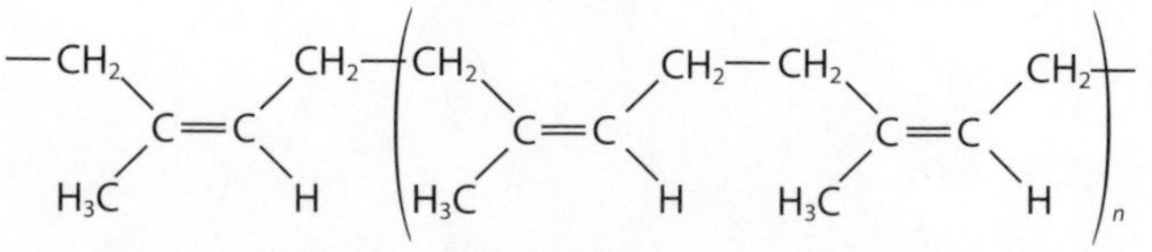

31. Sulfur is reacted with rubber to align rubber polymers and bond the polymer chains to one another.

33. Cross linking makes the polymer more rigid.

35. **(a)** Yes, styrene can undergo an addition reaction because it can add to the double bond in the ethene group.

(b) Yes, propene can undergo an addition reaction because it can add to the double bond.

(c) No, there are only single bonds, so nothing can add to any multiple bonds in ethane.

37. **(a)** Vinyl chloride   **(b)** Styrene

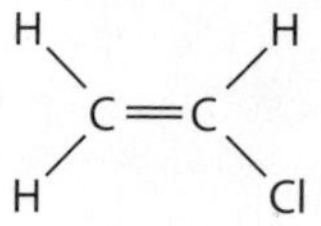

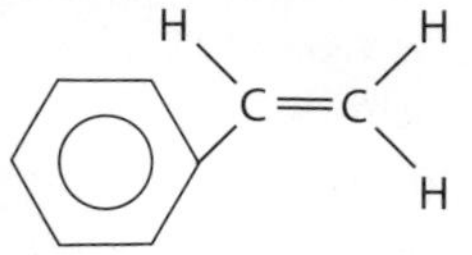

(c) Propylene

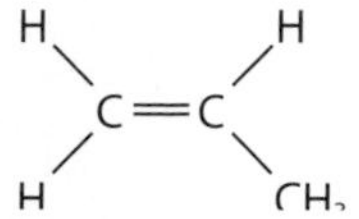

39.

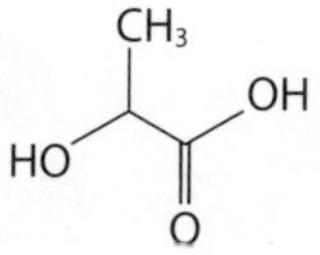

41. PET is polyethylene terephthalate

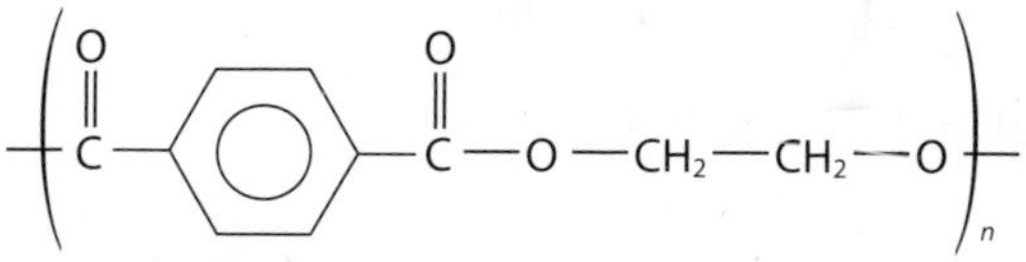

43. Nylon-6,6 is made from adipic acid and hexamethlyene diamine.

$$HOOCCH_2CH_2COOH + H_2NCH_2CH_2CH_2CH_2CH_2CH_2NH_2$$

45. Two examples mentioned in this chapter are starch and cellulose. Rayon, though prepared from cellulose, is distinct from cellulose. Other examples include DNA and RNA, which are discussed in Chapter 15.

47. The four parts to successful recycling are collection, sorting, reclamation, and end-use. Unfortunately, collected materials will not be sorted or reclaimed if the supporting infrastructure for processing doesn't exist.

49. Composites give greater strength, and they can be designed for a particular purpose. Yes, recycling can be more difficult because separation of fibers from the polymer during reprocessing may be difficult.

51. The recycling percentages for plastics are summarized in Figure 14.12. PET is the most widely recycled plastic in carpet fibers, fiberfill for jackets, sleeping bags, and tennis balls. The second most recycled plastic is high-density polyethylene, HDPE, in fencing and drain pipe.

53. Polymer composite materials are substances that have a polymer matrix with reinforcing fibers. Glass fiber-reinforced polyesters are used for boat hulls and car body panels. Graphite fiber-epoxy composites are used in fishing rods and tennis racquets.

55. $\dfrac{10,000 \text{ ethylene monomers}}{1 \text{ polyethylene polymer}}$

57. For this addition polymer, 400,000 monomers $\times$ 104 g/mol = 4×10^7 g/mol.

59. About 3% of refined petroleum is the starting material for the preparation of organic compounds, and approximately 80% of this amount is used to prepare polymers. Therefore, about 2.4% of refined petroleum is used to make polymers.

Chapter 15

1. Chirality describes objects that are nonsuperimposable mirror images of one another.

3. (a) Absolute stereochemical descriptor that compares substituent disposition at a chiral center to the disposition in natural L-glyceraldehyde.

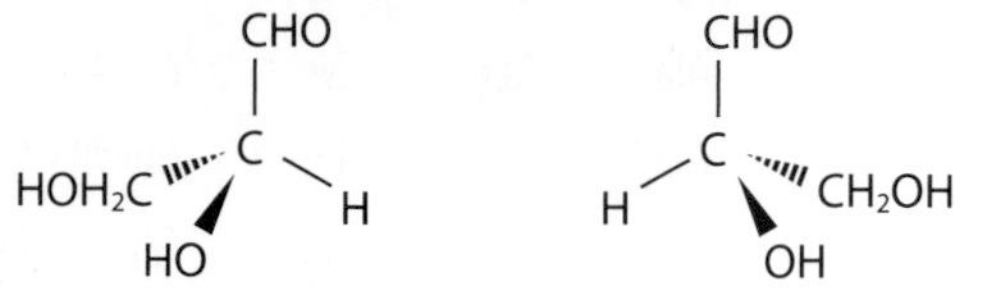

L-Glyceraldehyde   D-Glyceraldehyde

The wedge-shaped bonds indicate that the wide end of the bond is closer to the viewer.

(b) Absolute stereochemical descriptor that relates substituent disposition at a chiral center to the disposition in natural D-glyceraldehyde.

(c) Molecules that rotate the plane of polarization of plane-polarized light.

(d) A device used to measure the direction and extent of rotation produced by chiral molecules of the plane of polarization of plane-polarized light.

5. (a) same (b) enantiomers (c) different

7. (a) chiral

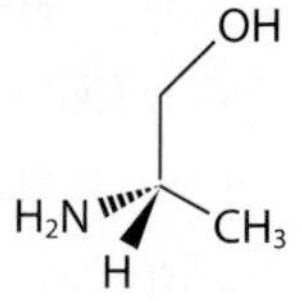

(b) achiral

(c) chiral

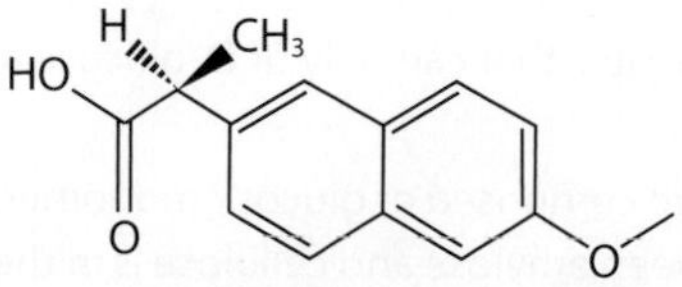

9. (a) A carbohydrate is a class of naturally occurring polyhydroxy aldehydes or ketones.

(b) A simple sugar is a monosaccharide. It cannot be broken down into simpler, smaller sugar molecules by hydrolysis (reaction with aqueous acid).

(c) A disaccharide is a carbohydrate that can be broken down into two monosaccharides when hydrolyzed.

(d) A polysaccharide is a carbohydrate built of many monosaccharides bonded together.

11. (a) Starches are polysaccharide polymers commonly built from D-glucose. (α-D-Glucose is shown.) Amylose is one form of starch consisting of straight chain polymers of about 60–300 α-D-glucose monomers. Amylopectin is another form of starch with molecules of up to 100,000 D-glucose monomers with a branch every 24 to 30 glucose units.

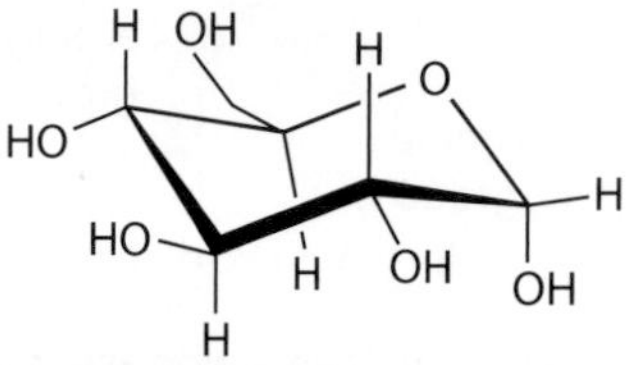

(b) Glycogen is an α-D-glucose polysaccharide with a side chain of indefinite size every 12 glucose units.

(c) Cellulose has only one form. It makes up cotton and the woody part of plant material. It is made of 2000–9000 D-glucose monomer units in the α-ring form. This is an unbranched structure.

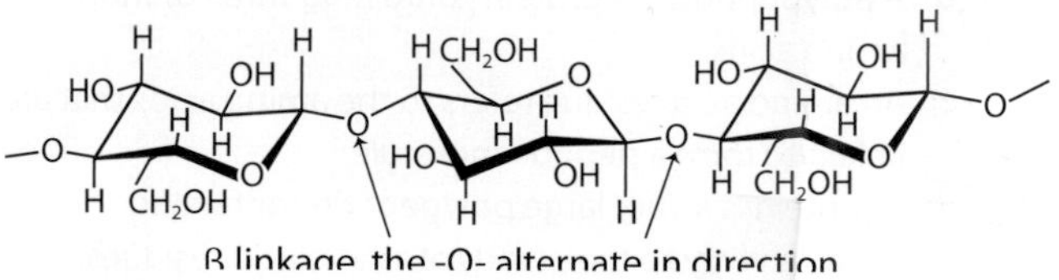

β linkage, the -O- alternate in direction

13. A saturated fatty acid is a carboxylic acid with only carbon-carbon single bonds in the carbon skeleton. The molecule has a long carbon chain and a carboxylic acid group, –COOH.

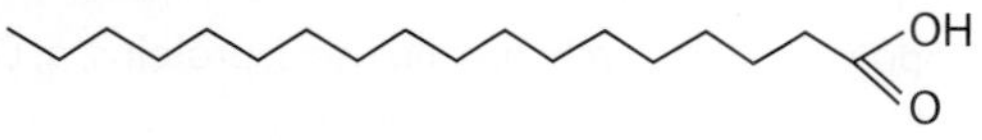

An unsaturated fatty acid has carbon-carbon double bonds in the carbon skeleton.

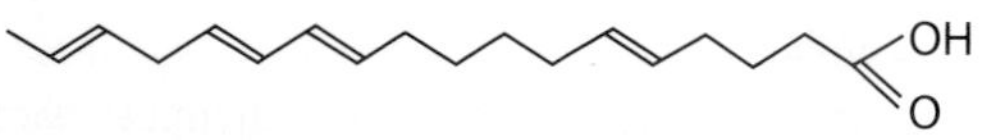

15. A *trans*-fatty acid has at least one carbon-carbon double bond with substituents on opposite sides (*trans*) of the double bond.

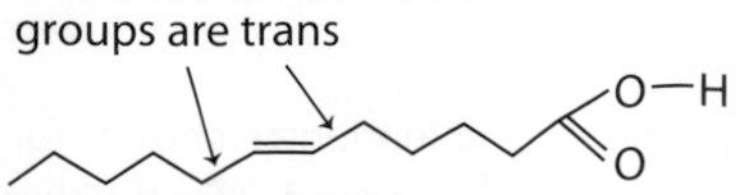

groups are trans

17. A soap is derived from natural sources, such as fats and oils, and the sodium or potassium salts of fatty acids. A detergent is an artificially prepared molecule, often with a sulfonate or quaternary ammonium head group.

19. They work because they stabilize tiny oil particles so they can remain in suspension in the aqueous layer. Without them the oil and water would separate into two layers because oil and water are not soluble in each other.

21. Water-in-oil emulsions give a greasy feel. The oil is the dominant part of the mixture.

23. An amino acid is a compound with an amino group and a carboxylic acid group attached to the same carbon atom.

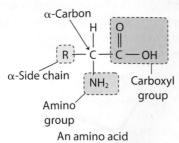

An amino acid

25. An essential amino acid is an amino acid that is not synthesized by the human body. An essential amino acid must be provided by the diet. Essential amino acids are valine, leucine, isoleucine, phenylalanine, methionine, threonine, tryptophan, and lysine.

27. (a) A dipeptide is a molecule formed by linking two amino acids.
    (b) A polypeptide is a protein containing three or more amino acids.
    (c) An amino acid residue refers to the amino acids that are linked to form a peptide molecule.
    (d) A protein is a very large polypeptide containing between fifty to thousands of amino acid residues.

29. Neuropeptides are biomolecules that transmit chemical messages along nerve pathways by connecting with receptors. Examples are methionine-enkephalin and leucine-enkephalin.

31. Biopolymers: a, starch; b, cellulose; f, proteins; g, DNA; h, RNA

33. (a) the spiral secondary protein structure
    (b) a common secondary protein structure where protein chains form parallel strands held together by hydrogen bonds

35. Enzymes are proteins that catalyze chemical reactions, and represent only a portion of all known proteins.

37. methionine and cysteine

39. phenylalanine, tryptophan, and tyrosine

41. alanine, proline, leucine, and isoleucine

43. peptide bond

45. (a) 24 possible combinations: 4 choices for the first position, 3 for the second, 2 for the third, and 1 for the fourth: $4 \times 3 \times 2 \times 1 = 24$. Order is important: ABCD is not equivalent to DCBA since the chain has a C-terminus and an N-terminus.

(b) The 24 possible combinations for the amino acids glycine, Gly; canine, Ala; serine, Ser; and cystine, Cys are shown in the following table.

1st	2nd	3rd	4th	Choice sequence
Ala	Gly	Ser	Cys	Ala-Gly-Ser-Cys
Ala	Gly	Cys	Ser	Ala-Gly-Cys-Ser
Ala	Ser	Gly	Cys	Ala-Ser-Gly-Cys
Ala	Ser	Cys	Gly	Ala-Ser-Cys-Gly
Ala	Cys	Ser	Gly	Ala-Cys-Ser-Gly
Ala	Cys	Gly	Ser	Ala-Cys-Gly-Ser
Gly	Ala	Ser	Cys	Gly-Ala-Ser-Cys
Gly	Ala	Cys	Ser	Gly-Ala-Cys-Ser
Gly	Ser	Ala	Cys	Gly-Ser-Ala-Cys
Gly	Ser	Cys	Ala	Gly-Ser-Cys-Ala
Gly	Cys	Ala	Ser	Gly-Cys-Ala-Ser
Gly	Cys	Ser	Ala	Gly-Cys-Ser-Ala
Ser	Ala	Cys	Gly	Ser-Ala-Cys-Gly
Ser	Ala	Gly	Gys	Ser-Ala-Gly-Cys
Ser	Gly	Cys	Ala	Ser-Gly-Cys-Ala
Ser	Gly	Ala	Cys	Ser-Gly-Ala-Cys
Ser	Cys	Gly	Ala	Ser-Cys-Gly-Ala
Ser	Cys	Ala	Gly	Ser-Cys-Ala-Gly
Cys	Ala	Ser	Gly	Cys-Ala-Ser-Gly
Cys	Ala	Gly	Ser	Cys-Ala-Gly-Ser
Cys	Gly	Ser	Ala	Cys-Gly-Ser-Ala
Cys	Gly	Ala	Ser	Cys-Gly-Ala-Ser
Cys	Ser	Gly	Ala	Cys-Ser-Gly-Ala
Cys	Ser	Ala	Gly	Cys-Ser-Ala-Gly

47. $5! = 5 \times 4 \times 3 \times 2 \times 1 = 120$ different peptides

49. threonylserylphenylalanine

51. Enzymes have active sites that can only fit D-glucose and not L-glucose.

53. Both biopolymers are composed of glucose monomers. The difference between amylose and cellulose is in the linkage between the glucose units.

55. Protein chains in hair are held together by hydrogen bonds, ionic bonds, and disulfide bonds. Hydrogen bonds form between H atoms attached to very electronegative atoms (N and O) and the electron-rich atoms with lone pairs. Ionic bonds result when carboxylate ions ($RCOO^-$) and the protonated amino groups $-NH_3^{1+}$ are attracted to one another. Disulfide bonds form between sulfur atoms in cysteine fragments in adjacent strands.

57. The disulfide linkage is a single bond between two sulfur atoms. The disulfide bonds between cysteine amino acid units hold parallel strands of hair protein in place.

59. DNA and RNA differ structurally because each contains a different sugar for the polymer chain, and they contain different bases. DNA contains the sugar a-2-deoxy-D-ribose, $C_5H_{10}O_4$, while RNA contains a-2-D-ribose, $C_5H_{10}O_5$. Both DNA and RNA contain adenine, guanine, and cytosine. The base thymine occurs only in DNA. The base uracil occurs only in RNA.

61. The monomers that polymerize to form DNA and RNA are nucleotides. The nucleotide monomers contain an organophosphate unit, a ribose unit, and a nitrogenous base. See Figure 15.19 in text.

63. Base pairs such as adenine-thymine and guanine-cytosine are complementary. There are two hydrogen bond sites between the A-T pair and three hydrogen bond sites between the G-C pair.

65. Information transmission from DNA strand: GCTGTAACCGAT

   (a) mRNA : CGACAUUGGCUA
   (b) tRNA : GCUGUAACCGAU
   (c) Amino acid sequence: Arg-His-Trp-Leu

67. (a) The three triplets GGG, AAC, and TTA are translated into mRNA as CCC-UUG-AAU.
   (b) After translation to tRNA, the tripeptide proline-leucine-asparagine is produced.

69. hydrogen bonding between specific bases that have complementary hydrogen bonding

71. DNA sequence = ...GTAGC... corresponds to the mRNA sequence ...CAUCG... .

73. mRNA sequence CU GU

75. Using Table 15.6, the following mRNA codes can be found:

   Proline: CCU, CCC, CCA or CCG
   Glutamic Acid: GAA or GAG
   Threonine: ACU, ACC, ACA or ACG

   Working backwards to obtain the DNA code, we must substitute A for U, G for C, C for G, and T for A. This gives the following possibilities for the DNA code of these amino acids:

   Proline: GGA, GGG, GGT, or GGC
   Glutamic Acid: CTT or CTC
   Threonine: TGA, TGG, TGT, or TGC

   So, there is more than one correct answer. Two possible correct answers are GGA-CTT-TGA-CTT or GGT-CTC-TGT-CTT

Answers to Applying Your Knowledge

Chapter 16

1. Digestion is the process of breaking large molecules in food into substances small enough to be absorbed by the body from the digestive tract.

3. (a) 9 kcal/g
   (b) 4 kcal/g
   (c) 4 kcal/g

5. The basal metabolic rate, BMR, is the rate at which the body uses energy to support the normal maintenance operations of the body. It is the minimum energy required for a person to stay alive.

7. A triglyceride is a triester formed from glycerol and three fatty acids.

9. (a) a disease that is associated with the buildup of fatty deposits on the inner walls of arteries
   (b) a lipid steroidal alcohol that contributes to the development of atherosclerosis
   (c) the yellowish deposit of cholesterol and lipid-containing material on artery walls that is a symptom of atherosclerosis
   (d) complex assemblages of lipids, cholesterol, and proteins that serve to transport water-soluble lipids in the bloodstream through the body
   (e) lipoproteins richer in lipid than in protein with a corresponding lower density
   (f) lipoproteins richer in protein than in lipid with a corresponding higher density

11. (a) nutrients the body needs in large amounts
    (b) nutrients needed by the body in small amounts

13. (a) plants; they are found in foods such as pasta, rice, potatoes, corn
    (b) produced in the liver and limited amounts are stored in liver tissue and muscle
    (c) plants

15. an organic compound essential to health; it is needed in small amounts in the diet

17. (c)

19. Because fat-soluble vitamins can accumulate in the body, large amounts of these vitamins can result in overdose.

21. E, B, A, D, C

23. It is a substance that dissolves in water to produce ions. The electrolyte balance refers to a condition of proper transfer of material through osmosis, normal nerve impulse transmission, normal acid-base balance, and extracellular volume.

25. phosphorus, calcium, magnesium, sodium, potassium, chlorine, and sulfur

27. Sequestrants are used to prevent metal ions from spoiling and adding unwanted flavors to food.

29. 30% of daily Calories

31. Milk should be promoted as a healthy food for the general public because it is an excellent source of calcium and protein. Milk should not be promoted as a healthy food because people who are lactose intolerant may experience serious adverse reactions and become ill when they consume milk.

33. Cholesterol is water insoluble because water is a small polar molecule that will have very weak attractions to the large, organic nonpolar cholesterol molecule. High levels of cholesterol are linked to atherosclerosis or hardening of the arteries. There are 27 carbon atoms in cholesterol, $C_{27}H_{45}OH$.

35. Women produce less estrogen after menopause. Estrogen inhibits calcium loss and bone erosion. Reduced estrogen levels result in more rapid calcium and bone loss. Osteoporosis can be minimized if adequate calcium intake occurs, especially from adolescence through young adulthood, and if estrogen replacements are taken after menopause.

37. Iodine deficiency leads to an enlargement of the thyroid. This condition is called "goiter." Potassium iodide, KI, is added to table salt to make iodized salt in order to prevent iodine deficiency.

39. There is no right answer for this question. Each answer will depend on the specific product.

41. 1100 Calories (estimated daily caloric need = 1760 Calories)

43. 20 minutes

45. 77 minutes for 100 g bread

47. $$\frac{240 \text{ mg cholestrol}}{100 \text{ mL blood}} \times \frac{473 \text{ mL}}{1 \text{ pint}} \times \frac{13 \text{ pints blood}}{\text{body}}$$
    $$= \frac{14{,}758 \text{ mg cholestrol}}{\text{body}}$$

49. 598 Calories total

51. 2750 Calories total, grams of fat = 92 g; grams of carbohydrate = 413 g; grams of protein = 69 g

53. Among nutrients, vitamin B_6 is not deemed to be of significant health concern.

Chapter 17

1. Malaria, pneumonia, bone infections, gonorrhea, gangrene, tuberculosis, and typhoid fever are bacterial diseases. Polio, AIDS, and Rubella (German measles) are viral diseases.

3. The United States Food and Drug Administration (FDA) has the responsibility for classifying drugs as either over-the-counter drugs or prescription drugs.

5. Three major classes of antibiotics are penicillins, cephalosporins, and tetracyclines.

7. Chemotherapy is the treatment of disease with chemical agents.

9. A retrovirus is a virus that uses RNA-directed synthesis of DNA instead of the usual DNA-directed synthesis of RNA.

11. (a) Vasodilators are used to treat heart disease and asthma. The vasodilator relaxes the walls of blood vessels and creates a wider passage for blood flow. This lowers blood pressure and reduces the amount of work that the heart must do to pump blood. It also enables a person to breathe more deeply and easily by dilating the bronchial tubes.

    (b) Alkylating agents are used to treat cancer. Alkylating agents react with the nitrogen bases of DNA in cancer cells and normal cells. Alkyl groups are added to the nitrogens in the bases. This has an effect on both cancer cells and normal cells, but cancer cells are usually dividing and duplicating DNA more often, so the cancer cells are impacted more.

    (c) Beta blockers are used to treat heart disease. Propranolol (Inderol) is used to treat angina, cardiac arrhythmia, and hypertension. Beta blockers act to keep epinephrine and norepinephrine from stimulating the heart.

Inderal, Propranolol

    (d) Antimetabolites are used to treat cancer. Anti-metabolites interfere with DNA synthesis, and cancer cells are more susceptible than normal cells because the cancer cells are generally replicating DNA more frequently than normal cells.

13. Histamines and neurotransmitters bind to receptor sites.

15. (a) Methotrexate is an antimetabolite cancer drug.
    (b) Chlorphenirimine (in Chlortrimeton) is an antihistamine.

17. Codeine is
    (a) an analgesic;   (b) an opioid; and
    (d) a controlled substance, a Schedule 2 drug.

19. Angina is chest pain on exertion and a symptom of heart disease.

21. A barbiturate, such as phenobarbital, is a depressant. Barbiturates bind to a receptor for GABA. This keeps channels for chloride ion, Cl^{1-}, transmission open. This inhibits transmission of nerve impulses. The physiological effects are a progression from sedation or relaxation, to sleep, to general anesthesia, to coma and death.

23. Heart muscle contracts more often and heart rate increases.

25. Estradiol has an alcohol group and a hydrogen on a carbon in the 5-membered ring, whereas ethinyl estradiol has both an alcohol group and an ethynyl group on that carbon.

27. (a) Dopamine helps control memory and emotion and regulate fine muscle movement.
    (b) Epinephrine is a neurotransmitter that produces increased blood pressure, dilation of blood vessels, and widening of the pupils of the eye.

29. Over-the-counter, prescription, unregulated non-medical drugs, controlled substances

31. (a) hallucinogen   (b) hallucinogen
    (c) hallucinogen   (d) hallucinogen   (e) depressant
    (f) antidepressant or stimulant

33. (a) Vasodilators treat angina. Vasodilators act to expand or dilate veins. This lowers resistance to blood flow and blood pressure. The heart needs to do less work to circulate the blood.
    (b) Diuretics treat hypertension. Diuretics reduce blood pressure and hypertension by stimulating excretion of sodium ion and urine production. Blood volume and pressure decrease when urine output increases.

35. Surgery, irradiation, chemotherapy

37. Hallucinogens cause a person to experience vivid illusions, fantasies, and hallucinations. Examples of hallucinogens are mescaline, PCP, and LSD.

39. (a) Morphine is a more effective pain killer than heroin and codeine.
    (b) Heroin is not a natural alkaloid.
    (c) Heroin is so addictive that it is not legal to sell or use it in the United States.

Chapter 18

1. (a) sedimentary rock: rock formed by deposition of dissolved or suspended substances
    (b) igneous rock: rock formed by solidification of molten rock

3. (a) Annealing is the process of heating and then slowly cooling a substance to make the substance less brittle and reduce strain in the solid.
    (b) Amorphous solids have no regular crystalline order.
    (c) Ceramics are materials generally made from clays and then hardened by heat.
    (d) Cement is a substance able to bond mineral fragments into a solid mass.

5. Magnesium is extracted from seawater. It is used for alloys for auto and aircraft parts.

7. Balanced reactions
    (a) $2 H_2O(l) + 2 e^- \longrightarrow H_2(g) + 2 OH^-(aq)$
    (b) $Mg^{2+}(aq) + 2 OH^-(aq) \longrightarrow Mg(OH)_2(s)$
    (c) $2 Cl^-(aq) \longrightarrow Cl_2(g) + 2 e^-$
    (d) $CaCO_3(s) \longrightarrow CaO(s) + CO_2(g)$

9. Coke, which is essentially carbon (C), iron oxide (Fe_2O_3), and limestone ($CaCO_3$) are mixed in a blast furnace to yield iron. The reducing agent is the carbon in coke.

11. Pig iron is the form of iron that comes directly from a blast furnace; it contains numerous impurities such as 4.5% carbon, which makes it brittle. Pig iron is used to make cast iron by pouring molten pig iron into molds. Cast iron also is brittle; it is used for engine blocks. Steel contains less carbon (about 1.3%) than pig iron or cast iron. Its composition can be varied to give it specific properties. Steel is used for heavy construction, automobile components, surgical instruments, tools, etc.

13. Oxygen is used to convert pig iron to steel. Oxygen gas is blown through molten pig iron to remove dissolved carbon by changing it to carbon dioxide and carbon monoxide.

15. (a) $2 Cu_2S(s) + 3 O_2(g) \longrightarrow 2 Cu_2O(s) + 2 SO_2(g)$
    (b) $2 FeS(s) + 3 O_2(g) \longrightarrow 2 FeO(s) + 2 SO_2(g)$

17. (a) approximately 140 million gallons
    (b) approximately 4.3 million gallons

19. Metals can be considered to have atomic nuclei and their core electrons at regularly spaced positions in the solid. The valence electrons can be shared between all of the electrons in the solid. These valence electrons are like a mobile sea, and they are free to flow throughout the solid. This electron mobility accounts for the electrical conductivity of metals and the thermal conductivity as well.

21. The dopants in an *n*-type semiconductor possess one more valance electron than a silicon atom. Therefore, elements in group 15 theoretically have the potential to be used. In practice, the most common dopants are P, As, and Sb.

23. Semiconductors have properties in common with metals and nonmetals. Therefore, they are found at the boundary between these types of elements in the periodic table. These elements are B, Ge, As, Sb, and Te.

25. The structure of an SiO_4 unit is tetrahedral.

27. Quartz contains both silicon and oxygen atoms held together by covalent bonds. Because the electrons in these bonds are held in place, they are unable to conduct electricity.

Chapter 19

1. Earth's carrying capacity depends on the amount of productive land, agricultural practices, biotechnology advances, the rate of degradation of farmland, global pollution, the amount of water for irrigation, and the level of an acceptable quality of life.

3. Soil is a mixture of mineral particles, organic matter, water, and air.

5. Soil has a structure made up of a series of layers called *horizons* that are loosely packed and permeable near the surface and gradually change to impermeable solid rock.

7. Limestone will neutralize acids in the soil and raise the pH.

9. carbon, hydrogen, and oxygen

11. (a) Metal ions from Group 1 and Group 2
    (b) The result is a more acidic soil, and the pH goes down.

13. (a) substances needed by green plants for healthy growth
    (b) Nonmineral nutrients are carbon, oxygen, and hydrogen. They are available from water and atmospheric carbon dioxide.
    (c) Mineral elemental nutrients are water-soluble substances that plants can only absorb as solutes through their roots.

15. A soil shortage of N, P, and K is more likely. The nutrients N, P, and K are primary elemental nutrients and are needed in greater amounts. They are more easily leached from the soil.

17. $N_2(g) + 3 H_2(g) \rightleftharpoons 2 NH_3(g)$. It is the synthetic source of ammonia fertilizer.

19. Chlorosis is a plant condition of low chlorophyll. It is caused by a deficiency of any one of the three nutrients magnesium, nitrogen, or iron. A symptom of chlorosis is the presence of leaves that are pale yellow instead of green.

21. In order, the numbers refer to the percentage, by weight, of the nutrients nitrogen, phosphorus (as P_2O_5), and potassium (as K_2O).

23. 12-12-2.5 indicates that, by weight, there is 12% nitrogen, 12% P_2O_5, and 2.5% potassium. Using example 19.1 as a guide to solving this problem, the amounts of these nutrients in a 50-lb. bag of the fertilizer are 6 lb. nitrogen, 2.6 lb. phosphorus, and 1 lb. potassium.

25. Phosphate rock is not very water soluble, so the phosphorus as phosphate is not readily available to plants. The superphosphate is more soluble, which makes phosphorus more available.

27. Straight fertilizers contain only one nutrient. Complete fertilizers contain all three primary nutrients; nitrogen, phosphorus, and potassium.

29. (a) pesticides kill and control pests;
    (b) insecticides kill insects
    (c) herbicides control and kill plants
    (d) fungicides control and kill fungi

31. The hydrogen gas used in the Haber process is derived from petroleum.

33. Ammonium nitrate is inexpensive and a good source of water-soluble nitrogen. Farmers benefit because cheap fertilizers keep food production costs down. Consumers benefit because food prices are low. The dark side of ammonium nitrate is that it is an explosive. Handling it can lead to industrial accidents. It can be used as an explosive by terrorists, as was done in 1995 when approximately a ton of ammonium nitrate was detonated next to the Federal Building in Oklahoma City.

35. DDT is fat soluble because "like dissolves like" and both are nonpolar. DDT poses problems because it is a carcinogen and does not degrade quickly. It is stored in the fat tissue of animals. This accumulation problem is exaggerated when one organism high in the food chain eats many other ones lower in the chain.

37. About 33% to 40% of food crop production is lost to pests each year worldwide. The monetary value of these losses is estimated at $20 billion per year.

39. A biodegradable pesticide is one that is quickly converted to harmless products by microorganisms and natural processes. Pyrethrins are an example of a class of biodegradable insecticides. An example of a pyrethrin is shown here.

Dimethrin

41. (a) Pesticides can increase crop yields and protect them when stored. They reduce the loss of crops to pests. Every bit of food saved is food that need not be replaced. Food supplies increase, and food is more plentiful at lower costs. Diseases carried by pests can be controlled or even eliminated. Pesticides can cause problems

of water and soil contamination if they are misused. Pesticide residues can contaminate crops and create long-term poisoning problems. Resistant strains of pests can develop and require higher levels of pesticides.

(b) Pesticides should be used early enough to require smaller amounts. They should be used only when absolutely necessary and when alternative methods do not exist.

43. Sustainable agriculture is a set of practices intended to improve profits, limit use of agrichemicals, and increase the use of environmentally friendly farming procedures.

45. Integrated pest management involves the use of disease-resistant plant varieties and biological controls such as predators or parasites to control pests.

47. (a) Genetically altered crops with specific genes inserted to produce plants with desirable properties

(b) Flavr Savr tomato, a weevil-resistant garden pea, and Bollgard cotton seed

49. Pheromones are insect sex attractants. They are used to lure insects into traps so insecticide spraying is not needed.

51. Yes, there is a conflict. Population growth will decrease the number of productive acres per person. The present 0.82 acre per person will decrease, and the efficiency of farming methods will need to improve in order to avoid famine and food shortages. This problem will be compounded because population centers usually are in the middle of prime agricultural land. The growth of urban areas will remove such land from production.

53.

Agrichemical	LD_{50}	
DDT	100 mg/kg	most toxic
primicarb	150 mg/kg	
carbaryl	250 mg/kg	
malathion	1000 mg/kg	
dimethrin	10,000 mg/kg	least toxic

CHEM

Index

H

Haber process, 404–405
  for ammonia production, 404–405f
Haemophilus influenza bacteria, 353
Hahn, Otto, 252
Hahnemann, Samuel, 375
Hair
  keratin in, 316–317
  permanent waves, 316–317
  rinses and conditioners, 306
Hairspray, 126
Half-lives, 246–247
  radioactive waste problem and, 257–260
Halides in Earth's crust, 380t
Hall, Charles, 183
Hall-Heroult process, 183
Hallucinogens, 366
  abuse of, 363
Halogens, 52–53
Halons, 132t
Handedness, 290–295
Hard water, 206–207
Harvard Center for Risk Analysis
  (HCRA), 10–11
Hazardous wastes
  radioactive waste, 257–260
  water and, 199–200
HCFCs (hydrochlorofluorocarbons),
  131–132
HDL (high-density lipoprotein), 332
  structure of, 332f
Health
  reducing agents and, 177
  See also Diseases; Drugs and medicines
Heart attacks, 371
  COX-2 inhibitors and, 363
  drugs for, 371
Heartburn, 167–168
Heart disease, 370–371
  cholesterol and, 331–332
  dietary fiber and, 331
  fatty acids and, 331
  as leading cause of death, 352t
Heat
  of combustion, 217, 219f
  of fusion, 97, 190–192
  proteins and, 332
  specific heat capacity of water, 190–192
  of vaporization, 191–192
  water, capacity of, 190–191
  water pollution and, 198t
Heat-transfer fluids for nuclear reactors, 256
Heavy oil, 267t
Helical protein structure, 313, 313f
Helium
  atoms of, 17f
  in dry air at sea level, 61t
  electron arrangements of, 43t, 45t
  properties of, 49t
  solubility of, 96
Hematite
  in Earth's crust, 380t
  iron oxide, 172, 173
Heme groups, 311
Hemoglobin, 311
  as quaternary protein structure, 314
  sickle cell anemia, 313
  structure of, 312f
  vitamin B$_6$ and, 334

Heptachlor, 210t
Heptane
  octane numbers of, 221t
  structural formula for, 227t
Herbicides, 409–410
  drinking water, maximum contaminant
    levels (MCL) for, 210t
Heroin, 361
  LD$_{50}$ (lethal dose-50%) of, 361t
  natural opiates and, 312
Heroult, Paul, 183
Heterogeneous matter, 16, 18–19f
Hevea brasiliensis tree, 278
Hexachlorobenzene, 210t
Hexane, 227t
HFCs (hydrofluorocarbons), 131–132
High blood pressure, 370
High-density lipoprotein (HDL), 332, 332f
High-density polyethylene (HDPE), 275, 275f
  polymer chains of, 276
  recycling of, 284–285
High-fructose corn syrup, 298
Histamine, 367
Histidine, 309t
  mRNA (messenger RNA) codes for, 323t
HIV/AIDS, 353, 355–357
  protease inhibitors and, 356
  thalidomide for treating, 295
Homeopathic medicine, 374–376
Homicide, 353
  as leading cause of death, 352t
Homogeneous matter, 16
  classification of, 18–19f
Honda hybrid automobiles, 182–183
Honey, 297–298
Horizons, 399f, 400
Hormones, 302
  proteins as, 313
  sex hormones, 302–303
  steroid hormones, 357–358
Hospitals, infections from, 355
Household products
  acidity of, 158f
  wastes and water, 200–201
  See also Cleaning supplies
Human genome, 320
Human Genome Project, 5, 320
Human-made risks, 11t
Humidity, 111
Humus, 399
Hybrid automobiles, 116, 182–183
Hydraulic fluids, 95
Hydrocarbons, 64
  and air pollution, 70–71
  alkanes, 85, 226–227
  alkenes, 86
  aromatic compounds, 233–234
  burning, products of, 138
  cyclic hydrocarbons, 233–235
  cycloalkanes, 233
  emission rates for, 71t
  formulas for, 226–227t
  multiple covalent bonds in, 86–87
  in photochemical smog, 66
  as primary pollutant, 67t
  rush hour, pollutant concentration
    during, 66f
  saturated hydrocarbons, 85
  single covalent bonds in, 85
  trees producing, 70

unsaturated hydrocarbons, 86–87
  See also Octane rating; Petroleum
Hydrochloric acid, 158, 160, 161t
  in cleaning supplies, 167
  in stomach, 167–168
Hydrochlorofluorocarbons (HCFCs), 131–132
Hydrocortisone, 235, 357
Hydrofluorocarbons (HFCs), 131–132
Hydrogen
  atomic number for, 38
  boiling points of hydrogen-containing
    compounds, 92f
  in dry air at sea level, 61t
  electron arrangements of, 43t, 45t
  energy and, 218–219, 219t
  heat of combustion of, 219t
  isotopes of, 39
  neutrons *vs.* protons in nuclei from,
    244–245t
  as plant nutrient, 401t
  properties of, 49t
  as reducing agent, 175t
  as submetal, 48–49
  symbol for, 21t
Hydrogenated fats, 300–301
Hydrogen bonding, 92
  in alcohols, 269
  boiling points and, 95
  of carboxylic acids, 271
  diagram of, 92f
  DNA (deoxyribonucleic acid) and, 320, 320f
  hair and, 316–317
  in helical protein structure, 313
  intermolecular hydrogen bonding, 313–314
  intramolecular hydrogen bonding, 313
  in nylon-6,6, 282f
  between thymine and adenine, 319
Hydrogen carbonate ion, 82t
Hydrogen chloride, 160
Hydrogen ions, 158
  acid-base reactions and, 160
  pH, relationship to, 164, 164f
Hydrogen peroxide
  liver, reaction with, 145f
  as oxidizing agent, 175t, 176
Hydrogen phosphate, 82t
Hydrogen spectrum, 42
Hydrogen sulfate, 82t
Hydrogen sulfite, 82t
Hydrolysis reactions, 315
Hydronium, 158
  pH and, 163–164
  in pure water, 162
  in strong acids, 160, 161t
Hydrophobic molecules, 305
Hydrophylic molecules, 305
Hydroscopic foods, 340–341
Hydrosphere, 378–380
  chemicals from, 381–383
Hydroxide, 82t, 158–159
  molarity of, 163
  pH, relationship to, 164, 164f
  in pure water, 162
Hypertension, 352t
Hypertensive renal disease, 352t
Hyperthyroidism, 252f
Hypertonic solution, 338
Hypervitaminosis A, 334
Hypochlorite, 82t
Hysterectomy, 358

I

Ibuprofen (Advil), *294, 363*
Ice, *15*
  crystallization of, *202–203*
  density of, *190*
  global warming and, *114*
  reversible state changes, *97–98*
  sublimation of, *97–98f*
  See also Snow/snow cover
Igneous rocks, *378*
Implosion, *254f*
Inderal, *371*
Indigestion, *168*
Industrial Revolution, *112*
  acid rain and, *192*
Industrial smog, *65*
Infectious diseases
  deaths from, *353*
  drugs for, *353–355*
  resistant to antibiotics, *354–355*
Influenza, *352t*
Infrared (IR) radiation, *106, 107–108*
  CFCs (chlorofluorocarbons) and, *111*
  energy of, *108, 120–121*
  greenhouse gases, efficiencies of
    absorption by, *111t*
Initiation stage of carcinogenesis, *372*
Inorganic compounds, *22*
Insecticides, *407–408*
  DDT (dichlorodiphenyltrichloroethane), *204*
  natural insecticides, *412–413*
  and water, *200t*
Insoluble substances, *96*
Insulators, nonmetals as, *49*
Insulin, amino acid sequence of, *311, 311f*
Integrated pest management (IPM), *412*
Intergovernmental Panel on Climate Change,
  United Nations, *113, 131*
Intermolecular forces, *91–92*
Intermolecular hydrogen bonding, *313–314*
International Desalination Association, *208*
International System of Units, *27, 214*
  prefixes for, *27t*
Internet, *9t*
Intramolecular hydrogen bonding, *313, 313f*
Invert sugar, *298*
Iodine, *252t, 337*
  solubility of, *96f*
  symbol for, *21t*
Ion-exchange columns/resin, *206, 207f*
Ionic bonds, *76, 78–79*
  diagram of, *90f*
Ionic compounds, *78–79*
  binary ionic compounds, *81–82*
  formulas for, *80–81*
  molecular compounds compared, *90–91*
  naming binary ionic compounds, *81–82*
  with polyatomic ions, *82–83*
  properties of, *91t*
Ionic solids, melting points of, *98t*
Ionizing smoke detectors, *252, 253f*
Ions
  common ions, list of, *80f*
  formation of, *51*
  See also Polyatomic ions
Iron, *336–337, 384–385*
  carbon monoxide and, *173–174*
  corrosion and, *184, 185f*

drinking water, maximum contaminant
  levels (MCL) for, *210t*
from Earth's crust, *383t*
as glass coloring, *392t*
in hard water, *206*
oxidation and, *173*
oxygen, reaction to, *142*
as plant nutrient, *401t, 402–403, 404t*
production of, *385f*
in pure water, *198t*
recycling of, *152*
symbol for, *21t*
Iron oxide, *172, 173*
Irradiation of foods, *4*
Irrigation and water use, *196*
Isoamyl acetate, *273*
Isoleucine, *309t*
  mRNA (messenger RNA) codes for, *323t*
Isomers, *228–229*
  cis isomers, *232–233*
  of octane, *229–230*
  optical isomers, *232, 291–292*
  stereoisomerism, *232–233*
  trans isomers, *232–233*
  See also Structural isomers
Isooctane
  energy and, *217–218*
  naming of, *229–230*
  octane numbers of, *221t, 222*
  octane rating and, *220–221*
Isopropyl alcohol, *229t, 235, 268t*
  as secondary alcohol, *267*
  uses for, *268*
Isotopes, *38–39*
  of radon, *249*
  See also Fission; Radioisotopes

J

Joint Institute of Nuclear Research, *249*
Joliot, Frédéric, *247*
Joliot, Irene Curie, *247*
Joules, *214*
Journal of the National Cancer Institute on skin
  cancers, *129*
Julian, Percy, *235*
Junk science, *5–6*

K

Kelsey, Frances, *294*
Kelvin scale, *109*
Keratin, *316–317*
Kerosene, *221t*
Ketones, *268*
Kettering, Charles, *7–8*
Kevlar vests, *281, 282*
Kidney disease, *352t*
Kilograms, *27*
Kilometers, *27*
Kinetic energy, *25*
Knocking properties, octane rating and,
  220–221
Krypton
  in dry air at sea level, *61t*
  in neon signs, *62*
Kyoto Conference/Protocol, *115–116*

L

Lactase, *316*
Lactic acid, *273t*
  enantiomers of, *292, 293f*
Lactose, *297–298*
  intolerance, *315–316*
  sucrose, sweetness relative to, *298t*
Lakes, *195f*
  acid rain and, *193–194*
Lambda (λ), *40–41*
Landfills
  for hazardous wastes, *199–200*
  water and runoff from, *199*
Lanolin, *304*
  in shampoos, *307*
Lanthanide series, *48*
Lanthanum, *390t*
Lard, *301t*
Latex, *278*
Lattices, *79, 387*
  doping process and, *388*
Lauric acid, *300t*
  genetic engineering of plants and, *414*
Lavoisier, Antoine, *33–34*
Law of conservation and matter, *33–34*
Law of conservation of energy, *343*
Law of definite proportions, *34*
LD$_{50}$ (lethal dose-50%), *361, 361t*
LDL (low-density lipoprotein), *332*
L-Dopa, *360*
Leaching effect, *401*
Lead
  automobile batteries, lead-acid, *181, 181f*
  drinking water, maximum contaminant
    levels (MCL) for, *210t*
  from Earth's crust, *383t*
  LD$_{50}$ (lethal dose-50%) of, *361t*
  one-mole quantity of, *140f*
  recycling of, *152*
  superconducting transition temperature
    for, *390t*
  symbol for, *21t*
  in water, *201*
Lead dioxide, *175t*
Leather, *199t*
Le Chatelier, Henri, *147*
Le Chatelier's principle, *147*
  acid-base buffers and, *166*
Legume nitrogen fixation, *402*
Lemon oil, *339*
L-enantiomers, *292–293*
Leprosy, *295*
Leucine, *309t*
  mRNA (messenger RNA) codes for, *323t*
Leucippus, *32*
Leu-enkephalin, *313f*
Levomethorphan, *368*
Levorotatory, *292*
Lewis, G. N., *43, 50, 76*
Lewis dot structures, *50*
  for alkanes, *85*
  atoms, symbols for, *50t*
  for covalent bonds, *83–84*
  sodium-chlorine interaction and, *78–79*
  for water, *88, 190*
Lexapro, *351, 352t*
Lexus hybrid automobiles, *182–183*
Librium, *364*

Section Summaries

1-1 The World of Chemistry (pp. 2–5)

From the food we eat to the technology we use, chemistry is all around us. Knowing the fundamentals of chemistry helps us better understand and analyze the world in which we live. While chemicals are sometimes associated with the adjective *toxic*, not all chemicals are harmful—many improve and can even save our lives. As a citizen of the world, it is important to understand how these chemicals interact and how they affect current and future social issues. None of humanity's questions can be fully addressed without some serious applied chemistry. Scientists must organize information, make predictions, and test these predictions through experimentation to learn more about the world.

1-2 DNA, Biochemistry, and Science (pp. 5–7)

Deoxyribonucleic acid (DNA) is present in cells throughout our bodies. It provides the equivalent of a fingerprint in identifying individuals. DNA analysis, one of the frontier areas of science, has become a very important tool of forensic science. It can help us understand genetics and diseases and can be implemented in new technology, such as genetic engineering. The study of DNA is one aspect of biochemistry, the natural science that unites the physical science of chemistry with the biological sciences, such as botany and genetics.

1-3 Air-Conditioning, the Ozone Hole, and Technology (pp. 7–9)

The development of refrigeration, a technology that changed the way humans eat, is rooted in chemistry. Through research and experimentation, chemists discovered a fluorine-based chemical that was neither toxic nor flammable. This process, the development of a new technology with a specific purpose, is an example of applied science. Technology, like pure and applied science, is a human activity. It is important to be able to critically evaluate these activities' societal impact. Only by staying informed can we be ready to adjust to changing times.

1-4 Fossil Fuel Use and Global Warming (pp. 9–10)

The world has become increasingly dependent on the use of fossil fuels such as crude oil and coal. Though agriculture, health care, climate control, and transportation depend largely on them, it is becoming increasingly apparent that the combustion of nonrenewable fossil fuels may be causing Earth to warm beyond what is natural and desirable. There is much debate over the scope and impact of global warming among scientists, but all citizens should become knowledgeable enough about the issue to make informed decisions about it.

1-5 Benefits/Risks and the Law (pp. 10–12)

It is important to consider how society weighs the benefits of new activities, chemicals, and technologies. Contemporary questions of security and medicine carry with them certain risks and benefits that need to be weighed. In any such analysis, several factors come into play. Scientists must take all of these factors into consideration as they develop new technologies, such as pesticides. No absolute answer can be provided to the question, "How safe is safe enough?" The determination of acceptable levels of risk requires value judgments that are difficult and complex. Risk assessment is the province of scientists, but risk management is a societal issue.

1-6 What Is Your Attitude Toward Chemistry? (pp. 12–13)

Before proceeding with the study of chemistry and its relationship to our society, you might want to examine your prejudices and attitudes about chemistry, science, and technology. Many feel fear or anxiety when approaching science, but understanding the basic concepts of chemistry and its positive and negative aspects will allow you to make informed decisions about personal problems and problems concerning society as a whole.

Key Terms

Chemistry: The study of matter and the changes it undergoes

Basic science: Pursuit of scientific knowledge without the goal of a practical application

Applied science: Science with a well-defined, short-term goal of solving a specific problem

Technology: The application of scientific knowledge in the context of industrial production, our economic system, and our societal goals

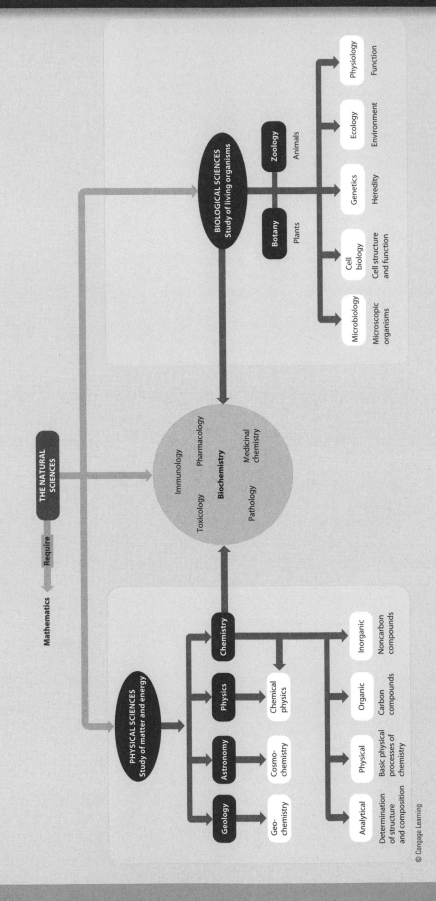

© Cengage Learning

Section Summaries

2-1 States of Matter, Mixtures, and Pure Substances (pp. 15–16)

Most samples of matter as they occur in nature are mixtures. Some, known as solutions, are homogeneous, and some are heterogeneous. Regardless of its state (solid, liquid, or gaseous), no amount of optical magnification will reveal a solution to be heterogeneous. Some mixtures, such as blood, appear to be homogeneous even if they are not. When a mixture is successfully separated into its components, it has been purified.

2-2 Elements: The Simplest Kind of Matter (pp. 16–17)

While most of the materials we encounter are complex mixtures, elements cannot be separated or broken down into any other kinds of matter. Elements, which exist as atoms, have unique and consistent properties by which they can be identified. Helium and mercury have distinctly different sets of properties, so they cannot be the same matter. A distinction between pure substances, elements, and atoms is the basis of the chemical model of matter.

2-3 Chemical Compounds: Atoms in Combination (pp. 17–18)

If a pure substance is not an element, it is a chemical compound. Water is a chemical compound of two hydrogen atoms and one oxygen atom. This combination forms one water molecule. Water is a pure substance because it has the same properties no matter where it comes from. Once elements are combined into compounds, the original properties of the elements are replaced by the characteristic properties of the compounds.

2-4 Classification of Matter (pp. 18–20)

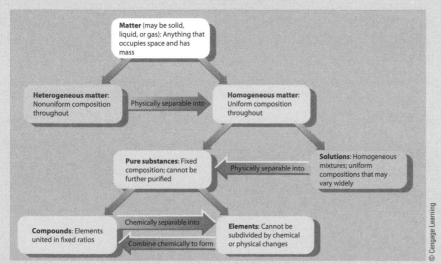

There are three basic reasons for studying pure substances: to understand how to utilize and apply their properties in everyday life, to help us deal intelligently with our environment, and out of simple curiosity. An essential part of basic research is investigation into the structure of matter—how atoms are connected in larger units of matter and how these units are arranged.

Key Terms

Macroscopic: Large enough to be seen, felt, and handled

Microscopic: Visible only with the aid of a microscope

Nanoscopic: In the range of the nanometer (0.000000001 m)

Heterogeneous mixture: Matter that is not uniform in composition

Homogeneous mixture: Matter that is uniform in composition

Solution: A homogeneous mixture that may be in the solid, liquid, or gaseous state

Pure substance: Matter with a uniform and fixed composition at the nanoscopic level

Element: A pure substance composed of only one kind of atom

Atom: The smallest particle of an element

Chemical compounds: Pure substances composed of atoms combined in definite, fixed ratios

Molecule: The smallest chemical unit of a compound

Diatomic molecules: Molecules composed of two atoms

Chemical formula: Written combination of element symbols that represents the atoms combined in a chemical compound

Subscripts: In chemical formulas, numbers written below the line (for example, 2 in H_2O) to show numbers or ratios of atoms in a compound

Structural formula: A chemical formula that illustrates the connections between atoms in molecules as straight lines

Molecular formula: A chemical formula that illustrates molecules with atomic symbols and subscripts

Organic compound: A derivative of the carbon and hydrogen compound

Inorganic compound: Any compound other than an organic compound

Key Terms

Chemical equation: A representation of a chemical reaction by the formulas of reactants and products

Balanced chemical equation: A chemical equation in which the total number of atoms of each kind is the same in reactants and products

Reactant: A substance that undergoes chemical change

Product: A substance produced as a result of chemical change

Coefficient: In a chemical equation, a number written before a formula to balance the equation

Mass: A measure of the quantity of matter in an object

Physical property: A property that can be observed without changing the identity of a substance

Density: Mass per unit volume

Chemical property: A property of matter that is observable in chemical reactions

Energy: The capacity for doing work or causing change

Potential energy: Energy in storage by virtue of position or arrangement

Kinetic energy: The energy of objects in motion

Quantitative: Describes information that is numerical

Qualitative: Describes information that is not numerical

Unit conversion: Converting a quantity expressed in one set of units to another

Equality: Two measurements that refer to the same quantity

Conversion factor: A ratio used to convert the units that describe a quantity

Section Summaries

2-5 The Chemical Elements (pp. 20–21)

We know much about the naturally occurring elements as well as the 18 or so elements that are not found anywhere in nature. Some elements only exist in nature as diatomic molecules. The varying properties of elements are largely determined by the composition of their atoms. Chemistry uses a special language and set of symbols to convey meaning. Some chemical symbols are single letters, while others are two letters. A single symbol is used to represent a single atom of that chemical.

2-6 Using Chemical Symbols (pp. 22–24)

Elemental symbols can be utilized in formulas that illustrate which and how many atoms are in a stable molecule. Structural formulas indicate which atoms are connected to which. To concisely represent chemical reactions, symbols and formulas are arranged in chemical equations. While changing a coefficient changes only the amount of an element or compound, changing a subscript changes its identity. Understanding formulas and equations is a fundamental aspect of chemistry.

2-7 Changes in Matter: Is It Physical or Chemical? (pp. 24–25)

In chemistry, the word *physical* is used to refer to processes that do not change chemical identities. Physical changes can often be measured numerically. By contrast, chemical reactions can alter a substance's identity and are not always observable. These reactions convert one or more substances (the reactants) to one or more entirely different substances (the products). As a noun, the word *chemical* can be applied to anything. As an adjective, it refers to this specific type of reaction. Some physical changes and almost all chemical reactions are accompanied by changes in energy.

2-8 The Quantitative Side of Science (pp. 25–28)

Numerical measurement is important to chemistry, as it establishes scientific facts and laws. Units are important to numerical measurement, as they tell us what precisely is being measured. Because they are built around multiples of ten, the units of the metric system are used widely in scientific measurement. The International System of Units (SI) is the current standard of the metric system.

Section Summaries

3-1 John Dalton's Atomic Theory (pp. 32–34)

In 1803, John Dalton established this atomic theory:

1. All matter is made up of indivisible and indestructible particles called atoms.
2. All atoms of a given element are identical, both in mass and in properties. By contrast, atoms of different elements have different masses and different properties.
3. Compounds form when atoms of different elements combine in ratios of small whole numbers.
4. Elements and compounds are composed of definite arrangements of atoms. Chemical change occurs when atomic arrays are rearranged.

This theory was used to define two scientific laws: Lavoisier's law of conservation of matter and Proust's law of definite proportions.

3-2 Structure of the Atom (pp. 34–36)

While there are more than 60 subatomic particles, three are most important. Protons and neutrons are found in the nucleus of the atom, while electrons orbit the nucleus. In 1899, Marie Curie discovered radioactivity, countering Dalton's idea that atoms are indivisible. Following Curie's discovery of alpha and beta particles and gamma rays, Ernest Rutherford found that the atom is mostly empty space, and its concentration of mass is mostly at its core.

3-3 Modern View of the Atom (pp. 36–39)

Thanks to the work of Rutherford and other scientists, we have a firm understanding of the atomic structure. Each element has a unique atomic number. Hydrogen has one proton in its nucleus, so its atomic number is 1. To find the number of neutrons in an element, one should subtract the atomic number from the mass number. The fundamental difference between isotopes is the different number of neutrons per atoms. Still, all isotopes have the same atomic number. Many isotopes have been produced artificially.

3-4 Building the Atomic Model: Where Are the Electrons in Atoms? (pp. 39–46)

In 1913, Niels Bohr introduced a new model of the hydrogen atom that incorporated the idea of orbits, or levels: When an electron absorbs with a quantum of energy, it moves from a ground state to an excited state. Each level can only hold a given number of electrons. Those in the highest occupied level are at the greatest stable distance from the nucleus. While Bohr's theory of a quantum number *n* remains valid, newer models have indicated that the electron field is comprised of a number of subshells and acts like an electrically charged cloud of energy.

3-5 Development of the Periodic Table (pp. 46–48)

In 1869, Dmitri Mendeleev wrote out all the known elements and their properties on cards and shuffled them. He realized that when elements were arranged by atomic weight, a trend in properties repeated itself several times. He arranged the element cards into groups that had similar properties and used the resulting periodic chart to

Key Terms

Law of conservation of matter: Matter is neither lost nor gained during a chemical reaction

Law of definite proportions: In a compound, the constituent elements are always present in a definite proportion by weight

Scientific theory: A set of ideas that seek to explain the world around us

Scientific model: Something that represents a part of the world around us

Radioactivity: Spontaneous decomposition of unstable atomic nuclei

Alpha (α) particle: A positively charged particle emitted by certain radioactive isotopes

Beta (β) particle: An electron ejected at high speed from the nuclei of certain radioactive isotopes

Gamma (γ) ray: High-energy electromagnetic radiation emitted from radioactive isotopes

Atomic number: The number of protons in the nucleus of an element

Mass number: The number of neutrons and protons in the nucleus of an atom

Isotopes: Atoms of an element with different mass numbers due to different numbers of neutrons

Atomic mass unit (amu): The unit for elements' relative atomic masses

Atomic weight: The average atomic mass of an element's isotopes weighted by percentage abundance

Continuous spectrum: A spectrum that contains radiation distributed over all wavelengths

Visible radiation: The portion of light that can be seen with the naked eye

Wavelength: The distance between similar points of two waves

Frequency: The number of waves that pass a fixed point in one unit of time

Energy: For light, energy is determined by the frequency and wavelength

Quantum: The smallest increment of energy

Ground state: The condition of an atom in which all electrons are in their normal, lowest energy level

Excited state: An unstable, higher energy state of an atom

Valence electrons: The outermost electrons in an atom

Shell: A principal energy level defined by a given value of n

Orbital: A region of three-dimensional space around an atom within which there is a significant probability that a given electron will be found

Subshell: A more specific energy level (orbital) within a given shell

Periodic law: When elements are arranged in the order of their atomic numbers, their chemical and physical properties show repeatable, or periodic, trends

Periodic table: An arrangement of elements by atomic number

Group: A vertical column of elements in the periodic table

Period: A horizontal row of elements in the periodic table

Metal: An element that conducts electric current

Nonmetal: An element that does not conduct electrical current

Insulator: A poor conductor of heat and electricity

Metalloid: An element with properties intermediate between those of metals and nonmetals

Semiconductor: A material with electrical conductivity intermediate between those of metals and insulators

Noble gas: An element in Group 18 of the periodic table

Ion: An atom or group of atoms with a positive or negative charge

Alkali metal: An element in Group 1 of the periodic table

predict the properties and places in the chart of as yet undiscovered elements. Thus, the periodic law and table were born.

3-6 The Modern Periodic Table (pp. 48–50)

The modern periodic table arranges elements by vertical groups of similar properties and horizontal periods related to energy levels for electrons. Most of the elements are metals, and 18 elements are nonmetals. The nonmetals are, except for hydrogen, found on the right side of the table. Nonmetals are insulators, while metalloids tend to be semiconductors.

3-7 Periodic Trends (pp. 50–52)

G. N. Lewis devised a system to use elements' symbols to represent their atomic nuclei. The outermost valence electrons are represented by dots placed around the symbol. The Lewis dot symbols are very useful in the discussion of bonding. We can use trends in size of atomic radii to predict trends in properties and radioactivity. The repeatable patterns of properties across the periods result from repeatable patterns in atomic structure.

3-8 Properties of Main-Group Elements (pp. 52–53)

Elements in a group generally react with other elements to form similar compounds. Certain groups are given specific names such as alkali metals, halogens, and noble gases. The elements of each of these groups share certain properties with the other elements in the group.

Concept Match

1. Periodic
2. Larger atoms
3. Two valence electrons
4. A noble gas
5. A transition metal
6. A main-group metal
7. Decrease in atomic radius
8. A halogen
9. An inner transition element
10. Valence electrons

a. Electron arrangement 2-8-2
b. Outermost occupied shell
c. Chromium (Cr)
d. Repeated pattern
e. At the bottom of a group
f. Praseodymium (Pr)
g. Eight valence electrons
h. Seven valence electrons
i. Across a period
j. Cesium

Answers: 1. d, 2. e, 3. a, 4. g, 5. c, 6. j, 7. i, 8. h, 9. f, 10. b

Cation: An atom or group of bonded atoms that has a positive charge

Alkaline earth metal: An element in Group 2 of the periodic table

Halogen: An element in Group 17 of the periodic table

Anion: An atom or group of bonded atoms that has a negative charge

Section Summaries

4-1 The Lower Atmospheric Regions and Their Composition (pp. 59–61)

A few vertical miles of gaseous chemicals comprise our atmosphere. Held in place by gravitational force, the atmosphere consists of the troposphere (closest to Earth), stratosphere, and ozone layer (farthest from Earth). The air we breathe contains over 15 chemicals, the most prevalent of which are nitrogen, oxygen, argon, and carbon dioxide.

4-2 Air: A Source of Pure Gases (pp. 61–62)

When air is fractionated, water vapor and carbon dioxide are removed to allow pure oxygen, nitrogen, and other gases to be separated and utilized. Pure oxygen is used in steelmaking and rocket production. Nitrogen is used in welding and medicine. Argon and other noble gases are used in colored "neon" signs.

4-3 Natural Versus Anthropogenic Air Pollution (pp. 63–64)

Nature pollutes the air on a massive scale with ash, mercury vapor, carbon dioxide, and other chemicals. We can do little to control these natural events. We can, however, control the anthropogenic pollution that noticeably affects air and water quality, especially in urban areas. Though it remains a problem, the U.S. government has taken important steps to curb pollution.

4-4 Air Pollutants: Particle Size Makes a Difference (p. 64)

Air pollutant particles range in size from fly ash particles to individual molecules and atoms. Aerosol particles are small enough to remain suspended in the atmosphere for long periods of time, and millions of tons of soot, dust, and smoke particles are emitted each year. These particles are harmful to humans as long as they remain in the air.

4-5 Smog (pp. 64–67)

The nomenclature *smog*, for the poisonous mixture of smoke, fog, air, and other chemicals, was first coined in 1911. There are certain circumstances that facilitate smog, such as windlessness and thermal inversion. Industrial smog and photochemical smog are both dangerous to humans, though they entail different chemicals and reactants.

4-6 Nitrogen Oxides (pp. 67–68)

About 97% of nitrogen oxides in the atmosphere are naturally produced. The 3% that is human-created, as well as a large contributor to photochemical smog, is emitted primarily by internal combustion engines. Nitrogen dioxide, a product of nitric oxide and atmospheric oxygen, is highly toxic and can form the secondary pollutant ozone.

4-7 Ozone as a Pollutant (pp. 67–69)

Ozone, an allotrope of molecular oxygen, consists of three oxygen atoms bound together. It has a pungent odor and is one of the most difficult pollutants to control, especially in urban areas.

Key Terms

Atmospheric pressure: The pressure exerted by the weight of the atmosphere at a given altitude

Troposphere: The region of Earth's atmosphere that extends from sea level up to about 10 km

Stratosphere: The region of Earth's atmosphere that extends from the troposphere to an altitude of about 50 km

Ozone layer: The stratospheric layer of gaseous ozone that protects life on Earth by filtering out most of the harmful ultraviolet radiation emitted by the Sun

Parts per million (ppm): 1 part out of every million parts

Parts per billion (ppb): 1 part out of every billion parts

Nitrogen fixation: Conversion of atmospheric nitrogen into soluble nitrogen compounds available as plant nutrients

Anthropogenic pollution: Pollution that is attributable to human activity

Hydrocarbon: A compound that contains only carbon and hydrogen atoms

Particulate: Either a solid particle or liquid droplet suspended in the atmosphere

Aerosol: A pollutant particle so small that it remains suspended in the atmosphere for a long period

Thermal inversion: An atmospheric condition in which warmer air is on top of cooler air

Industrial smog: Smog generally caused by industrial activity such as coal burning

Photochemical smog: Smog formed by the action of sunlight on photoreactive pollutants in the air

Key Terms

Primary pollutant: A pollutant emitted directly into the atmosphere

Secondary pollutant: A pollutant formed by chemical reactions in the atmosphere

Photodissociation: Decomposition of a reactant, caused by the energy of light

Allotrope: A different molecular form of an element

Global warming: Warming of Earth above a level that is desirable from a human viewpoint as a result of increased concentrations of one or more greenhouse gases

Section Summaries

4-8 Sulfur Dioxide: A Major Primary Pollutant (pp. 69–70)

Sulfur dioxide is produced when sulfur or sulfur-containing compounds are burned in air. Although volcanoes produce sulfur dioxide, humans are accountable for roughly 70% of its emission. SO_2 reacts with oxygen to form SO_3, which dissolves in aqueous aerosol particles to form acid rain.

4-9 Hydrocarbons and Air Pollution (pp. 70–71)

Methane gas and other hydrocarbon pollutants are produced by ruminant animals, the use of industrial solvents, bacteria, motor vehicle exhaust, and a number of other natural and human sources.

4-10 Carbon Monoxide (pp. 71–73)

At least ten times more carbon monoxide enters the atmosphere from natural sources than from all industrial and automotive sources combined. Contact with CO is normal and unavoidable. Still, CO is considered a dangerous air pollutant because it is colorless, odorless, and deadly in high concentrations.

4-11 A Look Ahead (p. 73)

Both legislative and technological efforts to reduce the concentrations of pollutants have shown positive results, but there is clearly still much to be done. It is only by being informed that we can address the world's current and future problems with pollution.

Concept Match

1. Gas with the highest percentage by volume in the atmosphere
2. Consisting mostly of very small water droplets
3. Abnormal temperature arrangement for air masses
4. A mixture of smoke, fog, air, and various other chemicals
5. A pollutant caused by reactions of other chemicals
6. Second most abundant gas in the atmosphere
7. A pollutant that is directly discharged into the atmosphere
8. A chemical species with an unpaired valence electron
9. Main ingredient in industrial smog
10. The product of a reaction between an oxygen atom and an oxygen molecule
11. An oxide of nitrogen produced by certain bacteria
12. The oxide of nitrogen produced by lightning, forest fires, and internal combustion engines
13. "Bad" ozone
14. An air pollutant that bonds to hemoglobin

a. Oxygen
b. Nitrogen
c. Aerosol
d. Secondary pollutant
e. SO_2
f. Smog
g. NO
h. Primary pollutant
i. Thermal inversion
j. CO
k. Ozone in the air we breathe
l. N_2O
m. Free radical
n. Ozone molecule

Answers: 1. b, 2. c, 3. i, 4. f, 5. d, 6. a, 7. h, 8. m, 9. e, 10. n, 11. g, 12. l, 13. k, 14. j

Section Summaries

5-1 Electronegativity and Bonding (pp. 77–78)

Atoms with high levels of electronegativity attract the valence electrons of other, electropositive, atoms. The transfer of valence electrons is referred to as an ionic bond, while the sharing of valence electrons is referred to as a covalent bond.

5-2 Ionic Bonds and Ionic Compounds (pp. 78–83)

When two atoms form an ionic bond to stabilize the electron octets of their outer shells, they become an ionic compound of a positively charged cation and a negatively charged anion, drastically changing their original properties. It is important to understand the naming system of ionic compounds and polyatomic ions.

5-3 Covalent Bonds (pp. 83–87)

Most chemical compounds are held together with single and multiple covalent bonds, which are illustrated by Lewis structures. Hydrocarbons represent an important and large class of molecules.

5-4 Shapes of Molecules (pp. 87–89)

The valence shell electron pair repulsion model is useful in representing three-dimensional molecules. Bonds might be linear, trigonal planar, tetrahedral, trigonal bipyramidal, or octahedral.

5-5 Polar and Nonpolar Bonding (pp. 89–90)

When there is an equal sharing of covalently bonded electrons, the bond is nonpolar. When covalently bonding electrons draw more toward one atom and away from the other, the bond is polar. The water molecule, for example, is polar.

5-6 Properties of Molecular and Ionic Compounds Compared (pp. 90–91)

Ionic compounds form brittle, crystalline solids with high melting and boiling points and are good conductors of heat and electricity when molten. Molecular compounds can be gases, liquids, or weak solids with low melting and boiling points and are never good conductors.

5-7 Intermolecular Forces (pp. 91–92)

Even though the attractive forces between separate molecules are relatively weak, they give matter exceedingly important properties. Changes in state and hydrogen bonding are two of these properties.

5-8 The States of Matter (pp. 92–93)

Matter in any state is composed of atoms, molecules, or ions in constant motion. When a gas is cooled to the point of condensation, its atoms slow and it becomes a liquid. When liquid is cooled, it solidifies. In contrast to solids, liquids and gases are fluid.

Key Terms

Octet rule: In forming bonds, main-group elements gain, lose, or share electrons to achieve a stable electron configuration with eight valence electrons

Electronegativity: A measure of the relative tendency of an atom to attract electrons to itself

Anion: An atom (or molecule) that has a negative charge

Cation: An atom (or molecule) that has a positive charge

Lewis dot symbol: A representation of a molecule, ion, or formula unit by atomic symbols with the valence electrons shown as dots

Ionic bond: The attraction between positive and negative ions

Ionic compound: A compound composed of positive and negative ions

Formula unit: In ionic compounds, the simplest ratio of oppositely charged ions that gives an electrically neutral unit

Lattice: A regular arrangement

Binary compound: A chemical compound composed of two elements

Polyatomic ion: A positive or negative ion composed of two or more atoms

Covalent bond: A bond in which two atoms share electrons

Bonding pair: A pair of electrons shared between two atoms

Lone pair: A pair of valence electrons that is not shared between two atoms but instead is associated with a single atom in the structure

Alkane: A hydrocarbon with carbon–carbon single bonds

Saturated hydrocarbon: A hydrocarbon that is an alkane

Double bond: A bond in which two pairs of electrons are shared between two atoms

Key Terms

Triple bond: A bond in which three pairs of electrons are shared between two atoms

Resonance structures: Two or more possible Lewis dot representations of a molecule or ion in which the only difference is the position of the valence electrons

Bond energy: The amount of energy required to break one mole of bonds between a specific pair of atoms

Alkene: A hydrocarbon with one or more carbon–carbon double bonds

Unsaturated hydrocarbon: A hydrocarbon that contains fewer than the maximum number of hydrogen atoms

Molecular geometry: The arrangement of atomic nuclei in a molecule

Electron pair geometry: The arrangement of bonds and lone pairs in a molecule

Nonpolar: Describes a bond or molecule with no or symmetrically oriented (and thus canceled) polar bonds

Polar: Describes a bond or molecule with positive and negative regions

Electrolyte: A compound that conducts electricity when melted or dissolved in water

Nonelectrolyte: A compound that does not conduct electricity when melted or dissolved in water

Intermolecular force: An attractive force that acts between molecules

Hydrogen bonding: Attraction between a hydrogen atom bonded to a highly electronegative atom and the lone pair on an electronegative atom in another or the same molecule

Condensation: The change of molecules from the gaseous state to the liquid state

Fluid: A state of matter that is capable of flowing; a gas or a liquid

Pressure: Force per unit area

Section Summaries

5-9 Gases (p. 94)

We are surrounded by gases. All gases share a set of common physical properties, including condensation, pressure, compressibility, and miscibility. Gaseous diffusion ensures that over time, any mixture of gases will become homogeneous.

5-10 Liquids (pp. 94–95)

Liquids are only slightly compressible, but their molecules remain mobile enough to flow. Raising the temperature of a liquid increases its volatility, bringing it closer to its boiling point. Higher molecular weight compounds have higher boiling points.

5-11 Solutions (pp. 95–96)

Solutions are any homogeneous mixtures, but liquid solutions are most common. Solubility is determined by the strengths of the forces of attraction between solute and solvent. Structurally similar compounds are more likely to be soluble.

5-12 Solids (pp. 96–97)

Particles exist in an orderly array in solids—atomic movement is restricted to vibration and sometimes rotation. Because their particles are so close together, solids maintain definite shapes. As their temperatures rise, solids can melt or even sublime.

5-13 Reversible State Changes (pp. 97–99)

The endothermic process of changing a solid to a liquid and then changing the liquid to a gas involves melting to the liquid state, heating of the liquid state to the boiling point, and vaporization of the liquid state to the gaseous state.

Miscibility: Ability to mix in all proportions

Volatile: Easily vaporized

Vapor pressure: The pressure above a liquid caused by molecules that have escaped from the liquid surface

Boiling point: The temperature at which the vapor pressure of a liquid equals the atmospheric pressure and boiling occurs

Normal boiling point: The boiling temperature if the atmospheric pressure is 1 atm

Solvent: In a solution, the substance present in the greater amount

Solute: In a solution, the substance dissolved in the solvent

Solubility: The maximum quantity of a solute that will dissolve in a given amount of a solvent at a specified temperature

Saturated solution: Solution that has dissolved all the solute that it can dissolve at a given temperature

Unsaturated solution: Solution in which less than the maximum amount of solute is dissolved at a given temperature

Insoluble: Describes a substance that will not dissolve in a given solvent

Melting point: The temperature at which a solid substance turns into a liquid

Crystallization (solidification): Formation of a crystalline solid from a melt or a solution

Sublimation: The process whereby molecules escape from the surface of a solid into the gas phase

Section Summaries

6-1 Atmospheric Carbon Dioxide Concentration over Time (pp. 104–106)

Although the concentration of carbon dioxide in our atmosphere is relatively low, evidence indicates that it has fluctuated significantly over the past 200,000 years. Over the past 200 years, higher CO_2 concentrations have correlated with higher than average temperatures, suggesting causation to some scientists.

6-2 What Is the Greenhouse Effect? (pp. 106–108)

The greenhouse effect is a well-known phenomenon. Radiation making its way from the Sun to Earth encounters many types of molecules in our atmosphere's stratosphere. Certain molecules, such as carbon dioxide, are better than others at absorbing radiation. The greenhouse effect is desirable up to a certain point, but excessive warming could upset Earth's comfortable range of temperature.

© Cengage Learning

6-3 Why Worry About Carbon Dioxide? (pp. 108–110)

Earth is warmer than can be explained simply by its distance from the Sun. This elevated temperature is attributable to Earth's balanced atmosphere. Over the last century, Earth has experienced a 25% increase in its atmospheric concentration of carbon dioxide. If indeed more CO_2 is the cause of higher temperatures, then Earth could experience a 1.5° to 6°C rise as early as 2050.

6-4 Other Greenhouse Gases (p. 111)

Chlorofluorocarbons owe their atmospheric presence to human activities, while other greenhouse gases such as methane and nitrous oxides do not. Most of these gases are present in low concentrations, but others, such as water vapor, have noticeable effects on temperature. The efficiency with which greenhouse gases absorb infrared radiation is not equal. Humans are fortunate that the concentrations of the most efficient absorbers are relatively low.

6-5 Sources of Greenhouse Gases (pp. 112–113)

The Industrial Revolution and internal combustion engine brought enormous appetites for fossil fuel. The combustion of coal, oil, gasoline, and natural gas produces both carbon dioxide and water vapor. Volcanic eruptions and other natural occurrences also generate greenhouse gases. Methane, another problematic greenhouse gas, is primarily released into the atmosphere from natural sources such as decaying vegetation and animal waste.

Key Terms

Greenhouse effect: A phenomenon of Earth's atmosphere by which solar radiation, trapped by Earth and reemitted from the surface as infrared radiation, is prevented from escaping by various gases in the atmosphere

Infrared (IR) radiation: Electromagnetic radiation whose wavelength is in the range between 2.5×10^{-5} m and 2.5×10^{-6} m

Chlorofluorocarbons (CFCs): A term used to refer collectively to compounds containing only carbon, chlorine, fluorine, and sometimes hydrogen

Radiational cooling: The phenomenon of enhanced loss of heat from Earth on a clear, cloudless night due to the absence of significant water vapor in the atmosphere

6-6 What Do We Know for Sure About Global Warming? (pp. 113–114)

There are some things about global warming that can be stated with increasing confidence:

1. Earth is definitely warming, and most scientists agree that humans are responsible.
2. The warming is almost certainly due to increased anthropogenic emissions of greenhouse gases.
3. A temperature increase of only 2°C would mean Earth would be warmer than it has been for the past 2 million years.
4. As the world's population increases, the rate and amount of greenhouse gases released will increase unless changes are made.

6-7 Consequences of Global Warming—Good or Bad? (pp. 114–115)

It is crucial that scientists and nonscientists alike consider the consequences of global warming. These consequences are controversial and difficult to predict. Most scientists who study the atmosphere agree upon a number of likely outcomes, though the extent to which each of the phenomena will manifest itself is debatable.

6-8 The Kyoto Conference Addresses Global Warming (pp. 115–116)

Many of the world's nations sent representatives to Kyoto, Japan, in 1997 to begin to address the issue of global warming. At this conference, 161 countries set greenhouse gas emissions goals for developed countries. In 2004, Russia ratified the Kyoto Protocol, making carbon dioxide reductions mandatory among 124 nations. Some countries have enacted more stringent emissions standards on top of the protocol. The United States refused to sign the Kyoto Protocol for economic reasons.

6-9 Possible Responses to Global Warming (pp. 116–117)

Assuming that the problem of global warming is real, scientists have begun to make suggestions about what can be done to either slow its progress or allow us to live with its consequences. Some of the actions suggested are obvious and practical, while others are unusual and impractical. Only time will tell if we respond appropriately.

Section Summaries

7-1 The Oxygen–Ozone Screen (pp. 120–122)

The Sun bombards Earth with electromagnetic radiation. Of this, 53% is infrared (the lowest in energy), 39% is visible, and 8% is ultraviolet (the highest in energy). Much of the ultraviolet (UV) radiation that streams from the Sun never reaches us—it is absorbed by two important screening molecules: oxygen and ozone. The energy required to break the oxygen–oxygen bonds of these molecules is supplied by UV radiation, which is nullified in this reaction.

7-2 Where Is the Ozone? (pp. 122–124)

The ozone layer is a band of higher than average ozone concentration within the stratosphere. In the absence of anthropogenic influences, the concentration of ozone has maintained a steady state for millions of years through the Chapman cycle. However, measured ozone concentrations are lower than can be accounted for by this natural cycle. While natural phenomena such as the generation of nitric oxide by microorganisms may be contributing to its thinning, anthropogenic influences have begun to perturb the ozone layer.

7-3 The Ozone Layer Is Disappearing (pp. 124–126)

Scientists began measuring stratospheric ozone concentrations over 80 years ago using relatively unsophisticated instrumentation. In tribute to G. M. B. Dobson, the Dobson unit (DU) was established to measure ozone concentration. Measurements first revealed a significant decline in stratospheric ozone in 1979, particularly during the months of September and November. By 1985, ozone decline was unmistakable, manifesting in an Antarctic ozone hole every year. F. Sherwood Roland, Mario Molina, and Paul Crutzen were awarded the 1995 Nobel Prize in Chemistry for discovering that chlorofluorocarbons were the key to the decline.

7-4 Why CFCs and Why at the Poles? (pp. 126–128)

After CFCs were discovered, their use as refrigerants, solvents, and propellants grew, mirroring the growth of modern, urbanized society. During the time CFCs were in common use, little effort was made to prevent them from escaping into the atmosphere. When released at lower altitudes, CFCs remained suspended and unreactive until natural weather patterns swept them into the stratosphere, where they were bombarded with UV radiation. This energy breaks CFCs down, generating highly reactive chlorine atoms that react cyclically with ozone molecules, converting them to oxygen molecules. The lower stratosphere above Antarctica is extremely cold and windless, which facilitates an eruption of ozone-depleting Cl atoms.

Key Terms

Rainout: The removal of pollutants from the atmosphere by natural precipitation

Steady state: A state in which the concentration of a chemical remains essentially constant even though the chemical may be undergoing numerous chemical reactions

Chapman cycle: The set of four reactions that represents the steady state formation and destruction of ozone in the stratosphere

Dobson unit (DU): A measure of stratospheric ozone concentration

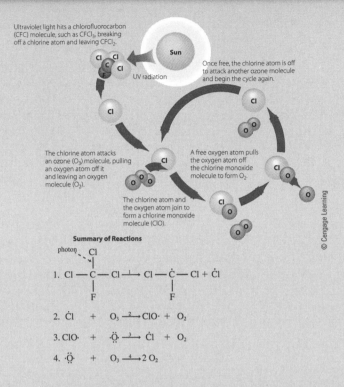

7-5 What Are the Implications of Increases in the Size of the Ozone Hole? (pp. 129–130)

There is evidence that the lowered ozone concentrations are already starting to have a negative effect on crop yields in the Southern Hemisphere. Skin cancer may become a serious issue if the ozone hole spreads over populated areas. A 1% decrease in ozone concentration has been predicted to cause a 2% increase in skin cancers, and some Chileans and Australians must already take special precautions to avoid overexposure.

7-6 Can We Do Anything About the Ozone Hole? (pp. 130–131)

Eighty percent of the chlorine and bromine in the stratosphere comes from manmade chemicals. In the United States, the use of CFCs as propellants in aerosol spray cans was banned in 1978, and their use as foaming agents was discontinued in 1990. The first international response to CFCs was the Montreal Protocol on Substances that Deplete the Ozone Layer in 1987. Over a hundred countries have banned the production of CFCs and other harmful chemicals entirely. While other countries have imposed taxes on CFC production, this has led to steep price hikes and black market smuggling.

7-7 What Will Replace CFCs? (pp. 131–132)

The demand for technology that used to use CFCs has not disappeared, so scientists have been working to find acceptable replacements. A number of researchers have tried replacing Cl atoms with H atoms (creating HCFCs and HFCs), which has met with some success and some failure. While they may still produce greenhouse gases, many replacements have much lower global warming potential than CFCs. The search for safe and effective alternatives continues as our knowledge continues to evolve.

Section Summaries

8-1 Balanced Chemical Equations and the "What?" Question (pp. 137–139)

Chemical equations represent what happens in chemical reactions at the nanoscopic level that we cannot see. Unbalanced equations are not faithful to reality. In balancing equations, we are applying the law of conservation of matter—the atoms in the reactants must all be present in the products. Only coefficients, not subscripts, can be changed to properly balance an equation. One cannot predict the products of a chemical reaction without prior knowledge of the reaction or knowledge of similar reactions.

8-2 The Mole and the "How Much?" Question (pp. 139–141)

The atomic weights of elements are relative, so to avoid the problems associated with difficult-to-remember individual masses, the mole, a counting unit consisting of a certain numbers of atoms, is necessary. One mole of anything is determined by Avogadro's number and is equal to the molar mass, the atomic weight in grams listed below the element symbol In the periodic table. To find the molar mass for a compound, one must multiply the number of each element by the atomic weight of the element and add them together. Moles allow us to determine the masses of reactants and products.

8-3 Rates and Reaction Pathways: The "How Fast?" Question (pp. 142–145)

Chemical reactions can occur in the blink of an eye or over hundreds of years. The exact pathway each atom takes from its position in the reactants to its position in the products is difficult to study but interesting. Activation energies and their resulting reaction rates help us understand these pathways. There are four strategies to controlling reaction rates: adjusting the temperature, adjusting the concentration of reactants, increasing contact between reactants, and adding a catalyst. In the presence of a catalyst, an alternate pathway with a lower activation energy is made available. Enzymes are biological catalysts.

8-4 Chemical Equilibrium and the "To What Extent?" Question (pp. 145–147)

In chemical equilibrium, a chemical reaction and its reverse are occurring at equal rates. Theoretically, all chemical reactions are reversible and therefore able to come to equilibrium. Reactions tend to reach equilibrium at a characteristic and predictable point. Chemicals do not always reach completion, but equilibrium can still be reached if they do not. Constant conditions are important to achieving equilibrium. If a reaction is altered midway, either forward or reverse movement may outpace the other for a period before the reaction returns to equilibrium. This is established by Le Chatelier's principle.

Key Terms

Mole: The counting unit used for atoms

Avogadro's number: 6.02×10^{23}; the number of objects in one mole

Molar mass: Mass in grams of one mole of any substance

Reaction rate: Amount of reactant converted to product in a specific amount of time

Activation energy (E_{act}): Quantity of energy needed for successful collision of reactants

Catalyst: A substance that increases the rate of a chemical reaction without being changed in identity

Chemical equilibrium: Condition in which a chemical reaction and its reverse are occurring at equal rates

Le Chatelier's principle: When a stress is applied to a system at equilibrium, the equilibrium shifts to relieve the stress

Exergonic: A chemical process in which there is a net energy release

Endergonic: A chemical process in which there is a net consumption of energy

Entropy: The disorder of matter

Enthalpy change: A measure of the quantity of heat transferred into or out of a system as it undergoes a chemical or physical change

Key Terms

Entropy change: A measure of the change in the disorder of a system as it undergoes chemical reaction or physical change

Thermodynamics: The science of energy and its transformations

First law of thermodynamics: Energy can be converted from one form to another but cannot be destroyed

Second law of thermodynamics: The total entropy of the universe is constantly increasing

Section Summaries

8-5 The Driving Forces and the "Why?" Question (pp. 147–151)

To examine the driving forces that account for chemical differences, we must consider two kinds of change—changes in energy as heat and changes in the amount of order that accompanies chemical reactions. Changes in energy are not related to reaction rates or equilibrium. In many reactions, potential energy is transformed into kinetic energy and released as heat. The opposite of this exergonic reaction is an endergonic reaction—one that requires energy to take place. Along with a decrease in energy, an increase in entropy lends to favorability. Understanding these concepts is important to understanding the first and second laws of thermodynamics, which summarize the universal conditions for changes in energy and entropy.

8-6 Recycling: New Metal for Old (pp. 151–152)

Many communities in the United States now recycle metals, paper, and plastic. Still, individuals are producing garbage at an alarming rate. Recycling counteracts the natural direction of increasing entropy and thus requires a great expenditure of energy in the collecting, transporting, and processing of materials. Generating less waste, by increasing the useful lifetime of our materials and diminishing excess use rather than consuming and recycling more, is the most efficient solution to the problem of waste. Still, recycling is better than dumping waste in landfills, especially in the case of toxic and abundant metals.

Concept Match

1. Number of atoms shown by the formula of $(NH_4)_3PO_4$
2. Mass of 1 mol of H_2O_2
3. Adding a catalyst
4. Cannot be destroyed
5. Mass of 1 mol of titanium
6. Balanced chemical equation
7. Lowering the temperature
8. Forward and reverse reactions proceeding at equal rates
9. Drives a reaction to completion
10. Loss of chemical potential energy
11. Number of moles of N_2 needed to produce 30 mol of ammonia, $N_2 + 3 H_2 \longrightarrow 2 NH_3$

a. Speeds up a reaction
b. 48 g
c. Same number of atoms of each element on both sides of the arrow
d. 34 g
e. Slows down a reaction
f. Dynamic equilibrium
g. Formation of gas that escapes
h. 20
i. 15
j. An exergonic reaction
k. Energy

Answers: 1. h, 2. d, 3. a, 4. k, 5. b, 6. c, 7. e, 8. f, 9. g, 10. j, 11. i

Section Summaries

9-1 Acids and Bases: Chemical Opposites (pp. 157–160)

We usually encounter acids and bases as solutions in water. In strong solutions, both can be extremely harmful to humans and should be handled with care, but not all acids and bases are dangerous—most that we encounter are dilute. All acidic water solutions have some chemical properties in common, as do all basic water solutions. Acidic solutions contain a higher concentration of H_3O^+ ions than OH^- ions, while the opposite is true for basic solutions. Acids and bases are often highly reactive with each other, and when exactly equivalent amounts of acid and base react, the result is a neutralization reaction. Every reaction in which a hydrogen ion is exchanged between reactants is an acid–base reaction, so water can react as either an acid or a base.

9-2 The Strengths of Acids and Bases (pp. 160–162)

The strength of an acid or base is determined by the extent of ionization in an aqueous solution. The greater the ionization, the stronger the acid or base. Many common acids react completely with water, and many establish equilibrium with water at a point where not all of the acid molecules have been converted to ions. These are typically weak, and the same holds true for bases. Anions play an Important role In solutions in which the acid concentration must remain constant.

9-3 Molarity and the pH Scale (pp. 162–165)

Concentrations of chemicals are expressed as molarity. A 1 molar solution contains 1 mol of solute per liter of solution. In neutral solutions such as water, the concentrations of hydronium ions and hydroxide ions are the same. In 1909, Danish biochemist S. P. L. Sørensen devised a method of measuring acidity—the pH scale. In a neutral solution, the pH is 7. A pH larger than 7 indicates a base, and a pH smaller than 7 indicates an acid. The further the pH from 7, the stronger the solution is. In the absence of a calculator, the bracketing method can be used to estimate a pH. The pH scale is nonlinear.

9-4 Acid–Base Buffers (pp. 166–167)

Buffer solutions contain a base that can react with an acid and an acid that can react with a base so that the pH of a solution remains close to its original value. These two components must not react with each other. Buffers are usually mixtures of a weak acid and its weakly basic anion, or a weak base and its weakly acidic cation. Buffers act to control pH and prevent a chemical disturbance while retaining the original conditions or structure. In buffered reactions, new components are never produced—products are components of the buffer. Buffers are very important to both technology and biology.

9-5 Corrosive Cleaners (p. 167)

Cleaners formulated with extremes in pH allow acidity or alkalinity to quickly attack unwanted dirt, grease, and stains. Strong acids, such as hydrochloric acid, can dissolve most mineral scale and iron stains. The pH of toilet-bowl cleaners is usually below 2. Because they contain strong acids, they can be quite harmful to skin and eyes on contact. On the other end of the pH scale are drain cleaners and oven cleaners, which have a pH of 12 or higher. Bases like sodium hydroxide are very good at dissolving drain clogs because they can cause rapid breaking of bonds in oils and greases. Oven cleaners are often strong bases propelled by aerosol. Care must be taken not to breathe these solutions, which are quite corrosive to nasal tissue, bronchial tubes, and lungs.

Key Terms

Acid–base indicator: A substance that changes color with changes in acidity or basicity of a solution

Hydronium ion: A hydrated proton, H_3O^+

Lewis structure

Acidic solution: A solution that contains a higher concentration of H_3O^+ ions than OH^- ions

Basic solution (alkaline solution): A solution that contains a higher concentration of OH^- ions than H_3O^+ ions

Salt: Ionic compound composed of the cation from a base and the anion from an acid

Neutralization reaction: Reaction of equivalent quantities of an acid and a base

Acid: Molecule or ion able to donate a hydrogen ion to a base

Base: Molecule or ion able to accept a hydrogen ion from an acid

Acid–base reaction: Reaction in which a hydrogen ion is exchanged between an acid and a base

Strong acid: Acid that is 100% ionized in aqueous solution

Weak acid: Acid that is only partially ionized in aqueous solution and establishes equilibrium with nonionized acid

Strong base: Base that is 100% ionized in aqueous solution

Weak base: Base that is only partially ionized in aqueous solution and establishes equilibrium with nonionized base

Concentration of a solution: The quantity of a solute dissolved in a specific quantity of a solvent or solution

Molarity: Number of moles of solute per liter of solution

pH: Negative logarithm of the hydronium ion concentration of a solution

Neutral solution: A solution in which the pH is 7

Buffer solution: Solution of an acid plus a base that controls pH by reacting with added base or acid

Buffer: Combination of an acid and a base that can control pH

Antacid: Base used to neutralize excess hydrochloric acid in the stomach

9-6 Heartburn: Why Reach for an Antacid? (pp. 167–168)

The walls of a human stomach contain thousands of cells that excrete hydrochloric acid, which kills microorganisms and aids in digestion. Heartburn occurs when the acid contents of the stomach back up into the esophagus and cause a burning sensation in the chest and throat. Antacids such as magnesium and aluminum hydroxides are bases used to neutralize the acid that causes heartburn. A new class of heartburn medication acts by preventing the secretion of acid altogether. Regular use of some antacids can cause constipation.

Concept Match

1. M	a. 7
2. pH of pure water	b. Molarity, moles of solute per liter of solution
3. Acidic pH	c. Hydrogen ion donor
4. Strong acid	d. Resists pH change
5. Strong base	e. Hydrogen ion acceptor
6. Acid definition	f. 10
7. Base definition	g. CH_3COOH, acetic acid
8. Weak acid	h. NaOH
9. Buffer	i. H_2SO_4
10. Basic pH	j. 3

Answers: 1. b, 2. a, 3. j, 4. i, 5. h, 6. c, 7. e, 8. g, 9. d, 10. f

Section Summaries

10-1 Oxidation and Reduction (pp. 172–175)

Oxidation got its name from the chemical changes that occur when oxygen combines with other elements or compounds. Many reactive metals are mined as oxides. A substance that has combined with oxygen has been oxidized. All combustion reactions are oxidations. The opposite of oxidation, the removal of oxygen, is reduction. Oxidation and reduction always occur together and have expanded in definition to consider electrons in addition to oxygen. If one reactant is oxidized, another must be reduced. Such reactions are known as oxidation–reduction reactions, or simply redox reactions.

10-2 Oxidizing Agents: They Bleach and They Disinfect (pp. 175–176)

When oxygen is involved in an oxidation, it is the oxidizing agent—the reactant—that causes an oxidation by gaining electrons. Oxidizing agents such as sodium hypochlorite, a water-soluble ionic compound, are fairly common. In laundering, the disinfectant NaOCl is used as a bleaching agent—it removes unwanted color by oxidizing colored chemical compounds to produce whiter clean clothes. A bleach disrupts the alternating single and double carbon–carbon bonds within a fabric that create color by absorbing visible light. A strong bleach can break bonds in molecules that compose the fabric itself, causing thin spots and holes. Another familiar oxidizing agent is hydrogen peroxide, which is used as an antiseptic and to bleach hair.

10-3 Reducing Agents: For Metallurgy and Good Health (pp. 176–178)

When metal reacts with another substance, it usually acts as a reducing agent. Reducing agents are crucial in the production of metals from their ores because most metals occur in nature only in chemical compounds. Hydrogen gas, like metals, is a reducing agent in virtually all of its reactions. Antioxidants are reducing agents that prevent oxidation by free radicals, which have the potential to cause cancer. Free radicals arise in the body naturally and are also produced in the presence of toxic substances like cigarette smoke.

10-4 Batteries (pp. 178–183)

A battery is any device that converts chemical energy to electrical energy. Essentially, it is a favorable oxidation–reduction reaction occurring inside a container that has two electrodes—an anode (−) and a cathode (+). To function, a battery takes advantage of the ease with which metals lose electrons. The electrons transferred between a metal and the ion of another metal can provide the electron flow, or current, in a battery. Throwaway batteries, or primary batteries, cannot be recharged. The alkaline battery is a common throwaway battery. Secondary batteries are reusable and allow the oxidation–reduction reactions at the electrodes to be reversed. Lead–acid automobile batteries are secondary batteries. Fuel cells produce energy from reactants that continuously flow into their compartments, while the chemical reaction products flow out of them.

Key Terms

Oxide: A compound of oxygen combined with another element

Combustion: Rapid oxidation that produces heat and (usually) light

Oxidation: The gain of oxygen, loss of hydrogen, or loss of electrons

Reduction: The loss of oxygen, gain of hydrogen, or gain of electrons

Oxidation–reduction reaction (redox reaction): A reaction in which one reactant is oxidized and another reactant is reduced

Oxidizing agent: A reactant that causes oxidation of another reactant

Bleaching agent: A chemical that removes unwanted color by oxidation of the colored chemical

Reducing agent: A reactant that causes reduction of another reactant

Coke: A mostly carbon product formed by heating coal in the absence of air to drive off volatile materials

Free radical: An atom or molecule that contains an unpaired electron

Electrochemical cell: A device in which a chemical reaction generates an electric current

Battery: A series of electrochemical cells that produces an electric current; commonly refers to any electrochemical, current-producing device

Electrode: A conducting material at which electrons enter and leave an electrochemical cell

Anode: Electrode at which electrons flow out of an electrochemical cell and oxidation takes place

Cathode: Electrode at which electrons flow into an electrochemical cell and reduction takes place

Primary battery: A battery that cannot be recharged because its reaction is not easily reversible

Secondary battery: A battery that can be recharged by reversing the flow of current, which reverses the current-producing reaction and regenerates the reactants

Electrolysis reaction: An oxidation-reduction reaction that requires the input of electrical energy.

Corrosion: The unwanted oxidation of metals during exposure to the environment

10-5 Electrolysis: Chemical Reactions Caused by Electron Flow (pp. 183–184)

In electrolysis reactions, electrons flow into an electrode, causing it to become negatively charged. Positive ions migrate toward that electrode and reduction takes place. In this type of cell, the electrodes need not be in separate compartments. Aluminum, the third most abundant element in Earth's crust, is produced through the Hall-Heroult process of electrolysis.

10-6 Corrosion: Unwanted Oxidation–Reduction (pp. 184–185)

More than $10 billion is lost each year to corrosion. Much of this is the rusting of iron and steel, although other metals corrode as well. The oxidizing agent in corrosion is usually oxygen. Whenever a strong reducing agent (the metal) and a strong oxidizing agent are in contact, a reaction is likely. Temperature, concentration, and the presence of salts affect the rate of corrosion. Rusting can be prevented by protective coatings such as paint and oil, which keep out salt-containing moisture.

Concept Match

1. Chemicals in an automobile lead storage battery
2. Reduction of Cu^{2+} ions produces this
3. Oxidizing agent used to purify drinking water
4. Oxidation
5. Reduction
6. Gain of hydrogen
7. Oxidized form of sulfur
8. Oxidized form of nitrogen
9. Allows ions to pass from one electrode compartment in a battery to the other
10. Fuel used in some fuel cells
11. Battery used in most automobiles

a. PbO_2, Pb, and H_2SO_4
b. Hydrogen
c. Loss of electron(s)
d. Reduction
e. SO_2
f. NO_2
g. Gain of electron(s)
h. Lead–acid battery
i. Chlorine
j. Copper metal
k. Salt bridge

Answers: 1. a, 2. j, 3. i, 4. c, 5. g, 6. d, 7. e, 8. f, 9. k, 10. b, 11. h

Section Summaries

11-1 The Unique Properties of Water (pp. 189–192)

Pure water is a liquid between 0°C and 100°C; the density of solid water is less than that of liquid water; the heat of fusion (melting) of ice is 80 cal/g; water has a relatively high heat capacity; water has a high heat of vaporization; water has a high surface tension; and water is an excellent solvent.

11-2 Acid Rainfall (pp. 192–194)

Any precipitation with a pH below 5.6 is considered to be acid rain, a phenomenon that has existed since 1900. When acid rain falls on natural areas, serious environmental problems can occur. Some lakes counter acid rain damage with limestone-lined bottoms, but industries and their regulatory agencies must address the problem as well.

11-3 How Can There Be a Shortage of Something as Abundant as Water? (pp. 194–197)

Because only 2.5% of water on Earth is freshwater, laws have been established by the Environmental Protection Agency and other agencies to regulate pollution and treatment. Still, aquifers and other wells are experiencing depletion, causing sinkholes and shortages of potable water.

11-4 What Is the Difference Between Clean Water and Polluted Water? (pp. 197–198)

Pollution, any condition that causes natural usefulness to diminish, may consist of pathogens, acids, toxic metal cations, or other synthetic and organic chemicals. The Clean Water Act of 1972 shifted responsibility for cleaning water from users to wastewater dischargers.

11-5 The Impact of Hazardous Industrial Wastes on Water Quality (pp. 199–200)

Before 1976, hazardous waste was dumped in landfills, contaminating ground and rainwater and costing millions of dollars in cleanup. Today, industrial producers must follow strict regulations in the transportation and disposal of hazardous waste.

11-6 Household Wastes That Affect Water Quality (pp. 200–201)

Personal waste can affect water supplies as much as industrial waste. Many common household products contain hazardous chemicals that require proper disposal. If these chemicals are poured down the sink or sent out with the trash, they can contaminate natural waters.

11-7 Toxic Elements Often Found in Water (pp. 201–202)

Mercury is widely used in industry, and once released into lakes and oceans it can contaminate edible fish. Lead, which accumulates in the body, has until recently been used in almost all plumbing fixtures. Arsenic occurs naturally in small amounts in many foods, but larger concentrations are deadly.

Key Terms

Heat of fusion: The heat required to melt a given quantity of a solid at its melting point

Specific heat capacity: The amount of heat required to raise the temperature of a 1 g sample of matter of a given size by 1°C

Heat of vaporization: The heat required to vaporize a given quantity of liquid at its boiling point

Surface tension: The amount of energy required to overcome the attraction for one another of molecules at the surface of a liquid

Acid rain: Rain that has a pH below 5.6

Aquifer: A layer of water-bearing porous rock or sediment held in place by impermeable rock

Potable: Describes water suitable for drinking

Groundwater recharge: Return to groundwater of water from treated sewage

Pathogen: A disease-causing microorganism

Hazardous waste: Industrial and household wastes responsible for water pollution

Biochemical oxygen demand (BOD): The amount of dissolved oxygen required by microorganisms that oxidize dissolved organic compounds to simple substances like CO_2 and H_2O

Biodegradable: Describes a substance that can be broken down into simple molecules by the action of microorganisms

Nonbiodegradable: Describes a substance that cannot be broken down by microorganisms and therefore persists in the environment

Key Terms

Denitrifying bacteria: Bacteria that convert ammonia or ammonium ion to nitrogen

Disinfection by-product: A compound formed by the action of disinfectants on substances found in water

Permeable: Allows substances to pass through

Semipermeable membrane: A membrane that allows water molecules but not ions or larger molecules to pass through

Osmosis: The flow of water molecules through a semipermeable membrane from a less-concentrated to a more-concentrated solution

Osmotic pressure: The external pressure required to prevent osmosis from taking place

Reverse osmosis: The application of pressure on a solution to cause water molecules to flow through a semipermeable membrane from a more-concentrated to a less-concentrated solution

Section Summaries

11-8 Measuring Water Pollution (p. 202)

The biochemical oxygen demand (BOD) is a measure of the quantity of dissolved organic matter in water. Highly polluted water often has a high concentration of organic material, which results in a larger BOD.

11-9 How Water Is Purified Naturally (pp. 202–204)

The water cycle offers a number of opportunities for nature to purify water. Distillation, crystallization, aeration, sedimentation, filtration, oxidation, and dilution all aid in purification. Nonbiodegradable substances are more difficult to deal with.

11-10 Water Purification Processes: Classical and Modern (pp. 204–205)

Sewage is treated using a variety of methods (ranging from least to most complex): primary, secondary, and tertiary wastewater treatment. There are various methods of tertiary water treatment, such as carbon black filtration and ammonia removal.

11-11 Softening Hard Water (pp. 206–207)

Hardness causes precipitates to form in boilers, causes soaps to form insoluble curds, and can impart a disagreeable taste to water. Removal of certain ions, as by a commercially available system, can soften water.

11-12 Chlorination and Ozone Treatment of Water (pp. 207–208)

Chlorine, a powerful oxidizing agent, is introduced as a gas into water to kill bacteria. Chlorination has reduced waterborne diseases significantly, but it can create carcinogenic disinfection by-products in industrial and city water supplies.

11-13 Freshwater from the Sea (pp. 208–209)

Technology has been developed for the conversion of seawater to freshwater. Solar distillation and reverse osmosis are used to desalinate water, though both methods are complex and costly.

11-14 Pure Drinking Water for the Home (pp. 209–211)

Many consumers have turned to bottled water (regulated by the FDA) and home water treatment devices to avoid the dangers of unclean water. Different treatment methods (distillation, reverse osmosis, and carbon filtration) produce different degrees of purity.

Section Summaries

12-1 Energy from Fuels (pp. 215–219)

Fuels are burned so that we may utilize the energy released from that process for some purpose. In chemical reactions, energy can be consumed or released or no net change may occur at all. Endothermic cleavages are represented in mathematical terms with a plus sign (+) by chemists. Exothermic bonding is represented in mathematical terms by a minus sign (−). Different types of fuels have different heats of combustion, but all hydrocarbons release about the same amount of heat per gram.

12-2 Petroleum (pp. 219–223)

Crude petroleum is a complex mixture of thousands of hydrocarbon compounds, and the actual composition of petroleum varies with the location in which it is found. Petroleum must be fractionated from crude oil, and its octane rating can be increased either by increasing the percentage of branched-chain and aromatic hydrocarbon fractions (catalytic re-forming) or by adding octane enhancers. Oxygenated and reformulated gasolines are sometimes used but may contaminate drinking water. After oil peaks, production will continue for several decades, but increasing costs and lower availability will favor the use of natural gas, coal, and alternative sources.

12-3 Natural Gas (p. 224)

Natural gas, a mixture of gases trapped with petroleum in Earth's crust, is the fastest-growing energy source in the United States. It is recoverable from oil and gas wells where the gases have migrated through the rock. About half the homes in the United States are heated by natural gas, but the United States has only about 5% of the known world reserves. Natural gas costs about one-third the price of gasoline and emits minimal amounts of carbon monoxide, hydrocarbons, and particulates. However, its disadvantage as a vehicle fuel is the requirement of a pressurized tank.

12-4 Coal (pp. 224–225)

Coal is a complex mixture of high molecular weight hydrocarbons that are about 85% carbon by mass. About 88% of our annual coal production is burned to produce electricity. Coal can be converted into a combustible gas or a liquefied fuel. In each case, environmental problems can be minimized but at additional costs per energy unit obtained from these fuels.

12-5 Methanol and Ethanol as Fuels (pp. 225–226)

Southern California has been testing methanol- and ethanol-powered cars since 1981. Flexible-fueled vehicles can operate on both standard gasoline and a blend of gasoline and either methanol or ethanol. Both ethanol and methanol burn more cleanly than gasoline, reducing air pollution. The technology for alcohol-powered vehicles has existed for many years. These fuels both cost and produce energy at half the rate of gasoline. Problems of distribution and storage are yet to be solved before purely alcohol-based fuels are viable.

Key Terms

Heat of combustion: The quantity of heat released when a fuel is burned

Fractional distillation: Separation of a mixture into fractions that differ in boiling points

Petroleum fraction: A mixture of hundreds of hydrocarbons with boiling points in a certain range that are obtained by fractional distillation of petroleum

Catalytic re-forming process: Process that increases octane rating of straight-run gasoline by converting straight-chain hydrocarbons to branched-chain hydrocarbons and aromatics

Catalytic cracking process: Process by which larger kerosene fractions are converted into hydrocarbons in the gasoline range

Oxygenated gasoline: A blend of gasoline that contains oxygenated organic compounds such as alcohols or ethers to cause the gasoline to burn more cleanly

Reformulated gasoline: A gasoline whose composition has been changed to reduce the percentage of olefins, aromatics, and sulfur and to add oxygenated compounds

Isomers: Two or more compounds with the same molecular formula but different arrangements of atoms

Structural isomers: Isomers that differ in the order in which the atoms are bonded together

Branched-chain isomers: Structural isomers of hydrocarbons that have carbon–carbon bonds in side chains

Straight-chain isomers: Structural isomers of hydrocarbons with no side chains

Key Terms

Alkyl groups: Alkanes with a hydrogen atom removed

Cis isomers: Stereoisomers with groups on the same side of a carbon–carbon double bond

Trans isomers: Stereoisomers with groups on opposite sides of a carbon–carbon double bond

Stereoisomerism: Describes isomers with the same molecular formulas and the same atom-to-atom bonding sequence but a different arrangement of the atoms in space

Alkyne: An organic compound containing a carbon–carbon triple bond

Cycloalkane: A saturated hydrocarbon with carbon atoms joined in a ring

Aromatic compound: A compound with one or more benzene rings or rings with benzene-like chemical properties

Alcohol: An organic compound containing a hydroxyl (OH) functional group

Functional group: An atom or group of atoms that is part of a larger molecule and that has a characteristic chemical reactivity

Ether: An organic compound with the general formula R—O—R′

Section Summaries

12-6 Classes of Hydrocarbons (pp. 226–230)

Alkanes (such as methane, the simplest alkane) contain carbon–carbon single bonds and are the backbone of organic chemistry. Every carbon in an alkane has a tetrahedral geometry. Chains of carbon atoms are usually represented with straight lines. Several different types of isomerism are possible for organic compounds. Branched-chain and straight-chain isomers are examples of structural isomers. Methyl and ethyl groups are types of alkyl groups.

12-7 Alkenes and Alkynes: Reactive Cousins of Alkanes (pp. 230–233)

Alkenes contain one or more carbon–carbon double bonds. In the alkene series, the possibility of locating the double bond between two different carbon atoms creates additional structural isomers. Some alkenes can also have *cis* and *trans* isomers, where isomers have the same molecular formulas and the same atom-to-atom bonding sequences, but the atoms differ in their arrangement in space. Alkynes contain one or more carbon–carbon triple bonds per molecule. *Cis* and *trans* isomers are not possible for alkynes.

12-8 The Cyclic Hydrocarbons (pp. 233–235)

In addition to straight and branched chains, hydrocarbons can form rings. Cycloalkanes are commonly represented by polygons in which each corner represents a carbon atom with two attached hydrogen atoms and the lines represent C—C bonds. In reality, the rings are not planar. Hydrocarbons containing benzene, benzene derivatives, and fused benzene rings are called aromatic compounds. These compounds are less reactive than alkenes and often have strong and pleasant odors. Their main feature is a six-carbon benzene ring.

12-9 Alcohols: Oxygen Comes on Board (pp. 235–236)

Alcohols are not strictly hydrocarbons since they also contain oxygen atoms. They have the general formula ROH, with R representing an alkyl group. Alcohols are essentially alkanes in which one of the hydrogen atoms has been replaced by the hydroxyl functional group. Methanol, also called methyl alcohol, is highly toxic. Ethanol, also called ethyl alcohol or grain alcohol, can be obtained by the fermentation of carbohydrates. It is popular both as a fuel and as an intoxicating agent. Diethyl ether and methyl propyl ether are common and well-known ethers.

Concept Match

1. Hydrocarbon	a. Benzene
2. Alkane	b. R—OH
3. Alkyl group	c. Major component of natural gas
4. Alkene	d. Contains only C and H
5. Alkyne	e. R—O—R′
6. Aromatic hydrocarbon	f. C_nH_{2n+2}
7. Methane	g. Mixture of CO and H_2
8. Alcohol	h. Contains C=C bond
9. Ether	i. Hydrocarbon that is missing a H atom
10. Synthesis gas	j. Contains C≡C bond

Answers: 1. d, 2. f, 3. i, 4. h, 5. j, 6. a, 7. c, 8. b, 9. e, 10. g

Section Summaries

Key Terms

13-1 The Discovery of Radioactivity (pp. 240–242)

While studying phosphorescence in 1896, Henri Becquerel discovered uranium radiation accidentally when it exposed some photographic plates. Three years later, Ernest Rutherford discovered the distinction between alpha and beta rays, and in 1900, Paul Villard identified gamma rays, which have great penetrating force.

13-2 Nuclear Reactions (pp. 242–244)

All of the elements heavier than bismuth, and a few lighter than bismuth, have natural radioactivity. Radioactivity is the result of a natural change of an isotope of one element into an isotope of a different element (a nuclear reaction). The total number of nucleons remains the same, but the identities of atoms can change. All matter is conserved.

13-3 The Stability of Atomic Nuclei (pp. 244–245)

The stability of nuclei is dependent on the relative numbers of protons and neutrons. The mass numbers of stable isotopes are always twice as large as (or greater than) the atomic number. Beta emission occurs in isotopes that have too many neutrons, but decay can also occur when there are too few neutrons and a positron is emitted. All isotopes of the elements beyond bismuth are unstable.

13-4 Activity and Rates of Nuclear Disintegrations (pp. 245–247)

The activity of a sample containing radioactive isotopes depends on the number of nuclei present and the rate at which they decay. Radioactive disintegrations are measured in *curies*. Every radioactive isotope has a characteristic half-life, and most of the naturally occurring radioactive isotopes are members of one of three decay series.

13-5 Artificial Nuclear Reactions (pp. 247–248)

In 1919, Rutherford was successful in producing the first artificial nuclear change by bombarding nitrogen with alpha particles. In this artificial transmutation, both product and nuclei are stable. Phosphorus-30 was the first radioactive isotope to be produced artificially. Today, more than 1000 other radioactive isotopes have been produced.

13-6 Transuranium Elements (pp. 248–249)

Neptunium was the first of the transuranium elements, discovered when uranium-239 was bombarded with a stream of high-energy deuterons. Neptunium was found to have a half-life of 2.33 days, after which it converted into a second new element—plutonium. A number of synthetic elements have been discovered since then.

13-7 Radon and Other Sources of Background Radiation (pp. 249–250)

While we sometimes encounter man-made radiation, natural background radiation pervades our lives. Radioactive radon exists as a gas and can contribute to lung cancer, especially among miners. Roughly 1 in 15 American homes is affected by radon contamination.

Nuclear reaction: A process by which an isotope of one element is transformed into an isotope of another element

Nucleon: A nuclear particle, such as a proton or neutron

Positron: A positively charged electron

Activity: The number of radioactive nuclei that disintegrate per unit of time

Half-life: The period required for one-half of a sample of a radioactive substance to undergo radioactive decay

Uranium series: The series of steps in the naturally occurring decay of uranium-238 to lead-206

Artificial transmutation: Experimental conversion of one element into another

Transuranium element: An artificial element with an atomic number greater than 92

Fission: The splitting of a heavy nucleus into smaller nuclei

Fusion: The combination of small nuclei into a heavier nucleus

Chain reaction: A process by which neutrons from one fission reaction can cause multiple fission reactions in nearby nuclei

Critical mass: The mass of fissionable material able to sustain a chain reaction

Binding energy: The force holding neutrons and protons together in a nucleus

Plasma: A high-temperature state of matter consisting of unbound nuclei and electrons

13-8 Useful Applications of Radioactivity (pp. 250–252)

Foods may be irradiated to retard the growth of organisms such as bacteria, molds, and yeasts. Carbon-14, a radioactive isotope that emits a beta particle and has a half-life of 5730 years, can be used to date materials up to 40,000 years old. Radiation is used in medical imaging both to diagnose and as therapy.

13-9 Energy from Nuclear Reactions (pp. 252–255)

A vast amount of energy is released both when heavy atomic nuclei are split and when light atomic nuclei are joined. Fission, which occurs when a neutron enters a heavy nucleus, has been utilized in both power plants and weapons of mass destruction. When a critical mass of fissionable material is brought together, an explosive chain reaction occurs.

13-10 Useful Nuclear Energy (pp. 256–261)

If the number of reacting thermal neutrons is controlled, the fission process can be maintained without leading to a chain reaction. Atomic reactors can thus be used to produce energy, but several vexing problems such as safety and waste remain. Controlled fusion reactions remain an ambitious goal.

Concept Match

1. Thermal neutrons
2. Most penetrating nuclear radiation
3. Least penetrating nuclear radiation
4. Discovered radioactivity
5. Fissionable atom
6. Co-discovered nuclear fission
7. Same as alpha particle
8. Reactor core meltdown
9. Stable isotope
10. Dangerous to the lungs
11. Source of H that may someday be used as a nuclear fuel
12. Positive electron
13. Useful in radio imaging
14. Nuclear disintegrations decrease by one-half
15. Loss from the nucleus causes an increase in atomic number
16. Dangerous by-product from controlled nuclear fission

a. Half-life
b. Plutonium-239
c. He nucleus
d. Lead-206
e. Radon-222
f. The oceans
g. Positron
h. Gamma ray
i. Becquerel
j. Technetium-99m
k. Alpha particle
l. Slow moving
m. Beta particle
n. Uranium-235
o. L. Meitner
p. Chernobyl

Answers: 1. l, 2. h, 3. k, 4. i, 5. n, 6. o, 7. c, 8. p, 9. d, 10. e, 11. f, 12. g, 13. j, 14. a, 15. m, 16. b

Section Summaries

14-1 Organic Chemicals (pp. 265–267)

Many of the organic chemicals used in the chemical industry are obtained from fossil fuels. The development of organic chemistry has led to cheap methods for the synthesis of both naturally occurring and new substances.

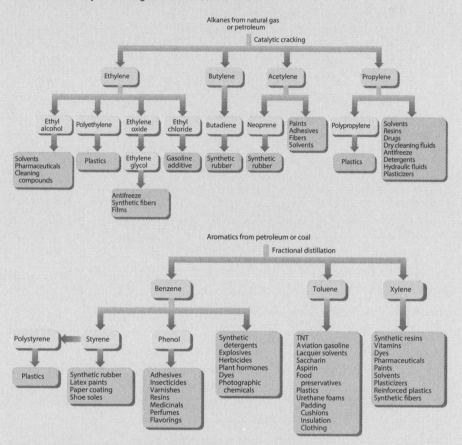

14-2 Alcohols and Their Oxidation Products (pp. 267–271)

Alcohols contain one or more hydroxyl functional groups bonded to carbon atoms and are a major class of organic compounds. Alcohols are classified according to the number of carbon atoms bonded directly to the —C—OH carbon as primary (one other C atom), secondary (two other C atoms), or tertiary (three other C atoms). Oxidation of alcohols may yield aldehydes, ketones, or carboxylic acids depending on the alcohol used and the extent of the oxidation.

14-3 Carboxylic Acids and Esters (pp. 271–273)

Carboxylic acids contain the —COOH functional group and are prepared by the oxidation of alcohols and aldehydes, reactions that occur quite easily. They are weak acids, are polar, and readily form hydrogen bonds. Formic and acetic acids are common carboxylic acids. When they react with alcohols in the presence of strong acids, esters are produced. The —OH of the carboxylic acid is replaced by the OR group from the

Key Terms

Organic chemistry: The chemistry of carbon compounds

Oxidation: For organic compounds, a process in which the number of bonds between carbon and oxygen is increased

Aldehyde: An organic compound containing a — C=O functional group | H

Ketone: An organic compound containing a carbonyl (C=O) functional group between two carbon atoms

Carboxylic acid: An organic compound containing a — C=O functional group | OH

Denatured alcohol: Ethanol with small added amounts of a toxic substance that cannot be removed easily by chemical or physical means

Ester: An organic compound containing a —COOR functional group

Polymer: A large molecule composed of repeating units derived from smaller molecules

Plastic: A polymer that can be molded into a variety of shapes

Monomers: The small molecules that combine to form polymers

Addition polymer: A polymer made from monomers joined directly with one another

Condensation polymer: A polymer made from monomers joined by a small molecule such as water

Natural rubber: Poly-*cis*-isoprene from the *Hevea brasiliensis* tree

Vulcanized rubber: Rubber with short chains of sulfur atoms that bond together the polymer of natural rubber

Key Terms

Condensation reaction: A reaction between two molecules in which a larger molecule is formed and a smaller molecule is eliminated

Primary amine: An organic compound containing an —NH_2 functional group

Amide: An organic compound containing a —$CONH_2$ or —$CONHR$ functional group

Reinforced plastics (composites): Plastics with fibers of another substance imbedded in a polymer matrix

Section Summaries

alcohol. In naming esters, the alkyl group from the alcohol is named first, followed by the name of the acid changed to end in –*ate*.

14-4 Synthetic Organic Polymers (pp. 273–282)

Synthetic organic polymers like plastic are ubiquitous in our daily lives—clothes, food packages, appliances, and cars all contain polymers. Approximately 80% of the organic chemical industry is devoted to the production of synthetic polymers, which are made by chemically joining together many monomers into one macromolecule. Polyethylene, which forms HDPE, is the world's most widely used polymer. Polymers of ethylene derivatives are used to make Styrofoam (polystyrene), bottles and fabrics (polypropylene), and pipes [poly(vinyl chloride)]. Latex and vulcanized rubber are utilized in a number of technologies, as are condensation polymers such as polyesters and polyamides (nylons).

14-5 New Polymer Materials (pp. 282–283)

Most plastics used today are composites. Reinforced plastics composed of polymer matrices have low densities and are stronger than steel on a weight basis. While any polymer can be used for the matrix material, glass fiber currently accounts for more than 90% of the fibrous material used in reinforced plastics because it is inexpensive, high in strength, and a good insulator. Graphite-polymer composites are used in a number of sporting goods and military aircrafts for their flexibility and strength.

14-6 Recycling Plastics (pp. 283–285)

Disposal of plastics has been the subject of considerable debate as municipalities face increasing problems in locating sufficient landfill space. Recycling programs for plastics developed much more slowly than those for metals, as plastics must be sorted by many different types—a costly process. Codes are stamped on plastic containers to help consumers identify their recyclable plastics. PET and HDPE are commonly recycled, while other plastics are not.

Concept Match

1. Aldehyde
2. Ketone
3. Carboxylic acid
4. Ester
5. Monomer
6. Present primary source of organic chemicals used as raw materials
7. Poly-*cis*-isoprene
8. Cross-linking via reaction with sulfur
9. Polyester
10. Nylon
11. Alcohol

a. Contains ROH
b. Polyamide
c. Vulcanization

d. Contains $R-\overset{\overset{\text{O}}{\|}}{C}-R'$

e. Contains $R-\overset{\overset{\text{O}}{\|}}{C}-H$

f. Contains $R-\overset{\overset{\text{O}}{\|}}{C}-OH$

g. Contains $R-\overset{\overset{\text{O}}{\|}}{C}-O-R'$

h. Petroleum
i. Formed from a di-alcohol and a diacid
j. Natural rubber
k. Building block for polymer

Answers: 1. e, 2. d, 3. f, 4. g, 5. k, 6. h, 7. j, 8. c, 9. i, 10. b, 11. a

Section Summaries

15-1 Handedness and Optical Isomerism (pp. 291–295)

A chiral molecule and its nonsuperimposable mirror image are called enantiomers. Both have the same boiling point and density, but they rotate a beam of plane-polarized light in opposite directions. Some enantiomers are found in nature, and some exist only in racemic mixtures. Thalidomide is a grave example of stark differences between enantiomers.

15-2 Carbohydrates (pp. 296–299)

Carbohydrates are divided into groups depending on how many monomers are combined by condensation polymerization. D-glucose, a monosaccharide, is found in fruit, blood, and living cells. Disaccharides such as sucrose, maltose, lactose, and artificial sweeteners are found in many types of food. Cellulose, a polysaccharide, is the most abundant organic compound on Earth.

15-3 Lipids (pp. 299–304)

Lipids vary widely in structure but never are polymers. Fats and oils are solid and liquid triglycerides, respectively. Steroids are four-ring lipids found in all plants and animals—sex hormones are steroids. Waxes are esters formed from long-chain fatty acids and long-chain alcohols.

15-4 Soaps, Detergents, and Shampoos (pp. 304–307)

Soap and synthetic detergent molecules consist of a long oil-soluble group and a water-soluble group. Shampoos are generally more complex, containing aniotic detergents. As with other detergents, dirt and oil is trapped in micelles and washed away.

15-5 Creams and Lotions (pp. 307–308)

Moisturizers lubricate dry skin, softening and smoothing it. Creams and lotions are emulsions, mixing emollients with water and other ingredients. Whether the oil or the water is suspended in the other, emulsions are examples of colloids.

15-6 Amino Acids (pp. 308–309)

Proteins are condensation polymers of amino acids. There are only 20 amino acids found in nature, and all but glycine follow a general formula. Glycene is achiral and does not have any enantiomers. Essential amino acids must be ingested from food.

15-7 Peptides and Proteins (pp. 309–311)

Dipeptides form when one molecule of water is eliminated between the carboxylic acid of one amino acid and the amine group of another. Simple and conjugated proteins are polypeptides. Proteins vary greatly in size, and enzymes often have small nonprotein parts called cofactors.

Key Terms

Biochemistry: The study of chemistry in living systems, including plants and animals

Chiral: Cannot be superimposed on its mirror image

Achiral: Can be superimposed on its mirror image

Enantiomers (optical isomers): A chiral molecule and its nonsuperimposable mirror-image molecule

Stereogenic center: A carbon atom with four different attached atoms or groups of atoms that gives rise to enantiomers

Configuration: A description of the specific arrangement of atoms in a chemical compound

Racemic mixture: A mixture of equal amounts of enantiomers

Teratogen: A chemical or factor that causes malformation of an embryo

Stereoisomers: Molecules with the same molecular formula but different three-dimensional structures

Carbohydrate: A biomolecule composed of simple sugars

Lipid: A class of biomolecules not soluble in water but soluble in organic solvents

Triglyceride: A triester of glycerol and fatty acids

Steroid: A lipid with a four-ring structure

Hormone: A chemical messenger secreted in the bloodstream by endocrine glands

Hydrophobic: Water fearing; does not dissolve in water

Hydrophilic: Water loving; dissolves in water

Emulsion: A stable mixture of water and an oily component

Emulsifying agent: A compound that has a water-soluble part as well as an oil-soluble part and that stabilizes an emulsion

Colloid: A particle larger than most molecules or ions dispersed in a solvent-like medium

Amino acid: A biomolecule containing an alpha-amino group and a carboxyl group

Essential amino acid: An amino acid that is not synthesized by the body and, therefore, must be obtained from the diet

Peptide bond: Amide group that connects two amino acids

Polypeptide: A condensation polymer of amino acids

Protein: A class of polypeptides that may also include non–amino acid parts

Primary protein structure: The amino acid sequence in a protein

Secondary protein structure: The shape of the backbone structure of a protein

Tertiary protein structure: The shape of an entire folded protein chain

Quaternary protein structure: The structure of a protein with more than one polypeptide

Denaturation: Destruction of proper functioning of a protein by changing its structure

Enzyme: A biomolecule that functions as a catalyst for biochemical reactions

Active site: The region of an enzyme where the catalytic chemistry occurs

Nucleic acids: Polymers of nucleotides that contain deoxyribose or ribose, nitrogen bases, and phosphate groups

Deoxyribonucleic acid (DNA): Nucleic acid that functions as a genetic information storage molecule

Ribonucleic acid (RNA): Nucleic acid that transmits genetic information and directs protein synthesis

Nucleotide: A biomolecule with a five-carbon sugar bonded to a nucleic acid base and a phosphate group

Gene: A segment of DNA that directs synthesis of proteins

Genome: The total sequence of base pairs in the DNA of a plant or animal

Replication: The process by which DNA is copied when a cell divides

15-8 Protein Structure and Function (pp. 311–316)

The order of the amino acid residues in a peptide or protein molecule is essential to its function. Primary, secondary, tertiary, and quaternary structures govern the shape and function of proteins. The structure of an enzyme's active site allows it to function as a catalyst.

15-9 Hair Protein and Permanent Waves (pp. 316–317)

Hair is composed principally of the protein keratin. Moisture affects hydrogen bonds between protein chains, and when their number and arrangement changes, hair changes. In "permanent waving," disulfide bonds between proteins are broken and reformed in a new shape.

15-10 Energy and Biochemical Systems (pp. 317–318)

The energy to sustain all but a few forms of life comes from the Sun. In the complex process of photosynthesis, carbon dioxide is reduced to make sugar, and water is oxidized to oxygen.

15-11 Nucleic Acids (pp. 318–323)

Genetic information is encoded in molecules called nucleic acids. Nucleotides polymerize to make DNA and RNA. DNA, the template for mRNA transcription and translation, is arranged in a double helix of polynucleotide pairs.

Concept Match

1. Energy "cash" in the living cell
2. Primary protein structure
3. Product of ATP hydrolysis
4. Saponification
5. D-Glucose
6. Enzymes
7. Carbohydrate stored in animals
8. Keratin
9. Polypeptides
10. DNA
11. Fibrous protein
12. Cysteine

a. ADP + energy
b. Amino acid common in hair
c. Structure determined by DNA and RNA
d. ATP
e. Hair protein
f. Proteins
g. Main sugar present in the blood
h. Polynucleotide
i. Biochemical catalysts
j. Glycogen
k. Collagen
l. Hydrolysis reaction in which a fat or oil produces glycerol and one or more fatty acids

Answers: 1. d, 2. c, 3. a, 4. l, 5. g, 6. i, 7. j, 8. e, 9. f, 10. h, 11. k, 12. b

Transcription: The process by which the information in DNA is read and used to synthesize RNA

Translation: The process for sequential ordering of amino acids that is directed by RNA during protein synthesis

Codon: A three-base sequence carried by mRNA that codes a specific amino acid in protein synthesis

Section Summaries

16-1 Digestion: It's Just Chemical Decomposition (pp. 329–330)

The chemical reactions of digestion begin when amylase, an enzyme in saliva, begins to break food down. Digestion continues in the stomach and is completed in the small intestine. Once digestion is finished, new molecules are used, recycled, or stored.

16-2 Sugar and Polysaccharides: Digestible and Indigestible (pp. 330–331)

Digestible carbohydrates include simple sugars, disaccharides, and some polysaccharides. Indigestible carbohydrates include cellulose and glucose. Indigestible polysaccharides are collectively referred to as dietary fiber, which can prevent constipation and decrease blood cholesterol.

16-3 Lipids: Mostly Fats and Oils (pp. 331–332)

We get fats from meat and fish, vegetables, vegetable oils, and dairy products. Fats can cause plaque to form on the inner walls of the arteries, instigating heart disease. LDLs are bad, transporting cholesterol throughout the arteries, while HDLs are good, transporting cholesterol to the liver.

16-4 Proteins in the Diet (pp. 332–333)

Meat, fish, eggs, cheese, and beans are high-protein foods. Dietary protein provides amino acids for new protein synthesis and nitrogen for synthesis of nitrogen-containing biomolecules. Protein–energy malnutrition is most common in children in developing nations.

16-5 Vitamins in the Diet (pp. 333–335)

Vitamins are organic compounds essential to our health in small amounts. The body does not synthesize vitamins but needs them to function as coenzymes. There are two kinds of vitamins—fat soluble (vitamins A, D, E, and K) and water soluble (vitamins B and C).

16-6 Minerals in the Diet (pp. 335–337)

Minerals are elements other than carbon, hydrogen, nitrogen, and oxygen that are needed for good health. Seven macronutrients make up about 4% of body weight: calcium, phosphorus, magnesium, sodium, potassium, chlorine, and sulfur.

16-7 Food Additives (pp. 337–341)

Food additives have little nutritive value. The FDA's GRAS List notes hundreds of generally safe additives. Additives include preservatives, antioxidants, sequestrants, flavors, flavor enhancers, colors, pH controllers, anticaking agents, stabilizers, and thickeners.

Key Terms

Nutrition: The science of diet and health

Nutrient: Chemical substance in food that provides energy and raw materials required by chemical reactions

Digestion: The breakdown of food into substances small enough to be absorbed from the digestive tract

Dietary fiber: Food polysaccharides that are not broken down by digestion

Lipoprotein: An assemblage of lipids, cholesterol, and water-soluble proteins

Low-density lipoprotein (LDL): A lipoprotein that transports cholesterol from the liver to tissues throughout the body

High-density lipoprotein (HDL): A lipoprotein that transports cholesterol from body tissues to the liver

Vitamin: Organic compound essential to health that must be supplied in small amounts by the diet

Coenzyme: A nonprotein molecule that makes enzyme function possible

Mineral: As a dietary nutrient, an element other than carbon, hydrogen, nitrogen, and oxygen that is needed for good health

Osteoporosis: Reduction of bone mass associated with calcium loss

Hypertonic solution: A solution with a concentration of solute higher than that found in its surroundings

Sequestrant: A chemical that ties up metal atoms or ions by forming bonds with them; used as a food additive

Basal metabolic rate (BMR): The minimum energy required to sustain an awake but resting body

16-8 Energy: Use It or Store It (pp. 341–343)

Even when resting, we are using a certain amount of energy (our basal metabolic rate). The human body is subject to the law of conservation of energy, so dieting requires expending more energy than is consumed. Glucose, which is burned throughout our bodies, is the major fuel molecule.

16-9 Our Daily Diet (pp. 343–344)

Nutrition Facts labels on food packages tell us much about the caloric, fat, and macronutrient content of our food, as well as % Daily Values. The macronutrients, vitamins, and minerals included on these labels are those deemed to be of greatest public health concern.

16-10 Some Daily Diet Arithmetic (pp. 345–346)

A good way to keep track of food Calories is to monitor your daily Caloric intake (food) and output (exercise). You can keep track of food Calories from fat by measuring the total grams of fat in your diet every day.

Concept Match

1. Chemical decomposition
2. Glucose
3. Calorie value per gram of fat
4. Calorie value per gram of protein
5. Leads to atherosclerosis
6. Result of deficient dietary iodine
7. LDL
8. HDL
9. Butylated hydroxyanisole
10. Nutrient and food coloring
11. Sodium benzoate
12. Flavor enhancer
13. Loss of calcium from bone
14. Result of deficient dietary iron
15. Physical activity level
16. Sequesters metals

a. Dietary saturated fat
b. Goiter
c. Anemia
d. Digestion
e. Sodium EDTA
f. Monosodium glutamate
g. 4 Cal
h. Beta-carotene
i. Major fuel molecule
j. Affects daily caloric need
k. Good cholesterol
l. Osteoporosis
m. Food preservative
n. 9 Cal
o. Bad cholesterol
p. Antioxidant food additive

Answers: 1. d, 2. i, 3. n, 4. g, 5. a, 6. b, 7. o, 8. k, 9. p, 10. h, 11. m, 12. f, 13. l, 14. c, 15. j, 16. e

Section Summaries

17-1 Medicines, Prescription Drugs, and Diseases: The Top Tens (pp. 350–353)

Americans spend billions of dollars on prescription and over-the-counter drugs. Many deadly infectious diseases have been neutralized since 1900. Meanwhile, HIV/AIDS has emerged and flourished in a devastating way.

17-2 Drugs for Infectious Diseases (pp. 353–355)

Modern chemotherapy, the treatment of diseases with chemical agents, was first discovered by Paul Ehrlich in 1904. Natural and synthetic antibiotics such as penicillin, the cephalosporins, and rifampin inhibit the growth of microorganisms.

17-3 AIDS: A Viral Disease (pp. 355–357)

HIV is a T-cell–destroying, AIDS-causing retrovirus consisting of an outer double lipid layer, an enzyme called reverse transcriptase, and RNA. A number of new and established drug cocktails are used to combat HIV, prolonging patients' lives indefinitely.

17-4 Steroid Hormones (pp. 357–358)

Hormones, which regulate biological processes by interacting with receptors, are produced by glands and secreted directly into the blood. Oral contraceptives and other synthetic analogs of female hormones can be used to prevent unwanted pregnancies.

17-5 Neurotransmitters (pp. 359–360)

Neurotransmitters carry nerve impulses from one nerve to another. The body has a variety of neurotransmitters. Norepinephrine controls coordination, emotion, and alertness; serotonin affects pain perception and mood; dopamine integrates fine muscular movement; and epinephrine produces the fight-or-flight response.

17-6 The Dose Makes the Poison (pp. 360–361)

A chemical substance may be lifesaving or deadly, depending on its dose. Because most substances can be poisonous in sufficiently large doses, toxicity is measured in laboratory animals before a substance is administered to humans. An LD_{50} is the dose found to be lethal to 50% of test animals under controlled conditions.

17-7 Painkillers of All Kinds (pp. 361–363)

Analgesics such as cocaine, morphine, and aspirin relieve pain. A number of pain-relieving alkaloids are derived from the unripened seed pods of opium poppies. Over-the-counter analgesics such as aspirin, acetaminophen, and ibuprofen can relieve mild aches and pains.

Key Terms

Over-the-counter drug: A medication that can be purchased without a prescription

Prescription drug: A medication that can be purchased only at the direction of a physician

Trade name: The manufacturer's name for a drug

Generic name: The generally accepted chemical name of a drug

Infectious disease: A disease caused by microorganisms

Antibiotic: A substance that inhibits the growth of microorganisms

Pathogenic: Capable of causing disease

Receptor: A molecule or portion of a molecule that interacts with another molecule to cause a change in biochemical activity

Analgesic: A drug that relieves pain

Antipyretic: A drug that reduces fever

Anti-inflammatory: A drug that reduces inflammation

Sedation: Induction of a relaxed state, usually by a medication

Hallucinogen: A drug that causes perceptions with no basis in the real world

Cardiovascular disease: Disease of the heart or blood vessels

Diuretic: A drug that causes excretion of fluid from the body

Carcinogen: An agent that causes cancer

Homeopathy: A system of medical treatment based on the use of minute quantities of natural substances that, in a healthy person, would produce symptoms of the same disease

Key Terms

Pharmacology: The branch of medicine concerned with the uses, effects, and modes of action of drugs

Section Summaries

17-8 Mood-Altering Drugs: Legal and Illegal (pp. 363–366)

Depressants such as barbiturates and tranquilizers dampen the central nervous system, stimulants such as methamphetamine excite the central nervous system, and hallucinogens such as mescaline activate serotonin receptors.

17-9 Colds, Allergies, and Other "Over-the-Counter" Conditions (pp. 366–368)

Seventy-five percent of illnesses are treated with over-the-counter products. Allergic symptoms occur when special nasal cells release histamines. The common cold is caused by a virus, and its symptoms can be countered with decongestants, antitussives, and expectorants. It is important to treat maladies with proper medications and dosages.

17-10 Preventive Maintenance: Sunscreens and Toothpaste (pp. 368–370)

Sunscreens of various SPFs can be applied to the skin to block cancer-causing ultraviolet rays. There is no safe tan—any change in the skin's natural color is a sign of damage. Brushing one's teeth with toothpaste and flossing regularly prevents plaque buildup, gingivitis, and cavities.

17-11 Heart Disease (pp. 370–371)

Cardiovascular disease, which results from any condition that decreases the flow of blood, and thus oxygen, to the heart, is the number one killer of Americans. The most common cause of heart disease is atherosclerosis. High blood pressure also contributes to heart disease. A number of drugs address each of these problems.

17-12 Cancer, Carcinogens, and Anticancer Drugs (pp. 371–374)

A cancer begins when a cell in the body multiplies without restraint, producing descendants that invade other tissue. Carcinogenesis is often a two-step process of initiation and promotion. Cancer is treated by surgery, irradiation, and chemotherapy.

17-13 Homeopathic Medicine (pp. 374–376)

Homeopathy is a system of alternative medicine developed by Samuel Hahnemann. To treat maladies, substances that cause similar symptoms are administered in very small doses. Homeopathy has no scientific basis.

Section Summaries

18-1 The Whole Earth (pp. 378–380)

Earth's hydrosphere supplies the saltwater and freshwater necessary to sustain life. The soluble salts in the oceans are a commercial source of magnesium, bromine, and sodium chloride (table salt). Three major types of rocks comprise Earth's crust: igneous, sedimentary, and metamorphic. The most abundant substances in rocks are silicates, and the more than 2000 kinds of known minerals fall into only a few major classes. The composition of the crust is not uniform—minerals have been redistributed throughout Earth's existence.

18-2 Chemicals from the Hydrosphere (pp. 381–383)

Sodium chloride is the major mineral in seawater, but other dissolved minerals are constantly being deposited into the oceans from rivers, undersea volcanoes, and thermal vents. After isolation, salt is purified. Much of it is destined for the chloralkali industry, which produces chlorine gas and sodium hydroxide. Magnesium, the lightest structural metal in common use, is a common component in alloys designed for light weight and great strength. Unlike salt, its extraction from seawater begins with the precipitation of insoluble magnesium hydroxide.

18-3 Metals and Their Ores (pp. 383–386)

Less reactive metals are often found as free agents, while the most reactive metals are present in nature only as soluble salts in the ocean and mineral springs. Iron is the fourth most abundant element in Earth's crust and the second most abundant metal. Our economy depends on iron and its alloys, particularly steel. To convert pig iron into carbon steel, excess carbon is burned out with oxygen, which is blown into the molten iron through a refractory tube. The properties of steel can be adjusted by the addition of alloy metals and the temperature and rate of its cooling. Copper both occurs in the elemental state and is extracted from copper sulfide ores through flotation and roasting.

18-4 Properties of Metals, Semiconductors, and Superconductors (pp. 386–390)

Nearly all metals are solids. Other shared properties include high electrical conductivity, high thermal conductivity, ductility, malleability, luster, and insolubility in water and other common solvents. Semiconducting elements sit between the excellent electrical conductivity of metals and the nonconductivity of nonmetals. The transistor, which revolutionized electronics, is built on the concept of semiconductivity. At certain low temperatures, metals reach superconducting transition temperatures, transforming them into unresisting superconductors. Solar cells convert the Sun's energy into usable electricity.

18-5 From Rocks to Glass, Ceramics, and Cement (pp. 390–394)

Earth's crust is largely held together by chemical bonds between silicon and oxygen, in either pure silica or silicate minerals. The linking of silicate chains forms clay, which is used in a number of technologies. When silica is melted, it forms the amorphous solid glass upon cooling. Common window glass is made by melting a mixture of silica with

Key Terms

Hydrosphere: All freshwater and saltwater that is part of planet Earth

Mineral: A naturally occurring, solid inorganic compound

Alloy: A mixture of two or more metals

Precipitation: Formation of a solid product during a chemical reaction between reactants in a solution

Slag: Mixture of nonmetallic waste products separated from a metal during its refining

Cast iron: Molded pig iron or other carbon–iron alloy

Steel: Malleable iron-based alloy with a relatively low percentage of carbon

Gangue: Unwanted substances mixed with a mineral during production of metal from ore

Roasting: In metallurgy, heating an ore in air, usually to produce an oxide

Semiconductor: A material with electrical conductivity intermediate between those of metals and insulators

Doping: In semiconduction, the addition of atoms of an element with extra or fewer valence electrons than the element of the semiconductor material

Transistor: A semiconductor device that controls electron flow in circuits

Superconductor: A substance that offers no resistance to electrical flow

Silica: Naturally occurring silicon dioxide

Silicate: A mineral composed of metal cations and silicon–oxygen anions

Glass: A hard, noncrystalline, transparent substance made by melting silicates with other substances

Key Terms

Annealing: Heating and cooling a metal, glass, or alloy in a manner that makes it less brittle

Ceramics: Materials and products made by firing mixtures of nonmetallic minerals

Cement: A material that bonds mineral fragments into a solid mass; any material that bonds other materials together

Section Summaries

sodium and calcium carbonates. Ceramic materials are commonly made of clay, sand, and feldspar. Though they are brittle, ceramics are utilized in many applications for their light weight and heat resistance. Cements bond mineral fragments into a single mass, which is used to make concrete.

Concept Match

1. Glass
2. SiO_4
3. Steel
4. Sodium hydroxide
5. Product of the blast furnace
6. $Fe^{3+} \longrightarrow Fe$
7. Valuable mixture of minerals
8. Ceramics and glass
9. Superconductor
10. Seashells
11. Nonmetal from seawater
12. Metal from the ocean
13. *p*-type semiconductor
14. Portland cement
15. $Fe \longrightarrow Fe^{3+}$

a. Reduction
b. Chlorine
c. Magnesium
d. Noncrystalline
e. Ore
f. In mortar and concrete
g. Pig iron
h. Product of chloralkali industry
i. No resistance to electrical flow
j. Building block of Earth's crust
k. B-enriched Si
l. Oxidation
m. Source of calcium carbonate
n. Alloy
o. Made of silicates

Answers: 1. d, 2. j, 3. n, 4. h, 5. g, 6. a, 7. e, 8. o, 9. i, 10. m, 11. b, 12. c, 13. k, 14. f, 15. l

Section Summaries

19-1 World Population Growth (pp. 396–399)

The world's current population is more than 3.5 times its size at the beginning of the 20th century and is more than double its size in 1960. However, largely attributable to a decline in women's fertility, growth is slowing. As the population increases, the carrying capacity of Earth comes into question. Improving food-use efficiency in the United States and around the world is the key to maintaining high quality of life with upward of 10 billion inhabitants.

19-2 What Is Soil? (pp. 399–401)

Soil is a mixture of mineral particles, organic matter, water, and air. Weathering processes in nature over thousands of years break rocks into the small mineral particles found in soil. Humus, which provides nutrients to plants, is important in maintaining soil's structure. Horizons within soil include the richly organic topsoil and the thick, mostly inorganic subsoil. Water can be absorbed onto the surface of or into the structure of a soil's particulate material. It can also occupy the pores ordinarily filled with air. It percolates away from the soil, but in doing so may leach important chemicals.

19-3 Nutrients (pp. 401–403)

At least 18 known elemental nutrients are required for normal green plant growth. The nonmineral nutrients (carbon, hydrogen, and oxygen) are absorbed from air and water. Mineral nutrients are absorbed through a plant's root system as solutes in water. Primary nutrients include nitrogen, phosphorus, and potassium. Secondary nutrients include calcium and magnesium, which are bound tightly by soil. Micronutrients such as iron are required only in very small amounts. Unless extensive cropping or other factors deplete the soil of these nutrients, sufficient quantities are usually available.

19-4 Fertilizers Supplement Natural Soils (pp. 403–407)

Fertilizers that contain only one nutrient, such as potassium chloride, are called straight fertilizers. Those containing a mixture of the three primary nutrients (nitrogen, phosphorus, and potassium) are called complete, or mixed, fertilizers. Quick-release fertilizers are water soluble as opposed to slow-release fertilizers, which require days or weeks to dissolve. Ammonia, the primary nitrogen fertilizer, is formed by the Haber process. Phosphorus is readily available in the form of phosphate rock, which can be transformed into the needed fertilizers.

19-5 Protecting Food Crops (pp. 407–411)

The natural enemies of crops include more than 80,000 viruses, bacteria, fungi, and algae; 30,000 weeds; 3000 nematodes; and about 10,000 species of plant-eating insects. About one-third of the world's food crops are lost to pests each year. Eighteen common classes of pesticides are fortified with more than 2600 active ingredients to battle pests. After DDT was banned, safer insecticides such as pirimicarb have been used widely to combat aphids and other insects. Selective and nonselective herbicides are used to kill plants, and since the banning of methyl bromide, a variety of organic compounds have been used as fungicides.

Key Terms

Humus: Decomposed organic matter in soil

Horizons: Layers within the soil

Topsoil: The layer of soil exposed to the air

Subsoil: The layer of soil just beneath the topsoil

Loam: Soil consisting of a friable mixture of varying proportions of clay, sand, and organic matter

Percolation: Movement of water through subsoil and rock formations

Leaching effect: The dissolving of soil material as water moves through the soil

Nonmineral nutrients: Carbon, hydrogen, and oxygen in soil

Mineral nutrients: Plant nutrients absorbed through roots as solutes in water

Primary nutrients: Major elements required by plants

Secondary nutrients: Elements required by plants in moderate amounts

Micronutrients: Elements required by plants in relatively small amounts

Nitrogenase: Bacterial enzyme that catalyzes nitrogen fixation

Chlorosis: Yellowing of plant leaves due to nutrient deficiency

Straight fertilizer: A fertilizer that contains only one nutrient

Complete (mixed) fertilizer: A fertilizer that contains the three primary nutrients (nitrogen, phosphorus, and potassium)

Quick-release fertilizer: A fertilizer that is water soluble

Slow-release fertilizer: A combination of plant nutrients that are slow to dissolve in water

Key Terms

Haber process: Industrial process for the reaction of nitrogen and hydrogen gases to form ammonia

Pesticide: A chemical used to kill pests

Insecticide: A pesticide that kills insects

Herbicide: A pesticide that kills weeds

Fungicide: A pesticide that kills fungi

Selective herbicide: An herbicide that kills only a particular group of plants

Nonselective herbicide: An herbicide that kills all plant life

Organic farming: Farming without synthetic chemical fertilizers and pesticides

Sustainable agriculture: An approach to agriculture whose broad goal is to provide food for the world's population while minimizing environmental impact

Integrated pest management (IPM): Limiting the use of pesticides by relying on a combination of disease-resistant crop varieties, natural predators or parasites, and selected use of pesticides

Transgenic crop: A crop that has been altered by the technique of genetic engineering

Section Summaries

19-6 Sustainable Agriculture (pp. 411–413)

Extensive use of synthetic fertilizers and pesticides, and development of higher yielding varieties of crops, have made it possible to increase crop yields dramatically. However, poor farming practices have resulted in soil erosion, loss of soil fertility, and pesticide-resistant pests. In response to these problems, organic farming and sustainable agriculture have emerged as popular alternatives to established farming practices. Integrated pest management and natural insecticides have also emerged as environmentally friendly and effective methods of pest control.

19-7 Agricultural Genetic Engineering (pp. 413–415)

The DNA of plants can be altered to produce faster growing, longer lasting, and more fruitful crops. Genetically engineered produce is common today—87 transgenic crop plants have been approved for commercial production. The most widely planted genetically modified food crops are soybeans, corn, rapeseed, and cotton. Genetically engineered crops remain controversial, as the potential risks and benefits to humans and the environment are not fully understood.

Significant Figures and Scientific Notation

Significant Figures

1. All nonzero digits are significant.

$$\overset{1\,2}{23}$$
Two significant figures

$$\overset{1\,2\,3\,4\,5}{15{,}699}$$
Five significant figures

2. Zeroes to the left of the first nonzero digit are not significant.

$$0.\overset{1\,2}{23}$$
Two significant figures

$$0.000\overset{1\,2}{23}$$
Two significant figures

3. Zeroes between nonzero digits are significant.

$$\overset{1\,2\,3\,4}{7077}$$
Four significant figures

$$\overset{1\,2\,3}{50.2}$$
Three significant figures

4. Zeroes at the end of a number that includes a decimal point are significant.

   These zeroes have been added to represent the correct number of significant figures.

$$\overset{1\,2\,3\,4\,5}{50.020}$$
Five significant figures

$$\overset{1\,2\,3}{7.00}$$
Three significant figures

For a number with zeros at the end and *no* decimal point, it is impossible to know how many of the digits are significant.

Calculations with Significant Figures

To find the correct number of significant figures for the answer to a calculation requires three steps: (1) Do the arithmetic. (2) Decide how many significant figures the answer can have. (3) Round off the answer.

In addition and subtraction an answer should have no more decimal places than the number in the calculation with the fewest decimal places. Consider the following addition:

$$\begin{array}{ll} 13.673 & \text{Three decimal places} \\ \underline{4.0} & \text{One decimal place} \\ \,[17.673] & \text{Three decimal places (unrounded answer)} \end{array}$$

The unrounded answer has two more decimal places than 4.0 and must therefore be rounded off to include only one decimal place as in 4.0. The rules for rounding off the answer are simple. If the numbers to be dropped are 5 or greater than 5, then 1 is added to the final digit that is retained. If the numbers to be dropped are less than 5, they are dropped and the final retained digit is unchanged.

In the preceding addition, 17.673 must be rounded off to have one decimal place, which means dropping the 73 and increasing the 6 by 1 to give 17.7.

For multiplication and division, the answer is limited to the number of digits in the number with the fewest significant figures. When, for example, 75 is multiplied by 0.05, the answer can have only one significant figure (as in 0.05); when 0.084 is divided by 0.00298, the answer can have only two significant figures (as in 0.084).

$$75 \times 0.05 = [3.75] \qquad \text{Rounded-off answer: 4}$$

$$\frac{0.084}{0.00298} = [28.18791946] \qquad \text{Rounded-off answer: 28}$$

Scientific Notation

Scientific notation, also known as exponential notation, is a way of representing large and small numbers as the product of two terms. The first term, the coefficient, is a number between 1 and 10. The second term, the exponential term, is 10 raised to a power—the exponent. For example,

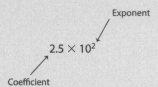

The exponent indicates the number of 10s by which the coefficient is multiplied to give the number represented in scientific notation:

$$2.5 \times 10^2 = 2.5 \times 10 \times 10 = 250$$

There are two great advantages to scientific notation. The first is that the very large and very small numbers often dealt with in the sciences are much less cumbersome in scientific notation. The second is that it removes the ambiguity in the number of significant figures in a number ending with zeroes. We can, for example, write

Two significant figures	2.5×10^2
Three significant figures	2.50×10^2
Four significant figures	2.500×10^2

All digits in the coefficient are always considered significant.

For numbers larger than 1, the exponent in scientific notation is a positive whole number, as illustrated above. For numbers less than 1, the exponent is a negative whole number that indicates the number of times the coefficient must be divided by 10 (or multiplied by 0.1) to give the number represented in scientific notation:

$$1.2 \times 10^{-2} = 1.2 \times \frac{1}{10} \times \frac{1}{10} = 0.012$$

There are two points to remember in converting a number into or out of scientific notation. (1) The value of the exponent is the number of places by which the decimal is shifted. (2) The coefficient should have only one digit before the decimal point. Look through the following examples and count the number of places the decimal points have been moved.

$10000 = 1 \times 10^4$	$12345 = 1.2345 \times 10^4$
$1000 = 1 \times 10^3$	$1234 = 1.234 \times 10^3$
$100 = 1 \times 10^2$	$123 = 1.23 \times 10^2$
$10 = 1 \times 10^1$	$12 = 1.2 \times 10^1$
$1 = 1 \times 10^0$	(Any number $\times\ 10^0 = $ the number itself.)
$1/10 = 1 \times 10^{-1}$	$0.12 = 1.2 \times 10^{-1}$
$1/100 = 1 \times 10^{-2}$	$0.012 = 1.2 \times 10^{-2}$
$1/1000 = 1 \times 10^{-3}$	$0.0012 = 1.2 \times 10^{-3}$
$1/10000 = 1 \times 10^{-4}$	$0.00012 = 1.2 \times 10^{-4}$

Calculation and Equation Aids

This card is designed to help you with conversion calculations and stoichiometric equations while working through the End-of-Chapter exercises.

1. Conversion between density, mass, and volume

Using this figure, you can obtain the following relations:

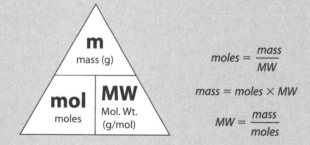

$$density = \frac{mass}{volume}$$

$$mass = density \times volume$$

$$volume = \frac{mass}{density}$$

2. Conversion between moles, grams, and molecular (or atomic) weight

Using this figure, you can obtain the following relations:

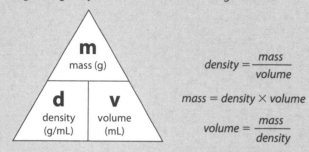

$$moles = \frac{mass}{MW}$$

$$mass = moles \times MW$$

$$MW = \frac{mass}{moles}$$

3. Determining the number of atoms or molecules in a sample of a pure substance

IMPORTANT: You must first determine the number of moles of the substance. See number 2 above for help in this.

$$moles \times \frac{6.022 \times 10^{23} \, atoms \, or \, molecules}{1 \, mole} = \# \, of \, atoms \, or \, molecules$$

4. Solution concentrations: Conversion between molarity, moles, and volume

Using this figure, you can obtain the following relations:

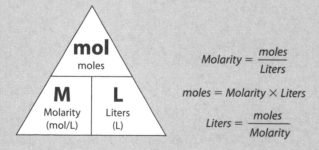

$$Molarity = \frac{moles}{Liters}$$

$$moles = Molarity \times Liters$$

$$Liters = \frac{moles}{Molarity}$$

5. Solution concentrations: Converting between percentage and parts per million or parts per billion

IMPORTANT: You need either the percentage or the concentration (ppm or ppb).

Parts per million: $\dfrac{percentage}{100} = \dfrac{ppm}{1,000,000}$

Parts per billion: $\dfrac{percentage}{100} = \dfrac{ppb}{1,000,000,000}$

6. Light: Frequency and wavelength of light

IMPORTANT: Sometimes it is necessary to first calculate the frequency. See number 7 below for help in this.

Using this figure, you can obtain the following relations:

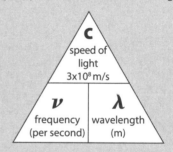

$$frequency = \frac{speed \, of \, light}{wavelength}$$

$$speed \, of \, light = frequency \times wavelength$$

$$wavelength = \frac{speed \, of \, light}{frequency}$$

7. Photons: Energy and frequency

Using this figure, you can obtain the following relations:

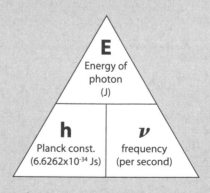

$$Energy \, of \, a \, photon = Planck \, Constant \times frequency$$

$$frequency = \frac{Energy}{Planck \, Constant}$$

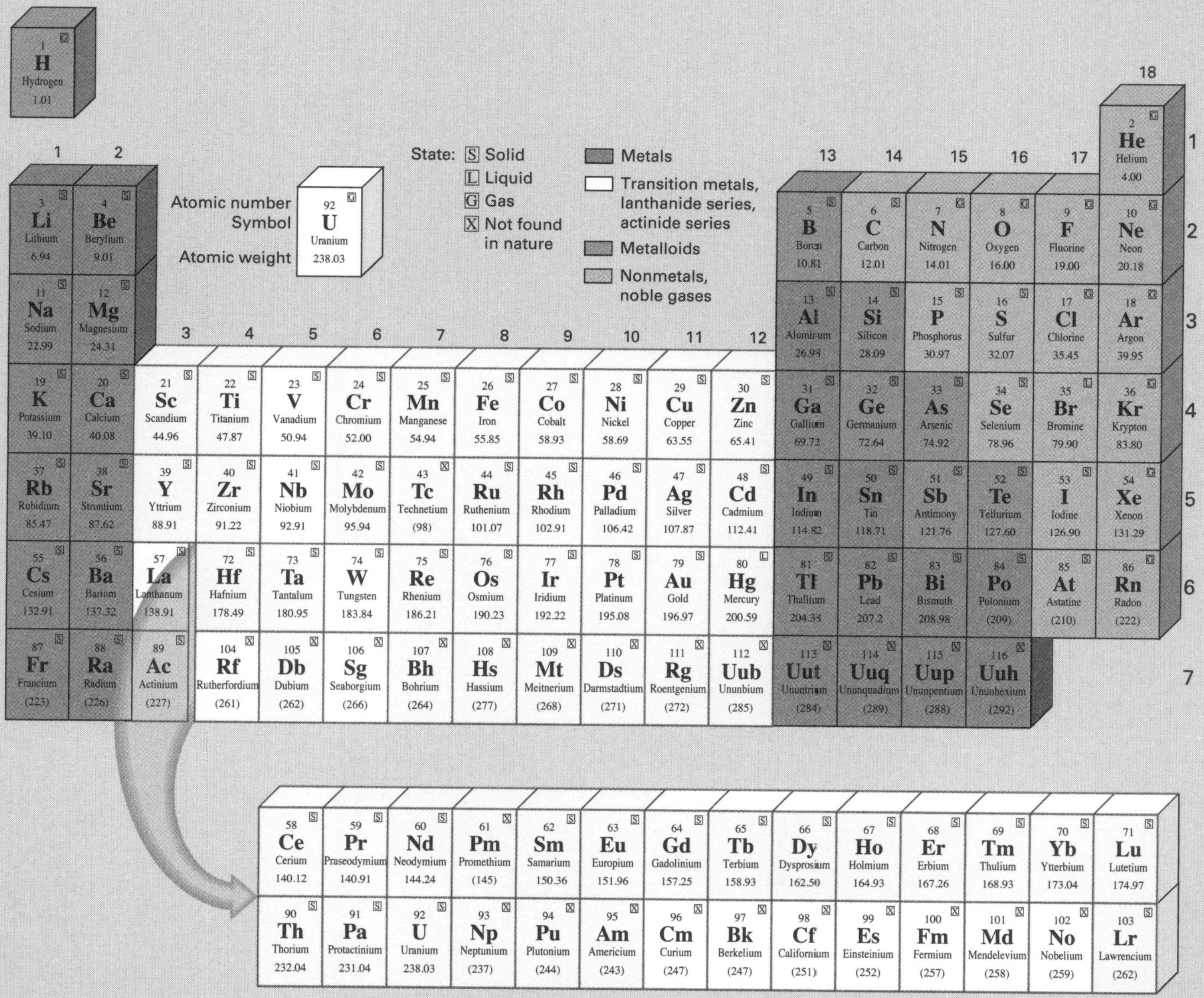
Atomic number
Symbol
Atomic weight
92 U Uranium 238.03
State: S Solid
L Liquid
G Gas
X Not found in nature
Metals
Transition metals, lanthanide series, actinide series
Metalloids
Nonmetals, noble gases
1 2 3 4 5 6 7 8 9 10 11 12 13 14 15 16 17 18
1 2 3 4 5 6 7
1 H Hydrogen 1.01
2 He Helium 4.00
3 Li Lithium 6.94
4 Be Beryllium 9.01
5 B Boron 10.81
6 C Carbon 12.01
7 N Nitrogen 14.01
8 O Oxygen 16.00
9 F Fluorine 19.00
10 Ne Neon 20.18
11 Na Sodium 22.99
12 Mg Magnesium 24.31
13 Al Aluminum 26.98
14 Si Silicon 28.09
15 P Phosphorus 30.97
16 S Sulfur 32.07
17 Cl Chlorine 35.45
18 Ar Argon 39.95
19 K Potassium 39.10
20 Ca Calcium 40.08
21 Sc Scandium 44.96
22 Ti Titanium 47.87
23 V Vanadium 50.94
24 Cr Chromium 52.00
25 Mn Manganese 54.94
26 Fe Iron 55.85
27 Co Cobalt 58.93
28 Ni Nickel 58.69
29 Cu Copper 63.55
30 Zn Zinc 65.41
31 Ga Gallium 69.72
32 Ge Germanium 72.64
33 As Arsenic 74.92
34 Se Selenium 78.96
35 Br Bromine 79.90
36 Kr Krypton 83.80
37 Rb Rubidium 85.47
38 Sr Strontium 87.62
39 Y Yttrium 88.91
40 Zr Zirconium 91.22
41 Nb Niobium 92.91
42 Mo Molybdenum 95.94
43 Tc Technetium (98)
44 Ru Ruthenium 101.07
45 Rh Rhodium 102.91
46 Pd Palladium 106.42
47 Ag Silver 107.87
48 Cd Cadmium 112.41
49 In Indium 114.82
50 Sn Tin 118.71
51 Sb Antimony 121.76
52 Te Tellurium 127.60
53 I Iodine 126.90
54 Xe Xenon 131.29
55 Cs Cesium 132.91
56 Ba Barium 137.32
57 La Lanthanum 138.91
72 Hf Hafnium 178.49
73 Ta Tantalum 180.95
74 W Tungsten 183.84
75 Re Rhenium 186.21
76 Os Osmium 190.23
77 Ir Iridium 192.22
78 Pt Platinum 195.08
79 Au Gold 196.97
80 Hg Mercury 200.59
81 Tl Thallium 204.33
82 Pb Lead 207.2
83 Bi Bismuth 208.98
84 Po Polonium (209)
85 At Astatine (210)
86 Rn Radon (222)
87 Fr Francium (223)
88 Ra Radium (226)
89 Ac Actinium (227)
104 Rf Rutherfordium (261)
105 Db Dubium (262)
106 Sg Seaborgium (266)
107 Bh Bohrium (264)
108 Hs Hassium (277)
109 Mt Meitnerium (268)
110 Ds Darmstadtium (271)
111 Rg Roentgenium (272)
112 Uub Ununbium (285)
113 Uut Ununtrium (284)
114 Uuq Ununquadium (289)
115 Uup Ununpentium (288)
116 Uuh Ununhexium (292)
58 Ce Cerium 140.12
59 Pr Praseodymium 140.91
60 Nd Neodymium 144.24
61 Pm Promethium (145)
62 Sm Samarium 150.36
63 Eu Europium 151.96
64 Gd Gadolinium 157.25
65 Tb Terbium 158.93
66 Dy Dysprosium 162.50
67 Ho Holmium 164.93
68 Er Erbium 167.26
69 Tm Thulium 168.93
70 Yb Ytterbium 173.04
71 Lu Lutetium 174.97
90 Th Thorium 232.04
91 Pa Protactinium 231.04
92 U Uranium 238.03
93 Np Neptunium (237)
94 Pu Plutonium (244)
95 Am Americium (243)
96 Cm Curium (247)
97 Bk Berkelium (247)
98 Cf Californium (251)
99 Es Einsteinium (252)
100 Fm Fermium (257)
101 Md Mendelevium (258)
102 No Nobelium (259)
103 Lr Lawrencium (262)

Key

Atomic number — Symbol — Atomic weight

| 92 | U | Uranium | 238.03 |

State: S Solid | L Liquid | G Gas | X Not found in nature

- Metals
- Transition metals, lanthanide series, actinide series
- Metalloids
- Nonmetals, noble gases

Periodic Table

1	H	Hydrogen	1.01

Group 1 / 2

3	Li	Lithium	6.94
4	Be	Beryllium	9.01
11	Na	Sodium	22.99
12	Mg	Magnesium	24.31
19	K	Potassium	39.10
20	Ca	Calcium	40.08
37	Rb	Rubidium	85.47
38	Sr	Strontium	87.62
55	Cs	Cesium	132.91
56	Ba	Barium	137.32
87	Fr	Francium	(223)
88	Ra	Radium	(226)

Groups 3–12 (transition metals)

21	Sc	Scandium	44.96
22	Ti	Titanium	47.87
23	V	Vanadium	50.94
24	Cr	Chromium	52.00
25	Mn	Manganese	54.94
26	Fe	Iron	55.85
27	Co	Cobalt	58.93
28	Ni	Nickel	58.69
29	Cu	Copper	63.55
30	Zn	Zinc	65.41
39	Y	Yttrium	88.91
40	Zr	Zirconium	91.22
41	Nb	Niobium	92.91
42	Mo	Molybdenum	95.94
43	Tc	Technetium	(98)
44	Ru	Ruthenium	101.07
45	Rh	Rhodium	102.91
46	Pd	Palladium	106.42
47	Ag	Silver	107.87
48	Cd	Cadmium	112.41
57	La	Lanthanum	138.91
72	Hf	Hafnium	178.49
73	Ta	Tantalum	180.95
74	W	Tungsten	183.84
75	Re	Rhenium	186.21
76	Os	Osmium	190.23
77	Ir	Iridium	192.22
78	Pt	Platinum	195.08
79	Au	Gold	196.97
80	Hg	Mercury	200.59
89	Ac	Actinium	(227)
104	Rf	Rutherfordium	(261)
105	Db	Dubnium	(262)
106	Sg	Seaborgium	(266)
107	Bh	Bohrium	(264)
108	Hs	Hassium	(277)
109	Mt	Meitnerium	(268)
110	Ds	Darmstadtium	(271)
111	Rg	Roentgenium	(272)
112	Uub	Ununbium	(285)

Groups 13–18

2	He	Helium	4.00
5	B	Boron	10.81
6	C	Carbon	12.01
7	N	Nitrogen	14.01
8	O	Oxygen	16.00
9	F	Fluorine	19.00
10	Ne	Neon	20.18
13	Al	Aluminum	26.98
14	Si	Silicon	28.09
15	P	Phosphorus	30.97
16	S	Sulfur	32.07
17	Cl	Chlorine	35.45
18	Ar	Argon	39.95
31	Ga	Gallium	69.72
32	Ge	Germanium	72.64
33	As	Arsenic	74.92
34	Se	Selenium	78.96
35	Br	Bromine	79.90
36	Kr	Krypton	83.80
49	In	Indium	114.82
50	Sn	Tin	118.71
51	Sb	Antimony	121.76
52	Te	Tellurium	127.60
53	I	Iodine	126.90
54	Xe	Xenon	131.29
81	Tl	Thallium	204.38
82	Pb	Lead	207.2
83	Bi	Bismuth	208.98
84	Po	Polonium	(209)
85	At	Astatine	(210)
86	Rn	Radon	(222)
113	Uut	Ununtrium	(284)
114	Uuq	Ununquadium	(289)
115	Uup	Ununpentium	(288)
116	Uuh	Ununhexium	(292)

Lanthanide series

58	Ce	Cerium	140.12
59	Pr	Praseodymium	140.91
60	Nd	Neodymium	144.24
61	Pm	Promethium	(145)
62	Sm	Samarium	150.36
63	Eu	Europium	151.96
64	Gd	Gadolinium	157.25
65	Tb	Terbium	158.93
66	Dy	Dysprosium	162.50
67	Ho	Holmium	164.93
68	Er	Erbium	167.26
69	Tm	Thulium	168.93
70	Yb	Ytterbium	173.04
71	Lu	Lutetium	174.97

Actinide series

90	Th	Thorium	232.04
91	Pa	Protactinium	231.04
92	U	Uranium	238.03
93	Np	Neptunium	(237)
94	Pu	Plutonium	(244)
95	Am	Americium	(243)
96	Cm	Curium	(247)
97	Bk	Berkelium	(247)
98	Cf	Californium	(251)
99	Es	Einsteinium	(252)
100	Fm	Fermium	(257)
101	Md	Mendelevium	(258)
102	No	Nobelium	(259)
103	Lr	Lawrencium	(262)